D0275226

PLANT BIOTECHNOLOGY AND TRANSGENIC PLANTS

BOOKS IN SOILS, PLANTS, AND THE ENVIRONMENT

Soil Biochemistry, Volume 1, edited by A. D. McLaren and G. H. Peterson
Soil Biochemistry, Volume 2, edited by A. D. McLaren and J. Skujiņš
Soil Biochemistry, Volume 3, edited by E. A. Paul and A. D. McLaren
Soil Biochemistry, Volume 4, edited by E. A. Paul and A. D. McLaren
Soil Biochemistry, Volume 5, edited by E. A. Paul and J. N. Ladd
Soil Biochemistry, Volume 6, edited by Jean-Marc Bollag and G. Stotzky
Soil Biochemistry, Volume 7, edited by G. Stotzky and Jean-Marc Bollag
Soil Biochemistry, Volume 8, edited by Jean-Marc Bollag and G. Stotzky
Soil Biochemistry, Volume 9, edited by G. Stotzky and Jean-Marc Bollag
Soil Biochemistry, Volume 10, edited by Jean-Marc Bollag and G. Stotzky

Organic Chemicals in the Soil Environment, Volumes 1 and 2, edited by C. A. I. Goring and J. W. Hamaker
Humic Substances in the Environment, M. Schnitzer and S. U. Khan
Microbial Life in the Soil: An Introduction, T. Hattori
Principles of Soil Chemistry, Kim H. Tan
Soil Analysis: Instrumental Techniques and Related Procedures, edited by Keith A. Smith
Soil Reclamation Processes: Microbiological Analyses and Applications, edited by Robert L. Tate III and Donald A. Klein
Symbiotic Nitrogen Fixation Technology, edited by Gerald H. Elkan
Soil–Water Interactions: Mechanisms and Applications, Shingo Iwata and Toshio Tabuchi with Benno P. Warkentin

Soil Analysis: Modern Instrumental Techniques, Second Edition, edited by Keith A. Smith

Soil Analysis: Physical Methods, edited by Keith A. Smith and Chris E. Mullins

Growth and Mineral Nutrition of Field Crops, N. K. Fageria, V. C. Baligar, and Charles Allan Jones

Semiarid Lands and Deserts: Soil Resource and Reclamation, edited by J. Skujiņš

Plant Roots: The Hidden Half, edited by Yoav Waisel, Amram Eshel, and Uzi Kafkafi

Plant Biochemical Regulators, edited by Harold W. Gausman

Maximizing Crop Yields, N. K. Fageria

Transgenic Plants: Fundamentals and Applications, edited by Andrew Hiatt

Soil Microbial Ecology: Applications in Agricultural and Environmental Management, edited by F. Blaine Metting, Jr.

Principles of Soil Chemistry: Second Edition, Kim H. Tan

Water Flow in Soils, edited by Tsuyoshi Miyazaki

Handbook of Plant and Crop Stress, edited by Mohammad Pessarakli

Genetic Improvement of Field Crops, edited by Gustavo A. Slafer

Agricultural Field Experiments: Design and Analysis, Roger G. Petersen

Environmental Soil Science, Kim H. Tan

Mechanisms of Plant Growth and Improved Productivity: Modern Approaches, edited by Amarjit S. Basra

Selenium in the Environment, edited by W. T. Frankenberger, Jr., and Sally Benson

Plant–Environment Interactions, edited by Robert E. Wilkinson

Handbook of Plant and Crop Physiology, edited by Mohammad Pessarakli

Handbook of Phytoalexin Metabolism and Action, edited by M. Daniel and R. P. Purkayastha

Soil–Water Interactions: Mechanisms and Applications, Second Edition, Revised and Expanded, Shingo Iwata, Toshio Tabuchi, and Benno P. Warkentin

Stored-Grain Ecosystems, edited by Digvir S. Jayas, Noel D. G. White, and William E. Muir

Agrochemicals from Natural Products, edited by C. R. A. Godfrey

Seed Development and Germination, edited by Jaime Kigel and Gad Galili

Nitrogen Fertilization in the Environment, edited by Peter Edward Bacon

Phytohormones in Soils: Microbial Production and Function, William T. Frankenberger, Jr., and Muhammad Arshad

Handbook of Weed Management Systems, edited by Albert E. Smith

Soil Sampling, Preparation, and Analysis, Kim H. Tan

Soil Erosion, Conservation, and Rehabilitation, edited by Menachem Agassi

Plant Roots: The Hidden Half, Second Edition, Revised and Expanded, edited by Yoav Waisel, Amram Eshel, and Uzi Kafkafi

Photoassimilate Distribution in Plants and Crops: Source–Sink Relationships, edited by Eli Zamski and Arthur A. Schaffer

Mass Spectrometry of Soils, edited by Thomas W. Boutton and Shinichi Yamasaki

Handbook of Photosynthesis, edited by Mohammad Pessarakli

Chemical and Isotopic Groundwater Hydrology: The Applied Approach, Second Edition, Revised and Expanded, Emanuel Mazor

Fauna in Soil Ecosystems: Recycling Processes, Nutrient Fluxes, and Agricultural Production, edited by Gero Benckiser

Soil and Plant Analysis in Sustainable Agriculture and Environment, edited by Teresa Hood and J. Benton Jones, Jr.

Seeds Handbook: Biology, Production, Processing, and Storage, B. B. Desai, P. M. Kotecha, and D. K. Salunkhe

Modern Soil Microbiology, edited by J. D. van Elsas, J. T. Trevors, and E. M. H. Wellington

Growth and Mineral Nutrition of Field Crops: Second Edition, N. K. Fageria, V. C. Baligar, and Charles Allan Jones

Fungal Pathogenesis in Plants and Crops: Molecular Biology and Host Defense Mechanisms, P. Vidhyasekaran

Plant Pathogen Detection and Disease Diagnosis, P. Narayanasamy

Agricultural Systems Modeling and Simulation, edited by Robert M. Peart and R. Bruce Curry

Agricultural Biotechnology, edited by Arie Altman

Plant–Microbe Interactions and Biological Control, edited by Greg J. Boland and L. David Kuykendall

Handbook of Soil Conditioners: Substances That Enhance the Physical Properties of Soil, edited by Arthur Wallace and Richard E. Terry

Environmental Chemistry of Selenium, edited by William T. Frankenberger, Jr., and Richard A. Engberg

Principles of Soil Chemistry: Third Edition, Revised and Expanded, Kim H. Tan

Sulfur in the Environment, edited by Douglas G. Maynard

Soil–Machine Interactions: A Finite Element Perspective, edited by Jie Shen and Radhey Lal Kushwaha

Mycotoxins in Agriculture and Food Safety, edited by Kaushal K. Sinha and Deepak Bhatnagar

Plant Amino Acids: Biochemistry and Biotechnology, edited by Bijay K. Singh

Handbook of Functional Plant Ecology, edited by Francisco I. Pugnaire and Fernando Valladares

Handbook of Plant and Crop Stress: Second Edition, Revised and Expanded, edited by Mohammad Pessarakli

Plant Responses to Environmental Stresses: From Phytohormones to Genome Reorganization, edited by H. R. Lerner

Handbook of Pest Management, edited by John R. Ruberson

Environmental Soil Science: Second Edition, Revised and Expanded, Kim H. Tan

Microbial Endophytes, edited by Charles W. Bacon and James F. White, Jr.

Plant–Environment Interactions: Second Edition, edited by Robert E. Wilkinson

Microbial Pest Control, Sushil K. Khetan

Soil and Environmental Analysis: Physical Methods, Second Edition, Revised and Expanded, edited by Keith A. Smith and Chris E. Mullins

The Rhizosphere: Biochemistry and Organic Substances at the Soil–Plant Interface, Roberto Pinton, Zeno Varanini, and Paolo Nannipieri

Woody Plants and Woody Plant Management: Ecology, Safety, and Environmental Impact, Rodney W. Bovey

Metals in the Environment: Analysis by Biodiversity, M. N. V. Prasad

Plant Pathogen Detection and Disease Diagnosis: Second Edition, Revised and Expanded, P. Narayanasamy

Handbook of Plant and Crop Physiology: Second Edition, Revised and Expanded, edited by Mohammad Pessarakli

Environmental Chemistry of Arsenic, edited by William T. Frankenberger, Jr.

Enzymes in the Environment: Activity, Ecology, and Applications, edited by Richard G. Burns and Richard P. Dick

Plant Roots: The Hidden Half, Third Edition, Revised and Expanded, edited by Yoav Waisel, Amram Eshel, and Uzi Kafkafi

Handbook of Plant Growth: pH as the Master Variable, edited by Zdenko Rengel

Biological Control of Crop Diseases, edited by Samuel S. Gnanamanickam

Pesticides in Agriculture and the Environment, edited by Willis B. Wheeler

Mathematical Models of Crop Growth and Yield, Allen R. Overman and Richard V. Scholtz III

Plant Biotechnology and Transgenic Plants, edited by Kirsi-Marja Oksman-Caldentey and Wolfgang H. Barz

Additional Volumes in Preparation

Handbook of Postharvest Technology, edited by A. Chakraverty, Arun S. Mujumdar, G. S. V. Raghavan, and H. S. Ramaswamy

Handbook of Soil Acidity, edited by Zdenko Rengel

PLANT BIOTECHNOLOGY AND TRANSGENIC PLANTS

EDITED BY

KIRSI-MARJA OKSMAN-CALDENTEY

VTT Technical Research Center of Finland
Espoo, Finland

WOLFGANG H. BARZ

Westphalian Wilhelm's University Munich
Munich, Germany

MARCEL DEKKER, INC.   NEW YORK · BASEL

ISBN: 0-8247-0794-X

This book is printed on acid-free paper.

Headquarters
Marcel Dekker, Inc.
270 Madison Avenue, New York, NY 10016
tel: 212-696-9000; fax: 212-685-4540

Eastern Hemisphere Distribution
Marcel Dekker AG
Hutgasse 4, Postfach 812, CH-4001 Basel, Switzerland
tel: 41-61-260-6300; fax: 41-61-260-6333

World Wide Web
http://www.dekker.com

The publisher offers discounts on this book when ordered in bulk quantities. For more information, write to Special Sales/Professional Marketing at the headquarters address above.

Current printing (last digit):
10 9 8 7 6 5 4 3 2 1

PRINTED IN THE UNITED STATES OF AMERICA

Preface

Biotechnology has become one of the most promising branches of science in recent years. Plant biotechnology is a rapidly developing field of plant research, and genetic engineering is an important tool in biotechnology. Biotechnology can provide, among other benefits, more nutritious and safer foods, better agronomical traits, and more effective pharmaceuticals or chemicals. Medicinal plants as well as other useful crops have great value for both the nonfood and the food industries. Plants can be used as such, or important metabolites can be isolated from them. Besides primary metabolites, plants produce drugs, pesticides, dyes, flavors, and fragrances. Plant breeding using either conventional or modern breeding methods has improved the quality or the agricultural properties of many crop plants.

Metabolic engineering of plants has shown its potential both in basic research and as a tool of modern plant breeding. Designer crops can already produce valuable enzymes, proteins, and antibodies. In many research laboratories and companies in Europe, the United States, and Japan, much work has been done in studying secondary metabolic pathways by isolating specific genes that regulate the function of some key enzymes. The output of this type of research has been tremendous due to the rapid development of plant molecular biology techniques. Furthermore, the large-scale production of useful plant material or metabolites in bioreactors shows new possibilities in plant biotechnology.

We intend in this book to provide the most up-to-date information on plant biotechnology and transgenic plant production. We review available methodologies for plant cell culture, transformation techniques for crop improvement, and strategies to yield high-value products. In addition, a wide spectrum of various applications of genetically engineered plants is presented. These activities are demonstrated in almost 30 chapters written by the most outstanding scientists in their respective field. We have kept in mind a broad range of readers in both academia and industry. We hope that this book also will be of interest to students of plant biology and biotechnology as well as to more experienced scientists who produce transgenic plants.

Kirsi-Marja Oksman-Caldentey
Wolfgang H. Barz

Contents

Preface   *iii*
Contributors   *ix*

1. Plant Biotechnology—An Emerging Field   1
    Wolfgang H. Barz and Kirsi-Marja Oksman-Caldentey

2. Plant-Derived Drugs and Extracts   23
    Yvonne Holm and Raimo Hiltunen

3. Industrial Strategies for the Discovery of Bioactive
    Compounds from Plants   45
    Helmut Kessmann and Bernhard Schnurr

4. Plant Cell and Tissue Culture Techniques Used in
    Plant Breeding   59
    Peter M. A. Tigerstedt and Anna-Maija Niskanen

5. Plant Cell Cultures as Producers of Secondary Compounds   77
    Kazuki Saito and Hajime Mizukami

6. Genetic Transformation of Plants and Their Cells   111
    Richard M. Twyman, Paul Christou, and Eva Stöger

7. Properties and Applications of Hairy Root Cultures   143
    Pauline M. Doran

8. Bioreactors for Plant Cell and Tissue Cultures   163
    Regine Eibl and Dieter Eibl

9. The Potential Contribution of Plant Biotechnology to
    Improving Food Quality   201
    D. G. Lindsay

10. Engineering Plant Biochemical Pathways for Improved
    Nutritional Quality   233
    *Michel Jacobs, Marc Vauterin, Eric Dewaele, and
    Adrian Craciun*

11. Transgenic Plants as Producers of Modified Starch and
    Other Carbohydrates   255
    Alan H. Schulman

12. Improving the Nutritional Quality and Functional Properties
    of Seed Proteins by Genetic Engineering   283
    Peter R. Shewry

13. Transgenic Plants as Sources of Modified Oils   305
    Sean J. Coughlan and Anthony J. Kinney

14. Flavors and Fragrances from Plants   323
    Holger Zorn and Ralf G. Berger

15. Fine Chemicals from Plants   347
    Michael Keil

16. Genetic Engineering of the Plant Cell Factory for Secondary
    Metabolite Production: Indole Alkaloid Production in
    Catharanthus roseus as a Model   373
    *Serap Whitmer, Robert van der Heijden, and
    Robert Verpoorte*

17. Transgenic Plants for Production of
    Immunotherapeutic Agents   405
    *James W. Larrick, Lloyd Yu, Sudhir Jaiswal, and
    Keith Wycoff*

18. Signal Transduction Elements   427
    Dierk Scheel

19. The Plant Cell Wall—Structural Aspects and
    Biotechnological Developments   445
    Bruno Moerschbacher

20. Lignin Genetic Engineering: A Way to Better Understand
    Lignification beyond Applied Objectives   477
    Alain-M. Boudet

Contents

21. Transgenic Plants Expressing Tolerance Toward
    Oxidative Stress   497
    Frank Van Breusegem and Dirk Inzé

22. Transgenic Plants with Increased Tolerance against
    Viral Pathogens   517
    Holger Jeske

23. Transgenic Plants with Enhanced Tolerance against
    Microbial Pathogens   549
    Raimund Tenhaken

24. Transgenic Crop Plants with Increased Tolerance to
    Insect Pests   571
    Danny J. Llewellyn and Thomas J. V. Higgins

25. Transgenic Herbicide-Resistant Crops—Advantages,
    Drawbacks, and Failsafes   597
    Jonathan Gressel

26. Plants and Environmental Stress Adaptation Strategies   635
    Hans J. Bohnert and John C. Cushman

27. Molecular Mechanisms that Control Plant Tolerance to
    Heavy Metals and Possible Roles in Manipulating
    Metal Accumulation   665
    *Stephan Clemens, Sébastien Thomine, and
    Julian I. Schroeder*

Index   *693*

Contributors

Wolfgang H. Barz   Institute of Plant Biochemistry and Biotechnology, Westphalian Wilhelm's University Munich, Munich, Germany

Ralf G. Berger   Institute of Biochemistry, University of Hannover, Hannover, Germany

Hans J. Bohnert   Department of Plant Biology, University of Illinois, Urbana, Illinois, U.S.A.

Alain-M. Boudet   Institute of Plant Biotechnology, UMR CNRS/UPS 5546, Castanet-Tolosan, France

Paul Christou   Molecular Biology Unit, John Innes Centre, Norwich, United Kingdom

Stephan Clemens   Leibniz Institute of Plant Biochemistry, Halle (Saale), Germany

Sean J. Coughlan   DuPont Nutrition and Health, Wilmington, Delaware, U.S.A.

Adrian Craciun   Department of Biotechnology, University of Brussels, Sint-Genesius-Rode, Belgium

John C. Cushman   Department of Biochemistry, University of Nevada, Reno, Nevada, U.S.A.

Eric Dewaele   Department of Biotechnology, University of Brussels, Sint-Genesius-Rode, Belgium

Pauline M. Doran   Department of Biotechnology, University of New South Wales, Sydney, Australia

Dieter Eibl   Department of Biochemistry, University of Applied Sciences Wädenswil, Wädenswil, Switzerland

Regine Eibl   Department of Biochemistry, University of Applied Sciences Wädenswil, Wädenswil, Switzerland

Jonathan Gressel   Department of Plant Sciences, Weizmann Institute of Science, Rehovot, Israel

Thomas J. V. Higgins   CSIRO Plant Industry, Canberra, Australia

Raimo Hiltunen   Department of Pharmacy, University of Helsinki, Helsinki, Finland

Yvonne Holm   Department of Pharmacy, University of Helsinki, Helsinki, Finland

Dirk Inzé   Department of Plant Systems Biology, Flanders Interuniversity Institute for Biotechnology, Ghent University, Ghent, Belgium

Michel Jacobs   Department of Biotechnology, University of Brussels, Sint-Genesius-Rode, Belgium

Sudhir Jaiswal   Planet Biotechnology Inc., Hayward, and Palo Alto Institute of Molecular Medicine, Mountain View, California, U.S.A.

Holger Jeske   Department of Molecular Biology and Plant Virology, University of Stuttgart, Stuttgart, Germany

Michael Keil   Boehringer Ingelheim Pharma KG, Ingelheim, Germany

Helmut Kessmann*   Discovery Partners International AG, Allschwil, Switzerland

Anthony J. Kinney   DuPont Nutrition and Health, Wilmington, Delaware, U.S.A.

James W. Larrick   Planet Biotechnology Inc., Hayward, and Palo Alto Institute of Molecular Medicine, Mountain View, California, U.S.A.

D. G. Lindsay   Institute of Food Research, Norwich, United Kingdom

Danny J. Llewellyn   CSIRO Plant Industry, Canberra, Australia

Hajime Mizukami   Faculty of Pharmaceutical Sciences, Nagoya City University, Nagoya, Japan

---

Current affiliation: Graffinity Pharmaceuticals AG, Heidelberg, Germany.

Bruno M. Moerschbacher   Institute of Plant Biochemistry and
Biotechnology, Westphalian Wilhelm's University Munich, Munich,
Germany

Anna-Maija Niskanen   Department of Applied Biology, University of
Helsinki, Helsinki, Finland

Kirsi-Marja Oksman-Caldentey   VTT Biotechnology, VTT Technical
Research Center of Finland, Espoo, Finland

Kazuki Saito   Research Center of Medicinal Resources, Graduate School
of Pharmaceutical Sciences, Chiba University, Chiba, Japan

Dierk Scheel   Department of Stress and Developmental Biology, Institute
of Plant Biochemistry, Halle (Saale), Germany

Bernhard Schnurr   Discovery Partners International AG, Allschwil,
Switzerland

Julian I. Schroeder   Division of Biology, University of California, San
Diego, La Jolla, California, U.S.A.

Alan H. Schulman   Institute of Biotechnology, University of Helsinki,
and MTT Research Finland, Helsinki, Finland

Peter R. Shewry   IACR–Long Ashton Research Station, Bristol, United
Kingdom

Eva Stöger   Molecular Biology Unit, John Innes Centre, Norwich,
United Kingdom

Raimund Tenhaken   Department of Plant Physiology, University of
Kaiserslautern, Kaiserslautern, Germany

Sébastien Thomine   Institute of Plant Sciences–CNRS, Gif-sur-Yvette,
France

Peter M. A. Tigerstedt   Department of Applied Biology, University of
Helsinki, Helsinki, Finland

Richard M. Twyman   Molecular Biology Unit, John Innes Centre,
Norwich, United Kingdom

Frank Van Breusegem   Department of Plant Systems Biology, Flanders
Interuniversity Institute for Biotechnology, Ghent University, Ghent,
Belgium

Robert van der Heijden   Leiden/Amsterdam Center of Drug Research,
Leiden, The Netherlands

Marc Vauterin   Department of Biology, University of Brussels, Sint-Genesius-Rode, Belgium

Robert Verpoorte   Leiden/Amsterdam Center of Drug Research, Leiden, The Netherlands

Serap Whitmer   Leiden/Amsterdam Center of Drug Research, Leiden, The Netherlands

Keith Wycoff   Planet Biotechnology Inc., Hayward, and Palo Alto Institute of Molecular Medicine, Mountain View, California, U.S.A.

Lloyd Yu   Planet Biotechnology Inc., Hayward, and Palo Alto Institute of Molecular Medicine, Mountain View, California, U.S.A.

Holger Zorn   Institute of Biochemistry, University of Hannover, Hannover, Germany

PLANT BIOTECHNOLOGY AND TRANSGENIC PLANTS

1

Plant Biotechnology—An Emerging Field

Wolfgang H. Barz
Institute of Plant Biochemistry and Biotechnology,
Westphalian Wilhelm's University Munich, Munich, Germany

Kirsi-Marja Oksman-Caldentey
VTT Biotechnology, VTT Technical Research Center of Finland,
Espoo, Finland

I. INTRODUCTION

Biotechnology is a scientific discipline with focus on the exploitation of metabolic properties of living organisms for the production of valuable products of a very different structural and organizational level for the benefit of men. The products can be the organisms themselves (i.e., biomass or parts of the organismic body), products of cellular or organismic metabolism (i.e., enzymes, metabolites), or products formed from endogenous or exogenous substrates with the help of single enzymes or complex metabolic routes. The organisms under question vary from microbes (bacteria, fungi) to animals and plants. In addition to intact organisms, isolated cells or enzyme preparations are employed in biotechnology. The possibility to submit the producing organisms or the cellular systems to technical and even industrial procedures has led to highly productive processes. The products of biotechnology are of importance for medicine, pharmaceutical sciences, agriculture, food production, chemistry, and numerous other disciplines.

   Biotechnology receives the necessary scientific and technical information from a considerable number of disciplines. Cell biology, morphology of the employed organisms, biochemistry, physiology, genetics, and various technical fields are major sources. In the last two decades, molecular biology and gene technology have substantially contributed to the spectrum of scientific disciplines forming biotechnology. As is always true for progress in natural sciences, it is especially true for biotechnology that more rapid development and gain of higher standards depend on the improvement of methods.

In the historical development of biotechnology, microbes have been used preferentially. They still offer an extremely rich potential for biotechnological application. Animal systems and their cells are also valuable systems, especially in view of the very costly products (i.e., antibodies, vaccines). Although much later in the chronological process, plant biotechnology has made an impressive development in gaining basic and applicable knowledge as well as in establishing production processes. It is therefore justified to speak of an emerging field. Major steps will be discussed in this chapter.

II. A LONG HISTORY TO REACH A HIGH STANDARD

In each ecosystem plants and other photosynthetically active organisms are responsible for primary production, which provides the energetic and nutritional basis for all subsequent trophic levels. The extremely high ability of plants to adapt to all kinds of environmental conditions and ecosystems has led to an extremely wide and differentiated spectrum of plants. Since ancient times higher plants have formed the main source of food for men, and therefore, concomitant with early phases of settlements and agriculture, men started to establish and improve crop plants. Archeological evidence has clearly shown how long well-known crop species (i.e., maize, cereals, legumes) have been grown, modified by selection, and thus improved in quality and yield. Plant breeding is indeed an old art that has been continuously developed in efficiency and scope. Quite typical for quality breeding of, for instance, cereals is the long procedure required (sometimes decades) to reach particular genotypes and to cross in specific genes or traits.

An interesting achievement in breeding of wheat is characterized by the term *green revolution*, in which (around 1950–1960) wheat genotypes from many different countries were used successfully on a very large scale to breed high-yielding and durable lines. For many countries such new varieties were a very great improvement for their agriculture.

Another important goal in breeding improved crop plants is the often achieved adaptation to unfavorable environmental conditions (i.e., heat, drought, salt, and other cues). Although good results have been obtained, such efforts will undoubtedly remain in the focus of future efforts. Better insight into the physiology, biochemistry, and chemical reactions as well as the gene regulation of the endogenous adaptation and defense mechanisms that plants can express will contribute to these objectives. Gene technology will be an essential component in these efforts.

Another characteristic feature of the long-term breeding of cereals, potatoes, or vegetables is the fact that during the long periods the shape and the outer appearance of the plants have changed so much that the original wild types were either lost or no longer easily identified as starting material. A typical example is corn. Modern agricultural crop plants are also bred for very uniform physical appearance, time of flowering, and maturity so that harvest by machines in an industrial manner is possible (examples are cotton, maize, and cereals). It is a feature of our high-yielding agriculture that all possible mechanical techniques are being employed.

Very precious treasures for future agriculture and for plant biotechnology are the gene banks and the International Breeding Centers, where great numbers of genotypes of crop plants are multiplied and carefully preserved for long periods of time. Such "pools of genes" represent the basis for sustainable development and allow future programs for improved adaptation of plants to human needs. Fortunately, the understanding has gained ground in recent years that in addition to crop plants all types of wild plants, in every ecosystem, must be preserved because of the genetic resources to be possibly exploited in the future.

An interesting development in itself, with a long history and remarkable contributions to culture and art, is the numerous and sometimes highly sophisticated ornamental plants produced in many countries. Beauty of color and flower shape were the guidelines in their breeding and selection. Rather early in this development the value of mutagenetic reagents was learned, and these ornamentals also served to shape the term of a mutant. Recent biochemical studies with, for example, snapdragon, tulip, chrysanthemum, or petunia and their flavonoid constituents clearly presented evidence that the various flower colors can contribute to identifing biosynthetic pathways.

In connection with flower pigments, which are secondary metabolites, it should be remembered that numerous other secondary constitutents of very different chemical structures are valuable pharmaceuticals. In many countries knowledge of plants as sources of drugs has been cherished for long times. Modern pharmacological and chemical studies have helped in the identification of the relevant compounds. Such investigations are still considered important objectives of plant biotechnology. In some cases extensive breeding programs have already achieved the selection and mass cultivation of high-yielding lines. In modern pharmacy, about 25% of drugs still contain active compounds from natural sources, which are primarily isolated from plants.

For a good number of years in the period from 1950 to 1980, plant biochemistry and plant biophysics concentrated on elucidation of the photosynthetic processes. The pathways of CO_2 assimilation as well as structure, energy transfer reactions, and membrane organization of chloroplasts and their thylakoids were objectives of primary interest. Chloroplast organization and molecular function of this organelle can be regarded as well-understood fields in plant biochemistry and physiology.

The last three decades of the 20th century were characterized by very comprehensive molecular analyses of chemical reactions, metabolic pathways, cellular organization, and adaptive responses to unfavorable environmental conditions in numerous plant systems. A very broad set of data has been accumulated so that plant biochemistry and closely related fields can now offer a good understanding of plants as multicellular organisms and highly adaptive systems. From a molecular point of view, the construction and the functioning of the different tissues and organs have become clear. Numerous experimental techniques have contributed to this development and some are typical plant-specific methods (i.e., cell culture techniques) with a very broad scope of application.

A fascinating field of modern plant biochemistry concerns the elucidation of the function and the molecular mechanisms of the various photoreceptor systems of higher plants. Red/far red receptors, blue light–absorbing cryptochromes, and ultraviolet (UV) light photoreceptors are essential components of plant development (1). These systems translate a light signal into physiological responses via gene activation. Quite remarkable, phosphorylated/unphosphorylated proteins are the essential components of the signal transduction system (1,2). Biotechnology will gain from this knowledge, and highly sensitive sensor systems could possibly be constructed.

In the history of plant sciences and biotechnology, the recent development of molecular biology and the introduction of gene technology deserve emphasis. Isolation, characterization, and functional determination of genes have become possible. Many plant genes were rather rapidly identified, and the number is increasing at enormous speed. Promoter analyses and identification of promoter binding proteins have decisively contributed to an understanding of the organization and function of plants as organisms consisting of multiple tissues and different organs. The phenomena of multigenes and multiple enzymes in one protein family were further revealed. Many different techniques in molecular biology and gene technology turned out to be extremely valuable. Recognition of the biology of *Agrobacterium tumefaciens* and application of its transferred DNA (T-DNA) system represented giant leaps forward. In general, because of these modern gene technological methods, plant biotechnology has grown into a new dimension with putative future possibilities that can hardly be overestimated.

In the following sections of this chapter several recent and future aspects of biotechnological relevance will be discussed.

III. PLANT TISSUE AND CELL CULTURES—A VERY VERSATILE SYSTEM

The present status of plant biotechnology cannot be evaluated without appreciation of the many possibilities and the potential of organ, tissue and cell suspension cultures. Plants of wide taxonomic origin have been subjected to culture under strictly aseptic conditions. Completely chemically defined media supplemented with growth regulators and phytohormones are the basis for the exploitation of this technique. Depending on the explant and the culture conditions cells either preserve their state of biochemical and morphological differentiation or return to a status of embryogenic, undifferentiated cells. The former situation can be used for organ cultures (e.g., pollen, anthers, flower buds, roots), whereas the latter leads to many callus and suspension types of cultures (3). For example, the cell culture technique has opened a facile route to haploid cells and plants, and such systems are of great importance for genetic and breeding studies.

Whenever a heterogeneous group of cells can be turned into a state of practically uniform cells, this much less complicated cellular system can then be exploited to study many problems. This has been performed with plant cell cultures for some 30 years now. Growth of cells in medium-size and large volumes has opened interesting applications for plant biotechnology. Numerous physiological, biochemical, genetic, and morphological results and data on cellular regulation stem from such investigations. Various primary and secondary metabolic routes have been elucidated with the help of cell culture systems. The typical sequence in pathway identification was first product and intermediate characterization, then enzyme studies, and finally isolation of genes. Furthermore, application of gene technology in the field of transgenic plants depends to some extent on the tissue and cell culture techniques (4).

Plants are characterized by *totipotency*, which means that each cell possesses and can express the total genetic potential to form a fully fertile and complete plant body. This fact, highly remarkable from a cell biological point of view, is the genetic basis for important and widely used applications of the cell culture technique. Differentiation of single cells or small aggregates of cells into embryos, tissues, and even plants allows the selection of interesting genotypes for several different fields of plant application (5).

The well-established procedures for mass regeneration of valuable specimens of ornamental and crop plants constitute an important business section in agriculture and gardening. Endangered plant species can be saved from extinction so that valuable gene pools will not disappear. Remarkable progress has been achieved in mass regeneration of trees from single plants or tissue pieces. This will undoubtedly be of further great benefit for forestry because several problems in tree multiplication can thus be circumvented (6).

Furthermore, it should be mentioned that plant cell suspension cultures possess a great potential for biotransformation reactions in which exogenously applied substrates are converted in sometimes high yields. Position and stereospecific hydroxylations, oxidations, reductions, and especially interesting glucosylation of very different substrates have been found (7). The plant cell culture technique has allowed the facile isolation of mutants from many plant species. The overwhelming importance of mutants for biochemical and genetic studies has been known for decades. Over the years, mutations from all areas of cellular metabolism have been selected and characterized. A good deal of our basic knowledge of the functioning and regulation of organisms and cells and their organelles stems from work with mutants. The various techniques of plated, suspended, or feeder cell–supported cell systems and even protoplasts have found wide applications (5). The normal rate of mutation and also increased levels of mutated cells induced by physical (UV light, high-energy irradiation) or chemical mutagens (many such compounds are known) have been used. The specific advantage of cell cultures for mutant selection is the possible isolation of single cells from a mass of unmutated ones. Heterotrophic, photomixotrophic, and photoautotrophic cells are available, and thus different areas of cell metabolism can be screened for mutations.

In the cell culture field, regulatory mutants (i.e., excessive accumulation of products of primary and secondary metabolism including visible pigments), uptake mutants (i.e., the normal cellular transport systems of nutrients into cells are invalidated), and resistance mutants (i.e., pronounced cellular tolerance against toxic compounds such as mycotoxins, pesticides, amino acid analogues, or salt) have especially been characterized. Various auxotrophic mutants in the field of growth regulators have also been of considerable value (3).

As an example, a series of studies using photoautotrophic cell suspension cultures and the highly toxic herbicide metribuzin blocking electron transport in photosystem II will be cited (8,9). A series of single, double, or even triple mutants of the D1 protein coded by the chloroplast *psbA* gene were selected and thoroughly characterized. The various lines allowed interesting insights into the mechanism of herbicide interference with the D1 protein.

In a discussion of plant mutants resistant to herbicides, the impressive results on herbicide-resistant crop plants require mention. Many of the mechanistic and metabolic aspects of herbicide resistance were first elucidated with cell cultures (10). Modern plant biotechnology has a wide choice of biochemical solutions for herbicide resistance by inactivation and detoxification reactions. Several major crop plant species (i.e., soybean, cotton, maize, rape) are presently cultivated to a large extent in the form of appropriately manipulated genotypes. This development is on the one hand regarded as a major advantage for agriculture but on the other hand as a subject of extensive and often very critical public debate.

IV. FROM GENES TO PATHWAYS TO BIOTECHNOLOGICAL APPLICATION

A landmark in our understanding of the structure, the organization, and the functioning of multicellular organisms is described by the extensive eukaryotic genome sequencing projects in the last decade. The genomes of the yeast *Saccharomyces cerevisiae*, the nematode *Caenorhabditis elegans*, and the fruit fly *Drosophila melanogaster* clearly revealed the genetic basis of the similarities and the differences of diverse multicellular organisms (11). This modest number should perhaps be compared with the 56 completed prokaryotic genome sequences (10 strains of archaea and 46 of bacteria) and the more than 200 in progress. In general, the number of genome sequencing projects is increasing rapidly.

The three eukaryotic genomes have a similar set of 10,000–15,000 different proteins, suggesting that this is the minimal complexity required by extremely diverse eukaryotes to execute development, essential metabolic pathways, and adequate responses to their environment. These available eukaryotic genome sequences thus also document basic lines of organismic evolution.

The recent completion and publication of the first complete genome sequence of a flowering plant, the brassica *Arabidopsis thaliana*, represents a further giant step forward (12). The genome of this model plant, dispersed over five chromosomes, documents for plant scientists a complete set of genes controlling developmental and growth patterns, primary and secondary metabolism, adaptative responses to environmental cues, and disease resistance. This full genomic sequence provides a means for analyzing gene function that is also important for other plant species, including commercial and agricultural crops. Plant biotechnology greatly benefits from the *Arabidopsis* genome project. The large set of identified genes and also the hitherto functionally unknown, predicted genes form the basis for more sophisticated plant genetic analysis and plant improvement by construction of plants better adapted to human needs.

The complete *Arabidopsis* genome appears to harbor 25,498 genes, of which 17,833 can presently be classified as predicted from careful sequence comparisons with genes from other organisms. Again, functional classification comprises altogether nine areas of metabolism (12) as known from the aforementioned other eukaryotic genome sequence projects (11). Thus 7665 genes (roughly 30%) remain to be functionally identified. In order to outline the amount of research still to be fulfilled with the *Arabidopsis* genome, it should be mentioned that altogether only some 10% of the genes have been characterized experimentally.

Although a detailed description of the *A. thaliana* genome cannot be given in this chapter, a few plant-specific aspects will be presented because they appear to be of importance for future plant biotechnological application, i.e., selection of specific lines, genetic modification, or transformation at sites of characteristic plant-specific properties.

A considerable number of the nuclear gene products (approximately 14%) are predicted to be targeted to the chloroplasts as indicated by appropriate signal peptide sequences. Such a value indicates the massive influx of nuclear-coded proteins into plastids. Protein kinases and the proteins containing a disease resistance protein marker as well as domains characteristic of pathogen recognition molecules are quite abundant in the *Arabidopsis* genome. The essential elements are domains (intracellular proteins with an amino terminal leucine zipper domain, a nucleotide binding site typical of small G proteins and leucine-rich repeats) that were already known from the *Arabidopsis RPS2* and *RPM1* genes as well as from other plant *R* genes (*R*, plant disease resistance genes) (13). These findings in the *Arabidopsis* genome as well as all other molecular data on plant mechanisms for recognizing and responding to pathogens (12,13) indicate that pathways transducing signals in response to pathogens and various other environmental factors are more essential elements in plants than in other eukaryotes.

Uptake, distribution, and compartmentalization of organic and inorganic nutrients; energy and signal transduction; and channeling of metabolites and end products are very essential elements in a plant's life. Membrane transport systems are especially decisive for a sessile organism composed of many different organs and tissues such as higher plants. Therefore, the comparatively large number of predicted membrane transport systems in the *Arabidopsis* genome appears understandable. Furthermore, it is not surprising that these transport systems are the well-known plant proton-coupled membrane potentials (in contrast to the animal and the fungal sodium-coupled systems). Proteins with sequence homologies to channel proteins and peptide transporters are further prominent components in the *Arabidopsis* genome. The importance of peptide transporters is further emphasized by the great number of *Arabidopsis* genes encoding Ser/Thr protein kinases;

thus, plant signaling pathways are presumably performed by the peptide–peptide phosphate mechanism (14). Future biotechnological applications of these documented plant transporters are, for example, construction of more facile cellular sequestration processes for biologically active or toxic xenobiotics (i.e., pesticides, parasitic toxins) into vacuoles.

Arabidopsis has over three times more transcription factors than identified in the genomes of the other eukaryotes. This great number should best be seen in connection with the expanded number of genes (also found in *Arabidopsis*) encoding proteins functioning in inducible metabolic pathways controlling defense and environmental interaction. Such routes are characteristic features of higher plants (15,16). Increased numbers of transcription factors are logically required to integrate gene function in response to the vast range of environmental factors that plants can perceive (17,18). Needless to say, these genes and their products represent very important tools for future biotechnology.

Finally, one aspect of the complex *Arabidopsis* genome analyses referring to signal transduction will be mentioned. The very high number of mitogen-activated protein (MAP) kinases in combination with the high number of PP2C protein phosphatases and biochemical evidence from inducible signal transduction studies support the assumption that plants operate signaling pathways with MAP kinase cascade moduls (19,20).

As mentioned before, the presence of genes encoding enzymes for pathways that are unique to vascular plants is of great importance for biotechnology. Thus, several hundred genes with probable roles in the synthesis and modification of cell wall polymers clearly emphasize the decisive role of cell walls in the life of plants. Cellulose synthetases and related enzymes involved in polysaccharide formation, polygalacturonases, pectate lyases, pectin esterases, β-1,3-glucanases, and numerous groups of polysaccharide hydrolases were among the most prominent enzymes indicated by the gene sequences of *Arabidopsis*. Again, this knowledge offers a wide range of experimental tools for either structural modification of cell walls in the living plant or in vitro studies with suitable substrates and isolated enzymes.

Furthermore, the considerable number of genes encoding peroxidases and diphenol oxidases (laccases) points at the importance of oxidative processes most likely in connection with lignin, suberin, and other polymers. Decisive reactions are thought to be cell wall stiffening and modification processes including cross-linking reactions of cell wall proteins (19). In connection with cell wall–located proteins, the large group of glycine-rich proteins (GRPs) may be used to show that plant molecular biology and biotechnology are quite often confronted with very complex metabolic systems. The plant GRPs possess a remarkable sequence homology with numerous animal proteins that are well known for their pronounced adhesive properties

(GRPs from shells), extreme mechanical flexibility (spider silk), or ability to resist high mechanical pressure (human skin collagen). These properties result from the specific amino acid sequence and certain repetitive motifs. In the case of plant GRPs, where again interesting and valuable properties for biotechnological application can be predicted from the sequences and functions, the characteristic elements are many glycine-rich motifs (i.e., GGGx or GGxxxGGx with x standing for tyrosine, histidine, or serine).

The gene sequences allow differentiation between two large groups of GRPs (21). Proteins with an N-terminal signal peptide are designed for apoplastic transport and cell wall localization. Protection of cells during antimicrobial defense and increased cell wall resistance toward enzymic digestion by microbial enzymes are logical functions. GRPs without N-terminal signal peptides are, in contrast, characterized by RNA-binding motifs, zinc finger domains or regions with oleosin character, i.e., proteins that stabilize oil droplets ("oleosomes") in the cytoplasm. In the last point, the ability of GRPs to form conformations with hydrophobic surfaces can be seen.

The real complexity in the GRP field results from the very different cues leading to their induction. Phytohormones, water stress, cold, wounding, light, nodulation, and pathogen attack have been demonstrated (22). Cytosolic compartments or matrix structures such as xylem, protoxylem, cell walls, epidermal cells, anthers, or root tips are the alternative expression sites. The putative function always appears to be to impregnate sensitive compartments with hydrophobic seals. A complete understanding of this complexity requires, in addition to the genes, identification of the various transcription factors and regulatory genes in order to open the GRP field for biotechnological application.

As a further illustration of surprising data obtained from the *Arabidopsis* genome project, the great number of genes encoding cytochrome P450 oxygenases will be mentioned. The P450 oxygenases represent a superfamily of heme-containing proteins that catalyze various types of hydroxylation reactions using NADPH and O_2. In plants these membrane-bound (endoplasmic reticulum) enzymes are known to be involved in pathways leading to various secondary metabolites as well as routes to plant growth regulators (23). Although of great importance in plant metabolism, the various plant P450s are poorly understood with only a very limited number characterized to any extent. In this context the very high number (~286) of *Arabidopsis* P450 genes must be seen in contrast to the 94 genes in *Drosophila*, the 73 genes in *C. elegans*, and just 3 genes in *S. cerevisiae*. Intensive analyses of plant P450 oxygenases will represent a major task in future years. Biotechnology will greatly benefit from such studies because hydroxylation-oxygenation pathways are already known as routes to valuable compounds. Further aspects of P450 will be discussed in connection with flavonoid and phytoalexin formation as well as xenobiochemical metabolism.

The great number of sequenced genes, especially in cases of isoenzymes encoded by multigene families, leads to an important question: Under which conditions of growth, tissue, and organ development or changing environmental conditions are such genes (selectively) transcribed? Knowledge of selective gene activation and changes therein under normal or adverse conditions is of great importance for our understanding of the complexity of multicellular organisms and for plant biotechnology.

A fascinating new technique (DNA microarray technology) allows the determination of RNA expression profiles on the genome level with many hundred genes at the same time. Samples of sequenced genes or characteristic gene fragments are immobilized as microspots on membranes or glass slides. These arrays are treated with mRNA preparations from the plant material under investigation. The process of specific DNA-mRNA hybridization can be followed or automatically recorded by various techniques of light emission or color formation (24). It is easy to predict that in the future plant sciences will benefit from DNA microarray technology to the same extent as already shown for medical and pharmaceutical applications (25).

Among other features, plants are characterized by their overwhelming number of structurally highly diverse secondary metabolites. These compounds (examples are alkaloids, terpenoids, flavonoids, and many other classes) are not essential for growth, energy conversion, and other primary metabolic pathways. They are, however, essential for interaction of the plant with its environment and other organisms; they are said to determine the "fitness" of a plant. Elucidation of many of their biosynthetic pathways, characterization of the enzymes involved, and cloning of the genes have been performed over many years. Detailed knowledge of the organ- or tissue-specific localization and integration of these compounds in developmental processes has been accumulated. Furthermore, many secondary metabolites of numerous different structural classes are well known for their biological (i.e., roles as attractants, repellents, defense compounds of plants to interact with other organisms) and physiological (organoleptic and other sensory properties and UV protection) characteristics as well as their medicinal and pharmaceutical value. Pharmaceuticals from plants still form a large portion of drugs in human therapy. Secondary metabolites will undoubtedly continue to be of great importance. Detecting, isolating, and producing biologically or pharmaceutically active secondary plant metabolites are high-priority objectives in many laboratories around the world (26). Such studies greatly benefit from the tremendous progress in analytical processes for valuable product recognition. The search for valuable plant secondary metabolites can make use of the large number of plants that have so far not been analyzed to any extent.

Another interesting aspect of the search for new secondary products is the fact that most likely all plants have a much greater genetic potential for the formation and accumulation of such products than actually expressed during normal growth conditions. Under stress (i.e., heat, drought, cold, high salt concentrations) and other difficult environmental conditions (i.e., UV irradiation, high light intensity) plants tend to form a much wider spectrum of secondary metabolites (27). Thus, numerous compounds (e.g., alkaloids, quinones, phenolics, lignans) are found as stress-related metabolites. Especially in response to pathogen (i.e., bacterial, fungal, viral) infection, a wide range of antimicrobial compounds called phytoalexins are inducibly formed de novo around infection sites. The large number of such phytoalexins indicates the reservoir of genetic information for secondary product formation that will be activated under particular circumstances (28).

The importance of phytoalexins as efficient antimicrobial defense compounds is elegantly demonstrated by the transfer of genes encoding key biosynthetic enzymes into plants that do normally not produce these compounds (29). The ability to synthesize the groundnut stilbene phytoalexin resveratrol has been expressed in tobacco, which resulted in much improved resistance of the transgenic plant toward established tobacco fungal parasites. This strategy to alter the spectrum of secondary metabolites in a plant by directing the flow of constitutive precursors into new products represents a valuable approach for modern plant biotechnology. Other examples, especially in the field of flavonoids and isoflavonoids, are feasible and are under investigation (30). With regard to pharmaceutical products, the value of transgenic plants has repeatedly been demonstrated (26).

A challenging field for plant biotechnology is the anthocyanin pigments in flowers. The introduction of additional hydroxyl functions in ring B by transfer of genes coding specific cytochrome P450 monooxygenases opens the possibility to create flowers with deeper (red-blue) color shades (31). Prerequisites are correct vacuolar pH conditions and copigmentation.

Plants kept under adverse conditions accumulate not only phytoalexins but also normal secondary metabolites, various of their biosynthetic intermediates, and many new compounds in sometimes high concentrations (32,33). In this context, a valuable technique for biotechnological application is connected with cell suspension cultures of the experimental plants in which secondary product accumulation and phytoalexin formation are stimulated or induced by treatment of cultures with microbial elicitors (34). These signal compounds of very different chemical structure (oligo- or polysaccharides of microbial cell wall structures, peptides or proteins of pathogens, as well as regulator compounds as glutathione, jasmonic acid, salicylic acid, or heavy metal ions as abiotic stress compounds) all tend to interfere with cell metabolism via signal transduction cascades to induce stress and defense responses (35). A large spectrum of secondary metabolites has been shown to accumulate (36). Because heterotrophic, photomixotrophic and photoautotrophic cell cultures can be used, the experimental possibilities are quite variable and wide (34). Elicitation of cell cultures has also been determined as a simple but efficient technique to detect new cytochrome P450 monooxygenases that are not expressed under normal conditions (37). The findings on new secondary products formed de novo under particular conditions support the interesting data from the *Arabidopsis* genome project showing that this plant as judged by sets of unexpected genes possesses (at least the genetic) potential to form secondary metabolites not yet isolated from *A. thaliana* (12).

In conclusion, the search for secondary products can use both new plants, not yet analyzed, and plants with known sets of these products because the genetic potential has not yet been fully exploited.

V. THE PLANT CELL ORGANELLES CONTAINING GENETIC INFORMATION

Plants possess three cellular compartments containing genetic information, namely the nucleus, the plastids, and the mitochondria. The genomes of these three compartments differ greatly in size and thus in number of heritable traits. The nucleus (size of the haploid genome $\sim 1.2 \times 10^8$ to 2.4×10^9 bp; $\sim 20,000-40,000$ genes) possesses a linear genome distributed over several chromosomes that normally occur as diploid sets of genes with the DNA material highly complexed with proteins (38). Identification and cloning of nuclear genes, their elimination or silencing, and introduction of foreign genes have almost become a routine procedure in numerous plant species. The highly sophisticated and efficient techniques of modern molecular biology that allow substantial modifications of nuclear genomes will be of utmost importance for plant biotechnology.

The mitochondria of plants carry circular genomes 200–2000 kb in length, differ in the number of genes ($\sim 50-70$), and even vary considerably between species and sometimes within one plant (39). Transformation of mitochondrial genomes is in its infancy.

The plastids harbor a circular double-stranded DNA molecule of 120–160 kb with about 130 genes. This genome has been found in all cellular types of plastids (i.e., proplastids, photosynthetically active chloroplasts, chromoplasts, and amyloplasts), and quite remarkably each chloroplast may contain up to 100 identical copies of the plastid genome. Given the fact that each leaf cell may possess as many as 100 chloroplasts, an exceptionally high degree of ploidy (up to approximately 10,000 plastid genomes) is the result for each cell. Successful attempts to engineer the chloroplast genome have so far been restricted to very few systems (i.e., *Chlamydomonas reinhardtii, Nicotiana tabacum, Arabidopsis thaliana*), but routine procedures with other crop plants suitable for biotechnological application are slowly emerging. Recent data on the stable genetic transformation of tomato (*Lycopersicum esculentum*) plastids and expression of a foreign protein in fruit represent a major step forward in technology (4). A key step in the chloroplast transformation experiments was the use of a specified region in the chloroplast genome as a component of transformation vectors in order to target transgenes by homologous recombination. The transplastomic tomato plants finally obtained were shown to transfer the foreign gene to the next generation via uniparentally maternal transmission.

This work also represents a significant breakthrough with regard to biotechnology because of the great advantages of transplastomic plants over conventional transgenic plants generated by transformation of the nuclear genome. Some advantages can be summarized as follows. Due to the polyploidy of the plastid genome, high levels of transgene expression and foreign protein accumulation (up to 40% of total cellular protein) can be expected. Because the chloroplast DNA lacks a compact chromatin structure, position effects of gene integration are most likely not involved. As mentioned earlier, transgene integration by homologous recombination provides an efficient integration system. Finally, as shown for the transplastomic tomato plants, most higher plants follow a strict uniparentally maternal inheritance pattern of chloroplasts, i.e., absence of pollen transmission of transgenes (4). Thus, the often criticized spread of transgenes from plants generated by nuclear transformation experiments can be avoided. This aspect will undoutedly be of major ecological importance. It is easy to predict that the availability of transplastomic plants offers a wide range of biotechnological applications. The new technology can be offered for the introduction of new biosynthetic pathways, resistance management of crop plants, and the use of plants as factories for biopharmaceuticals, proteins, enzymes, or peptides.

VI. METABOLISM OF XENOBIOCHEMICALS

Higher plants are often confronted with a wide range of exogenous organic compounds, of either natural or anthropogenic origin. Products in the latter category (especially prominent are herbicides, insecticides, and various other groups of pesticides) are intentionally applied to agricultural plants and thus are also introduced into the general biosphere. As expected from the very reason for their application, these environmental chemicals differ greatly in their biological activity or toxicity toward different plant species; this variability ranges from highly toxic to nontoxic because the plants' responses vary from very sensitive to highly resistant. This difference in itself allows important conclusions for biotechnology when the decisive mechanistic reason has been deciphered.

Great progress has been made in our understanding of the metabolism of these environmental chemicals in crop plants. In contrast to previous belief, plants have developed a pronounced potential for the metabolism of foreign compounds. Metabolism proceeds such that after uptake of foreign products, structural modifications (phase I: generation of functional groups such as —OH, —NH, or —SH) are introduced that finally allow transfer of hydrophilic metabolites (phase II: conjugation metabolism, formation of polar, water-soluble products by addition of glucosyl or amino acyl residues) into vacuolar long-term storage or peroxidative polymerization (phase III) of xenobiotic derivatives into polymeric structures such as lignin or cell wall–localized polyphenolic matrices.

Because plants cannot excrete organic waste or end products outside the plant body (as animals normally do), metabolic excretion aims at vacuoles or long-term durable polymers. A great variety of very different chemical structures can thus be changed to harmless metabolites (40). Complete degradation of the carbon skeleton of foreign products to CO_2 and water is very rare in plants. For plant biotechnology aiming at the generation of (crop) plants with a higher level of resistance toward xenobiochemicals, two enzyme systems are of special interest. Decisive hydroxylation reactions of phase I are catalyzed by cytochrome P450 monooxygenases. Numerous dealkylation, epoxidation, and hydroxylation reactions (at aromatic, heterocyclic, alicyclic, or aliphatic substrates) are the key introductory steps that convert toxic compounds into much less toxic or nontoxic metabolites. In addition to xenobiotic metabolism, P450 enzymes are involved in numerous reactions of primary (phytohormones) and secondary (e.g., flavonoid pigments, many phenylpropanoid compounds, terpenoids, alkaloids, phytoalexins) metabolism (23,28). The importance of cytochrome P450 oxygenases in plant metabolism can hardly be overestimated (41). The great number of P450 genes detected in the *Arabidopsis* genome (see earlier) adequately supports this statement. Furthermore, the well-characterized mammalian P450 enzymes and their documented decisive role in detoxification of drugs and other exogenous compounds have stimulated research in this field (23). Therefore, based on the knowledge that numerous xenobiochemicals are converted by P450 enzymes, clear identification of the relevant enzymes, determination of their substrate specificities, analysis of gene regulation (constitutive expression versus inducible formation), and cloning of the genes are now preferential objectives. Because cloning of P450 genes is often easier than isolation of the membrane-bound proteins and their biochemical characterization, numerous known gene sequences await functional identification (23,37).

It has clearly been shown that in *Helianthus tuberosum* a P450 enzyme was highly induced by exogenous chemicals (a phenomenon well known from animal systems). Upon heterologous expression in yeast, the enzyme converted a wide range of xenobiotics and herbicides to nonphytotoxic compounds (42). For plant biotechnology such genes are potential tools for the control of herbicide tolerance as well as soil and groundwater bioremediation. In accordance with this statement, a cytochrome P450 monooxygenase cDNA selected from a soybean P450 cDNA library was also shown to catalyze the oxidative metabolism of a range of herbicides and to enhance tolerance to such compounds in transgenic tobacco (43).

The preceding data would never have been obtained without the application of molecular biological techniques. Such procedures are of great importance for biotechnology in the search for other specific genes and their functional characterization. With the great number of genes obtained from the genome sequencing projects (e.g., *Arabidopsis*) or from the facile cloning of P450 genes, techniques for gene selection and functional determination become more and more of interest. In this context, new and elegant applications of the well-known technique to identify and characterize genes by constructing knockout mutants should be mentioned. Using T-DNA of *A. tumefaciens* as an insertional mutagen and PCR techniques with primers directed at the wanted gene(s), large collections of transformed *Arabidopsis* lines (or other plants if they can readily be transformed) have been made available for screening studies. In essence, any gene can thus be identified and the mutant plant analyzed for the resulting phenotype (44,45).

Highly lipophilic xenobiotics, especially those carrying conjugated double bonds, halogen substituents (Cl, Br) at aromatic or aliphatic structures, or nitro and nitroso groups are metabolized in plants by glutathione *S*-transferases (GSTs) (46). This highly complex set of isozymes is involved in the metabolism of endogenous substrates (protection against oxidative stress in respiratory and photosynthesis pathways, carrier systems for vacuolar transport of anthocyanin pigments and xenobiochemicals) as well as exogenous compounds (detoxification of herbicides and other foreign products, especially by nucleophilic attack of the S atom and displacement of the halogen or nitro substituent). The resulting peptide derivative may be processed further but will eventually be stored in vacuoles. The GSTs are homo- or heterodimers with the various subunits either expressed constitutively or formed inducibly upon treatment of plants with suitable substrates. Each distinct subunit is encoded by a different gene. Multiple homo- and heterodimers exist, and the isoenzymes show distinct but only partly overlapping substrate specificities. Intensive studies with maize, wheat, and soybean have shown that the constitutive expression or the manipulated overexpression of certain GST subunits represents a tool for promoting tolerance of crop plants toward specific agrochemicals (46,47).

The data collected so far on plant metabolism of xenobiochemicals and other foreign compounds clearly indicate that powerful techniques exist that provide interesting applications for plant improvement.

VII. CROP PLANTS AND RENEWABLE RESOURCES

With the *Arabidopsis* genome in hand, plant scientists are now eagerly looking for sequence data for crop plants such as rice and maize (48). In these cases the scientific challenge of genome sequencing is much bigger because these plants have genomes 4 to 25 times larger than the *Arabidopsis* genome. This results from the tendency of many plants to carry duplicate or multiple copies of large sections of DNA. In view of the economic importance of rice and maize as staple food for more than half of the world's population, the results of such projects will undoubtedly form the basis for better knowledge of the genetics of these plants. These efforts will eventually also lead to continued progress in improving the productivity and the quality of these crop plants. Thus, a challenging and fascinating chapter of plant biotechnology will be opened in a few years (48).

In general, the productivity of modern agricultural crop plants has been increased manyfold over the last decades. Adaptation of the various genotypes either to the often complex factors of the environment (i.e., soil, climate, temperature, water supply), to the specific prevailing agricultural conditions, or to pests and pathogens has been achieved very successfully at sometimes impressive speed. Furthermore, the different demands of markets and consumers with regard to product quality and fields of product application have been leading guides in the breeding programs. These programs were conducted by conventional techniques of crossing and selection, but more recently molecular biological procedures [e.g., restriction fragment length polymorphism (RFLP)] have also been introduced. In general, in addition to yield and quality, modern agricultural crop plants have been optimized for high consumption of fertilizers and water. This last aspect will have to be at least partly reversed because future agricultural practice in many countries will be confronted by a reduced water supply. Plants with appropriate mechanisms for low water management are a challenging scientific task in the future.

A few lines of foreseeable development in plant breeding and construction are certain. Plant breeding will more and more apply molecular biological and gene technological methods. The data from genome sequencing programs will be essential prerequisites. The diversification of lines within a given species will increase because of the diverse demands for product quality and product application. The overall productivity of our crop plants has to be greatly increased in order to feed the rapidly growing population.

A very interesting and scientifically important step into this modern field has been taken by the recent release of "Golden Rice." This transgenic rice supplies provitamin A and iron and is expected to reduce major micronutrient deficiencies in substantial populations where rice is the major diet (49). Iron deficiency (a health problem in many women) is compensated by several transgenes leading to better iron uptake and hydrolysis of phytate. Vitamin A (required to prevent eye problems and blindness) is provided by substantial levels of β-carotene accumulating in the rice grains due to four transgenes to allow carotinoid formation.

The wide field of renewable resources represents a further challenge for plant biotechnology and modern agriculture. Petrol oil and many mineral oil–derived chemicals as well as coal are to be replaced by plant biomass or plant-derived raw materials, various chemicals, biopolymers, and all sorts of high-molecular or low-molecular products formed by and isolated from plants. Such plant production requires little if any exhaustable energy resources.

Potato lines with structurally modified starch (changes in amylose/ amylopectin ratios), rape transgenic genotypes accumulating seed oil with other than the normal C16 and C18 fatty acids, or crop plants mainly storing fructans instead of sucrose in their roots are well-established suitable examples (50). From rape-derived "bio-diesel" as petrol for cars to highly sophisticated organic chemicals from suitably constructed plant lines, the design of new "industrial plants" opens wide possibilities for plant biotechnology on a practically unlimited scale.

VIII. CONCLUSIONS

Plant biotechnology has developed into a scientific discipline with substantial value in itself. In addition to the microbial and the animal systems, plants and their cells can be used with great benefit for biotechnological questions. This application will undoubtedly continue and most likely will increase in importance. This is especially mandatory because plants are the major and most important source of our nutrition. It is easy to predict that the use of transgenic plants will more and more become routine and a matter of course. The development that started a number of years ago is of so much value that there will be no way and no need to go back. All the biotechnological efforts have to be seen in the context of the pressure that the rapidly growing population exerts on the production of food and all materials that can be produced with plants.

REFERENCES

1. C Fankhauser, J Chory. Light control of plant development. Annu Rev Cell Dev Biol 13:203–229, 1997.
2. KC Yeh, JC Lagarias. Eukaryotic phytochromes: light-regulated serine/threonine protein kinases with histidine kinase ancestry. Proc Natl Acad Sci USA 95:13976–13981, 1998.
3. RLM Pierik. In-Vitro Culture of Higher Plants. 4th ed. Boston: Kluwer Academic Publishers, 1998.
4. S Ruf, M Hermann, IJ Berger, H Carrer, R Bock. Stable genetic transformation of tomato plastids and expression of a foreign protein in fruits. Nat Biotechnol 19:870–875, 2001.
5. R Endreß. Plant Cell Biotechnology. New York: Springer Verlag, 1994.
6. J Kleinschmit, A Meier-Dinkel. Biotechnology in forest tree improvement. In: R Rodriguez, R Sanchez Tames, DJ Durzan, eds. Plant Aging, Basic and Applied Approaches. New York: Plenum, 1990, pp 319–325.
7. T Suga, T Hirata. Biotransformation of exogenous substrates by plant cell cultures. Phytochemistry 29:2393–2406, 1990.
8. C Schwenger-Erger, J Thiemann, W Barz, U Johannaningmeier, D Naber. Metribuzin resistance in photoautotrophic *Chenopodium rubrum* cell cultures: characterization of double and triple mutations in the psbA gene. FEBS Lett 329:43–46, 1993.
9. C Schwenger-Erger, N Böhnisch, W Barz. A new psbA mutation yielding an amino-acid exchange at the lumen-exposed site of the D_1-protein. 7 Naturforsch 54c:909–914, 1999.
10. P Brandt. Transgene Pflanzen. Boston: Birkhäuser Verlag, 1995.
11. GM Rubin. Comparative genomics of the eukaryotes. Science 287:2204–2215, 2000.
12. The *Arabidopsis* genome initiative. Analysis of the genome sequence of the flowering plant *Arabidopsis thaliana*. Nature 408:796–815, 2000.
13. R Michelmore. Molecular approaches to manipulation of disease resistance genes. Annu Rev Phytopathol 15:393–427, 1995.
14. CA Ryan, G Pearce. Systemin: a polypeptide signal for plant defence genes. Annu Rev Cell Dev Biol 14:1–17, 1998.
15. AA Agrawal. Induced responses to herbivory and increased plant performance. Science 279:1201–1202, 1998.
16. SW Hutcheson. Current concepts of active defense in plants. Annu Rev Phytopathol 36:59–90, 1998.
17. P Reymond, EE Farmer. Jasmonate and salicylate as global signals for gene expression. Curr Opin Plant Biol 1:404–411, 1998.
18. A Stepanova, JR Ecker. Ethylene signalling: from mutants to molecules. Curr Opin Plant Biol 3:353–360, 2000.

19. O Otte, W Barz. The elicitor-induced oxidative burst in cultured chickpea cells drives the rapid insolubilization of two cell wall structural proteins. Planta 200: 238–246, 1996.
20. H Hirt. Multiple roles of MAP kinases in plant signal transduction. Trends Plant Sci 2:11–15, 1997.
21. G Sachetto-Martins, LO France, DE de Oliveira. Plant glycine-rich proteins: a family or just proteins with a common motif? Biochim Biophys Acta 1492:1–14, 2000.
22. H Cornels, Y Ichinose, W Barz. Characterization of cDNAs encoding two glycine-rich proteins in chickpea (*Cicer arietinum* L.): accumulation in response to fungal infection and other stress factors. Plant Sci 154:83–88, 2000.
23. PR Ortiz de Montellano. Cytochrome P450. Structure, Mechanism and Biochemistry. 2nd ed. New York: Plenum, 1995.
24. MB Eisen, PT Spellman, PO Brown, D. Botstein. Cluster analysis and display of genome-wide expression patterns. Proc Natl Acad Sci USA 95:14863–14868, 1998.
25. DS Latchman. How can we use our growing understanding of gene transcription to discover effective new medicines? Curr Opin Biotechnol 6:712–717, 1997.
26. KM Oksman-Caldentey, R Hiltunen. Transgenic crops for improved pharmaceutical products. Field Crops Res 45:57–69, 1996.
27. J Kuc. Phytoalexins, stress metabolites and disease resistance in plants. Annu Rev Phytopathol 33:275–297, 1995.
28. W Barz. Phytoalexins. In: H Hartleb, R Heitefuss, HH Hoppe, eds. Resistance of Crop Plants Against Fungi. Jena: G Fischer Verlag, 1997, pp 183–201.
29. R Hain, B Bieseler, H Kindl, G Schröder, R Stöcker. Expression of a stilbene synthase gene in *Nicotiana tabacum* results in synthesis of the phytoalexin resveratrol. Plant Mol Biol 15:325–335, 1990.
30. RA Dixon, ChL Steele. Flavonoids and isoflavonoids—a gold mine for metabolic engineering. Trends Plant Sci 4:394–400, 1999.
31. N de Vetten, J ter Hors, HP van Schaik, A De Boer, J Mol, R Koes. A cytochrome *b* (5) is required for full activity of flavonoid 3′,5′-hydroxylase, a cytochrome P450 involved in the formation of blue flower colors. Proc Natl Acad Sci USA 96:778–783, 1999.
32. U Jaques, H Keßmann, W. Barz. Accumulation of phenolic compounds and phytoalexins in sliced and elicitor-treated cotyledons of *Cicer arietinum* L. Z Naturforsch 42c:1171–1178, 1987.
33. T Vogt, M Ibdah, J Schmidt, V Wray, M Nimtz, D Strack. Light-induced betacyanin and flavonoid accumulatation in bladder cells of *Mesembryanthemum cristallinum*. Phytochemistry 52:583–592, 1999.
34. A Beimen, L Witte, W Barz. Growth characteristics and elicitor-induced reactions of photosynthetically active and heterotrophic cell suspension cultures of *Lycopersicon peruvianum* (Mill). Bot Acta 105:152–160, 1992.
35. O Otte, A Pachten, F Hein, W Barz. Early elicitor-induced events in chickpea cells. Functional links between oxidative burst, sequential occurrence of extra-

cellular alkalinization and acidification, K^+/H^+ exchange and defence related gene activation. Z Naturforsch 56c:65–76, 2001.

36. U Mühlenbeck, A Kortenbusch, W Barz. Formation of hydroxycinnamoylamides and alpha-hydroxyacetovanillone in cell cultures of *Solanum khasianum.* Phytochemistry 42:1573–1579, 1996.

37. S Overkamp, F Hein, W Barz. Cloning and characterization of eight cytochrome P450 cDNAs from chickpea (*Cicer arietinum* L.) cell suspension cultures. Plant Sci 155:101–108, 2000.

38. AD Bates, A Maxwell. DNA Topology. New York: Oxford University Press, 1993.

39. W Schuster, A Brennicke. The plant mitochondrial genome: physical structure, information content, RNA editing and gene migration to the nucleus. Annu Rev Plant Physiol Plant Mol Biol 45:61–78, 1994.

40. H Sandermann Jr. Higher plant metabolism of xenobiotics: the "green liver" concept. Pharmacogenetics 4:225–241, 1994.

41. Cl Chapple. Molecular-genetic analysis of plant cytochrome P450-dependent monooxygenases. Annu Rev Plant Physiol Plant Mol Biol 49:311–343, 1998.

42. T Robineau, Y Batard, S Nedelkina, F Cabello-Hurtado, M LeRet, O Sorokine, L Didierjeau, D Werk-Reichart. The chemically inducible plant cytochrome P450 CYP 76B1 actively metabolizes phenylureas and other xenobiotics. Plant Physiol 118:1049–1056, 1998.

43. B Simiszky, FT Corbin, ER Ward, TJ Fleischmann, RE Dewey. Expression of a soybean cytochrome P450 monooxygenase cDNA in yeast and tobacco enhances the metabolism of phenylurea herbicides. Proc Natl Acad Sci USA 96: 1750–1755, 1999.

44. PJ Krysan, JK Young, MR Sussman. T-DNA as an insertional mutagen in *Arabidopsis.* Plant Cell 2:2283–2290, 1999.

45. PJ Krysan, JC Young, F Tax, MR Sussman. Identification of transferred DNA insertions within *Arabidopsis* genes involved in signal transduction and ion transport. Proc Natl Acad Sci USA 93:8145–8150, 1996.

46. DP Dixon, I Cummins, DJ Cole, R Edwards. Glutathione-mediated detoxification systems in plants. Curr Opin Plant Biol 1:258–266, 1998.

47. KA Marrs. The functions and regulation of glutathione *S*-transferases in plants. Annu Rev Plant Physiol Plant Mol Biol 47:127–158, 1996.

48. D Adam. Now for the hard ones. Nature 408:792–793, 2000.

49. I Potrykus. Golden rice and beyond. Plant Physiol 125:1157–1161, 2001.

50. S Warwel, F Bruse, C Deimes, M Kunz, M Klaas-Rusch. Polymers and surfactants on the basis of renewable resources. Chemosphere 43:39–48, 2001.

2

Plant-Derived Drugs and Extracts

Yvonne Holm and Raimo Hiltunen
Department of Pharmacy, University of Helsinki, Helsinki, Finland

I. INTRODUCTION

The first medicines known to man were certainly made from locally grown wild plants. The knowledge about the actions of the plants was compiled by trial and error and passed down from generation to generation orally. This kind of traditional medicine, folk medicine, is still very much applied in many developing countries simply because they cannot afford expensive Western medicines. The study of traditional medicines, used in different parts of the world, by modern pharmacological methods is now a respected research area called ethnopharmacology. The indications for many plants used in traditional medicine have been verified by ethnopharmacological studies. For instance, hops have been used for centuries as a mild sedative in Europe, and in 1983 the active compound was identified as 2-methyl-3-buten-2-ol (1).

The next stage of development was to produce plant material for medicinal purposes by cultivation and it is still the most important way, although production by cell and tissue culture is gaining importance. Highly productive cultivars of the cultivated plants may be developed by breeding methods, e.g., crossing, and the plants are better developed owing to improved conditions of soil, pruning, and control of pests, fungi, etc. Gene modification has been used, for example, to increase the resistance of a plant to insects or plant diseases, but the use of genetically modified plants, e.g., soy, tomato, and corn, for production is very much debated. In field cultivation the harvesting can be performed by machines, and drying facilities are available for the correct treatment of the plant material. However, field cultivation is dependent on seasons and usually one to three harvests may be obtained in a year, whereas production of drugs by cell and tissue culture may be performed regardless of season.

Modern Western medicine (since late 19th century) is focused on single, isolated compounds or now (since the 1950s) preferably synthetic substances because they can easily be produced in large amounts. However, compounds are still isolated from plants if they cannot be easily synthesized; i.e., the synthesis comprises too many steps or the synthesis is economically unprofitable (the yield is low). For example, the total synthesis of taxol has been known since 1993, but it is still mainly isolated from its natural source, the yew, because the synthesis involves too many steps to be profitable. Semisynthesis, starting with a plant-derived compound, which is then modified either biologically or chemically, is also a much used means of producing desired drugs. If a plant contains several groups of biologically active compounds, it is not always possible to know which compound is responsible for the activity and then it is wiser to produce an extract. This is the case with purple coneflower (*Echinacea purpurea*), which contains both lipophilic and hydrophilic immunologically active compounds and is therefore used as a water-ethanol extract. Also, different groups of compounds in an extract may have synergistic effects. Plant extracts may also be standardized to contain a certain amount of one component or group of components known to possess biological activity. Plant extracts are used in phytotherapy, whereas extracted, purified compounds are used in official medicine. In this chapter some of the most important plant-based isolated drugs, extracts, or plant products and standardized extracts are presented.

II. DRUGS ISOLATED FROM PLANTS

A. Artemisinin

In the 1970s the antimalarial properties of extracts of annual or sweet wormwood (*Artemisia annua* L.), a traditional Chinese drug for fevers and malaria, were discovered and this led to the isolation of the active principle, artemisinin, which is also called *qinghaosu*. It is an endoperoxide found in the dried aerial parts of the plant in concentrations ranging from 0.01 to 0.86%, depending on the origin of the plant (2). Artemisinin is an antimalarial agent that is selectively toxic to various species of *Plasmodium* (*falciparum, vivax, ovale*) in vitro and in vivo, including chloroquine-resis-

tant strains. Synthetic efforts have yielded active derivatives, including α-
and β-artemethers, artether, and artesunate. Artemisinin and its derivatives
are sparingly soluble in water and lipidic phases and are therefore admin-
istered in aqueous or oily suspension, tablets, or suppositories. The use is
reserved to geographical areas with multiresistant *falciparum*, by prescrip-
tion (2).

artemisinin

B. Cardiac Glycosides

Two of the *Digitalis* genus, *D. purpurea* L. (purple foxglove) and *D. lanata*
Ehrh. (Grecian foxglove), are used for the extraction of digitoxin and digoxin. The purple foxglove contains 0.2–0.4% cardenolide glycosides and
is cultivated in the Netherlands for extraction of glycosides or, more rarely,
collected from natural habitats. The Grecian foxglove is cultivated in the
Netherlands and France. The leaves are collected and must be rapidly dried
at a temperature as low as possible. The Grecian foxglove is used industrially
for the extraction of digoxin and digitoxin as well as derivatives of its secondary glycosides, e.g., desacetyl-lanatoside C. Purple foxglove contains
about 30 glycosides, divided into three series: the A series, with digitoxi-
genin as aglycone (primary glycoside is purpurea glycoside A); the B series,
with gitoxigenin as aglycone; and the E series, with gitaloxigenin as aglycone. Varieties in which series A predominates (>50%) are preferred. The
constituents of the Grecian foxglove are called lanatosides and are divided
into five series; the A, B, and E series; the C series, with the primary glycoside lanatoside C; and the D series (2). The cardenolide glycosides are
still important drugs in the treatment of heart insufficiency, although synthetic drugs, e.g., β-blockers, are also used.

C. Opium

The all-time classic of plant-derived drugs is certainly opium. According to
the *European Pharmacopoeia*, raw opium is the air-dried latex obtained by
incision from the unripe capsules of opium poppy, *Papaver somniferum* L.

It contains not less than 10.0% morphine and not less than 2.0% codeine. Raw opium is intended only as starting material for the manufacture of galenical preparations (3).

The cultivation of the opium poppy may be divided into production of opium, mostly in India, and production of the straw for extraction of alkaloids, mostly in temperate climates. Legal opium cultivation is restricted to the following countries: Bulgaria, Greece, Iran, India, Russia, Turkey, and the former Yugoslavia. However, great amounts are illegally cultivated in Southeast Asia and Mexico (4). The opium production involves a great deal of manual work: cutting of the capsules to release the latex and later, when it has dried and turned brown, collecting (2). One capsule gives about 20 mg of opium (4). The latex is air dried and shaped into cakes of about 5 kg. In order to isolate alkaloids directly from the capsules, poppy varieties optimized for alkaloid production have been developed and cultivation methods optimized. The harvest may be conducted at complete maturity, which yields "straw," or before maturity, which yields "green poppy," rich in alkaloids. The capsules are dried and alkaloids may be extracted (2).

Morphine is the main alkaloid of opium and is known as a powerful analgesic that acts via the central nervous system. This means that it has many adverse effects as well as the risk of tolerance and dependence. Still, morphine is the drug of choice for severe pain, for instance, in cancer patients. Morphine is isolated from opium in a process that begins with mixing opium and calcium chloride to a thin paste, which releases the alkaloids from their salts and precipitates meconic acid. Morphine is solubilized as calcium morphinate. The next step is the purification of morphine by adding ammonium chloride, which makes morphine precipitate. The precipitate is collected, washed, and dried (2,4). The total synthesis of morphine was published in 1956, but it is still isolated from the plant (5). Over 90% of the morphine obtained is used for codeine production.

Codeine is present in small amounts in opium (about 2%) and can be extracted, but it is usually produced by semisynthesis, i.e., methylation, from morphine. Codeine is an antitussive agent (4). Dihydrocodeine is an analgesic (2).

Noscapin is one of the main alkaloids in opium (about 6%) and is obtained as a by-product in the isolation of morphine. It has antitussive properties (4).

Other semisynthetic derivatives of morphine are ethylmorphine, the 3-ethylether of morphine, and pholcodine, 3-morpholinyl-ethylmorphine, which are both antitussives. Diacetylmorphine (= heroin) has no use in therapeutics but has great abuse potential. Morphine antagonists are also prepared from morphine and include N-allylnormorphine (= nalorphine), which is a partial antagonist, and N-cyclopropylmethyl-14-hydroxynordi-

hydromorphinone (= naltrexone) and *N*-allyl-14-hydroxynordihydromor-
phinone (= naloxone), which are pure antagonists (2).

R = H: morphine
R = CH₃: codeine

D. β-Sitosterol

β-Sitosterol is one of several phytosterols that are widely distributed
throughout the plant kingdom. It is common in vegetables, grains, nuts,
seeds, and fruits and is one of the main sterols in fungi and algae (6). β-
Sitosterol may be obtained from wheat and rye germ oils, cottonseed, and
other seed oils, but it is obtained mainly as a by-product in the processing
of soybean oil and tall oil. The average Western diet provides 250–350 mg
of β-sitosterol daily (7).

β-Sitosterol, which chemically resembles cholesterol, inhibits the absorption of cholesterol. β-Sitosterol alone or combined with other phytosterols has been shown to reduce the plasma total cholesterol and low-density lipoprotein (LDL) levels in human subjects (8). It is also able to relieve
the symptoms of benign prostatic hyperplasia (BPH) and is used in herbal
remedies indicated for the treatment of this ailment (9). β-Sitosterol and its
saturated derivative, β-sitostanol, are used as cholesterol-lowering agents in
different kinds of functional foods.

sitosterol

E. Steroids

Although total synthesis of some medicinal steroids is applied commercially, there is also a great demand for natural products that can serve as a starting material for their semisynthesis. Medicinal steroids for which there is a demand are sex hormones (testosterone, estradiol, and progesterone), corticoids (cortisone acetate and betamethasone), oral contraceptives (norethisterone and mestranol), and diuretic steroids (spironolactone).

Hecogenin provides a practical starting material for the synthesis of corticosteroids. It is obtained commercially as the acetate in about 0.01% yield from sisal leaves (*Agave sisalana* Perr.). In East Africa, when removing the fiber, a hecogenin-containing "sisal concentrate" is obtained. From this the "juice" is separated and allowed to ferment for 7 days. The sludge contains about 80% of the hecogenin originally present in the leaves. It is processed by steam under pressure to complete the hydrolysis, filtered, and dried. The final concentrate contains about 12% hecogenin. Diosgenin, obtained from various *Dioscorea* species (yams), is suitable for the manufacture of oral contraceptives and sex hormones. It can also be used as a starting material for corticosteroid synthesis if microbiological fermentation is applied to introduce oxygen into the 11α-position of the pregnene nucleus. *Dioscorea* spp., wild or cultivated, contain steroidal saponins, which can be isolated by acid hydrolysis. Previous fermentation of the material for 4–10 days often gives a better yield. The water-insoluble sapogenin is then extracted with a suitable organic solvent (10). The soya beans (*Glycine max* Sieb. et Zucc.) contain appreciable amounts of stigmasterol and sitosterol. Stigmasterol provide a good source of progesterone and can replace diosgenin (6,10).

F. Taxol

The history of taxol began in 1964 in a screening program of the National Cancer Institute (NCI) in the United States, in which antineoplastic natural agents were tested. Taxol showed cytotoxic activity in vitro. The source of taxol was the Pacific yew, *Taxus brevifolia* Nutt., which contained 0.01% in the bark (11). Isolation was not possible without extinction of the species. Systematic screening of the *Taxus* genus resulted in selection and cultivation of cultivars whose needles constitute an exploitable source of taxol. In particular, *T.* x. *media* cv. *hicksii* has a taxol concentration of up to 0.06%. It is also quite feasible to prepare taxol by semisynthesis from baccatin III and 10-deacetylbaccatin III, which are structural analogues found in the leaves of the European yew, *Taxus baccata* L. Synthetic work has also led from the same 10-deacetylbaccatin III to esters, i.e., docetaxel, known under the trade name Taxotere®. Two approaches to the total synthesis of taxol were published in 1993, but the interest is purely academic because the process contains about 40 steps and is thus far from cost effective (2).

Many attempts to produce taxol from cell cultures of different *Taxus* species have been made. A quite successful experiment with *Taxus brevifolia* suspension cultures gave a maximum taxol level of 1.43 mg/L, which was 0.013% as a specific content (12). Taxol is a mitotic spindle poison with a very specific mode of action: it promotes the assembly of tubulin dimers into microtubules, which it stabilizes by inhibiting their depolymerization (2,11). Both taxol and docetaxel are indicated in the therapy of advanced ovarian and breast cancer when no other therapy is effective. Both substances are highly toxic and severe adverse effects, including neutropenia, peripheral neuropathy, cardiovascular problems, alopecia, nausea, and hypersensitivity to the solvent, are common (2).

taxol

G. Tropane Alkaloids

Tropane alkaloids, mainly atropine and scopolamine, are isolated mainly from two Australian species of *Duboisia*, *D. myoporoides* R. Br. and *D. leichhardtii* F. Mueller. They contain 0.6–5% alkaloids with either hyoscyamine or scopolamine as the main component. The isolation of the alkaloids starts with the powdered drug, which is extracted with a diluted mineral acid, for instance, 0.1 N sulfuric acid. Impurities are removed by shaking the acidic phase with an organic solvent, i.e., ethyl acetate, toluene, or chloroform, in which the impurities are dissolved. The water phase is made alkalic and the alkaloids are extracted using the same solvent (4). The tropane alkaloids are parasympatholytics and are applied, for instance, in ophthalmic preparations, to relieve smooth muscle spasms, and for motion sickness (scopolamine) (2).

H. Vinblastine and Vincristine

Vinblastine and vincristine are binary indole alkaloids isolated from the aerial parts of the Madagascan periwinkle (*Catharanthus roseus* G. Don syn. *Vinca rosea* L.). The plant contains 0.2–1% alkaloids, which form a very complex mixture in which about 95 components have been identified. Vincristine is found at a level of 0.0003%, and vinblastine is a little more abundant. In spite of the low concentration, both alkaloids are extracted from plant material. Many efforts have been made to produce these substances by cell culture, but so far no cost-effective method has been found. Both substances are antimitotic agents. Vincristine is used for the treatment of acute leukemia and in combination chemotherapy, and vinblastine is indicated in the treatment of Hodgkin's disease (2).

Two semisynthetic derivatives of vinblastine are marketed, namely vindesine and vinorelbine. Vindesine is prepared from vinblastine by formation of the hydrazide of 16-deacetylvinblastine (with hydrazine) and reduction of the acylhydrazide (by Raney nickel in methanol). An alternative method consists of forming the acylazide (by reaction with nitrous acid) and then the amide (by treatment with anhydrous ammonia). Vinorelbine is characterized by the replacement of the tryptamine moiety of the upper half (indole—CH_2—CH_2—N—) with a gramine-type moiety (indole—CH_2—N—), i.e., by the elimination of one carbon atom. The reaction goes via a bisiminium ion, by the Polonovski reaction on anhydrovinblastine, or through the bromoindolenine of anhydrovinblastine (2).

R = CH₃: vinblastine
R = CHO: vincristine

III. PLANT PRODUCTS AND EXTRACTS

A. Chamomile

At least three plants are known as chamomile: German or genuine chamomile (*Matricaria recutita* L. syn. *Matricaria chamomilla* L.), Roman or English chamomile [*Chamaemelum nobile* (L.) All. syn. *Anthemis nobilis* L.], and Moroccan chamomile (*Ormenis multicaulis* L.). It is important to distinguish between the different chamomiles because they have different chemical compositions and thus different effects. German chamomile is the most appreciated, most expensive, and most extensively investigated of these three. In addition to a blue essential oil (due to chamazulene), chamomile contains flavonoids, mostly apigenin, rutin, and their glucosides, which possess spasmolytic activity (13,14).

As is the case with all plants that are widely cultivated, there exist numerous cultivars and chemotypes of chamomile. According to the composition of the essential oil, six chemotypes may be distinguished:

1. A bisabolol oxide A type from Egypt, the Czech Republic, and Hungary.
2. A bisabolol oxide B type from Argentina.
3. A bisabolol type with up to 50% bisabolol from Spain. The cultivar Degumille® developed by Asta Pharma (formerly Degussa) belongs to this type.
4. A bisabolon oxide A type from Turkey and Bulgaria.
5. A matricine free or poor type (gives a green oil) from Egypt and Turkey.
6. A so-called uniform type, which contains almost equal amounts of bisabolol oxide A, B, and bisabolol (15).

There is a certain demand in the pharmaceutical industry for the chemotype containing α-bisabolol (often along with chamazulene), because α-bisabolol has anti-inflammatory, antimicrobial, and antipeptic activities and chamazulene is pain relieving, wound healing, and antispasmodic. In addition, chamomile oil is fungicidal (13). Chamomile flowers may be used for tea and extracts; tinctures and ointments are prepared for both internal and external use. The oil is also used in aromatherapy. Chamomile is the subject of a positive German monograph describing its use internally for gastrointestinal spasms and inflammations, externally for skin and mucous membrane inflammation (mouth) and bacterial skin disease, and as bath and irrigation therapy for inflammation of the genital and anal areas (16).

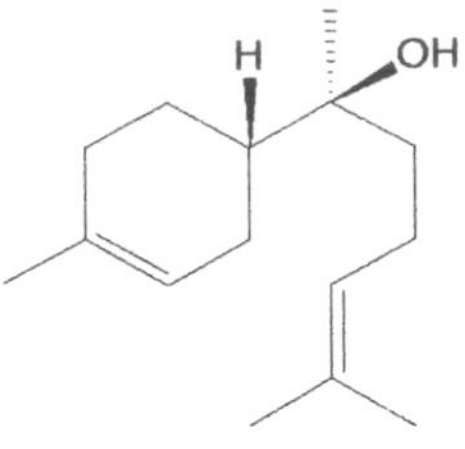

(-)-α-bisabolol

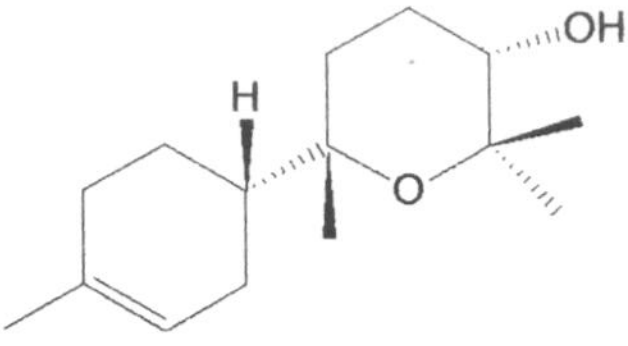

(-)-α-bisabolol oxide A

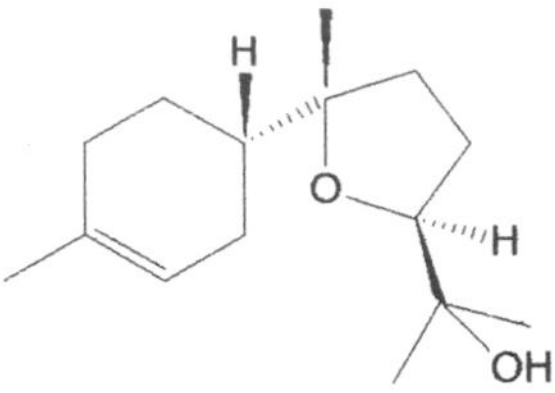

(-)-α-bisabolol oxide B

B. Echinacea

The roots and herbs of three species of *Echinacea* or coneflower, *E. angus-tifolia* DC, *E. pallida* (Nutt.) Nutt. and *E. purpurea* (L.) Moench. are widely used as immunostimulants. The chemical composition of *Echinacea* is quite complex and comprises eight groups of compounds: caffeic acid derivatives (echinacoside, cynarin, cichoric acid), flavonoids, essential oil, polyacety-lenes, alkylamides, alkaloids, polysaccharides, and other constituents, in-cluding phytosterols (17). Biological activities attributed to *Echinacea* are stimulation of phagocytosis, increase of respiratory activity, and increase of the mobility of leukocytes (18). Immunostimulatory principles have been demonstrated in both lipophilic and polar fractions of extracts of *Echinacea* species (13). Preparations containing *Echinacea* are mostly extracted using water and ethanol. They are not usually standardized (earlier often stan-dardized to echinacoside) but marker substances, cichoric acid in *E. pur-purea* and echinacoside or rutin in *E. angustifolia*, are determined. *Echina-cea* preparations have widespread use all over the world. They are registered drugs in Switzerland, Germany, Austria, Hungary, and Australia; herbal rem-edies in many European countries; and food supplements in the United States (13,19). The aboveground parts of *E. purpurea* are the subject of a positive German monograph on their use for the external treatment of hard-to-heal wounds, eczema, burns, herpes simplex, etc.; as a prophylactic in-ternal immunostimulant at the onset of cold and flu symptoms; and as an adjuvant for treatment of chronic respiratory infections, prostatitis, and uri-nary tract conditions. *E. purpurea* root, *E. angustifolia* root, and *E. angus-tifolia/E. pallida* aerial parts are not recommended because of lack of current clinical studies (13).

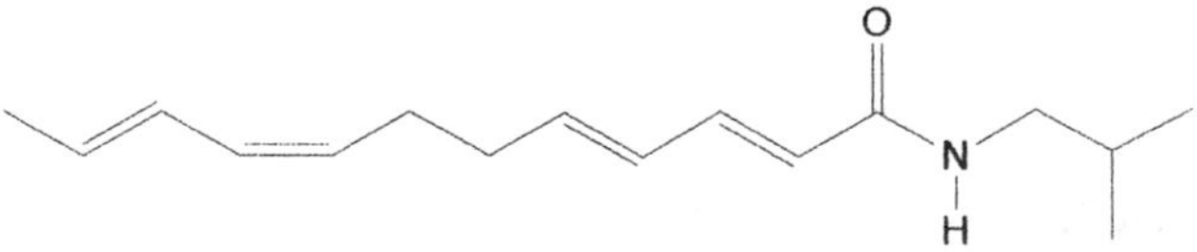

isobutylamine dodeca-(2*E*,4*E*,8*Z*,10*E*)-tetraenoate

isobutylamine undeca-(2*E*,4*Z*)-diene-8,10-diynoate

R^1 = rhamosyl, R^2 = glucosyl: echinacoside

C. Evening Primrose Oil

Evening primrose, *Oenothera biennis* L., originally a North American species, is cultivated in the United Kingdom and Canada for the production of seeds. The seeds contain up to 25% oil rich in polyunsaturated fatty acids. The major fatty acids are linoleic acid (65–80%), γ-linolenic acid (GLA, 8–14%), and oleic acid (6–11%) (2). The oil also contains smaller amounts of palmitic and stearic acid (20). Traditional breeding methods have been employed in order to develop cultivars with a high oil and GLA content. The oil is obtained by cold expression. Because of its high content of unsaturated fatty acids, it is difficult to preserve and is usually incorporated in soft gelatin capsules. Antioxidants, such as vitamin E, are often added to preserve fatty oils (2).

In Germany and the United Kingdom, evening primrose oil is approved for treatment of atopic eczema (4,13). Many other indications have been claimed but remain poorly backed up by clinical studies or controversial results. Therefore, evening primrose oil is usually considered a food supplement in most countries.

Other sources of GLA are borage oil (*Borago officinalis* L.), with a GLA content of 18–25%, and black currant seed oil (*Ribes nigrum* L.), with 16–17% GLA (13,21).

D. Garlic

Garlic (*Allium sativum* L.) has been known as a medicinal plant since ancient times. It is mentioned in Papyrus Ebers, which is dated to about 1550 BC. It is also a popular spice in spite of a sulfuric smell. The active compound in garlic is alliin, which is devoid of smell. When tissues are cut, alliin is degraded by an enzyme, alliinase, to allicin. Air oxidation of allicin leads to diallyldisulfide, which is the main constituent of garlic volatile oil. Present in garlic extracts are also allicin condensation products, 6Z- and 6E-ajoenes and cycloadducts of propenethial, vinylthiines (2,4).

Several pharmacological activities are attributed to garlic. Garlic extracts and the volatile oil are antibacterial and antifungal and have cholesterol-lowering properties and antihypertensive effects. The ajoenes have activity against platelet aggregation (2).

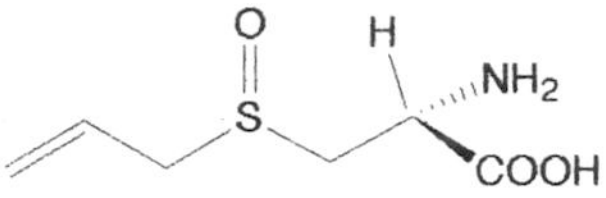

alliin
(S-allyl-L-cysteine sulphoxide)

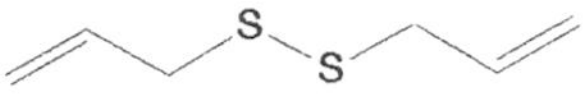

diallyl disulphide

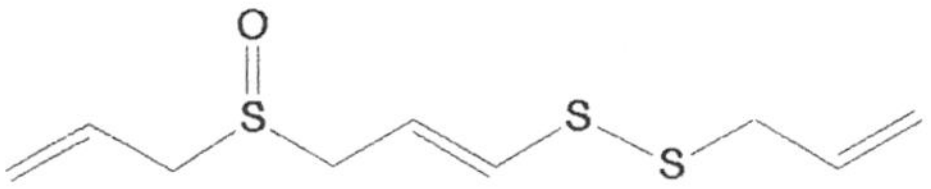

E-ajoene

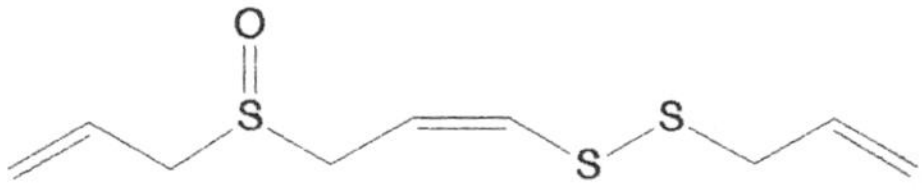

Z-ajoene

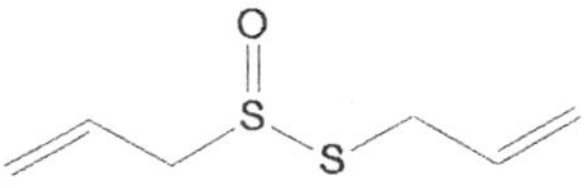

allicin

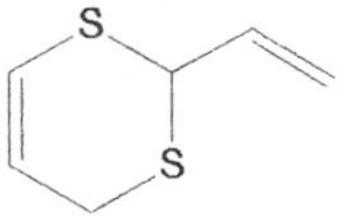

2-vinyl-4H-1,3-dithiin

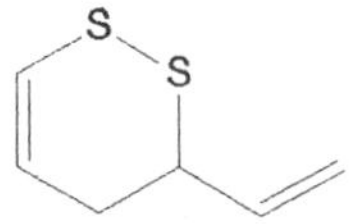

3-vinyl-4H-1,2-dithiin

There are numerous garlic preparations on the market, and they are produced in different ways:

1. Garlic powder (not stabilized). Garlic is dried in the sun or at 105°C, powdered, and usually dispensed in capsules.
2. Garlic powder (stabilized). Before drying, the garlic is treated with hot alcohol steam, which inactivates the enzymes. This means that products made in this way contain alliin.
3. Oil macerates. Fresh garlic is macerated in oil, for instance, in soybean oil, and the macerate is dispensed in soft gelatin capsules, which may be enteric coated.
4. Garlic juice may be expressed and sterilized.
5. Products containing volatile garlic oil are prepared by autolysis; i.e., garlic pieces are kept in water and the enzyme activity is allowed to go on. The process is continued by steam distillation, which yields about 0.2% oil without allicin but with di- and trisulfides. The oil is dispensed in gelatin capsules (4).
6. A special garlic product is the aged garlic extract, Kyolic® (Wakunaga, Japan), prepared by fermentation and standardized

with *S*-allyl cysteine. Some products may be standardized to alliin (19).

Garlic is the subject of a positive German monograph in which it is indicated for supportive dietary measures to reduce blood lipids and as a preventative for age-dependent vascular changes. The ESCOP monograph on garlic gives the same indications plus use for relief of coughs, colds, catarrh, and rhinitis (13).

E. Linseed Oil

Linseed oil is obtained from ripe, dried linseeds (*Linum usitatissimum* L.), which contain 35–45% oil. The oil is a source of α-linolenic acid (40–62%), linoleic acid (16–25%), and oleic acid (10–15%) (22). The oil was obtained earlier by hot expression of linseed meal and the press was adjusted to leave sufficient oil in the cake to make it suitable as cattle food. This method yielded a yellowish-brown drying oil, which on exposure to air gradually thickened, formed a hard varnish, and was used in paint (10). Now linseed oil is isolated by a special cold expression method that yields a yellow oil used as a food supplement.

The human body is able to produce a great variety of different types of fatty acids, with the exception of linoleic acid and α-linolenic acid, which are essential fatty acids (EFAs) and must be received from the diet. With a total EFA content higher than 70% and a total content of unsaturated fatty acids higher than 90%, linseed oil is a good source of both *n*-3 (α-linolenic acid) and *n*-6 (linoleic acid) polyunsaturated fatty acid precursors. Linoleic acid is the precursor of the *n*-6 series leading to γ-linolenic acid (GLA), dihomo-γ-linolenic acid (DGLA), and arachidonic acid (AA). α-Linolenic acid is the precursor of the *n*-3 series, including eicosapentaenoic acid (EPA) and docosahexaenoic acid (DHA). The conversion of EFAs into other fatty acids in the human body depends on the activity of the enzyme Δ-6-desaturase (23). Note that fish oils are natural sources rich in EPA and DHA.

F. Mint Oils and Menthol

Mint oils are some of the most commonly used essential oils all over the world. Essential oils are isolated by distillation procedures from the plant material, and this yields a 100% pure oil. The mint oils produced in the largest amounts are peppermint (*Mentha x piperita* L.), spearmint (*M. spicata* L.), and cornmint (*M. arvensis* L. var. *piperascens* Malinvaud). The United States is the major producer of peppermint (3200 tons in 1992), and spearmint and cornmint producers include Japan, Taiwan, and Brazil (13). Peppermint leaves contain 1–3%, spearmint about 0.7%, and cornmint 1–

2% essential oil of complex composition. Peppermint oil is characterized by menthol and menthone, spearmint by carvone, and cornmint by a high concentration of menthol (70–95%). The mint oils are used extensively as fragrance components in toothpastes, mouthwashes, gargles, soaps, detergents, creams, lotions, and perfumes; in flavoring chewing gums, candies, chocolates, and alcoholic beverages; and in medicines. Peppermint oil is used in cough mixtures as an expectorant, as a choleretic and antiseptic agent, and in enteric-coated capsules to treat irritable bowel syndrome. Cornmint oil is used mainly for the production of menthol. The oil is slowly cooled, which induces the crystallization of menthol. The dementholized oil still contains 30–45% menthol, and this is the commercial cornmint oil (2,13).

(-)-menthol

   The mint oils could in a way be regarded as standardized products because oils from different sources are mixed to meet composition requirements set by the International Standardization Organisation (ISO) or by different pharmacopoeias. For instance, the European pharmacopoeia sets the following requirements for the composition of peppermint oil: limonene 1.0–5.0%, cineole 3.5–14.0%, menthone 14.0–32.0%, menthofuran 1.0–9.0%, isomenthone 1.5–10.0%, menthyl acetate 2.8–10.0%, menthol 30.0–55.0%, pulegone not more than 4.0%, and carvone not more than 1.0%. In addition, the ratio of cineole content to limonene content is to be greater that 2 (3). Menthol may be produced by freezing cornmint oil, which gives the pure (−)-enantiomer plus dementholized oil, which is utilized to flavor toothpastes and chewing gum. Menthol can also be produced by semisynthesis starting from pinene or by total synthesis, which yields racemic menthol. Menthol is largely consumed by the tobacco industry and by the pharmaceutical industry in itch-relieving creams, cough mixtures, and preparations to relieve the symptoms of the common cold (2).

G. Saw Palmetto

As the life expectancy increases, all kinds of age-related health problems become more common. A problem of men older than 50 years is benign prostatic hyperplasia (BPH), i.e., enlargement of the prostata, which leads to different urinary tract symptoms, such as frequent urination, weak stream, hesitancy, intermittency, and incomplete emptying. A popular cure for these symptoms is an extract prepared from the fruits of saw palmetto, *Sabal serrulata* Hook. syn. *Serenoa repens* (Bartel.) Small. (2).

The chemical composition of the fruits still remains poorly studied. The fruits and seeds are rich in fatty oil with about 50% short-chain fatty acids, for instance, lauric acid. Commercially available are lipid and sterol hexane extracts, which contain fatty acids, linear and monounsaturated alkanes, aliphatic alcohols and their esters, phytosterols (among others sitosterol and its derivatives), and polyprenols (2). The pharmacological activities of the extract are inhibition of steroid 5α-reductase and 3α-reductase, decreased binding of dihydrotestosterone to androgen receptors, and inhibition of cyclooxygenase and lipoxygenase (2,13). In clinical trials, extracts of saw palmetto have been superior to placebo and comparable to finasteride, a synthetic drug used in BPH (24).

Lipidosterolic extracts of saw palmetto are the subject of a positive German monograph in which they are indicated for micturition problems in BPH. They are marketed as herbal remedies in many European countries. The status in the United States is undetermined (13).

IV. STANDARDIZED EXTRACTS

A. Ginger

Ginger root is mainly a spice but is used in Oriental traditional medicine and lately also in Western medicine. The root is rich in starch (60%) and contains proteins and fat. Secondary metabolites found in the root are essential oil (main component zingiberene) and pungent principles, i.e., gingerols, which occur alongside corresponding ketones (zingerone) and in the dried drug alongside dehydration products called shogaols. Shogaols are twice as pungent as gingerols (2,13).

Ginger oleoresin, which is widely used in the food industry, is prepared by organic solvent extraction (hexane, acetone, ether, alcohol). The solvent is removed under vacuum and the product is a viscous, semisolid extract (13). Powdered ginger root has also been used in the form of capsules to treat motion sickness (2). The newest indication for ginger root is to relieve rheumatic pain and help the mobility of joints. The active principles are believed to be gingerols, and extracts are standardized to a certain content of gingerols. Two such extracts, Zinax HMP-33 and EV.EXT™ 77 (ginger and galanga root), are marketed by Eurovita (Karslunde, Denmark). EV.EXT 77 is extracted by means of a patented extraction method based on the LipCell technology (Eurovita product information).

gingerols
(n=1-4,6,8,10)

B. Ginkgo

The leaves of the ginkgo or maidenhair tree (*Ginkgo biloba* L.) contain two groups of pharmacologically interesting compounds, flavonoids (0.5–1%) and terpenes (diterpenes up to 0.5% and sesquiterpenes). The flavonoids are represented by about 20 different flavonol glycosides, mostly with quercetin as aglycone. The diterpenes are also known as ginkgolides A, B, C, J (and M in the roots), and the sesquiterpene present is bilobalide (2). Ginkgolide B is an inhibitor of platelet-activating factor (PAF), and the ginkgo flavonoids are known as free radical scavengers. The ginkgo extract promotes vasodilatation and improves the blood flow in both arteries and capillaries (18). Ginkgo has been the subject of numerous clinical trials for "cerebral insufficiency," and some positive results have been published concerning the use of ginkgo in Alzheimer's disease (25). The *Ginkgo biloba* extracts on the market are concentrated (50:1), standardized products made by a multistep process. The process starts with dry, ground ginkgo leaves, which are extracted by an acetone-water mixture under partial vacuum. The organic solvent is then removed and the extract processed, dried, and standardized to a potency of 24% flavonoids and 6% terpenes. The product is available in both solid and liquid form. Ginkgo preparations are widely prescribed in Germany. In other countries they are mainly available as over-the-counter drugs or herbal remedies (18).

R^1= OH, R^2= R^3= H: ginkgolide A
R^1= OH, R^2= OH, R^3= H: ginkgolide B
R^1= R^2= R^3=OH: ginkgolide C

C. Ginseng

Since ancient times, ginseng roots have had a reputation in China as a tonic with revitalizing properties (2,13). Several species of ginseng are used: Asian ginseng (*Panax ginseng* C.A. Meyer), Japanese ginseng (*P. pseudoginseng* Wall.), and Western or American ginseng (*P. quinquefolius* L.). Sometimes Siberian ginseng is also mentioned, but this is an entirely different species, *Eleutherococcus senticosus* Maxim., which contain different substances. The ginseng root contain numerous compounds, but the ones considered active are the so-called ginsenosides (about 20). They are saponins, glycosides of tetracyclic aglycones of the dammarane series, i.e., a trihydroxylated type (protopanaxadiols) and a tetrahydroxylated type (protopanaxatriols). The ginsenosides are numbered: ginsenoside R_0, R_{a-1-2}, R_{b-1-3}, R_{c-f}, R_{g-1-2}, R_{h-1} (1). The different ginseng species differ in chemical composition. Ginseng is the subject of a positive German monograph. It is used as a tonic for invigoration for fatigue and reduced work capacity and concentration and during convalescence. It is generally referred to as an adaptogen (2,13). There are numerous ginseng products on the market, and at least the ginseng extract G115® from Pharmaton (Ridgefield, CT) is standardized. The manufacturer claims that "it is prepared using a highly sophisticated standardization process, ensuring a consistent level of active components" (Pharmaton product information).

ginsenoside R_{b-1}

D. Hawthorn

The hawthorn berries, leaves, and flowers are popular starting materials for phytotherapeutic preparations indicated for mild heart insufficiency in Europe. In the United States, the status of hawthorn is undetermined (13). According to the European pharmacopoeia, berries are collected from two species: *Crataegus monogyna* Jacq. (Lindm.) and *C. laevigata* (Poir.) DC (syn. *C. oxyacantha* L.) or their hybrids. A minimum content of 1.0% procyanidins is required. Leaves and flowers are collected from the two preceding species plus *C. pentagyna* Waldst. et Kit. ex Willd., *C. nigra* Waldst.

et Kit., and *C. azarolus* L. Here a minimum content of 1.5% flavonoids is required (26).

With so many accepted species, which have different chemical compositions, and different extraction procedures, it is necessary to standardize the products. Water, water-ethanol mixtures (30–70%), and methanol (dry extracts) are used as extraction solvents and extract different active groups of compounds. Water is good extractant of oligomeric proanthocyanidins (di- to hexameric), and ethanol extracts polymeric proanthocyanidins and triterpene acids (4). The detailed protocols of the extraction procedures seem to be confidential property of the respective companies (27). Products containing hawthorn leaves and flowers are generally standardized to flavonoids, calculated as hyperoside or vitexin, and products containing berries generally to oligomeric proanthocyanidins, calculated as catechin or epicatechin (28). In addition to standardized products, such as capsules, tablets, and drops, the pressed juice of hawthorn berries is used, especially in Germany (4).

E. St. John's Wort

The healing properties of St. John's wort or hypericum (*Hypericum perforatum* L.) were known to Dioscorides and Hippocrates, were forgotten for some time during the late 19th century, and are now being rediscovered (18). St. John's wort is a plant with a complex chemical composition containing naphthodianthrones, hypericin, isohypericin, and pseudohypericin; flavonoids, flavonols, flavones, glycosides, and biflavonoids (amentoflavone); phenols, phenolic acids, and prenylated derivatives of phloroglucinol (hyperforin); and tannins, essential oil, and other constituents, such as carotenoids, β-sitoserol, and different acids (2,20).

R = CH$_3$: hypericin
R = CH$_2$OH: pseudohypericin

hyperforin

The pharmacological activities attributed to hypericum extracts are numerous. Earlier studies claimed inhibition of monoamine oxidase (MAO), later found to be inhibition of MAO-A (13). Flavonoids and xanthones are found to inhibit catechol-O-methyltransferase (COMT), a hydroethanolic extract containing 0.15% hypericin interacted moderately with γ-aminobutyric acid A (GABA$_A$) and benzodiazepine receptors, and different hypericum extracts showed inhibition of serotonin uptake by postsynaptic receptors (28). Also, hyperforin, formerly known as the antibiotic compound of hypericum, was found to be a potent uptake inhibitor of serotonin, dopamine, noradrenaline, GABA, and L-glutamate (29). Clinical studies have shown the effectiveness of hypericum extracts compared with placebo and standard antidepressants in the treatment of mild to moderate depression (30).

Extracts of St. John's wort (for instance, HyperiFin™ by Finzelberg) are prepared using ethanol 60% (v/v) and methanol 80% (v/v) at temperatures of 50–80°C. The native extract is standardized to a hypericin content of not less than 0.3%. From this both dry extracts (for tablets and capsules) and liquid extracts (for drops and tonics) are prepared (Finzelberg product information). Supercritical carbon dioxide has been used to produce an extract of hypericum enriched in hyperforin (38.8%) for experimental purposes (31). Supercritical extraction is an effective method to produce plant extracts, but the use on a production scale is expensive. Pharmaton (Ridgefield, CT) has launched a product containing St. John's wort and standardized to hyperforin (Pharmaton product information).

V. CONCLUSIONS

Plants have been used as medicines or sources of medicines since time immemorial and will still be used. In our need for new antimicrobial, antiviral, and antitumor drugs we turn to nature because only a small percentage of the total plant species have been examined chemically and still fewer have been tested for biological activity. Large screening projects are going on and will certainly reveal new interesting molecules with desired activities. The program of the annual meeting of the Society for Medicinal Plant Research in 2000 contained lectures with the following titles: "Will science in the new millenium be different?," "Ethnobotany and ethnopharmacology: their role in future medicinal plant research," and "Understanding the molecular basis of action of natural products—a challenge for the new millenium" (32). These titles reflect the focus and direction of medicinal plant research in the near future.

On the cultivation front, it remains to be seen whether genetically modified organisms (GMOs) will be accepted for production of foodstuffs and medicinal plants. At the moment, opposition to GMOs is strong among the public, but this may change if researchers are able to show that genetically modified plants are safe to both consumers and the environment. Supercritical fluid extraction (SFE) will definitely be applied in the isolation of plant-derived drugs and the preparation of extracts when the method is out of its infancy, i.e., apparatus becomes cheaper, more readily available, and easy to operate. SFE is an extremely powerful isolation method, and as an extra bonus extraction using supercritical carbon dioxide leaves no solvent residues in the final product and is environmentally safe. Also, extraction with subcritical water may be a method of choice in the future.

The functional food market seems to expand daily and the line between food and drugs becomes more wavering in spite of different authorities' efforts to make a clear distinction between the two groups. In the future we might prevent diseases by eating functional food, cereals enriched with vitamins, yoghurt enriched with fibers, soy products providing phytoestrogens, green tea rich in flavonoids, etc., and treat diseases by replacing the gene (or genes) responsible for the particular disease. However, not all diseases can be treated by gene therapy; and there will always be a need for analgetics and antibiotics, among others, and nature will provide at least part of these drugs in the future.

REFERENCES

1. R Wohlfart, R Hänsel, H Schmidt. Nachweis sedativ-hypnotischer Wirkstoffe im Hopfen. 4. Mitt Planta Med 48:120–123, 1983.
2. J Bruneton. Pharmacognosy, Phytochemistry, Medicinal Plants. 2nd ed. Andover: Intercept, 1999, pp 157, 162–164, 207–209, 299–301, 329–331, 532–539, 623–627, 643–647, 735–742, 805–824, 929–948, 1016–1022.
3. European Pharmacopoeia 3rd ed. Strasbourg: Council of Europe, 1996, pp 1262–1263, 1299–1300.
4. E Steinegger, R Hänsel. Pharmakognosie. 5th ed. Berlin: Springer-Verlag, 1992, pp 500–515, 524–525, 529–533, 580–584, 633–637, 706.
5. H Auterhoff. Lehrbuch der Pharmazeutischen Chemie. Stuttgart: Wissenschaftliche Verlagsgesellschaft, 1968, p 483.
6. JPM Dewick. Medicinal Natural Products: A Biosynthetic Approach. New York: John Wiley & Sons, 1997, pp 228–239.
7. H Schilcher. Phytopharmaka zur Therapie von Prostata-Erkrankungen. Angew Phytother 2:14–16, 1981.
8. PJ Jones, DE MacDougall, F Ntanios, CA Vanstone. Dietary phytoserols as cholesterol-lowering agents in humans. Can J Pharmacol 75:217–227, 1997.
9. J Kraft. So kann dem Prostatiker geholfen werden. Artzl Prax 33:2167, 1981.
10. WC Evans. Trease and Evan's Pharmacognosy. 13th ed. London: Bailliére Tindall, 1989, pp 335–336, 482–486.

11. B Schneider. Taxol. Ein Arzneistoff aus der Rinde der Eibe. Dtsch Apoth Ztg 134:3389–3400, 1994.
12. JH Kim, JH Yun, YS Hwang, SY Byun, DI Kim. Production of taxol and related taxans in *Taxus brevifolia* cell cultures: effect of sugar. Biotechnol Lett 17:101–106, 1995.
13. AY Leung, S Foster. Encyclopedia of Common Natural Ingredients Used in Food, Drugs, and Cosmetics. 2nd ed. New York: John Wiley & Sons, 1996, pp 98–99, 145–148, 216–220, 235–237, 271–274, 277–281, 295–297, 368–372, 467–469.
14. M Lis-Balchin. The Chemistry and Bioactivity of Essential Oils. Park Corner: Amberwood Publishing, 1995, pp 37–39.
15. H Schilcher. Die Kamille. Stuttgart: Wissenschaftliche Verlagsgesellschaft, 1987, pp 97–98.
16. Monograph Matricariae flos, Bundesanzeiger, no 228, December 5, 1984.
17. R Bauer, H Wagner. Echinacea species as potential immunostimulatory drugs. In: H Wagner, NR Farnsworth, eds. Economic and Medicinal Plant Research. Vol 5. San Diego: Academic Press, 1991, pp 253–321.
18. VE Tyler. The Honest Herbal. 3rd ed. New York: Pharmaceutical Product Press, 1993, pp 115–117, 149–151, 275–276.
19. JG Bruhn, P Eneroth. Läkemedelsboken. Fakta och erfarenheter. Enskede: Boehringer Ingelheim, 1997, pp 98–109, 122–127.
20. CA Newall, LA Anderson, JD Phillipson. Herbal Medicines. A Guide for Health-Care Professionals. London: Pharmaceutical Press, 1996, pp 110–113, 250–252.
21. J Bruneton. Pharmacognosy, Phytochemistry, Medicinal Plants. 1st ed. London: Intercept, 1995, pp 105–106.
22. H Wagner. Pharmazeutische Biologie. 2. Drogen und ihre Inhaltstoffe. 3rd ed. Stuttgart: Gustav Fischer Verlag, 1985, p 300.
23. L Stryer. Biochemistry. 3rd ed. New York: WH Freeman, 1988, pp 467–493.
24. TJ Wilt, A Ishani, G Stark, R MacDonald, J Lau, C Mulrow. Saw palmetto extracts for treatment of benign prostatic hyperplasia: a systematic review. JAMA 280:1604–1609, 1998.
25. BS Oken, DM Storbach, JA Kaye. The efficacy of *Ginkgo biloba* on cognitive function in Alzheimer disease. Arch Neurol 55:1409–1415, 1998.
26. European Pharmacopoeia. 3rd ed. Supplement 2000. Strasbourg: Council of Europe, 1999, pp 780–783.
27. R Kaul. Der Weissdorn. Stuttgart: Wissenschaftliche Verlagsgesellschaft, 1998, pp 45–76.
28. ESCOP Monographs on the Medicinal Uses of Plant Drugs. Elburg: ESCOP Secretariat, 1996, Hyperici herba, 10 p.
29. SS Chatterjee, SK Bhattacharya, M Wonnemann, A Singer, WE Muller. Hyperforin as a possible antidepressant component of hypericum extracts (abstr). Life Sci 63:499–510, 1998.
30. K Linde, G Ramirez, CD Mulrow, A Pauls, W Weidenhammer, D Melchart. St. John's wort for depression—an overview and meta-analysis of randomised clinical trials. BMJ 313:253–258, 1996.

31. SK Bhattacharya, A Chakrabarti, SS Chatterjee. Activity profiles of two hy-
     perforin-containing extracts in behavioral models (abstr). Pharmacopsychiatry
     31(suppl 1):22, 1998.
32. Abstract Book of 48th Annual Meeting of the Society for Medicinal Plant
     Research and 6th International Congress on Ethnopharmacology, Zurich, Sep-
     tember 3–7, 2000.

3

Industrial Strategies for the Discovery of Bioactive Compounds from Plants

Helmut Kessmann* and Bernhard Schnurr
Discovery Partners International AG, Allschwil, Switzerland

I. INTRODUCTION

Plants have developed an enormous diversity of secondary products and most of them have evolved to exhibit a certain biological activity, which is useful for survival of the individual plant or species. Since 1000 BC, the Chinese and Indian literatures have described the medicinal use of plants and many recipes are still in use (1–9). Such information from traditional medicine has been used for decades to isolate medicinally active secondary products. These efforts have been very successful and a number of drugs are on the market as a result of this strategy, including plant-derived anticancer compounds such as taxol, docetaxel, and camptothecin. However, despite these successes, the advent of combinatorial chemistry led to decreased interest in natural products. Recently, this trend has changed again, and natural products, including those from plant sources, are now being used as a general source of chemical diversity, which may well complement the chemical structures amenable to total organic synthesis. As by far most plant species have not yet been studied for biologically active secondary products, there is great potential to discover new drugs or lead structures for further development. In addition, modification of secondary product formation by

Current affiliation: Graffinity Pharmaceuticals AG, Heidelberg, Germany.

45

molecular techniques and more extensive use of cell culture technologies may further increase the chances for successful drug discovery from plants.

There are in principle two approaches for drug discovery from plant sources: (1) sourcing using various strategies followed by "random" screening and (2) use of information from traditional medicine on biologically active plant extracts and isolation of the active secondary product(s). As strategy 2 is quite straightforward when the information is available, this chapter focuses on random screening approaches. As an introduction, current industrial strategies and paradigms used for drug discovery in general are described. The reader is referred not only to published papers but also to the Web sites of specialized companies in the particular fields. In addition, several databases, e.g., NAPRALERT, are available for on-line searches (or references) regarding all aspects of natural products.

II. CHANGING PARADIGMS: INDUSTRIALIZATION OF DISCOVERY

New chemical entities with a novel mode of action are urgently needed in the pharmaceutical industry for further improvement of the individual company product portfolio and to meet the expectations from the financial community. The number of drugs to be marketed by the major pharmaceutical companies needs to be tripled over the next years for companies to stay competitive and justify their current valuation (10).

The fundamental strategies in industrial drug discovery have changed over the past decades. Until the 1980s, basic research, low-throughput screening, and serendipity were the basis for successful drug discovery. During the 1980s "rational" design was developed as a new strategy. Based on protein structures, usually obtained by X-ray crystallography, chemical modulators were sought "in silico." Unfortunately, despite large investments by the pharmaceutical industry, this strategy did not fulfill its promises and new technologies were developed in the early 1990s. These strategies still dominate the drug discovery process in most companies (10).

The principal paradigm of drug discovery today is shown in Fig. 1. A key step in the discovery process is the identification of a biochemical target that is causally involved in a certain disease or a phenotype that one wishes to modify by drug application. Pharmaceutical companies focus on diseases for which a potential drug may lead to at least U.S. $100 million in yearly revenues. Diseases for which the market is smaller are usually not targeted, although recent legislation on such "orphan drugs" may change the situation and allow especially smaller companies to develop drugs for such often severe diseases. Natural compounds or derived structures may be particularly interesting for such orphan diseases.

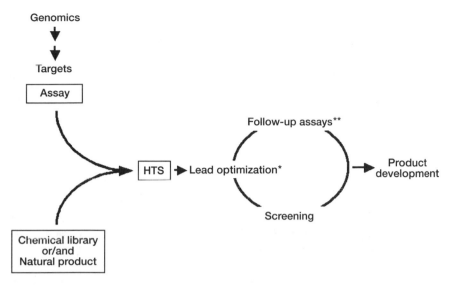

FIGURE 1 The current discovery scheme used in the pharmaceutical industry. For the discovery of new crop protection products, companies use, in addition to biochemical tests, assays directly with plants, insects, or plant diseases. *In case of natural products, this step would involve (re-)isolation of active ingredient and structural elucidation; the active structure may also be modified using traditional or combinatorial chemistry. **Follow-up assays include tests for specificity, initial efficacy studies in cell culture or animal models, as well as ADME (absorption, distribution, metabolism, and excretion tests).

Once a disease-specific target is identified by basic research or —more often—by genomic approaches, it is used to develop a biochemical assay for high-throughput-screening (HTS). The term HTS was created in the early 1990s to describe the screening of not only a few thousand but hundreds of thousands of chemical compounds in a short time.

In addition to new technologies for screening, the advent of combinatorial chemistry was a major breakthrough a decade ago (11–13). A large number of discrete compounds are being synthesized today using automated synthesis machines. However, not every chemical class can be synthesized using this approach, and additional chemical diversity is needed to increase the chance for discovery of novel chemical entities. Natural products are ideally suited to fill such "holes" in the diversity chemical space and also to serve as scaffolds for creation of natural compound–like libraries (7,8,14) (see also other chapters in this book).

There have been many technological breakthroughs, but the fundamental change is most likely the step from the very research-based approach to the "industrialization of discovery" (10,15–22). The genomic efforts offer a good example of this new paradigm. Although sequencing of the human and other genomes is a result of the synergy between innovations in automation and informatics, it is mainly a breakthrough in process management. The basic principle is that basic research is no longer used to discover, for example, the function of a certain gene but standardized technologies are used to analyze the sequence and function of all genes in the genome in a systematic, "industrial" process. Mass sequencing is done today in *factory-like setups* with dozens of to several hundred sequencers run in a 24-hour, 7-days-a-week fashion (15) (examples of companies pursuing this approach are Incyte Pharmaceuticals, Human Genome Sciences, Millennium, and recently PE Celara). It was expected that the human genome would be fully sequenced in the year 2000 and that by 2002 over 60 microbial genomes and several metazoan species would be available in company or public databases. This factory-like or industrialized process is now also applied to the next steps in drug discovery.

The development of combinatorial chemistry and HTS has moved the bottleneck in drug discovery to the next step: identification of important target genes causally involved in a certain disease. Again, high-throughput technologies are being developed to discover genes of interest in pharmaceutical companies as well as specialized biotechnology companies. One example of such a functional genomics approach has been published by CuraGen Corporation (New Haven, CT) in collaboration with the University of Washington, Seattle, and others (18). They have used the yeast two-hybrid system to discover protein-protein interaction in yeast. Most of the approximately 6000 yeast genes were expressed separately in two haploid yeast strains. One protein is fused to a DNA-binding domain of a transcription factor; the other is fused to the activator region of the same transcription factor. Upon mating, both proteins are expressed in the diploid yeast and only yeast where such a protein-protein interaction takes place will grow. Although this technology is well known, it had not been applied in such a systematic way for a entire genome. A total of 957 interactions were discovered involving 1004 yeast proteins. However, the fact that certain proteins interact in the yeast two-hybrid system does not necessarily mean that this interaction is relevant to the in vivo situation. Expression profiling during the yeast life cycle and study of the subcellular localization of the proteins will rule out some protein-protein interactions as not relevant to the in vivo situation (18).

The systematic study of the function of human genes is also under way using yeast, mice, zebra fish, *Caenorhabditis elegans, Drosophila melanogaster*, or other model species. In addition to analyzing phenotypes after overexpression or reduced expression of genes, expression profiling in diseased versus healthy tissue will help to identify new potential disease targets. In many cases such studies are performed by specialized companies and data are made available on a subscription-fee basis (see, e.g., www.exelixis.com, www.paradigmgenetics.com, www.curagen.com).

III. AUTOMATION FOR DRUG DISCOVERY

Until the mid-1990s, only a few thousand or 10,000 compounds were screened in a particular target. Today, large pharmaceutical companies or specialized drug discovery companies screen several hundred thousand compounds in a particular screening in a few days or weeks. This dramatic increase in screening capacity is due to advances in automation as well as to novel assay technologies (Table 1). The reader is also referred to specialized journals in this field such as *Drug Discovery Today*, *Journal of Biomolecular Screening*, and *Drug Discovery and Development*.

A modern industrial-scale drug discovery laboratory consists of the following (16,17,22) (see also *www.autoprt.co.uk*, *www.bohdan.com*, *www.shimadzu.com*):

A compound storage system where compounds can be automatically accessed for primary screening and follow-up tests. Compounds are stored in microtiter plates, often in dimethyl sulfoxide (DMSO).

A range of semiautomatic workstations or fully automated screening machines.

Sophisticated data management that allows on-line quality control of the screening process.

Assays are performed either in microtiter plates (96, 384, 1536, or 3456 plates) or in other very low volume screening devices. Because of the high degree of automation, very high throughput, and in particular the industrial process management, drug discovery has become a highly standardized and industrialized process, similar to the sequencing and functional genomic efforts described earlier.

In order to screen large numbers of compounds in a short time, the assay methods need to be fast, easy to perform, robust, and cost effective. Fluorescence-based methods are of particular interest because they offer high sensitivity at reasonable costs and high throughput. Fluorescence assay technologies, which have become widely used in the discovery process, are broadly divided into two classes: macroscopic fluorescence techniques (e.g., fluorescence intensity, fluorescence anisotropy or fluorescence polarization, time-resolved fluorescence technologies) and microscopic fluorescence technologies such as fluorescence correlation spectroscopy (FCS). The latter can

TABLE 1   Some Examples of Assay Technologies Applied Today Including Their
Potential Limitations in Screening of Natural Products

Commonly used assay technologies (examples)	Applied for target class (examples)	Possible problems in screening natural products
Fluorescence methods for noncellular assays including		Quenching, fluorescent ingredients, nonspecific binding
Fluorescence intensity	Broad	
Fluorescence polarization	Proteases, kinases, nuclear receptors, GPCR	
Time-resolved fluorescence	Broad	
Fluorescence correlation spectroscopy (FCS)	Broad	(Technology not yet widely used)
Isotropic methods		Nonspecific effects
Filtration assays		
Flashplate, SPA[a]	Broad	
Luminescence	Broad	
Ultraviolet-visible absorbance	Broad	Interference with assay signal
Cellular assays		
Quantification of expressed protein, e.g., ELISA	G protein–coupled receptors, nuclear receptors	Cytotoxic metabolites, general unspecific stress by metabolites
Reporter gene		

[a]SPA, scintillation proximity assay, a technology developed by Amersham.

be reduced down to nanoliter volumes and can even detect single molecule interactions (23,24) (see also www.evotec.com). Another method, homogeneous time-resolved fluorescence, is used for kinases, receptors, proteases, and other targets. As an example, in a protease assay one label (e.g., europium cryptate) is attached to one part of the protein substate, an allophycocyanin label to the other side. Excitation at 337 nm leads to an energy transfer from the cryptate to the allophycocyanin, which causes emission of a delayed fluorescence signal at 620 nm. Protease cleavage leads to a larger distance between the two labels, which results in a lower signal. This method is of particular interest in natural product screening because the interference with autofluorescence and quenching in such screening samples is limited (23,24). A similar concept has been developed for isotopic assays.

For most methods, standard plate readers are being used, although digital imaging will be the readout of choice for very large throughput (19). Other technologies currently being developed are using microfluidic devices (see, for example, *www.orchid.com*, *www.aclara.com*) (25,26).

A significant number of the potential drug targets are tested in insect or mammalian cell cultures, which are also run in 96-, 384-, 1536-plate or lower volume formats (27–29). An example is the screening for modulators of G protein–coupled receptors, the most important class of drug targets. If the interaction of a natural ligand causes an increase in the intracellular cyclic adenosine monophosphate (cAMP) concentration, such a receptor can be overexpressed in a mammalian cell line and the increase in cAMP quantified by the activity of a reporter gene linked to a cAMP-responsive element (CRE). If coupling of the G protein–coupled receptor is not via the cAMP pathway but through increase of intracellular calcium, dyes are used for quantification of calcium (27,29). Other targets are screened by quantification, e.g., by enzyme-linked immunosorbent assay (ELISA), of the expression of endogenous proteins in the selected cell culture after compound application.

For natural product screening, particularly for extracts, cellular screening systems are often difficult to use because of the presence of cytotoxic material in the extracts or other nonactive material, which interferes with the assay (e.g., by signal quenching). In most assays such undesired effects lead to false-negative results; i.e., if any biological activity was present in the extract, it is not discovered because of side effects.

IV. PLANT PRODUCTS IN THE INDUSTRIALIZED DRUG DISCOVERY PROCESS

In order to be successful in random screening, high chemical diversity needs to be applied. During the early to mid-1990s, the interest in natural products for screening decreased because combinatorial chemistry was believed to deliver large chemical libraries with the required diversity at much lower cost and at higher speed than the traditional synthesis or natural product screening. However, although the diversity that can be generated by synthetic chemistry is broad, many natural products cannot be synthesized in a cost-effective way, if at all. Therefore, natural compounds are used today (again) in many pharmaceutical or agrochemical companies to increase the diversity of their screening libraries or as scaffolds for the synthesis of natural compound–like libraries (1,5,7,8,14).

One obviously important factor for the discovery of novel secondary products is sourcing of the material (1,3,5–7,23). Several strategies have been developed, although it is not yet clear which of the strategies will lead to the most cost-effective means of discovery of new plant-derived drugs or drug leads. In addition, any "bioprospecting" must observe the rules of the United Nations Convention on Biological Diversity, which recognizes the right of countries over the biological resources within their boundaries. The source country should be involved in the discovery process and should benefit by technology transfer and commercial rights to novel products.

1. Use of information from traditional medicine, the most used or widely used strategy. Chinese medicine and Indian medicine have proved to be particularly valuable sources for plant species accumulating medicinally active compounds (1–9).
2. Sourcing from extreme and/or as yet untouched ecosystems (4,7–9).
3. Ecological observations. For example, individual plant species or lines not attacked by certain diseases may contain fungicidal compounds (6,7,23).
4. Chemotaxonomy, which is particularly useful for discovering species with a higher yield or modified lead structures (2,3,7,8).
5. Use of biologically activated plants, e.g., after infection, infestation by insects, or under extreme environmental conditions (7).
6. Use of cell cultures with different growth conditions and/or induction by microbial elicitors. This technology, which offers great commercial potential, is currently used by some specialized companies (30) (see, e.g., www.phytera.com).

In principle, two different approaches are applied today (Fig. 2):

1. The traditional approach of using crude extracts followed by purification of active principles after the discovery of activity in a particular screening system.
2. The alternative approach, which first separates the constituents of a plant extract and then applies such more or less pure products to the screening process.

Strategy 1 has been used since quinine and morphine were discovered more than hundred years ago (3,4,8). Large numbers of crude extracts can be prepared in a cost-effective manner, and such extracts are applied in cellular or noncellular screening protocols. The obvious advantage of this procedure is the low cost of extract preparation, which allows screening of many samples in a short time. Early discovery of known metabolites by thin-layer chromatography (TLC), liquid chromatography–mass spectrometry (LC-MS), or other technologies has eliminated the most important disadvantage of this technology: rediscovery of known metabolites. However, the major disadvantage is still the high investment in extraction and purifi-

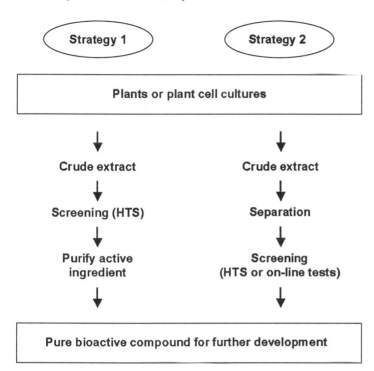

FIGURE 2 The two different approaches for discovery of novel discrete compounds from plant sources.

cation procedures, which are still performed at a low rate of automation. In addition, lack of standardization with every secondary product being present in different—and sometime insufficient—concentration remains a major bottleneck.

Strategy 2 is, in principle, a further development of the chemical screening approach originally developed by H. Zähner in close cooperation with the chemists W. Keller-Schierlein and A. Zeek during the 1970s (8). In the original procedures, compounds were first separated by TLC, isolated, and their structure determined. Purified compounds were applied for screening, or screening was performed directly on TLC plates.

This concept has now being further developed based on advances in high-performance liquid chromatography (HPLC), MS, and other analytical methods. Based on preparative HPLC, the company AnalytiCon AG (Berlin) and E. Merck KGA (Darmstadt) have developed a preparative HPLC (SepBox) that is able to separate gram quantities of crude extracts into hundreds of pure or almost pure compounds. The concentration of the compounds could be standardized using general detection methods such as evaporating light scattering (ELSD) or MS. Replicate samples can be eliminated prior to the screening by their MS profile, and all new structures can be applied to the screening process. However, with the further development of miniaturized screening devices and assay volumes in the very low microliter or even nanoliter range, each fraction from even a standard preparative HPLC separation can now be tested directly with sufficient material left for at least initial structural elucidation.

A different approach could be screening performed on line with the separation. Such an approach has been developed by the company Screentec in The Netherlands (www.screentec.nl).

Although very powerful, strategy 2 has not yet been implemented on a large scale. Reasons may be its novelty but also the very high investment and running costs. However, as soon as more efficient on-line screening protocols and cost-effective automation become available, this may change and create a new era of natural product discoveries.

V. OPPORTUNITIES FROM TRADITIONAL MEDICINE

A review on discovery of novel products from plants should also briefly address the important role of plant extracts, which have been used in medicine since the beginning of mankind and still play an important role in modern medicine. The key issue for commercialization of herbal-based medicines is standardization and consistency of material. Adulteration or even contamination is a potential risk. However, recent successes of plant-derived products, increasing interest in Chinese and Indian medicine, and favorable regulations for commercialization have created a fast-growing market for herbal-based medicines and nutraceuticals (Table 2). In addition to most major pharmaceutical companies, several small companies are specializing in such products (a list of Internet addresses is given at the end of the reference list).

The early recipes used in traditional medicine were often the source of pure bioactive compounds used in pharmaceutical products. First examples of active principles identified from plants include morphine from *Papaver somniferum*, quinine from *Cinchona* sp., and salicin (salicylic acid) from *Spirae ulmaria*. More recent examples are atropine from *Atropa belladonna*, vincristine and vinblastine from *Catharanthus roseus*, and cardiac glycosides from *Digitalis purpurea*. Another famous example is the anticancer drug taxol, introduced by Bristol-Myers-Squibb in 1993. The great commercial success of this compound led to some renewed interest in plant secondary products and additional drugs have been successfully launched, e.g., anticancer derivatives of camptothecin, originally isolated from the Chi-

TABLE 2 Selection of Currently Most Popular Botanical Medicines and Some Examples of New Products Being Reviewed by the Food and Drug Administration

Plant species	Activity or indication area
Ginkgo biloba	Stimulation of peripheral circulation; cognitive impairment[a]
Echinacea spp.	Immunostimulant
Allium sativum	Cholesterol reducing
Panax ginseng	General stimulant
Tanacetum parthenium	Migraine prophylaxis
Valeriana officinalis	Calmative, sleep inducing
Hypericum perforatum	Antidepressant[a]
Chinese herb formula	Menopausal hot flushes,[a] plantar warts,[a] eczema[a]

[a]New products being reviewed by the FDA. For references, see Refs. 2, 3, 5, and 7.

nese tree *Camptotheca acuminata* (1) or galanthamine from *Galanthus woronowiis*, a reversible cholinesterase inhibitor for treatment of Alzheimer's disease. There have been a number of excellent reviews on natural products as pharmaceutical compounds or lead structures for drug development (1–9) (see also other chapters in this book).

VI. OUTLOOK

Use of plant-based diversity has been historically successful in the discovery of new drugs, either for developing crude mixtures or as a source of discrete compounds. As there are estimated to be approximately 250,000 plant species in the world and probably only 10% have been tested for some type of biological activity, plants will continue to be a rich source of novel structures with desired biological activity. Such secondary products may also be used as scaffolds for further structure modification using combinatorial chemistry approaches. Strategies for discovery of novel products from plants have to be integrated into today's industrialized discovery process.

In order to be successful in using plant secondary products in an industrial discovery process, several approaches are possible:

1. Generation of a large number of extracts obtained under reproducible conditions and from diverse and reliable sources in a factory-like setup.
2. Generation of a large number of diverse *pure* natural products that can be submitted to the standard screening procedures.

3. Further development of on-line separation, screening, and structural elucidation.

As structural bioinformatics and virtual screening will become more and more important in the future, the structural elucidation of natural products should be as early as possible during the discovery process.

REFERENCES

1. AL Harvey. Medicines from nature: are natural products still relevant to drug discovery? Trends Pharmacol Sci 20(5):196–198, 1999.
2. K Ten Kate, SA Laird. The Commercial Use of Biodiversity: Access to Genetic Resources and Benefit Sharing. Earthscan Publications, 1999.
3. J Gwynn, PJ Hylands. Plants: a source of new medicines. Drug Discovery World 2:54–59, 2000.
4. NR Farnsworth. Ethnobotany and the Search for New Drugs. New York: John Wiley & Sons, 1994.
5. WR Strohl. The role of natural products in a modern drug discovery program. Drug Discovery Today 5:39–41, 2000.
6. A Harvey. Strategies for discovering drugs from previously unexplored natural products. Drug Discovery Today 5:294–299, 2000.
7. R Verpoorte. Exploration of nature's chemodiversity: the role of secondary products as leads in drug development. Drug Discovery Technol 3:232–238, 2000.
8. S Grabley, R Thierike, eds. Drug Discovery from Nature. Berlin: Springer, 2000.
9. J Josephson. Trapping traditional healers treasures. Mod Drug Discovery 5: 45–50, 2000.
10. J Drews. Human Disease—From Genetic Cause to Biochemical Effects. Berlin: Blackwell, 1997.
11. SR Wilson, AW Czarnick, eds. Combinatorial Chemistry: Synthesis and Application. New York: John Wiley & Sons, 1996.
12. AD Baxter. Combinatorial chemistry. Drug Discovery World 2:41–48, 2000.
13. T Studt. Raising the bar on combinatorial chemistry. Drug Discovery Dev 1: 24–29, 2000.
14. R Thierike, S Grabley, K Geschwill. Automation strategies in drug discovery. In: S Grabley, R Thierike eds. Drug Discovery from Nature. Berlin: Springer, 2000, pp 56–71.
15. S Bertels, S Frormann, G Jas, KU Bindseil. Synergistic use of combinatorial and natural product chemistry. In: S Grabley, R Thierike, eds. Drug Discovery from Nature. Berlin: Springer, 2000, pp 72–105.
16. J Boguslavsky. Millennium streamlines gene sequencing. Drug Discovery Dev 5:28–34, 2000.
17. D Kniaz. Drug discovery adopts factory model. Mod Drug Discovery 5:67–72, 2000.

18. P Uetz, L Giot, G Cagney, et al. A comprehensive analysis of protein-protein interactions in *Saccharomyces cerevisiae*. Nature 403:623–627, 2000.

19. G. Lennon. Genomics. Using expression databases for drug discovery. Drug Discovery World 2:18–24, 2000.

20. J Burbaum. Miniaturization technologies in HTS: how fast, how small, how soon? Drug Discovery Today 3:313–322, 1998.

21. K Powell. Target discovery and drug design: extracting the value from genomics. Drug Discovery World 2:25–30, 2000.

22. M Beggs, H Blok, A Diels. The high-throughput screening infrastructure: the right tools for the task. J Biomol Screening 4(3):143–149, 1999.

23. KJ Moore, S Turconi, S Ashman, M Ruediger, U Haupts, V Emerick, A Pope. Single molecule detection technologies in miniaturized high-throughput screening: fluorescence correlation spectroscopy. J Biomol Screening 4(6):335–353, 1999.

24. U Haupts, M Rüdiger, AJ Pope. Macroscopic versus microscopic fluorecence techniques in (ultra)-high-throughput screening. Drug Discovery Today HTS Suppl 1(1):3–9, 2000.

25. L Arbiola, M Chin, P Fuerst, R Schweitzer, MA Sills. Digital imaging as a detection method for fluorescent protease assay in 96-well and miniaturized assay plate formats. J Biomol Screening 4(3):121–127, 1999.

26. I Gibbons. Microfluidic arrays for high-throughput submicroliter assays using capillary electrophoresis. Drug Discovery Today HTS Suppl 1(1):33–37, 2000.

27. AM Maffia III, I Kariv, KR Oldenburg. Miniaturization of a mammalian cell–based assay: luciferase reporter gene readout in a 3 microliter 1536-format. J Biomol Screening 4(3):137–142, 1999.

28. LC Mattheakis, LD Ohler, Seeing the light: calcium imaging in cells for drug discovery. Drug Discovery Today HTS Suppl 1:15–19, 2000.

29. MA Verney, CM Suto. Discovery of a subtype-selective metabotropic glutamate receptor ligands using functional HTS assays. Discovery Today HTS Suppl 1: 20–26, 2000.

30. F DiCosmo, M Misawa. Plant Cell Culture Secondary Metabolism towards Industrial Application. Boca Raton, FL: CRC Press, 1996.

Interesting Web sites for companies specializing in natural products as pharmaceuticals or agrochemicals:

www.oxfordnaturalproducts.com
www.paracelsian.com
www.pharmaprint.com
www.herbtech.com
www.phytera.com
www.agraquest.com
www.phytopharm.com

4

Plant Cell and Tissue Culture Techniques Used in Plant Breeding

Peter M. A. Tigerstedt and Anna-Maija Niskanen
Department of Applied Biology, University of Helsinki,
Helsinki, Finland

I. INTRODUCTION

Plant breeding proper is applied science. It applies sciences, such as botany (taxonomy, morphology, physiology), genetics (cytology and ecological, population, quantitative, and molecular genetics), mathematics (biometry, statistics), chemistry (food, fiber, and fuel chemistry), and finally economics (agricultural, forest, and industrial economics).

Plant breeding is closely related to human prosperity, producing food, feed, fiber, and fuel. It is also closely related to human socioeconomic development, from subsistence farming through stages of societal development to industrial farming and forestry.

Plant breeding also has strong cultural heritage values in connection with the architecture of gardens, communities, and landscapes. Rural and urban landscapes both rely heavily on the use of cultivated plants that are the results of long-term plant domestication (man-made evolution spanning human generations) and short-term efficient plant breeding (genetic manipulations resulting in intellectual properties).

Depending on its socioeconomic status in the society, plant breeding is either a state-supported social undertaking, with its values interwoven into the state infrastructure, or a private enterprise, often with international coverage. Many forms of mixed structures in plant breeding are common, including state-subsidized breeding involving different kinds of nongovernmental organizations (NGOs). Generally, plant breeding can become a lucrative undertaking only in geographic regions that are optimal for the growth of certain food or nonfood crops and where the seed or plant market is large enough. Under more marginal conditions and in small cultivated areas, plant breeding is inevitably a state-supported social activity. Plant breeding typically includes agricultural crop plants, horticultural plants of various importance (vegetables, berries, fruits, flowers, ornamentals), and, during the last half-century, also industrial forest trees.

Plant biotechnology developed swiftly during the last quarter of the 20th century. With that, new tools have become available to plant breeding. It is the aim of this chapter to try to evaluate where, when, and how plant cell and tissue cultures can be adopted to practical plant breeding and also to plant production. It must be emphasized that plant biotechnology encompasses much more than merely cell and tissue culture techniques. Yet these techniques, widely adopted, have profound bearings on many of the phases in a plant-breeding program. These new techniques open a number of new vistas for plant breeders. These we will try to evaluate in the breeding context, keeping in mind what was said earlier of the many sciences adapted to a plant-breeding program.

We have reviewed two books that handle principles and prospects of plant breeding, namely *Plant Breeding—Principles and Prospects* (1) and *Principles of Plant Breeding* (2). From these two books and in addition from volume 44 of *Biotechnology in Agriculture and Forestry—Transgenic Trees* (3), we venture to look at cell and tissue culture techniques in the context of the following key plant and tree breeding items:

Plant Breeding—Principles and Prospects (1)
    Somatic hybridization
    Gene transfer to plants
    Micropropagation and somatic embryogenesis
    In vitro selection
    The need for a comprehensive plant breeding strategy
Principles of Plant Breeding (2)
    Breeding self-pollinated plants
    Breeding hybrid varieties of outcrossing plants
    Breeding clonally propagated plants
    Breeding hybrid varieties of selfing plants
Biotechnology in Agriculture and Forestry—Transgenic Trees (3)
        (articles therein have been reviewed to set following items):
    Cell and tissue culture versus other clonal propagation techniques

Forest tree population structures and the adoption of cloning
Ecological aspects in tree breeding
The clonal forestry option

We believe that the preceding book chapters and in addition our own experience in forest tree breeding will cover most aspects of how to use plant cell and tissue cultures. Taking this approach in the discussions, we know that repetitions are sometimes inevitable. We hope that the readers can tolerate this. Each item listed can be read independently; we do not refer to other items treated. There have been hundreds of articles and certainly tens of recent books that give the technological nitty-gritty of each cell and tissue culture technique that we are about to scrutinize. Therefore, we do not indulge in writing procedures or giving recipes. Rather, we venture to look at the new techniques from the viewpoint of their "comparative advantages" having in mind both socioeconomics and ecology. In doing so, we inevitably encounter restrictions on the use of certain culture methods. For instance, why use tissue culture for cloning if rooted cuttings does the job at a lower price? Why use somatic hybrids if sexual hybridization can be applied? Why approach somatic embryogenesis if seed propagation is easy? However, it is also true that many new methods used in plant biotechnology, e.g., transformations, protoplast fusions, or in vitro selections, do not function without adequate cell and tissue culture techniques. In such cases, we must look at how plant cell culture techniques are integrated in larger applications of biotechnology. In fact, the integration of new biotechnological methods in plant breeding programs is up front in our discussions. Finally, our view is global; we have used the Food and Agriculture Organization (FAO) 1995 study *World Agriculture: Towards 2010* (4) for setting the stage. We also wish to acknowledge the book *Plants, Genes and Agriculture* (5), whose illustrations and chapters have given us needed perspectives.

II. SOMATIC HYBRIDIZATION

Pelletier (6) introduces the use of somatic hybridization techniques in plant breeding as follows: "Somatic hybridization needs the 'isolation' of intact protoplasts, the interparental 'fusion' of these naked cells, the 'sustained divisions' of fusion products before or after their 'selection' and the 'regeneration' of plants.'"

The first time this was successfully done was in 1972, when Carlson and his group reported on parasexual plant hybridization in Solanaceae (7). Since then, the technique has been greatly refined and many protocols are available for producing fusion products. Basically, these protocols can be classified as chemical or electrochemical.

The somatic hybridization techniques should be used where sexual means do not function because of some form of incompatibility. Also, one must clearly define what characters may be useful to manipulate by fusion and how such hybrids or cybrids can be integrated in breeding programs. Generally, wide hybridization in plants is used in order to integrate or incorporate valuable traits between species or in some cases even between genera. It comes at the very beginning of a breeding program, generally to widen the scope of genetic variability, later selected for useful and balanced gene combinations. It can also be seen as a means of bridging between species, well exemplified in the bridging for disease resistance in barley (8). Such wide hybrids, whether sexual or parasexual, generally show distorted sexual balance, growth, and adaptation. Thus, they have to undergo a considerable period of "balancing" by either direct selection in populations or backcrossing to either parent or in some cases to a third parent. Therefore, such undertakings have recently been classified as "prebreeding" to underline their time requirement and separation from typical breeding programs. Looking at the economics of breeding programs, we have come across several cases in which plant breeders doubt that the extra economic burden of interspecific hybridization can be tolerated in plant breeding programs working in a competitive environment. It therefore seems more plausible that such activities should be carried out as direct extensions of national or international genetic resources units (gene banks), whereby the extra economic burdens are carried by national or international research centers, publicly funded.

At the end of the chapter on somatic hybridization, Pelletier points at the future possibilities. Thus, it can be stated that the technology has developed to a point at which plant regeneration from protoplasts is functioning in most of the major crop species, including the cereals. In conjunction with polyploidy, a number of new methods of transferring organelle genomes and combining cytoplasms with nuclei of widely differing species seems possible. One must keep in mind that from a plant breeder's point of view, the new techniques are not shortcuts to superior cultivars but they extend classical breeding programs. Still, these techniques must be used; they all widen the genetic variability that the plant breeder considers imperative for combining new genes for better cultivated plants in a future world that suffers from food, feed, fiber, and fuel shortages.

III.   GENE TRANSFER TO PLANTS

Through very intensive research, many shortcomings have been solved or diminished since 1993 and new techniques are developed almost monthly.

From the problems listed by Potrykus (9), we would like to repeat only those that have direct bearing on the use of cell and tissue cultures in plant breeding.

Cell walls are perfect physical barriers and must be removed to introduce foreign DNA in one way or another. Thus, protoplasts become the targets for introduction. Here protoplast regeneration back to embryonic tissues and ultimately to plants is the complication, functioning in some plant species but not in others. Potrykus mentions the existence of many obstacles that, as he says, "still limits efficient procedures for the recovery of sufficient numbers of independent transgenic plants which retain their varietal identity." However, he is optimistic that gene technology will become an integral procedure in plant breeding in the foreseeable future. In fact, many new techniques used in gene transfer have become more "gentle" to the living material, avoiding the use of protoplasts and thus eliminating the limiting procedure of protoplast regeneration to cells with walls and eventually to plants. We like to project this foreseeable future in plant breeding programs and ask, how does it affect tenures of plant breeding programs and ultimately their economics?

Gene transfer to many plant species is routine today, thanks to the development of commercial laboratory equipment and kits including all necessary chemicals and vehicles. Thus, in a plant breeding program, little time needs to be spent on technical matters. However, finding desirable single genes in donors that will work well in the recipient is problematic in a number of ways. The integration into the recipient genome is still today very much at random, often causing distortion of the growth and yield of the recipient. Just as in wide crosses, mentioned earlier, it may take many generations of genome balancing before the transgene becomes well integrated and the transgenic plant retains normal growth and development. This considerable time required should not be added to a conventional breeding program. It is now more than ever reasonable to separate gene mapping, gene tagging, and gene transfers to *genomics*, the manipulation of the genome. Just as in the research on genetic resources, it is of considerable importance to plant breeding, but it has to be excluded from a plant breeding program because of its ultimate target: genome research (not plant breeding). Commonly, this activity is now called *genetic enhancement* or *prebreeding*.

Cell and tissue culture techniques are, of course, of central importance in the regeneration of whole transgenic plants. Once a desirable gene is duly integrated in the recipient genome, cell cultures may again be used in cloning whole transgenic plants. It is obvious that the most diligent use of transgenic plants in producing new cultivars is in the domain of clonal plants, such as the potato. If, on the other hand, the transgenes must be integrated in autogamous plant cultivars, it will take a good deal of time and effort before the transgene is passed on to the whole seedlot. It will be even more difficult to integrate the transgene into allogamous cultivated plants because (A) backcrossing in most cases is out of the question due to inbreeding depressions and (B) the integration of the transgene in the whole population must inevitably cause linkage discrepancies as genes closely linked to the transgene are dragged along, eventually also causing inbreeding problems.

A large number of perennial woody plants, both gymnosperms and angiosperms, can now be genetically transformed (3). Going through this recent compilation, one finds that the step after gene transfer is again cell or tissue culture and regeneration, if possible from embryogenic cell lines. Most trees are highly allogamous and suffer from severe inbreeding depression. Therefore the only plausible path from the transgenic tree to plantations is through cloning of the ortet, thus introducing ramets into clonal tree plantations. In the case of, e.g., poplars and aspens (*Populus* sp.), which frequently form clones from root sprouts under natural conditions, this may be an ecologically sound means of tree breeding. But even in this case, one may run into problems of using only a few selected transgenic ortets, actually originating from a single batch of transgenes, all initiated from the same initial source of "transgenic infection" of, say, a leaf tissue. The random integration of the transgene into the genome may give rise to several "transgenic lines." However, these lines are genetically identical with the exception of the position of the transgene. Therefore they do not represent genetically diverse lines as understood by population genetics. Particularly in conifers, the adoption of clonal forestry of transgenes may have unpredictable ecological repercussions. One would need the transgene integrated in several different ortets (genotypes) and then grow a ramet mixture in order to retain genetic diversity in long-living tree plantations. It is the risks of host-parasite disequilibriums in large practical plantations that cause major concerns.

IV. MICROPROPAGATION AND SOMATIC EMBRYOGENESIS

Breeders of woody plants, including ornamentals, fruit trees, and forest trees, are especially inclined to adopt micropropagation in their breeding programs. Clearly, this technique has reached practical application and can in some cases considerably shorten the tail end of a breeding program: the production of material used in cultivation. In some of the most drastic cases, mutations found in natural populations can be directly micropropagated for usage, e.g., in propagating ornamental trees of special form or color. In Finland this technique has found direct application in propagating various birch clones (curly birch, laciniate forms, high-yielding single pair matings) and hybrid aspen clones. However, the advantages of micropropagation must be evaluated economically in comparison with other cloning techniques, such as root and shoot explants or direct rooting of cuttings. It appears in many cases that the more conventional rooting of cuttings is simply less expensive and therefore applicable to forest trees, where the magnitude of cloning is frequently in the million ramets range.

Economic comparisons of cloning techniques must be viewed in the light of final plant prices. For example, plants produced for forestry plantations must sell for a maximum $1 apiece as the landowner needs to plant stands of approximately 2000 plants per hectare, whereas ornamental rhododendron plants may sell for $20 apiece as the landowner considers only single plants and not stands. Thus, the extra price added through micropropagation is essential in case of a forestry plant but futile in the case of an expensive ornamental. The final method of cloning plants depends on the numbers required and unit prices.

The dream of every plant and tree breeder is the application of somatic embryogenesis to a breeding program. First, in this way one could avoid the tissue and plant aging phenomenon, which causes many problems particularly in woody species, including trees. It may cause plagiotrophic growth of tissues taken from mature trees as the basically totipotent cells are genetically fixed in developmental stages. Second, it may, in combination with cryopreservation, be used for keeping cell lines or tissues embryogenic and ready for multiplication once the same material has been through comparative field testing for yield or other special values for a number of years. Third, a somatically induced embryonic cell line or tissue can eventually be handled for the production of "artificial seed." This would be not only a seed but also a "clonal seed" that could be multiplied in great numbers. If the somatic tissue in addition has a desirable transgene, it would open great vistas for fast adoption of transgenic plants. Fourth, somatic embryogenesis could be used for "fixing heterosis" in clones. In this way one could circumvent the tedious construction of male sterility systems to produce hybrid seed. Basically, one would need only one heterotic plant for manufacturing large numbers of hybrid clonal seeds. There are several examples of vigorous growth in interspecific hybrids of trees. However, it has often been impossible in practice to utilize hybrid vigor because of difficulties in producing hybrid seed. This is the case in larch hybrids, such as *Larix decidua* × *L. gmelini japonica* (Fig. 1). To have practical importance, hybrid seed must be produced in hundreds of kilograms. This has not been possible by the seed orchard mode of breeding because of species differences in timing of flowering. Selection of superior larch hybrids, already present as mature trees, and the conversion of their somatic cells to an embryonic state would solve many present problems for practical forestry. Alternatively, cell lines

FIGURE 1   Shoot proliferation of larch hybrid (*Larix decidua* × *L. gmelini japonica*) through organogenesis in tissue culture. The culture was started from lateral buds of a 60-year-old tree in 1996. (Photo by A.-M. Niskanen 1997.)

or tissues could be kept cryopreserved until field testing indicates which hybrid genotypes to clone.

However, this section has to be ended with the statement that there is still a long way to go before things work on a practical scale. What you can manipulate in the research laboratory is a long way from practical application in a plant or tree breeding program. Eventually, however, with the large concentration on basic research at the moment, new useful applications will be adopted.

V.   IN VITRO SELECTION

The benefits of in vitro selection in plant cell cultures must be seen against the background of natural mutation rates, which form the basis of evolutionary development of plants. Typically, natural mutation rates are on the order of 1 mutant per 100,000–1,000,000 individuals/locus/generation. Thus, it takes a very large experimental plant population to find a spontaneous mutation for some desired character at a single locus. One could imagine an agricultural field of plants as an immensely large "petri dish." If the experimental plants were grown at a spacing of 25 cm, a hectare could contain 1,600,000 plants, which would be just over the critical 1/1,000,000. In addition, most mutations are recessives, so the chance of picking them up in a heterozygous state would be very small and only the dominant mutation would be easily distinguished. Of course, by using mutagenic agents this low natural mutation frequency can be much increased, but that would also introduce a situation in which lethals and sublethals would increase drastically.

The advantage of the field petri dish is, of course, that the selection is directly on the whole plant and not on single cells in suspensions or as protoplasts. There are many examples of successful selection of cells in culture for (1) biotic stress resistance (fungal disease) and (2) abiotic stress resistance (cold, aluminum tolerance, herbicide tolerance). However, very often, after regeneration to plants of the cell lines, this resistance is again lost or at least expressed together with abnormal plant development (10). Therefore the field petri dish may in many cases be the most appropriate unit for selection, for instance, if one desires to find tolerance of the kinds just mentioned. Merely growing the plant material under appropriate stress may be the best alternative. However, in vitro selection can be made much more effective than in vivo selection because the environmental error variation can be eliminated to a large extent. This superior precision, together with the whole sequence of plant transformation, including the use of genetic markers for finding the transgenes, has made in vitro selection in cell cultures an integral part of the production of transgenic plants. The uncertainty of retaining resistance after regeneration of selected cells or tissues and the risks of developing abnormal plants put in vitro selection in the prebreeding category, not to be directly integrated in a plant breeding program.

VI. THE NEED FOR A COMPREHENSIVE PLANT BREEDING STRATEGY

Man has moved from historical plant domestication, which took place as an integral part of human cultural development, to plant breeding, determined by Mendelian genetics, to "plant engineering," determined by the rapid recent development of molecular biology. It must be kept in mind that all cultivated plants have emerged in a historical perspective, adapting the "landraces" to human farming systems. This has taken thousands of years, the first signs of cultivated plants dating back more than 10,000 years. Only the past 100 years of plant breeding has been based on Mendelian genetics. During this period, yields of our most important crop plants, rice, maize, and wheat, have been doubled, doubled, and doubled again! When the "green revolution" came to Mexico less than 50 years ago, Mexican farmers grew 600–800 kg of wheat per hectare. With the integration of new wheat cultivars, resistant to lodging, into new farming systems, including irrigation, fertilization, and pest and disease management, yields rose to 6000–8000 kg/ha. A whole new farming system had to be built around the new wheat cultivars, the new ideotypes. It was shown that the miraculous increases in yield were due to the interaction of novel genotypes with novel environments, called genotype × environment interaction in statistical terms.

The green revolution caused threats to the environment from the ample use of water, fertilizers, and chemicals. It also caused genetic erosion, in which old landraces were replaced by a few new cultivars and many genotypes were lost. To counteract this development, genbanks were established all over the world.

At the moment, we are experimenting with even more precise and effective methods of changing our crop plants. We need to increase yields, we need to produce a better quality, and we need to keep cultivation risks down. Thus the new "greener green revolution" emphasizes the importance of sustainable yields, leaving the environment in a sustainable cultivation condition for new generations of humans.

The use of modern plant cell and tissue culture techniques in selection and ultimately in breeding new plants must be seen as a new promising and efficient tool in the hands of plant breeding. To make maximum use of the new technologies, they must be well integrated in plant breeding programs. New methods cannot replace conventional plant breeding, but they may enhance genetic improvement if properly integrated. At all times, the new methods must be evaluated for their comparative advantages over more conventional methods. If they are more effective, if they can cut expenses, and, above all, if new cultivars are superior in sustaining the environment in productive condition, then they should be adopted. However, to produce more but destroy less (of the environment) is not exclusively a plant breeding problem. New and more productive cultivars must be integrated in new and more environmentally benign farming systems in terms of the use of water, fertilizers, pesticides, insecticides, and mechanical implements. New efficient plant breeding must at all times consider its integration in such farming systems. No doubt, the question of sustainability in combination with higher yield levels in order to meet global human population increases and hunger contains a controversy. No new plant ideotype, whether produced by conventional plant breeding or by new biotechnology, can produce much more without more input of water and fertilizers. High-yielding new cultivars will thus inevitably cause environmental problems.

VII. BREEDING FOR SELF-POLLINATED PLANTS

Many of our major crop plants are autogamous, self-pollinators. The most outstanding are the grains (wheat, rice, barley, and oats) and the major legumes (soybean, bean, pea, and groundnut). It is projected that by 2010 the production of wheat and rice together will be 1194 million tonnes and thus make up about half of the world total cereal production of 2334 million tonnes (4).

Self-pollinators keep their natural populations in balance with ecology, pests, and disease by assembling plant populations that consist of pure-line mixtures. Occasional outbursts of cross-pollination between the lines create new genetic combinations. In this way self-pollinated species are minutely attuned to environmental conditions and can react immediately to changing conditions, such as disease outbreaks, by changing their population structure and line composition.

New autogamous cultivars must be superior, uniform, and stable to meet the requirements of the international "Plant Breeders Rights." Also, modern agricultural farming systems, including mechanized harvesting and postharvest technology, require uniformity of the crop. This is very much against ecological stability but has to be accepted in modern high-yielding and uniform-quality agriculture. Traditionally, breeders of self-pollinated crops have attained this uniformity by self-pollinating lines for about 6–10, generations after which almost total homozygosity has been reached and the crop is uniform enough to be registered as a new cultivar. How and where can cell and tissue cultures be used in such breeding programs?

As cell cultures and in vitro selection are an integral part of all gene transformation experiments, this is perhaps the outstanding usage of the technique. Here, however, it must be compared with conventional back-crossing techniques for transferring single genes. Clearly, if the desirable single gene transfers come from genes cloned from other unrelated organisms, sexually incompatible with the target receptor, then cell cultures and in vitro selection are the only alternative. However, if the desirable gene can be transferred by sexual hybridizing and backcrossing to the receptor cultivar, it may be less risky to use the conventional technique in spite of the fact that perhaps 5–10 generations of backcrossing are needed for the integration of the single gene. Using *Agrobacterium*-mediated or ballistic transfer may, of course, save time, as only the desired single gene is transferred to an otherwise homozygous line. However, it has been shown that single transgenes also normally need several sexual generations of integration in the receptor genome to function properly. This is particularly the case when the transgene comes from outside the receptor species' sexual range. Genes function differently, however, and the most economical mode of single gene transfer is therefore different from case to case. A single gene changing the lysine content in the storage protein of barley has been shown to lower total grain yields because of distortion of total storage proteins and consequent grain shriveling. Single genes that confer disease resistance, on the contrary, may have no detrimental effect on yield.

VIII. BREEDING HYBRID VARIETIES OF OUT-CROSSING PLANTS

Of our major crops, maize, rye, sunflower, rapeseed, sugar beet, and onion are originally allogamous, outcrossers. FAO projections to 2010 show that maize production alone will be an estimated 698 million tonnes or about 30% of total global cereal production (4). The dramatic yield increases that have evolved in maize started around 1930 with the utilization of heterosis or hybrid vigor (11). Since then, cytoplasmic male sterility techniques have been developed to produce hybrid seed at reasonable prices. At present one is working on another alternative means of "fixing heterosis," namely the transfer of apomixis to maize for asexual formation of seed. Virtually all the allogams just mentioned are the target of this new idea and astonishingly there is also an indication of hybrid vigor in typical inbreeders, such as rice and wheat, hybrid rice already having reached the market in China in the mid-1970s. It has been suggested that with seed costs falling it would now be profitable to grow hybrid rice on 70 million hectares worldwide, which is nearly half of the total global rice area of 145 million hectares (4).

   Apomixis would produce a "clonal seed" in which heterosis would be fixed, and growers could use their own seed for several generations and thus lower production costs. On the other hand, this is bad news for the seed industry, which experiments with various techniques that would force the grower to return to the seed shop each season. Protoplast and cell cultures may be used in many ways to produce apomicts. An ideal tool would be the use of somatic hybridization of protoplast to combine two genomes into a new apomictic polyploid (12). Plant cell culture of embryogenic cells (somatic embryogenesis) may be a useful complement to apomixis, perhaps even an alternative.

IX. BREEDING CLONALLY PROPAGATED PLANTS

Plants that naturally regenerate vegetatively, such as most tubers (potato, sweet potato, cassava), are particularly interesting plant breeding objects as their direct cloning is a rapid means of multiplication. Also, the plant breeder, contrary to using sexual seeds, can make use of the total genetic variation (broad sense heritability) that includes both additive and nonadditive sources of genetic variation. Clonally propagated plants do, of course, occasionally flower and set seed, and at this stage the breeder can select novel material with desirable trait combinations and then start cloning it for field testing. Plants that naturally regenerate vegetatively are usually allogamous in their sexual reproduction, so that new gene combinations can be formed in bursts of sexual regeneration alternating with the vegetative mode.

Not only tubers are clonally regenerating. Many trees have the option of root sprouting and thus, e.g., certain aspen and poplar species can form large natural clones, including thousands of ramets. Clonal forestry has become a realistic option in the breeding of certain forest tree species.

Traditionally, fruit breeding (apples, pears, plums, apricots, peaches, oranges, grapefruits) ends in cloning, either by grafting on rootstocks or rooting of cuttings. The same techniques have been used in propagating selected woody ornamentals, such as lilacs, rhododendrons, and roses.

In modern breeding of clonally propagated plants, traditional cloning techniques, as mentioned earlier, may be improved by some form of micropropagation. However, micropropagation may often be a more expensive method than the traditional rooting of cuttings. All depends on the numbers to be cloned and on the market price of the cloned product. In cloning forest trees, the numbers to be produced are in the millions of ramets. A forest plantation often has a stand density of about 2000 individuals per hectare. To be of economic importance, plantation forestry supplying wood for industrial fiber use must be applied on thousands of hectares. For 100,000 hectares of cultivated clonal forestry, 200 million ramets would be needed. The individual forest tree ramet to be planted must be produced at reasonable costs, perhaps about $1–2 apiece, which would make plant costs per hectare $2000–4000. This could be a tolerable cost for the landowner. When conventional forest tree seedlings are used, the plant costs per hectare are roughly half of the ramet prices. On the other hand, a single fruit tree or ornamental rhododendron plant may sell on the market for $10–20. The landowner is willing to pay this price because a fruit tree orchard may be planted at a density of 500 trees per orchard hectare and the annual yield sells for a regular high price. Likewise, the owner of a small garden has only a few fruit trees or rhododendron plants in her garden. Thus, the initial numbers of ramets to be cloned may stay much lower than in industrial forestry, but the piece price is 10 times higher. So the costs of cloning may also be much higher and new micropropagation methods may become realistic.

X. BREEDING HYBRID VARIETIES OF SELFING PLANTS

Hybrid rice is an amazing story. Chinese farmers first grew rice hybrids in the mid-1970s. In 1992–1993 it was grown on 19 million hectares in China and it was spreading to many other countries in Southeast Asia. In the Philippines, the International Rice Research Institute (IRRI) has recorded yields over 10 tonnes/ha and the upper limit may well be at 16 tonnes, approaching similar record yields of wheat in Mexico. The limiting factor in using hybrid rice has been the high seed costs. The problem is equivalent to the problem of using cloned trees in industrial forestry. In the 1970s yields of hybrid seed were 1–1.5 t/ha; the yield increased to 2.3 t in the 1980s and to 4.5 t in 1991–1992 with a top yield of 6.35 t/ha. Because of the decreasing hybrid seed cost, hybrid rice can now be grown profitably on nearly half of the total global rice area of 124 million ha (4). The more effective hybrid rice seed production has been based on photo- and thermosensitive genetic male sterility. Now there is a strong research thrust to develop apomictic hybrid seed production using transgenes and protoplast fusion. This would enable farmers to use their own hybrid seed for several rounds of production, which would virtually eliminate the seed cost problem. Clearly, the hybrid rice model could be adopted for other autogamous grains, foremost in the production of hybrid wheat.

XI. CELL AND TISSUE CULTURE IN TREE BREEDING

Because of the long generation cycle, tree breeding using conventional methods, such as seed orchards, is inevitably time consuming. From the first selection of superior phenotypes, in natural forests, to seed production in orchards may require 20–30 years in temperate and boreal regions and 10–20 years in tropical and subtropical regions. Such long-lasting programs are generally economically possible only when financed by the forest industry or by state resources. Very little conventional tree breeding is managed as truly profitable private forest tree breeding. It is here that cell and tissue culture (commonly micropropagation) may substantially shorten breeding programs, make them more efficient, and form the basis for private enterprises.

First, it may be possible within natural tree variation to find genotypes that carry valuable gene combinations directly without extra time for breeding. Such is the case in curly birch (*Betula pendula* var. *carelica*), which is particularly valuable for its ornamental wood structure. The inheritance of wood curliness is still unclear. It does not seem to obey Mendelian laws, and it may well be a growth anomaly caused by transposons. In this case the direct cloning of curly "genotypes" has proved quick and successful.

Second, in tree breeding a few individuals are easy to produce by artificial hand pollination. Examples are the hybrids in *Betula*, *Populus*, *Eucalyptus*, *Cryptomeria*, and *Larix*. However, from a few heterotic hybrid individuals to forests of hybrids is a long leap. The only reasonable technique here is some form of vegetative propagation, "fixing heterosis." It works well in the four first mentioned genera but so far not well enough in larch. Again, what works reasonably well in the research laboratory does not necessarily work well enough in a cost-critical practical application where plants for cultivation must be produced at competitive prices.

What about transgenic trees? What would the choice genes be? Herbicide resistance, drought tolerance, cold tolerance, lignin reduction, apical dominance? Whatever single gene would be transferred to a tree individual, it would still take a great effort to clone that transgenic tree to a large number of ramets. Here is the difference between an annual agricultural crop and a perennial tree with a long generation interval. A new transgene in an annual can be reproduced sexually until enough seed is available, but in case of a tree, whole plant cloning is the only reasonable alternative.

XII.   FOREST TREE POPULATION STRUCTURES AND THE ADOPTION OF CLONING

Almost all forest trees are allogamous cross-pollinators. Most of them are monoecious, but in some genera both monoecious and dioecious species may be found, as in the genus *Acer*. Other genera are exclusively dioecious, such as *Populus*. In many of the monoecious species there are self-incompatibility mechanisms that render progeny after selfing most unlikely. In most important conifers, selfing is possible, but many recessive lethal or sublethal genes lead to severe inbreeding depression. Thus, natural forest tree populations are as a rule highly heterozygous and differences between individuals within populations are generally large for a number of quantitative measurable traits, such as phenology, growth habit, and resistance to disease and pests.

Analyzed marker loci (isozymes, RFLP, RAPD, AFLP, ISSR) commonly indicate between 1 (fixation) and 10 alleles per locus. By retaining high heterozygosity and high variation between individuals in populations, perennial long-living trees buffer themselves against environmental variation (cold, heat, drought, etc.) and attacks by pests and disease (animal browsing, insects, fungi, viruses, etc.).

The individual cloning of such variable forest tree species for large-scale cultivation of industrially superior genotypes may be ecologically hazardous as it will inevitably reduce the genetic variation present in natural stands, thus causing instability particularly concerning disease and pest resistance. The whole tree cloning of transgenics may cause similar instability. The use in plantations of clonal mixtures has been discussed for years in forest genetic fora. There is no definite consensus on how an artificial clonal forest plantation should be built. Opinions vary from "just a few clones" (3–5) to "many clones" (10–30) per plantation. A plantation is an area of several hectares comprising more than perhaps 10,000 plants. However, it must be emphasized that some tree species have natural "clone structures." For instance, the European aspen, *Populus tremula*, and the North American trembling aspen, *Populus tremuloides*, both form clone structures that may cover more than a hectare a patch. These structures come from root sprouting of older trees. A recent unpublished study in Finland has shown that natural aspen stands may consist of between 2 and 10 clones. It suggests that clonal aspen or hybrid aspen plantations could be arranged in a "chessboard" structure, each clone grown in pure monoculture of say 1/4 ha or about 500 ramets of a selected ortet. A 1-ha plantation would consist of 4 clones and in larger plantations, clone numbers could increase to 10 and the same clone could be repeated in another location of the large plantation area. Trees that do not form clones under natural conditions, such as conifers, should, however, be grown as clonal mixtures, thus imitating natural stands with their genetic variation between individuals.

At this time there is not enough evidence from clonal plantations concerning the risks of growing clones in monoculture. Trees grow for tens of years, but insects and pests often have generation cycles of 1 year or less. Resistant clones may at any time be victims of new disease and pest strains that break resistance. This could be fatal to clonal forestry. We do not yet know what the risks are, but observations on agricultural plants and theoretical modeling suggest epidemics and instability in cultivated plant ecosystems if genetic uniformity is allowed. The classical example is the spread of the corn leaf blight (*Helminthosporium maydis*) in hybrid maize cultivations in the United States due to the use of a uniform male sterility cytoplasm (T-cytoplasm) in the production of hybrid seed (13).

XIII. ECOLOGICAL ASPECTS IN TREE BREEDING

Cell and tissue cultures adopted to tree breeding may greatly enhance genetic gains in characters such as wood quality (fiber length, lignin composition, specific gravity, bole straightness, crown form and branchiness, etc.) and wood volume (springwood/latewood ratio, growth initiation and cessation, nutrient uptake, drought and cold tolerance, etc.). Manipulation of wood quality and volume, all included in final yield, may, however, be contradictory to tree adaptation in a certain climate; additional yield may be at the cost of tree adaptation and fitness. During the past 40 years the world has experienced a considerable reduction of genetic diversity in agroecosystems with the advent of the "green revolution" with its high-yielding cultivars in rice, wheat, and maize, the three outstanding agricultural staple crops. Its consequences have been duly registered in pest and disease outbreaks.

Pressures on world forestry, particularly in developing countries that suffer shortages of fuelwood and where high population density causes forests to be turned into agriculture, there is a loss of tropical forests at about 15 million hectares per annum, 0.8% of the total tropical forest area (4). In developed industrial countries, increasing forest area is converted into plantation forestry, with a loss of natural forest biodiversity (stand variation in age classes and species mixtures). Forests are replaced by even-aged mono-cultures, easy to manage and high in production. Thus, trees have been converted to "cultivated plants." This is where we select species, seed sources, and tree breeding material and ultimately select clones for high production. It is here that transgenic trees may become cloned for mass production.

XIV. THE CLONAL FORESTRY OPTION

The industrial forestry sector has become aware of the risks involved, and there has been an encouraging change to more ecological thinking in the management of plantations. Where tree clones are used, their advantages are now weighted against increasing risks. There are several ways to avoid genetic narrowing in clonal forestry, the first being the use of clonal mixtures. But when planning to use cloned transgenic trees, one runs into problems. The transfer of a special gene to a certain location in the genome is still uncontrollable and therefore, if the same gene is to be transferred to several different genotypes, it ends up in different places in the genome, it functions differently, and it may influence adjacent genes differently. In annual crops this is not a particular problem, because the transgene can be backcrossed several times to different genotypes and thus the location is fixed and the gene duly integrated. Such backcrossing is prohibitive in trees simply due to their long generation interval but also due to their almost obligate cross-pollination with severe inbreeding depression. One is therefore forced to whole tree cloning when the desirable gene has been transferred.

Ultimately, one ends up having a single desirable transgenic clone that should go into plantation. There are, however, solutions to how such single clones could be used. First, they could be mixed in plantations with other nontransgenic clones according to regulations set for clonal forestry. The most valuable transgenic tree clone could thus be marked in the stand and left standing through stand thinnings until the harvesting of the most valuable end product at final stand rotation. Or a valuable transgenic clone could simply be planted among normal sexual tree seedlings, which would keep genetic variation at an acceptable "natural" level.

Whatever method is used in productive industrial forestry plantations, if the yield of the cloned trees is much superior to that of their wild relatives, there may be a possibility to grow industrial forests on a "plantation basis," leaving a larger share of natural forests as gene reservoirs. It is a fact that we are just now experiencing how forest trees are converted to cultivated plants. Simultaneously, the world is experiencing clear signs of global climatic change that may upset the adaptation of long-lived trees. Clonal forest stands have lost their ability to respond genetically to such change. Natural forests with high degrees of genetic variability have the capacity to respond dynamically to environmental change by genetic selection during natural regeneration. This is why natural forests must be managed and regenerated for a changing future environment.

REFERENCES

1. MD Hayward, NO Bosemark, I Romagosa, eds. Plant Breeding—Principles and Prospects. London: Chapman & Hall, 1993.
2. RW Allard. Principles of Plant Breeding. 2nd ed. New York: John Wiley & Sons, 1999.
3. YPS Bajaj. Biotechnology in Agriculture and Forestry 44: Transgenic Trees. New York: Springer-Verlag, 2000.
4. N Alexandratos. World Agriculture: Towards 2010. An FAO Study. New York: John Wiley & Sons, 1995.
5. MJ Chrispeels, DE Sadava. Plants, Genes and Agriculture. Boston: Jones & Bartlett, 1994.
6. G Pelletier. Somatic hybridization. In: MD Hayward, NO Bosemark, I Romagosa, eds. Plant Breeding—Principles and Prospects. London: Chapman & Hall, 1993, pp 93–106.
7. PS Carlson, HH Smith, RD Dearing. Parasexual plant hybridization. Proc Natl Acad Sci U S A 69:2292–2294, 1972.
8. M Veteläinen, E Nissilä, PMA Tigerstedt, R von Bothmer. Utilization of exotic germplasm in Nordic barley breeding and its consequences for adaptation. In: PMA Tigerstedt, ed. Adaptation in Plant Breeding. Dordrecht: Kluwer, 1997, pp 289–295.
9. I Potrykus. Gene transfers to plants: approaches and available techniques. In: MD Hayward, NO Bosemark, I Romagosa, eds. Plant Breeding—Principles and Prospects. London: Chapman & Hall, 1993, pp 126–137.
10. G Wenzel, B Foroughi-Wehr. In vitro selection. In: MD Hayward, NO Bosemark, I Romagosa, eds. Plant Breeding—Principles and Prospects. London: Chapman & Hall, 1993, pp 353–370.
11. AF Troyer. Breeding widely adapted, popular maize hybrids. In: PMA Tigerstedt, ed. Adaptation in Plant Breeding. Kluwer, 1997, pp 185–196.
12. APM den Nijs, GE van Dijk. Apomixis. In: MD Hayward, NO Bosemark, I Romagosa, eds. Plant Breeding—Principles and Prospects. London: Chapman & Hall, 1993, pp 229–245.
13. National Academy of Sciences. Genetic Vulnerability of Major Crops. Washington, DC: NAS, 1972.

5

Plant Cell Cultures as Producers of Secondary Compounds

Kazuki Saito
Graduate School of Pharmaceutical Sciences, Chiba University, Chiba, Japan

Hajime Mizukami
Faculty of Pharmaceutical Sciences, Nagoya City University, Nagoya, Japan

I. INTRODUCTION

Because plant secondary compounds are often produced only in small quantities in a particular type of cells of rare plant species, it is not always feasible to isolate secondary compounds from intact plants. Plant cell culture can be an alternative way to produce these compounds continuously under artificially controlled conditions. In particular, the production of pharmaceutically important plant metabolites has been a target for practical application of plant cell culture for the last few decades. Although not all attempts at practical production have been fully successful so far, several compounds, i.e., shikonin, berberine, and ginseng saponins, have been commercially produced from in vitro cell cultures.

In this chapter, we describe plant cell cultures for production of secondary metabolites, how cell cultures can be established, and which factors affect producibility of the metabolites and present several case studies of cell culture production of pigments and clinically used antineoplastic compounds.

77

II. CELL CULTURE SYSTEMS USED FOR PRODUCTION OF PHYTOCHEMICALS

A. Callus and Cell Suspension Cultures

Cell cultures are induced as callus tissues. Callus cultures are also usually the materials from which cell suspension cultures are obtained. In addition, selection of high-producing cell lines for a particular secondary metabolite is carried out using callus tissues of either small-aggregate or single-cell origin. Suspension cultures are established by transferring the callus tissue into liquid medium of the same composition as used for callus tissues and agitating the culture on a rotatory or reciprocal shaker. Suspension cultures generally comprise more homogeneous and less differentiated cell populations and grow more rapidly than their parent callus cultures. Furthermore, cell suspensions are suitable for continuous and/or chemostat culture and easy to feed various chemical factors during the culture. These properties make suspension cultures the material of choice for biochemical and molecular biological investigation of plant secondary metabolism. Scaling up from flask to bioreactor for the production of phytochemicals is always performed using suspension cultures.

B. Immobilized Cultures

Over the last two decades, immobilized culture systems have attracted much attention for efficient production of plant secondary metabolites. Cultured cells from high-density suspension cultures are trapped in an inert matrix such as calcium alginate gel beads, stainless steel, and foam particles. The immobilized entities are cultured in shaken flasks or aerated bioreactors. Alternatively, the cell-encapsulated beads can be packed into a column, which is percolated with nutrient medium. The major advantage of the immobilized culture system is that cell growth and secondary metabolite production can be separated by the precise manipulation of the chemical environment, allowing continuous or semicontinuous operation (1). However, establishing the immobilized culture system for large-scale production of phytochemicals is expensive. In addition, for efficient operation of the immobilized system, permeation of the product from the cells to the medium is necessary, which has not yet been fully achieved.

C. Organ Cultures

In spite of prolonged and concentrated efforts, many valuable phytochemicals such as morphinan alkaloids of *Papaver somniferum*, tropane alkaloids of various solanaceous species, and dimeric indole alkaloids of *Catharanthus roseus* cannot be produced by callus and cell suspension cultures. Because most of these compounds start to accumulate when the proper organs are regenerated from the cultured cells, production of these compounds in cultured cells requires decoupling of biochemical differentiation from morphological differentiation, which has so far been unsuccessful. This situation makes organ cultures a favored option. One major disadvantage of organ cultures is reduced productivity in bioreactors because the physical structure of shoots or roots results in various difficulties including handling problems at inoculation and shear of the organs during culture.

1. Shoot Cultures

Multiple shoots regenerated either from callus cultures or directly from explants including apical buds are cultured in solid or liquid medium. Shoot cultures have been considered appropriate when the target secondary metabolites are produced in aerial parts of plants. Monoterpenoid essential oil flavors, which are not produced in dedifferentiated callus or cell suspension cultures because of lack of oil-secretory tissues, have been reported to accumulate in shoot cultures (2,3). Production of a sesquiterpene lactone artemisinin that exhibits potent antimalarial activity in shoot cultures of *Artemisia annua* has also been actively investigated (4,5). A dimeric indole alkaloid anhydrovinblastine that is a direct precursor of antileukemic indole alkaloids, vinblastine and vincristine, accumulated in shoot cultures of *Catharanthus roseus* at a level similar to that in the leaves of intact plants (6). Vindoline and catharanthine, precursors of the dimeric indole alkaloids, were also produced in multiple shoot cultures of *C. roseus* (7). In a few cases, multiple shoots transformed with *Agrobacterium tumefaciens* were used to investigate secondary metabolite production (8).

It is interesting to note that providing the cultured shoots with environments similar to those of the intact plants sometimes results in enhancement of the particular secondary metabolism. For example, menthol production in *Mentha arvensis* shoot cultures increased with light illumination (3). Dimeric indole alkaloid production in *C. roseus* shoot cultures was also stimulated by near-ultraviolet light irradiation (9). Rooting was also reported to enhance artemisinin formation in cultured shoots of *A. annua* (10).

2. Root Culture

There are two types of root culture, untransformed root culture and hairy root culture, which are obtained by transformation with *Agrobacterium rhizogenes*. Production of phytochemicals by hairy root cultures has been intensively investigated and is reviewed in a separate chapter of this book. Hairy roots generally show more vigorous growth than untransformed roots. However, untransformed roots sometimes show vigorous growth to an extent similar to that of transformed roots when cultured in auxin-containing medium. Production of hyoscyamine and scopolamine, pharmacologically active tropane alkaloids, was more active in normal root cultures of *Duboisia myoporoides* than in the hairy roots (11). Scaling up of transformed *Atropa belladonna* root cultures was attained without any reduction in tropane alkaloid productivity by combining cutting treatment of seed roots with use of a stirred bioreactor with a stainless steel net (12). It should also be pointed out that the research on tropane alkaloid production in root cultures of solanaceous plants starting in the mid-1980s has fruited as molecular biological characterization and genetic engineering of tropane alkaloid biosynthesis (13).

D. Bioreactor Cultures

Irrespective of suspended cells, immobilized cells, or whole organs, it is necessary to establish efficient large-scale bioreactor systems for commercial production of secondary metabolites (14). Recent advances in this field are described in a separate chapter.

III. FACTORS AFFECTING SECONDARY METABOLITE PRODUCTION BY PLANT CELL CULTURES

A. Plant Growth Regulators

Effects of plant growth regulators, especially auxin and cytokinin, on secondary metabolism in cell cultures have been extensively investigated. It is well known that auxin is essential and cytokinin is preferable to induce cell dedifferentiation and to maintain cell proliferation in vitro. In has also been widely recognized that the concentration and balance of auxin and cytokinin affect organ regeneration from cultured cells. These growth regulators regulate secondary metabolism in in vitro–cultured cells probably through controlling cell differentiation. However, the effects of auxin and cytokinin are variable from species to species and from product to product, and the mechanism by which the plant growth regulator up- or down-regulates the particular secondary metabolism is not clear in most cases.

Gibberellin is usually not added to culture medium, and only a few reports describe its effect on natural product biosynthesis. Production of berberine in *Coptis japonica* cell cultures was increased by gibberellin (15). In contrast, gibberellin inhibited shikonin biosynthesis in *Lithospermum erythrorhizon* cell cultures (16).

For practical application of plant cell cultures to secondary metabolite production, it is desirable to culture the cells without phytohormones, especially when the product is used as a crude extract, for example, in the case of anthocyanin, because contamination of the phytohormones from the culture medium may influence human health. In mammalian cell cultures, hybridomas produced by fusion of antibody-producing cells having no proliferation activity with highly proliferative myeloma cells are used to produce monoclonal antibodies. A similar approach could be possible in plant cell cultures. Actually, protoplasts from petals *of Petunia hybrida* were fused with protoplasts from cultured crown gall tumor cells, and microcalli thus produced grew vigorously on hormone-free medium and formed anthocyanin characteristic of parent petals (17).

B. Medium Nutrients

Optimization of medium nutrients is important to increase the productivity of the particular secondary metabolites. There are a number of reports describing the effects of medium nutrients on secondary metabolism in plant cell cultures. Many of these investigations seem to indicate a negative correlation between cell proliferation and secondary metabolism. It might be possible that any manipulation for inhibiting cell growth leads to an increase in the productivity of secondary metabolites, leading to establishment of a two-stage culture system for production of phytochemicals where the cells are first cultured in the medium appropriate for maximum biomass production and then transferred to the growth-limiting medium for maximum productivity of secondary metabolites as established for shikonin production in *Lithospermum erythrorhizon* cell cultures (18).

One of the most important nutritional factors is the phosphate level in the medium. Since Nettlership and Slator (19) first reported in 1974 that use of a phosphate-free medium increased the alkaloid productivity of *Peganum harmala* cells, it has been recognized that reducing the phosphate concentration results in growth limitation and a concomitant increase in the level of secondary products.

Nitrogen is essential to support cell growth as a source of protein and nucleic acid synthesis, thus affecting secondary metabolism. Generally, culture medium contains N sources as NH_4^+ and NO_3^-, and both the concentration of total nitrogen and the ratio of NH_4^+ to NO_3^- regulate cell growth and secondary metabolism. In many cases, reducing the total nitrogen concentration in the medium leads to lower cell growth and higher product formation as typically reported for anthocyanin production by *Vitis vinifera* (20) cell cultures. However, cell growth and production of betacyanin in *Phytolacca americana* cell cultures increased with an elevated nitrogen supply (21). If used as a sole nitrogen source, NH_4^+ is often toxic to cell growth. Shikonin production in *L. erythrorhizon* cell suspension cultures was completely inhibited when the cells were cultured in NH_4^+-containing medium (22). It is important to find an optimum ratio of NH_4^+ to NO_3^- for attaining maximum production of secondary metabolites.

Sucrose is utilized most as a carbon source. In contrast to phosphate and nitrogen, an increase in the initial sucrose concentration in culture medium leads to an increase in secondary metabolite production. The enhancing effect of sucrose was most impressively shown in the case of rosmarinic acid formation in *Coleus blumei* cell suspension cultures, where the rosmarinic content increased sixfold in a medium containing 5% sucrose compared with that in the control medium (2% sucrose), reaching 12% of dry weight (23). This effect was not due to the higher osmotic pressure because addition of mannitol to low-sucrose medium did not increase rosmarinic acid production. In contrast, the stimulatory effect of sucrose on anthocyanin production in *Vitis vinifera* cell cultures was shown to be due to osmotic stress (24). The carbon-to-nitrogen ratio is also an important factor in secondary metabolism as shown by anthocyanin production in *Vitis* cell cultures (25).

Although less investigated compared with macronutrients, micronutrients are also expected to affect secondary metabolism. In fact, the shikonin content in *L. erythrorhizon* cell cultures increased drastically with an increasing Cu^{2+} level in the medium (22).

C. Elicitors

Elicitors are the active components in extracts of microbial and plant origin that induce defense responses when applied to plant tissues. The elicitors produced by microorganisms and plants are referred as *biotic elicitors*, while physical and chemical stresses such as ultraviolet (UV) irradiation, heat or cold shock, and heavy metals also induce a wide range of defense responses and are defined as *abiotic elicitors*. Abiotic elicitors are thought to induce the release of biotic elicitors from plant cell walls. It has been shown that elicitors are capable of not only inducing de novo formation of phytoalexins but also activating biosynthetic potentials of various constitutive metabolites in cultured plant cells. Production of sequiterpene gossypol in *Gossypium arboretum* was increased over 100-fold by elicitors prepared from *Verticillum dabliae* elicitors (26). Elicitor treatment increased the biosynthesis of the benzophenanthridine alkaloid sanguinarine 26-fold in *Papaver somniferum* cell cultures (27). Induction of isoflavonoid biosynthesis in *Pueraria lobata* cell cultures by either a biotic elicitor yeast extract or the abiotic elicitor $CuCl_2$ has also been extensively investigated, especially at the molecular level (28).

Elicitors provide important clues to understanding the molecular basis of the transducing pathway through which exogenous signals lead to secondary product biosynthesis, involving various signal compounds such as reactive oxygen species, jasmonic acid, Ca^{2+}, and phosphoinositides. Induction of secondary metabolism by elicitors in cell suspension cultures of

various plant species was correlated with earlier rapid and transient accumulation of jasmonic acid and its methyl ester methyl jasmonate, and jasmonic acid was proposed to be a key signal compound in the cellular process
of elicitation leading to the accumulation of various secondary metabolites
in the cultured plant cells (29). Production of various phytochemicals including rosmarinic acid (30), alkannin (31), taxol (32), shikonin (33), and
stilbene (34) has been reported to be induced by jasmonic acid or methyl
jasmonate. A cDNA encoding geranylgeranyl diphosphate synthase, which
catalyzes an important biosynthetic step leading to taxol, was cloned from
Taxus canadensis cell cultures pretreated with methyl jasmonate to induce
taxol biosynthesis (35). This suggests that jasmonic acid (or its methyl ester)
may be used as an inducer of secondary metabolism in cultured plant cells
not only for practical application but also for basic research.

D. Physical Factors

Physical factors controlling secondary metabolite production synthesis in
cultured plant cells include light, temperature, medium pH, aeration, cell
density, etc. The effect of light on natural product biosynthesis is quite varied. Light illumination usually induces chloroplast differentiation, which
sometimes leads to elevation of secondary metabolism. A lupine alkaloid
lupanine was produced only in the green callus of *Thermopsis lupinoides*
cultured under light illumination (36). Light illumination is often essential
to induce anthocyanin biosynthesis, although its biosynthesis is not localized
in chloroplasts. In contrast, biosynthesis of nicotine in tobacco cells (37)
and shikonin in *Lithospermum erythrorhizon* cell cultures is inhibited by
light illumination (38). From a biotechnological viewpoint, a light requirement for secondary metabolite formation is problematic because it is difficult
to provide adequate illumination without affecting temperature.

The optimal culture temperature and medium pH are usually between
20 and 25°C and between 5.6 and 6.0, respectively. Although these factors
are expected to affect secondary metabolism in cultured cells, they have not
received much attention so far. Aeration is also an important factor to regulate both cell growth and product yield, especially in bioreactor cultures,
and is discussed later in this chapter and in a separate chapter. Cell density,
which is mostly determined by cell inoculum size, is also an important factor
affecting product yield. High-density cultures were established for berberine
production by *Coptis japonica* cultures (39) and anthocyanin production by
Perilla frutescens cultures (40).

E. Biological Factors

One of the most important factors controlling secondary metabolite production is cell-to-cell variation. Within a population of cultured cells there is a

difference in metabolic behavior, especially in ability to synthesize particular metabolites even if the cell population was induced from the same piece of explant and cultured in the same physical and chemical environments. Although a molecular biological basis for such cellular variation in secondary metabolism has not yet been clarified, it has been well recognized that selective subculture of cell aggregates whose content of the secondary metabolite is higher than others will eventually result in the isolation of so-called high-producing cell lines for a particular secondary metabolite. Such selection is easy to perform when the target compound is a pigment (41) but still possible for colorless compounds by using a convenient cell-squash method (42) or a semiautomated immunoassay (43).

Another important biological factor is stability of the biosynthetic capability of cultured plant cells. Alkaloid-producing cell lines of *Catharanthus roseus* that were established by repeated selection lost their biosynthetic ability during subcultures; the indole alkaloid content decreased 70-fold over 8 years of subculture (44). A similar kind of biochemical instability was also reported for nicotine in *Nicotiana rustica* callus (45), cardenolides in *Digitalis purpurea* callus (46), and cinnamic acid in *Capsicum frutescens* cell suspension (47). These results indicate that it is important to subculture the cells under selection pressure or with occasional reselection.

IV. PRODUCTION OF PLANT PIGMENTS

A. Anthocyanin

Anthocyanin constitutes a major flavonoid pigment. It is ubiquitous in the plant kingdom and provides scarlet to blue colors in flowers, fruits, leaves, and storage organs. Chemically, they are based on a single aromatic molecule, that of delphinidin, and all are derived from this pigment by hydroxylation, methylation, or glycosylation (Fig. 1). Interest in anthocyanins as food colorants has been increasing not only because they are less toxic than synthetic red dyes but also because they exhibit significant antioxidant activity, which might protect against cardiovascular diseases and certain cancers (48).

There have been a number of publications describing anthocyanin production by plant cell cultures; some of the more recent ones are summarized in Table 1. Progress in sophisticated spectroscopic technology such as fast atom bombardment mass spectrometry (FAB-MS) and two-dimensional nuclear magnetic resonance (2D-NMR) together with high-performance liquid chromatography (HPLC) separation has led to elucidation of complex structures of anthocyanin as reported in cell suspension cultures of *Perilla* sp. (49), *Daucus carota* (50), *Ajuga reptans* (51), and *Ajuga pyramidalis* (52).

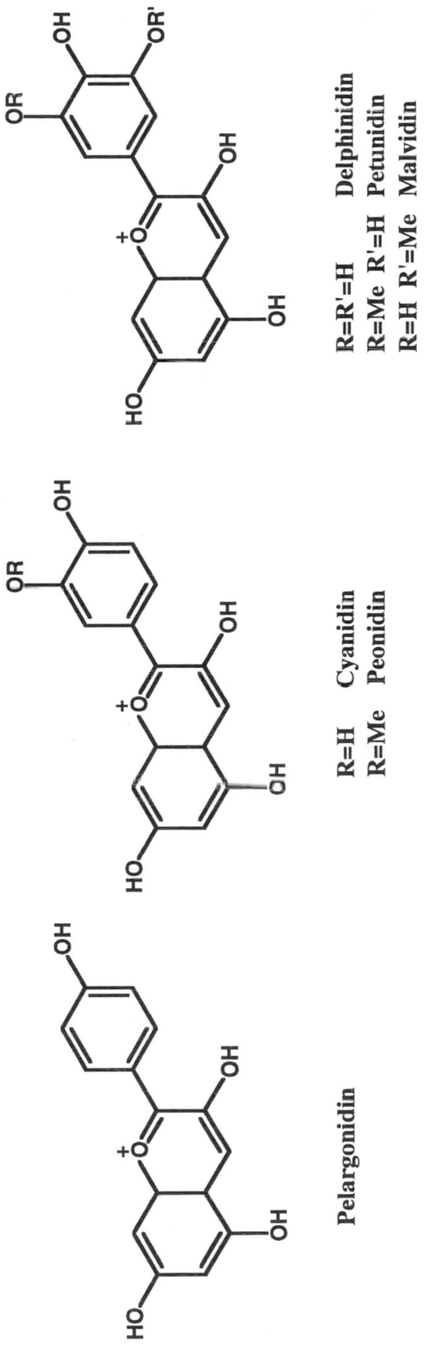

FIGURE 1   Chemical structures of six common anthocyanidins.

TABLE 1   Anthocyanin Production by Plant Cell Cultures Reported after 1990

Plant species	Culture system	Light	Productivity	Reference
Ajuga pyramidalis	Callus	Yes		52
Ajuga reptans	Callus/suspension	Yes	100 mg/L (s)[a]	76
				51
Aralia cordata	Callus	No	10.3% dw (c)	56
Daucus carota	Callus	Yes		50
			23.7% dw (c)	77
Fragaria ananasa	Callus	No	0.4% dw (c)	58
Hibiscus sabdariffa	Callus/suspension	Yes	3% dw (c)	69
Malus pumila	Callus/suspension	Yes	0.49% fw (c)	78
Oxalis linalis	Callus/suspension	Yes		79
Perilla frutescens	Callus/suspension	Yes	3.87 g/L (s)	40
Vitis hybrid	Callus/suspension	Yes	2.9 g/L (s)	63
Vitis vinifera	Callus/suspension	Yes	1.2 g/L (s)	54
				57

[a]c, in callus cultures; s, in suspension cultures.

Although it is suggested that polyacylated anthocyanins are stable in neutral aqueous solution and thus suitable for application as food colorants, the occurrence of such acylated anthocyanins in plant cell cultures has not been reported in many cases. However, this does not necessarily mean that plant cell cultures lack the ability to synthesize acylated anthocyanins but may mean that they have been overlooked because of the lability of the acylated pigments in the extraction solvent containing hydrochloric acid.

Because anthocyanin is relatively inexpensive, extensive optimization of anthocyanin production in cultured plant cells is necessary to reduce the production costs for commercial, including cell line selection, manipulation of the physical and chemical environment, process management of large-scale culture, and manipulation of the genome using genetic engineering.

1.   Cell Line Selection

Even when cell cultures accumulate anthocyanin, it is apparent that the cultures are mostly heterogeneous and the population of the pigmented cells is usually low. For example, only 10% of *Catharanthus roseus* cells in culture actively accumulated anthocyanin (53). However, because anthocyanins are colored compounds, it is relatively easy to perform visual cell aggregate selection by repeatedly subculturing red sectors of the callus tissues. Such cell aggregate selection was described most typically and systematically in

Euphorbia milli callus, whose anthocyanin levels increased sevenfold, reaching 13% of dry weight, and the capability was stable for 24 passages (41), and it has been applied to various anthocyanin-producing plant cell cultures. By small-aggregate cloning combined with HPLC analysis, a cell line of *Vitis vinifera* accumulating malvidin-3-glucoside, a main anthocyanin in most red wines, at 63% of total anthocyanins was established from an initial culture with a malvidin-3-glucoside level of 13% (54).

2. Physical Environment

One of the most important physical factors affecting anthocyanin biosynthesis is light. In many cases, strong light irradiation is required for anthocyanin production in cultured plant cells. However, for commercial exploitation of anthocyanin production by plant cell cultures, scaling up and efficient operation of photobioreactors with high light intensity are difficult and expensive, especially because light illumination always affects the temperature of the culture and thus generates the problem of cooling. Hiraoka et al. (55) succeeded in establishing a cell line of *Bupleurum falcatum* capable of producing anthocyanin in the dark by selectively subculturing red portions of the callus tissue that eventually appeared in the dark. This approach has been used to obtain culture strains accumulating anthocyanin in the dark of *Aralia cordata* (56), *Vitis vinifera* (57), and *Fragaria ananassa* (58). Bioreactor-cultured *Perilla frutescens* cells were able to accumulate significant amount of anthocyanin (about 10% dry weight) without light irradiation when aerated at 0.2 vvm but not at 0.1 vvm (59).

3. Chemical Factors

Effects of the plant growth regulators on anthocyanin production in cultured cells are apparently variable. In *Vitis vinifera* suspension cultures (54) and *Strobilanthes dyeriana* callus cultures (60), better growth and anthocyanin production were obtained with 1-naphthaleneacetic acid (NAA) than 2,4-dichlorophenoxyacetic acid (2,4-D). In contrast, 2,4-D favored anthocyanin production in *Hibiscus sabdariifa* (61), *Euphorbia milli* (62), *Vitis* hybrid (63), and *Fragaria ananassa* (64) cultured cells, whereas it was inhibitory at a higher concentration (63). Generally, cytokinins hardly affect anthocyanin production in cultured cells with a few exceptions; e.g., kinetin is superior to benzyladenine in grape cell suspension cultures (65). It was also shown that in strawberry suspension cultures the cyanidin-3-glucoside content increased and peonidin-3-glucose level decreased with increases in auxin and cytokinin levels (66).

Among various medium nutrients, nitrogen sources and phosphate are important factors affecting anthocyanin production by plant cell cultures. Reducing the nitrate concentration in the medium increased anthocyanin accumulation in cell cultures of *Vitis* hybrid (25) and *Catharanthus roseus* (67), whereas increasing the ratio of nitrate to ammonium was effective for anthocyanin production in cell suspension cultures of *Aralia cordata* (56) and *Euphorbia milli* (62). It has been reported that phosphate limitation is associated with growth limitation and a concomitant increase in anthocyanin level in the cultured cells of various plants, including *Vitis* hybrid (25), *Daucus carota* (68), *Hibiscus sabdariffa* (69), and *Vitis vinifera* (70).

4. Bioreactor Culture

There have been some reports describing successful scaling up of the cultures (up to a 500-L pilot scale) for anthocyanin production as summarized in Table 2. However, in some cases, including *Vitis vinifera* cells (57), anthocyanin productivity decreased in jar-fermenter cultures compared with suspension cultures. The CO_2 level in culture vessels, agitation method, and volumetric oxygen transfer are important factors controlling both biomass and anthocyanin production in a bioreactor.

5. Engineering Anthocyanin Production Using Recombinant DNA Technology

It is now possible to engineer plant cell metabolism genetically using a wide range of new techniques in molecular biology. The molecular biological aspects of anthocyanin biosynthesis have been extensively studied, and genes or cDNAs encoding almost all of the enzymes in anthocyanin biosynthesis have been cloned (71). In addition, some genes encoding transcriptional factors that regulate the expression of all or a subset of anthocyanin biosynthesis genes have been identified and isolated (72). Proper use of such regulatory genes may lead to a general strategy for switching on the entire pathway of anthocyanin biosynthesis. Expression of such transcriptional factors in cultured maize cells stimulated expression of the genes of flavonoid biosynthesis, resulting in accumulation of anthocyanin (73,74). Another interesting target for engineering anthocyanin production is its transport into central vacuoles. Hirasuna et al. reported that the up-regulation of anthocyanin production in grape cells in culture by reducing the nitrate concentration in the medium may be due to the involvement of a nitrate-sensitive ATPase in the accumulation of anthocyanin in vacuoles (63). The finding of a glutathione *S*-transferase involved in transport of cyanidin-3-glucoside into vacuoles (75) may provide a biochemical basis for engineering vacuolar targeting of anthocyanin in cultured plant cells.

B. Shikonin

Shikonin and its *S*-isomer alkannin are red naphthoquinone pigments accumulated as acyl esters in the cork layer of the roots of various boraginaceous

TABLE 2 Large-Scale Production of Anthocyanin by Bioreactor Cultures

Plant species	Bioreactor volume	Agitation/aeration	Light	Medium		Productivity	Reference
Vitis hybrid	30 L (20 L)[a]	Aeration 0.4 vvm	No	Modified MS[b] 2,4-D Kinetin Sucrose	0.01 mg/L 0.6 mg/L 0.292 M	12.5 mg/L/day (12 days)[c]	25
Perilla frutescens	2.6 L (2 L)	Aeration 0.1 vvm	Yes	LS 2,4-D BA Sucrose	1 μM 1 μM 3%	290 mg/L/day (10 days)	82
		Aeration 0.2 vvm	No			165 mg/L/day (10 days)	59
Aralia cordata	500 L (300 L)	Agitation 30 rpm Aeration 0.2 vvm with CO-enriched (0.3%) air	No	MS 2,4-D Kinetin Sucrose	1 mg/L 0.1 mg/L 3%	110 mg/L/day (12 days)	80
Euphorbia milli	30 L	Air flow rate 1.0 L/min	Yes	Modified B5[d] 2,4-D BA Sucrose	1 μM 0.01 μM 5%	50 mg/L/day (10 days)	81

[a] Working volume.
[b] Production medium was established based on Murashige and Skoog medium.
[c] Culture period.
[d] Production medium was constructed based on Gamborg's B5 medium.

species except for *Plagyobotrys arizonicus*, which accumulates alkannin in the cuticle layer of the leaves. Shikonin exhibits a wide range of pharmacological activities such as antimicrobial, anti-inflammatory, wound-healing, and antitumor actions. Investigations have also shown that shikonin inhibits topoisomerase-I activity in vitro and angiogenesis and induces apoptosis in the HL-60 human leukemia cell line (see Ref. 83 and literature cited therein).

1. Large-Scale Production of Shikonin Derivatives by *Lithospermum erythrorhizon* Cell Cultures

In 1974 Tabata et al. (38) first demonstrated that callus tissues of *L. erythrorhizon* produced the shikonin acyl esters (shikonin derivatives) as those in the roots. Visual selection of the cell aggregate led to the establishment of a high-producing cell line whose shikonin content increased 20-fold, reaching up to 1.2 mg/g fresh weight of the cells (84). By using this culture line, various factors controlling shikonin production in cultured cells of *L. erythrorhizon* were extensively investigated (Table 3). Media optimized for biomass production (growth medium) and for shikonin production (production medium) were separately established (22). Two-stage culture in a 750-L airlifted bioreactor yielded 1.4 to 2.3 g/L shikonin for 23 days (18). Thus, in 1983 Mitsui Petrochemical Industry declared commercial production of shikonin using *L. erythrorhizon* cell cultures, which became the first and is

TABLE 3 Physical and Chemical Factors Regulating Shikonin Production in *Lithospermum erythrorhizon* Cell Suspension Cultures

Factor	Reference
Up-regulating	
Sucrose	98
Cu^{2+}	22
Agar powder/agaropectin	99
Oligogalacturonide	90
Methyl jasmonate	33
Down-regulating	
Light	38
Lumiflavine	100
2,4-D	38
GA_3	16
NH^{4+}	22
Glutamine	101

still one of few examples of successful use for industrial production of a phytochemical.

2. Regulatory Mechanism of Shikonin Biosynthesis

L. erythrorhizon cell culture does not merely represent an industrial application of cell culture technology to phytochemical production but provides us with a model system suitable for investigating regulatory mechanisms of plant secondary metabolism.

After a preliminary experiment in which feeding radiolabeled precursors into an intact plant of *L. erythrorhizon* was unsuccessful because of extremely low incorporation of the tracers, shikonin was demonstrated to be biosynthesized through the prenylation of *p*-hydroxybenzoic acid (PHB) with geranyldiphosphate (GPP) yielding *m*-geranyl-*p*-hydroxybenzoic acid (GHB) by precursor feeding to *L. erythrorhizon* callus tissues (85). Since then, a series of investigations using *L. erythrorhizon* cell cultures has clarified a biosynthetic pathway leading to shikonin and its metabolically related compounds as shown in Fig. 2. GPP is synthesized in the mevalonate pathway in *L. erythrorhizon* cells (86), although it is postulated that the nonmevalonate pathway is involved in the formation of monoterpenes in plants.

The activity of GPP:PHB acid geranyltransferase (PHB geranyltransferase) catalyzing the formation of GHB from PHB was detected in a cell-free extract of *L. erythrorhizon* cells in culture (87). When shikonin production was repressed by culturing the cells under light illumination, the activity of this enzyme was strongly inhibited and instead PHB *O*-glucosyltransferase was activated, resulting in rapid accumulation of PHB *O* glucoside (88). The shikonin-nonproducing suspension cells cultured in LS medium produced dihydroechinofuran, which was probably derived from GHB via geranylhydroquinone (89). GHB formation was completely inhibited by white light but not by the other inhibitors of shikonin biosynthesis such as 2,4-D, glutamine, and gibberellin 3 (GA$_3$). Addition of oligogalacturonides to *L. erythrorhizon* cells cultured in MS medium in which shikonin biosynthesis was usually repressed rapidly and transiently induced PHB geranyltransferase activity followed by increases in the echinofuran level and then shikonin content (90). These results unambiguously indicate that the formation of GHB is an important step in the regulation of shikonin biosynthesis, especially by light. Neither purification of PHB geranyltransferase nor cloning of cDNA encoding this enzyme has been unsuccessful to date. Furthermore, the enzymes catalyzing biosynthetic steps after GHB formation have not yet been well characterized except for hydroxylation of the isoprenoid side chain of geranylhydroquinone (91) and conversion of deoxyshikonin to shikonin derivatives in cell-free extract of *L. erythrorhizon* cell

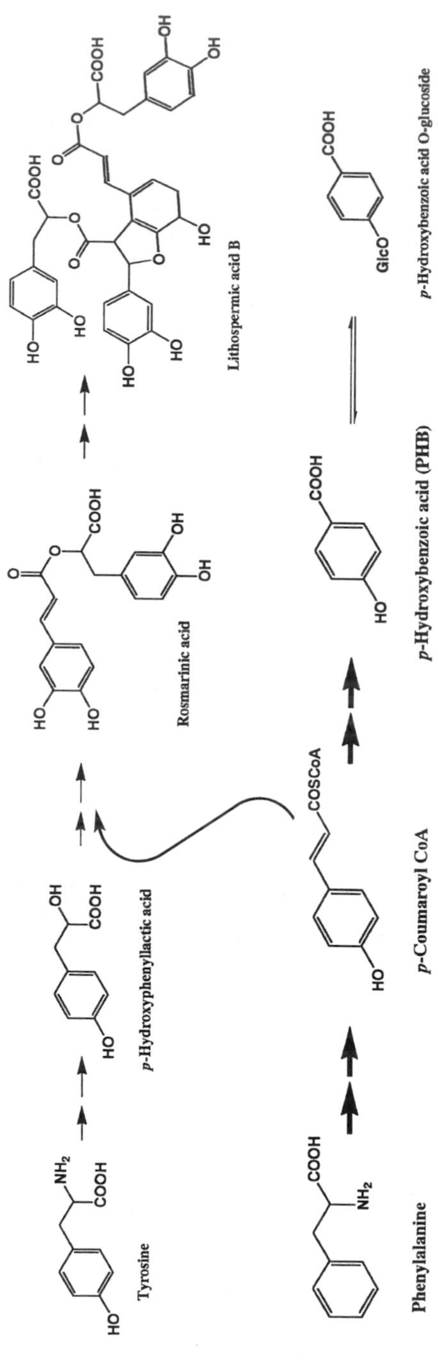

Tyrosine

p-Hydroxyphenyllactic acid

Rosmarinic acid

Lithospermic acid B

Phenylalanine

p-Coumaroyl CoA

p-Hydroxybenzoic acid (PHB)

p-Hydroxybenzoic acid O-glucoside

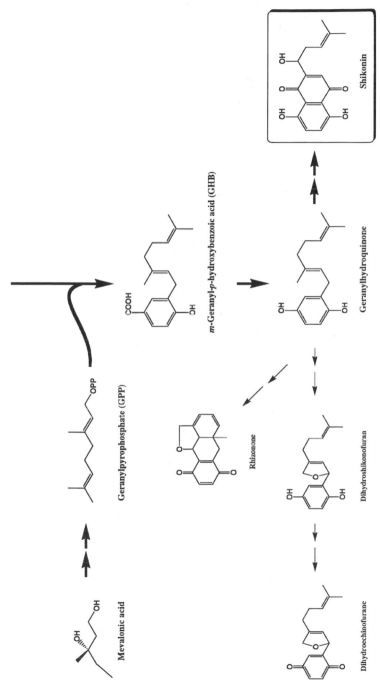

FIGURE 2   Biosynthetic pathway leading to shikonin and its metabolically related compounds.

cultures (92). These remain to be investigated to obtain further insight into regulatory mechanisms in shikonin biosynthesis.

3.   Intracellular Localization of Shikonin Biosynthesis

Based on electron microscopic and biochemical analyses (93,94), it has been postulated that a series of enzymatic reactions starting from prenylation of PHB by the prenyltransferase are located in the vesicles derived from rough endoplasmic reticulum (rER). The vesicles are suggested to fuse with the cell membrane and excrete shikonin pigments, which are accumulated on the cell wall as oil droplets.

4.   Engineering Shikonin Biosynthesis in Cultured Plant Cells

Shikonin biosynthesis in *L. erythrorhizon* cell cultures is suitable as a target for metabolic engineering because (1) the biosynthetic pathway is clearly elucidated and involved in a complex consisting of a shikimate-phenylpropanoid pathway and a mevalonate pathway, (2) we have various network factors in hand to either up-regulate or down-regulate shikonin biosynthesis, (3) various cell lines differing in shikonin biosynthesis capability are present, and (4) a stable transformation system using *Agrobacterium rhizogenes* (95) and a plant regeneration system from the callus tissues (96) have been established. However, the investigation in this direction has just started (97). Cloning and characterization of the structural and/or regulatory genes involved in shikonin biosynthesis will be a next step for further understanding and manipulating cellular events underlying shikonin biosynthesis.

V.   PRODUCTION OF CLINICALLY USED ANTINEOPLASTIC COMPOUNDS

A.   *Catharanthus* Alkaloids

The dimeric indole alkaloids, vincristine and vinblastine (Fig. 3), are clinically used as antileukemic agents, which are produced in *Catharanthus roseus* (periwinkle). These dimeric alkaloids are formed by condensation of vindoline and catharanthine, which are both derived from strictosidine as the common intermediate for all monoterpenoid indole alkaloids (Fig. 3). A number of studies have been performed to produce antineoplastic compounds in cell cultures of *C. roseus*. These include callus, cell suspension, and transformed hairy roots (102,103). Although substantial amounts of monomeric indole alkaloids can be produced in these cell cultures, production of the dimeric alkaloids is not fully successful.

    Feeding of biosynthetic precursors, secologanin and loganin, resulted in increased accumulation of ajmalicine and strictosidine (104). Cell differ-

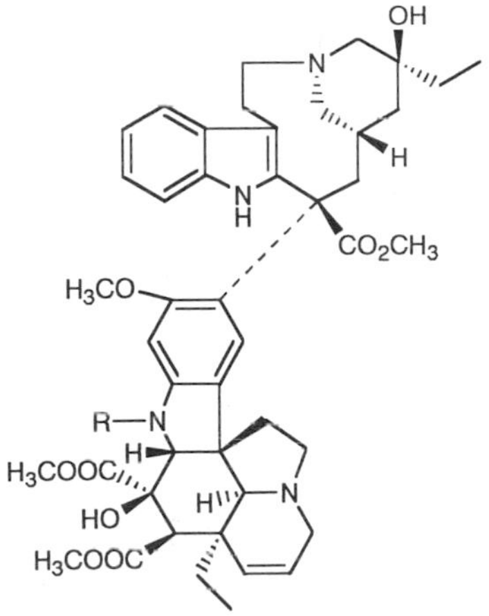

R = CHO vincristine
R = CH₃ vinblastine

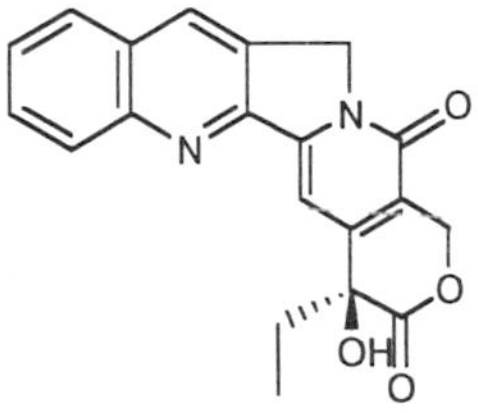

camptothecin

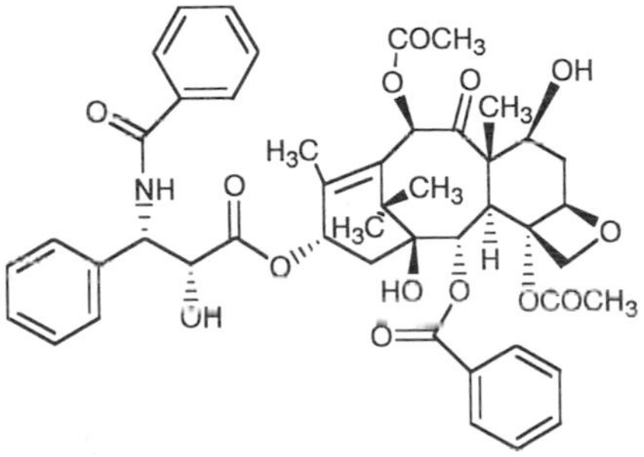

paclitaxel

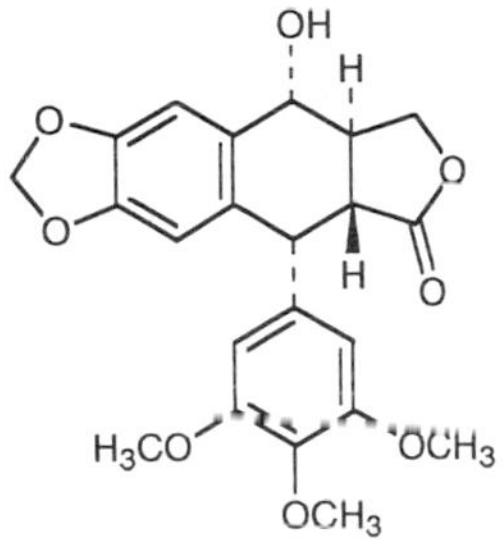

podophyllotoxin

FIGURE 3 Chemical structures of clinically used antineoplastic compounds from plants.

entiation also induced the production of the alkaloids (105). Since the genes encoding tryptophan decarboxylase and strictosidine synthase were isolated, the effects of overexpression of these two genes in *C. roseus* cell culture were elucidated (106). The cell lines with enhanced strictosidine synthase activity accumulated over 200 mg/L of the indole alkaloids. A recent study has partially clarified the signal transduction cascade of strictosidine synthase gene expression, which involves the octadecanoid pathway and protein phosphorylation (107). The specific transcriptional factor for inducible gene expression of tryptophan decarboxylase and strictosidine synthase has been isolated (108). These molecular factors for regulation of indole alkaloid biosynthesis are promising for future application to engineered cell culture production.

B. Camptothecin

Camptothecin is a pentacyclic quinoline alkaloid, which was isolated from *Camptotheca acuminata* in 1966 (109) (Fig. 3). It exhibits potent anticancer activity that is due to the strong inhibitory action on DNA topoisomerase I. Irinotecan is a semisynthetic derivative of camptothecin improved by decreasing the cytotoxicity of camptothecin and increasing the water solubility. Irinotecan is a prodrug that is metabolically activated by esterase to an ultimate form for anticancer action.

Although the final skeleton of camptothecin belongs to the group of quinoline alkaloids, it is synthesized via strictosidine, a common intermediate of monoterpenoid indole alkaloids. However, the biosynthetic pathway after strictosidine and strictosamide is not yet clear (110). Camptothecin has been found not only in *C. acuminata* (Nyssaceae) but also in several other plant species of different plant families, i.e., Icacinaceae, Rubiaceae, and Apocynaceae.

Because of its potent antitumor activity and the unique mechanism of action, a number of attempts at production by cell culture have been made (Table 4). These are mostly with *C. acuminata*. The first trial of production by cell culture of *C. acuminata* was performed in 1974 by Misawa's group (111). Although they established a rapidly growing cell suspension culture, the productivity was not satisfactory. A shoot culture instead of a cell suspension of *C. acuminata* produced a high amount (0.1% dry weight) of camptothecin, but the growth rate was low (112). High producibility (0.2% dry weight) was reported with a few cell lines of *C. acuminata* callus (113), although the growth rate of these lines was not measured. With *C. acuminata* cell cultures, the addition of yeast extracts and jasmonates increased camptothecin production (114).

Species in the genus *Ophiorriza* (Rubiaceae) are receiving growing attention as a target for cell cultures producing camptothecin instead of *C. acuminata*. Although a callus culture of *O. pumila* did not produce camptothecin, the regenerated aspetic plantlets accumulated camptothecin and its derivatives (115). More interestingly, a hairy root culture stimulated by infection of a pathogenic soil bacterium, *Agrobacterium rhizogenes*, showed rapid proliferation and also good camptothecin producibility (0.1% dry weight) (116,117). A substantial amount of camptothecin was excreted into the liquid medium. Because camptothecin is cytotoxic and exhibits inhibitory activity toward plant growth, there should be a particular mechanism of sequestering camptothecin into vacuoles or excretion to outside the cells.

TABLE 4 Camptothecin Production in Cell Culture

Species	Tissue	CPT content (% dry wt.)	Growth rate	Reference
Ophiorriza pumila	Hairy root	0.1	16/5 w	116,117
Camptotheca acuminata	Cell suspension	0.00025	10/3 w	111
C. acuminata	Cell suspension	0.004	16/4–5 w	123
C. acuminata	Shoot culture	0.1	15/4 w	112
C. acuminata	Callus	0.2	16/13 w	113
Nothapodytes foetida	Plantlets	0.0007	—	124

C. Paclitaxel (Taxol)

Paclitaxel (taxol) is a diterpene compound initially isolated from *Taxus brevifolia* (Pacific yew) in 1971 as a cytotoxic substance against leukemia cells (118). It is also found in all species belonging to the genus *Taxus*. Paclitaxel was discovered through an extensive program in the United States to screen plant extracts for discovery of new anticancer drugs among 110,000 compounds from 35,000 species (119). It is now clinically used for the treatment of ovarian and breast cancer. The mechanism of action is to promote the assembly of microtubules and inhibit the disassembly process.

For the clinical trials, the limited supply of paclitaxel only from the bark of *T. brevifolia* was problematic. Total synthesis of paclitaxel was achieved; however, the synthetic process involved long steps and a low yield and was not commercially feasible. The semisynthesis of paclitaxel from its precursors isolated from leaves and stems of *Taxus* cultivars is also established. However, plant cell culture has a strong long-term potential for production of paclitaxel or its precursors, in particular, by adopting advances in biosynthetic understanding and genetic engineering.

The production of paclitaxel and related compounds in plant cell culture has been investigated with several species of the genus *Taxus*, e.g., *T. brevifolia*, *T. baccata*, *T. chinensis*, *T. canadensis*, *T. cuspidata*, and *T. x media*. A number of trials of production in callus and cell suspension cultures of these *Taxus* plants have been reported as summarized by Wickremesinhe and Arteca (119). The highest producibility of paclitaxel was achieved upon inducion by methyl jasmonate with *T. media* cell culture (120). They found that the addition of 100 μM methyl jasmonate increased paclitaxel production up to 0.6% content in cells and 110 mg/L medium for 14 days. This producibility was fivefold higher than that of the control without addition of methyl jasmonate. The content of baccatin III was also induced 20-fold (0.2% of cell weight) with *T. baccata* culture. Jasmonate is known to play a role in a signal transduction system involved in the defense process in plant cells. This is a good example of how one can manipulate the production of secondary compounds in cell culture by a particular signal molecule.

D. Podophyllotoxin

Podophyllotoxin is a lignan compound derived from two phenyl propanoid units. The clinically used antitumor compounds etoposide and teniposide are semisynthetic derivatives of podophyllotoxin (Fig. 3). They act as inhibitors of the microtubule assembly for antitumor activity.

The rhizomes of *Podophyllum* species are the source for extraction of podophyllotoxin. However, the occurrence of these plants is limited and the growth of the plants is slow. Therefore, production by cell culture has been

TABLE 5 Podophyllotoxin Production in Cell Culture

Species	Tissue	Ptox content (% dry wt)	Growth rate	Comments	Reference
Linum album	Cell suspension	0.25	7/1 w		125
Juniperus chinensis	Callus	0.005–0.0⁻6	11/40 days	Enhanced by oligosaccharide elicitor and phenylalanine	122
Callitris drummondii	Cell suspension	0.02 (dark – 0.1 (light)	20–25%/3 w	Induced by light, mostly β-D-glucoside form	126
Podophyllum hexandrum	Cell suspension	0.1 (dark)– 0.03 (light)	2/2 w	Increased by feeding coniferyl alcohol as soluble form	126,127
Podophyllum peltatum	Callus				128
Linum flavum	Root culture				129

required. As shown in Table 5, cell cultures of several plant species have been established for the production of podophyllotoxin and its derivatives such as its glucoside, 5-methoxypodophyllotoxin, and desoxypodophyllotoxin. Production cell culture of *P. hexandrum* was increased by a factor of 6- to 30-fold by the addition of the precursor coniferyl alcohol, solubilized as a β-cyclodextrin complex or a a glucoside form, coniferin (121). Addition of a precursor, phenylalanine, and an elicitor, chito-oligosaccharide, also resulted in increased production of podophyllotoxin in *Juniperus chinensis* calli (122).

VI. FUTURE PROSPECTS

A number of attempts have been made for feasible production of secondary compounds in plant cell cultures. Although there are a few successful examples achieving high producibility in cell cultures, in most cases producibility was lower than that of differentiated particular cells of intact plants. These are presumably because of the lack of our understanding of molecular mechanisms of regulation of secondary compound production. However, as in some cases the molecular mechanism of signal transduction and regulation of gene expression of secondary metabolism is being revealed, these new findings can be directly applied for metabolic engineering in cell cultures. Entire DNA sequencing of the whole genome of *Arabidopsis thaliana*, a model species of higher plants, is being accomplished. Thus, in the near future it will be necessary to combine the molecular biological techniques with the cell culture methods more tightly.

ACKNOWLEDGMENTS

The author (K.S.) thanks Dr. H. Sudo for correcting the literature. The research in the authors' groups was supported in part by Grants-in-Aid for Scientific Research from the Ministry of Education, Science, Sports and Culture, Japan, by the Research for the Future Program (96I00302) of the Japan Society for the Promotion of Science, by CREST of Japan Science and Technology to K.S.

REFERENCES

1. RD Hall, MA Holden, MM Yeoman. The immobilization of higher plant cells. In: YPS Bajaj, ed. Immobilization of Plant Cells. New York: Springer-Verlag, 1988, pp 136–156.
2. ML Fauconnier, M Jaziri, M Marlier, J Roggenmans, JP Wathelet, G Lognay, M Severin, J Homes, K Shimomura. Essential oil production by *Arthemis nobilis* L. tissue culture. J Plant Physiol 141:759–761, 1993.

3. I Asai, K Yoshihira, T Omoto, N Sakui, K Shimomura. Growth and mono-terpene production in shoot culture and regenerates of *Mentha arvensis*. Plant Tissue Cult Lett 11:218–225, 1995.
4. NB Paniego, AE Maligne, AM Giulietti. *Artemisia annua* (Quing-Hao): In vitro culture and the production of artemisinin. In: YPS Bajaj, ed. Biotechnology in Agriculture and Forestry 24. Medicinal and Aromatic Plants V. New York: Springer-Verlag, 1993, pp 64–78.
5. CZ Liu, YC Wang, C Guo, F Ouyang, HC Ye, GF Li. Production of artemisinin by shoot cultures of *Artemisia annua* L. in a modified inner-loop mist bioreactor. Plant Sci 135:211–217, 1998.
6. T Endo, A Goodbody, M Misawa. Alkaloid production in root and shoot cultures of *Catharanthus roseus*. Planta Med 53:479–482, 1987.
7. K Hirata, M Horiuchi, T Ando, K Miyamoto, Y Miura. Vindoline and catharanthine production in multiple shoot cultures of *Catharanthus roseus*. J Ferment Bioeng 70:193–195, 1990.
8. A Spencer, JD Hamill, MJC Rhodes. Production of terpenes by differentiated shoot cultures of *Mentha citrata* transformed with *Agrobacterium tumefaciens* T37. Plant Cell Rep 8:601–604, 1990.
9. K Hirata, M Horiuchi, M Asada, T Ando, K Miyamoto, Y Miura. Stimulation of dimeric alkaloid production by near-ultra-violet light in multiple shoot cultures of *Catharanthus roseus*. J Ferment Bioeng 74:222–225, 1992.
10. JFS Ferreira, J Janick. Roots as an enhancing factor for the production of artemisinin in shoot cultures of *Artemisia annua*. Plant Cell Tissue Organ Cult 44:211–217, 1996.
11. H Deno, H Yamaga, T Emoto, T Yoshioka, Y Yamada, Y Fujita. Scopolamine production by root cultures of *Duboisia myoporoides*. II. Establishment of hairy root cultures by infection with *Agrobacterium rhizogenes*. J Plant Physiol 131:315–323, 1987.
12. KT Lee, T Suzuki, T Yamakawa, T Kodama, Y Igarashi, K Shimomura. Production of tropane alkaloids by transformed root cultures of *Atropa belladonna* in stirred bioreactors with a stainless steel net. Plant Cell Rep 18:567–571, 1999.
13. Y Yamada, M Tabata. Plant biotechnology of tropane alkaloids. Plant Biotechnol 14:1–10, 1997.
14. W Kreis, E Reinhard. The production of secondary metabolites by plant cells cultured in bioreactors. Planta Med 55:409–415, 1989.
15. Y Hara, TY Oshioka, T Morimoto, Y Fujita, Y Yamada. Enhancement of berberine production in suspension cultures of *Coptis japonica* by gibberellic acid treatment. J Plant Physiol 133:12–15, 1988.
16. N Yoshikawa, H Fukui, M Tabata. Effect of gibberellin A_3 on shikonin production in *Lithospermum erythrorhizon* callus cultures. Phytochemistry 25:621–622, 1986.
17. R Phillip, NJ Darrell. Pigment production in hormone-autonomous cell cultures following somatic fusion of petal and tumor protoplasts of *Petunia hybrida*. Plant Sci 83:95–102, 1992.

18. Y Fujita, M Tabata, A Nishi, Y Yamada. New medium and production of secondary compounds with two-staged culture medium. In: A Fujiwara, ed. Plant Tissue Culture 1982. Tokyo: Maruzen, 1982, pp 399–400.

19. L Nettleship, M Slaytor. Adaptation of *Peganum harmala* callus to alkaloid production. J Exp Bot 25:1114–1123, 1974.

20. CB Do, F Cormier. Effects of low nitrate and high sugar concentrations on anthocyanin content and composition of grape (*Vitis vinifera*) cell suspensions. Plant Cell Rep 9:500–504, 1991.

21. M Sakuta, T Takagi, A Komamine. Effects of nitrogen source on betacyanin accumulation and growth in suspension cultures of *Phytolacca americana*. Physiol Plant 71:459–463, 1987.

22. Y Fujita, Y Hara, C Suga, T Morimoto. Production of shikonin derivatives by cell suspension cultures of *Lithospermum erythrorhizon*. A new medium for the production of shikonin derivatives. Plant Cell Rep 1:61–63, 1981.

23. M Petersen, K Dombrowski, C Gertlowski, E Haeusler, B Kawatzki, J Meinhard, AW Alfermann. The use of plant cell cultures to study natural products biosynthesis. In: K Oono, T Hirabayashi, S Kikuchi, H Handa, K Kajiwara, eds. Plant Tissue Culture and Gene Manipulation for Breeding and Formation of Phytochemicals. Tsukuba: NAIR, 1992, pp 293–296.

24. CB Do, F Cormier. Accumulation of anthocyanins enhanced by a high osmotic potential in grape (*Vitis vinifera* L.) cell suspensions. Plant Cell Rep 9:143–146, 1990.

25. T Yamakawa, K Ishida, S Kato, T Kodama, Y Minoda. Formation and identification of anthocyanins in cultured cells of *Vitis* sp. Agric Biol Chem 47:997–1001, 1983.

26. PF Heinstein. Future approaches to the formation of secondary natural products in plant cell suspension cultures. J Nat Prod 48:1–9, 1985.

27. RT Tyler, U Eilert, COM Rijnders, IA Roewer, CK McNabb, WHW Kurz. Studies on benzophenanthridine alkaloid production in elicited cell cultures of *Papaver somniferum* L. In: WGW Kurz, ed. Primary and Secondary Metabolism of Plant Cell Cultures 11. Berlin: Springer-Verlag, 1989, pp 200–207.

28. U Sankawa, T Hakamatsuka, K Shinkai, M Yoshida, HH Park, Y Ebizuka. Changes of secondary metabolism by elicitor treatment in *Pueraria lobata* cell cultures. In: M Terzi, R Cella, A Falavigna, eds. Current Issues in Plant Molecular and Cellular Biology. Dordrecht: Kluwer Academic Publishers, 1995, pp 595–604.

29. H Glundelach, MJ Mueller, TM Kutchan, MH Zenk. Jasmonic acid is a signal transducer in elicitor-induced plant cell cultures. Proc Natl Acad Sci USA 89:2389–2393, 1992.

30. H Mizukami, Y Tabira, BE Ellis. Methyl jasmonate–induced rosmarinic acid biosynthesis in *Lithospermum erythrorhizon* cell suspension cultures. Plant Cell Rep 12:706–709, 1993.

31. H Urbanek, K Bergier, M Saniewski, J Patykowski. Effect of jasmonate and exogenous polysaccharides on production of alkannin pigments in suspension cultures of *Alkanna tinctoria*. Plant Cell Rep 15:637–641, 1996.

32. N Mirjalili, JC Linden. Methyl jasmonate induced production of taxol in suspension cultures of *Taxus cuspidata*: ethylene interaction and induction models. Biotechnol Prog 12:110–118, 1996.
33. K Yazaki, K Takeda, M Tabata. Effect of methyl jasmonate on shikonin and dihydroechinofuran production in *Lithospermum* cell cultures. Plant Cell Physiol 38:776–782, 1997.
34. S Krisa, F Larronde, H Budzinski, A Decendit, G Deffieux, JM Merillon. Stilbene production by *Vitis vinifera* cell susepnsion cultures: methyl jasmonate induction and ^{13}C biolabelling. J Nat Prod 62:1688–1690, 1999.
35. J Hefner, REB Ketchum, R Croteau. Cloning and functional expression of a cDNA encoding geranylgeranyl diphosphate synthase from *Taxus canadensis* and assessment of the role of this prenyltransferase in cells induced for taxol production. Arch Biochem Biophys 360:62–74, 1998.
36. K Saito, M Yamazaki, S Takamatsu, A Kawaguchi, I Murashi. Greening induced production of (+)-lupanine in tissue culture of *Thermopsis lupinoides*. Phytochemistry 28:2341–2344, 1989.
37. MC Hobbs, MM Yeoman. Effect of light on alkaloid accumulation in cell cultures of *Nicotiana* sp. J Exp Bot 22:1371–1378, 1991.
38. M Tabata, H Mizukami, N Hiraoka, M Konoshima. Pigment formation of callus cultures of *Lithospermum erythrorhizon*. Phytochemistry 13:927–932, 1974.
39. K Matsubara, S Kitani, T Yoshioka, T Morimoto, Y Fujita, Y Yamada. High density culture of *Coptis japonica* cells increases berberine production. J Chem Technol Biotechnol 46:61–69, 1989.
40. JJ Zhong, T Yoshida. High-density cultivation of *Perilla frutescens* cell suspensions for anthocyanin production: effects of sucrose concentration and inoculum size. Enzyme Microb Technol 17:1073–1079, 1995.
41. Y Yamamoto, R Mizuguchi, Y Yamada. Selection of a high and stable pigment-producing strain in cultured *Euphorbia milli* cells. Theor Appl Genet 61:113–116, 1982.
42. T Ogino, N Hiraka, M Tabata. Selection of high nicotine-producing cell lines of tobacco callus by single cell cloning. Phytochemistry 17:1907–1910, 1978.
43. MH Zenk, H El-Shagi, H Arens, J Stockigt, EW Weiler, B Deus, J Stoeckigt. Formation of the indole alkaloids serpentine and ajmalicine in cell suspension cultures of *Catharanthus roseus*. In: W Barz, E Reihard, MH Zenk, eds. Plant Tissue Culture and Its Bio-technological Application. Berlin: Springer-Verlag, 1977, pp 27–43.
44. B Deus-Neumann, MH Zenk. Instability of indole alkaloid production in *Catharanthus roseus* cell suspension cultures. Planta Med 50:427–431, 1984.
45. M Tabata, N Hiraoka. Variation in alkaloid production in *Nicotiana rustica* callus cultures. Physiol Plant 38:19–23, 1976.
46. M Hirotani, T Furuya. Restoration of cardenolide synthesis in redifferentiated shoots from callus cultures of *Digitalis purpurea*. Phytochemistry 16:610–611, 1977.
47. RR Holden, MA Holden, MM Yeoman. Variation in the secondary metabolism of cultured plant cells. In: RJ Robin, MJC Rhodes, eds. Manipulating Sec-

ondary Metabolism in Culture. Cambridge: Cambridge University Press, 1986, pp 15–29.

48. CA Rice-Evans, NJ Miller. Antioxidant activities of flavonoids as bioactive components of food. Biochem Soc Trans 24:790–794, 1996.

49. H Tamura, M Fujikawa, H Sugisawa. Production of phenyl-propanoids from cultured callus tissue of the leaves of Akachirimen-shiso (*Perilla* sp.). Agric Biol Chem 53:1971–1973, 1989.

50. WE Glaessgen, V Wray, D Strack, JW Metzger, HU Seitz. Anthocyanin from cell suspension cultures of *Daucus carota*. Phytochemistry 31:1593–1601, 1992.

51. N Terahara, A Callebaut, R Ohba, T Nagata, M Ohnishi-Kameyama, M Suzuki. Triacylated anthocyanins from *Ajuga reptans* flowers and cell cultures. Phytochemistry 42:199–203, 1996.

52. DL Madhavi, S Juthangkoon, K Lewen, MD Berber-Jimenez, MAL Smith. Characterization of anthocyanins from *Ajuga pyramidalis* Metallica Crispa cell cultures. J Agric Food Chem 44:1170–1176, 1996.

53. RD Hall, MM Yeoman. Intercellular and intracellular heterogeneity in secondary metabolite accumulation in *Catharanthus roseus* following cell line selection. J Exp Bot 38:1391–1398, 1987.

54. S Krisa, V Vitrac, A Decendit, F Larronde, G Deffieux, JM Merillon. Obtaining *Vitis vinifera* cell cultures producing higher amounts of malvidin-3-O-glucoside. Biotechnol Lett 212:497–500, 1999.

55. N Hiraoka, T Kodama, Y Tomita. Selection of *Bupleurum falcatum* callus line producing anthocyanins in darkness. J Nat Prod 49:470–474, 1986.

56. K Sakamoto, K Iida, K Sawamura, K Hajiro, Y Asada, T Yoshikawa, T Furuya. Anthocyanin production in cultured cells of *Aralia cordata* Thunb. Plant Cell Tissue Organ Cult 36:21–26, 1994.

57. F Cormier, F Brion, CB Do, C Moresoli. Development of process strategies for anthocyanin-based food colorant production using *Vitis vinifera* cell cultures. In: F DiCosmo, M Misawa, eds. Plant Cell Culture Secondary Metabolism: Toward Industrial Application. Boca Raton, FL: CRC Press, 1996, pp 167–185.

58. M Nakamura, Y Takeuchi, K Miyanaga, M Seki, S Furusaki. High anthocyanin accumulation in the dark by strawberry (*Fragaria ananassa*) callus. Biotechnol Lett 21:659–699, 1999.

59. JJ Zhong, M Yoshida, K Fujita, T Seki, T Yoshida. Enhancement of anthocyanin production by *Perilla frutescens* cells in a stirred bioreactor with internal light irradiation. J Ferment Bioeng 75:299–303, 1993.

60. SL Smith, GW Slywka, RJ Klueger. Anthocyanins of *Strobilanthes dyeriana* and their production in callus culture. J Nat Prod 44:609–610, 1981.

61. H Mizukami, K Tomita, H Ohashi, N Hiraoka. Anthocyanin production in callus cultures of roselle (*Hibiscus sabdariffa* L.) Plant Cell Rep 7:553–556, 1988.

62. Y Yamamoto, Y Kinoshita, S Watanabe, Y Yamada. Anthocyanin production in suspension cultures of high-producing cells of *Euphorbia milli*. Agric Biol Chem 53:417–423, 1989.

63. TJ Hirasuana, ML Shuler, VS Lackney, RM Spanswick. Enhanced anthocyanin production in grape cell cultures. Plant Sci 78:107–120, 1991.
64. T Mori, M Sakurai, JT Shigeta, K Yoshida, T Kondo. Formation of anthocyanins from cells cultured from different parts of strawberry plants. J Food Sci 58:788–792, 1993.
65. HV Meyer, J van Staden. The in vitro production of anthocyanin from callus cultures of *Oxalis linearis*. Plant Cell Tissue Organ Cult 40:55–58, 1995.
66. T Mori, M Sakurai, M Seki, S Furusaki. Use of auxin and cytokinin to regulate anthocyanin production and composition in suspension cultures of strawberry. J Sci Food Agric 65:271–276, 1994.
67. KH Knobloch, J Bast, J Berlin. Medium- and light-induced formation of serpentine and anthocyanins in cell suspension cultures of *Catharanthus roseus*. Phytochemistry 21:591–594, 1982.
68. DK Dougall, KW Weyrauch. Growth and anthocyanin production by carrot suspension cultures grown under chemostat conditions with phosphate as the limiting nutrient. Biotechnol Bioeng 22:337–352, 1980.
69. H Mizukami, M Nakamura, T Tomita, K Higuchi, H Ohashi. Effects of macronutrients on anthocyanin production in roselle (*Hibiscus sabdariffa* L.) cell cultures. Plant Tissue Cult Lett 8:14–29, 1991.
70. F Dedaldechamp, C Uhel, JJ Macheix. Enhancement of anthocyanin synthesis and dihydroflavonol reductase (DFR) activity in response to phosphate deprivation in grape cell cultures. Phytochemistry 40:1357–1360, 1995.
71. M Petersen, D Strack, U Matern. Biosynthesis of phenylpropanoids and related compounds. In: M Winl, ed. Biochemistry of Plant Secondary Metabolism. Sheffield: Sheffield Academic Press, 1999, pp 151–221.
72. G Consonni, F Geuna, G Gavazzi, C Tonelli. Molecular homology among members of the R gene family in maize. Plant J 3:335–346, 1993.
73. E Grotewold, M Chamberlin, M Snook, B Siame, L Butler, J Swenson, S Maddock, GS Clair, B Browen. Engineering secondary metabolism in maize cells by ectopic expression of transcription factors. Plant Cell 10:721–740, 1998.
74. W Bruce, O Folkert, C Garnaat, O Crasta, B Roth, B Bowen. Expression profiling of the maize flavonoid pathway genes controlled by estradiol-inducible transcription factors CRC and P. Plant Cell 12:65–79, 2000.
75. MR Alfenito, E Souer, CD Goodman, R Buell, J Mol, V Walbot. Functional complementation of anthocyanin sequestration in the vacuoles by widely divergent glutathione *S*-transferases. Plant Cell 10:1135–1149, 1998.
76. A Callebaut, AM Hendrickx, JC Motte. Anthocyanins in cell cultures of *Ajuga reptans*. Phytochemistry 29:2153–2158, 1990.
77. L Rajendran, G Suvarnalanta, GA Ravishankar, LV Venkataraman. Enhancement of anthocyanin production in callus cultures of *Daucus carota* L. under influence of fungal elicitors. Appl Microbiol Biotechnol 42:227–231, 1994.
78. T Komiya. Development of apple tissue culture techniques and their applications. Plant Tissue Cult Lett 9:69–73, 1992.
79. NR Crouch, LF van Staden, J van Staden, FE Drewes, HJ Meyer. Accumu-

lation of cyanidin-3-glucoside in callus and cell cultures of *Oxalis reclinata*. J Plant Physiol 142:109–111, 1993.
80.	Y Kobayashi, M Akita, K Sakamoto, H Liu, T Shigeoka, T Koyano, M Kawamura, T Furuya. Large-scale production of anthocyanin by *Aralia cordata* cell suspension cultures. Appl Microbiol Biotechnol 40:215–218, 1993.
81.	Y Yamamoto, Y Kinoshita, A Takahashi. Anthocyanin production of cultured *Euphorbia milli* cells. Plant Tissue Cult Lett 13:249–257, 1996.
82.	JJ Zhong, T Seki, SI Kinoshita, T Yoshida. Effect of light irradiation on anthocyanin production by suspended culture of *Perilla frutescens*. Biotechnol Bioeng 38:653–658, 1991.
83.	Y Yoon, YO Kim, NY Lim, WK Jeon, HJ Sung. Shikonin, an ingredient of *Lithospermum erythrorhizon* induced apoptosis in HL60 human premyelotic leukemia cell line. Planta Med 65:532–535, 1999.
84.	H Mizukami, M Konoshima, M Tabata. Variation in pigment production in *Lithospermum erythrorhizon* callus cultures. Phytochemistry 17:95–97, 1978.
85.	H Inouye, S Ueda, K Inoue, H Matsumura. Biosynthesis of shikonin in callus cultures of *Lithospermum erythrorhizon*. Phytochemistry 18:1301–1308, 1979.
86.	SM Li, S Hennig, L Heide. Shikonin: a geranyl diphosphate–derived plant hemiterpenoid formed via mevalonate pathway. Tetrahedron Lett 39:2721–2724, 1998.
87.	L Heide, M Tabata. Geranylpyrophosphate: *p*-hydroxybenzoate geranyltransferase activity in extracts of *Lithospermum erythrorhizon* cell cultures. Phytochemistry 26:1651–1655, 1987.
88.	L Heide, N Nishioka, H Fukui, M Tabata. Enzymatic regulation of shikonin biosynthesis in *Lithospermum erythrorhizon* cell cultures. Phytochemistry 28:1873–1877, 1989.
89.	H Fukui, M Tani, M Tabata. An unusual metabolite, dihydroechinofuran, released from cultured cells of *Lithospermum erythrorhizon*. Phytochemistry 31:519–521, 1992.
90.	M Tani, K Takeda, K Yazaki, M Tabata. Effects of oligogalacturonides on biosynthesis of shikonin in *Lithospermum erythrorhizon* cell cultures. Phytochemistry 32:1285–1290, 1993.
91.	H Yamamoto, K Inoue, SM Li, L Heide. Geranylhydroquinone 3′-hydroxylase, a cytochrome P-450 monooxygenase from *Lithospermum erythrorhizon* cell suspension cultures. Planta 210:312–317, 2000.
92.	T Okamoto, K Yazaki, M Tabata. Biosynthesis of shikonin derivatives from L-phenylalanine via deoxyshikonin in *Lithospermum* cell cultures and cell-free extracts. Phytochemistry 38:83–88, 1995.
93.	M Tsukada, M Tabata. Intracellular localization and secretion of naphthoquinone pigments in cell cultures of *Lithospermum erythrorhizon*. Planta Med 50:338–341, 1984.
94.	Y Yamaga, K Nakanishi, H Fukui, M Tabata. Intracellular localization of *p*-hydroxybenzoate geranyltransferase, a key enzyme involved in shikonin biosynthesis. Phytochemistry 32:633–636, 1993.

95. K Shimomura, H Sudo, H Suga, H Kamada. Shikonin production and secretion by hairy root cultures of *Lithospermum erythrorhizon*. Plant Cell Rep 10: 282–285, 1991.
96. HJ Yu, S Oh, MH Oh, DW Choi, YM Kwon, SG Kim. Plant regeneration from callus cultures of *Lithospermum erythrorhizon*. Plant Cell Rep 16:261–266, 1997.
97. S Sommer, A Koehle, K Yazaki, K Shimomura, A Bechthold, L Heide. Genetic engineering of shikonin biosynthesis. Hairy root cultures of *Lithospermum erythrorhizon* transformed with the bacterial *ubiC* gene. Plant Mol Biol 39:683–693, 1999.
98. H Mizukami, M Konoshima, M Tabata. Effect of nutritional factors on shikonin derivative formation in *Lithospermum erythrorhizon* callus cultures. Phytochemistry 16:1183–1186, 1977.
99. H Fukui, N Yoshikawa, M Tabata. Induction of shikonin formation by agar in *Lithospermum erythrorhizon* cell suspension cultures. Phytochemistry 22: 2451–2453, 1983.
100. M Tabata, K Yazaki, Y Nishioka, E Yoneda. Inhibition of shikonin biosynthesis by photodegradation products of FMN. Phytochemistry 32:1439–1442, 1993.
101. K Yazaki, H Fukui, H Kikuma, M Tabata. Regulation of shikonin production by glutamine in *Lithospermum erythrorhizon* cell cultures. Plant Cell Rep 6: 131–134, 1987.
102. R Verpoorte, R van der Heiden, PRH Moreno. Biosynthesis of terpenoid indole alkaloids in *Catharanthus roseus* cells. Alkaloids 49:221–299, 1997.
103. AJ Parr, ACJ Peerless, JD Hamill, NJ Walton, RJ Robins, MJC Rhodes. Alkaloid production by transformed root cultures of *Catharanthus roseus*. Plant Cell Rep 7:309–312, 1988.
104. PRH Moreno, R van der Heijden, R Verpoorte. Effect of terpenoid precursor feeding and elicitation of formation of indole alkaloids in cell suspension cultures of *Catharanthus roseus*. Plant Cell Rep 12:702–705, 1993.
105. OA Moreno-Valenzuela, RM Galaz-Avalos, Y Minero-Garcia, VM Loyola-Vargas. Effect of differentiation on the regulation of indole alkaloid production in *Catharanthus roseus* hairy roots. Plant Cell Rep 18:99–104, 1998.
106. C Canel, MI Lopes-Cardoso, S Whitmer, L van der Fits, G Pasquali, R van der Heiden, JHC Hoge, R Verpoorte. Effects of over-expression of strictosidine synthase and tryptophan decarboxylase on alkaloid production by cell cultures of *Catharanthus roseus*. Planta 205:414–419, 1998.
107. FLH Menke, S Parchamann, MJ Mueller, JW Kijne, J Memelink. Involvement of the octadecanoid pathway and protein phosphorylation in fungal elicitor–induced expression of terpenoid indole alkaloid biosynthetic genes in *Catharanthus roesus*. Plant Physiol 119:1289–1296, 1999
108. L van der Fits. Transcriptional regulation of stress-induced plant secondary metabolism. PhD dissertation, Leiden University, Leiden, The Netherlands, 2000.
109. ME Wall, MC Wani, CE Cooke, KH Palmer, AT McPhail, GA Sim. Plant antitumor agents. I. The isolation and structure of camptothecin, a novel al-

kaloidal leukemia and tumor inhibitor from *Camptotheca acuminata*. J Am Chem Soc 88:3888–3890, 1966.
110. CR Hutchinson. Camptothecin: chemistry, biogenesis and medicinal chemistry. Tetrahedron 37:1047–1065, 1981.
111. K Sakato, H Tanaka, N Mukai, M Misawa. Isolation and identification of camptothecin from cells of *Camptotheca acuminata* suspension cultures. Agric Biol Chem 38:217–218, 1974.
112. H Sudo, Y Hasegawa, Y Matsunaga. Patent Japan H2-70746, 173–179, 1991.
113. H Wiedenfeld, M Furmanowa, E Roeder, J Guzewska, W Gustowski. Camptothecin and 10-hydroxycamptothecin in callus and plantlets of *Camptotheca acuminata*. Plant Cell Tissue Organ Cult 49:213–218, 1997.
114. SH Song, SY Byun. Characterization of cell growth and camptothecin production in cell cultures of *Camptothecan acuminata*. J Microbiol Biotechnol 8:631–638, 1998.
115. M Kitajima, M Nakamura, H Takayama, K Saito, J Stöckigt, N Aimi. Constituents of regenerated plants of *Ophiorrhiza pumila*; formation of a new glycocamptothecin and predominant formation of (3*R*)-deoxypumiloside over (3*S*)-congener. Tetrahedron Lett 52:8997–9000, 1997.
116. K Saito, H Sudo, M Yamazaki, M Koseki-Nakamura, M Kitajima, H Takayama, N Aimi. Feasible production of camptothecin by hairy root culture of *Ophiorriza pumila*. Plant Cell Rep 20:267–271, 2001.
117. H Sudo, T Yamakawa, M Yamazaki, N Aimi, K Saito. Bioreactor production of camptothecin by hairy root cultures of *Ophiorrhiza pumila*. Biotech Lett 24: in press, 2002.
118. MC Wani, HL Taylor, ME Wall, P Coggon, AT McPhail. Plant antitumor agents. VI. The isolation and structure of taxol, a novel antileukemic and antitumor agent from *Taxus brevifolia*. J Am Chem Soc 93:2325–2327, 1971.
119. ERM Wickremesinhe, RN Arteca. XXI *Taxus* species (yew): in vitro culture, and the production of taxol and other secondary metabolites. In: YPS Bajaj, ed. Biotechnology in Agriculture and Forestry. Vol 41, Medicinal and Aromatic Plants X. Berlin: Springer-Verlag, 1998, pp 415–442.
120. Y Yukimune, H Tabata, Y Higashi, Y Hara. Methyl jasmonate–induced overproduction of paclitaxel and baccatin III in *Taxus* cell suspension cultures. Nat Biotechnol 14:1129–1132, 1996.
121. HJ Woerdenbag, W van Uden, HW Frijlink, CF Lerk, N Pras, TM Malingre. Increased podophyllotoxin production in *Podophyllum hexandrum* cell suspension cultures after feeding coniferyl alcohol as a β-cyclodextrin complex. Plant Cell Rep 9:97–100, 1990.
122. T Muranaka, M Miyata, K Ito, S Tachibana. Production of podophyllotoxin in *Juniperus chinensis* callus cultures treated with oligosaccharides and a biogenetic precursor. Phytochemistry 49:491–496, 1998.
123. AJ van Hengel, RM Buitelaar, HJ Wichers. VII *Camptotheca acuminata* Decne: in vitro culture and the production of camptothecin. In: YPS Bajaj, ed. Biotechnology in Agriculture and Forestry. Vol 28, Medicinal and Aromatic Plants II. Berlin: Springer-Verlag, 1994, pp 98–112.

124. G Roja, MR Heble. The quinoline alkaloids camptothecin and 9-methoxy-camptothecin from tissue cultures and mature trees of *Nothapodytes foetida*. Phytochemistry 36:65–66, 1994.

125. T Smollny, H Wichers, S Kalenberg, A Shahsavari, M Petersen, AW Alfermann. Accumulation of podophyllotoxin and related lignans in cell suspension cultures of *Linum album*. Phytochemistry 48:975–979, 1998.

126. W van Uden, N Pras, TM Maingre. The accumulation of podophyllotoxin-β-D-glucoside by cell suspension cultures derived from the conifer *Callitris drummondii*. Plant Cell Rep 9:257–260, 1990.

127. W van Uden, N Pras, JF Viser, TM Malingre. Detection and identification of podophyllotoxin produced by cell cultures derived from *Podophyllum hexandrum* royle. Plant Cell Rep 8:165–168, 1989.

128. PG Kadkade. Formation of podophyllotoxins by *Podophyllum peltatum* tissue cultures. Naturwissenschaften 68:481–482, 1981.

129. J Berlin, N Bedorf, C Mollenschott, V Wray, F Sasse, G Höfle. On the podophyllotoxins of root cultures of *Linum flavum*. Planta Med 204–206, 1988.

6

Genetic Transformation of Plants and Their Cells

Richard M. Twyman, Paul Christou, and Eva Stöger
Molecular Biotechnology Unit, John Innes Centre, Norwich, United Kingdom

I. INTRODUCTION

Plant transformation is an indispensable tool, both for the experimental investigation of gene function and for the improvement of plants either by enhancing existing traits or introducing new ones (1–3). It is now possible to introduce and express DNA stably in nearly 150 different plant species. Many aspects of plant physiology and biochemistry that cannot be addressed easily by any other experimental means can be investigated by the analysis of gene function and regulation in transgenic plants. This offers an unprecedented opportunity to study the molecular basis of important processes that have been intractable to conventional analysis, such as the complex signal transduction pathways and hierarchies of genetic regulation that underlie plant-microbe interactions, sexual reproduction, and development.

However, much of the effort in plant transformation research reflects expectations that the technology can rapidly produce plants with improved or novel traits for the benefit of mankind (2). These improvements would be difficult or impossible to achieve with conventional breeding alone. The fruits of this research are already available and include herbicide-tolerant plants; plants showing resistance to insect pests and viral, bacterial, and fungal diseases; plants with improved nutritional qualities; plants used as

111

bioreactors to produce valuable proteins such as antibodies and vaccines; and plants in which metabolic pathways have been engineered to produce valuable products such as speciality oils and drugs (4) (see Chapters 14–30 in this volume). This chapter reviews the technology available for gene transfer to plants, focusing particularly on recent advances, and discusses the constraints that must still be overcome to realize the full potential of plant transformation.

II. REQUIREMENTS FOR PLANT TRANSFORMATION

A. Overview

In general, plant transformation systems are based on the introduction of DNA into totipotent plant cells, followed by the regeneration of such cells into whole fertile plants. Two essential requirements for plant transformation are therefore an efficient method for introducing DNA into plant cells and the availability of cells or tissues that can easily and reproducibly regenerate whole plants. DNA can be introduced into isolated cells or protoplasts, explanted tissues, callus, or cell suspension cultures. However, the process is characteristically inefficient and only a proportion of cells in a target population will be transformed. These cells must be induced to proliferate at the expense of nontransformed cells, and this can be achieved by introducing a selectable marker gene and regenerating plants under the appropriate selective regime. Efficient DNA delivery, competence for regeneration, and a suitable selection system are therefore prerequisites for most plant transformation systems, although there has been recent development in the application of in planta transformation strategies, which circumvent the requirement for extensive tissue culture (discussed later). Other criteria that define an efficient transformation system are listed in Table 1.

B. DNA Transfer Methods

DNA transfer to plants was first attempted in the 1960s, although the lack of selectable markers and molecular tools to confirm transgene integration and expression made the outcome of such experiments unclear (5). A breakthrough came in the late 1970s with the elucidation of the mechanism of crown gall formation by *Agrobacterium tumefaciens* (6). The discovery that virulent strains of *A. tumefaciens* carried a large plasmid that conferred the ability to induce crown galls and that part of the plasmid (the T-DNA) was transferred to the plant genome of crown gall cells provided a natural gene transfer mechanism that could be exploited for plant transformation (7). Tobacco plants carrying recombinant T-DNA sequences were first generated in 1981, although the foreign genes were driven by their own promoters and

TABLE 1   Criteria for an Efficient Plant Transformation System

Essential prerequisites for plant transformation
  Efficient method of DNA transfer
  Availability of cells/tissues competent for regeneration (not required for in planta transformation strategies)
  Suitable selection system
Other criteria that define an efficient transformation system
  Reproducibly high transformation efficiency (number of transgenic plants recovered as a proportion of cells/explants originally transformed)
  Minimal culture time, to avoid somaclonal variation and sterility.
  High-frequency recovery of phenotypically normal, fertile transgenic plants
  Technically simple procedure
  Economical procedure
  Versatile (applicable to many species)
  Genotype-independent (applicable to all cultivars and varieties, including elite genotypes)

were not expressed in plant cells (8). The first transgenic tobacco plants expressing recombinant genes in integrated T-DNA sequences were reported in 1983 (9). The technique of *Agrobacterium*-mediated transformation has been developed and refined since then to become a widely used strategy for gene transfer to plants.

Although it is convenient and versatile, a major limitation of *Agrobacterium*-mediated transformation is its restricted host range, which until relatively recently excluded most monocotyledonous plants (3). The development of strategies to extend the range of plants susceptible to *Agrobacterium* infection is discussed in the following. A number of alternative plant transformation methods were developed to facilitate gene transfer to these recalcitrant species. These methods can be grouped under the term "direct DNA transfer" and include the transformation of protoplasts using polyethylene glycol (PEG) or electroporation, microinjection, the use of silicon carbide whiskers, and particle bombardment. So far, only direct DNA transfer to protoplasts and particle bombardment have gained widespread use (2). The development and application of *Agrobacterium*-mediated transformation, particle bombardment, protoplast transformation, and other transformation techniques is discussed in more detail below.

C. Cell and Tissue Culture and Plant Regeneration

Small explants of living plant tissue can be maintained on a simple nutrient medium. Transformation may be carried out on tissues dissected from seeds, leaves, stems, roots or buds because under the appropriate conditions these can be induced to dedifferentiate and proliferate to produce undifferentiated callus cultures. Different hormone treatments induce callus to form shoots and roots, allowing the regeneration of whole plants. Alternatively, the callus can be maintained and subcultured indefinitely or can be broken up in liquid medium to provide a cell suspension culture, which can yield individual cells and protoplasts. Depending on the transformation method, tissue explants, callus, dispersed cells, or protoplasts can be used as transformation targets (Fig. 1).

After transformation, cells are allowed to proliferate on selective medium to increase the amount of callus and kill nontransformed cells. Transgenic plants can then be regenerated by two methods: somatic embryogenesis or organogenesis. Somatic embryogenesis involves the formation of embryogenic callus direct from somatic tissues. This recapitulates the entire developmental pathway, including the embryonic stage. Organogenesis involves the direct growth of shoots from the callus of transformed tissues or in some cases direct growth from transformed explants without a callus stage. The shoots can be transferred to rooting medium and regenerated into plants or grafted onto seedling rootstock and propagated. In some species, only one regenerative process is possible under the conditions used for transformation. For example, transgenic rice and maize plants are generated predominantly by somatic embryogenesis, and transgenic cassava plants can be generated only by the organogenesis of shoots. In other species, such as banana and soybean, both processes are possible and the method of choice depends on the starting material and culture conditions and which process produces transgenic plants the most rapidly and with the greatest efficiency.

An alternative to these processes is the development of transgenic plants by true embryogenesis from transformed seeds, zygotes, or gametes that undergo diploidization. Such techniques do not require extensive tissue culture and are discussed along with in planta transformation strategies later.

D. Selectable Marker Genes

Most foreign genes introduced into plants do not confer a phenotype that can be conveniently used for selective propagation of transformed cells. For this reason, a selectable marker gene is introduced at the same time as the nonselectable foreign DNA. This confers upon transformed cells the ability to survive in the presence of a particular chemical, the selective agent, that is toxic to nontransformed cells (10,11). In direct DNA transfer methods, the selectable marker and nonselected transgene(s) may be linked on the same cointegrate vector or may be introduced on separate vectors (cotransformation). Both strategies are suitable because exogenous DNA, whether

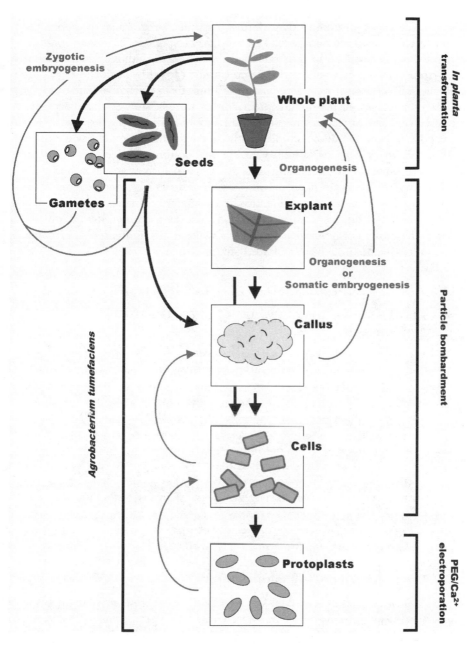

FIGURE 1 Strategies for the transformation of higher plants. The boxes show the various different targets for transformation and how they are obtained from whole plants (black arrows). The methods used to transform each of these targets are represented by the square brackets. The gray arrows show the routes used to obtain whole transgenic plants.

homogeneous or a mixture of different plasmids, predominantly integrates at a single locus (cointegration) (12). Similarly, separate T-DNAs introduced into the same cell tend to integrate at the same locus, although independent integration events have also been demonstrated (13). Ideal selection systems should kill nontransformed cells rapidly, but the selective agent (and its detoxification products) should affect neither the physiology nor the regenerative capacity of transformed cells and should be inactivated quickly. The selective regime should be simple and inexpensive to implement. Most plant selectable marker genes currently in use are dominant selectable markers, i.e., they function in the hemizygous state because they confer a heterologous property upon the plant cell for which there is no competing endogenous activity. In most cases, selection is toxic to nontransformed cells, but one system confers the ability to use mannose as a carbon source and restricts the growth of nontransformed cells without killing them (14). Popular selection systems in current use are listed in Table 2.

III. *AGROBACTERIUM*-MEDIATED TRANSFORMATION

A. *Agrobacterium tumefaciens:* A Natural Gene Transfer System

Agrobacterium tumefaciens is a Gram-negative soil bacterium responsible for crown gall disease, a neoplastic disease of many dicotyledonous plants characterized by the appearance of large tumors (galls) on the stems. Virulence is conferred by a large tumor-inducing plasmid (Ti plasmid) containing genes encoding plant hormones (auxins and cytokinins) and enzymes that catalyze the synthesis of amino acid derivatives termed opines (6). The plant hormones are responsible for the deregulated cell proliferation that accompanies crown gall growth, while the opines are secreted by the plant cells and used by the bacteria as food. These genes are contained on a specific region of the Ti plasmid, the T-DNA (transfer DNA), so called because it is transferred to the plant nuclear genome under the control of *vir* (virulence) genes carried elsewhere on the Ti plasmid (15). It is this natural gene transfer mechanism that is exploited for plant transformation.

B. Development of Ti-Plasmid Vectors

The earliest indication that T-DNA could be used as a plant transformation vector was the demonstration that DNA from an *Escherichia coli* plasmid (the Tn7 transposon) could be stably transferred to the plant genome by first incorporating it into the T-DNA (16). However, transgenic plants could not be recovered from the transformed cells, either by regeneration or by grafting onto normal plants, because the hormones encoded by the T-DNA oncogenes

TABLE 2 Plant Selectable Marker Genes in Current Use

Gene (product)	Source	Phenotype and other comments
aad (aminoglycoside adenyltransferase)	*Shigella flexneri*	Provides resistance to trimethoprim, streptomycin, spectinomycin, and sulfonamides. Used mainly for chloroplast transformation
ble (glycopeptide-binding protein)	*Streptalloteichus hindustantus*	Resistance to the glycopeptide antibiotics bleomycin and pheomycin (and the derivative Zeocin)
dhfr (dihydrofolate reductase)	Mouse	Resistance to methotrexate
sul1 (dihydropteroate synthase)	*E. coli*	Resistance to sulfonamides (Asulam)
epsps (enolpyruvylshikimate phosphate synthase)	*Petunia hybrida*	Resistance to the herbicide glyphosate
hpt (hygromycin phosphotransferase)	*Klebsiella*	Resistance to the aminoglycoside antibiotic hygromycin B. Popular selectable marker in rice
manA (mannose-6-phosphate isomerase, MIP)	*E. coli*	Ability to grow on mannose as sole carbon source
neo/nptII/aphII (neomycin phosphotransferase)	*E. coli*	Resistance to the aminoglycoside antibiotics neomycin, kanamycin, and geneticin (G148)
bar (phosphinothricin acetyltransferase)	*Streptomyces hygroscopicus*	Resistance to phosphinothricin (PPT), which is a component of the herbicides bialophos, Basta, and glufosinate

For details, see Refs. 10 and 11.

caused unregulated and disorganized callus growth (16). In rare cases, shoots were derived from such callus tissue, and analysis showed that much of the T-DNA (including the oncogenes) had been deleted from the genome (17). An important step in the development of T-DNA vectors was the realization that the only requirements for T-DNA transfer to the plant genome were the *vir* genes and the 24-bp direct repeat structures marking the left and right borders of the T-DNA. No genes within the T-DNA were necessary for transformation, and any sequence could be incorporated therein. This allowed the development of disarmed Ti plasmids (18) lacking all the oncogenes, facilitating T-DNA transfer to plant cells without causing neoplastic growth.

Once suitable selectable markers had been incorporated into the T-DNA (Table 2), Ti plasmids became very powerful gene delivery vectors. However, wild-type Ti plasmids were unsuitable for the task due to their large size, which made them difficult to manipulate in vitro (large plasmids have a tendency to fragment and they lack unique restriction sites for subcloning). An early strategy to overcome this problem was the development of intermediate vectors, where the T-DNA was subcloned into a standard *E. coli* plasmid vector, allowing in vitro manipulation by normal procedures, and then integrated into the T-DNA sequence of a disarmed Ti plasmid resident in *A. tumefaciens* by homologous recombination (19). This system was simple to use but relied on a complex series of conjugative interactions between *E. coli* and *A. tumefaciens*, requiring three different bacterial strains (triparental matings). However, because the *vir* genes act in trans to mobilize the T-DNA, it was soon discovered that the use of large natural Ti plasmids was unnecessary. Intermediate vectors have been largely superseded by binary vectors (20), in which the *vir* genes and the T-DNA are cloned on separate plasmids. These can be introduced into *A. tumefaciens* by conjugation with an *E. coli* donor or by freeze-thaw cycles or electroporation. Most contemporary *Agrobacterium*-mediated transformation systems employ binary vectors.

C. General Protocol for *Agrobacterium*-Mediated Transformation

At first, *Agrobacterium*-mediated transformation was achieved by cocultivating a virulent *A. tumefaciens* strain containing a recombinant Ti plasmid with plant protoplasts and then obtaining callus from the protoplasts from which fertile plants were regenerated. This strategy was widely replaced by a simpler method in which small discs were punched from the leaves of recipient species and incubated in medium containing *A. tumefaciens* prior to transfer to solid medium (21). Infection can be promoted under conditions that induce virulence (presence of 10–200 μM acetosyringone or α-hydroxyacetosyringone, acidic pH, and room temperature), although these need to be optimized for different species. After coculture for several days, the discs are transferred to a medium containing selective agents to eliminate nontransformed plant cells, antibiotics to kill the bacteria, and hormones to induce shoot growth. After a few weeks, shoots develop from transformed callus cells. These can be removed and transferred to rooting medium or grafted onto seedling rootstock. Most current protocols for the *Agrobacterium*-mediated transformation of solanaceous plants are variations on the leaf disc theme, although different tissue explants are suitable transformation targets in different species. Alternative methods are required for the transformation of monocots. Rapidly dividing embryonic cells (e.g., immature embryos or callus induced from scutellar tissue) are required for the transformation of rice (22) and other cereals. These are cocultured with *Agrobacterium* in the presence of acetosyringone (3).

D. Recent Advances—Expanding the *Agrobacterium* Host Range

Although versatile and efficient for many plants, *Agrobacterium*-mediated transformation was, until recently, limited to dicots and monocots of the orders Liliales and Arales (which excludes most of the agronomically important cereals). The range of plants amenable to genetic manipulation by direct DNA transfer is limited only by the availability of cells competent for regeneration, and the range of plants amenable to genetic manipulation by *A. tumefaciens* is further restricted to the species with cells that are both competent for regeneration and susceptible to infection by *A. tumefaciens*.

Driven in part by the potential financial benefits of transgenic crop plants, much research effort has been directed toward extending the *Agrobacterium* host range. In some cases careful optimization of transformation conditions is required. For example, the addition of a surfactant to the inoculation medium was responsible for the successful transformation of wheat embryos and callus (23). Such optimization can also dramatically increase the efficiency of transformation in susceptible plants. The duration of cocultivation was shown to be important for the transformation of citrange (24). The use of feeder cells can also significantly increase transformation efficiency; for example, Niu et al. (25) have shown that cocultivation on a tobacco cell feeder layer significantly increases the efficiency of *Agrobacterium* infection of peppermint. Artificial infiltration of the bacteria into plant tissues using a syringe or by vacuum infiltration can improve transformation efficiencies in tobacco and is routinely used for the transformation of *Arabidopsis*.

A further consideration is the fact that *A. tumefaciens* usually infects wounded cells in its natural hosts, being attracted to phenolic compounds such as acetosyringone that induce expression of the *vir* genes. Infection and T-DNA transfer are therefore stimulated by wounding tissues (e.g., by crushing or cutting) or by treating explants with acetosyringone. Several monocot species have been infected by pretreating the cells with the exudate from wounded dicot plants, such as potato (26). The transformation of recalcitrant species such as barley (27) and sunflower (28) has been achieved by prewounding embryonic or meristematic tissue, respectively, with metal particles or glass beads, while sonication has facilitated the uptake of *Agrobacterium* by soybean tissue (29). The preinduction of virulence genes may circumvent the necessity to induce a wounding response. Infection may be achieved by testing a range of different *Agrobacterium* strains, and hypervirulent strains such as ALG1 in which the Ti plasmid contains *vir* genes enhanced by mutation have proved especially useful. Virulence can also be increased by the use of superbinary vector systems wherein the *vir* component carries multiple copies of the virulence genes. Superbinary vector systems such as pTOK have facilitated the transformation of important monocots, including maize (30) and sugarcane (31) (see Table 3).

Finally, some crops are amenable to *Agrobacterium* infection but resist gene transfer in other ways. Comparative studies of T-DNA gene expression in tobacco and maize cells, for example, have suggested that maize inhibits

TABLE 3   Recent Advances in *Agrobacterium* Gene Transfer Technology for the Transformation of Cereals

Plant	*Agrobacterium* strain	Vector name	Strain/vector properties	Target tissue
Rice	EHA101 LBA4404	pIG121 pTOK233	Normal strain LBA4404 containing superbinary vector pTOK233 performed better than hypervirulent strain EHA101 containing the same superbinary vector	Immature embryos, callus, suspension cells
Maize	LBA4404	pTOK233	Superbinary vector	Immature embryos
Barley	AGL1	pDM805	Hypervirulent strain	Immature embryo explants
Wheat	C58 (ABI)	pMON18365	Binary vector	Immature embryos, embryogenic callus

Table shows *Agrobacterium* strain, vector name, and type and the tissue used as a transformation target.
Source: Data are taken from Refs. 22, 23, 27, 30, and 99.

the integration of T-DNA by suppressing gene expression (32). Other plants defend themselves against T-DNA integration by forming necrotic boundaries. In some species (e.g., grape), such defenses have been overcome by treatment with antioxidants (33).

E. Recent Advances—Increasing the Capacity of T-DNA

Another early problem with *Agrobacterium*-mediated transformation was the inability to transfer large DNA fragments to the plant genome. Although a maximum size limit for T-DNA gene transfer was not determined, the transfer of foreign DNAs larger than 30 kbp was not routinely achievable until about five years ago. This prevented the transfer of large genes or T-DNA sequences containing multiple genes. The problem was addressed by Hamilton and colleagues, who developed a novel binary vector system in which the T-DNA vector was based on a bacterial artificial chromosome and the *vir* component expressed higher levels of *virG* and *virE*. This bibac (binary BAC) vector facilitates the transfer of up to 150 kbp of DNA to the plant genome, allowing the introduction of gene libraries into plants and the simultaneous introduction of many genes on a cointegrate vector (34).

F. Conjugation Systems and Ri Plasmids

The mechanism of T-DNA transfer has many similarities with bacterial conjugation systems, and indeed it has been shown that broad host range bacterial plasmids will transfer from *Agrobacterium* to the plant genome under the control of their own mobilization genes (35). Another system with close similarities to the *A. tumefaciens* Ti plasmid is the Ri plasmid of *Agrobacterium rhizogenes*. This causes hairy root disease though the integration of its own T-DNA, carrying genes that activate endogenous plant hormones. Ri plasmids have also been exploited as gene transfer vectors and have the advantage that they do not induce cell proliferation in the host plant (i.e., they are naturally disarmed) (Table 3). *Agrobacterium* strains containing both Ti and Ri plasmids often transfer both T-DNAs to the host plant (36). The Ri T-DNA induces hairy root disease and acts as a marker for transformation, while the disarmed Ti plasmid carries the transgenes of interest. In this system, dominant markers are not required and regeneration is not prevented by the Ri T-DNA.

IV. PARTICLE BOMBARDMENT

In 1987, a novel method for plant transformation was introduced by John Sanford and colleagues (37). This was based on a device that used gunpowder to accelerate small tungsten particles to a velocity of approximately 400 m s^{-1}. The particles were coated with DNA and could penetrate plant cells without killing them. Initial experiments showed that this device was capable of delivering tobacco mosaic virus (TMV) RNA into onion epidermis cells (resulting in virus coat protein synthesis) and similarly DNA comprising the CaMV 35S promoter and the *cat* reporter gene could be delivered, resulting in high levels of transient CAT activity. Later experiments showed that particle bombardment was an efficient method for stable integrative transformation (38). A different device was developed by Agracetus Inc., which used a high-voltage electrical discharge to generate the accelerating force (Accell™) (39). The rather erratic gunpowder-based device was refined to a system based on high-pressure blasts of helium (40), and this is the only commercially available particle gun, which is marketed by Bio-Rad.

A. Advantages and Disadvantages of Particle Bombardment

Particle bombardment is simple both conceptually and in practice. Typically, plasmid DNA is prepared by standard methods and precipitated onto tungsten or gold particles using $CaCl_2$. Spermidine and PEG are included to protect the DNA during precipitation, and the particles are washed and suspended in ethanol before drying onto Mylar aluminized foil. This is fired against a retaining screen that allows the microprojectiles through, to strike the target tissue. Particle bombardment is widely used because it circumvents two major limitations of the *Agrobacterium* system. First, it is possible to achieve the transformation of any species and cell type by this method because DNA delivery is controlled entirely by physical rather than biological parameters. The range of plant species transformable by particle bombardment is therefore restricted only by the competence of cells for regeneration, and the technique is genotype independent and thus useful for the transformation of elite cultivars as well as model varieties. However, careful optimization is required to tailor the method for different species and different cell types and to achieve the highest efficiency transformation with the least cell damage. Important parameters include acceleration method, particle velocity (controlled by the discharge voltage and/or gas pressure), particle size, and the use of different materials (tungsten, gold) (reviewed in Ref. 41). It has also been shown that, for some species and/or tissues, osmotic pretreatment prior to bombardment increases the transformation efficiency (42). This can be achieved by partial drying or the addition of osmoticum (mannitol and/or sorbitol) to the culture medium. Optimization experiments are usually carried out by bombarding explants with a screenable marker gene such as *gusA* and assaying for transient expression (41). In general, stable transformation by any direct DNA transfer method occurs at a much lower frequency than transient transformation. A number of different reporter genes are used for transient expression analysis in plants as shown in Table 4.

Second, particle bombardment allows the stable and heritable introduction of many different genes at once using different plasmids, as these tend to concatemerize to form one DNA cluster that integrates at a single locus. Conversely, multiple transformation using the *Agrobacterium* system requires the cointegration of all the genes in the same T-DNA. Chen et al. (43) reported the cotransformation of rice with 14 separate plasmids containing various marker genes and showed data confirming the cointegration of at least 13 of the plasmids in one plant. Cotransformation has also been used to introduce up to four agronomically important genes into rice plants, producing plants showing resistance to a spectrum of insect pests and plants with pyramiding resistance against individual pests (44,45). Cointegration prevents different transgenes segregating at meiosis. This can be very important in breeding programs where plants carry two or more transgenes required to generate a single protein, e.g., in plants expressing recombinant human antibodies.

A potential disadvantage of the particle bombardment method is the cost of purchasing or hiring the bombardment device. However, a number of articles have been published providing instructions for the construction of alternative economical devices, such as the particle inflow gun based on flowing helium (46). Another disadvantage of particle bombardment is the tendency for DNA sequences introduced by this method to undergo complex rearrangements prior to or during integration. Transgene rearrangement is a pitfall of all direct DNA transfer methods but is perhaps more acute in particle bombardment because the forces involved may cause more DNA fragmentation than other methods and because bombarded plant cells may be induced to produce DNA degradation and repair enzymes in response to their injury (47). This may limit the usefulness of particle bombardment for the introduction of large DNA molecules. However, although an upper size limit has not been established, it has been possible to introduce YAC complementary DNA (cDNA) clones into plant cells by this method (48).

B. Recent Advances

Recent advances in particle bombardment technology include the clean DNA and agrolistic systems for limiting the amount of plasmid backbone sequence that enters the plant genome, as discussed in more detail below. Successful nuclear integration requires that the metal particle actually enters the nucleus (49), and recent experiments have shown that transgene integration may be facilitated by damage caused to DNA strands as the particle moves through

TABLE 4   Reporter Genes Used in Plants

Reporter gene (product)	Comments
gusA (β-glucuronidase) 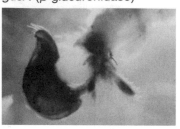	Source: *E. coli* Activity: catalyzes the hydrolysis of β-glucuronides Assays: nonisotropic; in vitro assays are colorimetric or fluorometric; also histochemical assay format using X-gluc Advantages: simple, sensitive, quantitative, many assay formats available, inexpensive Disadvantages: assays are destructive; enzyme is stable so unsuitable for studies of down-regulation
cat (chloramphenicol acetyltransferase)	Source: *E. coli* Tn9 Activity: catalyzes the transfer of acetyl groups from acetyl coenzyme A to chloramphenicol Assays: in vitro assays only, isotropic Advantages: simple to perform Disadvantages: low sensitivity, expensive, low resolution in vivo, reliance on isotopic assay format
luc (luciferase)	Source: The firefly *Photinus pyralis* Activity: light produced in the presence of luciferase, its substrate luciferin, oxygen, Mg^{2+}, and ATP Assays: nonisotopic bioluminescent assays in vitro and in vivo Advantages: sensitive, rapid turnover, quantitative Disadvantages: expensive detection equipment, limited reproducibility of some assay formats
Anthocyanin regulators	Source: *Zea mays* Activity: induces pigmentation Assays: visual screening for pigmented cells in vivo Advantages: simple, inexpensive, nondestructive Disadvantages: low sensitivity, not quantitative, background expression, adverse effects on transgenic plants

TABLE 4 Continued

Reporter gene (product)	Comments
GFP (green fluorescent protein) 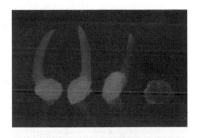	Source: the jellyfish *Aequorea victoria* Activity: intrinsic fluorescence under blue-UV light Assays: nonisotopic, in vivo assays in live plants Advantages: intrinsic activity (no substrate requirements), sensitivity, use in live plants Disadvantages: weak signal in some systems (this is being addressed through the use of modifed GFPs with stronger emission and emission at different wavelengths)

Examples of GFP and GUS activity are shown.
Source: Adapted from Refs. 100 and 101.

the nucleus. Occasionally, exogenous DNA integrates at two different sites separated by megabase pairs of DNA (although these still segregate as a single locus). Confocal analysis of the interphase nucleus following fluorescence in situ hybridization (FISH) to transgene sequences has shown that these sites may occupy the same region of the nucleus at interphase and that such integration patterns may reflect localized damage to DNA caused by the particles (author's unpublished data, in preparation).

V. TRANSFORMATION OF PROTOPLASTS

Protoplasts are plant cells from which the cell wall has been enzymatically or mechanically removed. As with animal cells, the contents of the protoplast cytoplasm are enclosed in a cell membrane, and the transformation of protoplasts can be achieved using many of the procedures routinely used to transfect cultured animal cells. Two procedures that are commonly used to introduce DNA into protoplasts include the uptake of naked DNA mediated by polyethylene glycol and a divalent cation (either Ca^{2+} or Mg^{2+}) (50,51) and electroporation (52), although other agents such as lipofectin have also been used (53). Plant protoplasts can also be transformed using *Agrobacterium*. Following transformation, protoplasts are placed on selective medium and allowed to regenerate new cell walls. The cells then proliferate and form a callus from which embryos or shoots and roots can be regenerated with appropriate hormonal treatments (54).

In principle, protoplasts from any plant species can be transformed, but the technology is limited by the ability of protoplasts to regenerate into whole plants, which is not possible for all species. Although economical and potentially a very powerful procedure, direct transformation of protoplasts is disadvantageous because of the long culture times involved. Not only does this mean that the transformation process itself is time consuming, but cells cultured for extensive periods either fail to regenerate or frequently regenerate plants that show full or partial sterility and other phenotypic abnormalities (somaclonal variation). Protoplast preparation, maintenance, and transformation is a skilled technique, which, compared with *Agrobacterium*-mediated transformation and particle bombardment procedures, requires a greater investment in training. Advances in other transformation techniques are making protoplast transformation obsolete.

VI. OTHER DIRECT TRANSFORMATION METHODS

Although the three transformation methods just discussed provide a route for the transformation of most plant species, there is value in the continuing exploration of novel transformation techniques. In some cases, it is possible to learn lessons from animal cell technology, where a number of alternative transformation systems have been established. Microinjection of DNA into animal eggs and zygotes, for example, is a routine procedure for generating transgenic animals that is applicable to all animal species. Microinjecting DNA into plant cells, while laborious and technically challenging, is advantageous in that it is the only current strategy available to study transformation events at a single-cell level. The injection of DNA into plant zygotes is being explored as a method for regenerating whole transgenic plants; however, it is currently highly inefficient. For example, isolated maize zygotes microinjected with DNA mimic embryonic development but abort at an early stage (55). Recently, the same technique has been applied to barley protoplasts, resulting in the recovery of numerous embryo-like structures but only two transgenic plants (56).

In 1986, Kurata et al. (57) devised a novel transformation strategy for animal cells in which DNA was taken up from solution after piercing the cell membrane with a finely focused laser beam. Although applicable to only a small number of cells, this technique achieved stable transformation efficiencies of 0.5%. A similar technique for plant transformation is available (58), although stable transformation has not been achieved.

Electroporation is used for the transformation of plant protoplasts but has been adapted for the transformation of walled plant cells in tissues. For example, maize explants have been transformed by electroporation either following partial enzymatic removal of the cell wall (59) or without treatment (60). This technique has also been successfully applied to rice and sugarcane explants (61,62). Pollen has also been transformed using electroporation (63).

Finally, a novel approach for the transformation of maize cells is to mix cells with silicon carbide whiskers. These penetrate the cells, allowing DNA uptake presumably through transient pores. This method is cheap, reproducible, and very simple to perform. Although initially used only for maize transformation (64), the technique has now been applied to other cereals (3,65).

VII. TRANSFORMATION WITHOUT TISSUE CULTURE

Most of the transformation procedures discussed so far have one major disadvantage—they rely on tissue culture for the regeneration of whole plants. As discussed, the tissue culture requirement adds significantly to the time and cost of producing transgenic plants and is perhaps the most serious limitation to the technology because it restricts the range of species amenable to genetic manipulation. Methods that circumvent tissue culture all together —in planta transformation systems—are therefore highly desirable, although until recently they were available only for the model dicot *Arabidopsis thaliana*. These techniques include seed transformation, in planta DNA inoculation, and flower dipping or infiltration, and all involve *Agrobacterium*-mediated transformation.

For seed transformation, *Arabidopsis* seeds are imbibed and cocultivated with *Agrobacterium tumefaciens* and then germinated (66). This is convenient in *Arabidopsis* because a large number of seeds are produced (10,000 per plant) and the seeds are very small. The procedure is relatively inefficient and is not very reproducible, but because the plant is small it is relatively easy to screen for transformants. A more efficient procedure is to inoculate plants with bacteria after severing the apical shoots (67). The bacteria infect the wounded meristematic tissue and about 5% of shoots emerging from the wound site are transgenic. Another method is to dip *Arabidopsis* flowers into bacterial suspension, or vacuum infiltrate the bacteria into the flowers, at around the time of fertilization (68,69). This generates plants with transformed and wild-type sectors of vegetative tissue, and these give rise to transgenic progeny at a low frequency (5–10 offspring per plant). Again, the low transformation frequency is acceptable because of the large number of seeds and the ease of screening.

In planta transformation techniques involving direct DNA transfer have also been developed. These techniques have been applied to a number of economically important crop plants, although they have yet to catch on be-cause they have low reproducibility. Generally, these techniques fall into four classes:

1. Introduction of DNA into meristematic tissue, e.g., by electro-poration, injection of DNA/lipofectin complexes, or particle bom-bardment, followed by the recovery of transgenic shoots—this has been attempted in rice (70).
2. Introduction of DNA into flowers around the time of fertilization, followed by the recovery of transgenic seeds—this has been at-tempted in cotton (71) and rice (72).
3. Imbibing seeds with a DNA solution—this has been used for soy-bean (29).
4. Introducing DNA into pollen or mixing pollen with DNA before applying to stigmata—this has been attempted with rice (72) and tobacco (73).

These techniques have met with variable levels of success.

VIII.   PLANT VIRUSES AS GENE TRANSFER VECTORS

Plant viruses do not integrate into the genome and therefore have little value as integrative transformation vectors. They also do not pass through the gametes, so their effects are not heritable through meiosis, although they can be transmitted to new stock by grafting. However, they have a number of attractive properties that suggest they could be developed as episomal gene transfer vectors. First, they naturally infect intact plant cells including those of species that have, so far, resisted other transformation methods. Viruses exist that infect all the commercially important crop plants, and in many cases even the naked nucleic acid is infectious if applied to, e.g., the leaf surface. Gene transfer by viral transduction is therefore much easier than *Agrobacterium*-mediated transformation, much simpler than protoplast transformation, and much cheaper than particle bombardment. Second, be-cause they multiply to high copy numbers in the plant cell, viruses can facilitate very high-level transgene expression. Third, many plant viruses cause systemic infections, i.e., they spread to every cell in the host plant, therefore obviating the need to regenerate transgenic plants from transformed cells. These properties have led to the development of a number of RNA virus–based expression vectors for recombinant protein synthesis, including those based on tobacco mosaic virus (74), potato virus X (75), and cowpea mosaic virus (76).

Plant viruses can also be used as a sensitive assay to detect T-DNA transfer to the plant nucleus. Since viruses replicate to such high copy numbers, viral infection symptoms can be used to confirm the presence of T-DNA in single cells of a given explant. This procedure, termed agroinfection, involves the transfer of a viral genome into the plant cell between T-DNA border repeats and was first used to confirm the genetic transformation of maize (77).

IX. ORGANELLE TRANSFORMATION

So far, only transformation of the nuclear genome has been considered. It can be desirable, however, to introduce DNA into organelle genomes as this increases the number of transgene copies in the cell and facilitates high-level transgene expression (78). The first reports of organelle transformation were serendipitous. For example, de Block et al. (79) reported the transformation of tobacco chloroplast DNA using *Agrobacterium*-mediated transformation of tobacco protoplasts. The rare organellar transformation event was discovered because of the maternal transmission of the transgene, confirmed by Southern blot analysis of chloroplast DNA. The targeted transformation of chloroplasts by particle bombardment was achieved in 1988 by Boynton and colleagues (80) using the unicellular green alga *Chlamydomonas rheinhardtii*. The same technique was used by Svab and Maliga (81) to transform tobacco chloroplast DNA. In each case, the particles penetrated the double membrane of the chloroplast so that a biological mechanism to pass through the membrane was unnecessary. More recently, PEG-mediated transformation of protoplasts has been used to introduce DNA into the tobacco chloroplast genome (82) and the two methods have been compared (83). Generally, only one of the many organelles in the cell is transformed, which means that homoplastomic cells must be obtained by intense selection. In most reports, the *aad* selectable marker gene has been used (Table 2). A recent strategy to track chloroplast transformation involves the fusion of this selectable marker to the gene for green fluorescent protein (84).

X. FUTURE PROSPECTS—CONTROLLED TRANSGENE INTEGRATION AND EXPRESSION

A. Overview

While the major goal of plant transformation technology over the last two decades has been to introduce and stably express transgenes in plants, the challenge for the next decade is to refine these techniques and introduce single transgene copies at defined sites without extraneous DNA sequences such as parts of the vector backbone or marker genes. The ultimate aim would be to produce transgenic plants with the transgene integrated at a known site, allowing the experimenter full and predictable control over transgene expression.

B. Clean DNA Transformation

A major disadvantage of all direct DNA transfer transformation methods is the tendency for exogenous DNA to undergo rearrangement and recombination events leading to the integration of multiple fragmented, chimeric, and rearranged transgene copies. Such clusters are prone to internal recombination and may therefore be unstable, as well as promoting transgene silencing. Kohli et al. (47) have shown that such rearrangements may be caused, in part, by recombinogenic elements in the vector backbone of the transformation plasmid. The backbone also contains bacterial markers and other sequences, such as the origin or replication, which are unnecessary for transformation.

A major breakthrough has been achieved in the last year in the production of transgenic plants without integrated vector backbone sequences. Fu et al. (85) describe a clean DNA technique involving the use of linear minimal transgene cassettes for particle bombardment. These cassettes lack all vector backbone elements but contain the DNA sequences essential for transgene expression: the promoter, open reading frame, and terminator. Transgenic rice plants regenerated from callus bombarded with such cassettes showed very simple integration patterns, with one or a few hybridizing bands on Southern blots. Comparative transformation experiments using whole plasmid DNA resulted in much more complex integration patterns, with multiple bands of different sizes (Fig. 2). Furthermore, release of a diagnostic fragment from the transgene showed that fewer rearrangement events had occurred in the plants carrying the linear cassettes. The progeny of these plants were examined for transgene expression, and there were no reported instances of transgene silencing. Conversely, whole plasmid transformation by particle bombardment routinely leads to silencing in about 20% of the resulting transgenic plants. This work also showed that linear cassettes promoted the same efficiency of cotransformation as normal plasmids, so the clean DNA strategy is a promising technique for the generation of transgenic plants carrying multiple transgenes.

Agrobacterium-mediated transformation was for a long time regarded as a naturally clean system, in which one to three intact T-DNA copies integrated into the genome, and only the DNA between the T-DNA border repeats was transferred. More recent data show, however, that DNA outside the border repeats is cotransferred to the plant genome at a very high frequency (>75% of transformants) (86). In many such cases, the backbone is

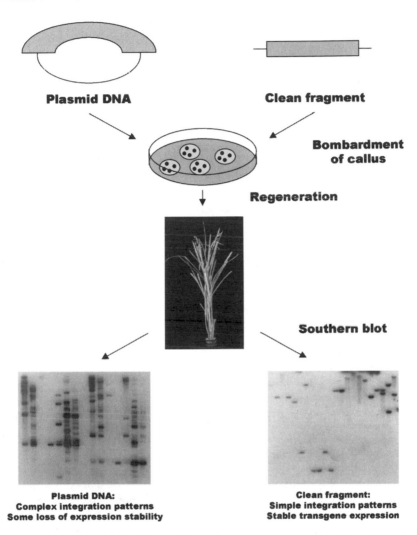

Plasmid DNA

Clean fragment

Bombardment of callus

Regeneration

Southern blot

Plasmid DNA:
Complex integration patterns
Some loss of expression stability

Clean fragment:
Simple integration patterns
Stable transgene expression

FIGURE 2   The clean DNA transformation system compared to particle bombardment with whole plasmid DNA. The transformation strategy is shown, and two representative Southern blots are compared to demonstrate the simpler integration patterns resulting from transformation with linear minimal cassettes. (Data from Ref. 85.)

linked to the T-DNA, suggesting that the VirD2 protein can skip over one T-DNA border, whereas in other cases the backbone segments are not linked to the T-DNA, suggesting an adventitious integration mechanism (87). Furthermore, T-DNA often integrates as multiple copies, predominantly inverted repeat structures, which are highly correlated with transgene silencing phenomena (88). Multimeric integration also occurs when two different T-DNAs are transferred to the same plant cell by different *Agrobacterium* strains, suggesting that multimer formation occurs before integration (89). This study showed a preference for linkage involving the right border repeat, although all other possible linkage combinations occurred at a lower frequency (89). Complex T-DNA integration patterns are more prevalent with hypervirulent *Agrobacterium* strains and/or superbinary vectors.

To address such problems, Chilton and colleagues have pioneered the technique of agrolistic transformation (90). Particle bombardment was used to introduce three plasmids into tobacco and maize cells, one containing a recombinant T-DNA and the others containing the *virD1* and *virD2* genes responsible for T-DNA mobilization. The resulting plants contain single integrated T-DNA sequences at high frequency, indicating that the Vir proteins could function correctly, although T-DNA sequences were also detected as part of the entire plasmid integrated into the plant genome. So, although this technique is useful for the generation of clean integration events, it does not eliminate backbone sequences completely. A possible strategy to limit the production of vector-containing plants is to flank the T-DNA with "killer genes" that confer a lethal phenotype on the transformant (91). Any plants carrying vector backbone sequences would be killed, provided the lethal genes were not disrupted or mutated upon integration and that segments of the vector backbone distal to the lethal genes did not integrate. A more refined technique devised by Gelvin (91) is to produce single-strand T-DNA in vitro using highly purified recombinant VirD2 protein. This development, analogous to the clean DNA strategy because backbone sequences are never introduced into the plant, may allow genuinely clean T-DNA insertions. However, the T-strands must be introduced into protoplasts by electroporation; thus, the procedure is conceptually no different from direct DNA transfer.

C.   Toward Marker-Free Transgenic Plants

There is some debate in the public domain about the potential dangers of selectable marker genes in transgenic plants. The permanent integration and expression of a selectable marker also prevents the same plant line from being retransformed with the same marker (i.e., if it was necessary to introduce further transgenes into an existing transgenic plant line). This issue has been tackled in a number of ways: (1) the use of genetic systems to eliminate marker genes from transgenic plants, (2) the development of strategies for marker gene segregation in breeding lines, and (3) the development of innocuous markers that do not confer antibiotic or herbicide resistance.

The first genetic system for marker elimination was reported by Goldsborough et al. (92) and involved the incorporation of transgenes within a maize *Ac* or *Ds* transposon, cloned within the T-DNA sequence. Two marker genes were used, and plants could be recovered that lacked one or other of the markers. In a subsequent system, the *Agrobacterium* isopentenyl transferase (*ipt*) gene was cloned in an *Ac* element within the T-DNA borders, while a second reporter gene carried in the T-DNA was outside the transposon (93). The *ipt* gene affects cytokinin metabolism so transformed plants showed a shooty phenotype. *Ac* elements often jump from site to site in the genome, but in about 15% of cases there is excision without subsequent reintegration. In these cases, the marker was removed and normal plant growth resumed as long as further copies of the T-DNA were not present. The excision events left the remainder of the T-DNA intact and did not affect expression of the linked reporter gene. This system is advantageous for a number of reasons: (1) the marker is innocuous, (2) the marker can be eliminated, and (3) there is automatic selection for single-copy insertions of the T-DNA.

In sexual populations, one method that could produce marker-free transgenic plants is to use cotransformation to introduce the nonselected gene and selectable marker at different loci and screen progeny in the R1 generation for segregants lacking the selectable marker gene. In direct DNA transfer cotransformation, there are occasional dispersive integration events resulting in transgene segregation and aberrant segregation ratios, although cointegration at a single locus predominates. The majority of T-DNA cotransformation events also result in cointegration, so that only a low frequency of marker-free plants is generated (13). The use of superbinary vector systems with two T-DNA plasmids may facilitate marker segregation, and segregation has been shown in 50% of the progeny of rapeseed and tobacco plants produced by *Agrobacterium*-mediated cotransformation with two binary plasmids (94).

The preceding strategy cannot be used for vegetatively propagated plants. An alternative strategy utilizes the bacteriophage site-specific recombinase system Cre-*loxP*. Tobacco plants were transformed with two marker genes, each flanked by *loxP* recombination target sites (95). The plants were then retransformed with the *cre* gene, and expression of Cre recombinase resulted in recombination between the *loxP* sites and excision of the original markers. One problem with this system is that transformation with the *cre* gene must itself involve an additional selectable marker. This requirement was circumvented by the transient expression of Cre, resulting in the production of a few marker-free plants.

D. Toward Precise Integration and Control of Expression

Another way in which Cre-*loxP*, and other site-specific recombination systems, can be exploited is to target transgenes to precise loci in the plant genome. Even in the absence of multiple and rearranged transgene copies, there is still much variation in transgene expression levels, and this has been attributed to so-called position effects reflecting the nature of the local molecular environment at the site of integration (e.g., the presence of regulatory elements, repetitive DNA, chromatin structure, and DNA methylation). In animal cells, a number of loci have been identified that are favorable to transgene expression because transgenes integrated there tend to be expressed faithfully and strongly. The controlled integration of DNA at such sites can be facilitated by flanking the transgene with *loxP* sites and placing a *loxP* site at the desired target site. A similar approach has been attempted in plants (96), but it appears that transgenes integrated at sites on several different chromosomes still display variable levels of expression within lines of identical transformants, suggesting that other factors are responsible for the variation (91). This agrees with data showing variable levels of transgene expression and different modes of transgene silencing involving different methylation patterns in a line of rice plants carrying the same single-copy, three-gene transgenic locus (97).

A different strategy to avoid position effects is to isolate the transgene from the surrounding molecular environment by flanking it with matrix attachment regions (MARs) that attach it to the nuclear matrix and isolate the transgene into an independent chromatin domain (98). The variability in transgene expression was reduced when reporter genes such as *gusA* were flanked by MARs, although as with the Cre-*loxP* system, there was still some variability that could be attributed to other factors (98).

XI. CONCLUDING REMARKS

It is now possible, at least in principle, to introduce any gene into any plant species. Problems of efficiency, cost, intellectual property, and other constraints remain but these are not insurmountable. The next few years will see further refinements in transformation technology, but it is likely that the focus will shift from technology development and optimization of the DNA transfer process itself to more challenging and fundamental territories, such as elucidating the factors that permit stable and predictable transgene expression, efficient transformation of organelles, and homologous recombination. Pathway engineering, involving the simultaneous introduction and concerted expression of multiple transgenes, will have to feature prominently in future strategies for plant transformation. Finally, the outputs from current genomics, proteomics, and metabolomics programs can be fully exploited only with sophisticated methods to introduce large pieces of DNA into the plant genome.

REFERENCES

1. G Hansen, MS Wright. Recent advances in the transformation of plants. Trends Plant Sci 4:226–231, 1999.
2. RG Birch. Plant transformation: problems and strategies for practical application. Annu Rev Plant Physiol Plant Mol Biol 48:297–326, 1997.
3. T Komari, Y Hiei, Y Ishida, T Kumashiro, T Kubo. Advances in cereal gene transfer. Curr Opin Plant Biol 1:161–165, 1998.
4. http://www.jic.bbsrc.ac.uk/science/molbio/index.htm
5. M Stroun, P Anker, P Charles, L Ledoux. Translocation of DNA of bacterial origin in *Lycopersicum esculentum* by ultracentrifugation in caesium chloride gradient. Nature 215:975–976, 1967.
6. N Van Larebeke, G Engler, M Holsters, S van den Elsacker, I Zaenen, RA Schilperoort, J Schell. Large plasmid in *Agrobacterium tumefaciens* essential for crown gall–inducing ability. Nature 252:169–170, 1974.
7. M-D Chilton, MH Drummond, DJ Merlo, D Sciaky, AL Montoya, MP Gordon, EW Nester. Stable incorporation of plasmid DNA into higher plant cells: the molecular basis of crown gall tumorigenesis. Cell 11:263–271, 1977.
8. L Otten, H de Greve, JP Hernalsteens, M van Montagu, O Schrieder, J Straub, J Schell. Mendelian transmission of genes introduced into plants by the Ti-plasmids of *Agrobacterium tumefaciens*. Mol Gen Genet 183:209–213, 1981.
9. L Herrera-Estrella, A Depicker, M van Montagu, J Schell. Expression of chimaeric genes transferred into plant cells using a Ti-plasmid–derived vector. Nature 303:209–213, 1983.
10. G Angenon, W Dillen, M van Montagu. Antibiotic resistance markers for plant transformation. In: SB Gelvin, RA Schilperoort, eds. Plant Molecular Biology Manual. 2nd ed. Dordrecht: Kluwer, 1994, pp 1–13.
11. A Wilmink, JJM Dons. Selective agents and marker genes for use in transformation of monocotyledonous plants. Plant Mol Biol Rep 11:165–185, 1993.
12. J Cooley, T Ford, P Christou. Molecular and genetic characterization of elite transgenic rice plants produced by electric-discharge particle acceleration. Theor Appl Genet 90:97–104, 1995.
13. T Komari, Y Hiei, Y Saito, N Murai, T Kumashiro. Vectors carrying two separate T-DNAs for cotransformation of higher plants mediated by *Agro-*

bacterium tumefaciens and segregation of transformants free from selectable markers. Plant J 10:165–174, 1996.

14.   M Joersbo, I Donaldson, J Kreiberg, SG Petersen, J Brunstedt, FT Okkels. Analysis of mannose selection used for transformation of sugar beet. Mol Breeding 4:111–117, 1998.

15.   J Zupan, P Zambryski. The *Agrobacterium* DNA transfer complex. Crit Rev Plant Sci 16:279–295, 1997.

16.   J Hernalsteens, F van Vliet, M de Beuckeleer, A Depicker, G Engler, M Lemmes, M Holsters, M van Montagu, J Schell. The *Agrobacterium tumefaciens* Ti plasmid as a host vector system for introducing foreign DNA in plants. Nature 287:654–656, 1980

17.   GJ Wullems, L Molendijk, G Ooms, RA Schilperoort. Retention of tumor markers in F1 progeny plants formed from in vitro induced octopine and nopaline tumor tissues. Cell 24:719–728, 1981.

18.   P Zambryski, H Joos, C Genetello, J Leemans, M van Montagu, J Schell. Ti plasmid vector for the introduction of DNA into plant cells without alteration of their normal regeneration capacity. EMBO J 2:2143–2150, 1983.

19.   AJM Matzke, M-D Chilton. Site-specific insertion of genes into T-DNA of the *Agrobacterium* tumour-inducing plasmid: an approach to genetic engineering of higher plant cells. J Mol Appl Genet 1:39–49, 1981.

20.   M Bevan. Binary *Agrobacterium* vectors for plant transformation. Nucleic Acids Res 12:8711–8721, 1984.

21.   RB Horsch, JE Fry, NL Hoffmann, D Eicholtz, SG Rogers, RT Fraley. A simple and general method for transferring genes into plants. Science 227: 1229–1231, 1985.

22.   Y Hiei, S Ohta, T Komari, T Kumashiro. Efficient transformation of rice (*Oryza sativa* L.) mediated by Agrobacterium and sequence analysis of the boundaries of the T-DNA. Plant J 6:271–282, 1994.

23.   M Cheng, JE Fry, SZ Pang, HP Zou, CM Hironaka, DR Duncan, TW Conner, YC Yang. Genetic transformation of wheat mediated by *Agrobacterium tumefaciens*. Plant Physiol 115:971–980, 1997.

24.   M Cervera, JA Pina, J Juarez, L Navarro, L Pena. *Agrobacterium*-mediated transformation of citrange: factors affecting transformation and regeneration. Plant Cell Rep 16:731–737, 1997.

25.   X Niu, X Li, P Veronesse, RA Bressan, SC Weller, PM Hasegawa. Factors affecting *Agrobacterium tumefaciens*–mediated transformation of peppermint. Plant Cell Rep 19:304–310, 2000.

26.   W Schafer, A Gorz, G Kahl. T-DNA integration and expression in a monocot crop plant after induction of *Agrobacterium*. Nature 327:529–531, 1987.

27.   S Tingay, D McElroy, R Kalla, S Fieg, M Wang, S Thornton, R Brettell. *Agrobacterium tumefaciens*–mediated barley transformation. Plant J 11:1369–1376, 1997.

28.   WS Grayburn, BA Vick. Transformation of sunflower (*Helianthus annuus* L.) following wounding with glass beads. Plant Cell Rep 14:285–289, 1995.

29.   HN Trick. Recent advances in soybean transformation. Plant Tissue Cult Biotechnol 3:9–26, 1997.

30. Y Ishida, H Saito, S Ohta, Y Hiei, T Komari, T Kumashiro. High efficiency transformation of maize (*Zea mays* L.) mediated by *Agrobacterium tumefaciens*. Nat Biotechnol 14:745–750, 1996.
31. AD Arencibia, ER Carmona, P Tellez, MT Chan, SM Yu, LE Trujillo, P Oramas. An efficient protocol for sugarcane (*Saccharum* spp. L.) transformation mediated by *Agrobacterium tumefaciens*. Transgenic Res 7:213–222, 1998.
32. SB Narasimhulu, X-B Deng, R Sarria, SB Gelvin. Early transcription of *Agrobacterium* T-DNA genes in tobacco and maize. Plant Cell 8:873–886, 1996.
33. G Hansen, M-D Chilton. Lessons in gene transfer to plants by a gifted microbe. Curr Top Microbiol Immunol 240:21–57, 1999.
34. CM Hamilton. A binary-BAC system for plant transformation with high-molecular-weight DNA. Gene 200:107–116, 1997.
35. V Buchanan-Wollaston, JE Passiatoire, F Channon. The *mob* and *oriT* functions of a bacterial plasmid promote its transfer to plants. Nature 328:172–175, 1987.
36. MA van Sluys, J Tempe, N Federoff. Studies on the introduction and mobility of the maize *Activator* element in *Arabidopsis thaliana* and *Daucus carota*. EMBO J 6:3881–3889, 1987.
37. TM Klein, ED Wolf, R Wu, JC Stanford. High velocity microprojectiles for delivering nucleic acids into living cells. Nature 327:70–73, 1987.
38. TM Klein, M Fromm, A Weissinger, D Tomes, S Schaaf, M Sletten, JC Stanford. Transfer of foreign genes into intact maize cells with high velocity microprojectiles. Proc Natl Acad Sci USA 85:4305–4309, 1988.
39. P Christou, D McCabe, WF Swain. Stable transformation of soybean callus by DNA-coated gold particles. Plant Physiol 87:671–674, 1988.
40. G-N Ye, H Daniell, JC Sanford. Optimization of delivery of foreign DNA into higher plant chloroplasts. Plant Mol Biol 15:809–819, 1990.
41. P Christou. Particle gun mediated transformation. Curr Opin Biotechnol 4:135–141, 1994.
42. P Vain, MD McMullen, JJ Finer. Osmotic treatment enhances particle bombardment–mediated transient and stable transformation of maize. Plant Cell Rep 12:84–88, 1993.
43. LL Chen, P Marmey, NJ Taylor, JP Brizard, C Espinoza, P de Cruz, H Huet, SP Zhang, A de Kochko, RN Beachy, CM Fauquet. Expression and inheritance of multiple transgenes in rice plants. Nat Biotechnol 16:1060–1064, 1998.
44. S Bano Maqbool, P Christou. Multiple traits of agronomic importance in transgenic indica rice plants: analysis of transgene integration patterns, expression levels and stability. Mol Breeding 5:471–480, 1999.
45. K Tang, P Tinjuangjun, Y Xu, X Sun, JA Gatchouse, PC Ronald, H Qi, X Lu, P Christou, A Kohli. Particle bombardment–mediated co-transformation of elite Chinese rice cultivars with genes conferring resistance to bacterial blight and sap sucking insect pests. Planta 208:552–563, 1999.
46. JJ Finer, P Vain, MW Jones, MD McMullen. Development of the particle inflow gun for DNA delivery to plant cells. Plant Cell Rep 11:323–328, 1992.

47. A Kohli, S Griffiths, N Palacios, RM Twyman, P Vain, DA Laurie, P Christou. Molecular characterization of transforming plasmid rearrangements in transgenic rice reveals a recombination hotspot in the CaMV 35S promoter and confirms the predominance of microhomology-mediated recombination. Plant J 17:591–601, 1999.
48. JM Vaneck, AD Blowers, ED Earle. Stable transformation of tomato cell cultures after bombardment with plasmid and YAC DNA. Plant Cell Rep 14: 299–304, 1995.
49. T Yamashita, A Iida, H Morikawa. Evidence that more than 90% of β-glucuronidase–expressing cells after particle bombardment directly receive the foreign gene in their nucleus. Plant Physiol 97:829–831, 1991.
50. FA Krens, L Molendijk, GJ Wullems, RA Schilperoort. In vitro transformation of plant protoplasts with Ti-plasmid DNA. Nature 296:72–74, 1982.
51. I Negrutiu, R Shillito, I Potrykus, G Biasini, F Sala. Hybrid genes in the analysis of transformation conditions. I. Setting up a simple method for direct gene transfer in plant protoplasts. Plant Mol Biol 8:363–373, 1987.
52. M Fromm, J Callis, LP Taylor, V Walbot. Electroporation of DNA and RNA into plant protoplasts. Methods Enzymol 153:351–366, 1987.
53. B Sporlein and HU Koop. Lipofectin—direct gene transfer to higher plants using cationic liposomes. Theor Appl Genet 83:1–5, 1991.
54. J Pazkowski, RD Shillito, M Saul, V Mandak, T Hohn, B Hohn, I Potrykus. Direct gene transfer to plants. EMBO J 3:2717–2722, 1984.
55. N Leduc, E Matthys-Rochon, M Rougier, L Mogensen, P Holm, JL Magnard, C Dumas. Isolated maize zygotes mimic in vivo early development and express microinjected genes when cultured in vitro. Dev Biol 10:190–203, 1996.
56. PB Holm, O Olsen, M Scnorf, H Brinch-Pedersen, S Knudsen. Transformation of barley by microinjection into isolated zygote protoplasts. Transgenic Res 9:21–32, 2000.
57. S Kurata, M Tsukakoshi, T Kasuya, Y Ikawa. The laser method for efficient introduction of foreign DNA into cultured cells. Exp Cell Res 162:372–378, 1986.
58. F Hoffman. Laser microbeams for the manipulation of plant cells and subcellular structures. Plant Sci 113:1–11, 1996.
59. CM Laursen, RA Krzyzek, CE Flick, PC Anderson, TM Spencer. Production of fertile transgenic maize by electroporation of suspension culture cells. Plant Mol Biol 24:51–61, 1994.
60. SM Pescitelli, K Sukhapinda. Stable transformation via electroporation into maize type II callus and regeneration of fertile transgenic plants. Plant Cell Rep 14:712–716, 1995.
61. XP Xu, BJ Li. Fertile transgenic indica rice plants obtained by electroporation of the seed embryo cells. Plant Cell Rep 13:237–242, 1994.
62. A Arencibla, PR Molina, G Delariva, G Selmanhousein. Production of transgenic sugarcane (Saccharum officinarum L) plants by intact cell electroporation. Plant Cell Rep 14:305–309, 1997.

63. CR Smith, JA Saunders, S van Wert, JP Cheng, BF Matthews. Expression of GUS and CAT activities using electrotransformed pollen. Plant Sci 104:49–58, 1994.
64. JA Thompson, PR Drayton, BR Frame, K Wang, JM Dunwell. Maize transformation utilizing silicon carbide whiskers—a review. Euphytica 85:75–80, 1995.
65. N Nagatani, H Honda, T Shimada, T Kobayashi. DNA delivery into rice cells and transformation using silicon carbide whiskers. Biotechnol Tech 11:471–473, 1997.
66. KA Feldmann, MD Marks. *Agrobacterium*-mediated transformation of germinating seeds of *Arabidopsis thaliana*—a non-tissue culture approach. Mol Gen Genet 208:1–9, 1987.
67. SS Chang, SK Park, BC Kim, BJ Kang, DU Kim, HG Nam. Stable genetic transformation of *Arabidopsis thaliana* by *Agrobacterium* inoculation in planta. Plant J 5:551–558, 1994.
68. SJ Clough, A Bent. Floral dip: a simplified method for *Agrobacterium*-mediated transformation of *Arabidopsis thaliana*. Plant J 16:735–743, 1998.
69. N Bechtold, J Ellis, G Pelletier. In planta *Agrobacterium*-mediated gene transfer by infiltration of adult *Arabidopsis thaliana* plants. C R Acad Sci Paris Life Sci 316:1194–1199, 1993.
70. SH Park, SRM Pinson, RH Smith. T-DNA integration into genomic DNA of rice following *Agrobacterium* inoculation of isolated shoot apices. Plant Mol Biol 32:1135–1148, 1996.
71. G Zhou, J Weng, Y Zheng, J Huang, S Qian, G Lui. Introduction of exogenous DNA into cotton embryos. Methods Enzymol 101:433–481, 1983.
72. P Langridge, R Brett-Schneider, P Lazzeri, H Lorz. Transformation of cereals via *Agrobacterium* and the pollen pathway: a critical assessment. Plant J 2:631–638, 1992.
73. A Touraev, E Stoger, V Voronin, E Heberle-Bors. Plant male germline transformation. Plant J 12:949–956, 1997.
74. MH Kumagai, J Donson, G Dellacioppa, D Harvey, K Hanley, LK Grill. Cytoplasmic inhibition of carotenoid biosynthesis with virus-derived RNA. Proc Natl Acad Sci USA 92:1679–1683, 1995.
75. DC Baulcombe, S Chapman, S Santa-Cruz. Jellyfish green fluorescent protein as a reporter for virus infections. Plant J 7:1045–1053, 1995.
76. K Dalsgaard, A Uttenthal, TD Jones, F Xu, A Merryweather, WDO Hamilton, JPM Langeveld, RS Boshuizen, S Kamstrup, GP Lomonossoff, C Porta, C Vela, JI Casal, RH Meloen, PB Rodgers. Plant-derived vaccine protects target animals against a viral disease. Nat Biotechnol 3:248–252, 1997.
77. J Escudero, G Neuhaus, M Schlappi, B Holn. T-DNA transfer in meristematic cells of maize provided with intracellular *Agrobacterium*. Plant J 10:355–360, 1996.
78. M Kota, H Daniell, S Varma, SF Garczynski, F Gould, WJ Moar. Overexpression of the *Bacillus thuringiensis* (Bt) Cry2Aa2 protein in chloroplasts confers resistance to plants against susceptible and Bt-resistant insects. Proc Natl Acad Sci USA 96:1840–1845, 1999.

79. M de Block, J Schell, M van Montagu. Chloroplast transformation by *Agrobacterium tumefaciens*. EMBO J 4:1367–1372, 1985.
80. JE Boynton, EH Gilham, EH Harris, JP Hosler, AM Johnson, AR Jones, BL Randolph-Anderson, D Robertson, TM Klein, KB Shark, JC Sanford. Chloroplast transformation in *Chlamydomonas* with high velocity microprojectiles. Science 240:1534–1538, 1988.
81. Z Svab, P Maliga. High frequency plastid transformation in tobacco by selection for the chimeric *aadA* gene. Proc Natl Acad Sci USA 90:913–917, 1993.
82. TJ Golds, P Maliga, HU Koop. Stable plastid transformation in PEG-treated protoplasts of *Nicotiana tabacum*. Biotechnology 11:95–97, 1993.
83. W Kofer, C Eibl, K Steinmuller, H-U Koop. PEG-mediated plastid transformation in higher plants. In Vitro Cell Dev Biol Plant 34:303–309, 1998.
84. MS Khan, P Maliga. Fluorescent antibiotic resistance marker for tracking plastid transformation in higher plants. Nat Biotechnol 17:910–915, 1999.
85. X Fu, LT Duc, S Fontana, BB Bong, P Tinjuangjun, D Sudhakar, RM Twyman, P Christou, A Kohli. Linear transgene constructs lacking vector backbone sequences generate low-copy-number transgenic plants with simple integration patterns. Transgenic Res 9:11–19, 2000.
86. B Martineau, TA Voelker, RA Sanders. On defining T-DNA. Plant Cell 6:1032–1033, 1994.
87. ME Kononov, B Bassuner, SB Gelvin. Integration of T-DNA binary vector backbone sequences into the tobacco genome: evidence for multiple complex patterns of integration. Plant J 11:945–957, 1997.
88. R Jorgensen, C Snyder, JDG Jones. T-DNA is organized predominantly in inverted repeat structures in plants transformed with *Agrobacterium tumefaciens* C58 derivatives. Mol Gen Genet 207:471–477, 1987.
89. M de Neve, S de Buck, A Jacobs, M van Montagu, A Depicker. T-DNA integration patterns in cotransformed plant cells suggest that T-DNA repeats originate from cointegration of separate T-DNAs. Plant J 11:15–29, 1997.
90. G Hansen, M-D Chilton. 'Agrolistic' transformation of plant cells: integration of T-DNA strands generated in planta. Proc Natl Acad Sci USA 93:14978–14983, 1996.
91. SB Gelvin. The introduction and expression of transgenes in plants. Curr Opin Biotechnol 9:227–232, 1998.
92. AP Goldsbrough, CN Lastrella, JI Yoder. Transposition mediated repositioning and subsequent elimination of marker genes from transgenic tomato. Biotechnology 11:1286–1292, 1993.
93. H Ebinuma, K Sugita, E Matsunaga, M Yamakodo. Selection of marker-free transgenic plants using the isopentenyl transferase gene. Proc Natl Acad Sci USA 94:2117–2121, 1997.
94. M Daley, VC Knauf, KR Summerfelt, JC Turner. Cotransformation with one *Agrobacterium tumefaciens* strain containing two binary plasmids as a method for producing marker-free transgenic plants. Plant Cell Rep 17:489–496, 1998.

95.   AP Gleave, DS Mitra, SR Mudge, BAM Morris. Selectable-marker-free trans-
      genic plants without sexual crossing: transient expression of Cre recombinase
      and use of a conditional lethal dominant gene. Plant Mol Biol 40:223–235,
      1999.
96.   H Albert, EC Dale, E Lee, DW Ow. Site-specific integration of DNA into
      wild type and mutant *lox* sites placed in the plant genome. Plant J 7:649–
      659, 1995.
97.   X Fu, A Kohli, RM Twyman, P Christou. Alternative silencing effects involve
      distinct types of non-spreading cytosine methylation at a three-gene single-
      copy transgenic locus in rice. Mol Gen Genet 263:106–118, 2000.
98.   L Mlynarova, LCP Keizer, WJ Stiekema, J-P Nap. Approaching the lower
      limits of transgene variability. Plant Cell 8:1589–1599, 1996.
99.   Y Hiei, T Komari, T Kubo. Transformation of rice mediated by *Agrobacte-
      rium tumefaciens*. Plant Mol Biol 35:205–218, 1997.
100. D McElroy, RIS Brettell. Foreign gene expression in transgenic cereals.
      Trends Biotechnol 12:62–68, 1994.
101. RM Twyman, CAB Whitelaw. Genetic engineering: animal cell technology.
      In: RE Spiers, ed. Encyclopedia of Cell Technology. New York: John Wiley
      & Sons, 2000, pp 737–819.

7

Properties and Applications of Hairy Root Cultures

Pauline M. Doran

Department of Biotechnology, University of New South Wales, Sydney, Australia

I. INTRODUCTION

Hairy roots are formed by genetic transformation of plant cells using *Agrobacterium rhizogenes*. Integration into the plant genome of T-DNA from the bacterial root-inducing (Ri) plasmid results in differentiation and growth of hairy roots at the infection site. Hairy roots can be excised, cleared of excess bacteria using antibiotics, and grown indefinitely in vitro by subculture of root tips in liquid medium. Practical techniques for initiation, culture, genetic manipulation, and molecular analysis of hairy roots are summarized in Hamill and Lidgett (1). Hundreds of plant species have been successfully transformed to hairy roots; lists of amenable species are provided in several publications (2–5).

For 15–20 years, hairy roots have been applied in a wide range of fundamental studies of plant biochemistry, molecular biology, and physiology, as well as for agricultural, horticultural, and large-scale tissue culture purposes. Several recent reviews describe current and potential uses of hairy root cultures in research and industry (4–8). The aim of this chapter is to outline some of the emerging and rapidly developing areas of hairy root research and application. The properties and culture characteristics of hairy

143

roots relevant to their scientific and commercial exploitation are summarized, and selected topics associated with organ coculture, foreign protein production, and the use of hairy roots in studies of phytoremediation and phytomining are reviewed.

II. PROPERTIES OF HAIRY ROOTS

There are several general features of hairy roots that confer significant technical advantages to them compared with untransformed roots or dedifferentiated plant cells. The attention given to hairy roots and their increasing adoption in scientific studies are due largely to properties such as

   Genotype and phenotype stability
   Autotrophy in plant hormones
   Fast growth
   High levels of secondary metabolites

It is important to realize, however, that not every hairy root culture displays these characteristics. In addition, as well as advantages, many researchers have experienced problems with hairy root initiation and maintenance. Some of the common difficulties encountered are outlined in the following paragraphs.

A. Genotype and Phenotype Stability

Like most differentiated plant tissues, hairy roots exhibit a high degree of chromosomal stability over prolonged culture periods (9). Stability has also been demonstrated in terms of growth characteristics, DNA analysis, gene expression, and secondary metabolite levels (10–13). Genotype and phenotype instability in hairy roots is therefore much less of a problem than in callus and suspended plant cell cultures, where somaclonal variations involving chromosome rearrangement and breakage, movement of transposable elements, and gene amplification and depletion can occur with relatively high frequency (14).

   The stability of hairy roots is an important advantage for both research and large-scale industrial applications. Nevertheless, cytological instability can sometimes occur, and there are several reports of variations in ploidy, chromosome number, and chromosome structure in hairy root cultures (15–17). Very high rates of chromosome elimination were observed in hairy roots of *Onobrychis viciaefolia* during 12 months of culture (18). It is possible that the altered karyotypes sometimes observed in hairy roots could arise from the presence of endopolyploid nuclei in the host cell genome (17) or be the result of localized callusing due, for example, to tissue damage. Callusing or loss of structural integrity is known to promote the development of polyploidy and aneuploidy in hairy root cultures (19). Minor structural rearrangements of chromosomes in hairy roots were considered most probably to arise from terminal deletions of DNA (17). Notwithstanding these observations, the frequency of chromosomal alteration in hairy roots is much lower than in cultures of dedifferentiated plant cells.

B. Autotrophy in Plant Hormones

Auxin metabolism is altered in plant cells after transformation with *A. rhizogenes*. Typically, the consequence for hairy root cultures is that exogenous growth regulators are not required in the medium; hairy roots are self-sufficient in plant hormone production. This is an advantage, as the medium for hairy root culture is simpler and cheaper than for suspended plant cells and untransformed roots, and regulatory hurdles associated with the use of synthetic growth regulators for production of food and pharmaceutical products are immediately overcome. Extensive empirical studies to identify the best combination of growth regulators for maintenance of hairy root cultures are also not required. Several reports describe the detrimental effects of exogenous plant growth regulators on hairy roots (19–21). Yet, some hairy root cultures have been found to grow better or produce higher levels of metabolites when growth regulators are applied (22–24); release of secondary metabolites into the medium may also be enhanced (24). Addition of gibberellic acid to hairy root cultures has had variable results, with reports of both positive (21,25–28) and negative (21,24) effects.

C. Fast Growth

Many hairy root cultures grow prolifically with doubling times of 1–2 days. These growth rates are similar to those of suspended plant cells and are much greater than typical values for untransformed roots in vitro. Despite this generalization, however, hairy root cultures display a wide range of growth rates and can also be very slow growing. Although obviously dependent on the environmental conditions employed, the ease with which hairy roots grow in culture also depends very much on the species. Examples of specific growth rates and doubling times measured in this laboratory for several hairy root cultures are listed in Table 1.

D. High Levels of Secondary Metabolites

Hairy roots are commonly associated with activated secondary pathways and high levels of secondary metabolites. Several studies have shown that morphological and structural organization of plant cells significantly enhances the formation of particular compounds (10,21); callus and cell suspensions

TABLE 1   Specific Growth Rates and Doubling Times for Several Species of Hairy Root

Species	Specific growth rate (day^{-1})	Doubling time (days)	Reference
Atropa belladonna	0.57	1.2	78
Nicotiana tabacum	0.31	2.2	62
Alyssum bertolonii	0.13	5.3	77
Alyssum tenium	0.12	5.8	77
Thlaspi caerulescens	0.11	6.3	67
Arabidopsis thaliana	0.10	6.9	79
Hyptis capitata	0.098	7.1	76
Solanum aviculare	0.09	7.7	26

derived from hairy roots produced much lower concentrations of secondary metabolites. The ability to synthesize valuable natural products in large-scale reactors at levels similar to those found in whole plants represents a major advantage for hairy roots compared with many suspended cell cultures. However, as hairy roots are not attached to other organs of the plant such as leaves, some differences can be expected in the range of products found in hairy roots compared with roots of intact plants. Because metabolites cannot be transported from hairy roots to alternative storage or biosynthetic sites for modification or turnover, hairy roots have been found to contain compounds not detected in the corresponding plant (29). Conversely, products that are synthesized in the roots of whole plants from precursor molecules translocated from the shoots are not normally produced in hairy root cultures.

E.   Species Resistant to Hairy Root Transformation

Although many dicotyledonous plants are susceptible to infection by *A. rhizogenes* and can be transformed to produce hairy roots, some species are resistant to either the transformation process or, if hairy roots develop at the infection site, to their subsequent culture after excision. Several difficult or recalcitrant species are listed by Mugnier (3); Papaveraceae and Rununculaceae plants were characterized by a particular lack of success for hairy root development. Species that have been subjected in this laboratory to many transformation attempts using several strains of *A. rhizogenes*, but which thus far have failed to produce sustainable hairy root cultures, include *Papaver somniferum*, *Castanospermum australe*, *Podophyllum hexandrum*, *Gossypium hirsutum*, *Hybanthus floribundus*, and *Berkheya coddii*. A common difficulty encountered during maintenance of some hairy roots is spontaneous callusing or loss of root morphology (30–32). This problem is exacerbated by any mild physical damage to the roots, e.g., during shake flask culture, which can accelerate callus formation.

To improve the frequency of transformation by *A. rhizogenes*, agents such as acetosyringone, which has been found to increase the activity of *Agrobacterium* virulence genes (33), may be employed. The practical outcome of acetosyringone treatment for hairy root initiation has been variable, however, with no change in transformation frequency (30) and slight negative effects (34) reported in some cases. Alternatively, exogenous growth regulators have been found to play an important role in hairy root induction from certain plants. For example, the transformation frequency of walnut (*Juglans regia*) was improved by applying IBA (indolebutyric acid) (35), pretreatment with NAA (α-naphthaleneacetic acid) significantly enhanced hairy root development on potato stems (36), and transformation of suspended plant cells to form hairy roots depended on the concentration of 2,4-D (2,4-dichlorophenoxyacetic acid) in the medium (37). The strain of *A. rhizogenes* used for the infection can also strongly influence the transformation frequency (34,36,38), as well as the properties of the resulting hairy root cultures (34,39). Other conditions, such as medium composition, pH, and the time allowed between wounding and bacterial infection, can also be important (34).

F. Clonal Variation

Considerable differences in root morphology, ploidy, growth rate, product levels, and excretion characteristics have been observed between individual hairy root clones initiated using the same materials and techniques but taken from independent infection sites (39–46). Different levels of expression of the T-DNA genes transferred to the plant cells may play a key role in generating variation between clones (41); for example, variations in growth rate, alkaloid levels, morphology, and ethylene production have been correlated with *rolC* gene expression levels in *Catharanthus roseus* hairy roots (46). In terms of bioprocess development, clonal variation provides the opportunity for selection of elite root lines with favorable production or culture characteristics.

III. COCULTURES USING HAIRY ROOTS

In plants, complete synthesis of many secondary metabolites requires the participation of both the roots and the leaves. For example, a precursor might be produced in the roots and then translocated to the shoots for conversion to a more valuable compound. If roots do not express in sufficient quantity the enzymes required for bioconversion of the precursor, it is very unlikely that hairy root cultures will be able to produce the desired final product. This is an important limitation associated with in vitro culture of single organs such as hairy roots, but may be overcome by coculturing the roots with shoots of the same or a different plant species.

As an example, hairy roots of *Atropa belladonna* produce high levels of hyoscyamine but do not express substantial quantities of the enzyme hyoscyamine 6β-hydroxylase (H6H), which is required to transform hyoscyamine to the more valuable pharmaceutical, scopolamine. Consequently, as indicated in Table 2, *A. belladonna* hairy roots in single-organ culture do not produce detectable levels of scopolamine. To demonstrate the concept of hairy root coculture, this situation was remedied in experiments using *A. belladonna* hairy roots cocultured with either *A. belladonna* shoots (47), which contain greater levels of H6H than the roots, or shoots of a *Duboisia* hybrid species used commercially for scopolamine production (48). The results in Table 2 indicate that coculture can be used to enhance significantly the production of metabolites such as scopolamine in vitro. Periodic damage of the hairy roots by gentle crushing increased the total scopolamine produced even further, by a factor of 5.4 compared with untreated cocultures (48), suggesting that greater release of hyoscyamine into the medium improved scopolamine formation. Cocultures have been carried out in single and dual shake flasks and bioreactor vessels (49).

Coculture of hairy roots and shoots provides the opportunity for precursors produced in the roots to be translocated through the medium to the shoots, where they may be transformed into the desired product. For bioconversions such as hyoscyamine to scopolamine that involve only a single enzyme, genetic manipulation of hairy roots to express the required enzyme is an attractive and feasible option and has been successfully carried out for improvement of scopolamine synthesis in hairy roots (31,50). However, the coculture approach has advantages when a large number of enzymes is required, when tight metabolic regulation of the enzyme array is necessary, when the enzymes must be compartmentalized in organelles for optimal activity, or if the enzymatic pathways are unknown. Interspecies and intergenus coculture also has the potential to allow in vitro production of completely new compounds that are not found in vivo because of the limited opportunity for exchange of metabolites between individual whole plants.

IV. PRODUCTION OF FOREIGN PROTEINS USING HAIRY ROOTS

Although most attention for foreign protein production has been given to microbial and animal cell cultures over the past 15–20 years, plants and

TABLE 2 Results from Coculture of A. belladonna Hairy Roots with Shoots of either A. belladonna or a Duboisia Hybrid

Parameter measured	Single-organ cultures			Coculture with A. belladonna shooty teratomas		Coculture with Duboisia shooty teratomas	
	A. belladonna hairy roots	A. belladonna shooty teratomas	Duboisia hybrid shooty teratomas	A. belladonna hairy roots	A. belladonna shooty teratomas	A. belladonna hairy roots	Duboisia hybrid shooty teratomas
Biomass (g dry weight)	3.0 ± 0.2	6.9 ± 0.3	3.4 ± 0.7	2.6 ± 0.2	5.6 ± 0.4	4.0 ± 0.5	3.0 ± 0.02
Hyoscyamine content (mg g^{-1} dry weight)	1.1 ± 0.03	None detected	Trace	0.21 ± 0.03	0.28 ± 0.05	0.7 ± 0.1	0.07 ± 0.0
Scopolamine content (mg g^{-1} dry weight)	None detected	None detected	None detected	None detected	0.84 ± 0.08	0.6 ± 0.2	1.9 ± 0.8
Total hyoscyamine (mg)	3.6	None detected	Trace	2.6		3.1	
Total scopolamine (mg)	None detected	None detected	None detected	5.0		8.1	

Coculture significantly enhanced the in vitro production of scopolamine compared with single-organ root or shoot cultures.
Source: Data from Refs. 47 and 48.

plant cells are now considered as viable and competitive expression systems for large-scale protein production. The development of transgenic plants and plant viral vectors for foreign protein synthesis has been reviewed extensively (51–56). There is increasing commercial activity aimed at the production of monoclonal antibodies, vaccines, and enzymes using plants, and several industrial processes using transgenic corn for synthesis of animal-derived proteins have already been developed (57–59).

Compared with whole plants grown in the field or glasshouse, plant tissue culture is a less well developed technology for producing commercially valuable foreign proteins. However, several transgenic plant cell cultures have been reported (60), either for investigation of large-scale, reactor-based protein production or as tools in fundamental studies of protein production by whole plants. So far, application of hairy roots as a production vehicle for foreign proteins has been limited. A. tumefaciens–mediated transfer of foreign genes into hairy roots was tested using green fluorescent protein (GFP) as a convenient indicator of cell transformation (61), demonstrating that hairy root clones previously selected for desirable traits could be genetically transformed using direct infection methods. In other work, hairy roots initiated from transgenic tobacco (Nicotiana tabacum) seedlings were used to produce fully assembled and functional murine immunoglobulin G_1 (IgG_1) monoclonal antibody in shake flask and bioreactor systems (62–64). Differences in growth and antibody secretion levels between eight hairy root clones allowed selection of elite root lines (63).

Typical results for biomass and IgG_1 antibody levels in tobacco hairy root and cell suspension cultures are compared in Table 3 (64). The pattern of antibody accumulation in the hairy roots was significantly different from

TABLE 3 Comparison of Maximum Biomass and Monoclonal Antibody Levels in Hairy Root and Suspended Cell Cultures of Transgenic *N. tabacum*

Parameter	Hairy root culture	Suspended cell culture
Maximum biomass (g dry weight)	0.51 ± 0.01	0.72 ± 0.04
Maximum antibody concentration in the biomass (mg g^{-1} dry weight)	0.71 ± 0.09	0.064 ± 0.02
Maximum percentage of antibody in the medium (%)	17 ± 1.2	48 ± 8
Maximum total antibody (mg)	0.24 ± 0.04	0.047 ± 0.01

Source: Data from Ref. 64.

that in the suspended cells: the hairy roots tended to secrete a relatively low proportion of antibody into the medium while retaining high intratissue levels, whereas the suspended cells contained low intracellular levels of antibody with about half the total antibody located in the medium. The tendency of hairy roots to retain protein in the biomass may be a disadvantage in terms of product purification and downstream processing; if the root biomass must be destroyed for antibody recovery, not only will the product be contaminated by a large number of endogenous proteins from the cells, but the roots cannot be used again in subsequent production cycles. On the other hand, plant culture medium has been shown to be a highly unstable environment for antibody molecules (65), so that facile secretion of product into the medium could be responsible for the low levels of total antibody recovered from plant cell suspensions. The results for maximum total antibody in Table 3 highlight the potential of hairy root systems for commercial antibody synthesis; hairy root cultures accumulated up to 5.1 times the antibody found with suspended plant cells. A maximum antibody titer of 18 mg L^{-1} or 1.8% total soluble protein has been measured in tobacco hairy root cultures (62).

A potential problem with transgenic plants cultivated as agricultural crops is long-term transgene instability, associated with either gene segregation or gene silencing. Variation in foreign gene expression within and between successive generations of corn has been reported (57,59) as well as loss of fertility and/or pollen transmission of the transgene (57). Accordingly, it is probable that breeding programs with continuous selection of high-producing lines will be necessary for consistent protein production in whole plants. Transgene stability is also of potential concern in hairy root cultures; however, several researchers have demonstrated stable long-term foreign protein synthesis in hairy roots. Monoclonal antibody levels in tobacco hairy roots were roughly constant after 19 months of culture (62), and production of GFP in *Hyoscyamus muticus* hairy roots continued for at least 30 weeks after transformation (61). In other work, stability of *gus* (β-glucuronidase) gene expression was demonstrated for *Lotus corniculatus* hairy roots cultured for 5 years under a variety of environmental conditions (66). Scopolamine production by hairy roots of transgenic *H. muticus* expressing the hyoscyamine 6β-hydroxylase gene from *H. niger* also remained high and stable after 2.5 years of culture (31).

V. HAIRY ROOTS IN PHYTOREMEDIATION AND PHYTOMINING STUDIES

The use of living plants to clean up polluted soils and waterways is a rapidly developing technology. Phytoremediation offers a range of advantages compared with existing remediation methods, including low cost, minimal site destruction and destabilization, low environmental impact, and favorable aesthetics. A related application of plants is phytomining. This emerging technology involves growing tolerant species on sites of mineralized soil or surface ore bodies, harvesting and drying the metal-rich crop, then ashing the biomass and treating the resulting bio-ore for metal recovery. The main advantage of phytomining compared with conventional mining is its low cost, allowing the exploitation of mineral deposits that are too metal-poor for direct mining operations.

Roots play a primary role in phytoremediation and phytomining, as they are the plant organs in direct contact with soil pollutants and heavy metals. Accordingly, there is a particular need to understand the biochemical and physiological functioning of roots in contaminated environments. Hairy root cultures are a convenient experimental system for such studies. In contrast to whole plants grown either in soil or hydroponically, hairy roots can be propagated indefinitely so that entire experimental programs can be carried out using tissues derived from the same plant, thus avoiding the effects of variability between individual specimens. Use of axenic conditions in hairy root culture prevents microbial symbiosis disguising the remediative activity of plant tissues, and better control over conditions at the roots can be exercised compared with soil cultivation. Separation of hairy roots from the leaves of plants also allows identification of the properties and functions of the roots without interference from translocation effects (67).

Several phytoremediation studies have been conducted using hairy root systems. Axenic cultures of *Catharanthus roseus* hairy roots were capable of removing 30 mg L^{-1} of 2,4,6-trinitrotoluene (TNT) from liquid solution within 5 days (68); disappearance of TNT from the medium was not due to simple accumulation within the biomass, as the roots were shown to facilitate TNT biotransformation into aminated nitrotoluenes and other soluble products. Hairy roots of *Armoracia rusticana*, *Atropa belladonna*, *Solanum aviculare*, and *S. nigrum* were found to metabolize polychlorinated biphenyls (PCBs); the most effective PCB transformations were correlated with an increase in total peroxidase activity in the roots (69). *Daucus carota* hairy roots have also been tested for transformation of phenols (70). Although chlorinated phenols were more toxic to root growth than phenol itself, both types of compound elicited peroxidase activity in the biomass. Peroxidase levels remained high in the hairy roots as phenol removal and metabolism took place.

As well as remediation of xenobiotic compounds, hairy roots have been applied to remove and concentrate heavy metals from liquid solutions. Cadmium accumulation has been investigated using several species, including *Nicotiana tabacum*, *Beta vulgaris* and *Calystegia sepium* (71), *Solanum nigrum* (72), *Rubia tinctorum* (73), and *Daucus carota* (74). In these studies, aspects of metal tolerance of the hairy roots, such as induction of phytochelatins (73,74), and stress responses such as ethylene production and lipid peroxidation (74) were quantified.

Hairy roots have been applied to investigate heavy metal uptake and detoxification by rare plant species capable of growing in high-metal environments and accumulating elevated levels of specific metal ions. These species, known as "hyperaccumulators," store heavy metals in their tissues at concentrations at least 100 times greater than those found in non-hyperaccumulator plants (75). About 400 hyperaccumulators have been identified, including 300 that hyperaccumulate nickel, 26 cobalt, 24 copper, 19 selenium, 16 zinc, 11 manganese, 1 thallium, and 1 cadmium (75). At the present time, the mechanisms of metal uptake by hyperaccumulating plants and the basis of their metal specificity are poorly understood. Hairy roots of several hyperaccumulators have been applied in metal uptake studies in liquid culture systems (67,76,77); these include *Alyssum bertolonii* and *Thlaspi caerulescens*, which were tested for hyperaccumulation of nickel and cadmium, respectively. As shown in Fig. 1, hairy roots of both hyperaccumulator species grew well at elevated metal concentrations that were toxic to hairy roots of the non-hyperaccumulator *Nicotiana tabacum*. The maximum Ni concentration measured in growing *A. bertolonii* hairy roots was 7200 μg g^{-1} dry weight (77), and the maximum Cd level in growing *T. caerulescens* was 10,600 μg g^{-1} dry weight (67); these metal contents are well above the threshold levels for plant hyperaccumulator status (1000 μg g^{-1} for Ni; 100 μg g^{-1} for Cd: Ref. 75). These results from hairy root cultures demonstrate that neither translocation of metal from roots to leaves nor interactions with soil microorganisms are necessary or responsible for metal hyperaccumulation in plants.

Hairy root cultures were used to investigate the role of the root cell wall and the distribution of metal between the apoplasm and symplasm of root cells during metal hyperaccumulation (67). As indicated in Fig. 2, there were significant differences in the patterns of metal uptake for the two species *A. bertolonii* and *T. caerulescens* compared with the non-hyperaccumulator *N. tabacum*. Over a period of 28 days, the proportion of Ni in the biomass that was retained in the cell wall fractions of *A. bertolonii* and *N. tabacum* hairy roots was relatively low, averaging only 17% for *A. bertolonii* and 24% for *N. tabacum* (Fig. 2a). In other words, most of the Ni taken up by both species quickly entered into the symplasm of the root cells. As Ni was toxic to *N. tabacum* but not to *A. bertolonii* (Fig. 1a), *A. bertolonii* must possess intracellular mechanisms for Ni detoxification that are not available to *N. tabacum*. In contrast, as shown in Fig. 2b, almost all of the Cd found in the *T. caerulescens* hairy roots was located in the cell walls during the first 7 days of contact with the metal, suggesting that association

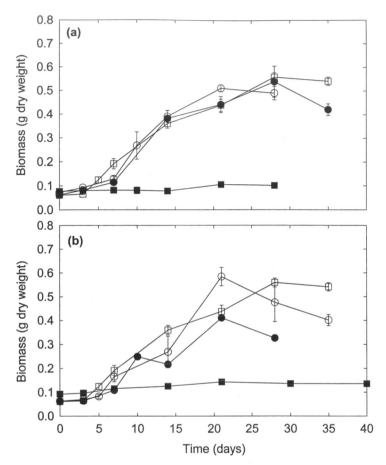

FIGURE 1 Growth of hairy roots of hyperaccumulator and non-hyperaccumulator species in medium with and without heavy metals. (a) Effect of Ni at an initial concentration of 20 ppm. (○) *A. bertolonii* without Ni; (●) *A. bertolonii* with Ni; (□) *N. tabacum* without Ni; and (■) *N. tabacum* with Ni. Nickel did not affect growth of *A. bertolonii* hairy roots but was very toxic to *N. tabacum*. (Data from Ref. 77.) (b) Effect of Cd at an initial concentration of 20 ppm. (○) *T. caerulescens* without Cd; (●) *T. caerulescens* with Cd; (□) *N. tabacum* without Cd; and (■) *N. tabacum* with Cd. Cadmium was particularly detrimental to growth of *N. tabacum* hairy roots. (Data from Ref. 67.)

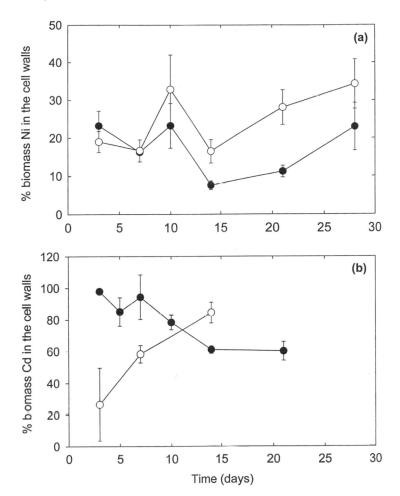

FIGURE 2   Percentage of heavy metal in the biomass associated with the cell walls. (a) Nickel was added to (●) *A. bertolonii* and (○) *N. tabacum* hairy roots at an initial concentration of 20 ppm. Most of the Ni in the biomass of both species was located in the symplasm of the cells. (b) Cadmium was added to (●) *T. caerulescens* and (○) *N. tabacum* hairy roots at an initial concentration of 20 ppm. Virtually all of the Cd in the *T. caerulescens* roots was retained in the cell walls for the first 7 days, whereas most of the Cd in *N. tabacum* roots was initially transferred into the symplasm. (Data from Ref. 67.)

with the cell walls is a primary mechanism of Cd accumulation in this species, at least during the early stages of metal exposure. The proportion of Cd in the walls decreased after 7 days as some Cd passed into the symplasm of the cells; however, most of the Cd (about 60%) continued to remain with the wall fraction.

The pattern of Cd uptake by *N. tabacum* hairy roots was the reverse of that with *T. caerulescens*. Initially, the cell walls accounted for only a minor proportion of the Cd in the *N. tabacum* biomass, as most of the Cd entered directly into the cells. Cell wall Cd levels in *N. tabacum* increased with time and eventually accounted for more than 80% of the biomass Cd; this increase was possibly due to cell lysis during the culture, as Cd had a severe detrimental effect on the growth (Fig. 1b) and integrity of *N. tabacum* roots. The ability of *T. caerulescens* hairy roots to hold virtually all of their Cd in the cell walls for several days could represent a critical defensive strategy, providing time for the development of intracellular mechanisms for Cd complexation or compartmentation. By the time Cd entered the symplasm of the *T. caerulescens* cells, the metal could be effectively detoxified so that growth of the roots was maintained (Fig. 1b).

VI. CONCLUSIONS

Application of hairy roots for genetic manipulation of plants and in studies of secondary metabolism and biosynthetic pathways has been an important focus over the last 10–15 years. The special properties of hairy roots also make them attractive for large-scale culture in bioreactors for the production of valuable natural products. More recently, hairy roots have been applied in new directions. They have been used as components of organ cocultures to expand the range of phytochemicals that can be synthesized in vitro, and as convenient research tools to investigate foreign protein production, phytoremediation, and phytomining using plants.

ACKNOWLEDGMENTS

The author's research was supported by the Australian Research Council (ARC).

REFERENCES

1. JD Hamill, AJ Lidgett. Hairy root cultures—opportunities and key protocols for studies in metabolic engineering. In: PM Doran, ed. Hairy Roots: Culture and Applications. Amsterdam: Harwood Academic, 1997, pp 1–29.
2. D Tepfer. Genetic transformation using *Agrobacterium rhizogenes*. Physiol Plant 79:140–146, 1990.

3. J Mugnier. Establishment of new axenic hairy root lines by inoculation with *Agrobacterium rhizogenes*. Plant Cell Rep 7:9–12, 1988.
4. B Canto-Canché, VM Loyola-Vargas. Chemicals from roots, hairy roots, and their application. In: F Shahidi, P Kolodziejczyk, JR Whitaker, A Lopez Munguia, G Fuller, eds. Chemicals via Higher Plant Bioengineering. Adv Exp Med Biol, vol 464. New York: Kluwer Academic/Plenum, 1999, pp 235–275.
5. A Giri, ML Narasu. Transgenic hairy roots: recent trends and applications. Biotechnol Adv 18:1–22, 2000.
6. PM Doran, ed. Hairy Roots: Culture and Applications. Amsterdam: Harwood Academic, 1997.
7. JV Shanks, J Morgan. Plant 'hairy root' culture. Curr Opin Biotechnol 10:151–155, 1999.
8. WR Curtis. Hairy roots, bioreactor growth. In: RE Spier, ed. Encyclopedia of Cell Technology, vol 2. New York: Wiley, 2000, pp 827–841.
9. ELH Aird, JD Hamill, MJC Rhodes. Cytogenetic analysis of hairy root cultures from a number of plant species transformed by *Agrobacterium rhizogenes*. Plant Cell Tissue Organ Cult 15:47–57, 1988.
10. HE Flores. Use of plant cells and organ culture in the production of biological chemicals. In: HM LeBaron, RO Mumma, RC Honeycutt, JH Duesing, eds. Biotechnology in Agricultural Chemistry. ACS Symp Ser 334. Washington, DC: American Chemical Society, 1987, pp 66–86.
11. IE Maldonado-Mendoza, T Ayora-Talavera, VM Loyola-Vargas. Establishment of hairy root cultures of *Datura stramonium*. Plant Cell Tissue Organ Cult 33: 321–329, 1993.
12. KH Lipp João, TA Brown. Long-term stability of root cultures of tomato transformed with *Agrobacterium rhizogenes* R1601. J Exp Bot 45:641–647, 1994.
13. R Ciau-Uitz, ML Miranda-Ham, J Coello-Coello, B Chí, LM Pacheco, VM Loyola-Vargas. Indole alkaloid production by transformed and non-transformed root cultures of *Catharanthus roseus*. In Vitro Cell Dev Biol Plant 30P:84–88, 1994.
14. PJ Larkin, WR Scowcroft. Somaclonal variation—a novel source of variability from cell cultures for plant improvement. Theor Appl Genet 60:197–214, 1981.
15. CH Hänisch ten Cate, K Sree Ramulu, P Dijkhuis, B de Groot. Genetic stability of cultured hairy roots induced by *Agrobacterium rhizogenes* on tuber discs of potato cv. Bintje. Plant Sci 49:217–222, 1987.
16. S Banerjee-Chattopadhyay, AM Schwemmin, DJ Schwemmin. A study of karyotypes and their alterations in cultured and *Agrobacterium* transformed roots of *Lycopersicon peruvianum* Mill. Theor Appl Genet 71:258–262, 1985.
17. G Ramsay, A Kumar. Transformation of *Vicia faba* cotyledon and stem tissues by *Agrobacterium rhizogenes*: infectivity and cytological studies. J Exp Bot 41:841–847, 1990.
18. ZQ Xu, JF Jia. The reduction of chromosome number and the loss of regeneration ability during subculture of hairy root cultures of *Onobrychis viciaefolia* transformed by *Agrobacterium rhizogenes* A_4. Plant Sci 120:107–112, 1996.
19. ELH Aird, JD Hamill, RJ Robins, MJC Rhodes. Chromosome stability in transformed hairy root cultures and the properties of variant lines of *Nicotiana*

rustica hairy roots. In: RJ Robins, MJC Rhodes, eds. Manipulating Secondary Metabolism in Culture. Cambridge: Cambridge University Press, 1988, pp 137–144.

20. TM Ermayanti, JA McComb, PA O'Brien. Cytological analysis of seedling roots, transformed root cultures and roots regenerated from callus of *Swainsona galegifolia* (Andr.) R. Br. J Exp Bot 44:375–380, 1993.

21. MJC Rhodes, AJ Parr, A Giulietti, ELH Aird. Influence of exogenous hormones on the growth and secondary metabolite formation in transformed root cultures. Plant Cell Tissue Organ Cult 38:143–151, 1994.

22. M Sauerwein, M Wink, K Shimomura. Influence of light and phytohormones on alkaloid production in transformed root cultures of *Hyoscyamus albus*. J Plant Physiol 140:147–152, 1992.

23. T Yoshikawa. Production of ginsenosides in ginseng hairy root cultures. In: PM Doran, ed. Hairy Roots: Culture and Applications. Amsterdam: Harwood Academic, 1997, pp 73–79.

24. L Vanhala, M Eeva, S Lapinjoki, R Hiltunen, K-M Oksman-Caldentey. Effect of growth regulators on transformed root cultures of *Hyoscyamus muticus*. J Plant Physiol 153:475–481, 1998.

25. H Ohkawa, H Kamada, H Sudo, H Harada. Effects of gibberellic acid on hairy root growth in *Datura innoxia*. J Plant Physiol 134:633–636, 1989.

26. MA Subroto, PM Doran. Production of steroidal alkaloids by hairy roots of *Solanum aviculare* and the effect of gibberellic acid. Plant Cell Tissue Organ Cult 38:93–102, 1994.

27. NB Paniego, AM Giulietti. Artemisinin production by *Artemisia annua* L. transformed organ cultures. Enzyme Microb Technol 18:526–530, 1996.

28. TC Smith, PJ Weathers, RD Cheetham. Effects of gibberellic acid on hairy root cultures of *Artemisia annua*: growth and artemisinin production. In Vitro Cell Dev Biol 33:75–79, 1997.

29. L Toivonen, J Balsevich, WGW Kurz. Indole alkaloid production by hairy root cultures of *Catharanthus roseus*. Plant Cell Tissue Organ Cult 18:79–93, 1989.

30. E De Vries-Uijtewaal, LJW Gilissen, E Flipse, K Sree Ramulu, B de Groot. Characterization of root clones obtained after transformation of monohaploid and diploid potato genotypes with hairy root inducing strains of *Agrobacterium*. Plant Sci 58:193–202, 1988.

31. K Jouhikainen, L Lindgren, T Jokelainen, R Hiltunen, TH Teeri, K-M Oksman-Caldentey. Enhancement of scopolamine production in *Hyoscyamus muticus* L. hairy root cultures by genetic engineering. Planta 208:545–551, 1999.

32. PM Santos, AC Figueiredo, MM Oliveira, JG Barroso, LG Pedro, SG Deans, AKM Younus, JJC Scheffer. Morphological stability of *Pimpinella anisum* hairy root cultures and time-course study of their essential oils. Biotechnol Lett 21:859–864, 1999.

33. SE Stachel, E Messens, M Van Montagu, P Zambryski. Identification of the signal molecules produced by wounded plant cells that activate T-DNA transfer in *Agrobacterium tumefaciens*. Nature 318:624–629, 1985.

34. L Vanhala, R Hiltunen, K-M Oksman-Caldentey. Virulence of different *Agrobacterium* strains on hairy root formation of *Hyoscyamus muticus*. Plant Cell Rep 14:236–240, 1995.
35. G Falasca, M Reverberi, P Lauri, E Caboni, A de Stradis, MM Altamura. How *Agrobacterium rhizogenes* triggers de novo root formation in a recalcitrant woody plant: an integrated histological, ultrastructural and molecular analysis. New Phytol 145:77–93, 2000.
36. A Dobigny, A Ambroise, R Haicour, C David, L Rossignol, D Sihachakr. Transformation of potato using mannopine and cucumopine strains of *Agrobacterium rhizogenes*. Plant Cell Tissue Organ Cult 40:225–230, 1995.
37. Z-Q Xu, J-F Jia, Z-D Hu. Enhancement by osmotic treatment of hairy root transformation of alfalfa suspension cultures, and chromosomal variation in the transformed tissues. Aust J Plant Physiol 24:345–351, 1997.
38. A Oostdam, JNM Mol, LHW van der Plas. Establishment of hairy root cultures of *Linum flavum* producing the lignan 5-methoxypodophyllotoxin. Plant Cell Rep 12:474–477, 1993.
39. M Zehra, S Banerjee, S Sharma, S Kumar. Influence of *Agrobacterium rhizogenes* strains on biomass and alkaloid productivity in hairy root lines of *Hyoscyamus muticus* and *H. albus*. Planta Med 65:60–63, 1999.
40. Y Mano, H Ohkawa, Y Yamada. Production of tropane alkaloids by hairy root cultures of *Duboisia leichhardtii* transformed by *Agrobacterium rhizogenes*. Plant Sci 59:191–201, 1989.
41. J Amselem, M Tepfer. Molecular basis for novel root phenotypes induced by *Agrobacterium rhizogenes* A4 on cucumber. Plant Mol Biol 19:421–432, 1992.
42. N Sevón, R Hiltunen, K-M Oksman-Caldentey. Somaclonal variation in transformed roots and protoplast-derived hairy root clones of *Hyoscyamus muticus*. Planta Med 64:37–41, 1998.
43. M Jaziri, J Homès, K Shimomura. An unusual root tip formation in hairy root culture of *Hyoscyamus muticus*. Plant Cell Rep 13:349–352, 1994.
44. T Aoki, H Matsumoto, Y Asako, Y Matsunaga, K Shimomura. Variation of alkaloid productivity among several clones of hairy roots and regenerated plants of *Atropa belladonna* transformed with *Agrobacterium rhizogenes* 15834. Plant Cell Rep 16:282–286, 1997.
45. R Bhadra, S Vani, JV Shanks. Production of indole alkaloids by selected hairy root lines of *Catharanthus roseus*. Biotechnol Bioeng 41:581–592, 1993.
46. J Palazón, RM Cusidó, J Gonzalo, M Bonfill, C Morales, MT Piñol. Relation between the amount of *rol*C gene product and indole alkaloid accumulation in *Catharanthus roseus* transformed root cultures. J Plant Physiol 153:712–718, 1998.
47. MA Subroto, KH Kwok, JD Hamill, PM Doran. Coculture of genetically transformed roots and shoots for synthesis, translocation, and biotransformation of secondary metabolites. Biotechnol Bioeng 49:481–494, 1996.
48. MGP Mahagamasekera, PM Doran. Intergeneric co-culture of genetically transformed organs for the production of scopolamine. Phytochemistry 47:17–25, 1998.

49. MA Subroto, MGP Mahagamasekera, KH Kwok, JD Hamill, PM Doran. Co-culture of hairy roots and shooty teratomas. In: PM Doran, ed. Hairy Roots: Culture and Applications. Amsterdam: Harwood Academic, 1997, pp 81–88.
50. T Hashimoto, D-J Yun, Y Yamada. Production of tropane alkaloids in genetically engineered root cultures. Phytochemistry 32:713–718, 1993.
51. JW Larrick, L Yu, J Chen, S Jaiswal, K Wycoff. Production of antibodies in transgenic plants. Res Immunol 149:603–608, 1998.
52. R Fischer, Y-C Liao, K Hoffmann, S Schillberg, N Emans. Molecular farming of recombinant antibodies in plants. Biol Chem 380:825–839, 1999.
53. CL Cramer, JG Boothe, KK Oishi. Transgenic plants for therapeutic proteins: linking upstream and downstream strategies. In: J Hammond, P McGarvey, V Yusibov, eds. Plant Biotechnology: New Products and Applications. Berlin: Springer-Verlag, 1999, pp 95–118.
54. DA Russell. Feasibility of antibody production in plants for human therapeutic use. In: J Hammond, P McGarvey, V Yusibov, eds. Plant Biotechnology: New Products and Applications. Berlin: Springer-Verlag, 1999, pp 119–138.
55. EE Hood, JM Jilka. Plant-based production of xenogenic proteins. Curr Opin Biotechnol 10:382–386, 1999.
56. JK-C Ma, ND Vine. Plant expression systems for the production of vaccines. In: J-P Kraehenbuhl, MR Neutra, eds. Defense of Mucosal Surfaces: Pathogenesis, Immunity and Vaccines. Berlin: Springer-Verlag, 1999, pp 275–292.
57. EE Hood, DR Witcher, S Maddock, T Meyer, C Baszczynski, M Bailey, P Flynn, J Register, L Marshall, D Bond, E Kulisek, A Kusnadi, R Evangelista, Z Nikolov, C Wooge, RJ Mehigh, R Hernan, WK Kappel, D Ritland, CP Li, JA Howard. Commercial production of avidin from transgenic maize: characterization of transformant, production, processing, extraction and purification. Mol Breed 3:291–306, 1997.
58. DR Witcher, EE Hood, D Peterson, M Bailey, D Bond, A Kusnadi, R Evangelista, Z Nikolov, C Wooge, R Mehigh, W Kappel, J Register, JA Howard. Commercial production of β-glucuronidase (GUS): a model system for the production of proteins in plants. Mol Breed 4:301–312, 1998.
59. G-Y Zhong, D Peterson, DE Delaney, M Bailey, DR Witcher, JC Register, D Bond, C-P Li, L Marshall, E Kulisek, D Ritland, T Meyer, EE Hood, JA Howard. Commercial production of aprotinin in transgenic maize seeds. Mol Breed 5:345–356, 1999.
60. PM Doran. Foreign protein production in plant tissue cultures. Curr Opin Biotechnol 11:199–204, 2000.
61. CD Merritt, S Raina, N Fedoroff, WR Curtis. Direct *Agrobacterium tumefaciens*–mediated transformation of *Hyoscyamus muticus* hairy roots using green fluorescent protein. Biotechnol Prog 15:278–282, 1999.
62. R Wongsamuth, PM Doran. Production of monoclonal antibodies by tobacco hairy roots. Biotechnol Bioeng 54:401–415, 1997.
63. R Wongsamuth, PM Doran. Hairy roots as an expression system for production of antibodies. In: PM Doran, ed. Hairy Roots: Culture and Applications. Amsterdam: Harwood Academic, 1997, pp 89–97.

64. JM Sharp, PM Doran. Effect of bacitracin on growth and monoclonal antibody production by tobacco hairy roots and cell suspensions. Biotechnol Bioprocess Eng 4:253–258, 1999.
65. W LaCount, G An, JM Lee. The effect of polyvinylpyrrolidone (PVP) on the heavy chain monoclonal antibody production from plant suspension cultures. Biotechnol Lett 19:93–96, 1997.
66. DE Cooke, KJ Webb. Stability of CaMV 35S-*gus* gene expression in (bird's foot trefoil) hairy root cultures under different growth conditions. Plant Cell Tissue Organ Cult 47:163–168, 1997.
67. TV Nedelkoska, PM Doran. Hyperaccumulation of cadmium by hairy roots of *Thlaspi caerulescens*. Biotechnol Bioeng 67:607–615, 2000.
68. JB Hughes, J Shanks, M Vanderford, J Lauritzen, R Bhadra. Transformation of TNT by aquatic plants and plant tissue cultures. Environ Sci Technol 31: 266–271, 1997.
69. M Mackova, T Macek, J Ocenaskova, J Burkhard, K Demnerova, J Pazlarova. Biodegradation of polychlorinated biphenyls by plant cells. Int Biodeterior Biodegrad 39:317–325, 1997.
70. M Pletsch, BS de Araujo, BV Charlwood. Novel biotechnological approaches in environmental remediation research. Biotechnol Adv 17:679–687, 1999.
71. L Metzger, I Fouchault, C Glad, R Prost, D Tepfer. Estimation of cadmium availability using transformed roots. Plant Soil 143:249–257, 1992.
72. T Macek, P Kotrba, M Suchova, F Skacel, K Demnerova, T Ruml. Accumulation of cadmium by hairy-root cultures of *Solanum nigrum*. Biotechnol Lett 16:621–624, 1994.
73. T Maitani, H Kubota, K Sato, M Takeda, K Yoshihira. Induction of phytochelatin (class III metallothionein) and incorporation of copper in transformed hairy roots of *Rubia tinctorum* exposed to cadmium. J Plant Physiol 147:743–748, 1996.
74. L Sanità di Toppi, M Lambardi, N Pecchioni, L Pazzagli, M Durante, R Gabrielli. Effects of cadmium stress on hairy roots of *Daucus carota*. J Plant Physiol 154:385–391, 1999.
75. RR Brooks, MF Chambers, LJ Nicks, BH Robinson. Phytomining. Trends Plant Sci 3:359–362, 1998.
76. TV Nedelkoska, PM Doran. Characteristics of heavy metal uptake by plant species with potential for phytoremediation and phytomining. Minerals Eng 13:549–561, 2000.
77. TV Nedelkoska, PM Doran. Hyperaccumulation of nickel by hairy roots of *Alyssum* species: comparison with whole regenerated plants. Biotechnol Prog 17:752–759, 2001.
78. K Kanokwaree, PM Doran. The extent to which external oxygen transfer limits growth in shake flask culture of hairy roots. Biotechnol Bioeng 55:520–526, 1997.
79. T Shiao, PM Doran. Root hairiness: effect on fluid flow and oxygen transfer in hairy root cultures. J Biotechnol 83:199–210, 2000.

8

Bioreactors for Plant Cell and Tissue Cultures

Regine Eibl and Dieter Eibl
Department of Biochemistry, University of Applied Sciences
Wädenswil, Wädenswil, Switzerland

I. INTRODUCTION

This chapter is concerned with existing bioreactors for different plant cell and tissue culture types producing bioactive compounds. It includes a general section on historical development.

Studies on bioreactor production of useful and valuable metabolites in plant cell and tissue cultures as well as mass propagation procedures on a large scale have been carried out since the end of the 1950s. These results and efforts initiated a large number of studies, reviews, and patents focused on the industrial application of this technology in the 1980s. Although there are some industrial applications (e.g., production of shikonin, taxol, ginseng biomass, berberine) and several new products have just reached semicommercial production levels, no great progress has been achieved for the last 10 years. This limited commercialization is a consequence of the economic feasibility of most processes based on plant cell and tissue cultures.

The bioreactor—the so-called heart of biotechnological production processes—has a key position here. Therefore, we give special attention to the most commonly used bioreactor types (stirred reactor, rotating drum reactor, airlift reactor, bubble column, packed bed reactor, trickle bed reactor), their modifications, their instrumentation, and the operational strategy for mass propagation of plant cell and tissue cultures. Furthermore, we describe a new generation of bioreactors, so-called low-cost reactors, available on the market today.

II. BACKGROUND

The first large-scale cultivations of plant cell and tissue cultures date back to 1959 (1). The researchers cultivated tobacco and vegetable cells in 10-L glass carboys, replaced by stainless steel tanks in 1960 (2). This stimulated a number of studies on mass cultivation processes using plant cell and tissue cultures as biocatalysts focused on the cultivation of tobacco cells (3,4). It culminated in the successful cultivation of *Nicotiana tabacum* cultures in a 20-m^3 stirred reactor in Japan in 1977 (5).

Other breakthroughs in the development of large-scale cultivations were the commercialization of the technical production of shikonin, the first production of secondary metabolites in 1983 (6), the production of ginseng biomass in 1988 (7), and the long-term cultivation of a number of cell lines in 50-m^3 vessels (8).

Table 1 summarizes commercially and semicommercially produced secondary metabolites and shows the leading position of Japan in the field. It shows a relatively small amount of pharmaceutically active agents that have great implications for the treatment of cancer and cardiac diseases, the so-called high-value products, and food additives as well as pigments for colors for cosmetics, foods and dyes for the textile industry.

III. GENERAL REQUIREMENTS FOR THE CULTIVATION OF PLANT CELLS IN BIOREACTORS

The majority of large-scale systems work with plant cell suspension cultures that can be grown in bioreactors in a manner similar to microbial fermentation (20).

As indicated in Table 2, some differences exist between microbial and plant cells concerning cell size and shape, growth behavior, and product formation, all of which influence cultivation process technology. Therefore, the following four general factors have to be considered for cultivation of plant cell suspension cultures in bioreactors (21,22):

1. Homogeneous and low-shear mixing for efficient nutrient transport without sedimentation and/or clumping as well as loss of cell viability
2. Optimal aeration with low shear stress

TABLE 1 Secondary Metabolites with Semicommercial and Commercial Production Levels

Metabolite	Species	Application	Manufacturer	Reference
Scopolamine	Duboisia sp.	Pharmacy (anticholinergicum)	Sumitomo Chemical Industries (Japan)	9
Podophyllotoxin	Podophyllum sp.	Pharmacy (antitumor agent)	Nippon Oil (Japan)	9
Protoberberines	Coptis japonica Thalictrum minus	Pharmacy (antibiotic) (anti-inflammatory agent)	Mitsui Petrochemical Industries (Japan)	10
Taxol (paclitaxel)	Taxus brevifolia Taxus chinensis	Pharmacy (antitumor agent)	ESCAgenetics (USA), Phyton Catalytic (USA/Germany), Nippon Oil (Japan)	11
Rosmarinic acid	Coleus blumei	Pharmacy (anti-inflammatory agent)	Nattermann (Germany)	12
Ginseng biomass containing ginsenosides	Panax ginseng	Pharmacy (dietary compound)	Nitto Denko (Japan)	13,14
Echinaceae polysaccharides	Echinacea purpurea Echinacea augustifolia	Pharmacy (anti-inflammatory agent, immunostimulant)	Diversa (Germany)	15
Shikonin	Lithospermum erythrorhizon	Pharmacy (antibiotic)	Mitsui Petrochemical Industries (Japan)	16,17
Geraniol	Geramineae sp.	Cosmetic (pigment) Pharmacy (essential oil)	Mitsui Petrochemical Industries (Japan)	9
Arbutin	Catharanthus roseus	Cosmetic (pigment, antisepticum)	Mitsui Petrochemical Industries (Japan)	9
Carthamin	Carthamus tinctorius	Cosmetic (pigment)	Kibun (Japan)	9
Vanillin	Vanilla planifolia	Food (flavor)	ESCAgenetics (USA)	11,18
Betacyanins	Beta vulgaris	Food (color)	Nippon Shinyaku (Japan)	9
	Euphorbia milli	Textile industry (dyes)	Nippon Paint (Japan)	11
Anthocyanins	Aralia cordata	Food (color)		19

TABLE 2   Comparison of Plant Cell Suspension and Microbial Culture

Feature	Microbial culture	Plant cell suspension culture
Culture characteristics		
Size	Small (diameter of 1– 10 μm)	Big (diameter of 40– 200 μm)
Individual cells, aggregates	Individual cells and aggregates	Frequently in aggregates
Growth rates	Rapid	Slow
Doubling time	Hours	Days
Shear sensitivity	Low	Moderate (high to low)
Stability	Stable	Unstable
Product accumulation	Often extracellular	Mostly intracellular
Process characteristics		
Culture medium	Often simple	Often complex
Inoculation density	Low	High (5–10%)
Temperature	Often 26–36°C	Species dependent
Aeration[a]	Often high (1–2 vvm)	Low (0.1–0.3 vvm)
Foaming	Often high	Sometimes foaming
Cultivation time	Days	Weeks

[a] vvm, volume per volume per minute.

3. Guarantee of the long-term sterility of the process as a practical consequence of slow growth rates
4. Introduction of light for heterotrophic, photomixotrophic, and photoautotrophic cultures for increasing the biosynthetic capacities

For more differentiated plant cell and tissue cultures such as organ cultures (root cultures, hairy root cultures, shoot and embryogenic cultures) it becomes necessary to realize increased cell-to-cell contact as well as a lower shear stress in the cultivation system. This is due to the morphology, the growth behavior, and the lower shear tolerance of these organ culture types.

IV.   INSTRUMENTATION OF BIOREACTORS FOR PLANT CELL AND TISSUE CULTURES

The optimum and specific process requirements for every growing plant cell and tissue culture line will usually be guaranteed by the bioreactor's configuration associated with reactor instrumentation. Process monitoring and control have become essential elements for control of bioreactor processes using plant cells as biocatalysts (23,24). Table 3 presents accessories that are typical parts of modern bioreactor equipment for plant cell and tissue system cultivations offered by a number of bioreactor companies today (Fig. 1).

The major physical process parameters influencing successful cultivation of plant cell and tissue culture are temperature, viscosity, gas flow rates, and foaming. Furthermore, for application of mechanically agitated reactors, agitator shaft power (estimated by wattmeter measures or torsion dynamometers) and impeller speed are also important parameters.

Temperature is one important parameter measured and controlled in all plant cell cultivation processes. Because plant cell cultures often show non-Newtonian characteristics (25), *on-line* measurement of viscosity can be quite difficult and *off-line* methods are often preferred. A common method for measuring the gas flow rates (air feed, exhaust gas) is to use air flow-meters such as rotameters, and liquid flow rates can be monitored with electromagnetic flowmeters or capacitance probes. Foaming, resulting from extracellular polysaccharides often excreted in the medium, is an ordinary problem in many cultivation processes. The foam can be detected by a capacitance or conductivity probe. The foam formation and foaming are controllable by mechanical foam destruction devices (foam breaker) such as single or rotating plates, ultrasonic irradiation, or addition of silicon- and polypropylene-based antifoam agents. The pressure monitoring is important during the in situ sterilization procedure before the reactor inoculation. Conductivity measurement is simple and typical for plant cell cultivation processes.

TABLE 3   Measurable and Controllable Process Parameters of Plant Cell and Tissue Culture Bioreactors

Accessories for plant cell and tissue cultures	Measured values	Controlled values
Standard accessories	Temperature Agitation speed Air flow rate Dissolved oxygen Foam level pH Pressure	Temperature Agitation speed Air flow rate
Additional accessories	Conductivity Viscosity Weight of vessel (volume)	Weight of vessel (volume)

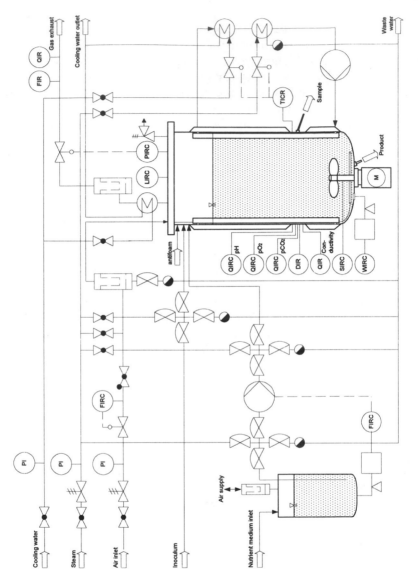

FIGURE 1 Plant cell bioreactor with instrumentation.

Although pH control is not always necessary in plant cell and tissue culture if a culture medium with an appropriate balance of NO_3 and NH_4 is used, the pH of the cultivation is commonly monitored with a steam-sterilizable pH electrode. Dissolved oxygen and sometimes other gaseous metabolites (e.g., CO_2) are useful, but the monitoring and utilization during long-term cultivations often become problematic. Low respiration rates of cultured plant cell and tissue cultures make implementation of *off*-gas analysis difficult (26).

V. GENERAL CLASSIFICATION OF BIOREACTOR TYPES

At present, a large number of bioreactor types are available. This is due to the rapid development of biotechnological processes resulting from progress in genetic engineering and the bioreactor business boom since the 1970s.

Devising a classification of all these bioreactor configurations becomes extremely difficult. Some reported studies have dealt with classification trials based on whether the reaction occurs, the bioreactor operation, the way in which the culture grows, the reactor mixing characteristics, the basis of the reactor structure, and the energy input (23,27–33).

However, the energy input classification dominates in the literature. It should also be noted that only aerobic reactor systems with high energy input will be considered using the classical energy input classification. But anaerobic cultures, immobilized cultures, and animal as well as plant cell and tissue cultures require a lower energy input for mass and heat transfer processes. It should also be borne in mind that they often occur as cell aggregates and show shear and pressure sensitivity. In addition, systems for cell immobilization become necessary. Aerobic reactors with mechanical, hydraulic, and pneumatic energy input (lower energy input or adapted energy input) are suitable reactor systems for such biocatalysts. Bed reactors and membrane reactors represent further commonly applied reactor variants. Figure 2 illustrates the general bioreactor classification with given examples of bioreactor types.

The power input for mass and heat transfer is controlled mechanically in mechanically agitated reactors. There are mechanically driven or agitated reactors with rotating agitators and with nonrotating agitators producing a stirring effect through a vibrating or tumbling motion. The third type of mechanically driven reactors works without an agitator while the whole reactor vessel rotates. The mechanical energy input by agitators or self-motion should ensure homogeneity of mass and temperature as well as gas dispersion inside the reactor.

Because of aseptic design requirements for mechanically driven reac-

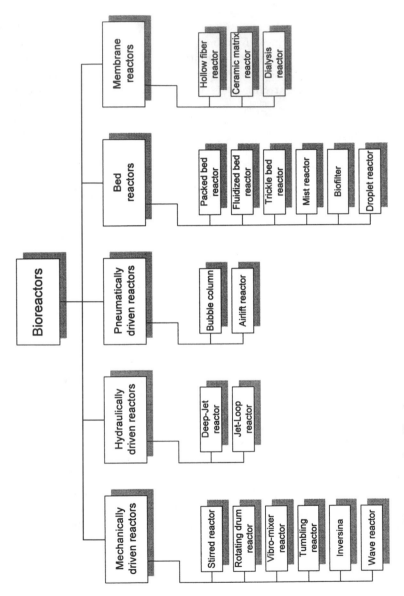

FIGURE 2 Bioreactor classification.

tors, one critical point is the seal between the reactor's inside and the rotating, tumbling, or vibrational shaft as well as the environment. The predominant current choice is a double mechanical seal for direct mechanical coupling of the shaft to the drive. Magnetic coupling has been recommended for alternative cases with high levels of containment. In addition, power transfer by magnetic drive is quite low (150 N m). Tumbling or vibrating stirrer shafts are sealed with a static seal such as a bellows or metal compensator (no mechanical seal). There is enormous know-how available on the design, scale-up, and operation of mechanically driven reactors. The most commonly used reactor type, applied in about 90% of all industrial processes, is the stirred reactor.

The working principles of hydraulically driven reactors are based on energy input produced by pumps. Special double-phase pumps guarantee fluid circulation in internal and external loops. Thereby the kinetic energy of the jet of liquid entering the medium after passing the nozzle (for example, slot nozzles, Venturi tubes, injectors, or ejectors) is used for gas distribution and liquid recirculation. Typical representatives of this reactor class are the deep-jet and the jet-loop reactor. The main advantages of hydraulically driven reactors are their simple design. They work without mechanical internals and seal shafts and show reduced shear forces, simple regulation of external loops, and the possibility to install heat exchangers in the external loop as well as to increase the number of loops.

The term pneumatically driven reactors covers reactors in which the energy input for mass and heat transfer takes place through a steam of gas or air. The effective differences in density between medium and gas lead to variations in fluid mixing and fluid dynamics influenced by viscosity, density difference, gas flow, and gas bubble size. The gas is usually injected by static gas distributors (diffuser stones, nozzles, perforated plates, diffuser rings) or dynamic gas distributors (slot nozzles, Venturi tubes, injectors or ejectors). Airlift reactors and bubble columns represent important pneumatically driven reactors. Although they are relatively simple in construction, sound design and operation are critical for optimal hydrodynamic behavior (31). If there are strong variations in biomass concentration, viscosity, surface tension, and ionic concentration, operational problems—for example, foaming, flotation, and bubble coalescence—will result.

Bed reactors are directly linked to the use of biocatalysts (enzymes, microorganism cells, plant cells, animal cells) occurring in the form of heterogeneous particles (e.g., flocs, cell aggregates, immobilized cells or enzymes); biofilms as well as biocatalysts require an immobilization matrix. Bed reactors with a continuous gas phase or continuous liquid phase are available. According to the dependence on the bed containing the biocatalyst and the flow rate inside the bed, we distinguish between packed bed reactors and fluidized bed reactors. The term packed bed reactor is used to define the reactor systems composed of a solid phase with packed or immobilized biocatalysts. The biocatalysts are in continuous or periodic contact with air and nutrient medium. Commonly used packed bed reactors are the trickle bed reactor and biofilter. With respect to the high packing density of bio-catalysts, the packed bed reactor is the bioreactor configuration with the highest achievable productivity per unit reactor volume. On the other hand, the height of the solid particles in a bed is influenced by the biocatalyst as well as the catalyst particles. These particles should be relatively incompressible and able to withstand their own weight in the bed without deforming and occluding liquid flow. Thus, crack formation and channeling inside the bed will be prevented. In fluidized bed reactors, the solid particles with the biocatalyst are maintained in fluidization by means of the circulation of the fluid phase (up-flow mode). Fluidization of the solid particles is reached when the fluid flow through the bed is high enough to compensate for their weight. The degree of internal mixing can vary to a great extent and depends on energy input. Configurations favoring complete liquid mixing are possible as well as configurations approaching plug flow. It is difficult to guarantee steady-state operation for fluidized bed systems with different biocatalyst densities and volumes.

Membrane reactors find application in procedures where separation is desired. Here membranes are used for immobilization of cells and enzymes, selective feeding and removal of medium components and metabolites, as well as separation of low- and high-molecular-weight nutrients. Applied membranes are dialysis, ultrafiltration, and microfiltration systems. The biocatalyst can be adsorbed in the form of a membrane layer or immobilized in the pores of the membrane. There are static membranes and dynamic membranes inside the reactor, external cross-flow systems, and special membrane reactors with medium circulation available. The most important advantage of membrane reactors is their recognition of specific culture conditions allowing biocatalyst usage without possible inactivation steps taking place. In membrane reactors the cells are retained by a membrane in a similar way to a natural cell (34). This ensures continuous or repeated operations. Principally, it is necessary to move the fluid on the pressure side to guarantee high filtration rates and long operational times.

Modern bioreactor configurations combine the working principles of a number of reactor basic configurations and are often equipped with external or internal reactor-separator systems (35,36). The eight main reactor types will be characterized briefly concerning the typical working principle in the following discussion.

A. Stirred Reactor

The ordinary stirred reactor (Fig. 3a) is equipped with baffles, air sparger and radial flow impellers, axial flow impellers, and impellers distributing the power input over a large fraction of the total reactor volume (e.g., Rushton, marine, or Intermig impeller). Its typical fermenter geometry (fermenter height/diameter ratio) must be 2:1 or 3:1. Mixing and mass dispersion are achieved by mechanical agitation. Different flow patterns and shear rates inside the vessel will be produced by different impeller shapes, sizes, and spacing (multiple stirrers) as well as installation of a coaxial draught tube (Fig. 3b). It is also possible to design and operate stirred reactors in a multistage mode.

B. Rotating Drum Reactor

Mass and energy inputs are realized by drum self-rotation in the rotating drum reactor (Fig. 3c). Systems with one reactor-chamber work with a long axis designed as a hollow shaft for air and gas exchange. Transport processes inside the reactor can be varied by the drum rotation rate. If there is the demand for low hydrodynamic shear stress conditions and high oxygen transfer rates for non-Newtonian cultures, the rotating drum reactor should be preferred to standard stirred reactors (37,38).

C. Bubble Column

The bubble column is another alternative to the stirred reactor that has no mechanical agitation and is structurally very simple (Fig. 4a). Mass and energy inputs are achieved only by pneumatic driving (gas sparging). Advantages of this reactor type are low capital cost and minimized problems of sterility based on lack of moving parts inside the reactor. Bubble columns can be divided into five main types of reactors on the basis of their structure: simple column reactors, multistage perforated plate column reactors as well as multistage column reactors with static mixers for repeated gas dispersion, and column reactors with nozzle aeration and tower reactors.

D. Airlift Reactor

As in bubble columns, mass and energy input in airlift reactors is accomplished pneumatically without mechanical agitation and associated with significantly lower shear levels than in stirred reactors. It should be pointed out that airlift reactors generally provide better mixing conditions than the bubble columns described. The use of a draught tube divides the flow in a riser and downcomer, and the density difference enables the liquid to circulate

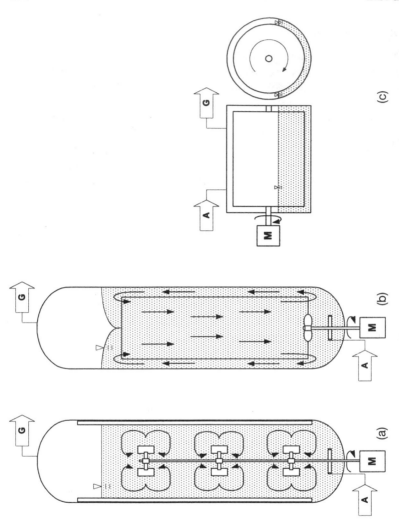

Figure 3 Mechanically driven reactors: (a) stirred reactor, (b) stirred reactor with draught tube, and (c) rotating drum reactor. A, air inlet; G, gas exhaust; M, motor.

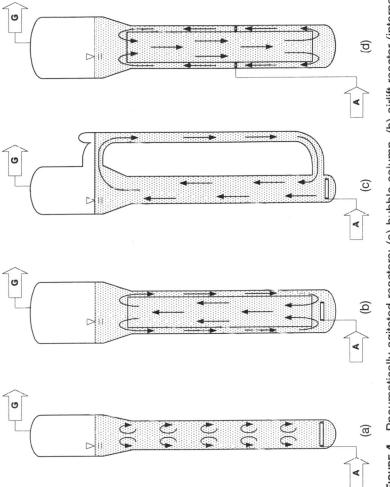

FIGURE 4 Pneumatically agitated reactors: (a) bubble column, (b) airlift reactor (internal loop), (c) airlift reactor (external loop), and (d) airlift reactor (draught tube). A, air inlet; G, gas exhaust.

with high turbulence. Three common airlift configurations, the internal loop airlift reactor, the external loop airlift reactor, and the draught tube airlift reactor, are shown in Fig. 4b–d. Airlift reactors of types c and d are also available as constructions using suitable foil material (39–41).

E.   Packed Bed Reactor

The typical packed bed reactor consists of a vertical column packed with cells and adsorbents such as polymeric beads (Fig. 5a). The nutrient medium can be fed either at the top or bottom of the tube and is circulated through the packed bed. To prevent gas pockets and flow channeling in the packed bed, air is sparged indirectly by aeration of a separately used storage vessel as well as a recycling medium vessel (42) for some applications.

F.   Fluidized Bed Reactor

When packed beds are operated in upflow mode, the bed expands at high liquid flow rates and follows the motion of the particles. This is the working principle of the ordinary fluidized bed reactor (Fig. 5d), where the particles are in a constant motion. A special fluidized bed reactor configuration with mixing characteristics similar to those of stirred reactors is depicted schematically in Fig. 5c. Considering the reactor geometry, the fluidized bed reactors with plug flow (Fig. 5d) are taller than fluidized bed reactors with complete backmixing (Fig. 5c).

G.   Trickle Bed Reactor

The trickle bed reactor is one variation of the packed bed configuration (Fig. 5b). One or a number of nozzles integrated in the headspace of the column spray the nutrient continuously or periodically onto the top of the packaging. The aeration takes place directly by introducing air at the base. A special version of the trickle bed reactor is the mist reactor. The nutrient medium is introduced as a mist phase produced by an injector or ultrasonic nozzle (43).

H.   Membrane Reactor

A few articles describe suitable membrane reactors for plant cell and tissue cultures. They are rarely used in comparison with the other bioreactor systems discussed (22,43). Typical membrane reactor configurations are depicted in Fig. 6. Figure 6a shows a membrane system with external cross-flow. The cells grow in a stirred reactor and will be retained. The membrane filter assembly is responsible for mass exchange, for example, a continuous separation of product in the course of the cultivation process. While the

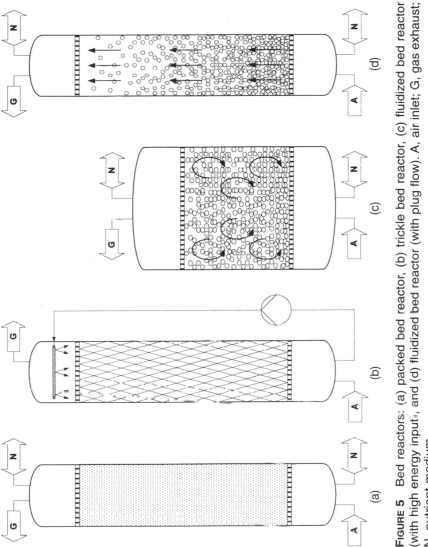

FIGURE 5 Bed reactors: (a) packed bed reactor, (b) trickle bed reactor, (c) fluidized bed reactor (with high energy input), and (d) fluidized bed reactor (with plug flow). A, air inlet; G, gas exhaust; N, nutrient medium.

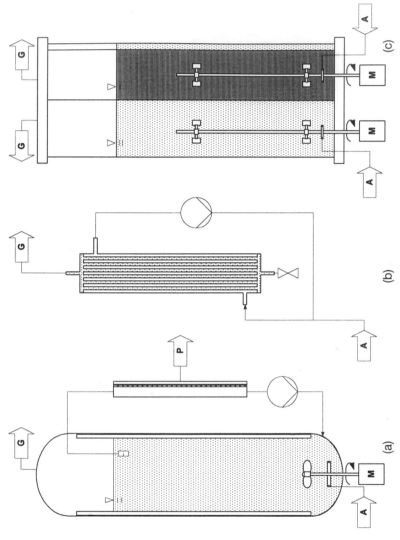

Figure 6 Membrane reactors: (a) membrane reactor with bypass membrane for perfusion mode, (b) typical membrane reactor for immobilized cells (hollow fiber reactor), and (c) membrane reactor with internal dialysis membrane (Bioengineering Ltd., Switzerland). A, air inlet; G, gas exhaust; M, motor; P, permeate.

product is leaving the reaction system on the permeate side, the culture medium passes the membrane using a bypass on the retentate side. The whole culture medium inside the bioreactor represents permeate. As shown in Fig. 6b, simultaneous nutrient and extracellular product exchange can take place in special membrane modules applied instead of cross-flow assemblies. Here the cells are immobilized on the inside of a hollow fiber module and the mass and gas exchange is realized by hollow fiber membranes. A membrane reactor from the company Bioengineering Ltd. is schematically illustrated in Fig. 6c. The inner and outer chambers are separated by a dialysis membrane. Depending on the molecular weight, medium components or products either diffuse through the membrane or are retained in their respective chamber (44,45).

VI. SUITABLE BIOREACTORS FOR PLANT CELL AND TISSUE CULTURES

An overview of the general working principles of bioreactors and their classification allow the conclusion that a variety of reactors exist. Table 4 gives an overview of the most commonly used bioreactor configurations for mass propagation depending on the cell culture type. As indicated in Table 4, an additional categorization into reactors for suspension cultures, hairy root cultures, and shoot and embryogenic cultures is useful.

A. Reactors for Plant Cell Suspensions

Stirred reactors, airlift reactors, rotating drum reactors, and fluidized bed reactors were adapted with only minor modifications and introduced to plant cell cultivation techniques with good results for a wide range of suspension cultures. The literature reveals the predominance of stirred reactors having volumes within a range of 75 m^3 for plant cell and tissue cultures (12,16, 25,46–49). They are generally equipped with aeration and mixing systems producing no or low shear damage.

In contrast to most animal cell cultures, a number of plant cell and tissue cultures are not damaged by air bubbles. Direct aeration by gas injection using spargers becomes possible. However, bubble-free aeration (membrane aeration, external aeration, indirect aeration using spin filters or other assemblies) can have advantages for cultures tending to foaming and flotation (79,84,85).

The primary impeller task is to guarantee mass and temperature homogeneity and gas dispersion in bioreactor cultivation processes based on plant cell suspension cultures. Gas dispersion plays a subordinate role. In contrast to reactors with high aeration rates showing optimal operation near

TABLE 4 Plant Cell and Tissue Culture Types and Bioreactors

Cell culture type	Reactor types	Products	Reference
Suspension culture	Stirred reactor	Ajmalicine, anthraquinone, berberine, biomass, rosmarinic acid, shikonin, vincristine	12,16,25,46–49
	Rotating drum reactor	Anthocyanin, biomass, shikonin	37,38,50
	Airlift reactor	Ajmalicine, artemisin, berberine, biomass, jatrorrhizine, saponin, vincristine, β-carotene	12,51–57
	Fluidized bed reactor	Betalaine, biomass	43,58–60
Hairy root culture	Stirred reactor	Ajmalicine, catharantine	61
	Modified stirred reactor	Ajmalicine, catharantine, ginsenosides, hyoscyamine, nicotine, shikonin	43,62–65
	Modified rotating drum reactor	Anthocyanin, hyoscyamine, scopolamine	66,67
	Modified airlift reactor	Ajmalicine, artemisin, biomass, catharantine	43,64,68
	Modified bubble column	Ajmalicine, catharantine, shikonin	64,65
	Modified packed bed reactor	Biomass	43,69
	Modified trickle bed reactor		
	Spray or droplet reactor	Betacyanin, betaxanthin, biomass, ginsenosides, hyoscyamine, nicotine, scopolamine, thiophenes	70–72
	Mist reactor	Betacyanin, betaxanthin, biomass, nicotine	73
Embryogenic culture	Stirred reactor	Alliin, coumarins, differentiated biomass	74–78
Shoot culture and embryogenic culture	Bubble column	Alliin, coumarins	79,80
	Aerated vessel	Differentiated biomass	81,82
	Airlift reactor	Coumarins	75,83

the flooding point situated on the top, plant cell and tissue culture bioreactors operate below this point. Suitable impellers are radial flow impellers, axial flow impellers, and impellers that distribute the power input over a large fraction of the total reactor volume (Fig. 7).

Radial flow impellers such as the Rushton impeller achieve high energy inputs. Therefore this impeller type is suitable for only limited duties in plant cell suspension cultivation procedures. Provided that a Rushton impeller should be applied, improved designs such as the Rushton impeller with concave blades should be preferred. This improved impeller type has a more uniform energy input and produces lower shear forces.

Large, slow-moving axial flow impellers providing good mixing at relatively low tip speeds up to 2.5 m sec^{-1} (e.g., marine impeller or special pitched blade impeller with rounded blades) are superior for plant cell and tissue cultures. Axial flow impellers are commonly used for more sensitive cultures requiring lower gas dispersion. There is a tendency to develop shear-sensitive improved axial flow impellers characterized by kinked, torsioned, and/or profiled large stirring paddles. Newly published design criteria for bioreactors that are really fashionable show tall vessels with multiple axial as well as radial flow impellers (e.g., multiple cross-arm paddle mixer, Intermig impeller and Sigma impeller). These impeller types also ensure intense and turbulent mixing in the top of the bioreactor. Improved asymmetrical impellers with the power input distributed over a large fraction of the total reactor volume and an axial displacement flow with superposed three-dimensional motions occurring additionally (e.g., multiple Seba impeller and Hint impeller) could be favorable for plant cell suspension cultures in tall bioreactors. It is commonly postulated that such improved asymmetrical impellers are suitable for fluids with medium to high viscosity. Furthermore, in those asymmetrical impellers, mixing zones with minimized mass transport as well as uncontrollable liquid swirling inside the impeller can be prevented (86) (H. Schindler, MAVAG Ltd., Switzerland, personal communication, 2000).

As demonstrated in the literature, impellers positioned near the vessel wall, for example, the spiral stirrer, helical ribbon impeller, and anchor impeller, are suitable impeller types for different plant cell lines (25,46–49), especially for cell broths with high viscosity. Major drawbacks are high costs of spiral and helical ribbon impellers resulting from the fabrication process and problematic cleanability. Alternatives are cut opened multiple helical ribbon impellers offered by various manufacturers. In addition, alternative systems were developed to avoid typical stirring movement, for example, the cell-lift fermenter. It is designed so that its cell-lift impeller (Fig. 8) acts as a fluid pump and aerator. Macrocirculation mixing is generated by a hollow central shaft with three rotating horizontal jet tubes in a low-shear

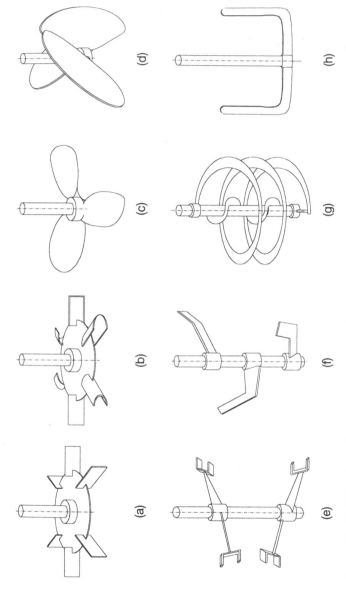

FIGURE 7 Suitable impeller types for plant cell suspensions: (a) Rushton impeller (concave blades), (b) Rushton impeller (pitched blade impeller (rounded blades), (c) marine impeller, (d) pitched blade impeller (rounded blades), (e) Intermig stirrer (Ekato, Germany), (f) Hint stirrer (Chema Balcke Dürr, Germany), (g) spiral impeller, and (h) anchor impeller.

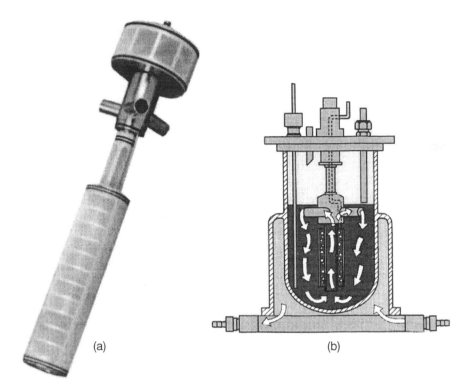

(a) (b)

FIGURE 8 Cell-lift fermenter (New Brunswick Scientific, USA, with kind permission of IG Instrumentengesellschaft, Zurich). (a) Cell-lift impeller and (b) reactor principle (88).

bulk movement of cells and nutrient medium. This reactor type was used successfully for cultivation of *Santalum album* L. cells producing phenolic compounds (87).

Airlift reactors and bubble columns have also been applied by a number of workers for plant cell suspension cultures (see Table 4). Although the airlift reactor is the system of choice for newly developed cell lines, it has been suggested by Doran (63) and Tanaka (89) that this reactor type can provide inadequate mixing related to low aeration rates and/or suspension broth viscosity. Another potential disadvantage of airlift technology is that it may not be possible to achieve cell growth as high as obtained in stirred reactors regularly (about 20 g dry weight L^{-1} (26)). One exception is the cultivation of *Berberis wilsonae* suspension cells in an 20-L airlift reactor, where Breuling et al. (51) described the highest growth rate reported to date for plant suspension cells (40 g dry weight L^{-1}). Tanaka et al. (37) developed a special rotating drum reactor that guarantees suspension homogeneity, a low-shear environment, and reduced wall growth. A few authors (42,58–60) recommend the application of fluidized bed reactors for suspension cultures, which also ensures cell immobilization. In conclusion, airlift reactors and stirred reactors are most often used in commercial as well as semicommercial plant cell cultivation.

B.   Reactors for Hairy Root Cultures

Suitable bioreactor performance for cultivation of hairy roots should consider the structural roots' integrity including physiology, morphology, and their rheological properties (Table 5). The growth behavior of the roots causes essential difficulties concerning the inoculation, harvesting, and sampling procedure in the course of the cultivation process. A special and enlarged inoculation assembly in the form of special inoculation vessels, cell bags, and inoculation tubes became necessary (72,77). A nonuniform biomass distribution in the reactor and insufficient mass transfer in the densely packed mass of growing roots result in cell necrosis and autolysis as well as loss of biosynthetic capability. It became clear that standard reactors are generally not suitable for hairy root cultures. In the only exception described in the literature, hairy roots of *Catharanthus trichophyllus* L. were successfully cultivated in a simple stirred reactor with a 14-L working volume (61) at low agitation rates (100 rpm) and a high aeration rate of 1 vvm. All other reported configurations (43,62–73) are modified standard reactors.

Usually, stainless steel or nylon meshes (basket) and polyurethane foam are introduced into the chosen standard reactor for roots to protect against mechanical and hydrodynamic shear (43). The basket makes it pos-

TABLE 5   Characteristics of Hairy Root Cultures (HRCs)

Advantages	Disadvantages
Stable production of pharmaceutical active compounds	Morphology and growth behavior of HRCs
Comparable or even higher growth rates than cell suspensions	Root development
Simplified cultivation process	Degree of branching
Less expensive and less complicated medium	Organized growth
Emerged growth possible (reduced medium volume and forced aeration)	Pressure sensitivity
	Shear sensitivity

sible to separate the growth space of the roots from the aeration and mixing area and to support the self-immobilization of the roots. Figure 9 illustrates such hairy root support systems applied for scopolamine- and hyoscyamine-producing hairy root cultures of *Hyoscyamus muticus* L. Laboratory rotating drum reactors were also adapted for hairy root cultivations. Such a modified rotating drum reactor named Inversina (Fig. 9c) produces an oloid movement with coupling of three basic movements (rotation, translation, inversion). The consequence is that the rhythmic movement and low shear stress did not inhibit the growth and biosynthetic capability of the cultivated roots. Biomass growth of about 45-fold was achieved in 28 days (77).

It could also be demonstrated that working trickle bed reactors in which the nutrient medium is sprayed or dispersed offer more ideal conditions for growth and productivity of hairy roots than ordinary submerged bioreactors. Although moisture is required by root tissue, if the medium film is too thick it limits nutrient and gas transfer to the tissue. The solution is a droplet or mist application cycle, which decreases the development of thick medium films on the roots. Laboratory trickle bed reactors with a support matrix usually with working volumes of 1 to 14 L have been suitable for mass propagations of *Hyoscyamus muticus* (77,90), *Nicotiana tabacum* (70), *Beta vulgaris* (73), and *Carthamus tinctorius* hairy roots (91,92).

Wilson (72) described the only designed large growth droplet bioreactor system with a volume of 500 L. The Wilson reactor consists of a sparge ring, a number of sample as well as addition ports, sterile filters for the air, a central inoculation port, two concentric rings of spray nozzles, a drain port for recycle of growth medium, a pressure relief valve, and the immobilization matrix. This immobilization assembly, consisting of a number of barbs forming chains, is located within the droplet reactor. A high-pressure diaphragm pump guarantees nutrient medium feeding. Measurable control parameters are temperature, pH value, pressure, oxygen, and weight of the growth vessel. The last parameter allows indirect root growth control. Wilson and co-workers were able to cultivate hairy roots of *Datura stramonium* in this bioreactor type. They showed the suitability and advantages of the designed immobilization matrix for even distribution of the roots and an easy harvesting procedure. A total biomass of 39.8 kg fresh weight was harvested after 40 days of cultivation (72).

C. Reactors for Embryogenic and Shoot Cultures

The characteristics of embryogenic cultures are quite different from those of shoot cultures, but they often seem to be similar to cell suspension cultures. In this case, the same bioreactor types suitable for suspension cultures can be applied for embryogenic cultures.

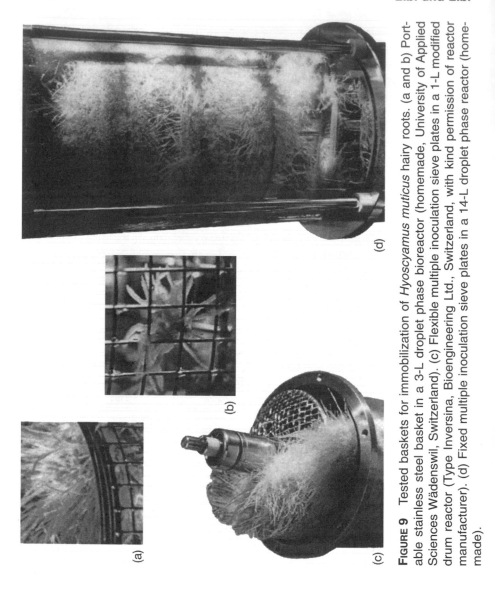

Figure 9 Tested baskets for immobilization of *Hyoscyamus muticus* hairy roots. (a and b) Portable stainless steel basket in a 3-L droplet phase bioreactor (homemade, University of Applied Sciences Wädenswil, Switzerland). (c) Flexible multiple inoculation sieve plates in a 1-L modified drum reactor (Type Inversina, Bioengineering Ltd., Switzerland, with kind permission of reactor manufacturer). (d) Fixed multiple inoculation sieve plates in a 14-L droplet phase reactor (homemade).

Introduction of light into the bioreactor becomes necessary for some embryogenic cultures as well as shoot cultures. The literature describes not only external illumination tubes that are installed around the reactor but also internal systems made from tubes and hollow fibers (77,79,93–95). Figure 10 shows the illumination cage of the 15-L reactor with an eccentric motion stirrer that has been used for cultivation of an embryogenic culture of *Allium sativum* (78). Another advantage of this reactor, in terms of oxygen transfer and the stirring principle, is that the motion of the multifunctional stirring system protects shear-sensitive cultures against damage and prevents the sedimentation of cell aggregates in the medium. The air enters the reactor through the cell-free interior of the cylindrical stirrer. The medium passes the metallic membrane of the stirrer and oxygenates without cells. Thus, the problems with flotation that arise with standard reactors are prevented.

Similar systems with bubble-free aeration achieved by silicone tubing inserted into the reactor and vibrating stirrers (vibromixers) instead of rotating stirrers have been proposed by Preil et al. and others (84–96) and Luttmann et al. (85).

Aerated vessels are also suitable large-scale reactors for shoot mass propagation. Akita et al. (82) reported their observations concerning mass propagation of shoots of *Stevia rebaudiana* L. in a 500-L reactor of this type (64.6 kg fresh weight after 4 weeks of cultivation).

VII. ECONOMIC FEASIBILITY AND TRENDS IN REACTOR DEVELOPMENT FOR PLANT CELL AND TISSUE CULTURES

The overview of suitable bioreactors for plant cell and tissue cultures indicates that the basic technology for production of plant-derived products in bioreactors is essentially in place. But the production of plant-derived products in bioreactors is often not economically viable compared with de novo synthesis. Exceptions are high-value products. Yoshioka and Fujita (97) and Goldstein (98) demonstrated that capital investment is an important instrument for reduction of secondary metabolite production costs. A possible key issue here is the bioreactor, with its high substantial costs based on requirements for aseptic bioprocess technology, sterilization in place (SIP), cleaning in place (CIP), and validation possibility including sophisticated instrumentation. In addition, costs for upstream processing and downstream processing as well as personal training have to be considered. The current price of a 100-L standard sterile reactor is estimated as $100,000 to $140,000. This price includes 30% costs for the bioreactor, 30% costs for instrumentation, 15% costs for piping, 10% costs for engineering, and 15% costs for qualification as well as the validation procedure. Increasing reactor size will result

FIGURE 10 Reactor with eccentric motion stirrer and internal illumination cage (3000 lux). (a) Cage and (b) reactor vessel. (Provided by the company Chema Balcke Dürr with their kind permission.)

in a proportional increase of costs. The costs concerning instrumentation, engineering, qualification, and validation will stay approximately at a constant level. This is underlined by the fact that the investment costs for a 500-L reactor are between $175,000 and $200,000 and the costs for a 10-m^3 reactor are between $400,000 and $450,000 (H. Schindler, MAVAG Ltd., Switzerland, personal communication, 2000).

A rethinking of the current reactor technology combined with a rational and simple reactor design using disposable materials is needed for a remarkable reduction of reactor costs influencing capital costs. This idea resulted in the development of a new bioreactor generation for plant cell and tissue cultures (disposable and low-cost bioreactors). At present, three types of such bioreactors for mass propagations of plant cell and tissue cultures have been proposed in the literature (26,99–101).

The so-called plastic-lined reactor with working volumes of 6.5 and 28.5 L based on an airlift reactor guarantees the mass propagation of *Hyoscyamus muticus* suspension cells. About 7 g dry weight of plant cells per liter of medium were grown in 13 days in the 28.5-L low-cost airlift reactor (99).

The wave reactor (Fig. 11) is a mechanically driven reactor system. This reactor system, available on the laboratory scale (working volume of 10 L) and pilot scale (working volume of 100 L), consists of three components: a rocker base unit, the disposable bioreactor chamber, and the measuring and control units. Here the energy input is caused by rocking the chamber forth and back, putting the cell culture and the medium in a wave movement. In this way, the surface of the medium is continuously renewed and bubble-free aeration can take place. Depending on the sensitivity of the cultivated cell line, the angle and amplitude of the rocking motion as well as the velocity can be varied. Specially designed sterile plastic cell bags that guarantee simple handling as well as optimal cell growth for hairy root cultures, suspensions, and embryogenic cultures have the function of the bioreactor chamber.

The wave laboratory reactor has shown a higher biomass increase for tropan alkaloids as well as for ginsenosides producing hairy root cultures (about 40% higher) and embryogenic cultures of *Allium sativum* (about 20% higher) compared with optimized stirred reactors, rotating drum reactors, and droplet phase reactors used in experiments for growth comparison (100).

Another low-cost system is the immersion reactor RITA of the French company CIRAD. The system has proved its efficiency for cultivation of embryogenic cultures of banana, coffee, citrus, oil palm, and rubber (101).

These low-cost reactors can thus enhance biomass productivity tremendously and may also have an economic advantage. It seems likely that

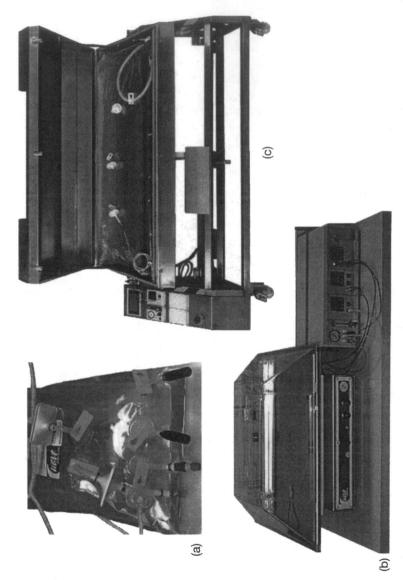

FIGURE 11 Wave reactor. (a) Cell bag, (b) laboratory reactor, and (c) pilot reactor. (With kind permission of Wave Biotech Ltd., Switzerland.)

such low-cost reactors represent interesting alternatives to other suitable standard bioreactors on the laboratory and pilot scale.

New low-cost as well as disposable reactor systems will be available on the market in the near future, such as a low-cost mist bioreactor for the production of bioactive compounds in transformed hairy root cultures (E. Wildi, ROOTec Ltd., Germany, personal communication, 2001).

VIII. OPERATIONAL STRATEGIES OF PLANT CELL AND TISSUE CULTURE BIOREACTORS

Apart from the choice of a suitable bioreactor system for cultivation of plant cell and tissue cultures, the operational strategy is of significance. Optimal results will be expected only from the combination of the most suitable reactor type and operational mode. Unfortunately, the possibilities of modern bioprocess engineering are not applied rigorously to mass propagation procedures based on plant cell and tissue cultures. This section gives a general survey of operational strategies of plant cell and tissue bioreactors. It is common knowledge that the mode of reactor operations can vary. Production methods have been developed based on batch, fed-batch, and continuous systems.

Cells (biocatalysts) and medium are added to a closed system (bioreactor) in a batch or discontinuous culture. In terms of the cultivation process, only the cultivation parameters—for example, temperature, aeration rate, and light—will be changed. Medium or biomass is not added or removed during total cultivation time. The culture is harvested after an appropriate period of cultivation, i.e., when growth has stopped because of lack of nutrient.

The medium is continuously fed into the bioreactor and biomass is continuously removed to keep the liquid level constant in continuous cultures. The steady-state level of the cell population, which is determined by the nutrient feed rate, is reached as long as the principle of the chemostat [growing biomass and dilution rate D (ratio of volumetric flow rate to reaction volume) $\leq$ specific growth rate] is followed. Theoretically, this method allows continuous exponential growth of the biocatalyst (20,24).

Compared with continuous cultures, batch systems have the advantages of simplicity, reliability, and flexibility as well as the drawback of a lower space-time yield. However, they can also cause significant disadvantages for systems with substrate as well as product inhibition in that the total substrate concentration has to be received at the beginning and the formed products have to stay inside the reactor during the total cultivation time. In contrast, continuous systems can define the substrate concentration through the regulation of the dilution rate as well as the contact time of the product by choosing the continuous reactor route. Drawbacks of continuous strategies are a high level of instrumentation, inflexibility, and susceptibility to failure related to infections by contaminants.

The fed-batch operation of a bioreactor is a compromise between an ordinary batch and a continuous operation. Fresh nutrient medium is added during the stationary phase of the biocatalyst growth in fed-batch operations, and as a result, the cell growth will continue. The volume of liquid medium is increasing during the cultivation time. Fed-batch cultivations are preferred for processes involving substrate inhibitions and biotransformations. The precursor represents the feeding medium in the course of biotransformation processes. It is also common practice to feed elicitors in fed-batch processes. In contrast to fed-batch processes, the nutrient medium is removed periodically and fresh medium is added during repeated fed-batch cultivations. The bioreactor is then left to run as a batch until the cycle is repeated. Such an operational route ensures prolongation of the process time as long as active biomass is available. A more difficult inoculum scale-up will not be necessary (102).

The two-stage batch mode is a modified fed-batch operation. In 1977 Zenk and co-workers successfully introduced a modified discontinuous procedure in which growth and product formation do not take place simultaneously, the so-called two-step procedure (20). The two-step or two-stage mode, which is also used commercially for shikonin production, works with two optimized media used separately. In reality, it is also possible to vary parameters such as oxygen supply, pH value, and light intensity to influence growth and product formation.

However, additional process strategies and reactors with external or internal (in situ) reactor-separator systems do exist. Special growth and product formation processes have been developed through reactor arrangements such as cascade operation (reactors in series) or through the combination of stirred tank and plug flow reactors. Here, depending on the growth and product formation phase, each bioreactor can be operated independently while using separate media or process parameters, for example temperature, pH, and oxygen supply. External or internal reactor-separator systems allow enrichment (increasing the cell concentration) or removal of biomass and metabolites in cultivation processes. It has been found that these assemblies are of special interest for slowly growing cultures where an increase of process efficiency can be achieved by high biomass concentrations as well as uninhibited product formation. Therefore, plant cell and tissue cultures have to be enriched and metabolites, medium components, or the complete medium have to be exchanged periodically or continuously inside the bioreactor. Typical cell culture configurations that meet these demands are membrane reactors and bioreactors with perfusion systems usually applied in the form of internal (spin filter) or external (centrifuge) cell retention devices (103). They guarantee the establishment of high-density cultures (e.g., commercial production of berberine). To our knowledge, other known technical reactor-separator systems are multiphase reactors, vacuum bioreactors, bioreactors with integrated adsorbers or ion exchangers, distillation and extraction bioreactors, and reactors with gas stripping (104).

IX. CONCLUSIONS

Plant cell and tissue cultures have been demonstrated to produce a wide range of secondary metabolites. A universal bioreactor suited for all plant cell and tissue culture types does not exist. Moreover, the most suitable bioreactor has to be estimated experimentally for every cell culture type and cell line. On the basis of known characteristics of the specific cell or tissue culture type, stirred reactors, rotating drum reactors, airlift reactors, bubble columns, fluidized bed reactors, packed bed reactors, and trickle bed reactors with working volumes from 0.75 L up to 75 m^3 in batch and fed-batch mode seem to be the most appropriate reactor configurations.

Hence, the production costs for products derived from plant cell and tissue cultures in these conventional bioreactors will be economically viable only for high-value products. Disposable, scalable, low-cost bioreactors were developed to overcome this problem. The types discussed here (plastic-lined reactor, wave reactor, RITA) provided advantages when used in production of secondary metabolites based on plant cell and tissue cultures. It is suggested that these new bioreactor types could improve the process efficiency for more valuable plant-derived products and result in a new wave of their industrial production.

REFERENCES

1. W Tulecke, LG Nickell. Production of large amounts of plant tissue by submerged culture. Science 130:863–864, 1959.
2. W Tulecke, LG Nickell. Methods, problems and results of growing plant cells under submerged culture. Trans N Y Acad Sci 22:196–204, 1960.
3. A Kato, Y Shimizu, S Nagai. Effect of initial kL_a on the growth of tobacco cells in batch culture. J Ferment Technol 53:744–751, 1975.
4. K Kato, S Kawazoe, Y Soh. Viscosity of the broth of tobacco cells in suspension culture. J Ferment Technol 56:224–228, 1978.
5. M Noguchi, T Matsumoto, Y Hirata, K Yamamoto, A Katsuyama, A Kato, S Azechi, K Kato. Improvements of growth rates of plant cell cultures. In: W Barz, E Reinhard, MH Zenk, eds. Plant Tissue Culture and Its Biotechnological Application. Berlin: Springer-Verlag, 1977, pp 85–94.

6. ME Curtin. Harvesting profitable products from plant tissue culture. Biotechnology 1:649–657, 1983.
7. T Yamamoto, K Ushiyama. Tissue cultured *Panax ginseng*'s chemical component and toxical safety. Nitto Tech Rep 26(2):61–70, 1988.
8. PR Westphal. Large scale production of new biologically active compounds in plant cell cultures. In: HJJ Nijkamp, LHW van der Plan, J van Aartrijk, eds. Progress in Plant Cellular and Molecular Biology. Dordrecht: Kluwer Academic, 1990, pp 601–609.
9. M Misawa. Plant Tissue Culture: An Alternative for Production of Useful Metabolites, FAO Agricultural Services Bulletin. Rome: FAO, 1994, pp 1–87.
10. F Sato, Y Yamada. High berberine producing cultures of *Coptis japonica* cells. Phytochemistry 23:281–285, 1984.
11. MAL Smith. Large scale production of secondary metabolites. In: M Terzi, ed. Current Issues in Plant Molecular and Cell Biology. New York: Kluwer Academic, 1995, pp 669–674.
12. B Ulbrich, W Wiesner, H Arens. Large-scale production of rosmarinic acid from plant cell cultures of *Coleus blumei* Benth. In: K H Neumann, W Barz, E Reinhard, eds. Primary and Secondary Metabolism of Plant Cell Cultures. Berlin: Springer, 1985, pp 293–303.
13. K Ushiyama, H Oda, Y Miyamoto. Large-scale culture of *Panax ginseng* roots. Proceedings of 6th International Congress of Plant Tissue and Cell Culture, IAPTC, Minneapolis, 1986, p 252.
14. K Hibino, K Ushiyama. Commercial production of ginseng by plant tissue culture technology. In: TJ Fu, G Singh, WR Curtis, eds. Plant Cell and Tissue Culture for the Production of Food Ingredients. New York: Kluwer Academic, 1999, pp 215–224.
15. E Ritterhaus, J Ulrich, K Westphal. Large scale production of plant cell cultures. Int Assoc Plant Tissue Cult Newslett 61:2–10, 1990.
16. Y Fujita, M Tabata. Secondary metabolites from plant cells. Proceedings of 6th International Congress of Plant Tissue and Cell Culture, IAPTC, Minneapolis, 1986, p 2.
17. GF Payne, V Bringi, C Prince, ML Shuler. Plant Cell and Tissue Culture in Liquid Systems. New York: Hanser, 1991, p 346.
18. O Sahi. Plant tissue culture. In: A Gabelman, ed. Bioprocess Production of Flavor Ingredients. New York: John Wiley & Sons, 1994, pp 239–267.
19. TJ Fu, G Singh, WR Curtis. Plant cell and tissue culture for food ingredient production. In: TJ Fu, G Singh, WR Curtis, eds. Plant Cell and Tissue Culture for the Production of Food Ingredients. New York: Kluwer Academic, 1999, pp 1–6.
20. D Hess. Biotechnologie der Pflanzen. Stuttgart: Verlag Eugen Ulmer, 1992, p 286.
21. AH Scragg, EJ Allan, F Leckie. Effect of shear on the viability of plant cell suspensions. Enzyme Microb Technol 10:361–367, 1988.
22. AH Scragg. Large scale plant cell culture: methods, applications and products. Curr Opin Biotechnol 3:105–109, 1992.

23. BH Lee. Fundamentals of Food Biotechnology. New York: VCH Publishers, 1996, pp 169–179.
24. R Endress. Plant Cell Biotechnology. Heidelberg: Springer Verlag, 1994, pp 51–63.
25. PM Kieran, PF MacLoughlin, DM Malone. Plant cell suspension cultures: some engineering considerations. J Biotechnol 59:39–52, 1997.
26. WR Curtis. Achieving economic feasibility for moderate-value food and flavor additives: a perspective on productivity and proposal for production technology cost reduction. In: TJ Fu, G Singh, WR Curtis, eds. Plant Cell and Tissue Culture for the Production of Food Ingredients. New York: Kluwer Academic, 1999, pp 225–236.
27. K Schügerl. Neue Bioreaktoren für aerobe Prozesse. Chem Ing Tech 52:951–965, 1980.
28. A Moser. Bioprozesstechnik. Vienna: Springer Verlag, 1981, pp 57–58.
29. A Einsele, RK Finn, W Samhaber. Mikrobiologische und biochemische Verfahrenstechnik. Weinheim: VCH, 1985, pp 123–128.
30. H Diekmann, H Metz. Grundlagen und Praxis der Biotechnologie. Jena: Gustav Fischer, 1991, pp 94–111.
31. H Voss. Fermentationstechnik und Aufarbeitung. In: H Ruttloff, ed. Lebensmittelbiotechnologie: Entwicklung und Aufarbeitung. Berlin: Akademie Verlag, 1991, pp 28–64.
32. J Pàca. Bioreaktoren. In: H Weide, J Pàca, WA Knorre. Biotechnologie. Jena: Gustav Fischer, 1987, pp 125–175.
33. F Menkel. Einführung in die Technik von Bioreaktoren. Munich: Oldenbourg Verlag, 1992, pp 13–15.
34. W Hartmeier. Immobilisierte Biokatalysatoren. Eine Einführung. Berlin: Springer, 1986, pp 74–83.
35. HJ Rehm, G Reed. Biotechnology. 2. Fundamentals on Biotechnology. Weinheim: VCH, 1985.
36. OM Neijssel, RR van der Meer, K.Ch.AM Luyben, eds. Proceedings of the 4th European Congress on Biotechnology. Amsterdam: Elsevier Science Publication, 1987.
37. H Tanaka, F Nishijima, M Suwa, T Iwamoto. Rotating drum fermentor for plant cell suspension cultures. Biotechnol Bioeng 25:2359–2370, 1983.
38. N Shibasaki, K Hirose, T Yonemoto, T Takadi. Suspension culture of *Nicotiana tabacum* in a rotary-drum bioreactor. J Chem Technol Biotechnol 53:359–363, 1992.
39. Bioengineering Ltd. Airlift Visual Safety Fermenter VSF. Wald, Switzerland, 1999.
40. Bioengineering Ltd. Loop Safety Fermenter LSF Wald, Switzerland, 1999.
41. M Rögner. Speeding up Structure Determination of Membrane Proteins. Bioforum 3: Reprint, 1999.
42. F Kargi, MZ Rosenberg. Plant cell bioreactors: present status and future trends. Biotechnol Prog 3:1–8, 1987.
43. H Wysokinska, A Chmiel. Transformed root cultures for biotechnology. Acta Biotechnol 17(2):131–159, 1997.

44. R Pörtner, H Märkl. Dialysis cultures. Appl Microbiol Biotechnol 50:403–414, 1998.
45. H Märkl, C Zenneck, PA Wilderer. Membranbioreaktoren für den mikrobiellen Abbau von BTX-Aromaten. Wasser Abwasser 132:414–415, 1991.
46. BS Hooker, JM Lee, G An. Cultivation of plant cells in a stirred vessel: effect of impeller design. Biotechnol Bioeng 35:296–304, 1990.
47. M Joliceur, C Chavarie, PJ Carreau, J Archambault. Development of a helical-ribbon impeller bioreactor for high-density plant cell suspension culture. Biotechnol Bioeng 39:511–521, 1992.
48. PM Kieran, DM Malone, PF MacLoughlin. Variation of aggregate size in plant cell suspension batch and semi-continuous cultures. Food and bioproducts processing. Trans Inst Chem Eng 71:40–46, 1993.
49. HJG ten Hoopen, WM van Gulik, JE Schlatman, PRH Moreno, JL Vinke, R Verpoorte. Ajmalicine production by cell cultures of Catharanthus roseus from shake flask to bioreactor. Plant Cell Tissue Organ Cult 38:85–91, 1994.
50. S Takashi, Y Fujita. Production of shikonin. In: A Komamine, M Misawa, F DiCismo. Plant Cell Culture in Japan. Tokyo: CMC, 1991, pp 72–78.
51. M Breuling, AW Alfermann, E Reinhard. Cultivation of cell cultures of Berberis wilsonae in 20 l airlift bioreactors. Plant Cell Rep 4:220–223, 1985.
52. PK Hegarty, NJ Smart, AH Scragg, MW Fowler. The aeration of Catharanthus roseus L.G. Don suspension cultures in airlift bioreactors: the inhibition effect at high aeration rates on culture growth. J Exp Bot 37:1911–1920, 1986.
53. U Fischer, UW Santore, W Hüsemann, W Barz, AW Alfermann. Semicontinuous cultivation of photoautotrophic cell suspension cultures in a 20 L airliftreactor. Plant Cell Tissue Organ Cult 38:123–134, 1994.
54. U Fischer, AW Alfermann. Cultivation of photoautotrophic plant cell suspensions in the bioreactor: influence of culture conditions. J Biotechnol. 41:19–28, 1995.
55. DP Fulzele, MR Hebele. Large-scale cultivation of Catharanthus roseus cells: production of ajmalicine in a 20-l airlift bioreactor. J Biotechnol 35:1–7, 1994.
56. T Matsushita, N Koga, K Ogawa, F Fujino, K Funatsu. High density culture of plant cells using an airlift column for production of valuable metabolites. In: DDY Ryu, S Furusaki, eds. Studies in Plant Science 4: Advances in Plant Biotechnology. Amsterdam: Elsevier, 1994, pp 339–353.
57. LL Villareal, C Arias, J Vega, J Feria-Valesco, T Ramirez, P Nicasis, G Rojas, R Quintero. Large scale cultivation of Solanum chrysotrichum cells: production of the antifungal saponin SC-1 in 10 l-airlift bioreactors. Plant Cell Rep 16:653–656, 1997.
58. B Dubuis, R Plüss, JL Romette, OM Kut, JE Prenosil, JR Bourne. Physical factors effecting the design and scale-up of fluidized-bed bioreactors for plant cell culture. In: AW Nienow, ed. Bioreactor and Bioprocess Fluid Dynamics. vol 5. London: MEP, 1993, pp 89–100.
59. B Dubuis, AM Kut, JE Prenosil. Pilot-scale cultures of Coffea arabica in a novel loop fluidised-bed reactor. Plant Cell Tissue Cult 43:171–183, 1995.

60. A Khlebnikov, B Dubuis, OM Kut, JE Prenosil. Growth and productivity of *Beta vulgaris* cell culture in fluidized-bed reactors. Bioprocess Eng 14:51–56, 1995.

61. E Davioud, C Kann, J Hamon, J Tempe, HP Husson. Production of indole alkaloids by in vitro cultures from *Catharanthus trichophyllus*. Phytochemistry 28:2675–2680, 1989.

62. MR Hilton, MJC Rhodes. Growth and hyoscyamine production of "hairy root" cultures of *Datura stramonium* in a modified stirred tank reactor. Appl Microbiol Biotechnol 33:132–138, 1990.

63. PM Doran. Design of reactors for plant cells and organs. Adv Biochem Eng 48:115–168, 1993.

64. AM Nuutila, L Toivonen, V Kauppinen. Bioreactor studies on hairy root cultures of *Catharanthus roseus*—comparison of three bioreactor types. Biotechnol Tech 8:61–66, 1994.

65. SJ Sim, HN Chang. Shikonin production by hairy roots of *Lithospermum erythrorhizon* in bioreactors with in situ separation. In: PM Doran, ed. Hairy Roots: Culture and Applications. Amsterdam: Harwood Academic Publishers, 1997, pp 219–225.

66. O Kondo, H Honda, M Taya, T Kobayashi. Comparison of growth properties of carrot hairy roots in various bioreactors. Appl Microbiol Biotechnol 32:291–294, 1989.

67. R Eibl. Pflanzenzellkulturtechnik an der Ingenieurschule Wädenswil. Bioworld 5:3–7, 1996.

68. C Liu, Y Wang, C Guo, F Quyang, L Ye, G Li. Enhanced production of artemisin by *Artemisia annua* L. hairy root cultures in a modified inner-loop airlift bioreactor. Bioprocess Eng 19:389–392, 1998.

69. EB Carvalho, WR Curtis. Characterization of fluid-flow resistance in root cultures with a convective flow tubular reactor. Biotechnol Bioeng 60:375–383, 1998.

70. PJ Whitney. Novel bioreactors for the growth of roots transformed by *Agrobacterium rhizogenes*. Enzyme Microb Technol 14:13–17, 1992.

71. SA McKelvey, SA Gehrig, KA Hollar, WR Curtis. Growth of plant root cultures in liquid- and gas-dispersed reactor environments. Biotechnol Prog 9:317–322, 1993.

72. PDG Wilson. The pilot-scale cultivation of transformed roots. In: PM Doran, ed. Hairy Roots: Culture and Applications. Amsterdam: Harwood Academic Publishers, 1997, pp 179–190.

73. PJ Wheathers, BE Wyslouzil, M Whipple. Laboratory-scale studies of nutrient mist reactors for culturing hairy roots. In: PM Doran, ed. Hairy Roots: Culture and Applications. Amsterdam: Harwood Academic Publishers, 1997, pp 191–199.

74. RHJ Kessel, AH Carr. The effect of dissolved oxygen concentration on growth and differentiation of carrot (*Daucus carota*) tissue cultures. J Exp Bot 23:996–1007, 1972.

75. DA Stuart, SG Strickland, KA Walker. Bioreactor production of alfalfa somatic embryos. HortScience 22:800–809, 1987.

76. DD Denchev, AI Kuklin, AH Scragg. Somatic embryo production in bioreactors. J Biotechnol 26:99–109, 1992.
77. R Eibl, D Hans, D Eibl. Vergleichende Untersuchungen zur Kultivierung pflanzlicher Zellen in Laborbioreaktoren. Tagungsband des 8 Heiligenstädter Kolloquiums "Technische Systeme für Biotechnologie und Umwelt," Heiligenstadt, 1996, pp 100–105.
78. R Eibl, D Hans, S Warlies, C Lettenbauer, D Eibl. Einsatz eines Taumelreaktorsystems mit interner Beleuchtung. Bioworld 2:10–12, 1999.
79. S Takayama, Y Arima, M Akita. Mass propagation of plants by fermenter culture techniques. Proceedings of VI International Congress of Plant Tissue Cell Culture, Minnesota, 1986, p 449.
80. M Ziv. Morphogenesis of gladiolus buds in bioreactors. Implications for scaled-up propagation of geophytes. In: HJJ Nijkamp, LHW van der Plan, J van Aartrijk, eds. Progress in Plant Cellular and Molecular Biology. Dordrecht: Kluwer Academic, 1990, pp 119–124.
81. PV Ammirato, DJ Styer. Strategies for large scale manipulation of somatic embryos in suspension culture. In: M Zaitlin, P Day, A Hollaender, eds. Biotechnology in Plant Science: Relevance to Agriculture in the Eighties. New York: Academic Press, 1985, pp 161–178.
82. M Akita, T Shigeoka, Y Koizumi, M Kawamura. Mass propagation of shoots of Stevia rebaudiana using a large scale bioreactor. Plant Cell Rep 13:180–183, 1994.
83. THH Chen, BG Thompson, DF Gerson. In vitro production of alfalfa somatic embryos in fermentation systems. J Ferment Technol 65:353–357, 1987.
84. W Preil. Application of bioreactors in plant propagation. In: PC Debergh, RH Zimmermann, eds. Micropropagation: Technology and Application. New York: Kluwer Academic, 1991, pp 425–445.
85. R Luttmann, P Florek, W Preil. Silicone-tubing aerated bioreactors for somatic embryo production. Plant Cell Tissue Organ Cult 39:157–170, 1994.
86. F Liepe, R Sperling, S Jembere. Rührwerke: Theoretische Grundlagen, Auslegung und Bewertung. Köthen: Eigenverlag Fachhochschule Köthen, 1998, p 27.
87. JV Valluri, WJ Treat, EJ Solts. Bioreactor culture of heterotrophic sandalwood (Santalum album L.) cell suspensions utilizing a cell-lift impeller. Plant Cell Rep 10:366–370, 1991.
88. New Brunswick Scientific. Prospekt Celligen Plus: The universal stirred-tank bioreactor system. New Brunswick Scientific Co., Edison, NJ, 1999.
89. H Tanaka. Oxygen transfer in broths of plant cells at high densities. Biotechnol Bioeng 24:425–452, 1982.
90. D Ramakrishnan, J Salim, WR Curtis. Inoculation and tissue distribution in pilot-scale plant root culture bioreactors. Biotechnol Tech 8:639–644, 1994.
91. AA DiIorio. Betacyanin production and efflux from transformed roots of Beta vulgaris in a nutrient mist bioreactor. PhD dissertation, Worcester Polytechnic Institute, 1991.
92. AA DiIorio, RD Cheatham, PJ Wheathers. Growth of transformed roots in a

nutrient mist bioreactor: reactor performance and evaluation. Appl Microbiol Biotechnol 37:457–462, 1992.

93. B Dubuis. Design and scale-up of bubble-free fluidised bed reactors for plant cell cultures. PhD dissertation, Swiss Federal Institute of Technology, Zurich, 1994.
94. Technische Universität Berlin. Entwicklung eines Photobioreaktors. Berlin: Fachgebiet Bioverfahrenstechnik, 1997.
95. Bioengineering Ltd. Illumination Unit (Photoreactor). Wald, Switzerland, 1999.
96. W Preil, P Florek, U Wix, A Beck. Towards mass propagation by use of bioreactors. Acta Hortic 226:99–105, 1988.
97. F Yoshioka, Y Fujita. Economic aspects of plant cell biotechnology. In: MSS Pais, F Mavituna, JM Novais, eds. Plant Cell Biotechnology. Berlin: Springer, 1988, pp 476–482.
98. WE Goldstein. Economic considerations for food ingridients produced by plant cell and tissue culture. In: TJ Fu, G Singh, WR Curtis, eds. Plant Cell and Tissue Culture for the Production of Food Ingredients. New York: Kluwer Academic, 1999, pp 195–213.
99. TY Hsiao, FT Bacani, EB Carvalho, WR Curtis. Development of a low capital investment reactor system: application for plant cell suspension culture. Biotechnol Prog 15:114–122, 1999.
100. R Fihl, C Lettenbauer, D Eibl, M Röll. Experiences in the application of the wave bioreactor. Bioforum 3:110–112, 1999.
101. CIRAD Ltd. Temporary immersion system RITA. France, 2000.
102. A Moser. Bioprocess Technology. New York: Springer-Verlag, 1988, pp 112–118.
103. M Seki, S Furusaki. Medium recycle as an operational strategy to increase plant secondary metabolite formation. In: TJ Fu, G Singh, WR Curtis, eds. Plant Cell and Tissue Culture for the Production of Food Ingredients. New York: Kluwer Academic, 1999, pp 157–163.
104. K Shimomura, H Sudo, H Saga, H Kamada. Shikonin production and secretion by hairy root cultures of *Lithospermum erythrorhizon*. Plant Cell Rep 10: 282–285, 1991.

9

The Potential Contribution of Plant Biotechnology to Improving Food Quality

D. G. Lindsay
Institute of Food Research, Norwich, United Kingdom

I. INTRODUCTION

The beginning of the agronomic revolution, some 10,000 years ago, was characterized by the collection of seeds from important staple plants for use in the next season. But a crop such as maize, which involves the cross between two genetically dissimilar wild plants, could not have survived in the environment without the intervention of man. This is because it has no natural process for dispersing its closely bound seeds.

The developments in fermentation to provide more diverse foods with longer storage capacity and its use in bread making were applications of biotechnology. Thus, biotechnology was applied throughout history in the production of food without any real scientific understanding of what was occurring.

It was not until the end of the 19th century that the principles governing inheritance influenced the process of plant breeding as we know it today. However, it is really only from the 1960s onward that high-yielding varieties with agronomically beneficial traits and other quality characteristics have been developed and introduced into the market. This has resulted from applications of the technologies described in other chapters and forms the beginning of the biotechnological revolution. The effect of this revolution is most dramatically observed in the increase in yield of rice (Fig. 1) that occurred throughout Southeast Asia with a continued increase in yields from 1968 until 1983

However, despite the application of these technologies to plant breeding, and the outstanding success they have had, problems remain that are best solved through the application of genetic engineering or are tractable only using this approach. Only whole chromosomes can be introduced into plants by "conventional" means, and in doing so both "good" and "bad" traits can be transferred. The whole process of developing a new variety can take years with conventional approaches.

Conventional plant breeding simply cannot ensure that a staple crop contains a beneficial constituent, such as an essential nutrient, if the biosynthetic machinery for that consituent is not expressed. Conventional plant breeding does not readily provide the basis for determining the effect of specific genetic changes on quality because there is no control over the specific biosynthetic pathways that enable any relationship to be assessed. Indeed, the current critics of the use of biotechnology ignore the fact that many plant products on the market are derived from "wide crosses," hybridizations in which genes are moved from one species or genus to another to create a variety of a plant that does not, and could not, exist in nature.

Developments in genetic techniques have enabled molecular genetic approaches to replace gene identification by standard biochemical approaches. As an example, *Arabidopsis* mutants have been characterized with altered production of a number of nutrients and phytochemicals that are important determinants of plant quality including carotenoids (1), flavonoids (2), tocopherols (3), and ascorbic acid (4). Information about their biosynthesis and the function of specific genes in controlling these quality deter-

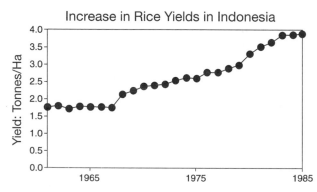

FIGURE 1 Increase in rice yields in Indonesia (1965–1985).

minants has been established. The increase in use of heterologous expression has allowed the functional cloning or characterization of the steps in the synthesis of many nutrients and phytochemicals to be established. The carotenoid pathway has been cloned by color complementation in *Escherichia coli* (5). The enzymes involved in iron uptake and biotin, thiamine, and vitamin E synthesis have been cloned or functionally characterized using heterologous expression systems (6–9).

Genetic manipulation provides the opportunity to look at the relationships between composition and quality in a systematic way. It is this application of the technology in conjunction with the genetic characterization of mutants that needs to be fully exploited to advance the effectiveness of conventional plant breeding. Once a gene or gene product, either in isolation or acting in tandem, has been determined to confer a specific beneficial characteristic, a tool exists to screen potential new varieties.

Modern gene technology, or genetic engineering, has really taken off only in the last decade or so. Large areas of genetically modified soybeans, corn, and canola have been successfully grown in the Western Hemisphere. In the United States alone in 1999, some 50% of all the soybeans planted were planted with GM-herbicide–resistant seeds. Although in this initial period the benefits have been primarily directed toward farmers, with the greatest focus on improved agronomic characteristics, there are specific applications of genetic engineering with unlimited potential benefits to consumers in the food and nonfood sectors. Many of these are described in more detail in other chapters.

This focus on producers' benefits has provided a reason, among many others, for opponents of the technology to argue forcefully against its use. Considerable efforts are being expended internationally to detect genetically manipulated crops for the enforcement of effective labeling. The development of genetic engineering is being severely restrained through sociopolitical concerns over its widespread adoption and safety. This is, in turn, leading to reticence in industry to make the long-term investment necessary to bring products to the market, given the strength of the opposition in the Western Hemisphere.

In all of these areas the effects on food composition that have occurred as a consequence of the genetic changes have not been well documented. But it is clear that compositional effects will have occurred from the application of conventional plant breeding. Plant breeding, thanks to modern genetic methodologies, will be less "hit or miss." It will be informed in the future about the consequences of a specific single-gene or multiple-gene alteration for the yield, the disease resistance, the nutritional value, and the flavor and texture of food plants. This will allow much more targeted screening of beneficial characteristics in the future, as well as the introduction of significantly improved nutrition, for the developing world communities whose dependence on a single plant crop frequently causes nutritional deficiency disease.

In practice however, it is unlikely that genetic engineering would ever supplant the more conventional biotechnological approaches entirely, especially in the context of improving the quality of our food supply. Both conventional and genetic biotechnologies will continue to be exploited to achieve this goal. There are very many commercial and technical obstacles to the development of crops with multiple gene transfers. Yet many of the benefits that consumers might wish to see in their plant foods would require such an approach. A seed company would have to be very convinced that either there were other benefits (e.g., resounding agronomic benefits) or the potential market size justified the investment before any such investment would be made. It is clear that some aspects of the quality of existing varieties of food crops could be significantly improved purely by conventional means, where multiple gene transfers are commonplace. Examples in which significant variations in the nutrient content of genotypes have been documented include a

> Twofold variation in calcium concentration in beans (10)
> Fourfold variation in β-catotene concentrations in broccoli (11)
> Fourfold variation in folates in beetroot (12)
> Two- to threefold variation in iron and zinc levels in maize (13)

II. PRIORITIES FOR THE FOOD SECTOR

In an international context, the priorities for the application of plant biotechnologies for the quality improvement of food vary depending on whether or not the community is in the developed or developing world.

A. The Developing World

Overwhelming proportions of the world's population are dependent on plants as their principal, if not exclusive, source of food. In such populations, the incidence of disease due to vitamin deficiencies is widespread. It has been estimated that over 100 million children worldwide are vitamin A deficient and that improving the vitamin A content of their food could prevent as many as 2 million deaths annually in young children (14). This is apart from the deficiencies in iodine intake, resulting in goiter, and from iron-deficient anemia, which are estimated to affect millions in the developing world. There is also an important need to improve the amino acid content of legume proteins that are deficient in essential sulfur amino acids.

There is evidence that the problem of vitamin A deficiency in particular has increased as a consequence of the application of traditional plant breeding. While the focus on improving disease resistance and yields of rice has rid many of these communities of hunger, it has had an impact on the local traditions. Less effort has been placed on growing alternative crops that may well have provided a source of vitamin A but are more difficult to make money from or do not produce consistent yields under local circumstances. This is particularly true of legume crops. With the focus only on this one aspect of basic food production, the problems of malnutrition have not disappeared. Indeed, the incidence of nutritional deficiencies may have increased.

1. Improvements in Protein Quality

This is an area of crucial importance to the majority of the world's population and is covered in greater detail in Chap. 12. Whatever strategies are adopted, there will need to be assurance that the introduction of any foreign protein into a plant will not increase the risks of an allergic response through the introduction of allergic epitopes. For example, the strategy of introducing a methionine-rich protein into legumes from the Brazil nut had to be abandoned because of the risks of an acute allergic response by some consumers (15). Apart from this problem, the sulfur-containing amino acid content of the legumes is not substantially raised. Suppression of synthesis of other methionine-rich proteins in the legume occurs (16). More information about the control of gene expression by the amino acid supply is required.

2. Reduction in Antinutritional Factors

The interest in reducing antinutritional factors in plants has been predominantly focused around improving the nutritional value of feeding stuffs. Phytates are present in many plant seeds and uptake of limit phosphorus or iron as well as other elements. The potential for introducing a phytase gene into feeding stuffs has been explored (17). However, there are other strategies that seem to be of greater overall value in human nutrition. Thioredoxin is thought to be an activator of the germination process in seeds (18). It is able to activate proteins to degradation by proteolysis and results in improved digestibility (19). It also has the potential advantage of being able to reduce allergenicity, presumably because of its capacity to break disulfide bonds by the action of the reduced thiol groups in the molecule and ensure that the tertiary structure of the protein is accessible to degradation by proteases (19). The insertion of the wheat thioredoxin gene into barley has produced a transgenic plant in which thioredoxin accounts for 7% of the total protein content in the barley and is a good source of sulphur amino acids (20).

B. The Developed World

In contrast to that in the developing world, the incidence of known nutritional deficiency disorders in the developed world is low. Even in the case of vegetarians, there appears to be no appreciable problem. Classical nutritional deficiency diseases have been avoided through the widespread fortification of food when it was realized that many staple crops contained insufficient concentrations of many essential vitamins and minerals. Fortification is also utilized to replace nutrients lost in the heat processing of staple foods and through oxidation. In addition, the consumption of nutritional supplements is becoming more widespread. Nonetheless, there is good evidence, as shown in Fig. 2, that even in a country such as the United States, where most nutritionally related disorders probably stem from overnutrition, there is inadequate intake of micronutrients compared with what would be found in the plasma if recommended dietary allowances (RDAs) were consumed.

For selected nutrients (iron, calcium, selenium, iodine, vitamin E, vitamin B_6, and vitamin A), the clinical and epidemiological evidence strongly

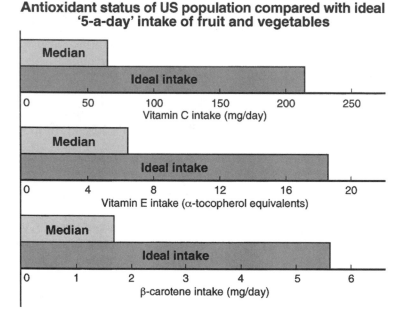

FIGURE 2   Plasma concentrations of selected antioxidant nutrients in the U.S. population.

supports the view that they play a significant role in maintaining health and are limiting in diets worldwide (21).

The RDAs were first developed as recommendations to alleviate nutritional deficiency diseases and have been periodically revised in the light of further data. A new set of values—to be known as dietary reference intakes—are expected to be published in 2000 soon by the U.S. Department of Agriculture (USDA). New values reflect the growing knowledge base in relation to the role of nutrients in optimizing health.

Diet is implicated as an important risk factor in the initiation or progression of many diseases of aging, particularly cardiovascular disease and cancer. One of the most striking and consistent observations, in terms of the relationship to health, has been the decrease in cancer risk associated with an increasing intake of fruit and vegetables in the diet (22). There appears to be no single explanation for this effect, but there is growing evidence for the role played by nutrients (vitamins, minerals) and other microconstituents of plants in minimizing the adverse effects of the processes leading to oxidative damage and degenerative diseases such as vascular disease and cancer. The basis for the protective effects of plant phytochemicals on carcinogenesis is the subject of intensive research internationally. This evidence needs to be taken into account especially in terms of reassessing overall nutritional needs throughout the human life span. Table 1 indicates the po-

TABLE 1 Recommended Dietary Allowances (RDAs) of Nutrients and Minerals Compared with Their Maximum Safe Intakes

Nutrient	Adult RDA	Upper safe intake ($\times$RDA)	Reference
Calcium	1200 mg	2	62
Iron	15 mg	5	63
Iodine	150 μg	13	64
Selenium	70 μg	13	65
Vitamin C	60 mg	16	66
Vitamin B$_6$	2 mg	125	67
Folate	200 μg	50	68
Biotin	30–100 μg	300	69
Vitamin E	10 mg TE	100	70
Vitamin A	1 mg RE	5 (retinol)	71
		100 (β-carotene)	
Vitamin K	80 μg	375	71

Source: Ref. 72.

tentially large increase in intakes of certain nutrients that might bring additional health benefits without posing any additional health risks to the general population.

It might be argued that because fruit and vegetables are known to reduce the risk of the development of cancer and cardiovascular disease, the only action that is required is to encourage the greater consumption of these foods. Under such circumstances, the need to consider any specific enhancement of their functional benefits is pointless. This argument ignores a number of facts:

> There is growing evidence that the augmentation of certain nutrients will have health benefits because RDAs of nutrients are based on requirements for healthy growth, not on optimal nutritional needs for the prevention of long-term disease.
> Consumer choice is influenced by socioeconomic and cultural constraints. Freedom of choice ensures that foods are selected that are enjoyable and available, but these may not always be healthy. Enhancement of levels of nutrients in specific staple crops, or their introduction if absent, could meet consumer health needs without the need to change dietary habits radically, which is rarely a practicable option.
> The consumption of fruit and vegetables may not provide optimal protection against the risk of disease.
> Plant food composition is constantly changing as new varieties are marketed. Because the exact protective mechanisms that are induced by eating fruits and vegetables are not known, it is not possible to link composition to specific functional mechanisms. No criteria could be applied to the development of new plant varieties, in terms of their composition, other than that they remain the same. This is an unsatisfactory basis to proceed, from the perspectives of both the plant breeding sector and the consumer. Phytochemicals or vitamins vary by an order of magnitude or more in the gene pool. For the prevention of disease it could help to know in which direction this pool should be altered and with what likely consequences.

There needs to be a systematic examination of the factors that determine the biosynthesis in the plant of phytochemicals, including nutrients, that are bioactive and protective as well as the factors that determine their turnover and catabolism. Genetic engineering provides a tool from which to expand knowledge more rapidly than could be achieved in any other way.

III. APPLICATION OF GENETIC ENGINEERING TO FOOD QUALITY IMPROVEMENT

Potential strategies for the enhancement of specific metabolites could target on

1. Overexpression of enzymes that control the final steps in the biosynthesis of a metabolite
2. Overexpression of rate-limiting enzymes
3. Silencing of genes whose expression causes the metabolite to be degraded
4. Increased expression of genes that are not subject to metabolic feedback control
5. Increasing the number of plastids in a plant
6. Increasing metabolic flux into the pathway of interest
7. Expression in storage organs using site-specific promoters

The strategy that has had the greatest success at present is the first one, especially in conjunction with the last strategy. However, this presupposes that the metabolite of interest is the final one in a particular pathway. In practice, if a substantial increase in the concentration of a metabolite is required, the use of specific promoters directing the synthesis to a particular organelle normally used for storage purposes, or where the plant normally synthesizes the metabolite, is essential. Failure to do so could cause toxicity in the plant by interfering with the production or function of other essential metabolites.

Because of extensive synteny among cereal crops, it is likely that information on the function of a gene in one cereal crop will be of direct relevance to its function in another. The differences in genome size between cereals are predominantly due to amplification of interspersed repetitive sequences, and knowledge of gene order and sequence in one cereal will enable the isolation and characterization of genes in another cereal. Even the characterization of the genes in *Arabidopsis*, although only distantly related to cereals, has enabled functions and traits of similar genes in other food plants to be located and confirmed.

No strategies have yet been applied where multiple gene insertions are necessary to produce the metabolite or where plastid numbers have been increased. However, rapid accumulation of sequence data of both chromosomal DNA and expressed sequence tags of plants and other species is providing rapid advances in knowledge of the genetic makeup and functions of several plants, and it is expected that these other possibilities will soon be feasible.

IV. PLANT SECONDARY METABOLITES AS DETERMINANTS OF QUALITY

There is considerable evidence that plant secondary metabolites are of crucial importance in determinining the quality characteristics of fruit and vegetables. Color changes with maturity are characterized by the synthesis of
chlorophylls and specific carotenoids and flavonoids (anthocyanins). Flavor
and aroma are often determined by complex chemical changes centered on
the production of terpenoids and aldehydes resulting from the action of
lipoxygenases as well as changes in the composition of complex phenolics
that can produce the ideal balance between astringency and bitterness in
beer or wine. Texture is determined by the structure of plant cell walls and
the enzymic changes in their structure that can be induced during the ripening and storage process. Fortunately, a number of the pathways leading
to the production of key phytochemicals of health interest are also important
in determining these other quality characteristics.

  The very diverse secondary metabolites in plants, which could have
an important impact on food quality, are biosynthesized via two important
pathways—the phenylpropanoid (PP) and the isoprenoid (IP) pathways—
with some common links. However, the way in which these pathways are
utilized to produce specific metabolites in food plants and how these relate
to the quality parameters described are at present insufficiently understood
to undertake little more than a hit-and-miss approach to genetic manipulation.

V. THE PHENYLPROPANOID PATHWAY

Phenylpropanoids (PPs) provide the basic units from which a wide variety
of plant metabolites are synthesized. From these basic units a series of important branch pathways result in the synthesis of flavonoids and isoflavonoids, coumarins, soluble phenolic esters, and cell wall polymers (Fig. 3).
The precise branch points are not known in all cases.

  The functions of the PP derivatives are as diverse as their structures;
they serve as pigments, phytoalexins, ultraviolet protectants, insect repellents, and signaling molecules in plant-microbial interactions; in the regulation of auxin transport; and in the stimulation of pollen germination. They
also produce structural support for plants, acting as the building blocks for
the biosynthesis of suberin, lignin, and other cell wall components (see
Chaps. 11 and 22). Given this wide diversity in structure and function, there
are major differences in temporal and spatial distribution of these metabolites
in the development of the plant and between different plant organs and cell
types.

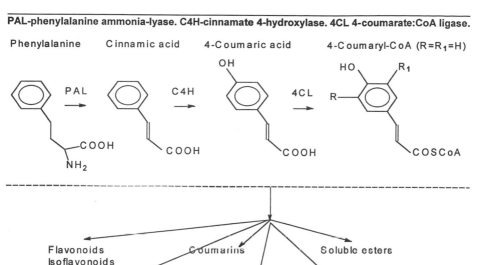

FIGURE 3   General phenylpropanoid metabolism.

The potential health interest in the majority of these metabolites lies in the fact that many of the simple and complex phenols that are present in foods have been shown to be powerful antioxidants in vitro (23) and demonstrate a wide range of potential benefits in animal experiments. They play an important role in the determination of flavor, color, and taste.

Clearly, given the importance of these metabolites to the plant, any proposed genetic modification must pay careful attention to a possible adverse impact on the plants' responses to environmental stress. For this reason, most of the successful attempts at up-regulating the production of a specific metabolite in the pathway have been where the gene is involved in the control of the end step in the biosynthesis rather than at earlier stages in the biosynthetic process. Careful attention also has to be given to the choice of promoters for correct temporal and spatial expression. For the strategy to succeed, promoters must direct the biosynthesis to edible storage organs or organelles. Up-regulation of expression in all parts of a plant often leads to adverse consequences (24).

A great deal is known about the biochemical events that occur in PP metabolism. An array of mutants are available in which steps in the pathway are blocked, especially in the pathways to flavonoid pigmentation. Complementary DNA (cDNA) and/or gene sequences are also available for the early steps in PP biosynthesis as well as in flavonoid biosynthesis (25).

PAL deaminates phenylalanine to cinnamate, which is subsequently hydroxylated to 4-hydroxycinnamic acid through the action of C4H. PAL is present constituitively, together with C4H and 4CL, in varying activities throughout plant development. The evidence suggests that these enzymes are highly regulated and exist in individual isoforms that may well be species and cell dependent.

Although PAL overexpression is likely to play a key role in increasing flux into total phenylpropanoid metabolism, it is probably too nonspecific and many biological functions would be affected. It has been found that overexpression of PAL in tobacco increases its resistance to microbial attack but decreases its resistance to insect larvae (26). Similar problems are likely to be found with the overexpression of other enzymes early in the pathway. The free acids that are synthesized through this pathway rarely accumulate to high levels within the plant and are usually conjugated to sugars, cell wall carbohydrates, or organic acids.

The texture of fruit and vegetables is closely related to the chemical and physical properties of the cell walls. The end point determinant of texture is related to the overall mechanical properties of the plant, which are determined by the structural characteristics of the cell wall, and the way in which these determine cell-to-cell adhesion (27). An important factor in tissue softening is the separation of individual cells. Fruit ripening generally results in the biochemical dissolution of the middle lamella pectic polysaccharides, which bind cells together. Thermal treatment can also cause β-eliminative degradation of the pectic polymers involved in cell-cell adhesion.

Ripening and thermally induced cell separation occur predominantly in nonlignified, thin-walled tissues. Tissues that fail to soften are frequently the result of secondary thickening and associated lignification. However, some parenchyma-rich plant tissues with thin, nonlignified tissues soften only slowly on cooking. In any studies of the texture of fruit and vegetables it will be important to understand the factors that control the processes behind maturation within both the middle lamellar and the xylem tissues. The role of specific steps in the biosynthesis of these structures in fruit and vegetable textural characteristics is described further in Chap. 11.

A. Flavonoids and Isoflavonoids

Flavonoids, whose basic structure is shown in Fig. 4, form an important class of metabolites of the PP pathway. They show protective effects against cancer progression in experimental animals. In vivo and in vitro studies have shown a range of potentially beneficial effects such as antioxidant properties, and they are effective vasodilators and platelet disaggregators (28). They are present in significant amounts in many food plants. They also illustrate ex-

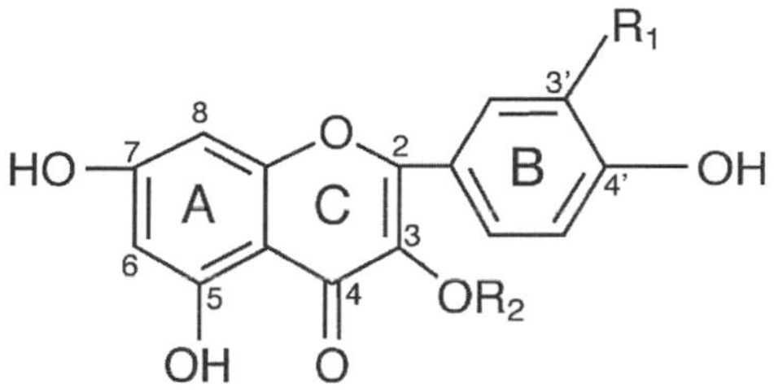

FIGURE 4 Basic structure of flavonoids.

tremely well the complex issues that need to be considered in focusing on specific metabolites in achieving the aim of improving food quality.

All flavonoids are derived from a chalcone precursor, the product of condensation of 4-coumaryl coenzyme A (CoA) (derived from phenylalanine produced in the shikimate pathway), and three molecules of malonyl-CoA (derived from acetyl CoA through the action of a carboxylase) involving the enzyme chalcone synthase (CHS) (Fig. 5). The CHS enzymes form a family of closely related multigene enzymes that produce a variety of secondary metabolites (29). This is achieved through variations in the number of rounds of condensation and differences in the folding pattern of the intermediate polyketide. Stilbene synthases from a variety of sources have exactly the same substrate specificity as CHS but the intermediate polyketide is folded differently, resulting in decarboxylation to form the stilbene (30). The stilbene resveratrol, which is found in wine, has been shown to possess anticancer activities in various cancer models (31). Resveratrol can be engineered successfully by genetic engineering (32).

The stereospecific cyclization of the chalcone, catalyzed by chalcone isomerase (CHI), provides a 2S-flavanone with the typical flavonoid skeleton. Chalcone reductase (CHR) isomerizes both 6'-hydroxy and 6'-deoxychalcone to 5-hydroxy and 5-deoxyflavanone, respectively. CHR and CHS are coinduced in elicited or infected cells, whereas CHS alone is expressed in the arial portions of uninfected plants. Consequently, it is likely to be of greater importance in terms of its ability to influence food quality parameters. 2',4',4'-Trihydroxychalcone (the product of CHS and CHR) is a branch point for the synthesis in legumes of different flavonoids. Thus, chalcone isomerase (CHI) converts the chalcone to 4,7-dihydroxyflavanone and then to the isoflavones. Another enzyme, chalcone O-methyltransferase (CHOMT), forms a chalcone that is a potent inducer of *Rhizobium* nodulation genes. Over 50 regiospecific O-methyltransferase sequences are known, which adds to the diversity of the flavonoids, isoflavanoids, and hydroxycinnamates found in plants (33). The possibility exists of protein-

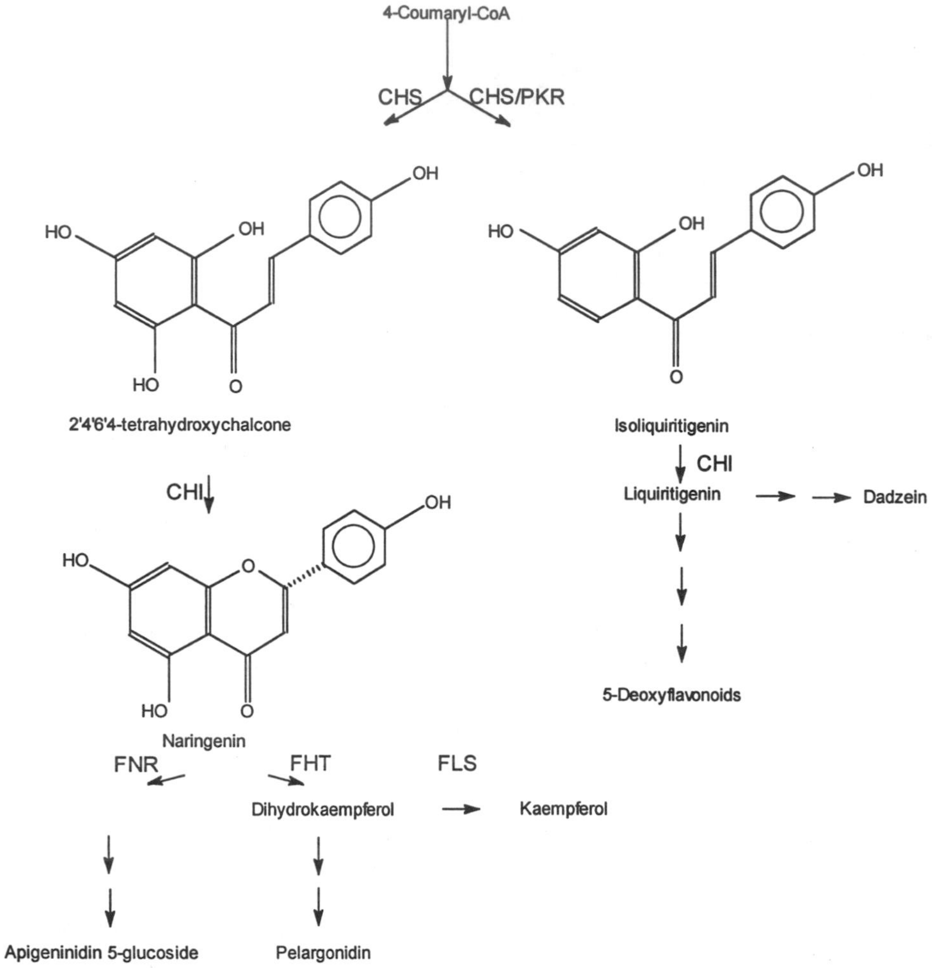

CHI-chalcone isomerase. PKR-Polyketide reductase. CHS-chalcone synthase. CHR-chalcone isomerase. FNR-Flavanone 4-reductase. FHT-Flavanone 3-hydroxylase. FLS-Flavanol synthase.

FIGURE 5 Biosynthesis of chalcones.

protein reactions between all of these enzymes, the outcome dependent on many factors controlling their expression.

    In addition, there are enzymes that are crucial for determining the broad class of flavonoids that are produced (chalcone isomerase for flavanone production, flavanone reductase for flavone production, and isoflavone synthase for the isoflavones). The highly species-specific nature of many of the biosynthetic steps makes it very difficult to predict the outcome of over-expressing any one enzyme (34). Instability has been observed in the expression of a transgene that can be exacerbated by environmental factors. Crosses between wild-type and transgenic plants in *Petunia* was exacerbated by environmental factors that caused methylation-mediated expression of the transgene. This result is in contrast to the manipulation of *Arabidopsis* and tobacco with the regulatory gene controlling pigmentation in maize. The gene controls the expression of several steps such as CHS and CHI. In all cases, there was a substantial increase in anthocyanin production (34,35). Nonetheless, the situation is not straightforward and severe oxidative stress symptoms have been observed with plants overexpressing a transcription factor gene.

Whereas very extensive information is available on the biosynthesis of flavonoids in terms of their induction, regulation, and tissue-specific expression in a wide range of species, very little is known about the post-translational regulation of flavonoid biosynthesis. The simple flavonoids with a single hydroxyl in the B-ring can be extensively modified by hydroxylation, methylation, glycosylation, acylation, and a number of other modifications.

In terms of food quality characteristics, the enzymes that occur at the end of the biosynthetic pathway are likely to be of the greatest interest. The hydroxylases that are responsible for the conversion of kaempferol into quercetin and subsequently myricetin or pelargonidin into cyanidin and delphinidin result in an increase in color, antioxidant activity, and potential nutritional value. Similarly the acyl- and methyltransferases increase pigment stability and may increase nutritional value. The glycosyltransferases have an effect on taste (bitterness). Similarly, astringency is determined by the concentration and type of tannins, hydrolyzable tannins, and flavan-3-ols.

Competing with these reactions are the enzymes that result in a loss of quality of the product, many of which involve some reaction with the phenolics present in the plant. Enzymatic browning is one of the main oxidative reactions leading to an undesirable loss of color, flavor, and nutrients in some products (e.g., browning of apples or cut salads) and an increase of quality in others (e.g., prunes, dates, and figs). The principal classes of enzymes involved are the peroxidases (PODs) and polyphenoloxidases (PPOs). PPO activity is found in all higher plant tissues and organs. A requirement of the processing of all frozen crops is that they be inactivated by blanching, before freezing, to prevent off-flavors developing as a result of their action and other quality deficits. No satisfactory genetic manipulation approach to reducing their effects has been devised, and this may prove difficult given the important role they play in the overall physiology of the plant (36).

Different plant families produce different chemical classes of PPs. Isoflavanoids are found predominantly in the Leguminosae, whereas the Solanaceae produce sesquiterpenes. At present it is not possible to transfer the complete sesquiterpene phytoalexin of Solanaceae, such as tobacco, to a legume (or vice versa) as the biochemistry of the complex enzymatic pathways is not fully understood.

Isoflavones are less widely distributed and are found in highest concentrations in the soybean and food products manufactured from it. Genistein and daidzein are estrogenic and show a wide range of physiological responses as a consequence, including delaying the menstrual cycle in women and a series of beneficial effects on cardiovascular function and bone loss (37). Extensive screening of soy varieties is taking place to identify high- and low-isoflavone genotypes. Interestingly, the selection for agronomically beneficial alleles also appears to select for the varieties with higher isoflavone levels, providing an example in which the goals of farmers and consumers may be congruent.

The branch pathway for the formation of the isoflavones is almost completely characterized and has several common links with that for anthocyanins (Fig. 6). However, the first reaction specific for isoflavone synthesis is unique. The 2-hydroxyisoflavanone synthase (2-HIS), which causes a 2-hydroxylation, couples to an aryl migration of the B-ring of a flavanone. The cloning of 2-HIS offers the possibility of intoducing isoflavone synthesis into other food crops (38).

B.   The Isoprenoid Pathway

The IP pathway is responsible for the biosynthesis of a vast range of compounds that play a crucial role in maintaining membrane fluidity (sterols), electron transport (ubiquinone), glycosylation of proteins (dolichol), and the regulation of cellular development (gibberellins etc.). Terpenoids synthesized via the IP pathway are also used for more specialized purposes including defense (terpene-derived phytoalexins and volatile signals), pollinator attractants (monoterpenes), and phytoprotectants (carotenoids). As far as food quality is concerned, among the most important characteristics determined by the pathway are flavor and aroma as well as color. Antioxidant benefits are also likely from the carotenoid pathway as well as the capacity of terpenes to protect against the adverse effects of procarcinogens by inducing phase II enzymes.

The key metabolite from which this diverse range of chemicals is synthesized is isopentenyldiphosphate (IPP). This is formed by two independent pathways (39) (Fig. 7). The cytoplasmic acetate-mevalonate pathway appears principally responsible for the synthesis of sterols and

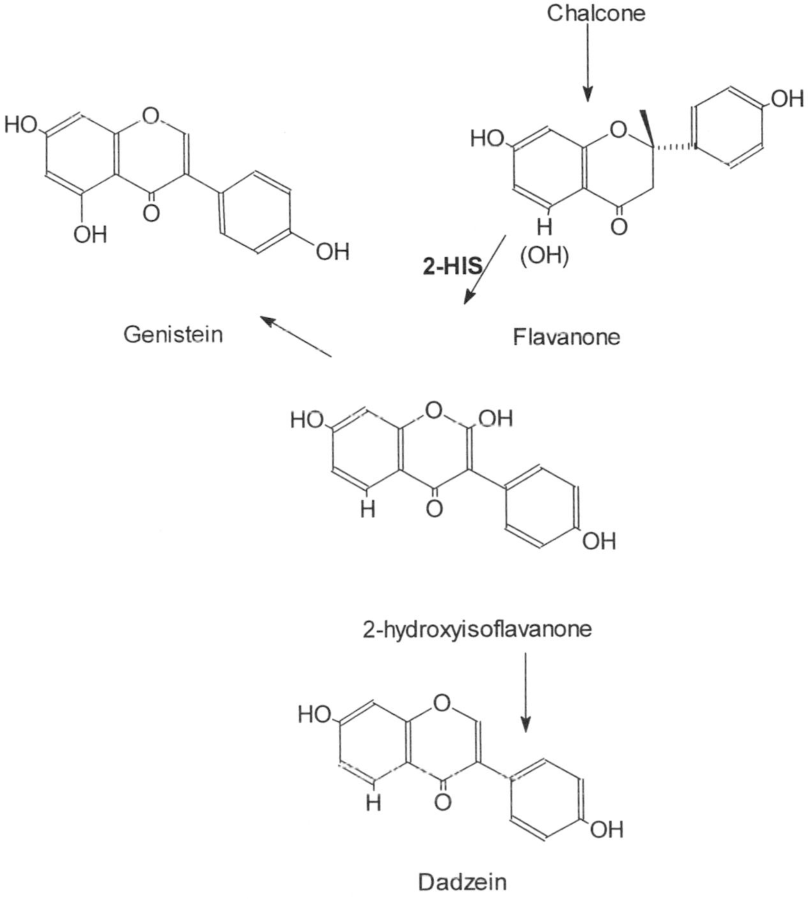

FIGURE 6 Biosynthesis of isoflavones.

sesquiterpenoids. The second pathway—the GAP-pyruvate pathway—is responsible for the synthesis of carotenoids and mono- and diterpenes and is localized within the plastid. An exchange of cytoplasmic and plasticidal metabolites has been found in *Chamomilla recutita* that utilizes both pathways (40). The extent to which this occurs could well depend on the tissue type and developmental phase, and IPP may move between organelles depending on numerous factors. Elucidation of the enzymology and regulation of the GAP-pyruvate pathway for IPP synthesis is a prerequisite to understanding and rationally manipulating terpenoid production in plants.

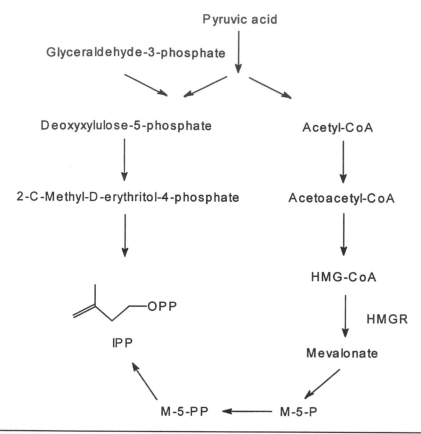

HMG-CoA-3-hydroxy-3-methylglutaryl-coenzyme-A. HMGR-HMG-reductase.
M-5-P-Mevalonate-5-phosphate. M-5-PP-Mevalonate-5-phosphate.

FIGURE 7   Biosynthesis of isopentenyl pyrophosphate (IPP).

Multiple isoforms exist of the enzymes that are involved in terpenoid biosynthesis. These isoforms appear to direct subtle control over the biosynthesis of metabolites and can direct synthesis into specific organelles (Fig. 8). One of the most important enzymes is 3-hydroxy-3-methylglutaryl-CoA reductase (HMGR), which is highly regulated and exists in many different isoforms that vary between species and tissues and during development.

A common aspect of the regulation of terpenoid biosynthesis is the coordinated modulation of many enzymes to affect necessary changes in flux of metabolites through the different branches of terpenoid metabolism

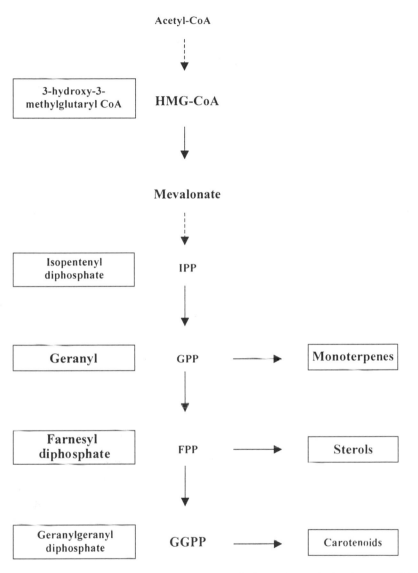

FIGURE 8   Synthesis of secondary metabolites through the isoprenoid pathway.

(41). Cholesterol synthesis in mammals requires the coordinated synthesis of HMG-CoA synthase, HGMR, and FPP synthase. However, the degradative turnover of these enzymes appears to be regulated independently of each other (42). This example of the effect of multiple enzymes to change flux in different pathways of terpenoid metabolism has been demonstrated in plants where exposure to specific metabolites can alter the production of either the steroid, glycoalkaloid, or sesquiterpene phytoalexin pathways, depending on the elicitor.

1. Sterol Biosynthesis

The nutritional interest in this class of phytochemicals derives from the fact that sterols have a similar structure to cholesterol and some sterols have a capacity to reduce the levels of total plasma cholesterol and low-density lipoprotein (LDL)-cholesterol (43,44). There has been a dramatic reduction in the morbidity and mortality from cardiovascular diseases with the use of hypolipidemic drugs (statins), and the interest in plant sterols lies in their potential to act as a natural, preventative dietary product.

The sterols are thought to act through their ability to limit the uptake of cholesterol from the diet as well as their potential to increase cholesterol excretion. This knowledge has led to the development of specific foods with cholesterol-reducing claims. These are already marketed in some countries as Benecol® and Take Control®.

Sterol biosynthesis begins with the conversion of farnesyl diphosphate into squalene, which determines the channeling of the isoprenoid pathway into the branches that produce phytosterols (Fig. 9). The sterol pathway involves a sequence of more than 30 enzyme-catalyzed reactions, all of which are membrane linked. Nothing is known about the catabolism of plant sterols (45).

Up-regulation of HMG-CoA in transgenic tobacco, corn, and tomatoes results in the accumulation of cycloartenol only, rather than sterols. Targeting of enzymes upstream from cycloartenol results in a decrease in the total amount of plant sterols with potentially deleterious effects on membrane stability in the plant and the protection against infection. However, pentacyclic triterpenes (such as α- and β-amyrin) are formed by another route from squalene oxide, and inhibition of this route could increase free sterol concentrations. Potential strategies for sterol production are, however, likely to focus on overexpression of terminal enzymes in the pathway (45).

As far as the regulation of enzymes downstream from cycloartenol is concerned, the methylation of cycloartenol is a critical rate-limiting step. Inhibition of two of these enzymes, COI and OBT 14DM, leads to the accumulation of unusual steroids rather than natural Δ^5-sterols (46).

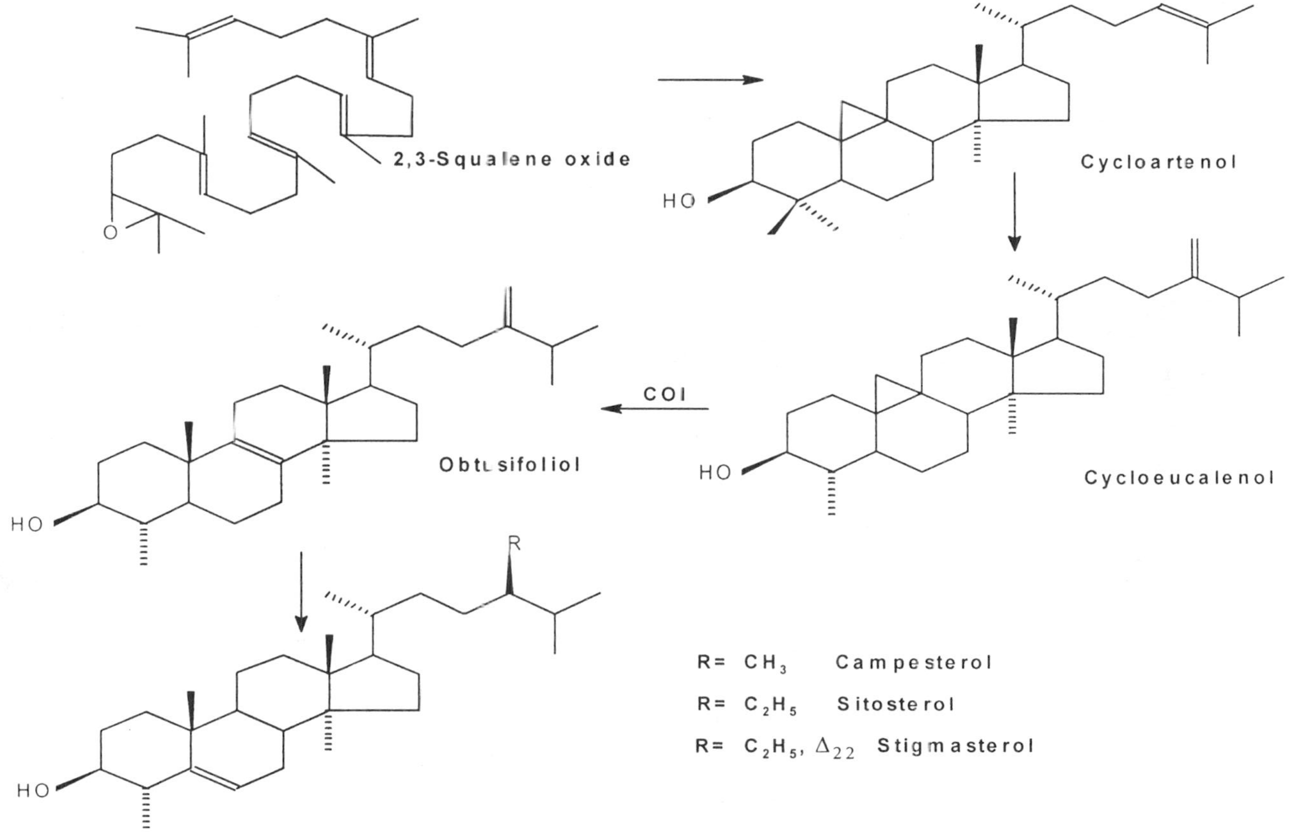

COI-cycloeucalenol obtusifoliol isomerase. OBT14DM-obtusifoliol 14-demethylase

FIGURE 9   Biosynthesis of plant sterols in photosynthetic phyla.

From the food quality perspective, the principal interest is in the biosynthesis of sterol esters, which give a more consistent cholesterol-lowering effect than the free sterols. The sitostanol esters (47) were the first shown to be consistently effective in lowering cholesterol, but sterol esters appear to be equally effective. Ideally, stages in the final phases of biosynthesis need to be studied to ensure that all the sterols are esterified and thereby ensure that they are more bioavailable (see Sec. VI).

2. Carotenoid Biosynthesis

Carotenoids are also derived from the general isoprenoid pathway. The first committed step in the carotenoid pathway is the head-to-head condensation of two geranylgeranyl diphosphate (GGPP) molecules to produce phytoene. The three enzymes responsible for converting GGPP into β-carotene are shown in Fig. 10. β-Carotene is of particular interest because it functions as a metabolically well-regulated precursor of vitamin A. Its importance to the health of children in the developing world has been discussed in Sec. II.A.

Although rice synthesizes GGPP, the enzymes required to convert it into β-carotene are not all present. In a remarkable feat of genetic engineering, three genes encoding these enzymes have been introduced into the endosperm of rice (48). Phytoene synthase and phytoene desaturase were introduced using a construct that did not have a selectable marker. This was introduced through the simultaneous introduction of another construct, which carried the third gene of interest (lycopene β-cyclase) as well as a selectable antibiotic resistance gene. This cotransformation strategy should

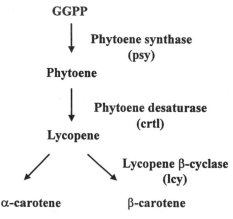

FIGURE 10   Biosynthesis of β-carotene.

enable a segregation of the antibiotic resistance gene away from the phytoene synthase and desaturase genes. This would eliminate one of the major issues that are raised when the safety of genetically manipulated foods is discussed. Fortunately, it has already been demonstrated that these plants will still produce β-carotene because these authors have demonstrated that plants engineered with standard transformation procedures to express only the first two enzymes in the pathway do not accumulate lycopene as would be expected. It seems that the enzymes needed to convert lycopene into β-carotene are constitutively expressed in normal rice endosperm or are induced when lycopene is formed.

VI. RELATIONSHIP OF STRUCTURE TO NUTRITIONAL QUALITY (BIOAVAILABILITY)

The overall content of a given nutrient in a food is not always a useful indicator of its nutritional value as not all of the nutrient present is absorbed. Nutritionists must concern themselves with understanding the proportion of an available nutrient that is digested, absorbed, and ultimately utilized. In the case of nutrients or phytochemicals whose beneficial effects are directed toward inhibiting degenerative diseases, it is important to know whether or not the nutrient is reaching the particular target organ and in a form that is active. Otherwise, the claims concerning the health benefits of that chemical would not be justified, especially as it is difficult to demonstrate benefits from long-term human studies.

The diet plays an important role in the uptake of specific nutrients and phytochemicals. Those that are lipophilic are absorbed much more readily from a lipid-rich diet. Frying tomatoes in oil dramatically improves the uptake of lycopene compared with the consumption of fresh tomatoes (49). Raw carrots, which have high levels of provitamin A carotenoids, are poorer sources of β-carotene than gently cooked carrots (50). The bioavailability of certain trace elements is increased by cooking or processing, such as the increased bioavailability of iron in canned spinach (51).

The chemical form of the phytochemical present in food is very important in determining the uptake through the gastrointestinal tract (Fig. 11). Quercetin-β-glucoside is more easily absorbed than the aglycone quercetin. Isorhamnetin-β-glucoside, which is chemically similar to quercetin, differing only by a single methoxyl group, is much more readily absorbed. Flavonoid rutinosides (rhamnosyl 1–6 glucosides) are not easily absorbed (52). Sterol esters are more bioavailable than the free sterols (53,54).

The chemical form of the phytochemical is of profound importance when considering the biological relevance of specific chemicals and their levels in the diet. Although some phenols might be better antioxidants than

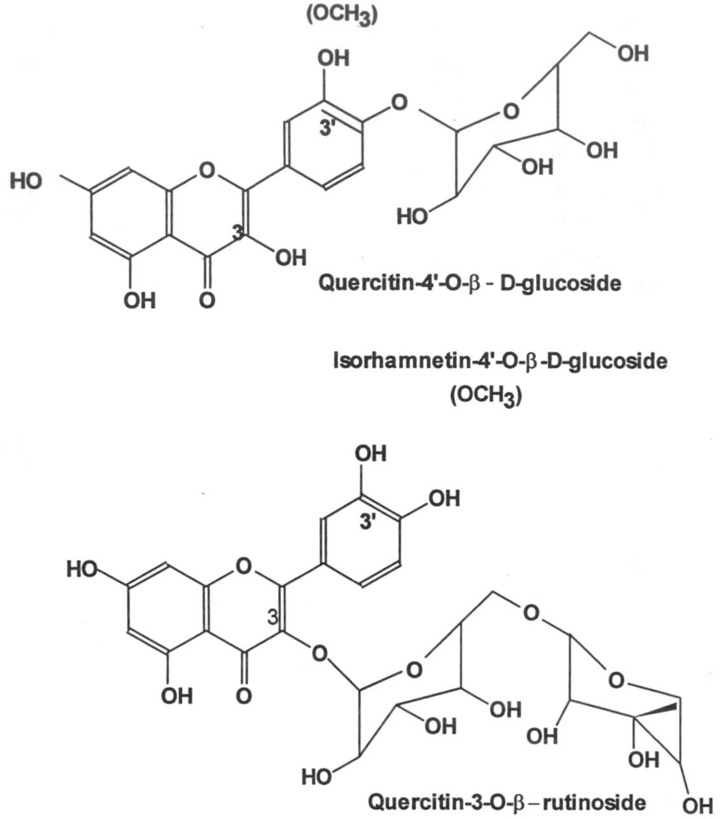

FIGURE 11   Structural features affecting the absorption of flavonoids.

others when tested in in vitro systems, this is of little significance in terms of health relevance. What matters is whether the compounds are easily absorbed, are not quickly degraded in tissues, and are able to reach the target sites. Flavonoids that are not absorbed undergo extensive degradation by gut microorganisms and may play only a limited role in preventing oxidative damage in the colon.

VII.   SAFETY AND REGULATORY ISSUES

The safety issues related to new foods created by biotechnology were, until comparatively recently, reasonably clear. In general, food plants produced by modern plant breeding techniques were subject to no specific controls. However, the discovery that a new variety of potato (Lenape) with good agronomic characteristics contained higher than average levels of the potentially toxic alkaloid solanine was sufficient grounds to withdraw it from commercial use. Similarly, in Sweden the variety Magnum Bonum was found to contain potentially toxic levels of glycoalkaloids and was withdrawn (55). In a number of countries, this led to voluntary codes of practice being adopted with the plant breeding industry to ensure that if there was more than a 10–20% variation in composition, the product would be referred to food safety agencies for evaluation.

As far as plants subject to genetic engineering are concerned, the principal concerns have been focused around the potential allergenicity of the modified plant and the use of antibiotic resistance genes for the selection of new strains. The possible variations in overall composition and the position to be adopted in terms of the need for safety evaluation have been considered by a number of international bodies.

In 1991 the United Nations Food and Agriculture Organization (FAO) and the World Health Organization (WHO) (56) convened an expert panel to consider strategies for assessing the safety of foods produced by biotechnology. The expert panel recommended that genetically engineered foods should be evaluated for safety by comparison with that of their natural antecedents. If the composition was similar to that of the traditional food and this food had a safe history of use, then the genetically modified food would be regarded as safe.

This led to the concept of "substantial equivalence" being developed following a meeting of the Organization for Economic Cooperation and Development (57). This concept assumes that if a genetically modified food can be characterized as substantially equivalent, it can be assumed to pose no new health risks and can be marketed without the need to undertake extensive toxicological and nutritional studies to determine its safety in use.

The principle of substantial equivalence has been adopted into the European Union (EU) Regulation on Novel Foods and Novel Food Ingredients (58). The Regulation excludes from its controls foods and food ingredients obtained through traditional propagating or breeding practices and which have a history of safe use. Genetically modified plants are considered as "novel" under the terms of the Regulation. However, the detailed safety evaluation provisions of the Regulation do not apply to foods produced by genetic engineering "if on the basis of the scientific evidence available they are substantially equivalent to existing foods with regard to their composition, nutritional value, metabolism, intended use, and the level of undesirable substances present." The Regulation regards food as novel if the characteristics of the food differ from the conventional food having regard to the accepted limits of natural variation of such characteristics. However, the principle of substantial equivalence is vague and difficult to define in many cases.

The U.S. attitude to regulation has so far been to regard safety as an issue that relates to the characteristics of the food and not to the process(es) that leads to it. Novel food products, of which products produced by genetic engineering are included in the definition, are not subject to any specific approval on safety grounds if the constituents of the food are the same as or substantially similar to those of substances currently found in other foods.

It is clear that it is never going to be possible to argue that a genetically modified plant is safe any more than to argue that a plant produced by conventional plant breeding is safe. The very concept can be addressed only in the context of a history of safe use as a human food. Clearly, the overwhelming evidence supports the view that health benefits arise as a consequence of the regular consumption of a variety of fruits and vegetables, few if any of which have any close compositional relationship to the wild types from which they were bred. Similarly, their production, storage, and distribution have depended on the use of a wide range of chemical fertilizers and pesticides. These chemicals are extensively tested for safety before approval is given for their marketing and use, but this has not removed the widely held view among consumers that "organic products" are better for your health. There is no evidence to support this view, and any adverse health effects that there might be as a consequence of the use of pesticides appear to be outweighed by the beneficial effects from the consumption of fruit and vegetables. What determines "safety" is the overall effect of consumption over a period, not the effects of a specific chemical that might be present.

A. Food Chemical Risk Assessment

The historic approaches to assessing the safety in use of a chemical in the food supply continue in spite of the fact that a growing body of evidence suggests that this is an inappropriate way to assess the safety of a chemical that may be present in the diet in only trace amounts (59,60). Animal feeding studies of fundamentally benign foods cannot be used to determine a dose-response relationship as in the case of a single chemical. Their usefulness is severely limited because of the differing diets of animals and the need to ensure that test and controls do not differ because of differences in weight or nutritional balance.

Many of the phytochemicals that are present in plants might be judged to pose an unacceptable risk if they were subject to the same approval procedures as are applied for synthetic chemicals (61). The value of doing so is highly questionable. It seems clear that any possible adverse effects of these phytochemicals are neutralized by other protective phytochemical factors in the plant.

It is possible that the standardized toxicological bioassays do not reflect the mechanisms that are operating in vivo through low-dose exposure to these compounds in the environment of the food. The past application of toxicology to food chemical risk assessment provided a generalized basis for protecting the population. Such a risk assessement process was strongly influenced by the need to ensure that there were no risks to consumers from the ingestion of food chemicals that were to be used in the production and processing of food. Thus, the only factor to consider was the inherent toxicological properties of the compound in question. The inherent toxicological properties of the compounds already in food and the influence they might exert in altering the toxicity of the added chemical were not considered. Neither was it deemed of interest to determine whether, or under what circumstances, the same chemicals might exert some health benefit as opposed to risk.

Although the safety in use of any food produced by genetic manipulation would be based on nutritionally based experimentation, there is likelihood of an insistence on applying traditional animal toxicological bioassays that have been designed to assess risk and not to make risk-benefit assessments. This is especially so if the food is judged not to be compositionally identical to the "traditional food." The assumption that chemicals, be they natural or synthetic, must pose a health risk of some kind, no matter what the dose, should be challenged. Risks can arise when exposure to chemicals is high. But it does not follow that there is some degree of residual risk when exposures are low.

It is beholden on scientists to demonstrate that this is so through rationally based arguments that have a biological validity. Risk evaluation using animal bioassays invariably applies rules that overemphasize the likely risks. The process highlights low-level risks and is costly in terms of both manpower and the use of experimental animals. The process might be justified if it resulted in providing consumers with total security over the safety in use of a food or of a chemical, but this is not the outcome. The very use of the word "risk" implies something unacceptable to many consumers. The application of extensive and rigorous approval procedures has done little to provide consumers with a sense of security in consuming food produced by using chemicals or other technologies.

None of the present methods for assessing safety take into account the fact that food is a complex mixture of chemicals, each of which has the potential to alter the overall risk-benefit ratio of consuming that food. Assessing the "risk" of chemicals in isolation and ignoring any assessment of "benefit" does not inspire confidence in the scientific process. Lack of confidence in the scientific basis for regulating chemicals produces a lack of confidence in the regulatory process itself. Unfortunately although it is accepted that animal toxicological bioassays will play a diminishing role in establishing the "wholesomeness" of a food, there are presently no alternative or viable substitutes. There is an urgent need to develop procedures for the safety evaluation of chemicals in food that are more realistic and that are developed from an understanding of the mechanisms that are occurring at usual exposure levels rather than on an empirical description of the phenomena and a focus entirely on perceived risk. Failure to do so will provide arguments for the opponents of any uses of genetic modification to impede progress toward its use in the development of even healthier plant foods.

VIII. CONCLUSIONS

There is little doubt that in terms of global health and nutrition needs, there is an urgent need to develop new plant crops with enhanced levels of iron, vitamin A, iodine, and the essential amino acids. This development will need to take place with plant breeders working closely with nutritionists. The issues that will need to be considered include the magnitude of any compositional change that is possible, its potential impact on health, the effect on crop yields and other agronomic issues, the role of the environment versus genetic manipulation in increasing levels, consumer acceptance, and a thorough analysis of any possible risks. Such an approach is already being adopted through the activities of the international agricultural research centers.

For the 10% of the world where there is overnutrition, the major focus of the plant breeding community will have to be on demonstrating the importance of their science to benefits that consumers perceive to be worthwhile. This is almost certainly to require a focus on health and nutrition issues. Any other application of genetic manipulation in food production is unlikely to be as effective in overcoming the current antagonism to its use. Again, this will require plant breeding institutes to work as part of multidisciplinary teams focused on specific targets that will improve the public health.

REFERENCES

1. B Pogson, K McDonald, M Truong, G Britton, D DellaPenna. *Arabidopsis* carotenoid mutants demonstrate that lutein is not essential for photosynthesis in higher plants. Plant Cell 8:1627–1639, 1996.
2. BW Shirley WL Kubasek, G Storz, E Brygemann, M Koornneff, FM Ansubel, HM Goodman. Analysis of *Arabidopsis* mutants deficient in flavonoid biosynthesis. Plant J 8:659–671, 1995.

3. SR Norris, TR Barette, D DellaPenna. Genetic dissection of carotenoid synthesis in *Arabidopsis* defines plastoquinone as an essential component of phytoene desaturase. Plant Cell 7:2139–2148, 1995.

4. PL Conklin JE Pallanca, RL Last, N Smirnoff. L-Ascorbic acid metabolism in the ascorbate-deficient *Arabidopsis* mutant vtc1. Plant Physiol 115:1277–1285, 1997.

5. FX Cunningham Jr, E Gantt. Genes and enzymes of carotenoid biosynthesis in plants. Annu Rev Plant Physiol Plant Mol Biol 49:557–583, 1998.

6. SR Norris, X Shen, D DellaPenna. Complementation of the *Arabidopsis* pds1 mutation with the gene encoding *p*-hydroxyphenylpyruvate dioxygenase. Plant Physiol 117:1317–1323, 1998.

7. P Baldet, C Alben, R Douce. Biotin synthesis in higher plants: purification and characterisation of bioB gene product equivalent from *Arabidopsis thaliana* in *Escherichia coli* and its subcellular localisation in pea leaf cells. FEBS Lett 419:206–210, 1997.

8. FC Belanger, T Leustek, B Chu, AL Kriz. Evidence for the thiamine biosynthetic pathway in higher plants and its developmental regulation. Plant Mol Biol 29:809–821, 1995.

9. D Eide, M Broderius, J Fett, ML Guerinot. A novel iron-regulated metal transporter from plants identified by functional expression in yeast. Proc Natl Acad Sci USA 93:5624–5628, 1996.

10. JM Quintana, HC Harrison, J Nienhuis, JP Palta, MA Grusak. Variation in calcium concentration among sixty S1 families and four cultivars of snap bean (*Phaseolus vulgaris* L.). J Am Soc Hortic Sci 121:789–793, 1996.

11. I Schonhof, A Krumbein. Gehalt an wertgebenden Inhaltsstoffen verschiedener Brokkolitypen (*Brassica oleracea* var *italica* Plenck). Gartenbauwissenschaft 61:281–288, 1996.

12. M Wang, IL Goldman. Phenotypic variation in free folic acid content among F1 hybrids and open-pollenated cultivars of red beets. J Am Soc Hortic Sci 121:1040–1042, 1996.

13. International Food Policy Research Institute. Agricultural strategies for micronutrients. http://www.cgair.org/ifpri/themes/grp06.htm.

14. World Health Organisation. Global Prevalence of Vitamin A Deficiencies. Micronutrient Deficiency Information Systems. Working Paper No 2. Geneva: WHO, 1995.

15. JA Nordlee, SL Taylor, JA Townshend, LA Thomas, LR Beach. Transgenic soybeans containing Brazil nut 2S storage protein. Issues regarding allergenicity. In: G Eisenbrand, H Aulepp, DD Dayan, PS Elias, W Grunow, J Ring, J Schlatter, eds. Food Allergies and Intolerances. Weinheim: VCH Verlagsgesellschaft, 1996, pp 196–202.

16. K Muntz, V Christov, R Jung, G Saalbach, I Saalbach, D Waddell, T Pickhardt, O Schieder. Genetic engineering of high methionine proteins in grain legumes. In: WJ Cram, LJ deKok, I Stuhlen, C Brunhold, H Rennenberg, eds. Sulphur Metabolism in Higher Plants. Molecular, Ecophysiological and Nutritional Aspects. Leiden: Backhuys Publishers, 1997, pp 71–86.

17. J Pen, TC Verwoerd, PA van Paridon, RF Beudeker, PJM van den Elzen, K Geerse, JD van der Kils, HAJ Versteeg, AJJ van Ooyen, A Hoekema. Phytase-containing transgenic seeds as a novel feed additive for improved phosphorous utilisation. Biotechnology 11:811–814, 1993.

18. JH Wong, I Besse, RM Lozano, JA Jiao, BC Lee, A Peters, K Kobrehel, BB Buchanan. New evidence for a role for thioredoxin h in germination and seedling development. Plant Physiol 2:403, 1996.

19. BB Buchanan C Adamidi, RM Lozano, BC Yee, M Momma, K Kobrehel, R Ermel, OL Frick. Thioredoxin-linked mitigation of wheat allergies. Plant Physiol 114:1614, 1997.

20. BB Buchanan. Thioredoxin: a photosynthetic regulatory protein finds application in food improvement. J Sci Food Agric 82:45–52, 2001.

21. PA Lachance. Overview of key nutrients: micronutrient aspects. Nutr Rev 56: S34–S39, 1998.

22. G Block, B Patterson, A Subar. Fruit, vegetables and cancer prevention: a review of the epidemiological evidence. Nutr Cancer 18:1–29, 1992.

23. CA Rice-Evans, NJ Miller, G Paganga. Antioxidant properties of phenolic compounds. Trends Plant Sci 2:152–159, 1997.

24. PM Mullineaux, GP Creissen. Opportunities for the genetic manipulation of antioxidants in plant foods. Biochem Soc Trans 24:829–835, 1996.

25. RA Dixon, CL Steele. Flavonoids and isoflavonoids—a gold mine for metabolic engineering. Trends Plant Sci 4:394–400, 1999.

26. GW Felton, KL North, JL Bi, SV Wesley, DV Huhman, MC Mathews, JB Murphy, C Lamb, RA Dixon. Inverse relationship between systemic resistence of plants to microorganisms and to insect herbivory. Curr Biol 9:317–320, 1999.

27. KW Waldron, AC Smith, AJ Parr, A Ng, ML Parker. New approaches to understanding and controlling cell separation in relation to fruit and vegetable texture. Trends Food Sci Technol 8:213–220, 1997.

28. MGL Hertog, PCH Hollman. Potential health effects of the dietary flavanol quercitin. Eur J Clin Nutr 50:63–71, 1996.

29. RA Dixon, CJ Lamb, S Masoud, VJH Sewalt, NL Paiva. Metabolic prospects for crop improvement through the genetic manipulation of phenylpropanoid biosynthesis and defense responses—a review. Gene 179:61–71, 1996.

30. J Schröder. A family of plant-specific polyketide synthases: facts and predictions. Trends Plant Sci 2:373–378, 1997.

31. M Jang, GO Udeani, KV Slowing, CF Thomas, CWW Beecher, HHS Fong, NR Farnsworth, RG Mehta, et al. Cancer chemopreventive action of resveratrol, a natural product derived from grapes. Science 275:218–220, 1997.

32. R Hain, B Bieseler, H Kindle, G Schröder, R Stöcker. Expression of a stilbene synthase gene in Nicotiana tabacum results in synthesis of the phytoalexin resveratrol. Plant Mol Biol 15:325–335, 1990.

33. RK Ibrahim, A Bruneau, B Bantignies. Plant O-methyl transferases: molecular analysis, common signature and classification. Plant Mol Biol 36:1–10, 1998.

34. G Folkmann. Control of pigmentation in natural and transgenic plants. Curr Opin Biotechnol 4:159–165, 1993.

35. AM Lloyd, V Walbot, RW Davies. Arabidopsis and Nicotiana anthocyanin

production activated by maize regulator-R and regulator-C1. Science 258: 1773–1775, 1992.

36. MJ Amiot, A Fluriet, V Cheymer, J Nicholas. Phenolic compounds and oxidative mechanisms in fruit and vegetables. In: FA Tomas-Barberan, RJ Robins eds. Phytochemistry of Fruit and Vegetables. Oxford: Clarendon Press, 1997, pp 51–85.

37. KDR Setchell, A Cassidy. Dietary isoflavones: biological effects and relevance to human health. J Nutr 129:758S–767S, 1999.

38. CL Steele, M Gijzen, D Outob, RA Dixon. Molecular characterization of the enzyme catalysing the aryl migration of isoflavonoid synthesis in soybean. Arch Biochem Biophys 367:147–150, 1999.

39. HK Lichtenhaler, M Rohmer, J Schwender. Two independent pathways for isopentenyl diphosphate and isoprenoid biosynthesis in higher plants. Physiol Plant 101:643–652, 1997.

40. K-P Adam, R Thiel, J Zapp. Incorporation of 1-[1-^{13}C] deoxy-D-xylose in chamomile sesquiterpenes. Arch Biochem Biophys 369:127–132, 1999.

41. D McCaskill, R Croteau. Some caveats for bioengineering terpenoid metabolism in plants. Trends Biotechnol 16:349–355, 1998.

42. DJ Wilkin, SY Kutsunai, PA Edwards. Isolation and sequence of the human farnesyl pyrophosphate synthetase cDNA. J Biol Chem 265:4607–4614, 1990.

43. X Pelletier, S Belbraouet, D Mirabel, F Mordet, JL Perrin, X Pages, G Debry. A diet modestly enriched in phytosterols lowers plasma cholesterol concentrations in normocholesterolemic humans. Ann Nutr Metab 39:291–295, 1995.

44. PJ Jones, FY Ntanios, M Raeini-Sarjaz, CA Vanstone. Cholesterol-lowering efficacy of a sitostanol-containing phytosterol mixture with a prudent diet in hyperlipidemic men. Am J Clin Nutr 69:1144–1150, 1999.

45. MA Hartmann. Plant sterols and the membrane environment. Trends Plant Sci 3:170–175, 1998.

46. A Rahier, M Tataon. Fungicides as tools in studying postsqualene sterol synthesis in plants. Pestic Biochem Physiol 57:1–27, 1997.

47. TA Miettinen, H Gylling. Regulation of cholesterol metabolism by dietary plant sterols. Curr Opin Lipidol 10:9–14, 1998.

48. X Ye, S Al-Babali, A Kloti, J Zhang, P Lucca, P Beyer, I Potrykus. Engineering the provitamin A (β-carotene) biosynthetic pathway into (carotenoid-free) rice endosperm. Science 287:303–305, 2000.

49. C Gärtner, W Stahl, H Sies. Increased lycopene bioavailability from tomato paste as compared to fresh tomatoes. Am J Clin Nutr 66:116–122, 1997.

50. CL Rock, JL Lovalvo, C Emenhiser, MT Ruffin, SW Flatt, SJ Schwartz. Bioavailability of beta-carotene is lower in raw than in processed carrots and spinach in women. J Nutr 128:913–916, 1998.

51. K Lee, FM Clydesdale. Effect of thermal processing on endogenous and added iron in canned spinach. J Food Sci 46:1064–1067, 1981.

52. AA Aziz, CA Edwards, MEJ Lean, A Crozier. Absorption and excretion of conjugated flavonols, including quercitin-4'-O-β-glucoside and isorhamnetin-4'-O-β-glucoside by human volunteers after the consumption of onions. Free Radic Res 29:257–269, 1998.

53. TA Miettinen. Stanol esters in the treatment of hypocholesterolemia. Eur Heart J Suppl 1:S50–S57, 1999.
54. HFJ Hendricks, JA Westrate, T Van Vliet, GW Meijer. Spreads enriched with three different levels of vegetable oil sterols and the degree of cholesterol lowering in normocholesterolemic and mildly hypercholesterolemic subjects. Eur J Clin Nutr 53:319–327, 1999.
55. K-E Hellenas, C Branzell, H Johnsson, P Slanina. High levels of glycoalkaloids in Swedish potato variety Magnum-Bonum. J Sci Food Agric 68:249–255, 1995.
56. WHO. Strategies for Assessing the Safety of Foods Produced by Biotechnology. Geneva: WHO, 1991.
57. OECD. Safety Evaluation of Foods Derived by Modern Biotechnology. Paris: OECD, 1993.
58. EU. Regulation No 258/97 of the European Parliament and the Council concerning novel foods and novel food ingredients. Off J Eur Communities 43: 1–7, 1997.
59. SA Miller. Novel foods: safety and nutrition. Food Technol 46:114–117, 1992.
60. DG Lindsay. Pesticide residues in food: risks or benefits? Pest Outlook 8:6–10, 1997.
61. LS Gold, TH Stone, NB Manley, BN Ames. Rodent carcinogenesis: setting priorities. Science 258:261–265, 1992.
62. CD Arnaud, SD Sanchez. In: EE Ziegler, LJ Filer Jr, eds. Present Knowledge in Nutrition. Washington, DC: International Life Sciences Institute, 1996, pp 245–255.
63. JJ Winzerling, LH Law. Comparative nutrition of iron and copper. Annu Rev Nutr 17:501–526, 1997.
64. F Delange. The disorders induced by iodine deficiency. Thyroid 4:107–128, 1994.
65. OA Levander, RF Burk. Selenium. In: EE Ziegler, LJ Filer Jr, eds. Present Knowledge in Nutrition. Washington, DC: International Life Sciences Institute, 1996, pp 320–328.
66. HE Sauberlich. Pharmacology of vitamin C. Annu Rev Nutr 14:371–391, 1994.
67. JE Leklem. Vitamin B6. In: EE Ziegler, LJ Filer Jr, eds. Present Knowledge in Nutrition. Washington, DC: International Life Sciences Institute, 1996, pp 174–183.
68. CE Butterworth Jr, A Bendich. Folic acid and the prevention of birth defects. Annu Rev Nutr 16: 73–97, 1996.
69. DM Mock. Biotin. In: EE Ziegler, LJ Filer Jr, eds. Present Knowledge in Nutrition. Washington, DC: International Life Sciences Institute, 1996, pp 220–235.
70. MG Traber, H Sies. Vitamin E in humans: demand and delivery. Annu Rev Nutr 16:321–347, 1996.
71. CL Rock. Carotenoids: biology and treatment. Pharmacol Ther 75:185–197, 1997.
72. MA Grusak, D DellaPenna. Improving the nutrient composition of plants to enhance human health. Annu Rev Plant Physiol Plant Mol Biol 50:133–161, 1999.

10

Engineering Plant Biochemical Pathways for Improved Nutritional Quality

Michel Jacobs, Marc Vauterin, Eric Dewaele, and Adrian Craciun
Department of Biotechnology, Free University of Brussels, Saint-Genesius-Rode, Belgium

I. INTRODUCTION

Considering recent controversies about genetically modified crops and genetically modified food, one frequently asked question concerns the real need for such modifications, especially if we refer to the first commercialized traits such as herbicide or insect resistance and the possible adverse effects of their utilization on the environment. Creating crops by transgenesis with an improved nutritional value may represent a more positive example in the public perception and would result in better public acceptance of biotechnology. It can also be emphasized that such an approach holds a great deal of promise in terms of attaining more appropriate levels of quality food components, which can be particularly crucial for the developing world.

Focusing on food nutritional quality, a series of targets can be identified. A major goal of plant genetic engineering consists of improving the content of essential amino acids of edible crop organs to provide better food for humans and a balanced composition of feeds for domestic animals. Most of this chapter will be devoted to the progress that has been made to improve the amino acid composition of plants and in particular the lysine content. Important results have also been reported on the qualitative aspects of a human diet, in particular vitamins, micronutrients, and phytochemicals relevant to human health. In time, plant biotechnology methods could lead to the production of feeds in which no addition of vitamins or minerals would be required. To prevent deficiencies, the daily intake of such nutrients has to reach minimal levels that have been defined on the basis of the work of nutritionists. Unfortunately, a more or less acute lack of many of these nutrients exists in various countries of the developing world. The diet of human populations is often essentially composed of a major staple food that does not fulfill the nutritional norms and may lead to severe illness and impair growth and development, especially in children.

We intend to underline strategies that have been developed to improve the nutritive value of food by increasing the level of one or another of these nutrients and amino acids. As such products are built up by the metabolic machinery of the plant cell, we will briefly mention a series of approaches that can be applied to redirect the metabolic flux toward increasing the level of a desired product. Then we will envisage how these methods can be applied to modify the amino acid composition of plant organs through metabolic engineering. We will focus on the main limiting amino acids in terms of nutritional impact, essentially lysine and threonine in cereals and sulfur-containing amino acids in legumes. Finally, we will describe methods of improvement beneficial to human nutrition and health with examples concerning increases of vitamin A and E levels and accumulation of deficient micronutrients such as iron.

II. STRATEGIES TO MANIPULATE PLANT METABOLIC PATHWAYS TOWARD THE PRODUCTION OF A DEFINED COMPOUND OF NUTRITIONAL VALUE

The definition of possible targets for engineering a primary metabolic pathway implies a sufficient knowledge of the whole pathway in terms of genes and enzymes involved and in particular identifying the critical regulatory points in the metabolic route.

In the past, such studies have been developed for a series of amino acids as exemplified by excellent reviews in books and journals (1–4). This acquired knowledge of the factors controlling the carbon flux through the pathway, such as identification of rate-limiting steps, enzyme regulation by feedback inhibition or repression versus induction, and the presence of competitive branches in the pathway, makes it possible to establish adequate strategies for manipulating the expression level of key enzymes in the pathway with the goal of provoking changes in flux toward the synthesis of a product. We can now modify the expression of a gene controlling a rate-limiting enzyme by insertion in the plant genome of the corresponding homologous or heterologous gene under the control of a strong promoter, expecting that a higher level of the encoded enzyme activity will be obtained. The transgene can also be modified in such a way that it now encodes an enzyme with a higher affinity for the substrate used in the branch leading to the desired compound. A mutated gene coding for a feedback-insensitive form of the enzyme could be obtained and introduced by transgenesis. This has already been achieved for both plant and microbial genes. Competitive branches of the pathway can be, at least partially, blocked by means of an antisense or knockout approach. Silencing the genes that exert their effects at or beyond branch nodes is thus an effective way to increase the flux toward the production of the desired metabolite by avoiding or at least decreasing a wasteful flow of intermediary products in another pathway. The lack of success in accumulating a metabolite can be also related to the fact that its catabolic rate is increased. Here also, antisense and down-regulation approaches may allow us to decrease the effects of such catabolic degradation (5).

These methods that rely on engineering of at least some of the genes of the pathway of interest have now been extended to the regulatory network controlling the overall pathway. The goal is to identify master control points that can influence the expression of a series of structural genes involved in metabolic conversions in the pathway to take advantage of this global effect on the synthesis of the product of interest. Modifying the expression of a single transcription factor may affect the expression level of a battery of biosynthetic genes involved in individual branches of a complex biosynthetic pathway. Some examples showing the potential of this approach are already available in the case of secondary metabolism (6–8) and also amino acid biosynthesis (9). The increasing availability of regulatory genes and the dissection of signal transduction pathways influencing primary metabolism make this type of methodology particularly attractive for the biotechnologist.

Finally, we also have to take into account the complexity of the biosynthesis we wish to orientate toward a defined direction. In that context, the development of functional genomics will provide us with invaluable information about the changes at the RNA, protein, and metabolite levels that occur when we modify defined steps of a specific pathway. It will also allow us to monitor the introduced changes so that desired alterations remain compatible with the growth and development of the crop. Moreover, we have to mention the possibility of engineering not only nuclear genes but also organelle genomes, in particular in the case of plant chloroplastic DNA. Here we can take advantage of the impressive increase in the expression of an inserted gene due to the large number of copies present in each leaf cell of the plastid transformed plant (10). As many steps of amino acid biosynthesis have been located inside the chloroplast, this represents a new avenue full of promise for overexpression of amino acid metabolism.

All the forecast benefits of metabolic engineering may still be limited or annulled by the complexity of the regulation controlling many plant biosynthetic pathways that still have to be worked out before we can obtain the production of the desired metabolites systematically and reliably. Besides manipulating gene expression of a pathway, the compartmentation of the metabolites in different cellular compartments and their transport inside the cell and to the different plant organs can totally alter the results of manipulating the properties of one key enzyme in the pathway. In particular, better knowledge of how quantifying and directing metabolic flux obtained by nuclear magnetic resonance spectroscopy appears essential (11).

III. MANIPULATING ESSENTIAL AMINO ACID METABOLIC PATHWAYS TO INCREASE THE NUTRITIONAL VALUE OF CROPS

The nutritional quality of crop plants is determined by their content of essential amino acids provided as proteins in foods for humans or in feed for monogastric animals. Plant proteins are often deficient in some of the 10 essential amino acids that are required in the human and animal diet. In general, cereals are deficient in lysine and legumes in the sulfur amino acids, methionine and cysteine.

Several approaches have been presented and used with the goal of increasing the relative content of essential amino acids in plants, in particular the aspartate-derived amino acids, lysine and threonine.

One of the earliest approaches consists of screening genotypes and induced mutants to identify plants with higher lysine levels in the seed. This method was first applied successfully to maize and led to the high-lysine opaque-2 type (12) . Later, similar types of mutants were reported in barley and *Sorghum* (see review in Ref. 13). The increase in total lysine content was due to a shift in the spectrum of storage proteins to the benefit of lysine-rich albumins and globulins instead of lysine-poor prolamines. However, this type of mutation was accompanied by a drastic reduction in yield and protein content.

The introduction of gene transfer technologies opened new ways to alter the amino acid composition of the seed proteins, in cereals and legumes in particular. In this case, genes encoding storage proteins naturally rich in essential amino acids or engineered genes leading to the production of "synthetic" proteins displaying a high content of defined essential amino acids were expressed in transgenic plants under the control of suitable promoters. In order to modify the nutritional value significantly, the foreign proteins must constitute a high percentage of the total proteins, so that in addition to high levels of transcription and translation, the stability of the additional proteins is crucial. As underlined by Galili et al. (14), although numerous attempts founded on such strategies have been reported, a significant, reliable accumulation of a stable nutritionally valuable protein still has to be convincingly demonstrated. In that context, it is worthwhile to mention the work of Chakraborty et al. (15) on the expression in potato of an *AmA-1* gene coding for a seed-specific albumin from *Amaranthus hypochondriacus* with a well-balanced amino acid composition. The authors reported that when this gene was put under the control of a granule-bound starch synthase promoter, a striking increase in most essential amino acids present in potato tubers was observed. Unexpectedly, the growth and production of tubers as well as the total protein content of this accumulation organ were significantly increased.

An alternative method consists of producing plants containing elevated levels of specific amino acids by manipulating their metabolism. This can be achieved through alteration of key enzymes from those biosynthetic pathways by mutation and transformation. We have focused on the aspartate-derived amino acid biosynthetic pathway producing the four essential amino acids lysine, threonine, isoleucine, and methionine. Considerable research, aided by genetic engineering, has been devoted to this pathway, its enzymes, and their regulation with the aim of optimizing amino acid content (16).

A. Biochemical Regulation of Lysine and Threonine Biosynthesis

Higher plants synthesize lysine, threonine, isoleucine, and methionine from the common precursor aspartate via a complex branched pathway (Fig. 1) (17). Understanding the inner regulation taking place during this process is essential with respect to knowledge of basic plant metabolism but also to possibly modifying the nutritional value of crops.

Close examination of the lysine and threonine biosynthesis revealed the existence of several enzymatic steps feedback controlled by these two end products. Essentially, two enzymes play key regulatory roles in the pathway: aspartate kinase (AK), the first enzyme of the overall pathway, which is retroinhibited by either lysine or threonine, and dihydrodipicolinate synthase (DHDPS), the first enzyme of the lysine-specific branch, which is severely inhibited by lysine. The main limiting step in lysine production appears to occur at the DHDPS level via the feedback control exerted by lysine (K_i between 5 and 20 μM). Most of the biosynthetic enzymes of the pathway have been localized in the plastids on the basis of biochemical data (17) and molecular evidence from genes encoding the majority of the bio-

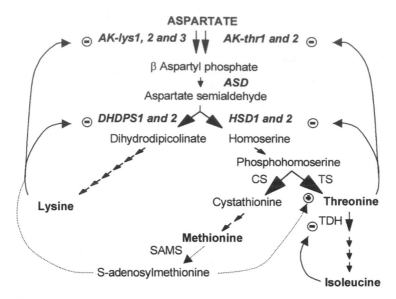

FIGURE 1   The aspartate family biosynthetic pathway. Enzyme abbreviations: AK-lys, aspartate kinase sensitive to lysine feedback inhibition; AK-thr, aspartate kinase sensitive to threonine feedback inhibition: ASD, aspartic semialdehyde dehydrogenase; HSD, homoserine dehydrogenase; DHDPS, dihydrodipicolinate synthase; TS, threonine synthase; TDH, threonine dehydratase; CS, cystathione-gamma-synthase; SAMS, S-adenosyl methionine synthase. Curved arrows indicate regulation loops by the end products. The (−) symbols indicate feedback inhibition and the (+) stimulation of enzyme activity.

synthetic enzymes, in which putative chloroplastic transit peptides have been identified.

B.   Mutants Overproducing Free Lysine or Threonine

We will now focus on an approach based, as previously mentioned, on the deregulation of biosynthetic pathways of such essential amino acids, in particular in the case of lysine and threonine.

In various species, mutants accumulating lysine or threonine in the soluble amino acid pool have been isolated by selecting for growth in the presence of the lysine amino acid analogue S-2-(aminoethyl)-1-cysteine (AEC) or toxic combinations of lysine and threonine in the culture medium (18). Although threonine overproduction was detected in all tissues analyzed and in particular in seeds, the increase of free lysine was noticeable only in leaves and calli. Moreover, an aberrant phenotype characterized by reduction of the foliar surface, absence of shoot elongation, and sterility was observed when high levels of free lysine accumulated (25% and more of the total free amino acid pool) (19). The accumulation of both amino acids in leaves is also dependent on the developmental stage, reaching a maximum at shoot elongation.

The biochemical characterization of these mutants has shown that we are dealing with deregulation at the level of the two key enzymes, DHDPS in the lysine overproducer mutant and AK in the threonine overproducer. These mutated forms are both fully insensitive to feedback inhibition by the respective amino acid (19).

C. Key Genes of the Lysine and Threonine Pathways

1. Aspartate Kinase Genes

At least five forms of aspartokinase have been identified in plants and may be classified on the basis of their differential feedback properties. One gene coding for a threonine-sensitive form (ak_1-hsd_1) was isolated by heterologous hybridization in *Arabidopsis thaliana*. The deduced amino acid sequence revealed the presence of a second region corresponding to the enzyme activity of homoserine dehydrogenase (HSD), the first committed enzyme in the branch of the pathway that leads to threonine synthesis. We are thus dealing with a gene coding for a bifunctional protein with an AK activity at the NH_2-side and an HSD activity at the COOH-terminus (20). The expression of the *A. thaliana* ak_1-hsd_1 gus gene was elevated in an array of young plant tissues containing actively growing cells (meristems, young leaves, cortical and vascular stem tissues, anthers, gynoecium, developing seeds). During the development of the embryo its expression appeared coordinated with the initiation and onset of storage protein synthesis (21). A second gene showing a high identity (82%) with the ak_1-hsd_1 nucleotide sequence was identified in a bacterial artificial chromosome (BAC) clone [European Molecular Biology Laboratory (EMBL bank)] and localized as ak_2-hdh_2 on chromosome 4 of the *A. thaliana* genome (V. Frankard, personnal communication).

As the major part of the AK activity is sensitive to feedback inhibition by lysine, attempts have been made to clone the corresponding genes. Degenerated primers corresponding to conserved motifs between lysine-sensitive bacterial AKs were used to clone two genes (*ak-lys1* and *ak-lys2*) encoding monofunctional AKs in *A. thaliana* (22,23). The presence of two nuclear genes both encoding AK-lys proteins and both targeted to the chloroplast, as shown by the presence of a transit peptide, but with only 70% identity at the deduced amino acid level raises questions about their respective roles in plant development. Therefore, *A. thaliana* and *Nicotiana tabacum* have been transformed with the 5' upstream region of either *ak-lys* gene fused with the *uidA* reporter gene encoding glucuronidase (GUS). In *A. thaliana*, a comparison of the GUS patterns of *ak-lys1* and *ak-lys2* revealed strongly predominant expression of *ak-lys2* over *ak-lys1*. Throughout the vegetative phases, *ak-lys2-GUS* expression reached a high level, especially in the vasculature, whereas *ak-lys1-GUS* plants showed a markedly reduced intensity of the histochemical staining. In the reproductive phase, both genes were well expressed in flowers, *ak-lys2* was the only one expressed in the fruits, and no staining could be detected in seeds in both cases. The physical position of *ak-lys1* and *ak-lys2* could be assigned to chromosome 5 at two different loci at a distance of 2 cM.

The *Arabidopsis ak-lys* gene family is thus composed of at least two different members. We identified a third *ak-lys* gene located on chromosome 3 of *Arabidopsis* using the Blast program to screen databases. The deduced *ak-lys3* amino acid sequence showed higher identity with the *ak-lys2* (82.8%) than with the *ak-lys1* (68%) peptide sequence. The *ak-lys3* gene was reported to be highly expressed in leaf vein tissues (24). The isolation and characterization of *A. thaliana* mutants displaying an AK-HSD isozyme less sensitive to threonine inhibition or an AK-lys isozyme less sensitive to lysine feedback inhibition (25,26) will make it possible to identify at the nucleotide level mutations that lead to threonine accumulation. After incorporation into the corresponding copy DNA (cDNA), these mutated plant genes can be respectively expressed in an appropriate construct to obtain threonine accumulation in specific tissues, especially in seeds.

2. Cloning and Characterization of Wild-Type and Mutant Genes Encoding Dihydrodipicolinate Synthase

A mutant of *Nicotiana sylvestris* (RAEC-1) was shown to overproduce lysine because of a mutation in the DHDPS gene that causes the DHDPS enzyme to be insensitive to the normal feedback inhibition of lysine. The dhdps-r1 mutation was identified as a substitution of two nucleotides changing asparagine in isoleucine in a conserved region of the protein (27). In maize, a series of single amino acid substitutions were found to eliminate lysine inhibition of DHDPS (28). DHDPS-encoding sequences were cloned by functional complementation in a bacterial DHDPS-deficient strain. This first isolated clone made it possible to obtain and sequence a full-length *Arabidopsis* DHDPS cDNA (29). Constructs derived from the *Arabidopsis* cDNA allowed the generation through ethyl methane sulfonate in vitro mutagenesis of clones encoding fully insensitive forms of the DHDPS protein (30). Furthermore, the clones successfully isolated by functional complementation in a DHDPS-deficient *E. coli* mutant were all found to encode insensitive enzyme forms, which means that complementation selects for insensitive DHDPS plant enzymes. In soybean, three mutants were constructed containing specific amino acid substitutions that lead to lysine-desensitized DHDPS (31).

In a further step, the *Arabidopsis* promoter has been isolated and fused with the reporter gene gus to study the transcription properties of this upstream sequence. Expression of *GUS* was detected in meristems and vascular tissues of roots, in vascular tissues of stems and leaves, and in the meristems of young shoots. In flowers, high expression was found in the carpels, pollen grain, and young embryos but not in endosperm of mature seeds. No lysine-induced repression of the *dhdps* gene could be detected (32).

An *Arabidopsis* genomic sequence encoding a second DHDPS enzyme was identified by screening the EMBL database. The *dhdps-2* coding sequence shows 84% identity with the nucleotide sequence of *dhdps-1*. The genomic *dhdps-2* sequence contains three exons and two introns, whereas only one intron and two exons are present in *dhdps-1* (Fig. 2). Comparison of the promoter regions of *dhdps-1* and *dhdps-2* did not reveal boxes with any significant conservation. The two *Arabidopsis* genes were localized on

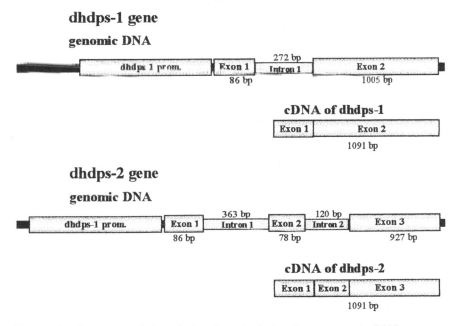

FIGURE 2 Structure of the *dhdps-1* and *dhdps-2* genes and cDNAs. prom, promoter; bp, base pair.

chromosome 2 for *dhdps-2* and chromosome 3 for *dhdps-1*. That *dhdps-2* encodes a functional protein was shown by the growth on a nonsupplemented medium of an *E. coli* DHDPS-deficient strain transformed with the *dhdps-2* apoprotein coding sequence. Activity tests performed in the presence of increasing concentrations of lysine proved that the DHDPS-2 enzyme is also strongly feedback inhibited, with a 50% loss of activity at 30 μM lysine.

Via promoter-GUS fusion, expression of the *dhdps-2* gene was observed during the whole developmental cycle and appears to be quite similar to that observed for the *dhdps-1* gene, although the *dhdps-2* gene is in general a little more expressed. DHDPS-2 activity was strongly detected in the vasculature of stems and leaves, in carpels, and in developing seeds (33).

D. Improving Lysine and Threonine Accumulation via Transfer of Bacterial Genes

The availability of bacterial and plant genes encoding feedback-insensitive enzymes allowed redirecting the expression of these genes in plants and in particular in storage organs. Initial experiments made use of constructs monitored by the strong cauliflower mosaic virus (CAMV) promoter harboring a bacterial *dapA* gene that encodes a DHDPS partially or fully feedback insensitive and mutated alleles of the *E. coli lysC* gene encoding a feedback-insensitive AK (see Ref. 16 for review). Transgenic lines have been obtained in tobacco (34–36), potato (37), barley (38), *Arabidopsis* (39), and oilseed rape and soybean (40).

As previously mentioned for *Nicotiana sylvestris* mutants, expression in tobacco of the bacterial AK led to accumulation of free threonine not only in vegetative tissues and in tubers in the case of potato but also in reproductive organs and seeds. In the case of the bacterial DHDPS, the lysine overproduction, although present in vegetative tissues, was not expressed in mature seeds. Constitutive overexpression of the bacterial DHDPS gene was also accompanied by phenotypical alteration of the transgenic plants, as observed with the selected *N. sylvestris* mutants, at least when a high lysine level was reached (36). Targeting the expression of these bacterial genes to sink organs such as seeds was then proposed as a way to alleviate such deleterious effects. Expression of the chimeric AK and DHDPS gene under the control of the β-phaseolin promoter obtained from the bean gene coding for this seed protein was thus evaluated. Transgenic tobacco seeds showed an important increase in free threonine but no significant change in the amount of lysine in mature seeds, although the activity of bacterial feedback-insensitive DHDPS was clearly higher than the DHDPS activity measured in nontransformed plants.

This lack of lysine accumulation in tobacco seeds was then ascribed to enhanced lysine catabolism as shown by higher activities of the major lysine catabolism enzyme lysine-ketoglutarate reductase (LKR) (41). We have to underline that the lysine content present in mature seeds of transgenic plants appears to be the result both of increased accumulation due to high activity of a feedback-insensitive DHDPS and of the rate of lysine catabolism. According to the balance between enhanced biosynthesis and induced catabolic degradation, the accumulation of lysine in seeds may vary among plant species (Table 1). This is best shown by the results obtained by Falco et al. (40) and Mazur et al. (42) when similar bacterial *ak* and *dhdps* genes were transformed into oilseed rape and soybean. Alone or combined with AK, expression of the bacterial DHDPS resulted in a dramatic increase of free lysine (10–100 times the level of nontransformed plants) with a significant effect on the total lysine content, which increased more than twice. In these last cases, accumulation of catabolic products such as saccharopine and α-aminoadipic acid was eventually observed but only at a minor level. However, in maize, when the bacterial AK and DHDPS were expressed with an endosperm or an embryo-specific promoter, lysine accumulation was detected only in the embryo but was sufficient to raise the overall lysine concentration in seed by 50 to 100% (42). Bacterial enzymes can thus be expressed with success in vegetative and sink organs such as seeds leading to deep modifications in lysine metabolism.

E. Improving Lysine Content Using Plant Gene Transfer

The cloning of mutated plant genes provided similar tools for metabolic engineering of plants. With the goal of increasing lysine production, several

TABLE 1 Levels of Lysine Overproduction in Seeds of Some Transgenic Crops (Fold Increase Compared with Untransformed Control) As Refers to Amino Acids

| Plant species | Inserted genes | Lysine | | Reference |
		Free aa	Total aa	
Brassica napus	Corynebacterium dhdps	3.1–140	1–1.75	39
Brassica napus	Corynebacterium dhdps + E. coli ak	4.7–38	1–2.0	39
Glycine max	Corynebacterium dhdps	12–25	1.25	39
Glycine max	Corynebacterium dhdps + E. coli ak	335	0.9–4.7	39
Hordeum sativum	E. coli dhdps	2	1.05	37

chimeric constructs harboring the *dhdps-r1* mutated allele (as described in Sec. III.C.2) have been developed under the control of the 35S CAMV promoter as well as tuber- or seed-specific promoters and transferred in a model species (*Nicotiana plumbaginifolia*) and crops such as *Solanum tuberosum* and *Sorghum bicolor*.

In *N. plumbaginifolia*, overexpression of the mutated *dhdps-r1* gene led to significant lysine overproduction in vegetative tissues, flowers, and immature seeds. Thus, the insensitivity of the mutated enzyme to feedback inhibition is critical for accumulating lysine, but the level of the enzyme also plays an important role in the increased lysine synthesis, (Fig. 3A and B). Transformants with high DHDPS activities are also characterized by high free lysine levels. The K_m and V_{max} of DHDPS thus represent determinant factors in the monitoring of lysine biosynthesis (E. Haryanto, personnal communication). Constitutive overproduction of this insensitive DHDPS was

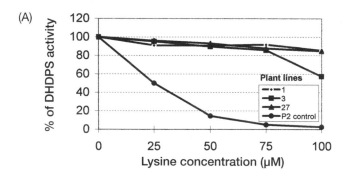

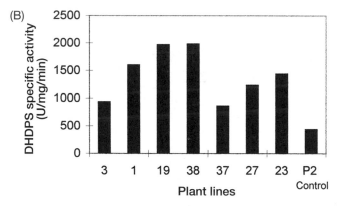

FIGURE 3   (A) Effect of lysine on DHDPS activity in leaf extracts of transgenic (1,3,27) and wild-type (P2) *Nicotiana plumbaginifolia* plants. (B) DHDPS activity in leaves from transgenic lines and wild-type P2 plants.

accompanied by abnormal phenotypes (multiple shoots, linear leaves, partial sterility) at least when high free lysine contents were reached in the transgenic plants, as already mentioned for lysine overproducer mutants of *Nicotiana sylvestris* and transgenic tobacco plants.

In potato, the same gene was introduced under the control of a patatin promoter to confine the lysine overproduction to tubers. In 23 analyzed tranformants, the range of lysine content varied from 0.9 to 13.7% of the total free amino acids while the control variety displayed a mean value of 2.6%. Thus, transformation of potato resulted in a fivefold increase in free lysine content without any noticeable modifications in tuber structure and production.

The properties of transgenic sorghum plants expressing the mutated *dhdps-r1* gene of *N. sylvestris* under the control of the 35S CAMV promoter were also determined. Transgenic plants were produced by introduction of the gene into *Sorghum* immature embryos by microprojectile bombardment. These embryos were then selected on the lysine analogue 2-aminoethyl-L-cysteine (AEC) from callus initiation up to regenerated roots. As a result of the expression of the gene, an active DHDPS enzyme insensitive to feedback inhibition was produced in the primary transformants and their progenies. In leaves of transgenic plants, a low but significant increase of lysine (1.5 to 2.5 times more than the control) could be associated with the expression level of the ectopic *dhdps-r1* gene. *Sorghum* seeds also synthesized higher levels of free lysine than observed in the original cultivar during the first phase of their development, from 15 to 25 days after pollination (Fig. 4). However, toward maturity the lysine content of transgenic seeds decreased and was almost comparable to the control value (44). This evaluation during maturation can be ascribed to enhanced lysine catabolism, as already mentioned for transgenic tobacco plants. A similar observation has been reported in corn, in which the expression of the deregulated bacterial *dhdps* gene under the control of an endosperm-specific promoter did not lead to any lysine accumulation (41). The presence of LKR activity has been demonstrated in immature endosperm in maize (43) and also in *Sorghum* seeds in our laboratory (44).

In conclusion, these results show clearly that although we are dealing with a complex, highly regulated biosynthetic pathway leading from aspartate to four essential amino acids, the overexpression of a single gene encoding a feedback-insensitive form of the key DHDPS enzyme exerts a significant effect on the carbon flux through the aspartate pathway toward lysine. This accumulation of lysine was in some species accompanied by phenotypical alterations, which should be avoided when this method is applied to crops. Organ-specific overproduction of lysine and threonine

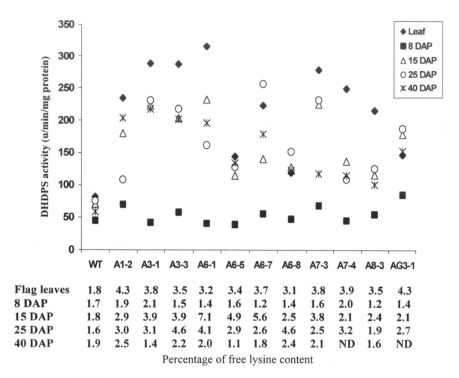

	WT	A1-2	A3-1	A3-3	A6-1	A6-5	A6-7	A6-8	A7-3	A7-4	A8-3	AG3-1
Flag leaves	1.8	4.3	3.8	3.5	3.2	3.4	3.7	3.1	3.8	3.9	3.5	4.3
8 DAP	1.7	1.9	2.1	1.5	1.4	1.6	1.2	1.4	1.6	2.0	1.2	1.4
15 DAP	1.8	2.9	3.9	3.9	7.1	4.9	5.6	2.5	3.8	2.1	2.4	2.1
25 DAP	1.6	3.0	3.1	4.6	4.1	2.9	2.6	4.6	2.5	3.2	1.9	2.7
40 DAP	1.9	2.5	1.4	2.2	2.0	1.1	1.8	2.4	2.1	ND	1.6	ND

Percentage of free lysine content

FIGURE 4 DHDPS specific activity measured in flag leaves and seeds of *Sorghum*, harvested on wild-type plants (WT) and T2 progenies of transformants (from A 1–2 to A G-3) at different developmental stages. Seeds were collected at 8, 15, 25, and 40 days after pollination (DAP) and were analyzed for DHDPS activity. Values at the bottom represent the percentage of lysine in the free amino acid pool. Each value is the mean of three independent assays from three individual T2 plants in each line.

through the use of specific promoters can overcome the limitations due to the constitutive expression of deregulated genes in transgenic plants.

Although lysine synthesis can be increased in plants, the level of free lysine in mature seeds may be determined not only by the efficiency of the biosynthetic pathway (in terms of activity of a feedback-insensitive DHDPS) but also by the stability of the amino acid accumulated in the free pool. In this case, the increased lysine catabolism due to the induction of LKR in endosperm tissues is a limiting factor leading to the reduction of free lysine accumulation (45). The genes encoding the two first catabolic enzymes, LKR and saccharopine dehydrogenase, have been cloned from *Arabidopsis* and maize (46,47). Therefore, by using an antisense or cosuppression approach, it might be possible to reduce the expression of LKR and the buildup of lysine degradation products. Another complementary approach will be to produce transgenic plants, in particular cereals, that altogether overproduce free lysine and express genes encoding lysine-rich proteins in seeds. This would provide a sink for stable accumulation of the increased lysine supply as bound amino acid in the endosperm lysine-rich proteins, avoiding the induction of strong catabolic activity.

F. Manipulating Plant Micronutrients to Benefit Human Health and Nutrition

Plant foods also contribute to ensuring sufficient essential vitamins and minerals for the human diet. Such micronutrients are often poorly represented in staple crops, and attempts have been developed to increase the synthesis of these compounds by isolating genes required for their synthesis and over-expressing them under the control of pertinent promoters.

Lack of vitamin A and vitamin E and iron deficiency are common problems in developing as well as in developed countries, involved in a series of illnesses, particularly in children.

In the major staple food rice, provitamin A is not synthesized in the endosperm, which represents the main part of the seed. The latest precursor of the provitamin A carotenoid is geranylgeranyl diphosphate (GGPP). Theoretically, the synthesis of β-carotene from this early intermediate would require the expression of four additional enzymes: phytoene synthase, phytoene desaturase, carotene desaturase, and lycogene β-cyclase. A phytoene synthase gene originating from daffodil was associated in a vector with the sequence coding for a bacterial phytoene desaturase that is capable of introducing the required double bonds. Lycopene β-cyclase from daffodil was carried by another vector, and cotransformed rice plants were selected. The relevant genes were placed under the control of an endosperm-specific glutelin promotor or the constitutive CAMV 35S promoter in the case of phytoene desaturase (48). Transgenic rice harboring this combination of genes produced seeds with a yellow endosperm indicating carotenoid formation. Biochemical analysis confirmed the presence of provitamin A at a level of at least 2 μg/g in homozygous lines, which should ensure the necessary daily amount of vitamin A in a typical Asian rice diet (300 g of rice per day). The grain known as golden rice is expected to be beneficial to people suffering from vitamin A deficiency.

In the case of vitamin E, the example illustrates the power of a genomics-based approach. The final step of the pathway is of special interest here because the methylation of γ-tocopherol to α-tocopherol, the biologically active form of the vitamin E, is performed by γ-tocopherol methyltransferase (γ-TMT), whose activity is a limiting factor in many seeds. Therefore, the objective was to clone the corresponding gene and to overexpress it in seeds. This was made possible by a genomics-based approach by identifying probable orthologs of the γ-TMT of *Arabidopsis* using the protein sequence of a cyanobacterium *Synechocystis*, which also synthesizes α-tocopherol (49). In transgenic *Arabidopsis* plants overexpressing the γ-TMT enzyme, more than 95% of the total tocopherol pool was in the form of α-tocopherol whereas the wild-type seeds contain practically only γ-tocopherol. Similar manipulations can now be used to improve the vitamin E level of important crops such as rapeseed, maize, and soybean.

Iron deficiency is also a significant nutritional problem for 30% of the world population. Rice endosperm contains only low amounts of iron, which, combined with a high content of phytate (an inhibitor of iron resorption in the intestine), causes a dietary iron deficiency in populations with a rice-based diet. One approach has been to express under the control of glutelin promoter a soybean gene coding for a ferritin protein used as a storage form of iron (50). Transgenic seeds have been shown to accumulate three times more iron than seeds from a control line. With this modification can also be associated the reduction of phytate concentration in cooked rice via a transgene coding for a thermotolerant phytase from *Aspergillus fumigatus* (51). Both vitamin A and iron-rich traits can be combined by sexual crossing, and such material is available for noncommercial use to developing countries.

The low availability of phosphorus (P) in many soils is an important factor limiting agricultural production worldwide. In an approach for developing novel crop varieties more efficient in the use of P, Lopéz-Bucio et al. (52) engineered plants able to produce more organic acids and in particular citrate by overexpressing a bacterial citrate synthase coding sequence in tobacco. Citrate-overproducing plants yield more leaf and fruit biomass when grown under P-limiting conditions and require less P fertilizer to achieve optimal growth.

In conclusion, we think that in time genetic engineering may lead to varieties carrying all the additional vitamins and minerals required in foods and feeds.

IV. CONCLUSIONS AND PERSPECTIVES FOR THE FUTURE

Transgenic crops with improved nutritional value have already been produced by overexpressing homologous, heterologous, and eventually mutated genes involved in key steps of the metabolism of amino acids and vitamins. In the case of lysine and threonine, overproduction of these essential amino acids has been achieved by the expression of feedback-insensitive AK and DHDPS enzymes. We stress that the overexpression of a single gene encoding a feedback-desensitized DHDPS exerts a significant effect on the flux through the aspartate-derived amino acid pathway toward lysine. With regard to metabolic engineering, this example is quite notable although direct extrapolation to other metabolic pathways will not necessarily meet with success. Rigidity of the metabolic network, functional redundancy of genes, and posttranscriptional control processes can alter the expected results and lead to undesirable side effects. The desired metabolite may be degraded, limiting its accumulation, as exemplified for lysine in seeds of at least some plant species.

Another constraint concerns the number of genes that have to be transferred in the recipient plant genome to reach or/and optimize the accumulation of a valuable product. Coordinated alterations of rate-limiting enzymes associated with the overexpression of master regulatory genes will probably contribute to increasing the metabolic flux toward the desired metabolite. However, the expression of multiple genes involved in a complex biosynthetic pathway will require the development of vector systems and methods ensuring the integration and stable expression of gene batteries in the plant genome (53,54).

Further progress in enhancing the nutritional quality of crops will also rely on emerging technologies, and we can anticipate decisive advances through functional genomics, DNA microarray technology, and metabolic flux analysis to obtain more insight into metabolic networks and their regulation in plants. It is important to obtain more information on metabolic sites of synthesis and storage and on metabolite transport between organs to learn how to exploit the possibilities offered by biochemical engineering and to apply them to metabolic processes relevant to the nutritional quality of crops.

ACKNOWLEDGMENTS

Work and grants in the laboratory of M.J. are supported by the Fund for Scientific Research—Flanders (FWO), the Biotech (BIO4-CT97-2182) program, and the Life Quality (QLK-1-1999-00765) program of the European Union.

REFERENCES

1. RA Azevedo, P Arruda, WL Turner, PJ Lea. The biosynthesis and metabolism of the aspartate derived amino acids in higher plants. Phytochemistry 46:395–419, 1997.

2. G Galili. Regulation of lysine and threonine synthesis. Plant Cell 7:899–906, 1995.
3. BF Matthews. Lysine, threonine and methionine biosynthesis. In: BK Singh, ed. Plant Amino Acids: Biochemistry and Biotechnology. New York: Marcel Dekker, 1999, pp 205–225.
4. R Amir, G Galili. Regulation of lysine and threonine metabolism. In: JK Setlow, ed. Genetic Engineering. New York: Plenum, 1998, 21, pp 57–77.
5. AJ Kinney. Manipulating flux through plant metabolic pathways. Curr Opin Plant Biol 1:173–178, 1998.
6. E Grotewold, M Chamberlin, M Snook, B Siame, L Buttler, J Swenson, S Maddock, G St Clair, B Bowen. Engineering secondary metabolism in maize cells by ectopic expression of transcription factors. Plant Cell 10:721–740, 1998.
7. L Tamagnone, A Merida, A Parr, S Mackay, FA Culianez-Macia, K Roberts, C Martin. The AmMYB308 and AmBYB330 transcription factors from *Antirrhinum* regulate phenylpropanoid and lignin biosynthesis in transgenic tobacco. Plant Cell 10:135–154, 1998.
8. W Bruce, O Folkerts, C Garnaat, O Crasta, B Roth, B Bowen. Expression profiling of the maize flavonoid pathway genes controlled by estradiol-inducible transcription factors CRC and P. Plant Cell 12:65–79, 2000.
9. J Bender, GR Fink. A Myb homologue, ATR1, activates tryptophan gene expression in *Arabidopsis*. Proc Natl Acad Sci USA 95:5655–5660, 1998.
10. L Bogorad. Engineering chloroplasts: an alternative site for foreign genes, proteins, reactions and products. Tibtech 18:257–263, 2000.
11. JKM Roberts. NMR adventures in the metabolic labyrinth within plants. Trends Plant Sci 5:30–34, 2000.
12. ET Mertz, LS Bates, OE Nelson. Mutant gene that changes protein composition and increases lysine content of maize endosperm. Science 145:279–280, 1964.
13. J Habben, BA Larkins. Improving protein quality in seeds. In: J Kiegel, G Galili. Seed Development and Germination. New York: Marcel Dekker, 1995, pp 791–809.
14. G Galili, O Shaul, A Perl, H Karchin. Synthesis and accumulation of the essential amino acids lysine and threonine in seeds. In: J Kigel, G Galili. Seed Development and Germination. New York: Marcel Dekker, 1995, pp 811–832.
15. S Chakraborty, N Chakraborty, A Datta. Increased nutritive value of transgenic potato by expressing a nonallergenic seed albumin gene from *Amaranthus hypochondriacus*. Proc Natl Acad Sci USA 97:3724–3729, 2000.
16. G Galili, BA Larkins. Enhancing the content of the essential amino acids lysine and threonine in plants. In: BK Singh, ed. Plant Amino Acids. Biochemistry and Biotechnology. New York: Marcel Dekker, 1998, pp 487–507.
17. JK Bryan. Advances in the biochemistry of amino acid biosynthesis. In: BJ Miflin, PJ Lea, eds. The Biochemistry of Plants: Amino Acids and Derivatives. Vol 5. New York: Academic Press, 1990, pp 161–196.
18. I Negrutiu, A Cattoir-Reynaerts, I Verbruggen, M Jacobs. Lysine overproducer mutants with an altered dihydrodipicolinate synthase from protoplast culture

of *Nicotiana sylvestris* (Spegazzini and Comes). Theor Appl Genet 68:11–20, 1984.

19. V Frankard, M Ghislain, M Jacobs. Two feedback-insensitive enzymes of the aspartate pathway in *Nicotiana sylvestris*. Plant Physiol 99:1285–1293, 1992.
20. M Ghislain, V Frankard, D Vandenbossche, BF Matthews, M Jacobs. Molecular analysis of the aspartate kinase–homoserine dehydrogenase gene from *Arabidopsis thaliana*. Plant Mol Biol 24:835–851, 1994.
21. JX Zhu-Shimoni, S Lev-Yadun, B Matthews, G Galili. Expression of an aspartate kinase homoserine dehydrogenase gene is subject to specific spatial and temporal regulation in vegetative tissues, flowers, and developing seeds. Plant Physiol 113:695–706, 1997.
22. V Frankard, M Vauterin, M Jacobs. Molecular characterization of an *Arabidopsis thaliana* cDNA coding for a monofunctional aspartate kinase. Plant Mol Biol 34:233–242, 1997.
23. G Tang, JX Zhu-Shimoni, R Amir, IB Zchori, G Galili. Cloning and expression of an *Arabidopsis thaliana* cDNA encoding a monofunctional aspartate kinase homologous to the lysine-sensitive enzyme of *Escherichia coli*. Plant Mol Biol 34:287–293, 1997.
24. Y Yoshioka, S Kurei, Y Machida. Identification of a monofunctional aspartokinase gene of *Arabidopsis thaliana* with spatially and temporally regulated expression. Genes Genet Sys 76:189–198, 2001.
25. B Heremans, M Jacobs. Threonine accumulation in a mutant of *Arabidopsis thaliana* (L.) Heynh. with an altered aspartate kinase. J Plant Physiol 146:249–257, 1995.
26. B Heremans, M Jacobs. A mutant of *Arabidopsis thaliana* (L.) Heynh with modified control of aspartate kinase by threonine. Biochem Genet 35:139–153, 1997.
27. M Ghislain, V Frankard, M Jacobs. A dinucleotide mutation in dihydrodipicolinate synthase of *Nicotiana sylvestris* leads to lysine overproduction. Plant J 8:733–743, 1995.
28. JM Shaver, DC Bittel, JM Sellner, DA Frisch, DA Domers, BG Gengenbach. Single-amino acid substitutions eliminate lysine inhibition of maize dihydrodipicolinate synthase. Proc Natl Acad Sci USA 93:1962–1966, 1996.
29. M Vauterin, M Jacobs. Isolation of a poplar and an *Arabidopsis thaliana* dihydrodipicolinate synthase cDNA clone. Plant Mol Biol 25:545–550, 1994.
30. M Vauterin, V Frankard, M Jacobs. Functional rescue of a bacterial dapA auxotroph with a plant cDNA library selects for mutant clones encoding a feedback-insensitive dihydrodipicolinate synthase. Plant J 21:239–248, 2000.
31. GW Silk, BF Matthews. Soybean DapA mutations encoding lysine-insensitive dihydrodipicolinate synthase. Plant Mol Biol 33:931–933, 1997.
32. M Vauterin, V Frankard, M Jacobs. The *Arabidopsis thaliana dhdps* gene encoding dihydrodipicolinate synthase, key enzyme of lysine biosynthesis, is expressed in a cell-specific manner. Plant Mol Biol 39:695–708, 1999.
33. A Craciun, M Jacobs, M Vauterin. *Arabidopsis* loss-of-function mutant in the lysine pathway points out complex regulation mechanisms. FEBS Lett 487:234–238, 2000.

34. O Shaul, G Galili. Increased lysine synthesis in transgenic tobacco plants expressing a bacterial dihydrodipicolinate synthase in their chloroplasts. Plant J 2:203–209, 1992.

35. O Shaul, G Galili. Threonine overproduction in transgenic tobacco plants expressing a mutant desensitized aspartate kinase from *Escherichia coli.* Plant Physiol 100:1157–1163, 1992.

36. O Shaul, G Galili. Concerted regulation of lysine and threonine synthesis in tobacco plants expressing bacterial feedback-insensitive aspartate kinase and dihydrodipicolinate synthase. Plant Mol Biol 23:759–768, 1993.

37. A Perl, O Shaul, G Galili. Regulation of lysine synthesis in transgenic potato plants expressing a bacterial dihydrodipicolinate synthase in their chloroplasts. Plant Mol Biol 19:815–823, 1992.

38. H Brinch-Pedersen, G Galili, S Knudsen, PB Holm. Engineering of the aspartate family biosynthetic pathway in barley (*Hordeum vulgare* L.) by transformation with heterologous genes encoding feed-back–insensitive aspartate kinase and dihydrodipicolinate synthase. Plant Mol Biol 32:611–620, 1996.

39. I Ben-Tzvi Tzchori, A Perl, G Galili. Lysine and threonine metabolism are subject to complex patterns of regulation in *Arabidopsis.* Plant Mol Biol 32: 727–734, 1996.

40. SC Falco, T Guida, M Locke, J Mauvais, C Sanders, RT Ward, P Webber. Transgenic canola and soybeen seeds with increased lysine. Biotechnology 13: 577–582, 1995.

41. H Karchi, O Shaul, G Galili. Lysine synthesis and catabolism are coordinately regulated during tobacco seed development. Proc Natl Acad Sci USA 91:2577–2581, 1994.

42. B Mazur, E Krebbers, S Tingey. Gene discovery and product development for grain quality traits. Science 285:372–375, 1999.

43. P Arruda, WJ Silva. Lysine-ketoglutaric acid reductase activity in maize: its possible role on lysine metabolism of developing endosperm. Phytochemistry 22:206–208, 1983.

44. Y Tadesse. Genetic transformation of *Sorghum* towards improving nutritional quality. PhD dissertation, Vrije Universiteit Brussel, 2000.

45. P Arruda, EL Kemper, F Papes, A Leite. Regulation of lysine catabolism in higher plants. Trends Plant Sci 5:324–330, 2000.

46. S Epelbaum, R McDevitt, SC Falco. Lysine-ketoglutarate reductase and saccharopine dehydrogenase from *Arabidopsis thaliana*: nucleotide sequence and characterization. Plant Mol Biol 35:735–748, 1997.

47. G Tang, D Miron, JX Zhu-Shimoni, G Galili. Regulation of lysine catabolism through lysine-ketoglutarate reductase and saccharopine dehydrogenase in *Arabidopsis.* Plant Cell 9:1305–1316, 1997.

48. X Ye, S Al-Babili, A Klöti, J Zhang, P Lucca, P Beyer, I Potrykus. Engineering the provitamin A (β-carotene) biosynthetic pathway into (carotenoid-free) rice endosperm. Science 287:303–305, 2000.

49. D Shintani, D DellaPenna. Elevating the vitamin E content of plants through metabolic engineering. Science 282:2098–2100, 1998.

50. F Goto, T Yoshihara, N Shigemoto, S Toki, F Takaiwa. Iron fortification of rice seed by the soybean ferritin gene. Nat Biotechnol 17:282–286, 1999.
51. I Potrykus. Vitamin-A and iron-enriched rices may hold key to combating blindness and malnutrition: a biotechnology advance. Nature 17(suppl):37, 1999.
52. J López-Bucio, O Martínez de la Vega, A Guevara-García, L Herrera-Estrella. Enhanced phosphorus uptake in transgenic tobacco plants that overproduce citrate. Nat Biotechnol 18:450–453, 2000.
53. WD Rees, SM Hay. The biosynthesis of threonine by mammalian cells: expression of a complete bacterial biosynthetic pathway in an animal cell. Biochem J 309:999–1007, 1995.
54. L Chen, P Marmey, NJ Taylor, JP Brizard, C Espinoza, P D'Cruz, H Huet, S Zhang, A de Kochko, RN Beachy, CM Fauquet. Expression and inheritance of multiple transgenes in rice plants. Nat Biotechnol 16:1060–1064, 1998.

11

Transgenic Plants as Producers of Modified Starch and Other Carbohydrates

Alan H. Schulman

Institute of Biotechnology, University of Helsinki, and
MTT Agrifood Research Finland, Helsinki, Finland

I. INTRODUCTION

Carbohydrates are a ubiquitous part of human and animal diets. Their abundance and omnipresence are difficult to overestimate. Carbohydrates provide clothing and shelter as the cellulose of cotton and wood, a means of communications as the cellulose of paper, dietary nutrients and fiber as starch and β-glucans, and the basis for most beverages ranging from fruit juice and soft drinks to cognac. The starch of most plant-derived foods and beverages is derived from either seeds or tubers. The properties of these are determined largely by their carbohydrate and protein components.

Humans throughout time, from the earliest hunter-gatherers to today's consumer, have chosen specific foodstuffs and processed them for consumption, generally unknowingly, in a way reflecting the functionality of their carbohydrate components. Starch, as the major edible carbohydrate component of foods, a major industrial feedstock, and the most abundant edible biopolymer, attracts the greatest attention as well regarding its functional properties and the possibilities for their modification. Besides their importance for food and beverage, plant carbohydrates are increasingly used in many industrial sectors including the production of biodegradable plastics.

They can be seen as a "green" alternative to hydrocarbon-based polymers. Carbohydrates contribute to many of the properties of novel "functional foods," ranging from sweetening to flavor enhancement, fat replacement, dietary fiber supplementation, texture preservation, and edible packaging (1). This chapter will concentrate on the synthesis and transgenic modification of α-glucans, in particular starch, and fructans in plants and leave aside the β-glucans including cellulose.

II. APPROACHES TO STARCH AND CARBOHYDRATE MODIFICATION

The various applications require starches with different properties, whether as raw starch granules, as gelatinized (cooked) starch, or as various hydrosylates. Suitable starches can be found by selecting either the appropriate plant source (tuber vs. grain), the particular plant variety, or the specific postharvest treatment. Postharvest, chemical or enzymatic alteration ("modification") of starch structure and hence properties has begun to be replaced with alternative strategies. Conventional breeding is one approach. If germplasm with the desired properties exists, easily scored and closely linked markers are needed for introgression of the requisite genes into a breeding line with good agronomic characteristics. Furthermore, a rapid screening method is needed for the trait.

Often, insufficient variation in storage carbohydrate properties can be found in the existing germplasm pool for a given crop. In this case, another option is to create the exotic germplasm by transgenic methods. There are basically four approaches to transgenic modification. The first is to change the quantity of carbohydrates in a seed or storage organ. Altering the relationship between the export strength of the carbohydrate source (generally, leaves) and the import strength of the carbohydrate sink (tubers, seeds) not only generally affects carbohydrate quantity and ultimately harvest yield but also may affect quality and downstream applications. Modulation of the expression levels or introduction of the biosynthetic enzymes, or introduction of novel forms with different catalytic properties, for an existing pathway such as starch biosynthesis is a broad second category of approaches. Another is the expression of degradative, glucolytic or glycolytic activities to alter the structure of a native carbohydrate. Lastly, an entirely novel pathway of branch can be introduced based on the metabolites present in the target tissue. These approaches will be examined in turn. Because starch is the major storage carbohydrate in plants, we shall concentrate on its biosynthesis, properties, and their modification.

III. STARCH STRUCTURE

A. Amylose and Amylopectin, the Two Components of Starch

Virtually all of the starch used by humans is storage starch, that which accumulates in seeds and tubers, and our knowledge of starch structure is mainly based on studies of this. Starch consists of two components, amylose and amylopectin, both of which are largely α-1,4-linked glucan polymers. Starch structure has been subject to many general reviews (2–4). Generally, amylose constitutes 20–30% of the total and contains rare, α-1,6 branches, whereas amylopectin, which has frequent α-1,6 branches, makes up the rest.

Amylopectin consists of linear, α-1,4-linked glucan chains frequently branched by α-1,6 bonds. The average chain length (degree of polymerization, DP) in amylopectin is on average 21–25 glucoses which, because of the frequent branching, yields a weight-average molecular weight (M_w) for amylopectin about 300 times larger than that of amylose. The branch points are not randomly distributed within amylopectin but rather are clustered. The chains of amylopectin are classified as the C-chain, the "core" chain containing the only reducing glucose in the molecule; the B-chains, which branch from the C-chain; and the A-chains, which are defined as the outermost branches of the molecule. Progressive amylolytic digestion followed by chromatography has shown that the B-chains are distributed in several size classes. Chains of DP 15–20 are found in the linear portions of clusters, whereas chains of DP 45–60 extend between clusters. A third form of starch, the highly branched "phytoglycogen," appears in certain mutants and is now thought to be an intermediate in the biosynthesis of amylopectin (5,6). The general properties of these components are presented in Table 1.

B. Starch Granules

Starch is packed into granules whose shape and size are characteristic of each plant species (7). Starch granules in potato tubers, for example, are smooth, irregular, and 15–75 μm in diameter, whereas those from maize are polyhedral and 5–20 μm. The starches of the Triticeae cereals including barley and wheat are present in a bimodal distribution of large A-granules (15–30 μm) and small B-granules (2–5 μm). Starch granule form confers properties to the starches that are important in various applications, from paper coating to brewing. In malting, the A- and B-granules are amylolytically digested in different manners, the A-granules by pinholing followed by internal digestion and the B-granules by surface erosion (8). This leads to uneven conversion during malting, a problem in brewing. The existence

TABLE 1 Starch Components and Their Properties

Property	Amylose	Amylopectin	Phytoglycogen
Degree of branching	Branches rare	Branched	Highly branched
Solubility in water	Insoluble	Insoluble	Soluble
Limiting viscosity number [η], mL (mg)$^{-1}$	240–390	185–188	
Iodine affinity, mg I_2 (100 mg)$^{-1}$ starch	19–19.9	0.33–1.5	
Molecular size	10^5–10^6	10^7–10^8	10^7
Glucose residues (DP)	~10^3	~10^5–~10^6	~10^5
Average chain length (CL)	100–1800	18–25	10–14
A-chains/B-chains		1.0–1.5	0.6–1.1
Iodine coloration (λ_{max})	660	530–550	430–450
β-amylolysis limit (%)	~70	~55	~30–45
α-amylolysis limit (%)	~100	~90	~80

Source: Refs. 2, 124, and 125.

of mutants affecting starch granule size distribution or morphology in various plants including barley (9,10) and pea (11) indicates that engineering of crops for specific granule size distributions is in principle possible.

The properties of starch, and the ultimate effects of biotechnologically induced changes in starch biosynthesis, largely reside in the organization of the starch granule. Work with advanced physical techniques and biochemical studies extending back almost 20 years have produced much insight into granule structure (5,12–18), the generally accepted view of which is presented in Fig. 1. At the lowest level of organization, paired, adjacent amylopectin chains form double helices. The helices are arranged as clusters, and the clusters in turn form crystallites. The linear, helical regions of the α-1,4 glucan chains form, in a radial direction, crystalline lamellae. These alternate at a periodicity of 9 nm with amorphous regions containing the α-1,6 branch points. Sets of these alternating crystalline and amorphous regions form semicrystalline zones hundreds of nanometers wide that alternate with broad amorphous zones. Together, the semicrystalline and amorphous bands form concentric shells termed growth rings. Interspersed within the amylopectin structure is the amylose component of the granule. In addition, lipids are tightly associated with the helix cores and proteins, in particular granule-bound starch synthase (GBSS), are tightly bound to the starch. The emergent properties of starch granules when treated with enzymes, solvents, or heat are greatly affected by the organization of the granule and by the bound lipids (19,20). These properties are related to the activities of the starch synthetic enzymes in complex ways, complicating rational approaches to specific targets in starch functionality. The genetic, developmental, and biochemical variations among plants producing storage starch result in a wide range of final properties of the starch (Table 1).

IV. STARCH DEPOSITION

In storage tissues, starch is synthesized within amyloplasts, which are derived as are chloroplasts from proplastids. Starch is also synthesized diurnally for transient assimilate storage in leaf chloroplasts. Starch biosynthesis is part of the complex process of tuberization, conversion of a stem into storage tissue, in potato and other crops producing storage tubers. In the cereals, starch is deposited in the starchy endosperm, whereas in most dicotyledonous plants it accumulates in fleshy cotyledons. Within developing endosperm, starch granules appear within a day of the onset of cellularization and continue until the grain dries (21–23). Tubers, however, have no sharp end point for starch biosynthesis.

A. Source of Photosynthate for Starch Biosynthesis

Starch biosynthesis with its key enzymes and metabolites is diagrammed in Fig. 2. Photosynthate is generally supplied as sucrose via the phloem of the maternal tissues. Both source and sink strength are critical to starch yield in storage organs. In some plants, breakdown and resynthesis of sucrose appear necessary to maintain a sucrose gradient and thus sink strength, although this is not the case in others such as barley. The sucrose taken into the endosperm is subsequently converted into UDPglucose by sucrose synthase (UDPglucose:D-fructose-2-glucosyltransferase, EC 2.4.1.13):

$$\text{sucrose} + \text{UDP} \rightarrow \text{UDPglucose} + \text{D-fructose}$$

This is a reversible reaction but, under the conditions found in storage tissues, the breakdown of sucrose is favored.

In many plants, sucrose synthase activity appears to be important to overall sink strength and hence yield. Antisense-mediated reductions in sucrose synthase levels in transgenic tomato (24) and potato (25) reduce overall starch biosynthesis, as do mutations to the sucrose synthase genes such as found in the maize *sh1* and *sus1* mutants (26). The UDPglucose product of sucrose synthase is then converted to glucose-1-phosphate by UDPglucose pyrophosphorylase (UTPglucose-1-phosphate uridylyltransferase, EC 2.7. 7.9). The UDPglucose pyrophosphorylase enzyme has been purified (27) and the gene encoding it cloned (28) from barley as well as from other plants. Glucose-1-phosphate is further processed to ADPglucose, the specific nu-

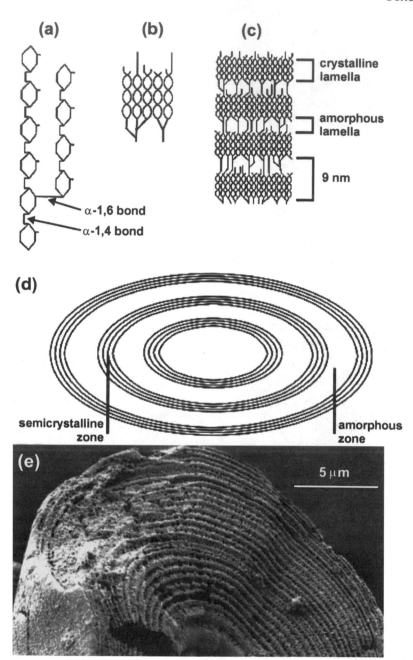

cleotide sugar that serves as the substrate for the starch synthases. This is catalyzed by the enzyme ADPglucose pyrophosphorylase (AGP, glucose-1-phosphate adenylyltransferase, EC 2.7.7.27) in the reaction

$$\text{ATP} + \alpha\text{-D-glucose-1-phosphate} \rightarrow \text{pyrophosphate} + \text{ADPglucose}$$

B. The Committed Pathway of Starch Biosynthesis

The conversion of glucose-1-phosphate to ADPglucose by AGP is considered the first specific, or committed, step in starch biosynthesis. The AGP enzyme has been extensively studied and reviewed since the 1960s (29,30) and also is the target for engineering of the pathway as discussed in the following. The enzyme in all tissues is a heterotretramer of two regulatory (small) and two catalytic (large) subunits (31). In most tissues, it is allosterically regulated, activated by 3-phosphoglycerate but inhibited by orthophosphate. Due in part to its regulation and also to the severely shrunken phenotypes of mutants of AGP (32), it has been seen as the major control point for the flow of carbon into starch. Flux analyses, however, contradict this interpretation (33,34).

Until recently, it was universally held that AGP is nuclear encoded but localized in the plastids in all tissues, photosynthetic and storage. However, at least for maize and barley endosperm (30), a combination of investigations on the *bt-1* mutant (35,36), studies of isolated amyloplasts (37), and messenger RNA (mRNA) transcriptional analyses (38,39) has shown that up to 95% of the cereal AGP is cytosolic (Fig. 2). A reasonable explanation for the difference between chloroplasts and amyloplasts regarding AGP localization rests on chloroplasts being sources of energy whereas amyloplasts are sinks. If AGP were restricted to amyloplasts, the ATP would have to be

←————————————————————————————————————

FIGURE 1 Current view of starch structure and its successive stages of organization within the granule. (a) Segment of amylopectin indicating the two bond types. (b) An amylopectin cluster showing the double helices formed between adjacent chains. (c) Helices are packed into crystalline lamellae spaced at intervals of 9 nm, interspersed with amorphous regions containing the parallel branch points. (d) The interspersed crystalline and amorphous lamellae form concentric semicrystalline zones several hundreds of nanometers wide. These zones are separated by amorphous zones lacking orderly packing of amylopectin helices. A pair of semicrystalline and amorphous zones form a growth ring in the granule. (e) Scanning electron micrograph of a potato starch granule showing the growth rings. The granule has been digested with α-amylase to remove partially the amorphous zones, which are more easily hydrolyzed. (From Ref. 45.)

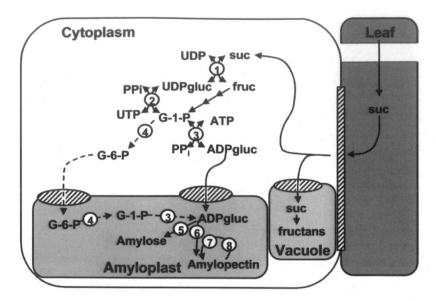

FIGURE 2   Schematic diagram of the currently accepted pathway for starch biosynthesis in storage organs. Photosynthate is transported as sucrose from source leaves through the phloem to the storage organ. It is then moved as sucrose into the storage tissues or cleaved in some plants by a cell wall invertase to glucose and fructose (not shown) to be resynthesized as sucrose by sucrose phosphate synthase in the cytoplasm. Storage as fructans in the vacuole represents an alternative to starch biosynthesis in temperate grasses. The key enzymes of starch biosynthesis are (1) sucrose synthase (SucSyn), (2) UDPglucose pyrophosphorylase (UGP), (3) ADPglucose pyrophosphorylase (AGP), (4) phosphoglucomutase (PGM), (5) granule-bound starch synthase (GBSS), (6) soluble starch synthase (SS), (7) starch branching enzyme (SBE), and (8) debranching enzyme (DBE). Not all alternative shunts in the pathway are shown. Fructose can be converted to glucose-1-phosphate via fructokinase (FK), phosphoglucoisomerase (PGI), and phosphoglucomutase (PGM). The relative proportions of ADPglucose synthesized in the cytoplasm and amyloplast vary from species to species. Translocators are shown as ovals on the organelle membranes.

imported and then converted to PPi and AGPglucose. This is energetically less favorable than movement of ADPglucose into the plastid and transport of ADP outward in return. Nevertheless, some AGP is plastidic even in the cereals, where the majority is cytoplasmic; the relative roles of the two forms remain to be established. Furthermore, in potato tubers it appears that the majority of the carbon moves as glucose-6-phosphate into the amyloplasts

(40), where it is subsequently converted to glucose-1-phosphate and then to starch. The details of the pathway in any particular plant are important regarding the possibilities of modifying starch quantity or quality. Starch content as well as starch quality can be affected by modulating the activity and properties of the AGP present in storage tissues.

C. Synthesis of Amylose

Amylose consists of glucose subunits linked by α-1,4 bonds into linear chains, with occasional α-1,6 branch points connecting additional α-1,4-linked chains onto a backbone (41). In barley, the average chain length is 1800 glucose units (42), but it may vary in the cereals between 1000 and 4400 glucose moieties, yielding a molecular weight of between 1.6×10^5 and 7.1×10^5 (43). In most normal starches, amylose makes up 20–30% of the total by weight. This is reduced to virtually none in the *waxy* mutants. The general features of amylose and the other main component of starch, amylopectin, are well established (44,45). The α-1,4 links in both amylose and amylopectin (see later) are made by the starch synthases (EC 2.4.1.21). The enzyme occurs in multiple forms, but all forms use ADPglucose as the glucose donor to the growing chain. In the storage organs, namely endosperm, cotyledons, and tubers, amylose is synthesized by the form called granule-bound starch synthase I (GBSS or GBSSI).

The "waxy" starches, perhaps the most common example of a modified carbohydrate created through both breeding and transgenic biotechnology, virtually completely lack amylose because of the absence of the GBSSI (46–48). The other forms of starch synthase are unable to compensate in such mutants, called *waxy* ("glutinous" in rice) because of the resulting property of the starch (the gene for GBSSI thereby being *wx*). In nonstorage tissues, however, amylose continues to be synthesized in *waxy* mutants, a form of the starch synthase called GBSSII carrying out the task in the cases examined (49). The gene or transcript for GBSSI has been cloned from many sources; alignments of these sequences revealed that these are highly conserved (50). The α-1,6 branch points in amylose are not synthesized by GBSSI or GBSSII but may derive from the action of a starch branching enzyme (SBE, see next for amylopectin) or from a branched oligosaccharide as the starch-synthetic substrate, with the poorly branched product subsequently elongated by GBSS. The substrate for amylose biosynthesis remains controversial and in vivo may be either amylopectin chains or soluble malto-oligosaccharides, both, or neither (45).

D. Synthesis of Amylopectin

Amylopectin is considerably more complex as a molecule, and its biosynthesis is commensurately more intriguing. The linear, α-1,4-linked portion of the polymer is produced by the soluble starch synthases (EC 2.4.1.21), which catalyze growth of the α-1,4 glucan chain by addition of glucose residues from ADPglucose. Historically, these enzyme forms received their name because they are not bound tightly to the starch granule, in contrast to the granule-bound starch synthase or GBSS (51). More recently, it has become clear that all forms of starch synthase are to some extent partitioned onto the starch granules or somehow become trapped in the growing, insoluble granule, so the original distinction is not very useful (52).

The amylopectin-synthesizing starch synthases are found in multiple forms in virtually all plants examined (reviewed in Ref. 45). Alignment of the proteins encoded by the sequenced soluble synthase form divides them into three main groups: SSI, SSII, and SSIII (53,54). Investigations of mutants and transgenics lacking or reduced in the activity of one of the SS forms indicate that each plays a specific, or at least preferential, role in amylopectin synthesis (9,11,55,56). These efforts have been complemented by expression of specific forms in *E. coli* and analyses of the α-glucan products made in the bacteria (57). From such experiments, SSII appears to synthesize α-1,4 chains of intermediate length (11), whereas the SSI form in barley (J. Tanskanen and A. H. Schulman, unpublished) appears to be involved in initiation of new chains (56). Potatoes expressing antisense to SSII (58), consistent with this view, have reduced relative abundance of chains of DP 18–50. The complexity of amylose biosynthesis from the perspective of engineering the pathway lies not only in the multiplicity of forms but also in their overlapping roles. Although one form may, because of its kinetics, be responsible for producing chains of a certain size class, in a mutant or transgenic plant where this form is absent another form may substitute but only partially or with identical results. The combination of overlapping roles and pleiotropism can lead to novel or unpredicted amylopectin structures in engineered starches (58,59).

The soluble synthases cover half of the story of amylopectin biosynthesis, however. The starch branching enzymes (SBEs, α-1,4-glucan, α-1,4-glucan-6-glucosyl transferase, EC 2.4.1.18, Q-enzyme) are responsible for producing the α-1,6 branches on the amylopectin molecule, which can then be further extended by the soluble starch synthases (44,60). Because it is the branching of amylopectin that confers its specific functional properties and behavior in food and beverage production, the SBEs have attracted considerable interest for the genetic tailoring of starch. The SBEs are transferases rather than synthases, detaching an α-1,4 -linked oligoglucan from the end of an amylopectin chain and moving it into an α-1,6 position elsewhere in the molecule. Nevertheless, they stimulate soluble starch synthases by increasing the effective substrate concentration determined by the number of nonreducing α-glucan ends in the amylopectin. As with the starch synthases, multiple isoforms have been identified that show organ (usually leaf or storage tissue) or temporal specificity in their expression patterns (61,62).

The various forms show differences as well in the length of chains transferred, which has implications for engineering of starch. These forms have been characterized as A or B types by their distinct properties (63,64). Antisense work in potato indicates that SBE A is responsible for transferring shorter chains than SBE B because average chain length increases in its absence (65). The well-known *amylose-extender* (*ae*) mutants illustrate the profound effect SBE has on starch properties. Rather than containing an increased amount of amylose as would be produced by the GBSS enzyme, these plants are in fact defective in amylopectin branching (47,66,67).

Over the last several years, a revolution in thinking about amylopectin biosynthesis has taken place with the introduction of the *preamylopectin trimming* model (5,6). The model addresses the question of how the non-random distribution of branch points typical of amylopectin may arise. It also helps to explain why mutants lacking a debranching enzyme such as the *sugary1* of maize (68) or a similar one in the alga *Chlamydomonas* (6) and *Arabidopsis* (69) contain a highly branched α-glucan referred to as phytoglycogen. In the model, SBEs and debranching enzymes (DBEs) carry out discontinuous steps of synthesis and amylolysis so that excess branches added by the SBE are removed. Crystallization of the product removes it from the cycle and fixes the structure as, in essence, a partially debranched glycogen.

An alternative, the *soluble glucan recycling* model, has been proposed (69). In this hypothesis, DBE plays only a subsidiary role in forming amylopectin, helping to turn over branched, soluble oligoglucans. This hypothesis explains the occurrence of phytoglycogen in DBE mutants but does not take the clustered branching of amylopectin into account. The validity of the two models is currently difficult to test. Furthermore, the actual in vivo functions of the soluble starch synthases, SBEs, and debranching enzymes still remain to be disentangled from the pleiotropic effects seen in mutants and antisense experiments.

V. TRANSGENIC MODIFICATION OF CARBOHYDRATE BIOSYNTHESIS

Efforts to alter carbohydrate biosynthesis extend back to about 1990 and have proceeded hand in hand with the use of overexpression and antisense inhibition to unravel carbohydrate biosynthesis in plants. The approaches can be divided into those that seek to alter starch quantity through affecting the strength of the carbon source or sink, those that attempt to convert starch to simple sugars, those with the goal of altering the amylose/amylopectin ratio, those trying to alter amylopectin structure, and lastly those that seek to produce novel carbohydrates through the introduction of new biosynthetic activities. Overviews of these efforts have been made several times from various perspectives (70–73). The key point is that grain or tuber quality and end use are related to the structures of the starch and protein components. These structures can be modified transgenically if suitable natural mutants are not available.

A. Alteration of Starch Quantity

Storage organs constitute net consumers or sinks for photosynthetically produced carbon, whereas leaves are the sources. Generally, source-sink balances are regulated by sugar levels (hexoses as well as sucrose) and by stress (74). Willmitzer and his colleagues demonstrated (75,76) that source strength in tobacco is inhibited by accumulation of sugar in the leaves. Expression of a yeast invertase in the cell wall of tobacco cleaved the sucrose normally loaded into the phloem and blocked its export, mimicking a very weak sink. This work was repeated later in transgenic potato plants (77), and photosynthesis was shown to be inhibited by sugar accumulation in the leaves. Using a parallel approach, sink strength was investigated by the same team (25). Sucrose synthase was demonstrated, through its removal in plants expressing sucrose synthase antisense under the strong 35S CaMV promoter, to play a crucial role in determining the sink strength of a potato tuber. Similar results were obtained by inhibiting the next step on the starch biosynthetic pathway, glucose-1-phosphate synthesis, through the expression of pyrophosphatase and concomitant reduction in pyrophosphate (PPi) content (78).

Following the pathway further, the accumulation of both starch and protein was inhibited by antisense knockdown of AGPase levels in potato tubers (79). Instead, the tubers accumulated up to 30% of their dry weight as sucrose and 8% as glucose, resulting in their increased fresh weight but decreased dry weight as well as pleiotropic effects on the transcription of other genes on the starch synthetic pathway. Generally, the practical goal is to increase the accumulation of starch in tubers or grains rather than to block it. Low-starch, high-sugar potatoes would be quite poor for a major market sector, chips, crisps, and fries, because sugar accumulation results in discoloration of chips or slices during frying. The postharvest accumulation of sugar in tubers has been limited by transgenic inhibition of UDPglucose pyrophosphorylase activity (80). An alternative, more effective approach was taken more recently by the expression of a tobacco invertase inhibitor in tubers (81). This reduced conversion of starch to soluble sugars by up to 75%, which appears to be at levels sufficient for the practical improvement of potato processing.

Increased sink strength has been engineered through expression of a yeast invertase in the tuber, promoting cleavage of sucrose and hence a stronger translocation gradient to the tuber (82). One strategy to improve potato tubers is to increase starch levels. Besides increasing total yield in dry weight, a higher starch content is correlated with a decrease in fat uptake during frying and therefore a more healthful product. In an attempt to do this, a mutant *E. coli* AGP form has been expressed in tubers as a translational fusion to a ribulose bisphosphate carboxylase transit peptide and driven by a patatin promoter (83). Other efforts using the same AGP form failed to increase starch content because of associated higher turnover of starch into sugars (84,85). As reviewed elsewhere (86), efforts to date to increase yield and hence starch biosynthesis through the manipulation of single enzyme levels have not been very predictable. An example of the capacity of carbohydrate metabolism to yield surprising results is the effect of expressing viral movement proteins in tobacco and potato (87). The protein MP17 of potato leaf roll virus increased soluble sugar and starch amounts in source leaves but did not affect photosynthesis in the leaf blade because of sequestration of the sugars in the vacuole. Although such experiments do not provide a ready recipe for engineering sugar or starch accumulation in plants, much can be learned about carbohydrate metabolism in the meantime, and in the end, effective quantitative manipulations may become straightforward.

B. Production of Simple Sugars in Storage Organs

As an outcome of analyses of sugar and starch metabolism and source-sink interactions, know-how has developed on the manipulation or production of simple sugars in storage organs. Work on cold sweetening of potatoes, a problem discussed in the previous section, examined the role of acid invertase in the process (88). Although the experiments showed that invertases do not control the conversion of starch to sugars during storage, they do determine the hexose-to-sucrose ratios. Transgenic expression of soluble invertase could thus be used as a strategy to produce hexoses in vivo. Tomatoes and most fruits, in contrast to potato tubers, accumulate sugar rather than starch. In experiments with goals opposite to what was attempted in potato (89), natural invertase levels in tomato fruits were reduced by an antisense strategy. Sucrose levels increased and hexose levels decreased in the antisense fruits, accompanied by a 30% reduction in fruit size.

In very promising newer work (90), an alternative approach to the production of hexose, in this case fructose, in potato tubers has been taken. Rather than introduce single enzymatic activities or reduce existing ones, a fusion coding for α-amylase from *Bacillus stearothermophilus* and glucose isomerase from *Thermus thermophilus*, both thermostable, was expressed in transgenic tubers under control of the GBSS promoter. The complex was not enzymatically active during tuber development. Instead, production of fructose and glucose was achieved by crushing the tubers and heating for 45 minutes to 65°C. A parallel approach, demonstrated in transgenic tobacco, was reported by a different group (91) slightly thereafter. In related work, a heat-tolerant β-glucanase has been transferred to two malting varieties of barley (92). This approach should improve malting quality through reduction of the content of β-glucans in wort, the source of filtration problems and of cloudiness in beer. The native β-glucanases do not withstand well the heating of the mashing process. These experiments clearly demonstrate the potential of in planta starch modification and of transferring an industrial process into the farmer's field to create a novel product.

An example of an unexpected effect on carbohydrate synthesis or turnover from a transgene was shown through overexpression of wheat thioredoxin h in barley endosperm (93). Thioredoxin h has been known to be important in germination for mobilization of storage protein in the endosperm. The thioredoxin must first be reduced, and the NADPH needed for this can be produced through hydrolysis of starch in the endosperm. Thus, it is perhaps satisfying but nonetheless surprising that overexpressed thioredoxin h should lead to a fourfold increase in α-1,6-debranching (pullulanase) activity in germinating grains. Although the authors do not present the glucan profile, this approach should greatly alter the limit dextrin profile of germinating grain and have an impact on malting.

The disaccharide trehalose, known for many years to be produced primarily by fungi and some insects, attracted interest because of its potential use as an osmoprotectant or stress-mitigating agent. Efforts were therefore made to engineer its expression in tobacco and potato through the introduction of trehalose-6-phosphate synthase (*otsA*) and trehalose-6-phosphate phosphatase (*otsB*) genes from *E. coli* (94). Although only very low levels of trehalose (0.11 mg g^{-1} fresh weight) could be obtained in this way, it was discovered that this poor yield is due at least in part to the presence of native trehalase activity not only in the transgenic regenerants but also in the control plants. It later became clear that the enzymatic machinery for synthesizing trehalose is in fact universal among the angiosperms, although in most plants trehalase blocks the accumulation of the sugar (95). This would not have been realized if control experiments with the trehalase inhibitor validamycin A had not been carried out, and it illustrates that metabolic engineering in plants is still very much of an adventure.

C. Alteration of the Amylose Complement in Starch

Some of the earliest efforts at qualitatively altering starch biosynthesis were directed at the amylose-to-amylopectin ratio. This was because the abundance of natural *waxy* mutants showed that amylose could be eliminated by suppression of GBSS activity and because low-amylose starches had certain processing advantages. The first successful creation of a low-amylose (*amf*) potato was achieved in the group of Jacobsen and Visser through mutagenesis rather than transformation (96). This was followed by antisense expression of GBSS by the same group (97,98). Often, antisense suppression of endogenous genes succeeds even without full sequence identity in the transgene. In other experiments, the GBSS of cassava, bearing only 74% identity to potato GBSS, was able in some regenerant lines to inhibit native GBSS synthesis completely (99). Glutinous, or amylose-deficient, rice is important in the diet of Japan, but a range of amylose contents may offer broader uses in foods. An antisense approach to GBSS suppression in rice yielded transgenic lines varying in their amylose content from slight reduction to complete absence (100). In plants more recalcitrant to transformation such as wheat, the more traditional approach of combining mutants by crossing has so far been more effective in achieving low-amylose lines (101).

D. Alteration of Amylopectin Structure

A major goal in many laboratories has been the transgenic tailoring of amylopectin structure. The reason for this is that much of starch functionality in cooking, baking, and extrusion is determined by the degree and pattern of branching in amylopectin. Linear chains readily form interchain hydrogen bonds, producing crystalline regions in starch granules and falling out of solution in the process called retrogradation in gelatinized, cooked starch. Lower levels of crystallinity result in more stable, but more wettable, gels. In malting and fermentation, digestibility by amylase is also directly linked to amylopectin structure. Exoglucanases, in particular β-amylase, digest inward from the nonreducing ends of α-1,4-glucan chains in starch. These enzymes are blocked by α-1,6 bonds, yielding a "limit" dextrin. In malting, the endoglucanase α-amylase is also present, breaking the α-1,4 bonds within the glucan chain. Because α-amylase does not cleave terminal α-1,4 bonds or those near α-1,6 bonds, α-limit dextrins remain after digestion. Hence, the processing benefits of increased yield of monosaccharides and disaccharides in starch hydrolysis and of tailoring of starch behavior during cooking have driven interest in using transgenic approaches.

One of the first efforts in which starch structure was altered in a transgenic plant involved expression of the *E. coli* glycogen synthase (*glgA*) in potato (102). Glycogen synthase carries out the same reaction as plant starch synthases, transferring glucoses into α-1,4 glucan chains. Total starch content in the tubers declined, and amylose was reduced from 23% to 8–9%. The short chains (A + B1) in the amylopectin increased from 66% of the total chains detectable following hydrolysis to 85%, while the long (B2 + B3) chains decreased from 33% to 15%. In a complementary effort, the *amf* low-amylose potato, which had been developed earlier by mutation breeding (96), was transformed with the gene for the *E. coli* glycogen branching enzyme (*glgB*) (103). As in the previous example, this enzyme carries out the same reaction as the corresponding starch branching enzyme of the plant, although the final product in bacteria is highly branched glycogen. Up to 25% more branches were made in the transgenic amylopectin and average chain length dropped, associated with more short chains of DP < 16.

For certain applications, it would be highly useful to obtain virtually pure amylose directly from the plant rather than through chemical fractionation of starch. So-called high-amylose or amylose-extender cereals have long been known (104,105), but, as described earlier, this is due not to synthesis of more of the product of GBSS ("true" amylose) but rather to less SBE activity and hence a less branched, more amylosic, amylopectin (47,66,67). Taking a cue from these mutants, a group at Unilever was able to produce potato tubers virtually lacking normal amylopectin but containing apparent amylose levels as high as found in any commercial cereal (65,106). Potato starch is phosphorylated in the amylose fraction, the phosphorylation conferring increased solubility, and this transgenic starch contained fivefold higher phosphorus contents than normal. Rather than knocking out the branching enzyme activity to alter amylopectin structure, the Kossmann laboratory expressed a chimeric antisense construct against the genes of both major soluble starch synthases, SSIII and SSII, in potato (58). Total starch synthase activity was reduced up to 90%, but amylose production was normal and amylopectin not eliminated. Instead, the amylopectin contained more chains of DP < 15, fewer of 15 to 80 glucose units, and more very long chains. The effect of removing one or the other SS form was not consistent with eliminating both simultaneously, indicating a complex interaction between the SS forms during amylopectin biosynthesis.

In a more direct approach, the Kossmann group was able to modify the amount of starch phosphorylation in a transgenic tuber (107). The group began by isolating proteins bound to starch and raising antibodies to them, with the expectation that starch-bound proteins would in some way be involved in starch biosynthesis. A gene for one of these, a protein of ~160 kDa, was cloned by screening a complementary DNA (cDNA) expression library with the antiserum. This protein, named R1, bears no resemblance to any previously characterized enzyme of starch biosynthesis. When the level of this protein is reduced in antisense-transformed potatoes, the level of starch phosphorylation is likewise lowered to 10–50% of normal. Glucose-6-phosphate was commensurately reduced, and the effect on starch phosphorus levels was seen at both C-3 and C-6 positions to an equal degree. When R1 was expressed in *E. coli*, it led to phosphorylation of the bacterial glycogen. Coincidentally, cold-induced sweetening, discussed earlier, was decreased through a secondary effect on starch digestibility. Curiously, in the antisense transgenic plants, leaves accumulated starch in excess of normal. It remains to be seen whether the gene can be used to phosphorylate starch in plants particularly. The authors reported that similar sequences are expressed in rice and *Arabidopsis*, yet these starches are not normally phosphorylated. If activation of phosphorylation becomes possible in the cereals, an important new class of starches will be available to the marketplace.

E. Production of Nonstarch Carbohydrates

The interconvertibility of many of the sugar metabolites on the starch biosynthetic pathway in leaves and storage organs by native and exogenous enzymes indicates that, in principle, many new carbohydrates could be synthesized in transgenic plants. An early attempt at this was the production of cyclodextrins in potato tubers (108). Cyclodextrins are rings comprising six to eight glucose units produced by bacterial cyclodextrin glucosyltransferases from a starch substrate. This group at Calgene expressed a *Klebsiella* cyclodextrin glucosyltransferase in potato tubers driven by a patatin promoter. They were able to produce both six-unit (α-) and seven-unit β-cyclodextrins by this approach, although the yield was exceptionally low, 0.001–0.01% of the starch being converted to cyclodextrins.

A more promising effort was made to produce mannitol in transgenic tobacco (109). Sugar alcohols or polyols such as mannitol and sorbitol are found in diverse plant species, where they are believed to confer osmoregulatory and stress-ameliorating functions, as well as in bacterial, fungi, and mammals. An *E. coli* gene for mannitol-1-phosphate dehydrogenase (*mtlD*) was expressed in tobacco and drove production of mannitol in excess of 6 μmol $(g)^{-1}$ of fresh weight in the leaves and in roots. In further work by the same group (110), specific targeting of this enzyme to tobacco chloroplasts led to accumulation of up to 100 mM mannitol in the plastids of one transgenic line, which was otherwise phenotypically normal. The mannitol increased resistance to oxidative stress induced by methyl viologen, apparently through improved scavenging of hydroxyl radicals. Useful as this may be for plant improvement, no one has yet attempted commercial production and harvesting of mannitol in this manner.

Perhaps the greatest attention has been paid to the biosynthesis of fructans in transgenic plants. Fructans are fructose polymers localized, unlike starch, in vacuoles rather than plastids. They are synthesized by disproportionation, whereby a fructosyl residue is first transferred from one sucrose to another to make the shortest fructan, gluc-fruc-fruc. The process proceeds by further fructosyl transfers from sucrose as well as by transfers between fructans. Mature fructans are found with a wide variety of branching patterns. In temperate grasses such as barley, fructans are an alternative to starch for carbon storage (111). This accumulation of fructans appears to contribute to yield stability under conditions unfavorable for starch biosynthesis because the fructans can later be converted to starch when conditions improve. They also accumulate early in endosperm development but are turned over to support synthesis of starch as the grain matures. If starch biosynthesis is reduced by cold temperatures or blocked such as in the *shx* mutant of barley (112), fructans rather than starch may persist or accumulate. Aspects of fructan biosynthesis have been summarized (113).

The focus on fructan engineering derives from its use as a potential pro- or prebiotic, antitumorigenic component of the human diet (114,115). The most common sources of fructans, in particular inulin, have been the Jerusalem artichoke (*Helianthus tuberosus*) and chicory. However, initial efforts at engineering production of fructans in transgenic tobacco employed the bacterial *SacB* gene, encoding levan sucrase, from *Bacillus subtilis*. The transformed plants accumulated 3–8% fructan of the levan type found in the bacterium (116). In a second effort with a bacterial transgene (117), the Willmitzer group expressed levan sucrase from *Erwinia amylovora* in the transgenic potato line previously engineered to be starch free with an antisense AGP. When the levan sucrase was targeted to the vacuole, 12 to 19% of the tuber dry weight was present as levan. However, yield was not increased relative to the parent line lacking starch.

The first fructan synthetic enzyme to be cloned from a plant was sucrose-fructan 6-fructosyltransferase (6-SFT) from barley (118). The group then expressed this clone in tobacco and in chicory. Chicory normally produces fructan of a different type than those, the graminans and phleins, found in barley (119). The transgenic tobacco was able to synthesize the trisaccharide kestose as well as unbranched fructans of the phlein type. Chicory, normally making inulin, gained the ability to make graminan fructans, in particular the tetrasaccharide bifurcose, which is the main form in barley leaves. In further experiments (120), chicory was transformed with a gene for an enzyme from onion, fructan:fructan 6G-fructosyltransferase (6G-FFT), a key enzyme in the synthesis of the inulins found in the Liliales, which had been cloned by screening with a 6-SFT probe from barley. Expression of the onion gene in chicory led to synthesis of the expected onion-type branched fructans as well as of linear inulin.

Following similar lines with potatoes, a clone encoding a 6-SFT–like enzyme was first isolated from a cDNA library of the globe artichoke (*Cynara scolymus*). When transformed into potato, the transgenic tubers produced high levels of 1-kestose along with nystose and traces of fructosyl-nystose. In subsequent work (121), the same group expressed both the sucrose:sucrose 1-fructosyltransferase (1-SST) and the fructan:fructan 1-fructosyltransferase from *C. scolymus* in transgenic potatoes. The tubers produced up to 5% of their dry weight as high-molecular-weight inulins of a range identical to those found in the native artichoke, and some fructan was also detected in leaves. The tuber fructans were synthesized partly at the expense of starch production. An obvious choice as the transgene host for fructan biosynthesis is the sugar beet, which stores sucrose, the very substrate needed, rather than starch, for fructans. When the gene for 1-SST, cloned from *H. tuberosus* together with that for 1-fructan:fructan fructosyl transferase (1-FFT) (122), was transferred into sugar beet, the expected small fructans up to DP 4 were obtained (123). Remarkably, more than 90% of the sucrose of the beet was converted into fructans, and no deleterious effects on plant growth were observed under greenhouse conditions.

VI. CONCLUSIONS

Plants are well suited as producers of modified starch and novel carbohydrates. Photosynthesis supplies a sucrose feedstock to carbohydrate storage organs, and the many intermediate steps in the conversion of sucrose to starch represent potential branch points at which the sugar may be shunted to new products. Storage starch, being insoluble, is not physiologically active, and the many mutations affecting starch synthesis demonstrate that a wide variety of structures can be tolerated by the plant. Tubers and storage roots (beets) are not required for propagation, so the storage starch in these organs need not be accessible to the plant for turnover. Cereal grain starch is, however, important in germination; therefore, modification or replacement of this carbon source must take physiological needs into account.

To date, the main efforts in carbohydrate engineering in plants have been directed to alterations in starch yield, to increasing or decreasing the effective amylose content, and to changing the degree of branching in amylopectin. Antisense approaches have been highly effective in bringing about qualitative changes in starch, although the type of starch produced has not been fully predictable. Great progress has been made over the last several years in understanding amylopectin biosynthesis. However, the intricacies of the interactions between the various isoforms of starch synthase, starch branching enzyme, and debranching enzymes continue to dog attempts at rational starch design based on the functional properties desired in the final product. Part of the difficulty lies as well in the elaborate nature of the starch structure itself, our limited ability to determine the structure fully as can be done for proteins or nucleic acids, and the complex relationship between gelatinization and retrogradation thermodynamics and rheology and starch structure. The most success in producing novel carbohydrates has been achieved by the transfer of fructan biosynthesis from plants where it is common to other crops, particularly sugar beet and potato, where these carbohydrates are not normally found. Fructans attract increasing interest as functional foods.

Ultimately, the position of starch and other plant carbohydrates as "green," greenhouse-neutral replacements for petrochemicals offers great potential for the farming of crops containing specialized storage products. In addition to nonfood uses, applications ranging from fat substitutes to fiber (as "resistant starch") in novel foods promise to create new markets for plant carbohydrates and new demand for their creation. At present, however, the rejection by the public of genetic engineering in general, widespread in Europe and growing in North America and elsewhere despite the environmental benefits it can bring to agriculture, is discouraging growth in the production of transgenic carbohydrates. It remains to be seen whether modified starch and carbohydrates produced in transgenic plants but destined for nonfood products can escape such pressures.

REFERENCES

1. P Chinachoti. Carbohydrates: functionality of foods. Am J Clin Nutr 61:922S–929S, 1995.
2. DJ Manners. Starch. In: DJ Manners, ed. Biochemistry of Storage Carbohydrates in Green Plants. London: Academic Press, 1985, pp 149–203.
3. DJ Manners. Some aspects of the structure of starch and glycogen. Denpun Kagaku 36:311–323, 1989.
4. W Manners, DD Muir. Structure and chemistry of the starch granule. In: W Manners, DD Muir, eds. The Biochemistry of Plants. Vol 3. New York: Academic Press, 1980, pp 321–369.
5. S Ball, H-P Guan, M James, A Myers, P Keeling, G Mouille, A Buléon, P Colonna, J Preiss. From glycogen to amylopectin: a model for the biogenesis of the plant starch granule. Cell 86:349–352, 1996.
6. G Mouille, ML Maddelein, N Libessart, P Talaga, A Decq, B Delrue, S Ball. Preamylopectin processing—a mandatory step for starch biosynthesis in plants. Plant Cell 8:1353–1366, 1996.
7. J-L Jane, TK Kasemsuwan, S Leas, H Zobel, JF Robyt. Anthology of starch granule morphology by scanning electron microscopy. Starch/Stärke 46:121–129, 1994.
8. AML McDonald, JR Stark, WR Morrison, RP Ellis. The composition of starch granules from developing barley genotypes. J Cereal Sci 13:93–112, 1991.

9. AH Schulman, RF Tester, H Ahokas, WR Morrison. The effect of the shrunken endosperm mutation *shx* on starch granule development in barley seeds. J Cereal Sci 19:49–55, 1994.

10. RF Tester, WR Morrison, AH Schulman. Swelling and gelatinization of cereal starches. V. Risø mutants of Bomi and Carlsberg II barley cultivars. J Cereal Sci 17:1–9, 1993.

11. J Craig, JR Lloyd, K Tomlinson, L Barber, A Edwards, TL Wang, C Martin, CL Hedley, AM Smith. Mutations in the gene encoding starch synthase II profoundly alter amylopectin structure in pea embryos. Plant Cell 10:413–426, 1998.

12. A Buléon, C Gérard, C Riekel, R Vuong, H Chanzy. Details of the crystalline ultrastructure of C-starch granules revealed by synchrotron microfocus mapping. Macromolecules 31:6605–6610, 1998.

13. DJ Gallant, B Bouchet, PM Baldwin. Microscopy of starch: evidence of a new level of granule organization. Carbohydr Polym 32:177–191, 1997.

14. GT Oostergetel, EJF van Bruggen. The crystalline domains in potato starch granules are arranged in a helical fashion. Carbohydr Polym 21:7–12, 1993.

15. TY Bogracheva, VJ Morris, SG Ring, CL Hedley. The granular structure of C-type pea starch and its role in gelatinization. Biopolymers 45:323–332, 1998.

16. J Jane, JJ Shen. Internal structure of the potato starch granules revealed by chemical gelatinization. Carbohydr Res 247:279–290, 1993.

17. S Hizukuri. Relationship between the distribution of the chain-length of amylopectin and the crystalline structure of the starch granules. Carbohydr Res 141:295–306, 1985.

18. S Hizukuri, Y Maehara. Fine structure of wheat amylopectin: the mode of A to B chain binding. Carbohydr Res 206:145–159, 1990.

19. P Colonna, V Leloup, A Buléon. Limiting factors of starch hydrolysis. Eur J Clin Nutr 46:S17–S32, 1992.

20. RF Tester, SJJ Debon, MD Sommerville. Annealing of maize starch. Carbohydr Polym 42:287–299, 2000.

21. MD Bennett, JB Smith, I Barclay. Early seed development in the Triticeae. Philos Trans R Soc Lond Ser B 272:199–277, 1975.

22. O-A Olsen. Endosperm developments. Plant Cell 10:485–488, 1998.

23. RC Brown, BE Lemmon, O-A Olsen. Endosperm development in barley: microtubule involvement in the morphogenetic pathway. Plant Cell 6:1241–1252, 1994.

24. M-A D'Aousta, S Yellea, B Nguyen-Quoc. Antisense inhibition of tomato fruit sucrose synthase decreases fruit setting and the sucrose unloading capacity of young fruit. Plant Cell 11:2407–2418, 1999.

25. R Zrenner, M Salanoubat, L Willmitzer, U Sonnewald. Evidence of the crucial role of sucrose synthase for sink strength using transgenic potato plants (*Solanum tuberosum* L.). Plant J 7:97–107, 1995.

26. PS Chourey, EW Taliercio, SJ Carlson, YL Ruan. Genetic evidence that the two isozymes of sucrose synthase present in developing maize endosperm are

critical, one for cell wall integrity and the other for starch biosynthesis. Mol Gen Genet 259:88–96, 1998.

27. L Elling, MR Kula. Purification of UDP-glucose pyrophosphorylase from germinated barley (malt). J Biotechnol 34:157–173, 1994.

28. K Eimert, P Villand, A Kilian, LA Kleczkowski. Cloning and characterization of several cDNAs for UDP-glucose pyrophosphorylase from barley. Gene 170:227–232, 1996.

29. J Preiss. ADPglucose pyrophosphorylase: basic science and applications in biotechnology. Biotechnol Annu Rev 2:259–279, 1996.

30. LA Kleczkowski. Back to the drawing board: redefining starch synthesis in cereals. Trends Plant Sci 1:363–364, 1996.

31. TW Greene, LC Hannah. Maize endosperm ADP-glucose pyrophosphorylase SHRUNKEN2 and BRITTLE2 subunit interactions. Plant Cell 10:1295–1306, 1998.

32. MJ Giroux, LC Hannah. ADP-glucose pyrophosphorylase in *shrunken-2* and *brittle-2* mutants of maize. Mol Gen Genet 243:400–408, 1994.

33. GW Singletary, R Banisadr, PL Keeling. Influence of gene dosage on carbohydrate synthesis and enzymatic activities in endosperm of starch-deficient mutants of maize. Plant Physiol 113:293–304, 1997.

34. K Denyer, J Foster, AM Smith. The contributions of adenosine 5'-diphosphoglucose pyrophosphorylase and starch-branching enzyme to the control of starch synthesis in developing pea embryos. Planta 197:57–62, 1995.

35. JC Shannon, FM Pien, H Cao, KC Liu. Brittle-1, an adenylate translocator, facilitates transfer of extraplastidial synthesized ADP-glucose into amyloplasts of maize endosperms. Plant Physiol 117:1235–1252, 1998.

36. JC Shannon, F-M Pien, K-C Liu. Nucleotides and nucleotide sugars in developing maize endosperms. Plant Physiol 110:835–843, 1996.

37. K Denyer, F Dunlap, T Thorbjørnsen, P Keeling, AM Smith. The major form of ADP-glucose pyrophosphorylase in maize endosperm is extra-plastidial. Plant Physiol 112:779–785, 1996.

38. T Thorbjørnsen, P Villand, K Denyer, O-A Olsen, AM Smith. Distinct forms of ADPglucose pyrophosphorylase occur inside and outside the amyloplasts in barley endosperm. Plant J 10:243–250, 1996.

39. T Thorbjørnsen, P Villand, LA Kleczkowski, O-A Olsen. A single gene encodes two different transcripts for the ADP-glucose pyrophosphorylase small subunit from barley (*Hordeum vulgare*). Biochem J 313:133–138, 1996.

40. E Tauberger, AR Fernie, M Emmermann, A Renz, J Kossmann, L Willmitzer, RN Trethewey. Antisense inhibition of plastidial phosphoglucomutase provides compelling evidence that potato tuber amyloplasts import carbon from the cytosol in the form of glucose-6-phosphate. Plant J 23:43–53, 2000.

41. RL Whistler, JR Daniel. Molecular structure of starch. In: RL Whistler, JR Daniel, eds. Starch: Chemistry and Technology. 2nd ed. New York: Academic Press, 1984, pp 153–182.

42. Y Kano, N Kunitake, T Karakawa, H Taniguchi, M Nakamura. Structural changes in starch molecules during the malting of barley. Agric Biol Chem 45:1969–1975, 1981.

43. NL Kent, AD Evers. Technology of Cereals. 4th ed. Oxford: Elsevier Science, 1994.
44. C Martin, AM Smith. Starch biosynthesis. Plant Cell 7:971–985, 1995.
45. AM Smith. Making starch. Curr Opin Plant Biol 2:223–229, 1999.
46. T Nakamura, M Yamamori, H Hirano, S Hidaka, T Nagamine. Production of *waxy* (amylose-free) wheats. Mol Gen Genet 248:253–259, 1995.
47. Y-J Wang, P White, L Pollak, J Jane. Characterization of starch structures of 17 maize endosperm mutant genotypes with Oh43 inbred line background. Cereal Chem 70:171–179, 1993.
48. ZY Wang, FQ Zheng, GZ Shen, JP Gao, DP Snustad, MG Li, JL Zhang, MM Hong. The amylose content in rice endosperm is related to the post-transcriptional regulation of the *waxy* gene. Plant J 7:613–622, 1995.
49. PL Vrinten, T Nakamura. Wheat granule-bound starch synthase I and II are encoded by separate genes that are expressed in different tissues. Plant Physiol 122:255–264, 2000.
50. RJ Mason-Gamer, CF Weil, EA Kellogg. Granule-bound starch synthase: structure, function, and phylogenetic utility. Mol Biol Evol 15:1658–1673, 1998.
51. RB Frydman, CE Cardini. Biosynthesis of phytoglycogen from adenosine diphosphate D-glucose in sweet corn. Biochem Biophys Res Commun 14: 353–357, 1964.
52. C Mu-Forster, R Huang, JR Powers, RW Harriman, M Knight, GW Singletary, PL Keeling, BP Wasserman. Physical association of starch biosynthetic enzymes with starch granules of maize endosperm. Plant Physiol 111:821–829, 1996.
53. J Marshall, C Sidebottom, M Debet, C Martin, AM Smith, A Edwards. Identification of the major starch synthase in the soluble fraction of potato tubers. Plant Cell 8:1121–1135, 1996.
54. Z Li, G Mouille, B Kosar-Hashemi, S Rahman, B Clarke, KR Gale, R Appels, MK Morell. The structure and expression of the wheat starch synthase III gene. Motifs in the expressed gene define the lineage of the starch synthase III gene family. Plant Physiol 123:613–624, 2000.
55. M Gao, J Wanat, PS Stinard, MG James, AM Myers. Characterization of *dull1*, a maize gene coding for a novel starch synthase. Plant Cell 10:399–412, 1998.
56. AH Schulman, S Tomooka, A Suzuki, P Myllärinen, S Hizukuri. Structural analysis of starch from normal and *shx* (shrunken endosperm) barley (*Hordeum vulgare* L.). Carbohydr Res 275:361–369, 1995.
57. A Edwards, A Borthakur, S Bornemann, J Venail, K Denyer, D Waite, D Fulton, A Smith, C Martin. Specificity of starch synthase isoforms from potato. Eur J Biochem 266:724–736, 1999.
58. JR Lloyd, V Landschutze, J Kossmann. Simultaneous antisense inhibition of two starch-synthase isoforms in potato tubers leads to accumulation of grossly modified amylopectin. Biochem J 338:515–521, 1999.
59. A Edwards, DC Fulton, CM Hylton, SA Jobling, M Gidley, U Rössner, C Martin, AM Smith. A combined reduction in the activity of starch synthases

II and III of potato has novel effects on the starch of tubers. Plant J 17:251–261, 1999.

60. J Preiss, MN Sivak. Biochemistry, molecular biology and regulation of starch synthesis. Genet Eng (N Y) 20:177–223, 1998.

61. C Sun, P Sathish, S Ahlandsberg, A Deiber, C Jansson. Identification of four starch branching enzymes in barley endosperm: partial purification of forms I, IIa and IIb. New Phytol 137:215–222, 1997.

62. C Sun, P Sathish, S Ahlandsberg, C Jansson. The two genes encoding starch-branching enzymes IIa and IIb are differentially expressed in barley. Plant Physiol 118:37–49, 1998.

63. MK Morell, A Blennow, B Kosar-Hashemi, MS Samuel. Differential expression and properties of starch branching enzyme isoforms in developing wheat endosperm. Plant Physiol 113:201–208, 1997.

64. RA Burton, JD Bewley, AM Smith, MK Bhattacharyya, H Tatge, S Ring, V Bull, WDO Hamilton, C Martin. Starch branching enzymes belonging to distinct enzyme families are differentially expressed during pea embryo development. Plant J 7:3–15, 1995.

65. SA Jobling, GP Schwall, RJ Westcott, CM Sidebottom, M Debet, MJ Gidley, R Jeffcoat, R Safford. A minor form of starch branching enzyme in potato (*Solanum tuberosum* L.) tubers has a major effect on starch structure: cloning and characterisation of multiple forms of SBE A. Plant J 18:163–171, 1999.

66. K Mizuno, T Kawasaki, H Shimada, H Satoh, E Kobayashi, S Okumura, Y Arai, T Baba. Alteration of the structural properties of starch components by the lack of an isoform of starch branching enzyme in rice seeds. J Biol Chem 268:19084–19091, 1993.

67. KN Kim, DK Fisher, M Gao, MJ Guiltinan. Molecular cloning and characterization of the Amylose-Extender gene encoding starch branching enzyme IIB in maize. Plant Mol Biol 38:945–956, 1998.

68. MG James, DS Robertson, AM Myers. Characterization of the maize gene *sugary1*, a determinant of starch composition in kernels. Plant Cell 7:417–429, 1995.

69. SC Zeeman, T Umemoto, WL Lue, P Au-Yeung, C Martin, AM Smith, J Chen. A mutant of *Arabidopsis* lacking a chloroplastic isoamylase accumulates both starch and phytoglycogen. Plant Cell 10:1699–1712, 1998.

70. G Palmgren. Transgenic plants: environmentally safe factories of the future. Trends Genet 13:348, 1997.

71. DS Brar, T Ohtani, H Uchimiya. Genetically engineered plants for quality improvement. Biotech Genet Eng Rev 13:167–169, 1996.

72. B Mazur, E Krebbers, S Tingey. Gene discovery and product development for grain quality traits. Science 285:372–375, 1999.

73. J Kossmann, J Lloyd. Understanding and influencing starch biochemistry. Crit Rev Biochem Mol Biol 35:141–196, 2000.

74. T Roitsch. Source-sink regulation by sugar and stress. Curr Opin Plant Biol 2:198–206, 1999.

75. U Sonnewald, M Brauer, A von Schaewen, M Stitt, L Willmitzer. Transgenic tobacco plants expressing yeast-derived invertase in either the cytosol, vac-

uole, or apoplast: a powerful tool for studying sucrose metabolism and sink/ source interactions. Plant J 1:95–106, 1991.

76. A von Schaewen, M Stitt, R Schmidt, U Sonnewald, L Willmitzer. Expression of a yeast-derived invertase in the cell wall of tobacco and *Arabidopsis* plants leads to accumulation of carbohydrate and inhibition of photosynthesis and strongly influences growth and phenotype of transgenic tobacco plants. EMBO J 9:3033–3044, 1990.

77. D Bussis, D Heineke, U Sonnewald, L Willmitzer, K Raschke, HW Heldt. Solute accumulation and decreased photosynthesis in leaves of potato plants expressing yeast-derived invertase either in the apoplast, vacuole or cytosol. Planta 202:126–136, 1997.

78. T Jelitto, U Sonnewald, L Willmitzer, M Hajirezcai, M Stitt. Inorganic pyrophosphate content and metabolites in potato and tobacco plants expressing *E. coli* pyrophosphatase in their cytosol. Planta 188:238–244, 1992.

79. B Müller-Röber, U Sonnewald, L Willmitzer. Inhibition of the ADP-glucose pyrophosphorylase in transgenic potatoes leads to sugar-storing tubers and influences tuber formation and expression of tuber storage protein genes. EMBO J 11:1229–1238, 1992.

80. AY Borovkov, PE McClean, JR Sowokinos, SH Ruud, GA Secord. Effect of expression of UDP-glucose pyrophosphorylase ribozyme and antisense RNAs on the enzyme activity and carbohydrate composition of field-grown transgenic potato plants. J Plant Physiol 147:644–652, 1996.

81. S Greiner, T Rausch, U Sonnewald, K Herbers. Ectopic expression of a tobacco invertase inhibitor homolog prevents cold-induced sweetening of potato tubers. Nat Biotechnol 17:708–711, 1999.

82. U Sonnewald, MR Hajirezaei, J Kossmann, A Heyer, RN Trethewey, L Willmitzer. Increased potato tuber size resulting from apoplastic expression of a yeast invertase. Nat Biotechnol 15:794–797, 1997.

83. DM Stark, KP Timmerman, GF Barry, J Preiss, GM Kishore. Regulation of the amount of starch in plant tissues by ADP glucose pyrophosphorylase. Science 258:287–292, 1992.

84. LJ Sweetlove, MM Burrel, T ap Rees. Characterization of transgenic potato (*Solanum tuberosum*) tubers with increased ADPglucose pyrophosphorylase. Biochem J 320:487–492, 1996.

85. LJ Sweetlove, MM Burrel, T ap Rees. Starch metabolism in tubers of transgenic potato (*Solanum tuberosum*) with increased ADPglucose pyrophosphorylase. Biochem J 320:493–498, 1996.

86. K Herbers, U Sonnewald. Molecular determinants of sink strength. Curr Opin Plant Biol 1:207–216, 1998.

87. K Herbers, E Tacke, M Hazirezaei, KP Krause, M Melzer, W Rohde, U Sonnewald. Expression of a luteoviral movement protein in transgenic plants leads to carbohydrate accumulation and reduced photosynthetic capacity in source leaves. Plant J 12:1045–1056, 1997.

88. R Zrenner, K Schuler, U Sonnewald. Soluble acid invertase determines the hexose-to-sucrose ratio in cold-stored potato tubers. Planta 198:246–252, 1996.

89. EM Klann, AB Hall, AB Bennnet. Antisense acid invertase (TIV1) gene alters soluble sugar composition and size in transgenic tomato fruit. Plant Physiol 112:1321–1330, 1996.

90. A Beaujean, C Ducrocq-Assaf, RS Sangwan, G Lilius, L Bülow, BS Sangwan-Norreel. Engineering direct fructose production in processed potato tubers by expressing a bifunctional alpha-amylase/glucose isomerase gene complex. Biotechnol Bioeng 70:9–16, 2000.

91. R Montalvo-Rodriguez, C Haseltine, K Huess-LaRossa, T Clemente, J Soto, P Staswick, P Blum. Autohydrolysis of plant polysaccharides using transgenic hyperthermophilic enzymes. Biotechnol Bioeng 70:151–159, 2000.

92. AM Nuutila, A Ritala, RW Skadsen, L Mannonen, V Kauppinen. Expression of fungal thermotolerant endo-1,4-beta-glucanase in transgenic barley seeds during germination. Plant Mol Biol 41:777–783, 1999.

93. MJ Cho, JH Wong, C Marx, W Jiang, PG Lemaux, BB Buchanan. Overexpression of thioredoxin h leads to enhanced activity of starch debranching enzyme (pullulanase) in barley grain. Proc Natl Acad Sci USA 96:14641–14646, 1999.

94. OJ Goddijn, TC Verwoerd, E Voogd, RW Krutwagen, PT de Graaf, K van Dun, J Poels, AS Ponstein, B Damm, J Pen. Inhibition of trehalase activity enhances trehalose accumulation in transgenic plants. Plant Physiol 113:181–190, 1997.

95. O Goddijn, S Smeekens. Sensing trehalose biosynthesis in plants. Plant J 14:143–146, 1998.

96. JHM Hovenkamp-Hermelink, E Jacoben, AS Ponstein, RGF Visser, GH Vos-Scheperkeuter, EW Bijmolt, JN de Vries, B Witholt, WJ Feenstra. Isolation of an amylose-free mutation of the potato (*Solanum tuberosum* L.). Theor Appl Genet 75:217–221, 1987.

97. AG Kuipers, WJ Soppe, E Jacobsen, RG Visser. Field evaluation of transgenic potato plants expressing an antisense granule-bound starch synthase gene: increase of the antisense effect during tuber growth. Plant Mol Biol 26:1759–1773, 1994.

98. AG Kuipers, WJ Soppe, E Jacobsen, RG Visser. Factors affecting the inhibition by antisense RNA of granule-bound starch synthase gene expression in potato. Mol Gen Genet 246:745–755, 1995.

99. SN Salehuzzaman, E Jacobsen, RG Visser. Isolation and characterization of a cDNA encoding granule-bound starch synthase in cassava (*Manihot esculenta* Crantz) and its antisense expression in potato. Plant Mol Biol 23:947–962, 1993.

100. R Terada, M Nakajima, M Isshiki, RJ Okagaki, SR Wessler, K Shimamoto. Antisense *waxy* genes with highly active promoters effectively suppress *waxy* gene expression in transgenic rice. Plant Cell Physiol 41:881–888, 2000.

101. C Kiribuchiotobe, T Nagamine, T Yanagisawa, M Ohnishi, I Yamaguchi. Production of hexaploid wheats with waxy endosperm character. Cereal Chem 74:72–74, 1998.

102. CK Shewmaker, CD Boyer, DP Wiesenborn, BD Thompson, MR Boersig, JV Oakes, DM Stalker. Expression of *Escherichia coli* glycogen synthase in the

tubers of transgenic potatoes (*Solanum tuberosum*) results in a highly branched starch. Plant Physiol 104:1159–1166, 1994.

103. AJ Kortstee, A Vermeesch, M., BJ de Vries, E Jacobsen, RG Visser. Expression of the *Escherichia coli* branching enzyme in tubers of amylose-free potato leads to an increased branching degree of the amylopectin. Plant J 10: 83–90, 1996.

104. NR Merritt. A new strain of barley with starch of high amylose content. J Inst Brew 73:583–585, 1967.

105. A-C Salomonsson, B Sundberg. Amylose content and chain profile of amylopectin from normal, high amylose, and waxy barleys. Starch/Stärke 46:325–328, 1994.

106. GP Schwall, R Safford, RJ Westcott, R Jeffcoat, A Tayal, YC Shi, MJ Gidley, SA Jobling. Production of very-high-amylose potato starch by inhibition of SBE A and B. Nat Biotechnol 18:551–554, 2000.

107. R Lorberth, G Ritte, L Willmitzer, J Kossmann. Inhibition of a starch-granule-bound protein leads to modified starch and repression of cold sweetening. Nat Biotechnol 16:473–477, 1998.

108. JV Oakes, CK Shewmaker, DM Stalker. Production of cyclodextrins, a novel carbohydrate, in the tubers of transgenic potato plants. Biotechnology 9:982–986, 1991.

109. MC Tarczynski, RG Jensen, HJ Bohnert. Expression of a bacterial *mtlD* gene in transgenic tobacco leads to production and accumulation of mannitol. Proc Natl Acad Sci USA 89:2600–2604, 1992.

110. B Shen, RG Jensen, HJ Bohnert. Increased resistance to oxidative stress in transgenic plants by targeting mannitol biosynthesis to chloroplasts. Plant Physiol 113:1177–1183, 1997.

111. CJ Pollock, AJ Cairns. Fructan metabolism in grasses and cereals. Annu Rev Plant Physiol Plant Mol Biol 42:77–101, 1991.

112. J Tyynelä, M Stitt, A Lönneborg, S Smeekens, AH Schulman. Metabolism of starch synthesis in developing grains of the *shx* shrunken mutant of barley (*Hordeum vulgare*). Physiol Plant 93:77–84, 1995.

113. AG Heyer, JR Lloyd, J Kossman. Production of modified polymeric carbohydrates. Curr Opin Biotechnol 10:169–174, 1999.

114. E Menne, N Guggenbuhl, MB Roberfroid. Fn-type chicory inulin hydrolysate has a prebiotic effect in humans. J Nutr 130:1197–1199, 2000.

115. MB Roberfroid. Chicory fructooligosaccharides and the gastrointestinal tract. Nutrition 16:677–679, 2000.

116. MJ Ebskamp, IM van der Meer, BA Spronk, PJ Weisbeek, SC Smeekens. Accumulation of fructose polymers in transgenic tobacco. Biotechnology 12: 272–275, 1994.

117. M Röber, K Geider, B Müller-Röber, L Willmitzer. Synthesis of fructans in tubers of transgenic starch-deficient potato plants does not result in an increased allocation of carbohydrates. Planta 199:528–536, 1996.

118. N Sprenger, K Bortlik, A Brandt, T Boller, A Wiemken. Purification, cloning, and functional expression of sucrose:fructan 6-fructosyltransferase, a key en-

zyme of fructan synthesis in barley. Proc Natl Acad Sci USA 400:355–358, 1995.

119. N Sprenger, L Schellenbaum, K van Dun, T Boller, A Wiemken. Fructan synthesis in transgenic tobacco and chicory plants expressing barley sucrose: fructan 6-fructosyltransferase. FEBS Lett 400:355–358, 1997.

120. I Vijn, A van Dijken, N Sprenger, K van Dun, P Weisbeek, A Wiemken, S Smeekens. Fructan of the inulin neoseries is synthesized in transgenic chicory plants (*Chicorium intybus* L.) harbouring onion (*Allium cepa* L.) fructan: fructan 6G-fructosyltransferase. Plant J 11:387–398, 1997.

121. EM Hellwege, S Czapla, A Jahnke, L Willmitzer, AG Heyer. Transgenic potato (*Solanum tuberosum*) tubers synthesize the full spectrum of inulin molecules naturally occurring in globe artichoke (*Cynara scolymus*) roots. Proc Natl Acad Sci USA 97:8699–8704, 2000.

122. IM van der Meer, AJ Koops, JC Hakkert, AJ van Tunen. Cloning of the fructan biosynthesis pathway of Jerusalem artichoke. Plant J 15:489–500, 1998.

123. R Sévenier, RD Hall, IM van der Meer, HJ Hakkert, AJ van Tunen, AJ Koops. High level fructan accumulation in a transgenic sugar beet. Nat Biotechnol 16:843–846, 1998.

124. AW MacGregor, GB Fincher. Carbohydrates of the barley grain. In: AW MacGregor, GB Fincher, eds. Barley: Chemistry and Technology. St. Paul, MN: American Association of Cereal Chemists, 1993, pp 73–130.

125. S-H Yun, NK Matheson. Structures of the amylopectins of waxy, normal, amylose-extender, and wx:ae genotypes and of the phytoglycogen of maize. Carbohydr Res 243:307–321, 1993.

12

Improving the Nutritional Quality and Functional Properties of Seed Proteins by Genetic Engineering

Peter R. Shewry
IACR–Long Ashton Research Station, Bristol, United Kingdom

I. INTRODUCTION

The protein content of seeds varies widely, from about 10–12% in cereals to 20–25% in pulses and oilseeds and up to 50% in soybean. It can be calculated that the total protein yield of the major seed crops amounts to over 350 million tonnes per annum, with cereals accounting for about 70% of the total (1). Seeds therefore form the major source of protein for the nutrition of humans and livestock. In addition, seed proteins are largely responsible for the processing properties of many seeds including the use of wheat for making bread and other processed foods such as pasta and noodles. The impact of seed proteins on the nutritional and processing quality of seeds has provided a major stimulus to their study over a period exceeding 250 years since Beccari (2) first described the isolation of wheat gluten.

Seed proteins also provided an attractive model system to study gene structure and expression, as their synthesis is tightly regulated with high levels occurring in specific seed tissues and at specific stages of development. Consequently, complementary DNAs (cDNAs) and genes for seed proteins were among the earliest to be cloned from plants (3), providing a basis for longer term strategic studies of the structures and functional (i.e., processing) properties of the encoded proteins. This has led over the past few years to success in improving aspects of seed quality by genetic engineering, although no improved varieties have so far reached the level of commercial production.

Furthermore, the capacity of seeds to synthesize and store high levels of protein, combined with well-established systems for seed production, harvesting, storage, and fractionation, has led to the use of seeds as bioreactors to produce novel proteins and peptides, notably high value products for biomedical applications. These form the subject of other chapters in this volume and I will, therefore, focus on the manipulation of seed proteins to improve their properties for traditional end uses (i.e., food and feed). However, before proceeding with this I will briefly discuss the structures of seed proteins and their impact on quality.

II. SEED PROTEINS

Seed proteins can be broadly classified into two groups, storage and nonstorage. Storage proteins can be defined as proteins that have no biological role except to provide a store of carbon, nitrogen, and sulfur that is mobilized during germination to support seedling growth. The amounts of storage proteins present vary between seeds of different species, accounting for about half of the total seed proteins in cereals but a higher proportion in more protein-rich oilseeds and legumes. Furthermore, the amounts also vary with the nutritional status of the plant, with high levels of nutrient availability resulting in increased accumulation. Storage proteins are invariably deposited in discrete protein bodies, where they form dense masses and do not interfere with essential cellular processes. They usually accumulate in cotyledonary and/or endosperm tissues, although storage in the perisperm or hypocotyl may also occur (4).

Storage proteins are usually classified into four groups on the basis of their solubility properties and sedimentation coefficients (S_{20w}) following a system and using a nomenclature standardized by Osborne (5).

III. GLOBULIN STORAGE PROTEINS

The most widespread storage proteins are globulins, which are defined by their solubility in dilute salt solutions. These comprise 7–8S and 11–12S types that are often called vicilins and legumes, respectively, reflecting their characteristic presence in legume seeds. The 7–8S globulins are typically trimeric molecules with three subunits of M_r about 40–50,000 forming a protein of M_r about 150–190,000 (6). However, in some species posttranslational proteolysis and/or glycosylation of the subunit chains results in a much more complex spectrum of components. For example, pea contains unprocessed subunits of M_r about 50,000 and a range of smaller components (M_r 12,000, 13,000, 16,000, 19,000, 25,000, 30,000, 33,000, and 34,000), the latter all being derived from M_r 47,000 precursor subunits (6). The 7–8S globulin storage proteins occur in many but not all legumes, in other dicotyledonous plants such as cotton and cocoa (7), and in the embryos and aleurone cells of cereals (8).

The 11–12S globulins differ from the 7–8S globulins in being hexameric, typically comprising six subunits of M_r about 60,000. Posttranslational proteolysis results in cleavage of the subunits into acidic (also called α or A) chains of M_r about 40,000 and basic (β, B) chains of M_r about 20,000, which remain associated by a single disulfide bond (6). Further proteolysis does not occur and glycosylation is rare. The 11–12S globulins are widely distributed in legumes and other dicotyledonous seeds (e.g., composites, crucifers, cucurbits) with related proteins forming the major storage protein fraction in oats and rice and minor components in wheat (7,9,10).

Although they have little sequence homology, the 7–8S and 11–12S globulins have similar three-dimensional structures (11,12), which may facilitate packing in the protein bodies, and appear to be derived from a common ancestral protein (12–14).

IV. PROLAMINS

Prolamins are characterized by their insolubility in water or dilute salt solutions but solubility in alcohol-water mixtures. They are major storage proteins in most cereal species (except oats and rice, where they are minor components) and vary widely in their structures and properties (15–19). The individual components are characterized by high contents of proline and glutamine (from about 30 to 70% of the total residues, respectively) and the presence of repeated amino acid sequences, while some individual subunits may be assembled into oligomers or polymers stabilized by interchain disulfide bonds.

V. 2S ALBUMINS

These are typically low-M_r proteins derived from single polypeptide chains that are posttranslationally cleaved to give subunits of M_r about 4–5000 and 8–9000 that remain associated by two interchain disulfide bonds (20). They are widespread in dicotyledonous plants (e.g., legumes, crucifers, composites), where they occur together with 7–8S and/or 11–12S globulins.

VI. NUTRITIONAL QUALITY OF SEED PROTEINS

Animals are able to synthesize only about half of the 20 amino acids commonly found in proteins, the remainder being required in their diet and hence termed "essential." If only one of these is limiting, the remaining amino acids cannot be utilized but are broken down and excreted, leading to poor feed conversion and environmental pollution. This is not usually a problem with ruminants, in which the rumen bacteria provide all 20 amino acids, but is a problem for nonruminants such as pigs and poultry.

Comparison of the amino acid compositions of seeds with the requirements of essential amino acids for humans and nonruminant livestock shows that both legumes and cereals are deficient in one or more amino acids. Furthermore, these deficiencies are determined by the amino acid compositions of the storage protein fractions.

The prolamins of wheat, barley, and maize contain low levels of lysine (≈ 1 mol % or less), and this is the first limiting amino acid in all three species. The prolamins (zeins) of maize also contain little or no tryptophan, which is the second limiting amino acid in this species, and all three also contain low levels of threonine. In contrast, the 7S and 11S globulins of legumes are severely deficient in cysteine and methionine, resulting in deficiencies in these amino acids in the whole seeds. Because cereal and legume seeds are largely complementary in their contents of essential amino acids, they are often used in appropriate mixtures to provide a balanced diet of essential amino acids. Nevertheless, there is considerable interest in producing high-lysine cereals and high-sulfur legumes to provide more flexibility in producing low-cost diets for livestock.

VII. HIGH-LYSINE CEREALS

Interest in improving the nutritional quality of cereals was stimulated by the discovery in the 1960s and 1970s of a range of high-lysine mutants in maize, sorghum, and barley. The first of these were the *opaque2* (21) and *floury2* (22) mutants of maize, followed by the *lys* gene of Hiproly barley (23) and *h1* mutant of sorghum (24). These are all spontaneously occurring mutant genes; a range of other spontaneous high-lysine genes were subsequently identified in maize (17), and a range of chemical and physical mutagens were used to generate further high-lysine genes in sorghum (25) and barley (26,27). Despite initial optimism, only one of these mutant genes, *opaque2* of maize, has been successfully exploited in plant breeding programs, using genetic modifiers to convert the soft starchy endosperm to a normal type (17). Although the other mutants have failed to be exploited because of associated deleterious effects on yield, analysis of their composition and molecular basis has contributed to our understanding of the regulation of seed protein biosynthesis and allowed us to make more informed decisions on strategies for improving quality.

The high-lysine phenotype of most mutant lines is due, at least in part, to a reduction in the proportion of the lysine-poor prolamins and increases in other more lysine-rich proteins. It is, therefore, tempting to suggest that high-lysine cereals could be engineered by down-regulation of prolamin gene expression using antisense or cosuppression technology, as described for 2S albumins (napins) and globulins in transgenic oilseed rape (28,29). In this work, antisense technology was used to down-regulate either napin or 12S globulin gene expression in developing seeds, resulting in compensatory increases in the remaining protein group. However, in all high-lysine cereals the decreased prolamin content is associated with reduced starch synthesis and hence lower yield, and this has proved difficult to separate from the high-lysine character by plant breeding. Consequently, reduction in prolamins is not generally considered to be a valid strategy for engineering high-lysine cereals. Instead, three different strategies have been proposed.

A.   Manipulation of the Amino Acid Composition of Prolamins

Wallace et al. (30) designed modified maize prolamins (zeins) containing single and double substitutions of lysine for neutral amino acids; insertions of oligopeptides (five to eight residues) rich in lysine, threonine, and tryptophan; and insertion of an M_r 17,000 peptide derived from the SV40 VP2 protein. All except for the latter (included as a control) were synthesized and aggregated into dense particles when the messenger RNA (mRNA) was injected into egg cells of the frog *Xenopus*, indicating that the proteins should also be stable in transgenic cereals. However, the drawback of this strategy is that zeins account for about half of the total protein in the maize seed and appear to be encoded by up to 100 genes (17). Consequently, it would be necessary to express the mutant zeins under control of a highly active promoter and possibly also to combine this with down-regulation of the endogenous wild-type zeins. A similar strategy could be applied to other cereals.

B.   Expression of Specific Lysine-Rich Proteins

Detailed analyses of the high-lysine barley mutant Hiproly demonstrated that about half of the increase in lysine resulted from elevated levels of four specific lysine-rich proteins, which together account for about 17% of the total grain lysine, compared with about 7% in normal lines (31). These proteins are β-amylase ($\simeq$5.0 g % Lys), protein Z (now known to be a serpin proteinase inhibitor) ($\approx$7.1 g % Lys), and chymotrypsin inhibitors CI-1 ($\approx$9.5 g % Lys) and CI-2 ($\approx$11.5 g % Lys) (31).

CI-1 and CI-2 were both purified from Hiproly, providing a basis for molecular cloning. This demonstrated that both proteins occur in two iso-forms (32) with the major form of CI-2 comprising 84 residues with an M_r of 9,380. This form contains seven lysine residues (8.3 mol %) but no cys-teine residues (and hence no disulfide bonds) (33). The three-dimensional structure of the protein shows a wedge-shaped disc of about 28 $\times$ 27 $\times$ 19 Å with a single α-helix of 3.6 turns and four strands of β-sheet with a left-handed twist (Fig. 1). The reactive (inhibitory) site (Met-59) of CI-2 is present on an exposed nine-residue loop (Gly-54 to Arg-62). Perutz and co-workers (34,35) have shown that it is possible to mutate or extend this loop region, inserting up to 10 glutamine residues at Met-59 in order to explore the role of glutamine repeats in neurodegenerative diseases. More recently,

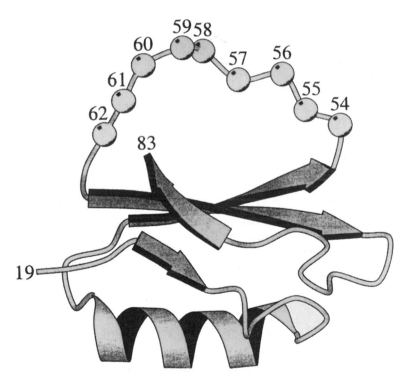

FIGURE 1   Ribbon diagram of the three-dimensional structure of the barley chymotrypsin inhibitor CI-2, showing the residues in the reactive (inhibitory) loop. (From Ref. 34, based on Ref. 82.)

Roesler and Rao (36) have described the design and expression in *E. coli* of five mutant forms of CI-2 containing 20–25 mol % lysine, based mainly on mutation of surface-exposed residues. Although all the mutants appeared to have conformations similar to that of the wild-type protein, as determined by circular dichroism spectroscopy, they had lower thermodynamic stabilities. Nevertheless, at least one of the mutants was considered to be suitable for expression in transgenic plants.

The thionins are another group of lysine-rich seed proteins (37). Rao et al. (38) designed high-lysine analogues of barley α-hordothionine, a 45-residue protein containing five lysines (11 mol %). These included an analogue containing seven additional lysines (i.e., a total of 27 mol %), which was synthesized and shown to exhibit antifungal activity similar to that of the wild-type protein. The authors have not, however, attempted to express any of the mutant proteins in transgenic plants, and such material would perhaps be unlikely to be accepted by consumers or regulatory authorities because of the wide biological activity of thionins (37).

Pea legumin contains about 5 mol % lysine (39), which is similar to the nutritional requirements for humans (5.5 g %) (Table 1). Sindhu et al. (40) reported that pea legumin was stably accumulated in the seeds of transgenic rice plants but did not determine the effect on the lysine content of the grain.

An alternative approach to using naturally occurring high-lysine proteins, or mutant forms of these, is to design high-quality proteins ab initio. Keeler et al. (41) demonstrated the feasibility of this concept by designing low-M_r proteins (3600 to 6700) with an α-helical coiled coil structure and containing up to 43 mol % lysine accompanied by high levels of methionine and in some cases also tryptophan. The proteins were initially synthesized in *E. coli* and one protein containing 31 mol % lysine and 20 mol % methionine was then expressed in tobacco seeds under control of the bean phaseolin and soybean β-conglycinin promoters. Increases in lysine of up to 0.8 mol %, compared with a wild-type level of 2.56 mol %, were observed in the primary transformants and were inherited through three generations. No impact on seed methionine levels was reported, and the work has not so far been extended to cereals or other crop plants.

C. Increased Accumulation of Free Lysine

The most spectacular success in increasing the lysine content of cereals has been achieved by increasing the amount of free rather than protein-bound lysine. This is based on eliminating the feedback regulation that normally limits the pool of free lysine by transforming with genes for feedback-insensitive enzymes from microorganisms. Two enzymes are important in this

TABLE 1 The Essential Amino Acid Contents of Soybean, Maize, Wheat, and Barley Seeds and Their Major Storage Protein Fractions Compared with the WHO Recommended Levels[a]

Amino acid	Soybean			Maize			Wheat			Barley		WHO
	Total seed	7S conglycinin	12S glycinin	Total seed	Ethanol-soluble zein	Ethanol-insoluble zein	Total seed	Gliadins	Glutenins	Total seed	Hordein	
Cysteine	1.3	0.3	1.4	3.1	1.7	5.0	2.6	3.0	2.4	2.9	1.9	⎫ 3.5
Methionine	1.3	0.2	1.8	2.0	1.6	7.9	1.3	1.7	1.4	1.7	1.1	⎭
Lysine	6.4	7.0	4.9	3.5	0	0.1	2.0	0.6	2.3	3.1	0.8	5.5
Isoleucine	4.5	6.4	4.7	3.6	4.4	2.5	3.6	4.3	3.7	3.6	0.4	4.0
Leucine	7.8	10.2	7.2	11.6	23.0	13.9	6.7	6.8	6.6	7.2	7.8	7.0
Phenylalanine	4.9	7.4	5.7	4.9	7.1	3.6	5.1	5.5	4.8	5.5	9.3	⎫ 6.0
Tyrosine	3.1	3.6	4.1	2.3	6.5	6.3	2.6	2.6	3.6	2.7	4.3	⎭
Threonine	3.9	2.8	3.4	3.9	3.1	3.7	2.7	2.2	3.1	3.3	1.7	4.0
Tryptophan	1.3	0.3	1.6	0.9	nd	nd	1.1	0.7	2.1	2.0	nd	1.0
Valine	4.8	5.1	5.1	4.9	3.7	4.6	3.7	4.1	4.2	4.6	3.5	5.0
Histidine	2.5	1.7	2.2	3.2	1.2	4.8	2.2	2.2	2.3	1.9	1.8	—
Notes	[b]	[c]	[d]	[e]	[f]	[g]	[b]	[g]	[g]	[b]	[h]	[i]

[a] nd, not determined; 16 g N is generally regarded as equivalent to 100 g protein and g/16 g N as percentage by weight.

[b] FAO (83) expressed as g/16 g N.

[c] Koshiyama (84) expressed as g/100 g protein.

[d] Bradley et al. (85) expressed as g/100 g protein.

[e] Ewart (86) expressed as g/100 g recovered amino acids.

[f] Misra and Mertz (87) expressed as g/100 g protein. The ethanol-soluble zein fraction contains mainly α-zeins and the ethanol-insoluble fraction β-, γ-, and δ-zeins. Tryptophan is known to be absent from α-zeins, β-zeins, and M_r 27,000 γ-zeins and present at about 1 mol % or less in M_r 16,000 γ-zeins and δ-zeins (17).

[g] Ewart (88) expressed as g/100 g recovered amino acids.

[h] Singh and Sastry (89) expressed as g/16 g N. Data for alcohol-soluble hordeins only.

[i] WHO recommended levels (90) expressed as g/100 g protein. Histidine is regarded as essential for human children but not adults and is not listed in the WHO recommendations.

respect: aspartate kinase (AK), which catalyzes the first reaction in the pathway, and dihydrodipicolinate synthase (DHDPS), which controls the branch point leading to lysine (Fig. 2). Falco et al. (42) initially showed that a twofold increase in total lysine occurred in seeds of oilseed rape (canola) expressing a DHDPS gene from *Corynebacterium* and a fivefold increase occurred in soybean seeds expressing the same enzyme together with an AK gene from *E. coli*. More recent work, described in a patent application (43), has shown that expression of the DHDPS gene in maize under control of the embryo + aleurone-specific globulin-1 promoter resulted in increases in free lysine from 1.4 to 15–27% of total free amino acids, representing increases in total lysine from 2.3 to 3.6–5.3%. In contrast, no effect was observed when the same gene was expressed under control of the starchy endosperm-specific glutelin-2 promoter, which was thought to be due to increased lysine catabolism. It was concluded that down-regulation of lysine catabolism would be required in order to achieve accumulation of lysine in the maize starchy endosperm.

VIII. HIGH-METHIONINE LEGUMES

Legume seeds are deficient in both cysteine and methionine. However, animals are capable of converting methionine to cysteine but not vice versa.

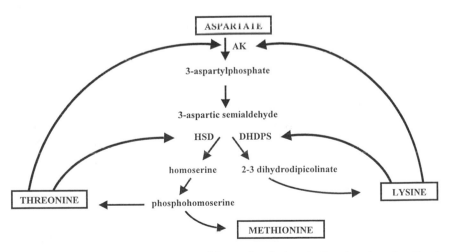

FIGURE 2 The pathway of lysine biosynthesis in plants showing feedback regulatory steps as arrows. AK, aspartate kinase; DHDPS, dihydrodipicolinate synthase.

Consequently, there is interest in increasing the methionine content of sulfur-deficient legume seeds.

A range of sulfur-rich proteins has been characterized from plant seeds including the β-zeins (4.4 mol % Cys, 11.4 mol % Met), γ-zeins (over 7 mol % Cys), and δ-zeins (23–27 mol % Met) of maize (17). In addition, Chui and Falco (44) have described the molecular cloning of a protein that appears to be related to δ-zeins but contains over 40 mol % methionine. Similarly, the M_r 10,000 prolamins of rice contain about 10 mol % Cys and 20 mol % Met (45).

The methionine-rich δ-zein gene has been used to increase the methionine content of maize seeds by up to 30% (46), but neither this nor any other sulfur-rich cereal prolamin genes have so far been used in transgenic legumes.

Instead, the main interest in improving the sulfur amino acid content of legume seeds has been focused on the 2S albumins. Youle and Huang (47) demonstrated that the 2S albumin fraction from Brazil nut (*Bertholletia excelsa*) contained over 17 mol % Met and 13 mol % Cys, and subsequent studies showed the presence of at least six closely related components (48,49). Methionine-rich 2S albumins have since been reported in sunflower, cotton, and amaranthus and cysteine-rich components in quinoa and pea (20).

Expression of the Brazil nut albumins has resulted in significant increases in the methionine content of seeds of oilseed rape, tobacco (by up to 30%) (50,51), *Arabidopsis* (by 20%) (52), narbon bean (*Vicia narbonensis*) (by up to threefold) (53), and soybean (R. Jung, personal communication). However, it is now known that the Brazil nut protein is allergenic to humans (54,55) and the commercial development of seeds expressing this protein has therefore been suspended.

Although a number of other (i.e., methionine-poor) 2S albumins are also allergenic (e.g., from mustards, castor bean, cotton), there is no evidence that this applies to the methionine-rich albumin (SFA8) of sunflower, which contains 16 Met out of 103 total residues (but see note added in proof). Molvig et al. (56) reported that the expression of this protein in seeds of lupin resulted in an increase in methionine by 94%, although there was no effect on total seed sulfur, the increase being at the expense of cysteine (reduced by 12%) and sulfate. Feeding trials with rats showed that the nutritional quality of the seeds was improved significantly, with increases in live weight gain, true protein digestibility, biological value, and net protein utilization. These results demonstrate that the methionine content of legume seeds can be increased by genetic engineering but that the extent of this may ultimately be limited by the availability of sulfur within the seed. Further increases may require the manipulation of the sulfur economy of the plant to increase the transport of sulfur into the developing grain. This will require further basic studies to identify the underlying mechanisms and their regulation.

Other approaches can also be proposed to increase the methionine content of legume seeds, some of which have been tested in seeds of model species or other crop species. Saalbach et al. (53,57,58) showed that a methionine-enriched 7S globulin from *Vicia* was stably accumulated at levels of 1.5 to 2.2% of the total protein in seeds of transgenic tobacco plants, whereas methionine-enriched 11S globulin from the same species did not accumulate and was subsequently shown to be degraded (59). Similarly, Utsumi and co-workers (60,61) have shown that wild-type and methionine-enriched 11S globulins of soybean accumulate stably in seeds of transgenic rice. De Clercq et al. (62) constructed modified 2S albumins from *Arabidopsis* and chimeras comprising parts of the Brazil nut and *Arabidopsis* albumins and expressed them in seeds of tobacco, *Arabidopsis*, and oilseed rape. Down-regulation of the sulfur-poor 11S globulins (cruciferins) in oilseed rape has also been shown to result in compensatory increases in more sulfur-rich proteins (29). Similarly, Kjemtrup et al. (63) demonstrated that mutant phytohemagglutinins of bean containing four additional methionine residues were stably accumulated in seeds of transgenic tobacco although the level was not reported. However, none of these approaches has yet been successfully applied to grain legumes.

IX. SEED PROTEINS AND FUNCTIONALITY

Although some seeds are consumed by humans with little or no processing (e.g., boiled rice), the vast majority are processed by the food industry into a wide range of products. The ability to process seeds is determined by their functional properties, which in turn depend, to a large extent, on the seed proteins. For example, Kinsella (64) has identified 10 functional properties that are conferred by soybean proteins used in food systems: solubility, water absorption and binding, viscosity, gelation, cohesion-adhesion, elasticity, emulsification, fat absorption, flavor binding, and foaming. Similarly, the viscoelastic and cohesive properties conferred by wheat gluten proteins allow flour to be processed into bread, pasta and noodles, other baked goods, and various other food products. Because these are biophysical properties determined by the structures and/or interactions of seed components, they are difficult to define in molecular terms and to manipulate. Nevertheless, good progress has been made with the two most important seed protein systems exploited in food processing, the gluten proteins of wheat and soybean globulins.

X. GEL FORMATION AND EMULSIFICATION PROPERTIES OF SOYBEAN GLOBULINS

Gelation forms the basis for making tofu, which is an important food product in the Far East. Gels result from the formation of an ordered protein network that is capable of holding water. In contrast, emulsification properties are important for stabilizing oil-water mixtures in products such as mayonnaise, dressings, and spreads, and protein emulsifiers usually have amphipathic properties that allow them to orientate at the oil-water interface.

Utsumi and co-workers have explored the molecular basis for the gelation and emulsification properties of soybean 12S globulins (glycinins), using protein engineering and expression in *E. coli*. A range of mutations and additions, and deletions of single residues and short sequences were made, focusing on five variable regions identified on the basis of sequence comparisons (65,66). Several mutant proteins were purified and their gelation and emulsification properties determined. These contained deletions of 11 or 8 residues from the protein N- and C-termini, respectively; the addition of blocks of four methionines into two of the variable regions; and the mutation of two cysteine residues involved in disulfide bond formation. All mutations showed greater emulifying activity, which was increased twofold in two of the mutants, while four of them formed harder gels. Furthermore, the presence of additional methionine residues in two of the mutants raised the possibility of simultaneously improving the functional and nutritional properties. The wild-type glycinins and one of the methionine-containing mutants have been expressed in transgenic tobacco at levels up to 4% of the total protein. The proteins were correctly processed, assembled into hexamers, and accumulated in protein bodies, although about half appeared to be partially degraded (67). More recent work from the same group has shown that the normal and methionine-containing glycinins can be expressed in transgenic rice seeds at levels up to about 5% of the total protein (60). The protein was colocated in protein bodies with the structurally related "glutelin" storage proteins of rice and assembled into 7S (trimeric) and 11S (hexameric) forms similar to those observed in developing soybean seeds. Some of the glycinin also appeared to form hybrid oligomers with the rice glutenin. Effects on the composition and digestibility of the grain were reported (61) (see earlier) but not on the functional properties.

XI. THE VISCOELASTICITY OF WHEAT GLUTEN

The gluten proteins of wheat correspond to the prolamin storage proteins and account for about half of the total nitrogen in the mature grain. They are initially deposited in protein bodies in the cells of the developing starchy endosperm, but these deposits coalesce during the later stages of grain maturation to give a continuous matrix surrounding the starch granules. When white flour (i.e., starchy endosperm cells) is mixed with water to form dough, the gluten proteins form a continuous network that confers cohesive and viscoelastic properties. This network is crucial for processing; in particular, it is expanded by fermentation of leavened bread to give a light porous structure that is then "fixed" on baking. Highly elastic (strong) wheats are required for making both bread and pasta, and poor processing quality often results from low elasticity.

Wheat gluten is a complex mixture of proteins that are classified into two broad groups (68). The glutenins consist of individual subunits that form high-M_r (above 1×10^6) polymers stabilized by interchain disulfide bonds. They are the major determinants of gluten elasticity, appearing to form an elastic network. In contrast, the gliadins are monomeric proteins that interact with each other and with the glutenin polymers by strong noncovalent interactions (notably hydrogen bonds). They appear to be the major determinants of gluten viscosity, acting as plasticizers.

Evidence from genetic (69) and biochemical studies (70,71) indicates that one group of glutenin proteins is particularly important in determining the elasticity (i.e., strength) of gluten. These high-molecular-weight (HMW) subunits of glutenin comprise between three and five individual proteins in hexaploid bread wheat (69) and usually one or two proteins in tetraploid pasta wheat (72). Each expressed subunit accounts for about 2% of the total protein (73,74), and quantitative differences in the total amount of subunit protein associated with variation in the number of expressed genes (three, four, or five in bread wheat) appear to contribute to differences in dough strength between cultivars. However, differences in dough strength are also associated with allelic variation in expressed proteins (particularly those encoded by genes on the D genome of bread wheat). These are presumed to result from subtle differences in subunit structure and interactions.

The demonstration of a relationship between subunit gene expression, protein amount, and dough strength has provided a rationale for attempts to improve gluten elasticity by introducing additional genes for HMW subunits. This has, so far, been achieved for bread wheat in four laboratories, which have introduced genes for wild-type HMW subunits encoded by chromosomes 1A and 1D of bread wheat or a chimeric gene produced by combining parts of two subunits encoded by chromosome 1D (75–79). Similarly, He et al. (80) transformed pasta wheat with the same genes from chromosomes 1A and 1D of bread wheat. In all cases, the transgenes were controlled by their own promoters and expression levels up to or above those of the endogenous genes were observed.

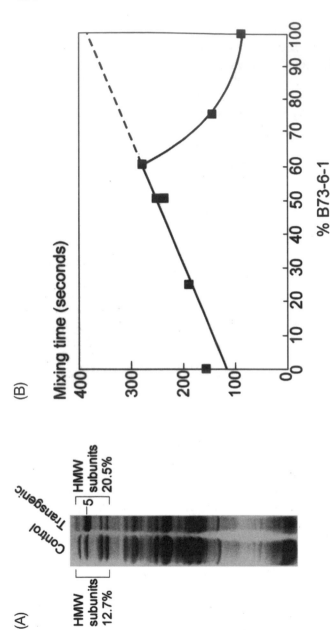

FIGURE 3 (A) Comparison of the sodium dodecyl sulfate–polyacrylamide gel electrophoresis (SDS-PAGE) profiles of total seed proteins from the background line L88-6 (control) and a transgenic line (B73-6-1) expressing additional gene copies for subunit 5. This results in an increase in the proportion of total HMW subunits from 12.7 to 20.5% and of subunit 5 from 2.7 to 10.7%. (B) The mixing time determined using a small-scale (2 g) Mixograph is a measure of dough strength. Blending of flour from the transgenic line B73-6-1 with flour from the normal bread wheat cultivar Banks results in increased mixing time (i.e., dough strength) up to about 60%, after which the dough becomes "overstrong" and the mixing time falls. (From Ref. 81.)

Expression of HMW subunit transgenes resulted in increased dough strength in a poor quality bread wheat line containing only two subunits (77) and also in pasta wheat (80) when the transgene levels were similar to those of the endogenous genes. However, very high level expression of a chromosome 1D encoded subunit in a bread wheat line expressing five subunits (81) or in pasta wheat (80) resulted in unusual mixing characteristics that were similar in some respects to those of "overstrong" varieties. In both cases the flour could be blended to increase the strength of a poor quality bread wheat flour (Fig. 3), indicating that such lines may be of commercial value for fortification of weak flours.

This work demonstrates that transformation with HMW subunits can be used to manipulate the functional properties of wheat flour, either to improve its quality for traditional end uses (bread, pasta, noodles) or to confer new properties for food (e.g., ingredients) or nonfood (e.g., packaging, films) applications. However, a deeper understanding of the relationships between the amounts, structures, and functional properties of proteins is required to allow changes to be specifically targeted. Furthermore, attention should be paid to other groups of glutenin subunits and gliadins in addition to the HMW subunits.

XII. CONCLUSIONS

In the present chapter I have focused on four aspects of seed composition that determine the quality for major end uses: the nutritional quality of cereals (lysine content) and legumes (methionine content) and the processing properties of wheat gluten proteins and soybean glycinin. These examples illustrate the range of approaches that can be taken to improve quality, including the expression of genes encoding novel proteins from other plant species, the up-regulation (by inserting additional gene copies) or downregulation (by antisense or cosuppression) of endogenous genes, and the construction and expression of mutant or ab initio designed genes. It is not surprising that amino acid composition has been one of the early successes as this can be readily defined at the molecular level. Similarly, wheat gluten and legume globulin proteins are probably the best understood seed proteins in terms of the molecular basis for their functional properties in food systems.

There are, of course, many other protein quality targets that can be addressed in the future, and there is no doubt that diet and health will have a major impact on the selection of these. Examples are the removal of proteins that are toxic, allergenic, or involved in intolerance (e.g., coeliac disease) or the introduction of novel proteins with beneficial properties. Such manipulations will be of direct benefit to the consumer and may consequently help to increase the acceptability of transgenic crops and food.

ACKNOWLEDGMENTS

IACR receives grant-aided support from the Biotechnology and Biological Sciences Research Council of the United Kingdom.

Note added in proof: Since this article was prepared it has been reported that the methionine-rich sunflower seed albumin SFA may be an allergen (JD Kelly, SL Hefle. 2S methionine-rich protein (SSA) from sunflower seeds is an IgE binding protein. Allergy 55:556–559, 2000).

REFERENCES

1. PR Shewry. Seed proteins. In: M Black, JD Bewley, eds. Seed Technology and its Biological Basis. Sheffield, UK: Sheffield Academic Press, 2000, pp 42–84.
2. JB Beccari. De Frumento. De Bononiensi Scientiarum et Artium Instituto atque Academia Commentarii, II Part I, 1745, pp 122–127.
3. CJ Leaver. Genome Organization and Expression in Plants. New York: Plenum, 1980, p 607.
4. JD Bewley, M Black. Physiology and Biochemistry of Seeds in Relation to Germination. 1. Development, Germination, and Growth. Berlin: Springer-Verlag, 1978.
5. TB Osborne. The Vegetable Proteins. 2nd ed. London: Longmans, Green, 1924, p 154.
6. R Casey. Distribution and some properties of seed globulins. In: PR Shewry, R Casey, eds. Seed Proteins. Boston: Kluwer Academic Publishers, 1999, pp 159–169.
7. M Delseny, M Raynal. Globulin storage proteins in crucifers and non-legume dicotyledonous families. In: PR Shewry, R Casey, eds. Seed Proteins. Boston: Kluwer Academic Publishers, 1999, pp 427–452.
8. AL Kriz. 7S globulins of cereals. In: PR Shewry, R Casey, eds. Seed Proteins. Boston: Kluwer Academic Publishers, 1999, pp 477–498.
9. MA Shotwell. Oat globulins. In: PR Shewry, R Casey, eds. Seed Proteins. Boston: Kluwer Academic Publishers, 1999, pp 389–400.
10. F Takaiwa, M Ogawa, TW Okita. Rice glutelins. In: PR Shewry, R Casey, eds. Seed Proteins. Boston: Kluwer Academic Publishers, 1999, pp 401–425.
11. S Utsumi, AB Gidamis, Y Takenaka, N Maruyama, M Adachi, B Mikami. Crystallization and X-ray analysis of normal and modified recombinant soybean proglycinins. Three-dimensional structure of normal proglycinin at 6 Å resolution. In: N Parris, A Kato, LK Creamer, J Pearce, eds. Macromolecular Interactions in Food Technology. Washington, DC: American Chemical Society, 1996, pp 257–270.

12. MC Lawrence. Structural relationships of 7S and 11S globulins. In: PR Shewry, R Casey, eds. Seed Proteins. Boston: Kluwer Academic Publishers, 1999, pp 517–541.

13. AD Shutov, H Bäumlein. Origin and evolution of seed globulins. In: PR Shewry, R Casey, eds. Seed Proteins. Boston: Kluwer Academic Publishers, 1999, pp 543–561.

14. JM Dunwell. Cupins: a new superfamily of functionally diverse proteins that include germins and plant storage proteins. Biotechnol and Genet Eng Rev 5: 1–32, 1998.

15. PR Shewry, AS Tatham. The characteristics, structures and evolutionary relationships of prolamins. In: PR Shewry, R Casey, eds. Seed Proteins. Boston: Kluwer Academic Publishers, 1999, pp 11–33.

16. PR Shewry, AS Tatham, NG Halford. The prolamins of the Triticeae. In: PR Shewry, R Casey, eds. Seed Proteins. Boston: Kluwer Academic Publishers, 1999, pp 35–78.

17. CE Coleman, BA Larkins. Prolamins of maize. In: PR Shewry, R Casey, eds. Seed Proteins. Boston: Kluwer Academic Publishers, 1999, pp 109–139.

18. A Leite, GC Neto, AL Vettore, JA Yunes, P Arruda. The prolamins of sorghum, *Coix* and millets. In: PR Shewry, R Casey, eds. Seed Proteins. Boston: Kluwer Academic Publishers, 1999, pp 141–157.

19. PR Shewry. Avenins: the prolamins of oats. In: PR Shewry, R Casey, eds. Seed Proteins. Boston: Kluwer Academic Publishers, 1999, pp 79–92.

20. PR Shewry, MJ Pandya. The 2S albumins storage proteins. In: PR Shewry, R Casey, eds. Seed Proteins. Boston: Kluwer Academic Publishers, 1999, pp 563–586.

21. ET Mertz, LS Bates, OE Nelson. Mutant gene that changes protein composition and increases lysine content of maize endosperm. Science 145:279–280, 1964.

22. OE Nelson, ET Mertz, LS Bates. Second mutant gene affecting the amino acid pattern of maize endosperm proteins. Science 150:1469–1470, 1965.

23. L Munck, KE Karlsson, A Hagberg, BO Eggum. Gene for improved nutritional value in barley seed protein. Science 168:985–987, 1970.

24. R Singh, JD Axtell. High lysine mutant gene (*hl*) that improves protein quality and biological value of grain sorghum. Crop Sci 3:535–539, 1973.

25. JD Axtell, SW Van Scoyoc, PJ Christensen, G Ejeta. Current status of protein quality improvement in grain sorghum. In Seed Protein Improvement in Cereals and Grain Legumes. Vol 1. Vienna: IAEA, 1979, pp 357–365.

26. H Doll. Barley seed proteins and possibilities for their improvement. In: W Gottschalk, HP Muller, eds. Seed Proteins: Biochemistry, Genetics, Nutritive Value. The Hague: Martinus Nijhof, 1983, pp 205–223.

27. PR Shewry, MS Williamson, M Kreis. Effects of mutant genes on the synthesis of storage components in developing barley endosperms. In: H Thomas, D Grierson, eds. Mutant Genes That Affect Plant Development. Cambridge: Cambridge University Press, 1987, pp 95–118.

28. J Kohno-Murase, M Murase, H Ichikawa, J Imamura. Effects of an antisense napin gene on seed storage compounds in transgenic *Brassica napus* seeds. Plant Mol Biol 26:1115–1124, 1994.

29. J Kohno-Murase, M Murase, H Ichikawa, J Imamura. Improvement in the quality of seed storage protein by transformation of *Brassica napus* with an antisense gene for cruciferin. Theor Appl Genet 91:627–631, 1995.

30. JC Wallace, G Galili, EE Kawata, RE Cuellar, MA Shotwell, BA Larkins. Aggregation of lysine-containing zeins into protein bodies in *Xenopus* oocytes. Science 240:662–664, 1988.

31. J Hejgaard, S Boisen. High lysine proteins in Hiproly barley breeding: identification, nutritional significance and new screening methods. Hereditas 93: 311–320, 1980.

32. PR Shewry. Barley seed proteins. In: J MacGregor, R Bhatty, eds. Barley: Chemistry and Technology. St. Paul, MN: AACC, pp 131–197.

33. MS Williamson, J Forde, B Buxton, M Kreis. Nucleotide sequence of barley chymotrypsin inhibitor-2 (CI-2) and its expression in normal and high-lysine barley. Eur J Biochem 165:99–106, 1987.

34. K Stott, M Blackburn, PJG Butler, M Perutz. Incorporation of glutamine repeats makes protein oligomerize: implications for neurodegenerative diseases. Proc Natl Acad Sci USA 92:6509–6513, 1995.

35. MF Perutz. Glutamine repeats and inherited neurodegenerative diseases: molecular aspects. Curr Opin Struct Biol 6:848–858, 1996.

36. KR Roesler, AG Rao. Conformation and stability of barley chymotrypsin inhibitor-2 (CI-2) mutants containing multiple lysine substitutions. Protein Eng 12:967–973, 1999.

37. F García-Olmedo. Thionins. In: PR Shewry, R Casey, eds. Seed Proteins. Boston: Kluwer Academic Publishers, 1999, pp 709–726.

38. AG Rao, M Hassan, JC Hempel. Structure-function validation of high lysine analogs of α-hordothionin designed by protein modelling. Protein Eng 7:1485–1493, 1994.

39. E Derbyshire, DJ Wright, D Boulter. Legumin and vicilin, storage proteins of legume seeds. Phytochemistry 15:3–24, 1976.

40. AS Sindhu, Z Sheng, N Murai. The pea seed storage protein legumin was synthesized, processed, and accumulated stably in transgenic rice endosperm. Plant Sci 130:189–196, 1997.

41. SJ Keeler, CL Maloney, PY Webber, C Patterson, LT Hirata, SC Falco, JA Rice. Expression of de novo high-lysine α-helical coiled-coil proteins may significantly increase the accumulated levels of lysine in mature seeds of transgenic tobacco plants. Plant Mol Biol 34:15–29, 1997.

42. SC Falco, T Guida, M Locke, J Mauvais, C Sanders, RT Ward, P Webber. Transgenic canola and soybean seeds with increased lysine. Biotechnology 13: 577–582, 1995.

43. SC Falco. Chimeric genes and methods for increasing the lysine and threonine content of the seeds of plants. U.S. Patent 5,773,691, 1998.

44. C-F Chui, SC Falco. A new methionine-rich seed storage protein from maize. Plant Physiol 107:291, 1995.

45. DG Muench, M Ogawa, TW Okita. The prolamins of rice. In: PR Shewry, R Casey, eds. Seed Proteins. Boston: Kluwer Academic Publishers, 1999, pp 93–108.

46. J Anthony, W Brown, D Buhr, G Ronhovde, D Genovesi, T Lane, R Yingling, K Aves, M Rosato, P Anderson. Transgenic maize with elevated 10 kD zein and methionine. In: WJ Cram, LJ De Kok, I Stulen, C Brunold, H Rennenberg, eds. Sulphur Metabolism in Higher Plants. Leiden, The Netherlands: Backhuys Publishers, 1997, pp 295–297.
47. RJ Youle, AHC Huang. Occurrence of low molecular weight and high cysteine containing albumin storage proteins in oilseeds of diverse species. Am J Bot 68:44–48, 1981.
48. C Ampe, J van Damme, LAB de Castro, MJAM Sampaio, M van Montagu, J Vanderkerckhove. The amino-acid sequence of the 2S sulfur-rich proteins from seeds of Brazil nut (*Bertholletia excelsa* H.B.K.). Eur J Biochem 159:597–604, 1986.
49. SB Altenbach, KW Pearson, FW Leun, SSM Sun. Cloning and sequence analysis of a cDNA encoding a Brazil nut protein exceptionally rich in methionine. Plant Mol Biol 8:239–250, 1987.
50. SB Altenbach, KW Pearson, G Meecker, LC Staraci, SSM Sun. Enhancement of the methionine content of seed proteins by the expression of a chimeric gene encoding a methionine-rich protein in transgenic plants. Plant Mol Biol 13:513–522, 1989.
51. SB Altenbach, C-C Kuo, LC Staraci, KW Pearson, C Wainwright, A Georgescu, J Townsend. Accumulation of a Brazil nut albumin in seeds of transgenic canola results in enhanced levels of seed protein methionine. Plant Mol Biol 18:235–245, 1992.
52. A de S Conceição, A Van Vliet, E Krebbers. Unexpectedly high expression levels of a chimeric 2S albumin seed protein transgene from a tandem array construct. Plant Mol Biol 26:1001–1005, 1994.
53. I Saalbach, T Pickardt, F Machemehl, G Saalbach, O Schieder, K Müntz. A chimeric gene encoding the methionine-rich 2S albumins of the Brazil nut (*Bertholletia excelsa* H.B.K.) is stably expressed and inherited in transgenic grain legumes. Mol Gen Genet 242:226–236, 1994.
54. VMM Melo, J Xavier-Filho, MS Lima, A Prouvost-Danon. Allergenicity and tolerance to proteins from brazil nut (*Bertholletia excelsa* H.B.K.) Food Agric Immunol 6:185–195, 1994.
55. JA Nordlee, SL Taylor, JA Townsend, LA Thomas, RK Bush. Identification of Brazil nut allergen in transgenic soybeans. N Engl J Med 334:688–692, 1996.
56. L Molvig, LM Tabe, BO Eggum, AM Moore, S Graig, D Spencer, TJV Higgins. Enhanced methionine levels and increased nutritive value of seeds of transgenic lupins (*Lupinus angustifolius* L.) expressing a sunflower seed albumin gene. Proc Natl Acad Sci USA 94:8393–8398, 1997.
57. G Saalbach, R Jung, G Kunze, R Manteuffel, I Saalbach, K Müntz. Expression of modified legume storage protein genes in different systems and studies on intracellular targeting of *Vicia faba* legumin in yeast. Proceedings of the 49th Nottingham Easter School: Genetic Engineering of Crop Plants. London: Butterworth, 1990, pp 151–158.

58. G Saalbach, V Christov, R Jung, I Saalbach, R Manteuffel, G Kunze, K Bram-barov, K Müntz. Stable expression of vicilin from *Vicia faba* with eight addi-tional single methionine residues but failure of accumulation of legumin with attached peptide segment in tobacco seeds. Mol Breed 1:245–258, 1995.

59. R Jung, MP Scott, Y-W Nam, TW Beaman, R Bassüner, I Saalbach, K Müntz, NC Nielsen. Processing and assembly of 11S seed storage globulins in vitro and in trangenic tobacco seeds: specificity of limited proteolysis and its role in 11S globulin assembly. Plant Cell 10:343–357, 1998.

60. T Katsube, N Kurisaka, M Ogawa, N Maruyama, R Ohtsuka, S Utsumi, F Takaiwa. Accumulation of soybean glycinin and its assembly with the glutelins in rice. Plant Physiol 120:1063–1073, 1999.

61. K Momma, W Hashimoto, S Ozawa, S Kawai, T Katsube, F Takaiwa, M Kito, S Utsumi, K Murata. Quality and safety evaluation of genetically engineered rice with soybean glycinin: analyses of the grain composition and digestibility of glycinin in transgenic rice. Biosci Biotechnol Biochim 63:314–318, 1999.

62. A De Clercq, M Vandewiele, J Van Damme, P Guerche, M Van Montagu, J Vanderkerckhove, E Krebbers. Stable accumulation of modified 2S albumin seed storage proteins with higher methionine contents in transgenic plants. Plant Physiol 94:970–979, 1990.

63. S Kjemtrup, EM Herman, MJ Chrispeels. Correct post-translational modifica-tion and stable vacuolar accumulation of phytohemagglutinin engineering to contain multiple methionine residues. Eur J Biochem 226:385–391, 1994.

64. JE Kinsella. Functional properties of soy proteins. J Am Oil Chem Soc 56: 242–258, 1979.

65. S Utsumi. Plant food protein engineering. Adv Food Nutr Res 36:89–208, 1992.

66. T Katsube, N Maruyama, F Takaiwa, S Utsumi. Food protein engineering of soybean proteins and the development of soy-rice. In: PR Shewry, JA Napier, PJ Davis, eds. Engineering Crop Plants for Industrial End Uses. London: Port-land Press, 1998, pp 65–76.

67. F Takaiwa, T Katsube, S Kitagawa, T Hisaga, M Kito, S Utsumi. High level accumulation of soybean glycinin in vacuole-derived protein bodies in the en-dosperm tissue of transgenic tobacco seed. Plant Sci 111:39–49, 1995.

68. PR Shewry, AS Tatham, F Barro, P Barcelo, P Lazzeri. Biotechnology of bread-making: unravelling and manipulating the multi-protein gluten complex. Bio-technology 13:1185–1190, 1995.

69. PI Payne. Genetics of wheat storage proteins and the effect of allelic variation on breadmaking quality. Annu Rev Plant Physiol 38:141–153, 1987.

70. PR Shewry, NG Halford, AS Tatham. The high molecular weight subunits of wheat, barley and rye: genetics, molecular biology, chemistry and role in wheat gluten structure and functionality. In: BJ Miflin, ed. Oxford Surveys of Plant Molecular and Cell Biology. Vol 6. Oxford: Oxford University Press, 1989, pp 163–219.

71. PR Shewry, NG Halford, AS Tatham. The high molecular weight subunits of wheat glutenin. J Cereal Sci 15:105–120, 1992.

72. G Branlard, JC Autran, P Monneveaux. High molecular weight glutenin subunits of durum wheat (*T. durum*). Theor Appl Genet 78:353–358, 1989.
73. W Seilmeier, H-D Belitz, H Wieser. Separation and quantitative determination of high-molecular-weight subunits of glutenin from different wheat varieties and genetic variants of the variety Sicco. Z Lebensm Unters Forsch 192:124–129, 1991.
74. NG Halford, JM Field, H Blair, P Urwin, K Moore, L Robert, R Thompson, RB Flavell, AS Tatham, PR Shewry. Analysis of HMW glutenin subunits encoded by chromosome 1A of bread wheat (*Triticum aestivum* L.) indicates quantitative effects on grain quality. Theor Appl Genet 83:373–378, 1992.
75. AE Blechl, OD Anderson. Expression of a novel high-molecular-weight glutenin subunit gene in transgenic wheat. Nat Biotechnol 14:875–879, 1996.
76. E Altpeter, V Vasil, E Stoeger, IK Vasil. Accelerated production of transgenic wheat (*Triticum aestivum* L.) plants. Plant Cell Rep 16:12–17, 1996.
77. F Barro, L Rooke, F Békés, P Gras, AS Tatham, R Fido, PA Lazzeri, PR Shewry, P Barcelo. Transformation of wheat with high molecular weight subunit genes results in improved functional properties. Nat Biotechnol 15:1295–1299, 1997
78. AE Blechl, HQ Le, OD Anderson. Engineering changes in wheat flour by genetic transformation. J Plant Physiol 152:703–707, 1998.
79. ML Alvarez, S Guelman, NG Halford, S Lustig, MI Reggiardo, N Ryabushkina, P Shewry, J Stein, RH Vallejos. Silencing of HMW glutenins in transgenic wheat expressing extra HMW subunits. Theor Appl Genet 100:319–327, 2000.
80. GY He, L Rooke, S Steele, F Békés, P Gras, AS Tatham, R Fido, P Barcelo, PR Shewry, PA Lazzeri. Transformation of pasta wheat (*Triticum turgidum* L. var. *durum*) with high-molecular-weight glutenin subunit genes and modification of dough functionality. Mol Breed 5:377–386, 1999.
81. L Rooke, F Békés, R Fido, F Barro, P Gras, AS Tatham, P Barcelo, P Lazzeri, PR Shewry. Overexpression of a gluten protein in transgenic wheat results in highly elastic dough. J Cereal Sci 30:115–120, 1999.
82. CA McPhalen, I Svendsen, I Jonassen, MNG James. Crystal and molecular structure of chymotrypsin inhibitor 2 from barley seeds in complex with subtilisin Novo. Proc Natl Acad Sci USA 82:7242–7246, 1985.
83. FAO Amino Acid Content of Foods and Biological Data on Proteins. Rome: FAO, 1970.
84. I Koshiyama. Chemical and physical properties of a 7S protein in soybean globulins. Cereal Chem 45:394–404, 1968.
85. RA Badley, D Atkinson, H Hauser, D Oldani, JP Green, JM Stubbs. The structure, physical and chemical properties of the soybean protein glycinin. Biochim Biophys Acta 412:214–228, 1975.
86. JAD Ewart. Amino acid analyses of glutenins and gliadins. J Sci Food Agric 18:111–116, 1967.
87. PS Misra, ET Mertz. Studies on corn proteins. IX. Comparison of the amino acid composition of Landry-Moureaux and Paulis-Wall endosperm fractions. Cereal Chem 5:699–704, 1976.

88. JAD Ewart. Amino acid analyses of cereal flour proteins. J Sci Food Agric 18: 548–552, 1967.

89. U Singh, LVS Sastry. Studies on the proteins of the mutants of barley grain. 2. Fractionation and characterization of the alcohol-soluble proteins. J Agric Food Chem 25:912–917, 1977.

90. FAO Energy and Protein Requirements. FAO Nutritional Meeting Report Series No 52, WHO Technical Report Series No 552, Rome, 1973.

13

Transgenic Plants as Sources of Modified Oils

Sean J. Coughlan and Anthony J. Kinney
DuPont Nutrition and Health,
Wilmington, Delaware, U.S.A.

I. INTRODUCTION

The importance of oils and fats in human nutrition is well documented (1). Lipids form a vital component of many cell constituents and are an important source of energy. Oils and fats also contribute significantly as a functional ingredient in improving the sensory characteristics of numerous processed food products.

About 70% of edible oils are derived from plant sources (about 50 million tonnes/annum), and there are three major groups of oil-producing crops. These are temperate annual oilseeds (soy, rapeseed, sunflower, and peanut), about 60% of the total vegetable oil production; perennial tropical crops (oil palm, coconut, and babassu nut), about 25% of total oil production; and crops such as cotton and corn where the embryo is a by-product of processing. This latter group accounts for about 10% of total vegetable oil production. The remaining 5% of total vegetable oils are derived from miscellaneous niche crops such as olive (3%), linseed (1%), and sesame (1%) (2).

Many oils are also used for nonfood applications (about 2% of total production). The plant oils most commonly used for industrial purposes include coconut, castor, linseed, and soy. The predominant source of industrial fatty acids, however, is tall oil, a by-product of the wood pulp and paper mill industry (2).

The utility of a given oil in a food or industrial application is determined primarily by acyl composition of the storage triacylglycerol (TAG). For most edible oils the acyl composition of the TAG is qualitatively the same as that of the membrane lipids. That is, the same five acyl groups palmitate (16:0), stearate (18:0), oleate (18:1), linoleate (18:2), and linolenate (18:3) occur in both lipid classes, often in the same molar ratios. The degree of fatty acid desaturation determines both the melting range and the thermal stability of the oil (3). Industrial oils such as castor, which is rich in hydroxy-fatty acids, often contain fatty acids with functional groups other than methylene-interrupted double bonds. These functional groups determine the reactivity and cross-linking ability of oils used for such applications as paints and coatings. The number of different fatty acids used for industrial purposes, however, is currently quite small compared with the number available (over 300 different types) from nature (4).

Thus, a review of plant seed oil modification is really a review of the fatty acids that make up the oil. Here we discuss the potential for extending the range of domesticated crops, such as soybean and rape, for novel industrial and food purposes by producing new and useful fatty acids in these crops.

II. SHORT- AND MEDIUM-CHAIN SATURATED FATTY ACIDS

There are basically two classes of oilseed species with high levels of saturated fatty acids in their oil. The classes are divided by the presence or absence of saturated fats at the sn-2 position of their TAG. Coconut oil is the best known example of an oil with >90% of its acyl chains saturated (C8:0 to C16:0, predominantly C12:0) and thus containing saturated fats at the sn-2 position. In this species both the fatty acid elongation machinery and the acyl transferase (LAPAAT) that transfers acyl chains to the sn-2 position are modified (5).

Other examples of these types of oils can be found in many plants including the Lauraceae, Myristicaceae, and Lythraceae, which have predominantly medium-chain (C10:0 to C14:0) acyl chains in all three sn TAG positions in their seed oil (5).

In most oils saturates are not found at the sn-2 position of TAG because normal LPAAT cannot use saturated acyl chains as substrates (5). Thus, the second class of oilseed accumulates saturated acyl chains only up to about 60% of the total oil. In these oils the specificity of LPAAT has excluded

16:0 and 18:0 acyl chains from the *sn*-2 position of TAG. Examples of these oils are cocoa butter (30% 16:0, 30% 18:0) and oil palm mesocarp (40% 16:0).

The major biochemical determinant for seed oils containing medium-chain (C8:0 to C16:0) fatty acids is now known to be the presence in the developing seed of a specialized acyl-acyl carrier protein (ACP) thioesterase that can redirect the common fatty acid synthase complex to produce medium-chain fatty acids, which are then incorporated into TAG formation. This has been elegantly shown by the work of the Calgene group, who first purified and cloned a medium-chain fatty-acyl-ACP thioesterase from developing California bay seeds (6). Overexpression of this enzyme in seeds of both *Arabidopsis* and canola redirected fatty acid accumulation in the TAG of both seeds to accumulate up to 40% laurate (7).

Since then, complementary DNAs (cDNAs) encoding other medium-chain thioesterases have been cloned from a variety of plants whose seeds are enriched in medium-chain fatty acids, most notably *Cuphea* sp., coconut, and elm (8). It is now clear that all of these are members of the *fat B* gene subfamily. The production of medium-chain fatty acids (C8:0 to C14:0) in developing seeds of, for example, *Cuphea* is catalyzed by the seed-specific expression of these novel *fat B* enzymes (9).

Although the specialized *fat B* enzymes are necessary and sufficient for medium-chain fatty acid production, as with plants with unusual desaturases, other modifications are required to optimize short-chain production in vivo. In *Cuphea* sp. it has been shown that the fatty acid synthase was also modified to produce shorter chain fatty acids (10). These workers suggested that, in *Cuphea wrightii*, the short-chain β-ketoacylsynthetase (*kas A*) somehow increased the efficiency of the medium-chain alkyl-ACP thioesterases. This hypothesis was supported by the work of Dehesh et al. (11), who isolated embryo-specific *kas A* cDNAs from C8:0- and C10:0-producing *Cuphea* species and functionally tested the encoded polypeptides in seeds of transgenic canola. In transgenic seed extracts, the medium-chain elongation reaction had increased resistance to cerulenin, and in double constructs with the C8:0/C10:0-specific thioesterases, the medium-chain fatty acid production was increased relative to TE-only transformants (12).

A schematic of the pathway of triglyceride biosynthesis is shown in Fig. 1, showing the critical enzymatic steps that affect fatty acid and hence TAG quality.

III. LONG CHAIN SATURATED FATTY ACIDS

Two major targets for fatty acid modification of edible oils are increased monounsaturates (18:1) with a concomitant decrease in polyunsaturates

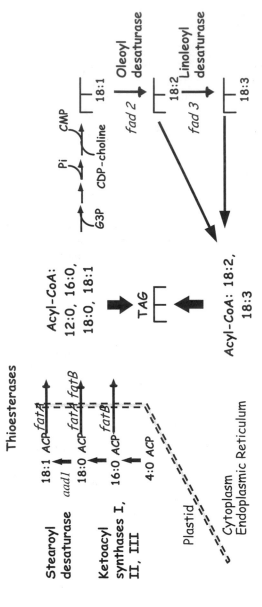

FIGURE 1 A simplified pathway of triglyceride biosynthesis. Fatty acid biosynthesis from C2 to C16:0, C18:1 occurs in the plastid. Critical steps affecting oil quality are the stearoyl desaturase and the thioesterases. The remaining steps in TAG biosynthesis are located in the smooth endoplasmic reticulum. The oleoyl and linoleoyl desaturases are the most critical extraplastidic steps affecting oil quality.

(18:2, 18:3), combined with a reduction in total saturates (16:0, 18:0). This would yield more a chemically stable oil (high 18:1) with reduced total saturated fat content. Another important target has been to increase the total 18:0 and 18:1 content of the plant oil so that it has an acceptable solid fat functionality for use in margarine and other confectionery applications but without the perceived deleterious health effects of partially hydrogenated oils rich in *trans*-fatty acids.

It has been shown that the *fat B* class of acyl:ACP thioesterases control the release of 16:0 into the cytoplasm, making it available for TAG biosynthesis. Thus, inactivation of *fat B* should give a low 16:0 phenotype. Conversely, overexpression of *fat B* should give a high 16:0 phenotype. The insertion of the first double bond into fatty acyl chains is a plastidic reaction catalyzed by a soluble desaturase enzyme, stearoyl-ACP desaturase (AAD1). Inactivation of the *aad 1* gene should result in a high 18:0-coenzyme A (CoA) pool in the cytoplasm and a resultant high 18:0 content of seed TAG. Oleoyl-CoA is also incorporated into membrane phosphatidylcholine (PC), where it is desaturated to linoleoyl-PtdCho by a membrane-bound δ-12 desaturase encoded by a *fad 2* gene (13). Inactivation of the *fad 2* gene should give a high 18:1 phenotype.

High-oleic mutants of corn, peanut canola, and sunflower have been described, with an 18:1 ranging from 60 to 90%. The high-oleic sunflower is particularly noteworthy because it is a dominant mutation in a structural *fad 2* gene (14). Interestingly, no good high-oleic mutant of soybean has yet been found by chemical mutagenesis despite intensive screening by both private and public breeders. In contrast to most high-oleic mutants, which are primarily recessive mutations, transgenic soy lines in which the seed-specific expression of *fad 2* has been inactivated result in a consistently high oleic (>80%) content of the seed oil. This phenotype is not affected by environment and has no yield penalty (15).

Low-saturate mutants of soy also exist, primarily due to altered *fat B* activity (16), but again they are recessive and are the result of mutations in several genes. Transgenic soy lines in which the *fat B* thioesterase activities are suppressed have a 50% reduction in total 16:0 and are dominant (15).

Again, both high-16:0 (inactive *kas 2*) and high-stearate (reduced *aad 1* activity) mutants of soy exist. The high-16:0 mutants (>40% 16:0 in seed oil) are relatively stable across different environments, although the mutants are recessive (17). Transgenic soybean plants overexpressing *fat B* also produce >40% 16:0 in the seed oil but as a dominant trait (15).

The high (20–30% 18:0) stearate mutant of soy, by contrast, has severe phenotypic consequences associated with the trait including poor germinability and loss of vigor (17). These characteristics were overcome in both transgenic soy and canola in which one or more endogenous seed-specific *aad 1* genes were inactivated in the transgenic seeds. The resultant oil in both cases was enriched in stearate (35–40% 18:0) and the vegetative tissue of the transgenic plants was normal. In both cases, however, the seed germination was impaired, possibly because of the presence of 18:0 in the membrane phospholipids as well as the TAG (15,18). Sunflower, by contrast, has a high-stearate mutant (25% 18:0) that appears to be normal in all other respects (19). Further, transgenic sunflowers that contain 35% 18:0 in the TAG (seed-specific suppression of *aad 1*) germinated normally, although the seed oil content was reduced (S. Coughlan, unpublished observations).

There is also a report of a specialized thioesterase involved in the production of stearate in the seeds of mangosteen (*Garcinia mangostana*) (20). Stearate is the predominant fatty acid (45–50% of total fatty acids) in this seed. When the mangosteen thioesterase gene (*fat A*) was expressed in canola seeds, stearate levels of 20% were observed (20).

IV. MONOUNSATURATED FATTY ACIDS

Oleic acid ($\Delta 9$ 18:1) is a component of most common plant oils, ranging from about 20% in soybean to over 80% in some mutant varieties of canola and sunflower (21). The $\Delta 9$ desaturase enzyme described earlier normally inserts a double bond at the $\Delta 9$ position of stearoyl-ACP to form oleoyl-ACP. There are, however, a number of naturally occurring oils that contain monounsaturated fatty acids with double bonds in positions other than the n-9 carbon from the carboxyl group (4). For, example, seeds of the Umbelliferae species such as carrot and coriander contain oils rich in petroselenic acid ($\Delta 6$ 18:1). This unusual monounsaturate is the result of the activity of a seed-specific plastidial $\Delta 4$ desaturase that converts palmitoyl-ACP to $\Delta 4$ hexadecanoyl-ACP, which is then elongated to petroselenoyl-ACP (22). Thus, this soluble desaturase has both a different substrate (chain length) specificity and different regiospecificity from the canonical stearoyl-ACP desaturase, yet based on its primary amino acid sequence it is clearly a member of the same gene family (23).

This gene family (*aad 1*) includes a number of other members that vary from the standard $\Delta 9$ 18:0-ACP in their substrate specificity, regiospecificity, or both. Additional examples include the $\Delta 6$ 18:0-ACP desaturase from *Thunbergia alata* (24) and $\Delta 9$ 16:0-ACP desaturases from *Doxantha* spp. and from *Asclepia syriaca* (25), all of which have >70% amino acid sequence similarity to $\Delta 9$ 18:0-ACP desaturases. It has been shown that mutations in as little as five amino acids can produce changes in both substrate specificity and regiospecificity of an acyl-ACP desaturase, for example, converting a $\Delta 6$ 16:0-ACP desaturase into a $\Delta 9$ 18:0-ACP desaturase (26).

Despite this strong conservation of primary sequence in the *aad 1* gene family, it is necessary but not sufficient to express the coriander Δ4 palmitoyl-ACP desaturase in a standard oilseed (e.g., soybean, sunflower) in order to produce large amounts of petroselenic acid instead of oleic acid (27). This is because the Umbelliferae species producing this fatty acid have a modified fatty acid biosynthetic pathway adapted to produce this fatty acid. Adaptations include a modified *kas A* gene product resulting in a condensing enzyme with specificity for Δ4 hexadecanoyl-ACP and a modified *fat B* gene encoding a petroselenyl-ACP thioesterase, plus other additional variants including acyltransferases, ferredoxins, and possibly ACPs (28,29). These observations demonstrate both the complexity of the evolutionary process that has resulted in divergent oils and the technical challenges of producing novel fatty acids in oilseed plants.

Very long chain monounsaturated fatty acids (VLCFAs, 20–26 carbons) are found in the storage oils of some plants, such as members of the Cruciferae family, and in epicuticular and storage wax esters. Most unsaturated VLCFAs are the result of the elongation of oleoyl-CoA by a membrane-bound elongase complex (30).

Waxes, esters of long-chain alcohols and fatty acids, are abundant in plants as a cell wall component to give hydrophobicity, resistance to fungal attack, and light absorption and reflection (31). Jojoba (*Simmondsia chinensis*) is the only angiosperm known to accumulate liquid waxes in its seeds as an energy store. Up to 60% of the seed dry weight consists of linear wax esters of ω-9 monounsaturated C20, C22, and C24 fatty acids and alcohols. TAG is absent (32,33). From earlier biochemical data it was inferred that an acyl-CoA–based chain elongation was responsible for the long-chain acyl chains of jojoba wax. This was confirmed by the Calgene group's purification and cDNA cloning of all three components, a jojoba β-ketoacyl-CoA synthase, a fatty acyl reductase, and a wax synthase proposed as necessary for wax synthesis (34–36). In transgenic *Arabidopsis* seed expressing all three of these cDNAs, large quantities of short-chain liquid wax accumulated, representing up to 70% of the seed oil (36). It has thus been demonstrated that not only are these three enzyme activities necessary and sufficient for seed wax formation but also this pathway can efficiently divert carbon from TAG in vivo.

An *Arabidopsis* gene (*fae 1*) related to the jojoba condensing enzyme (β-ketoacyl-CoA synthase) has been cloned by insertional mutagenesis (37). *Fae 1* was shown to complement *Arabidopsis* mutants that have reduced 20-C fatty acids and increased oleic acid and to produce VLCFAs when expressed in tobacco (38). It seems probable based on these observations that a single condensing enzyme controls the elongation of C18 to C20, C22 acyl CoAs in VLCFA-producing seed.

A *fae 1* homologue has also been cloned from *Limnanthes* and has been expressed in soybean seeds (39). The result is the formation of C20 and C22 saturated fatty acids at the expense of palmitic acid (16:0). When coexpressed with the *Limnanthes* Δ5 desaturase, 20:1 is formed at about 20% relative abundance. These results illustrate that the specificity of the elongation step of acyl-CoAs appears to be determined entirely by the *fae 1* gene product and that the *Limnanthes fae 1* gene product has specificity for saturated fatty acids, unlike the monounsaturated specificity of the *Arabidopsis fae 1*.

V. POLYUNSATURATED FATTY ACIDS

The 18-carbon polyunsaturated fatty acids (PUFAs) linoleic and α-linolenic acids are essential components of plant membranes (18:2, 18:3) (40). PUFAs are also substrates for the lipoxygenase-mediated production of various volatile compounds including methyl jasmonate (41). This is an area of active research because the volatiles produced are important both as signaling agents in response to pathogen attack (42) and as flavoring components in the area of food and fragrance chemistry (43). The PUFAs α-linoleic and linolenic acids are widely distributed in nature. In addition, γ-linolenic acid (GLA) is found in the seeds of a few plant species (borage, evening primrose, black currant) as well as cyanobacteria and fungi (4). Longer chain n-6 PUFAs such as arachidonic acid (ARA 20:4n-6) are found in microorganisms such as the fungus *Morteriella alpina* and the marine diatom *Porphyridiom cruentum* (4). The related n-3 LC-PUFAS eicosapentanoic acid (EPA 20:5n-3) and docosahexanoic acid (DHA 22:6n-3) are found in marine microorganisms and in fish oil (4).

Both linoleic and α-linolenic acids are essential fatty acids in the human diet because the Δ12 and n-3 desaturase activities necessary to convert oleic acid to these fatty acids are lacking in most mammalian microsomes. The dietary PUFAs (18:2, 18:3) are used primarily in mammals as precursors for the eicosanoids including prostaglandins and leukotrienes (44). Because of the known beneficial effects associated with the intake of LC-PUFAs in the diet of both infants and adults (44) and the limited natural sources commercially available, there has been much interest lately in the possibility of producing ARA, EPA, and DHA in plant seed oils, where they do not normally occur (4).

In higher eukaryotes, the major long-chain PUFA arachidonate (ARA 20:4n-6) is derived from linoleic acid (18:2, and EPA (20:5n-3) and DHA (22:6n-3) from α-linolenic acid (ALA). The metabolic pathways for long-chain PUFA formation have been known for some time (30), but until recently none of the proteins had been identified (see Figure 2).

Simplified pathway of plant long chain - polyunsaturated fatty acid biosynthesis

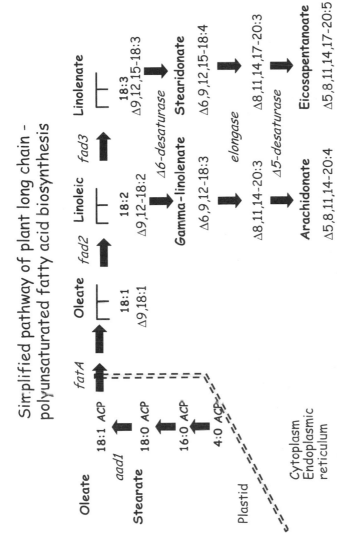

FIGURE 2 Higher plant pathway of long-chain PUFA formation from linoleic and α-linolenic acids. A simplified schematic showing the extraplastidic pathway of arachidonate and eicosa-pentaenoate from stearate. The minimum amounts of enzyme activities are shown with no cross talk between the two pathways.

The pathways of very long chain PUFA synthesis in fungi and marine algae involve a series of sequential reductions and elongations that resemble the production of long-chain fatty acids such as erucic acid in plants like *Arabidopsis* and rapeseed (30). With the advent of modern genomic techniques and the existence of known desaturase and elongase protein sequences, it was possible to search cDNA EST databases of PUFA-containing organisms using conserved desaturase and elongase domains. In this way, Δ5 and Δ6 desaturases and the 18:3 n-6 elongase from the ARA-accumulating filamentous fungus *Morteriella alpina* were identified (45–48). Using similar approaches, orthologues of the Δ5 desaturase and the 18:3n-6 elongase have been identified from *C. elegans* (49,50) and human cDNA EST databases respectively (51,52). Rapid functional identification of putative cDNA clones was carried out by expression in yeast (45–52). This technique enabled identification of individual positive cDNA clones and led to the possibility of producing these very long chain PUFAs in the oil of transgenic plants. Earlier work on the expression of a δ-6 desaturase from a cyanobacterium in transgenic tobacco showed no accumuation of GLA or OTA in seeds (53). In contrast, coexpression of *M. alpina* Δ6 and Δ12 desaturases in canola seed resulted in a GLA content of about 40% of the seed oil fatty acids (54). Similarly, seed-specific expression of the borage Δ6 desaturase in transgenic soy seed resulted in a similar total GLA content in the seed oil (S. Coughlan, unpublished results). The differences between canola and soy seed may be related to the higher amount of 18:2 naturally present in soy oil (50%) compared with canola oil (20%). With the cloning of all of the individual components of the eukaryotic pathway of LC-PUFA formation, the technical challenges remaining are those of metabolic pathway engineering. For example, it will be necessary to co-express up to six different enzymes simultaneously in a developing oilseed in order to produce DHA from α-linolenic acid in transgenic plants. Finally, it should be noted that some marine microorganisms (bacteria and diatoms) have a completely different pathway of LC-PUFA formation using polyketide synthases (55).

VI. VARIATIONS ON THE METHYLENE-INTERRUPTED DOUBLE BOND

As discussed earlier, the polyunsaturated fatty acids of plant membrane phospholipids and of the most common plant oils are predominantly linoleic and α-linolenic acids. These are 18-carbon fatty acids with two or three methylene-interrupted double bonds. A wide variety of plants have oils containing fatty acids with other types of functional groups or non–methylene-interrupted bonds, and many of these fatty acids are the result of the activity of diverged members of the phospholipid Δ12 desaturase (*fad 2*) gene family (5).

The first diverged *fad-2* genes to be identified were the genes encoding 12-hydroxylases from *Ricinus communis* and *Lesquerella fendleri* (56,57). These enzymes catalyze the introduction of hydroxy groups into fatty acids. Heterologous expression of these diverged *fad 2* cDNAs in transgenic *Arabidopsis* seeds resulted in the production of up to 20% hydroxylated fatty acids in the seed oil (58). Also observed were a concomitant increase in oleic acid, which is a common by-product of heterologous expression of diverged *fad 2* cDNAs, and an accumulation of the novel fatty acid in membrane phosphatidylcholine. One interpretation of these observations is that the novel fatty acid is not being properly channeled from the PC to TAG after the functional group has been inserted. Thus, the whole cycle is slowed down and there is a backup of oleoyl-CoA waiting to be incorporated into PC for desaturation. Thus, it is likely that additional diverged enzymes exist in the species producing these unusual fatty acids to ensure the appropriate channeling from the membrane phospholipid into the storage lipid. To obtain levels of these unusual fatty acids in transgenic oilseeds comparable to those in the native plant, it will be necessary to understand the metabolic pathway and to coexpress the additional enzyme activities into the transgenic oilseed together with the diverged *fad 2*. Candidates for these activities include novel phospholipase(s), acyl CoA synthases, and the acyltransferases including the newly described PDAT (phospholipid:diacylglycerol acyltransferase) class (59).

The *Lesquerella* hydroxylase is actually a bifunctional enzyme with some omega-6 desaturase activity (56), and only a few amino acid substitutions are necessary and sufficient to change the hydroxylase to desaturase activities of the enzyme (60). These similarities in primary amino acid sequence between *fad 2* desaturases and *fad 2* hydroxylases led to homology-based searching for other *fad 2*–related genes in oilseed species containing unusual fatty acids in their oil. This approach led to the identification of *fad 2*–related epoxygenases from *Vernonia galamensis* (61) and *Crepis palaestrina* (62) and a fatty acid acetylenase from *Crepis alpina* (62). These enzymes catalyze the formation of epoxy groups and acetylenic triple bonds, respectively, into fatty acids.

More recently, *fad 2*–related conjugase cDNAs from *Momordica charantia*, *Impatiens balsamica*, and *Calendula officinalis* have been identified (63,64). Conjugated fatty acids are polyunsaturated fatty acids with non–methylene-interrupted double bonds. *Momordica* is rich in α-eleostearic acid (18:3 Δ9*cis*, 11*trans*, 13*trans*). *Impatiens* is rich in α-parinaric acid (18:4 Δ9*cis*, 11*trans*, 13*trans*, 15*cis*) and *Calendula* in calendic acid (18:3 Δ8*trans*, 10*trans*, 12*cis*).

Finally, the plant sphingolipid desaturases (65) represent a possible ancestral fusion of an N-terminal cytochrome b_5 with a Δ6-desaturase first described in borage (*Boragio officinalis*) seeds. Borage seeds, in common with a few other plants such as evening primrose (*Oenothera biennis*) and black currant (*Ribes nigrum*), contain oils rich in γ-linolenic acid (18:3 Δ6,9,12). Given the primary sequence homology between the ubiquitous plant sphingolipid desaturases and the borage Δ6 desaturase, it seems probable that the borage-type Δ6 desaturases represent enzymes that have diverged in substrate specificity from the Δ8/Δ6 sphingolipid desaturase family. Curiously, the borage Δ6 desaturase is extremely effective at producing up to 50% GLA in transgenic soy seed (S. Coughlan, unpublished results) but not in transgenic *Arabidopsis* or canola seeds (66).

VII. COMMERCIAL PRODUCTION OF NEW OILS FROM TRANSGENIC PLANTS

In all the preceding examples in which fatty acid metabolism has been manipulated in transgenic plants, the abundance of novel fatty acid content in the resulting seed oil varies depending on the type of fatty acid produced. Very significant amounts of short-chain, monounsaturated and polyunsaturated fatty acids have been produced in transgenic plants. In contrast, fatty acids containing unusual functional groups are produced in transgenic plants in amounts considerably less than that of exotic plants that normally produce them. The differences in the relative accumulation of different fatty acids in transgenic plant oils are not yet fully understood. Thus, a key technical goal that remains is the ability to produce any given novel fatty acid as a large proportion of the seed oil.

Additional hurdles to the commercial production of novel oils includes any possible negative impact of the transgenes on seed yield and the costs associated with gaining worldwide regulatory approval to grow and produce the new crop (currently running into many millions of dollars). The ability to sell these new oils for a premium over commodity oil that is large enough to cover these costs is essential to convert the many technical achievements described in this chapter into commercial successes.

REFERENCES

1. G Taubes. The soft science of dietary fat. Science 291:2536–2545, 2001.
2. G Robbelen, RK Downey, A Ashri. Oil Crops of the World. New York: McGraw-Hill, 1989, pp 1–21.
3. WS Singleton. Properties of the liquid state. In: KS Markley, ed. Fatty Acids. Part 1. 2nd ed. Malabar: Robert E Krieger, 1983, pp 499–607.

4. FD Gunstone, JL Harwood, FB Padley. The Lipid Handbook. 2nd ed. London: Chapman & Hall, 1995, pp 47–146.
5. T Voelker, AJ Kinney. Variations in the biosynthesis of seed-storage lipids. Annu Rev Plant Physiol Mol Biol 52:335–361, 2001.
6. MR Pollard, L Anderson, C Fan, DJ Hawkins, HM Davies. A specific acyl-ACP thioesterase implicated in medium chain fatty acid production in immature cotyledons of *Umbellularia californica*. Arch Biochem Biophys 284:306–312, 1991.
7. TA Voelker, AC Worrell, L Anderon, J Bleibaum, C Fan, DJ Hawkins, SF Radke, HM Davies. Fatty acid biosynthesis redirected to medium chains in transgenic oilseed plants. Science 257:72–74, 1992.
8. A Jones, HM Davies, TA Voelker. Palmitoyl-acyl carrier protein (ACP) thioesterase and the evolutionary origin of plant acyl-ACP thioesterases. Plant Cell 7:359–371, 1995.
9. K Dehesh, A Jones, DS Knutzon, TA Voelker. Production of high levels of 8:0 and 10:0 fatty acids in transgenic canola by over-expression of Ch FatB, a thioesterase cDNA from *Cuphea hookeriana*. Plant J 9:167–172, 1996.
10. MB Slabaugh, JM Leonard, SJ Knapp. Condensing enzymes from *Cuphea wrightii* associated with medium chain fatty acid biosynthesis. Plant J 13:611–620, 1998.
11. K Dehesh, P Edwards, J Fillati, MB Slabaugh, J Byrne. KAS IV: a 3-ketoacyl-ACP synthase from *Cuphea* sp. is a medium chain specific condensing enzyme. Plant J 15:383–390, 1998.
12. JM Leonard, SJ Knapp, MB Slabaugh. A *Cuphea* B-ketoacyl-ACP synthase shifts the synthesis of fatty acids towards shorter chains in *Arabidopsis* seeds expressing *Cuphea FatB* thioesterases. Plant J 13:621–628, 1998.
13. J Okuley, J Lightener, K Feldman, N Yadav, E Lark, J Browse. *Arabidopsis FAD2* gene encodes a protein that is essential for polyunsaturated lipid biosynthesis. Plant Cell 6:147–158, 1994.
14. V Hongtrahul, MB Slabaugh, SJ Knapp. A seed specific delta-12 oleate desaturase gene is duplicated, rearranged, and weakly expressed in high oleic acid sunflower lines. Crop Sci 38:1245–1249, 1998.
15. AJ Kinney. Plants as industrial chemical factories—new oils from genetically engineered soybeans. Fett/Lipid 100:173–176, 1998.
16. WR Fehr, GA Welke, EG Hammond, DN Duvick, SR Cianzio. Inheritance of reduced palmitic acid content in seed oil of soybean. Crop Sci 31:88–89, 1991.
17. DM Bubeck, WR Fehr, EG Hammond. Inheritance of palmitic and stearic acid mutants of soybean. Crop Sci 29:652–656.
18. DS Knutzon, GA Thompson, SE Radke, WB Johnson, VC Knauf, JC Kridl. Modification of *Brassica* seed oil by antisense expression of a stearoyl-acyl carrier protein desaturase gene. Proc Natl Acad Sci USA 89:2624–2628, 1992.
19. S Cantisan, E Martinez-Force, A Alvarez-Ortega, R Garces. Lipid characterization in vegetative tissues of high saturated fatty acid sunflower mutants. J Agric Food Chem 47:78–82, 1999.
20. DJ Hawkins, JC Kridl. Characterization of acyl-ACP thioesterases of mango-

steen (*Garcinia mangostana*) seed and high levels of stearate production in transgenic canola. Plant J 13:743–752, 1998.

21. AJ Kinney. Genetic modification of the storage lipids of plants. Curr Opin Biotechnol 5:144–151, 1994.

22. EB Cahoon, JB Ohlrogge. Metabolic evidence for the involvement of a delta-4 palmitoyl-ACP desaturase in petroselenic acid synthesis in coriander endosperm and transgenic tobacco cells. Plant Physiol 104:827–837, 1994.

23. J Shanklin, EB Cahoon. Desaturation and related modifications of fatty acids. Annu Rev Plant Physiol Mol Biol 49:611–641, 1998.

24. EB Cahoon, AM Cranmer, J Shanklin, JB Ohlrogge. Δ-6 hexadecenoic acid is synthesized by the activity of a soluble Δ-6 palmitoyl-acyl carrier protein desaturase in *Thunbergia alata* endosperm. J Biol Chem 269:27519–27526, 1994.

25. EB Cahoon, S Coughlan, J Shanklin. Characterization of a structurally and functionally diverged ACP desaturase from milkweed seed. Plant Mol Biol 33: 1105–1110, 1997.

26. EB Cahoon, Y Lindqvist, G Schneider, J Shanklin. Redesign of soluble fatty acid desaturases from plants for altered substrate specificity and double bond position. Proc Natl Acad Sci USA 94:4872–4877, 1997.

27. EB Cahoon, J Shanklin, JB Ohlrogge. Expression of a coriander desaturase results in petroselenic acid production in transgenic tobacco. Proc Natl Acad Sci USA 89:11184–11188, 1992.

28. MC Suh, DJ Schultz, JB Ohlrogge. Isoforms of acyl carrier protein involved in seed-specific fatty acid biosynthesis. Plant J 17:679–688, 1999.

29. DJ Schulz, MC Suh, JB Ohlrogge. Stearoyl-acyl carrier protein and unusual acyl-acyl carrier protein desaturase activities are differentially influenced by ferredoxin. Plant Physiol 124:681–692, 2000.

30. JL Harwood. Recent advances in the biosynthesis of plant fatty acids. Biochim Biophys Acta 130:7–56, 1996.

31. D Post-Beitenmiller. Biochemistry and molecular biology of wax production in plants. Annu Rev Plant Physiol Mol Biol 47:405–430, 1996.

32. JB Ohlrogge, MR Pollard, PK Stumpf. Studies on the biosynthesis of waxes by developing jojoba seed tissue. Lipids 13:203–210, 1977.

33. MR Pollard, T McKeon, LM Gupta, PK Sumpf. Studies on biosynthesis of waxes by developing jojoba seed. II. The demonstration of wax biosynthesis by cell free homogenates. Lipids 14:651–662, 1979.

34. MW Lassner, K Lardizabal, JG Metz. A jojoba B-ketoacyl-CoA synthase cDNA complements the canola fatty acid elongation mutation in transgenic plants. Plant Cell 8:281–292, 1996.

35. JG Metz, MR Pollard, L Anderson, TR Hayes, MW Lassner. Purification of a jojoba embryo fatty acyl-coenzyme A reductase and expression of its cDNA in high erucic acid rapeseed. Plant Physiol 122:635–644, 2000.

36. KD Lardizabal, JG Metz, T Sakamoto, WC Hutton, MR Pollard, MW Lassner. Purification of a jojoba embryo wax synthase, cloning of its cDNA, and production of high levels of wax in seeds of transgenic *Arabidopsis*. Plant Physiol 122:645–655, 2000.

37. DW James, E Lime, J Keller, I Ploy, E Ralson, HK Dooner. Directed tagging of the *Arabidopsis FAE1* gene with the maize transposon activator. Plant Cell 7:309–319, 1995.
38. AA Millar, L Kunst. VLCFA biosynthesis is controlled through the expression and specificity of the condensing enzyme. Plant J 12:121–131, 1997.
39. EB Cahoon, EF Marilla, KL Stecca, SE Hall, DT Taylor, AJ Kinney. Production of fatty acid components of meadowfoam oil in somatic soybean embryos. Plant Physiol 124:243–251, 2000.
40. M McConn, J Browse. Polyunsaturated membranes are required for photosynthetic competence in a mutant of *Arabidopsis*. Plant J 15:521–530, 1998.
41. M McConn, RA Creelman, E Bell, JE Mullet, J Browse. Jasmonate is essential for insect resistance in *Arabidopsis*. Proc Natl Acad Sci USA 94:5473–5377, 1997.
42. P Vijayan, J Shockley, CA Levesque, RJ Cook, J Browse. A role for jasmonate in pathogen defense of *Arabidopsis*. Proc Natl Acad Sci USA 95:7209–7214, 1998.
43. I Gill, R Valivety. Polyunsaturated fatty acids, part 2: biotransformations and biotechnological applications. Trends Biotechnol 15:470–478, 1997.
44. I Gill, R Valivety. Polyunsaturated fatty acids, part 1: occurrence, biological activities and applications. Trends Biotechnol 15:401–409, 1997.
45. YS Huang, S Chaudhary, JM Thurmond, EG Bobik Jr, L Yuan, GM Chan, SJ Kirchner, P Mukerji, DS Knutzon. Cloning of delta 12- and delta 6-desaturases from *Morteriella alpina* and recombinant production of gamma-linolenic acid in *Saccharomyces cerevisiae*. Lipids 34:649–659, 1999.
46. LV Michaelson, CM Lazarus, G Griffiths, JA Napier, AK Stobart. Isolation of a delta 5-fatty acid desaturase gene from *Morteriella alpina*. J Biol Chem 273: 19055–19059, 1998.
47. DC Knutzon, JM Thurmond, YS Huang, S Chaudhary, EG Bobik Jr, GC Chan, SJ Kirchner, P Mukerji. Identification of a Δ5-desaturase from *Morteriella alpina* by heterologous expression in bakers yeast and canola. J Biol Chem 273:29360–29366, 1998.
48. JM Parker-Barnes, T Das, E Bobik, AE Leonard, JM Thurmond, LT Chang, YS Huang, P Mukerji. Identification and characterization of an enzyme involved in the elongation of n-6 and n-3 polyunsaturated fatty acids. Proc Natl Acad Sci USA 97:8284–8289, 2000.
49. LV Michaelson, JA Napier, M Lewis, G Griffith, CM Lazarus, AK Stobart. Functional identification of a fatty acid d-5 desaturase gene from *Caenorhabditis elegans*. FEBS Lett 439:215–218, 1998.
50. F Beaudoin, LV Michaelson, SJ Hey, MJ Lewis, PR Shewrey, O Sayanova, JA Napier. Heterologous reconstitution in yeast of the polyunsaturated fatty acid biosynthetic pathway. Proc Natl Acad Sci USA 97:6421–6426, 2000.
51. AE Leonard, B Kelder, EG Bobik, LT Chuang, JM Parker-Barnes, JM Thurmond, PE Kroeger, JJ Kopchik, YS Huang, P Mukerji. CDNA cloning and characterization of human Δ5-desaturase involved in the biosynthesis of arachidonic acid. Biochem J 347:719–724, 2000.

52. AE Leonard, EG Bobik, J Dorado, PE Kroeger, LT Chuang, JM Thurmond, JM Parker-Barnes, T Das, YS Huang, P Mukerji. Cloning of a human cDNA encoding a novel enzyme involved in the elongation of long-chain polyunsaturated fatty acids. Biochem J 360:765–770, 2000.

53. AS Reddy, TL Thomas. Expression of a cyanobacterial d6-desaturase gene results in gamma-linolenic acid production in transgenic plants. Nat Biotechnol 14:639–642, 1996.

54. JW Liu, YS Huang, S DeMichele, M Bergana, E Bobik Jr, C Hastilow, LT Chuang, P Mukerji, D Knutzon. Evaluation of the seed oils from a canola plant genetically transformed to produce high levels of gamma-linolenic acid. In: YS Huang, VA Ziboh, eds. Gamma-Linolenic Acid: Recent Advances in Biotechnology and Clinical Applications. Champaign, IL: AOCS Press, 2001, pp 61–71.

55. JG Metz, P Roessler, D Facciotti, C Levering, F Dittrich, M Lassner, R Valentine, K Lardizabal, F Domergue, A Yamada, K Yazawa, V Knauf, J Browse. Production of polyunsaturated fatty acids by polyketide synthases in both prokaryotes and eukaryotes. Science 293:290–293, 2001.

56. FN van de Loo, P Broun, S Turner, CR Somerville. An oleate 12-hydroxylase from *Ricinus communis* L is a fatty acyl desaturase homolog. Proc Natl Acad Sci USA 92:6743–6747, 1995.

57. P Broun, S Boddupalli, C Somerville. A bifunctional oleate 12-hydroxylase: desaturase from *Lesquerella fendleri*. Plant J 13:201–210, 1998.

58. P Broun, C Somerville. Accumulation of ricinoleic, lesquerolic and densipolic acids in seeds of transgenic *Arabidopsis* plants that express a fatty acyl hydroxylase cDNA from castor bean. Plant Physiol 113:933–942, 1997.

59. A Dahlqvist, U Stahl, M Lenman, A Bafor, M Lee, L Sandager, H Ronne, S Stymne. Phospholipid:diacylglycerol acyltransferase: an enzyme that catalyzes the acyl-CoA independent formation of triacylglycerol in yeast and plants. Proc Natl Acad Sci USA 97:6487–6492, 2000.

60. P Broun, J Shanklin, E Whittle, C Somerville. Catalytic plasticity of fatty acid modification enzymes underlying chemical diversity of plant lipids. Science 282:1315–1317, 1998.

61. AJ Kinney, WD Hitz, S Knowlton, EB Cahoon. Re-engineering oilseed crops to produce industrially useful fatty acids. In: J Sanchez, E Cerda-Olmedo, E Martinez-Force, eds. Advances in Lipid Research. Sevilla: Secr Publ Univ Sevilla, 1998, pp 623–628.

62. M Lee, M Lenman, A Banas, M Bafor, S Singh, M Schweizer, R Nilsson, C Liljenberg, A Dahlqvist, PO Gummeson, S Sjodahl, A Green, S Stymne. Identification of non-heme diiron proteins that catalyze triple bond and epoxy group formation. Science 280:915–918, 1998.

63. EB Cahoon, TJ Carlson, KG Ripp, BJ Schweiger, GA Cook, SE Hall, AJ Kinney. Biosynthetic origin of conjugated double bonds: production of high-value drying oils in transgenic soybean embryos. Proc Natl Acad Sci USA 96: 12935–12940, 1999.

64. EB Cahoon, KG Ripp, SE Hall, AJ Kinney. Formation of conjugated d-8, d-

10-double bonds by d-12-oleic acid desaturase–related enzymes. Biosynthetic origin of calendic acid. J Biol Chem 276:2637–2643, 2001.

65. P Sperling, U Zahringer, E Heinz. A sphingolipid desaturase from higher plants. Identification of a new cytochrome *b*5 fusion protein. J Biol Chem 273:28590–28596, 1998.

66. DS Kuntzon, VC Knauf. Manipulating seed oils for polyunsaturated fatty acid content. In: J Harwood, ed. Plant Lipid Biosynthesis: Fundamental and Agricultural Applications. Cambridge: Cambridge University Press, 1998, pp 287–304.

14

Flavors and Fragrances from Plants

Holger Zorn and Ralf G. Berger
Institute of Biochemistry, University of Hannover, Hannover, Germany

I. INTRODUCTION

The sensation of odor is triggered by highly complex mixtures of small, rather hydrophobic molecules from many chemical classes that occur in trace concentrations and are detected by receptor cells of the olfactory epithelium inside the nasal cavity. The nonvolatile chemical messengers of the sense of taste interact with reporters located on the tongue and impart four basic impressions only: sweet, sour, salty, and bitter. In scientific Anglo-Saxon usage, all sensory (odor, taste, color and texture) attributes of food have been classed under the general term "flavor." Fragrances, as used in perfumes, cosmetics, and toiletries, are distinguished from volatile flavors mainly by the different range of application.

According to European food legislation, aroma compounds derived from a natural source by physical means are classified as "natural"; their synthetic counterparts are "nature-identical," and compounds without a natural prototype are "artificial." Most of the natural flavors currently processed by the flavor and fragrance industry are originally derived from plant sources. Naturalness of flavors started to become increasingly important particularly for the food market about two decades ago. Today, the U.S. demand for natural flavors accounts for about 70–80% of all flavor-added products. Europe also favors natural flavors, and a widespread growth in Asia over the next few years was predicted (1). This rising demand cannot be covered by means of traditional flavor recovery processes, which mainly revert to extraction and distillation of field-grown plant material (2). Thus, this gap of supply represents one major push for plant biotechnology for the production of natural flavors. Although the specific fields of research overlap and interact to a wide extent, an attempt to structure plant aroma biotechnology is presented in Fig. 1.

Recombinant DNA technology, although discussed controversially in public, raises multifaceted expectations. Since the introduction of the first genetically engineered whole food, the FLAVRSAVR™ tomato, in 1994, recombinant DNA technology has developed rapidly. The majority of genetic engineering imparted pest or pesticide resistance to the plant, whereas improved flavor quality has been aspired to only recently.

Notwithstanding the long breeding history of aromatic fruit, we still lack essential information on the molecular, cellular, and physiological events that control the processes of flavor genesis. With the help of recombinant DNA technology, the genetic information responsible for flavor formation in plants can be characterized and isolated (e.g., Ref. 3). Afterward it may be either transferred into a suitable microbial host strain or used specifically to modify food plants. The understanding of these fundamental biochemical principles is indispensable for the development of competitive

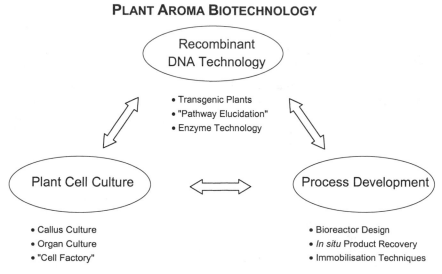

FIGURE 1   Research fields of plant aroma biotechnology.

industrial processes through manipulation of precursors, enzymes, and horticultural and storage conditions. Leahy and Roderick (4) and Takeoka (5) published comprehensive articles with numerous references on general and on fruit- and vegetable-specific principles of flavor genesis from nonvolatile precursors. The employment of stable isotope-labeled precursors to trace their integration in the target compounds or the modification of enzyme activities expanded the methodology for elucidation of biosynthetic pathways leading to fruit flavors (6). Last but not least, the recent advances in analytical chemistry vitally contribute to deepen our knowledge of biochemical pathways of aroma genesis and furthermore permit the detection of new powerful aroma compounds as potential target molecules for biotechnological processes (e.g., Ref. 7). Usually, only a few constituents of a complex plant flavor determine the overall flavor profile. Aroma dilution analysis combined with coupled gas chromatography–olfactometry aids in detecting these so-called character impact compounds. Only trace concentrations of an impact molecule are required to modify a flavor or fragrance composition.

The great potential of *plant cell and tissue cultures* for the production of food ingredients was emphasized by Fu et al. (8). Tremendous progress in the basic understanding of plant metabolic pathways and regulatory mechanisms and in the development of lucrative high-yielding cell lines has been made. In particular, the short culture periods in comparison with the whole plant and the facile accessibility of the genome are attractive merits. An exciting field of application is the bioconversion of externally applied substrates, which is facilitated by the saprophytic nature of most plant cell culture systems. The refinement of abundant agrochemicals from industrial waste streams to high-priced flavor compounds is within reach, as is the precise direction of the flavor profile by addition of potential aroma precursors. The use of plant cell cultures for flavor production has been reviewed by Scragg (9). As the overall secondary metabolite profile of callus and organ cultures may differ significantly from that of the parent plant, exhaustive application and safety considerations and, where necessary, safety testing have to be undertaken before the commercial use of cell culture extracts as food additives (10).

Far-reaching *process developments* are necessary to translate ideas from recombinant DNA technology and cell culture systems into industrial practice. Detailed economic considerations regarding the difficult interplay between technical and business factors may be found in the report of Goldstein (11). Process developments may comprise bioreactor design or the provision of an external accumulation site. The immobilization of cells or enzymes on solid supports may both increase production of secondary metabolites decisively and simplify product purification.

II. RECOMBINANT DNA TECHNOLOGY

A. Genetically Engineered Food Plants

In contrast to numerous efforts to establish pest- or pesticide-resistant field crops on the world market, few genetically engineered plants that provide benefits such as improved flavor properties or enhanced vitamin supply have been developed up to now (Table 1). A major breakthrough in consumer acceptance of genetically modified food plants could possibly be achieved by the work of Ye et al. (12), who transferred the entire provitamin A (β,β-carotene) biosynthetic pathway into the endosperm of rice. In a single *Agrobacterium*-mediated transformation, two plant genes from daffodil (*Narcissus pseudonarcissus*), phytoene synthase and lycopene β-cyclase, and a bacterial phytoene desaturase from *Erwinia uredovora* were introduced into the rice genome. The resulting genetically engineered "golden rice" is capable of forming β,β-carotene from endogenous geranylgeranyl diphosphate and thus may counteract vitamin A deficiency, which is a serious health problem in many third-world countries (12). Carotenes may act as precursors not only of vitamins but also of flavors. Common flavor impact compounds such as the C_{13}-norisoprenoids α-ionone, β-ionone, and grasshopper ketone can be structurally traced back to carotenoid progenitors (13,14).

Overexpression of a yeast $\Delta9$ desaturase gene in tomato (*Lycopersicon esculentum* Mill.) increased not only monoenic fatty acids but also polyunsaturated fatty acids in tomato fruits. The changes of the fatty acid profile were accompanied by changes in certain flavor compounds derived from enzymatic fatty acid degradation. Especially the linolenic acid peroxidation products (Z)-3-hexenal and (Z)-3-hexen-1-ol, derived from (Z)-3-hexenal by alcohol dehydrogenase, were increased in transgenic fruit (15). An approach to the directed modification of the flavor profile of tomato fruit was reported by Speirs et al. (16). Tomato plants were transformed with gene constructs containing a tomato alcohol dehydrogenase complementary DNA (cDNA) in a sense orientation relative to the tomato polygalacturonase promoter to provide fruit-ripening specific expression of the cDNA. The transformed plants displayed enhanced alcohol dehydrogenase (ADH) activities in the ripening fruit, which influenced the balance between some of the aldehydes and the corresponding alcohols associated with flavor perception. In particular, hexanol and Z-3-hexenol levels were increased in fruit with enhanced ADH activity. In a preliminary taste trial, the transgenic tomato fruit exhibited a more intense "ripe fruit" flavor (16).

A transgenic method for producing plant foodstuffs such as fruits (e.g., melons), vegetables, or seeds with a modified sweet component was patented by Tomes (17). A gene that encodes the sweet protein brazzein was linked to a promoter capable of directing the expression of brazzein in a favored

TABLE 1 Genetically Engineered Plants with Modified Aroma Profile

Plant species	Genetic modification	Altered properties	Ref.
Lycopersicon esculentum	Overexpression of a yeast Δ-9 desaturase gene	Levels of (Z)-3-hexenal and (Z)-3-hexen-1-ol increased	15
Lycopersicon esculentum	Insertion of a tomato alcohol dehydrogenase cDNA in sense direction	Hexanol and Z-3-hexenol levels increased	16
Cucumis melo	Transformation with a 1-aminocyclopropane-1-carboxylic acid oxidase antisense gene	Reduction of ethylene synthesis, delayed ripening	18
Cucumis melo	Expression of a brazzein-encoding gene	Enhanced sweet flavor component	17

plant organ and thus enhancing the sweet component of the flavor. *S*-Linalool synthase from *Clarkia breweri* (Onagraceae), an annual plant native to California, was the subject of another patent specification. By expressing the respective nucleic acid sequence in appropriate host plants, enhancement of their scent production has been attempted (18).

1-Aminocyclopropane-1-carboxylic acid oxidase (ACO) is an enzyme involved in the biosynthesis of ethylene, a plant growth regulator initiating fruit ripening. To improve the storage and handling characteristics of cantaloupe charentais melon (*Cucumis melo* var. *cantalupensis*, Naud. cv Védrandais), a cantaloupe melon line was transformed with an ACO antisense gene. A strong reduction of ethylene synthesis and, as a consequence, delayed ripening were achieved. The total quantity of volatiles detected in the antisense fruit was only 20 to 40% of that in the control fruit, indicating that biosynthesis of flavors is strongly controlled by ethylene. Exogenous ethylene was able to restore a qualitative and quantitative aroma profile very similar to that of the control fruit without antisense ACO (19).

Ultimately aiming at the modification of the essential oil composition of aromatic plants by genetic transformation, Faure et al. (20) developed an efficient in vitro shoot regeneration method from spearmint (*Mentha spicata* L.) and peppermint (*Mentha × piperita*) leaf disks. On the basis of these results, Diemer et al. (21,22) established an *Agrobacterium tumefaciens*–mediated transformation procedure for peppermint, spearmint, and cornmint (*Mentha arvensis* L.). The stable integration and expression of two reporter genes was confirmed by polymerase chain reaction (PCR), Southern blot hybridization, reverse transcription PCR (RT-PCR), and a histoenzymatic assay (21,22).

B. Plant Enzymes and Genomics

Although microbial enzymes are already widely applied in industrial flavor production (e.g., Refs. 23,24), the failure in isolating and operating active and stable multienzyme complexes stable in vitro has still hampered broad technical implementation of plant enzymes. Based on up-to-date RNA-DNA techniques, dissection of the complex developmental process of flavor genesis becomes feasible. The identification of flavor related-genes and their corresponding proteins as well as the growing understanding of molecular regulation of gene expression now adds new tools to flavor biotechnology and thus establishes a basis for profitable exploitation.

The genes and peptides of an entire pathway toward the formation of volatile aliphatic and aromatic esters in strawberry fruit (*Fragaria* sp.) have been disclosed by Aharoni et al. (25). DNA sequences that encode strawberry fruit–specific aminotransferase, pyruvate decarboxylase, thiolase, alcohol dehydrogenase, or acyltransferase were cloned and characterized. Further acyltransferases and esterases involved in aroma genesis were isolated from apple (*Malus domestica* Borkh.), mango (*Magnifera indica* L.), and banana (*Musa* sp.). These nucleic acid or protein sequences may be used in expression systems or to modify plants with the goal of producing natural or synthetic flavors. In this context, a novel strawberry acyltransferase was identified by use of cDNA microarrays combined with appropriate statistical analyses. Such microarray assays allow systematic studies of the expression profiles of large subsets of genes in given tissues under specific physiologic and environmental conditions (26,27). Another approach to isolate ripening-related genes from strawberry fruit utilized the differential screening of a high-quality cDNA library, whereby altogether 26 ripening-related cDNAs were identified (28).

Complementary DNA clones encoding 4-coumarate-coenzymeA (CoA) ligase (4CL) (EC 6.2.1.12) and caffeic acid *O*-methyltransferase (EC 2.1.1.6) (29,30), key enzymes of phenylpropanoid metabolism, were isolated from a cDNA library constructed from messenger RNA (mRNA) of a kinetin-treated cell suspension culture of vanilla (*Vanilla planifolia* Andr.). Based on studies using inhibitors of 4CL, it was concluded that down-regulation of this enzyme by the antisense technique would result in a redirection of the flow of phenylpropanoid precursors from lignin biosynthesis into flavor compounds. These attempts hold much promise because vanillin not only is the impact compound of the world's most popular flavor, but also has antioxidant properties in food. The complex interrelations of phenylpropanoid biosynthesis were reviewed by Dixon et al. (31), who presented examples of genetic engineering of, e.g., tobacco (*Nicotiana tabacum* L.), alfalfa (*Medicago sativa* L.), and soybean (*Glycine max* (L.) Merr.) plants and cell cultures to alter pathway flux.

Several plant enzymes involved in the early steps of terpenoid biosynthesis have been characterized, including acetoacetyl-CoA thiolase (EC 2.3.1.9), mevalonate kinase (EC 2.7.1.36), isopentenyl diphosphate isomerase (EC 5.3.3.2), and 3-hydroxy-3-methylglutaryl-CoA synthase (EC 4.1.3.5) (32). Cyclases convert linear isoprenoid diphosphates such as geranyl diphosphate, farnesyl diphosphate, and geranylgeranyl diphosphate into a variety of mono- and polycyclic hydrocarbons and alcohols. A number of terpene cyclases from plants, representing soluble, magnesium-containing enzymes, have been cloned and expressed in *Escherichia coli*. Thus, there is now the potential to engineer plants to produce specific cyclic terpenes for use in the flavor and fragrance industries (33).

Cloning and expression of the monoterpene synthases (−)-4-*S*-limonene synthase from spearmint, (+)-bornyl diphosphate synthase from sage (*Salvia officinalis* L.), and (−)-pinene synthase from grand fir (*Abies gran-*

dis) in *E. coli* enabled Schwab et al. (34) to gain insight into the mechanistic procedures of the reaction sequence toward cyclic monoterpenes. An overview of the enzymology and regulation of essential oil biosynthesis with detailed description of the reaction mechanisms leading to the basic isoprenoid skeletons is given in Ref. 35.

Natural (Z)-3-hexenol (leaf alcohol), traditionally isolated from mint terpene fractions, is formed from linoleic and linolenic acid via the lipoxygenase pathway. Lipoxygenases have been isolated, cloned, and characterized from various plant sources (for a review see Ref. 36). The enzymatic genesis of the 13-hydroperoxy linoleic and linolenic acid is followed by the hydroperoxide lyase (HPO-lyase)–catalyzed cleavage to (Z)-3-hexenal, which is subsequently reduced to the corresponding alcohol. As in a reconstituted production system the activity of hydroperoxide lyase proved to be the rate-limiting factor, the gene coding for this enzyme was cloned from banana and heterologously expressed in yeast cells to yield a highly active lyase material (37). Another plant HOP-lyase was purified 300-fold from tomatoes (38). Whereas only 13-hydroperoxides from linoleic acid and α-linolenic acid were cleaved by the tomato enzyme, HOP-lyase from alfalfa (*Medicago sativa* L.) also accepted 9-hydroperoxides as substrates to form the respective volatile C9-aldehydes (39). The specific activity for 9-hydroperoxy fatty acids was about 50% of the activity for the 13-isomers.

The characteristic flavor of onion occurs when the enzyme alliinase (EC 4.4.1.4) hydrolyzes *S*-alk(en)yl-L-cysteine sulfoxides (ACSOs) to form pyruvate, ammonia, and sulfur-containing volatiles. Physical characterization of alliinase and molecular analysis of the respective cDNA revealed that two genes and thus two protein subunits were expressed in onion bulb tissue. Although these genes for alliinase are highly homologous in their DNA sequence, there are differences in the proteins that they code for, probably due to varying degrees of glycosylation (40).

Flavorless glycosides represent one accumulation form of aroma substances in fruit and in many other plant tissues. In addition to developments in the analysis of these polar plant constituents, current attention is focused on biotechnological methods for flavor release and flavor enhancement through enzymatic hydrolysis of the glycosidically bound aroma precursors. The liberation of the "bound" aroma portion offers a tool for the production of natural flavors from otherwise waste materials such as peelings, skins, and stems. Because the practical application of endogenous and exogenous glycosidases is limited by their low activities at neutral pH values and strong inhibition by glucose, the construction of chimeric genes with improved hydrolytic properties and their overexpression in different hosts are the subjects of actual research (Refs. 13,14 and references therein).

An industrially relevant route for the production of a "natural" topnote flavor of concord grapes (*Vitis labrusca* "concord"), methylanthranilate (MA), is the peroxidase-catalyzed *N*-demethylation of methyl *N*-methylanthranilate (MNMA). Comparison of different commercial peroxidase preparations showed soybean peroxidase to be the most effective biocatalyst for the *N*-demethylation of MNMA to MA. Upon complete conversion of MNMA, the yield of soybean peroxidase–catalyzed MA amounted to 82% within 10 minutes at 70°C and pH 4 (41).

Current research is focused on the improvement of cacao plants (*Theobroma cacao* L.) for the food industry. The efficiency of breeding programs could be increased if genetically based maps were available and markers associated with major quantitative trait loci for quality and productivity could be identified. Even though the target organism in this case is not genetically modified, genomics acts as an invaluable research tool (42).

III. PLANT CELL, TISSUE, AND ORGAN CULTURES

The term *tissue culture* is applied to any nondifferentiated cell culture grown on solid or, as suspension culture, in liquid medium. As the propagation of the cultures is strictly based on mitotic events, the entire genome and, thus, the full potential to form flavors and fragrances are maintained in each cultured cell. However, organized cultures often exhibit an enhanced capacity to form volatile flavors when compared with that of unorganized cell suspension cultures. Commonly used are the hairy root cultures induced by the transformation of aseptic plantlets with *Agrobacterium rhizogenes*. Independent of the type of cell culture used, usually simultaneous application of different strategies for the improvement of yields is necessary to reach time-dependent volumetric yields [mg product (L × day)$^{-1}$] sufficient for industrial commercialization. These strategies include the selection of stable, high-yielding cell lines, variation of medium components and gas phase composition, precursor feeding, use of elicitors, in situ product removal, and immobilization techniques. Although individual strategies may result in enhanced secondary metabolite formation, often several strategies have to be combined to give a synergistic response (43,44).

Numerous attempts to produce flavor and aroma compounds by plant cell, tissue, and organ cultures have been described (9,24,45,46). To avoid recapitulations, only references that were not cited in the preceding reviews were considered for the present section.

A. Tissue Cultures

Capsaicin, (Fig. 2), the major pungent principle of chilli pepper (*Capsicum frutescens* L.), may be extracted from callus cultures. By developing cell

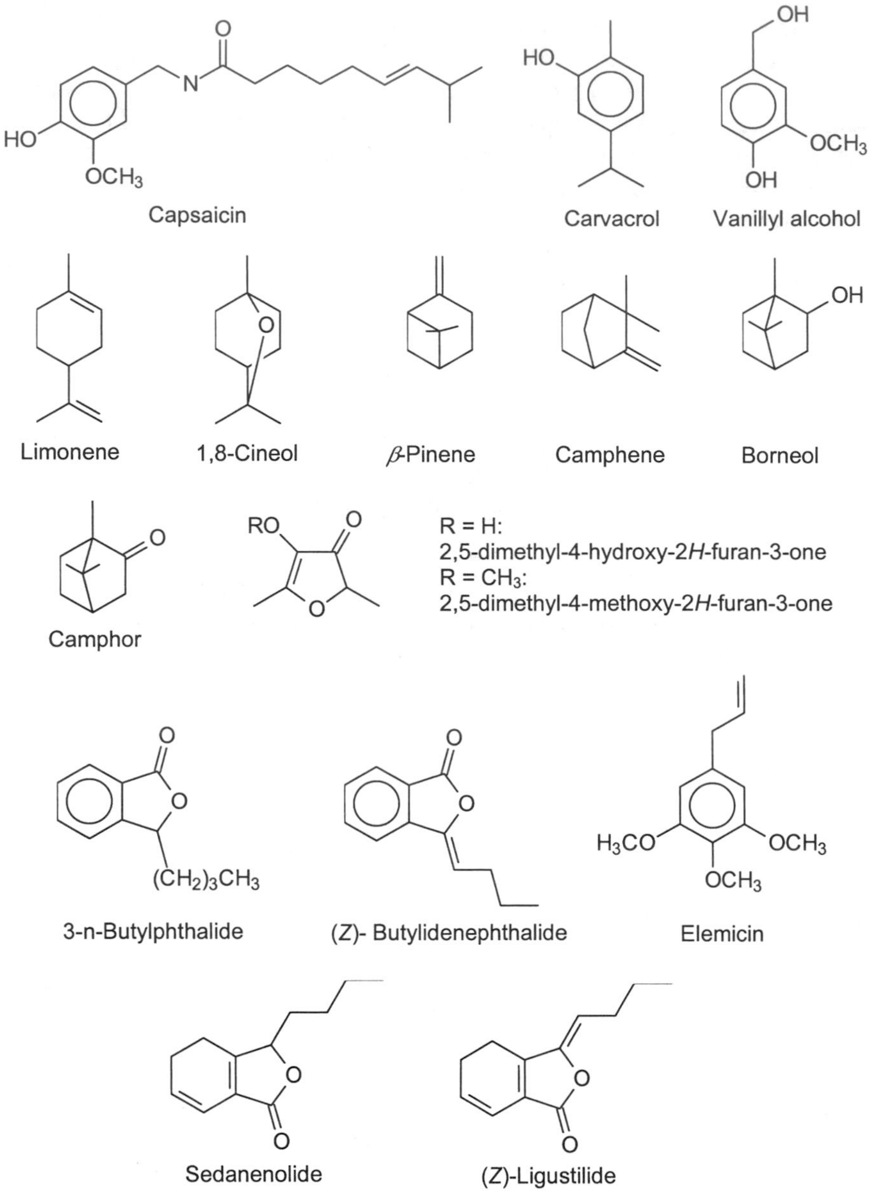

FIGURE 2   Flavors and fragrances from plant cell cultures.

lines resistant to p-fluorophenylalanine (PFP), capsaicin yield could be increased up to 45% over normal cell lines (80 μg capsaicin/g fresh weight). The activity profile of phenylalanine ammonia lyase, the enzyme responsible for conversion of phenylalanine to (E)-cinnamic acid, exhibited no correlation with the capsaicin content in both control and PFP-resistant cells (47).

In the flavor and fragrance industry there is an enormous demand for essential oils, and many of the more than 3000 different essential oils have been utilized in the creation of fragrances. Consequently, there has been extensive research on the production of essential oils by plant cell cultures.

Only very low yields of volatile oil were achieved with callus cultures of pot marjoram (*Origanum vulgare* L.) (48). In contrast to the composition of essential oil from plants cultivated in the field, microdistillation of calli revealed only three major essential oil constituents, one of which was identified as carvacrol (Fig. 2).

In contrast to expectations, the source tissue of the explant may crucially affect the production of volatile compounds of the resulting tissue cultures. Whereas embryogenic cell lines of sweet orange (*Citrus sinensis* (L.) Osbeck) emitted a sweet, fruity aroma, nonembryogenic cell lines derived from immature juice vesicles failed to form the characteristic flavor constituents of sweet orange. Seventeen aroma active compounds were identified from embryogenic cell lines, altogether amounting to 420 mg volatiles/kg fresh weight tissue (49).

A frequently underestimated aspect of culturing plant cells is the influence of the light regime. For the callus cells of grapefruit (*Citrus paradisi* Maof.), lemon (*C. limon* (L.) Burm.), and lime (*C. aurantiifolia* (Christm. et Panz.) Swingle) a photomixotrophic state was indispensable for the generation of monoterpenes (50). The chlorophyll content and accumulation of isoprenoid-derived volatiles were positively correlated in each case. After optimization of the growth medium's phytoeffector composition and the light regime, about 40 volatile mono- and sesquiterpene hydrocarbons, oxygenated terpenes, and aliphatic aldehydes could be recovered from grapefruit callus cultures. The best yielding callus contained about 186 mg aroma active compounds per kg wet weight, representing about 5% of the volatiles found in peel tissue of the whole fruit.

Similarly, the accumulation of chlorophyll and volatile oligoisoprenoids by white diosma (*Coleonema album* Thunb.) photomixotrophic cell cultures was favoured by high light intensities, an extended photoperiod, and elevated concentrations of phytoeffectors. Total volatiles, including e.g. limonene and phellandrenes, accumulated to approximately 73 mg per kg wet weight (51).

Few differences in the growth patterns were observed between dark- and light-grown vanilla (*Vanilla planifolia*) cultures. However, the light conditions did affect the production of compounds associated with the vanillin pathway, particularly of 4-hydroxy-3-methoxybenzyl alcohol (vanillyl alcohol) (Fig. 2) (52).

Still a serious problem in establishing plant callus and suspension cultures is the risk of endogenous microbial infections, for which remedy can be found in the application of an appropriate antibiotic. On the other hand, microbial infection may even contribute to flavor biosynthesis. Elicitors, components of biological origin involved in inter- or intraspecies interactions of plants, may be of significance for the formation of flavor constituents. In cell cultures of rosemary (*Rosmarinus officinalis* L.) emanating cineole and β-pinene (Fig. 2), persistent contamination with *Pseudomonas mallei* was reported (53). Unintended elicitation may be proposed to explain this observation.

Two important character impact compounds of strawberry flavor, the furanones 2,5-dimethyl-4-hydroxy-2*H*-furan-3-one (DMHF) and 2,5-dimethyl-4-methoxy-2*H*-furan-3-one (mesifuran), were synthesized by strawberry tissue cultures (*Fragaria* × *ananassa*, cv. Elsanta) after they were treated with *Methylobacterium extorquens* (Fig. 2). Untreated (sterile) cell cultures and the bacteria alone were not capable of forming DMHF or mesifuran. A biosynthetic pathway in the strawberry callus–*Methylobacterium* system for the two furanones was proposed as follows: endogenous strawberry 1,2-propanediol is oxidized to 2-hydroxypropanal (lactaldehyde) by the *Methylobacterium* species. The microbially derived lactaldehyde could be further condensed with dihydroxyacetone phosphate by the strawberry cells to form the furanone progenitor 6-deoxy-D-fructose-1-phosphate (54). Photomixotrophic cell cultures of parsley (*Petroselinum crispum* (Mill.) Nym.) accumulated volatiles solely after treatment with the autoclaved homogenate of fungal cells. Homogenates of the wood-destroying basidiomycetes *Polyporus umbellatus* and *Tyromyces sambuceus* induced a spicy, celery-like odor, triggered by elemicin (5-allyl-1,2,3-trimethoxybenzene), 3-*n*-butylphthalide, (*Z*)- and (*E*)-butylidenephthalide, sedanenolide, and (*Z*)-ligustilide (Fig. 2) (55).

Considerable amounts of odorous mono- and sesquiterpenes have been recovered from in vitro callus cultures of Brazilian snapdragon (*Otacanthus coeruleus*) (Scrophulariaceae), an ornamental pot plant originating from east Brazil. Interestingly, the amount of essential oil extracted from the nutrient media was often higher than the amount extracted from the cell cultures. High-sucrose treatments, especially from 40 g L^{-1} on, increased the oil quantities found in the medium, probably due to increased osmotic stress. Up to 0.34% total oil content was reached in cultures plus medium, which was more than in the respective plants (56). With the aim of obtaining high levels of essential oils, cell cultures of two genotypes of rosemary were established.

The compositions of the growth media for both callus formation and regeneration of plants crucially influenced the monoterpenoid profile. Concentrations of calcium ions, sucrose, and plant growth regulators significantly affected the yields of camphene, 1,8-cineole, linalool, camphor, borneol, and bornyl acetate (57) (Fig. 2).

B. Organ Cultures

Hairy root cultures of anise (*Pimpinella anisum* L.) were grown in different nutrient media in darkness and under periodic light conditions (58,59). The composition of the essential oils obtained from hairy root cultures differed significantly from that of the fruits. Whereas the major components of the essential oils from the hairy root cultures were the anethole precursor (*E*)-epoxypseudoisoeugenyl 2-methylbutanoate, zingiberene, β-bisabolene, geijerene, and pregeijerene, the terpene spectrum of the fruits was dominated by (*E*)-anethole (Fig. 3). The highest essential oil yield obtained from hairy root cultures was 0.1%, which was comparable to that obtained from the roots of the parent plant and, when considering hairy roots on a dry weight basis, also that of the fruits. With regard to potential biotechnological exploitation, the morphological stability of the "rooty" phenotype is of great importance. In one of four growth media tested, the hairy root cultures revealed high morphological stability with no dedifferentiation or greening even after more than 2 years of culture (60). Encouraged by these results,

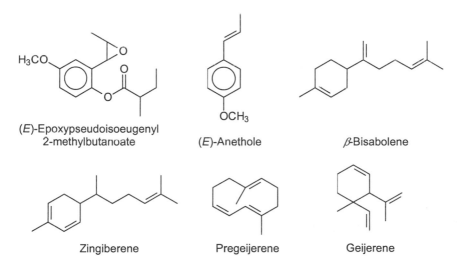

(*E*)-Epoxypseudoisoeugenyl
2-methylbutanoate

(*E*)-Anethole

β-Bisabolene

Zingiberene

Pregeijerene

Geijerene

FIGURE 3 Flavors from hairy root cultures and fruits of anise (*Pimpinella anisum* L.).

Matsuda et al. (61) produced mutant hairy roots of musk melon (*Cucumis melo* L.) with an altered metabolism of essential oils by means of T-DNA insertion mutagenesis. From more than 6500 clones, five fragrant hairy root clones were obtained and the clone emitting the strongest fruity flavor of ripe fresh melons was selected for further investigations. Extraction of the volatile compounds and identification by means of gas-liquid chromatography–mass spectrometry (GLC-MS) revealed the presence of (Z)-3-hexenol, (E)-2-hexenal, 1-nonanol, and (Z)-6-nonenol, which also shape the flavor of melon fruits. Aroma emanation was successfully maintained in the hairy roots when they were subcultured repeatedly for more than 3 years. Preliminary scale-up experiments using a 4-L jar fermenter showed an overall profile of the extracted essential oils that was very similar to that of hairy roots cultured on a routine laboratory scale. When compared with that of the fresh ripe melon fruit, the yield of aroma compounds in this scale-up approach was approximately 6.5-fold higher, indicating possible biotechnological exploitation. Plant sources of flavor components from cell cultures are summarized in Table 2.

C. Biotransformation by Plant Cell Cultures

As with microbial conversions, cultured plant cells can be employed as "cell factories" to conduct bioconversions of exogenous substrates (Fig. 4). Looking at flavor effectiveness and availability, monoterpenes are of outstanding interest. A comprehensive overview of biotransformations of monoterpenoid alcohols, aldehydes, ketones, and oxides by plant and microbial cell cultures was given by Shin (62). The conversion of monoterpenes, steroids, and indole alkaloids using numerous cell cultures was summarized by Hamada and Furuya (63). Special attention in these reviews was dedicated to the regiospecificity and stereospecificity of the biochemical reactions as well as to immobilization techniques for cells or enzymes.

Immobilized and free cells of kangaroo apple (*Solanum aviculare*) and of a yam species (*Dioscorea deltoidea*) were utilized for the oxidation of (−)-limonene to (Z)- and (E)-carveol and to carvone. Depending on the immobilization medium, either carvone or (Z)- and (E)-carveol were formed predominantly (64).

Suspended cells of chilli pepper (*Capsicum frutescens*) accumulated vanilla flavor metabolites such as vanillin, vanillic acid, and ferulic acid when fed with isoeugenol. Increased biotransformation rates of isoeugenol could be achieved by immobilizing cells with sodium alginate and applying fungal elicitors. Product yields up to 23 μg mL^{-1} were reached by the simultaneous addition of β-cyclodextrin and isoeugenol (2.5 mM) (65). Further attempts to optimize the biotechnological production of vanilla flavor compounds have been reviewed (66).

TABLE 2 Possible Plant Sources of Flavor Components from Cell Cultures

Common name	Botanical name	Principal metabolites	Ref.
Paprika	*Capsicum frutescens*	Capsaicin	47
Oregano	*Origanum vulgare*	Carvacrol	48
Sweet orange	*Citrus sinensis*	3-hydroxy-2-butanone/ethyl acetate/ acetaldehyde	49
Grapefruit	*Citrus paradisi*	Limonene/nootkatone/citronellal	50
White diosma	*Coleonema album*	Limonene/phellandrenes	51
Vanilla	*Vanilla planifolia*	Vanillin/vanillyl alcohol	52
Rosemary	*Rosmarinus officinalis*	β-Pinene/camphene/cineole/linalool/ camphor	53,57
Strawberry	*Fragaria × ananassa*	2,5-Dimethyl-4-hydroxy-2*H*-furan-3-one/ 2,5-dimethyl-4-methoxy-2*H*-furan-3- one	54
Brazilian snapdragon	*Otacanthus coeruleus*	Menthone/valencene	56
Parsley	*Petroselinum crispum*	3-*n*-Butylphthalide/butylidenephthalide/ elemicin/sedanenolide/ligustilide	55
Anise	*Pimpinella anisum*	Epoxypseudoisoeugenyl 2-methlbutyrate/zingiberene/ β-bisabolene/geijerene/pregeijerene	58,59
Musk melon	*Cucumis melo*	(*Z*)-3-Hexenol/(*E*)-2-hexenal/1-nonanol/ (*Z*)-6-nonenol	61

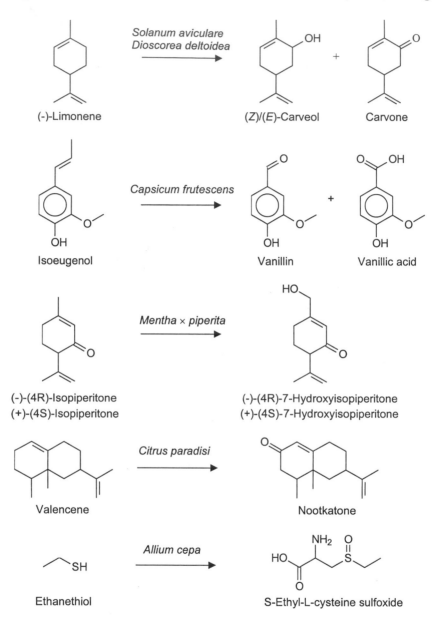

FIGURE 4 Bioconversion of exogenous substrates by plant cell cultures.

The potential of peppermint (*Mentha* × *piperita*) cell suspension cultures to synthesize menthol has been investigated intensively. Studies showed that peppermint cells possess extensive hydroxylation activity toward terpenes. After application of the biosynthetic precursors (−)-(4*R*)- and (+)-(4*S*)-isopiperitones to cell suspension cultures, metabolism yielded the corresponding 7-hydroxyisopiperitones, which were concomitantly converted into the respective glucopyranosides (67).

The flavor synthesis of various *Allium* tissue cultures (onion, garlic, and chive) was less productive when compared with the respective plants (68). This was attributed to the low aroma precursor concentrations inside the cells rather than to the C-S lyase activity. Based on the knowledge of the well-characterized biosynthetic pathways to the flavor precursors (+)-*S*-alk(en)yl-L-cysteine sulfoxides (ACSOs) in *Allium* species, attempts were made to alter the flavor profile of onion root cultures by feeding of aroma precursors. Addition of cysteine, glutathione, and methionine increased the yields of methyl- and propenylcysteine sulfoxides, and the ratio of the propenyl to the methyl form was shifted significantly depending on the amount and type of the precursor used. *S*-Ethyl-L-cysteine sulfoxide, which is not a naturally occurring compound, was produced by application of ethanethiol to the root cultures, indicating the possible use of such cultures for the production of novel homologous metabolites (69).

Incubation of suspended, stationary phase cells of grapefruit with exogenous valencene led to the intermediary formation of the 2-hydroxy derivative, followed by conversion to the 2-oxo compound, nootkatone. The transformation rate was 68% in 24 hours, and when the concentration of about 0.7 mg L^{-1} was reached, it was maintained for another 48 hours without noticeable change (50).

IV. PROCESS DEVELOPMENTS

To increase production rates of secondary metabolites, different in situ extraction procedures have been applied to plant cell and tissue cultures. The extraction phases can either be water-immiscible liquids, such as *n*-hexadecane or Miglyol™, or solid adsorbents such as the hydrophobic polystyrene-divinylbenzene resins (e.g., Amberlite XAD). Zeolites have also been successfully applied for the specific accumulation of essential oil compounds from liquid fermentation media (70). By provision of an external accumulation site, further metabolism and degradation of products in the medium as well as cytotoxic product concentrations may be impeded. Moreover, and especially for highly volatile flavors, the danger of physicochemical losses of product is minimized by addition of an appropriate adsorbent. In combination with further optimization attempts, such as immobilization or elicitation, the productivity of a cell culture process can be vitally enhanced (for reviews see Refs. 24,71). Superior production rates of cell culture systems compared with field-grown material are sought by numerous efforts to improve bioreactor design. Current apparatus proposals for suspension cultures as well as for organized cultures were summarized by Scragg (9). Design equations for oxygen transfer rates in large-scale root culture reactors have been discussed by Tescione et al. (72).

The tomato flavor enzyme system (lipoxygenase and hydroperoxide lyase) was harnessed as a crude enzyme preparation in a hollow-fiber reactor to produce hexanal from linoleic acid. At exogenous substrate (linoleic acid) concentrations of 16 mmol L^{-1}, hexanal production rates of about 5 μg min^{-1} were achieved. The reactor system proved to be stable over an operation period of 5 days, indicating that flavor production with immobilized membrane-associated enzymes in a hollow-fiber reactor is a promising technique. It allows retention of the enzyme system and substrate, with concomitant removal of the product (73).

For the synthesis of extracellular metabolites the immobilization of intact plant cells provides several processing advantages, such as protecting cells from mechanical stress, enhanced productivity, high stability, and facilitated product recovery (10). Several immobilization agents (polyurethane foam, carrageenan, alginate, pectate, polyphenyleneoxide) for cell cultures of *Solanum aviculare* and *Dioscorea deltoidea* did not affect the biotransformation course of $(-)$-limonene to (Z)- and (E)-carveol and to carvone but significantly changed product ratios (64).

V. CONCLUSIONS

The ever increasing consumer demand for "naturalness" of foodstuffs and cosmetics along with the enormous potential of plants in biosynthesis and bioconversion represents an attractive basis for plant aroma biotechnology. The potential market for biotechnology-derived flavors and fragrances is estimated to be 10% or more of the overall flavor market, which was expected to amount to $10 billion in 2000, tendency ascending (1). Indeed, up to now, the lion's share thereof accounts for microbially catalyzed processes (74). Contrary to the successful generation of high-value pharmaceuticals, plant cell cultures still need to be shown to be competitive with field plants or microorganisms in the production of flavor compounds on an industrial scale. To reach this aim, the productivity of the cell culture systems will need to be improved significantly. Most promising starting points are given by recombinant DNA technologies (75). The genetic information responsible for fruit and herbal flavor formation can be isolated, characterized, and afterward either transferred into a suitable microbial expression system or used specifically to modify food plants. A steadily increasing number of publications and patent applications on plant genes and transgenic plants indicates rapid scientific progress as well as industrial interest. After all, the short- and middle-term success of these technologies will, beyond technological and economic considerations, depend crucially on the acceptance of the novel products by the consumer. Extensive evaluation and communication of scientific facts will be indispensable to deal with widespread concerns before achievements in basic science will turn into marketable flavors and flavor-enriched products.

REFERENCES

1. R Shamel, A Udis-Kessler. The sweet taste of success. Chem Ind 10(1):27–28, 2000.
2. T Münch, B Müller. Bioengineering challenges of natural flavor production. In: P Schieberle, KH Engel, eds. Frontiers of Flavour Science. Garching: Deutsche Forschungsanstalt für Lebensmittelchemie, 2000, pp 343–347.
3. D Havkin-Frenkel, A Podstolski, E Witkowska, P Molecki, M Mikolajczyk. Vanillin biosynthetic pathways. In: TJ Fu, G Singh, WA Curtis, eds. Plant Cell and Tissue Culture for the Production of Food Ingredients. New York: Kluwer Academic/Plenum Publishers, 1999, pp 35–43.
4. MM Leahy, RG Roderick. Fruit flavor biogenesis. In: R Teranishi, EL Wick, I Hornstein, eds. Flavor Chemistry. New York: Kluwer Academic/Plenum Publishers, 1999, pp 275–286.
5. G Takeoka. Flavor chemistry of vegetables. In: R Teranishi, EL Wick, I Hornstein, eds. Flavor Chemistry. New York: Kluwer Academic/Plenum Publishers, 1999, pp 287–304.
6. A Mosandl, S Fuchs, M Wüst, T Beck. Biosynthetic studies on monoterpenoids using stable isotope labelling and enantioselective analysis. In: P Schieberle, KH Engel, eds. Frontiers of Flavour Science. Garching: Deutsche Forschungsanstalt für Lebensmittelchemie, 2000, pp 433–439.
7. S Widder, CS Lüntzel, T Dittner, W Pickenhagen. 3-Mercapto-2-methylpentan-1-ol, a new powerful aroma compound. J Agric Food Chem 48:418–423, 2000.
8. TJ Fu, G Singh, WR Curtis. An introduction. In: TJ Fu et al., eds. Plant Cell and Tissue Culture for the Production of Food Ingredients. New York: Kluwer Academic/Plenum Publishers, 1999, pp 1–6.
9. AH Scragg. The production of aromas by plant cell cultures. In: RG Berger, ed. Biotechnology of Aroma Compounds. Berlin: Springer, 1997, pp 239–263.
10. TJ Fu. Plant cell and tissue culture for food ingredient production: safety considerations. In: TJ Fu et al., eds. Plant Cell and Tissue Culture for the Production of Food Ingredients. New York: Kluwer Academic/Plenum Publishers, 1999, pp 237–250.
11. WE Goldstein. Economic considerations for food ingredients produced by plant cell and tissue culture. In: TJ Fu et al., eds. Plant Cell and Tissue Culture for

the Production of Food Ingredients. New York: Kluwer Academic/Plenum Publishers, 1999, pp 195–213.

12. X Ye, S Al-Babili, A Klöti, J Zhang, PBP Lucca, I Potrykus. Engineering the provitamin A (β-carotene) biosynthetic pathway into (carotenoid-free) rice endosperm. Science 287:303–305, 2000.

13. P Winterhalter, GK Skouroumounis. Glycoconjugated aroma compounds: occurrence, role and biotechnological transformation. In: RG Berger, ed. Biotechnology of Aroma Compounds. Berlin: Springer, 1997, pp 73–106.

14. P Winterhalter. Carotenoid-derived aroma compounds: biogenetic and biotechnological aspects. In: GR Takeoka, R Teranishi, PJ Williams, A Kobayashi, eds. Biotechnology for Improved Foods and Flavors. ACS Symposium Series. Washington, DC: American Chemical Society, 1996, pp 295–308.

15. C Wang, CK Chin, CT Ho, CF Hwang, JJ Polashock, CE Martin. Changes of fatty acids and fatty acid–derived flavor compounds by expressing the yeast delta-9 desaturase gene in tomato. J Agric Food Chem 44:3399–3402, 1996.

16. J Speirs, E Lee, K Holt, K Yong-Duk, NS Scott, B Loveys, W Schuch. Genetic manipulation of alcohol dehydrogenase levels in ripening tomato fruit affects the balance of some flavor aldehydes and alcohols. Plant Physiol 117:1047–1058, 1998.

17. DT Tomes. Methods and compositions for production of plant foodstuffs with enhanced sweet component of flavor. Patent WO9742333, 1997.

18. E Pichersky. Use of linalool synthase in genetic engineering of scent production. Patent WO9715584, 1997.

19. AD Bauchot, DS Mottram, P John. Aroma formation in cantaloupe charentais melon. In: P Schieberle, KH Engel, eds. Frontiers of Flavour Science. Garching: Deutsche Forschungsanstalt für Lebensmittelchemie, 2000, pp 463–468.

20. O Faure, F Diemer, S Moja, F Jullien. Mannitol and thidiazuron improve in vitro shoot regeneration from spearmint and peppermint leaf disks. Plant Cell Tissue Organ Cult 52:209–212, 1998.

21. F Diemer, JC Caissard, S Moja, F Jullien. Agrobacterium tumefaciens–mediated transformation of Mentha spicata and Mentha arvensis. Plant Cell Tissue Organ Cult 57:75–78, 1999.

22. F Diemer, F Jullien, O Faure, S Moja, M Colson, E Matthys-Rochon, JC Caissard. High efficiency transformation of peppermint (Mentha × piperita L.) with Agrobacterium tumefaciens. Plant Sci 136:101–108, 1998.

23. P Schreier. Enzymes and flavour biotechnology. In: RG Berger, ed. Biotechnology of Aroma Compounds. Berlin: Springer, 1997, pp 51–72.

24. RG Berger. Aroma Biotechnology. Berlin: Springer, 1995, pp 92–115.

25. A Aharoni, HA Verhoeven, J Luecker, AP O'Conell, AJ Van Tunen. Fruit flavour related genes and the use thereof. Patent WO0032789, 2000.

26. A Aharoni, LCP Keizer, HJ Bouwmeester, Z Sun, M Alvarez-Huerta, HA Verhoeven, J Blaas, AMML van Houwelingen, RCH De Vos, H van der Voet, RC Jansen, M Guis, J Mol, RW Davis, M Schena, AJ van Zunen, AP O'Connell. Identification of the SAAT gene involved in strawberry flavor biogenesis by use of DNA microarrays. Plant Cell 12:647–662, 2000.

27. B Lemieux, A Aharoni, M Schena. Overview of DNA chip technology. Mol Breed 4:277–289, 1998.

28. K Manning. Isolation of a set of ripening-related genes from strawberry: their identification and possible relationship to fruit quality traits. Planta 205:622–631, 1998.

29. ZT Xue, PE Brodelius. Kinetin-induced caffeic acid O-methyltransferases in cell suspension cultures of *Vanilla planifolia* Andr. and isolation of caffeic acid O-methyltransferase cDNAs. Plant Physiol Biochem 36:779–788, 1998.

30. PE Brodelius, ZT Xue. Isolation and characterization of a cDNA from cell suspension cultures of *Vanilla planifolia* encoding 4-coumarate: coenzyme A ligase. Plant Physiol Biochem 35:497–506, 1997.

31. RA Dixon, PJ Howles, C Lamb, K Korth, XZ He, VJH Sewalt, S Rasmussen. Plant secondary metabolism. In: TJ Fu et al., eds. Plant Cell and Tissue Culture for the Production of Food Ingredients. New York: Kluwer Academic/Plenum Publishers, 1999, pp 7–22.

32. R van der Heijden, AE Schulte, AC Ramos Valdivia, R Verpoorte. Characterization of some isoprenoid-biosynthetic enzymes from plant cell cultures. In: K Kieslich, CP van der Beek, JAM de Bont, WJJ van den Tweel, eds. New Frontiers in Screening for Microbial Biocatalysts. Amsterdam: Elsevier Science, 1998, pp 177–184.

33. MH Beale, AL Phillips. The terpenoid pathway—closing the loop. Biochem Soc Trans 27:A17, 1999.

34. W Schwab, DC Williams, R Croteau. On the mechanism of monoterpene synthases: stereochemical aspects of (−)-4S-limonene synthase, (+)-bornyl diphosphate synthase and (−)-pinene synthase. In: P Schieberle, KH Engel, eds. Frontiers of Flavour Science. Garching: Deutsche Forschungsanstalt für Lebensmittelchemie, 2000, pp 445–451.

35. D McCaskill, R Croteau. Prospects for the bioengineering of isoprenoid biosynthesis. In: RG Berger, ed. Advances in Biochemical Engineering Biotechnology. Berlin: Springer, 1997, pp 107–146.

36. R Casey, SI West, D Hardy, DS Robinson, Z Wu, RK Hughes. New frontiers in food enzymology: recombinant lipoxygenases. Trends Food Sci Technol 10:297–302, 1999.

37. AHA Muheim, B Schilling, K Lerch. The impact of recombinant DNA-technology on the flavour and fragrance industry. R Soc Chem 214:11–20, 1997.

38. CNSP Suurmeijer, M Perez-Gilabert, DJ van Unen, HTWM van der Hijden, GA Veldink, JFG Vliegenthart. Purification, stabilization and characterization of tomato fatty acid hydroperoxide lyase. Phytochemistry 53:177–185, 2000.

39. MA Noordermeer, GA Veldink, JFG Vliegenhart. Alfalfa contains substantial 9-hydroperoxide lyase activity and a 3Z:2E-enal isomerase. FEBS Lett 443:201–204, 1999.

40. SA Clark, ML Shaw, D Every, JE Lancaster. Physical characterization of alliinase, the flavor generating enzyme in onions. J Food Biochem 22:91–103, 1998.

41. MJH van Haandel, FCE Saraber, MG Boersma, C Laane, Y Fleming, H Weenen, IMCM Rietjens. Characterization of different commercial soybean

peroxidase preparations and use of the enzyme for N-demethylation of methyl N-methylanthranilate to produce the food flavor methylanthranilate. J Agric Food Chem 48:1949–1954, 2000.

42. RD Pridmore, D Crouzillat, C Walker, S Foley, R Zink, MC Zwahlen, H Brüssow, V Petiard, B Mollet. Genomics, molecular genetics and the food industry. J Biotechnol 78:251–258, 2000.

43. ML Shuler. Overview of yield improvement strategies for secondary metabolite production in plant cell culture. In: TJ Fu et al., eds. Plant Cell and Tissue Culture for the Production of Food Ingredients. New York: Kluwer Academic/Plenum Publishers, 1999, pp 75–83.

44. R Verpoorte, R van der Heijden, HJG ten Hoopen, J Memelink. Novel approaches to improve plant secondary metabolite production. In: TJ Fu et al., eds. Plant Cell and Tissue Culture for the Production of Food Ingredients. New York: Kluwer Academic/Plenum Publishers, 1999, pp 85–100.

45. H Dörneburg, D Knorr. Generation of colors and fragrancies in plant cell and tissue cultures. Crit Rev Plant Sci 15:141–168, 1996.

46. CG Kumar, A Sharma, SK Kanawjia. Prospects for bioproduction of food flavours. Indian Food Ind 17:98–111, 1998.

47. T Sudhakar Johnson, R Sarada, GA Ravishankar. Capsaicin formation in p-fluorophenylalanine resistant and normal cell cultures of *capsicum frutescens* and activity of phenylalanine ammonia lyase. J Biosci 23:209–212, 1998.

48. KP Svoboda, RP Finch, E Cariou, SG Deans. Production of volatile oils in tissue culture of *Origanum vulgare* and *Tanacetum vulgare*. Acta Hortic 390: 147–151, 1995.

49. RP Niedz, MG Moshonas, B Peterson, JP Shapiro, PE Shaw. Analysis of sweet orange (*Citrus sinensis* L. Osbeck) callus cultures for volatile compounds by gas chromatography with mass selective detector. Plant Cell Tissue Organ Cult 51:181–185, 1997.

50. G Reil, RG Berger. Accumulation of chlorophyll and essential oils in photomixotrophic cell cultures of *Citrus* sp. Z Naturforsch 51c:657–666, 1996.

51. G Reil, RG Berger. Variation of chlorophyll and essential oils in photomixotrophic cell cultures of *Coleonema album* (Thunb.). J Plant Physiol 150:160–166, 1997.

52. D Havkin-Frenkel, A Podstolski, D Knorr. Effect of light on vanillin precursors formation by in vitro cultures of *Vanilla planifolia*. Plant Cell Tissue Organ Cult 45:133–136, 1996.

53. A Shervington, R Darwish, F Afifi. Volatile oil produced by microbial infected *Rosmarinus officinalis* callus and suspension cultures. Alex J Pharm Sci 11: 81–84, 1997.

54. I Zabetakis. Enhancement of flavour biosynthesis from strawberry (*Fragaria* × *ananassa*) callus cultures by *Methylobacterium* species. Plant Cell Tissue Organ Cult 50:179–183, 1997.

55. G Reil, RG Berger. Elicitation of volatile compounds in photomixotrophic cell culture of *Petroselinum crispum*. Plant Cell Tissue Organ Cult 46:131–136, 1996.

56. AC Ronse, H de Pooter, A van De Vyver, MP de Proft. *Otacanthus* species: in vitro culture, plant propagation, and the production of essential oil. In: YPS Bajaj, ed. Biotechnology in Agriculture and Forestry. Berlin: Springer, 1998, pp 305–319.

57. AA Tawfik, PE Read, SL Cuppett. *Rosmarinus officinalis* L. (rosemary): in vitro culture, regeneration of plants, and the level of essential oil and mono-terpenoid constituents. In: YPS Bajaj, ed. Biotechnology in Agriculture and Forestry. Berlin: Springer, 1998, pp 349–365.

58. PM Santos, AC Figueiredo, MM Oliveira, JG Barosso, LG Pedro, SG Deans, AKM Younus, JJC Scheffer. Essential oils from hairy root cultures and from fruits and roots of *Pimpinella anisum*. Phytochemistry 48:455–460, 1998.

59. N Andarwulan, K Shetty. Phenolic content in differentiated tissue cultures of untransformed and *Agrobacterium*-transformed roots of anise (*Pimpinella anisum* L.). J Agric Food Chem 47:1776–1780, 1999.

60. PM Santos, AC Figueiredo, MM Oliveira, JG Barosso, LG Pedro, SG Deans, AKM Younus, JJC Scheffer. Morphological stability of *Pimpinella anisum* hairy root cultures and time-course study of their essential oils. Biotechnol Lett 21:859–864, 1999.

61. Y Matsuda, H Toyoda, A Sawabe, K Maeda, N Shimizu, N Fujita, T Fujita, T Nonomura, S Ouchi. A hairy root culture of melon produces aroma compounds. J Agric Food Chem 48:1417–1420, 2000.

62. SW Shin. Biotransformation of exogenous monoterpenoids by plant cell culture. Korean J Pharmacogn 26: 227–238, 1995.

63. H Hamada, T Furuya. The selectivity by plant biotransformation. In: TJ Fu et al., eds. Plant Cell and Tissue Culture for the Production of Food Ingredients. New York: Kluwer Academic/Plenum Publishers, 1999, pp 113–120.

64. T Vanek, I Valterova, R Vankova, T Vaisar. Biotransformation of (−)-limonene using *Solanum aviculare* and *Dioscorea deltoidea* immobilized plant cells. Biotechnol Lett 21:625–628, 1999.

65. S Ramachandra Rao, GA Ravishankar. Biotransformation of isoeugenol to vanilla flavour metabolites and capsaicin in suspended and immobilized cell cultures of *Capsicum frutescens*: study of the influence of β-cyclodextrin and fungal elicitor. Process Biochem 35:341–348, 1999.

66. S Ramachandra Rao, GA Ravishankar. Vanilla flavour: production by conventional and biotechnical routes. J Sci Food Agric 80:289–304, 2000.

67. SH Park, KS Kim, Y Suzuki, SU Kim. Metabolism of isopiperitones in cell suspension culture of *Mentha piperita*. Phytochemistry 44:623–626, 1997.

68. F Mellouki, A Vannereau, L Cosson. Les précurseurs d'arome dans des cultures cellulaires d'*Allium*. Acta Bot Gall 143:131–136, 1996.

69. CL Prince, ML Shuler, Y Yamada. Altering flavor profiles in onion (*Allium cepa* L.) root cultures through directed biosynthesis. Biotechnol Prog 13:506–510, 1997.

70. W Treffenfeld, H Vollmer, A Preuß, RG Berger, U Krings, E Latza. Verfahren zur selektiven Anreicherung und Trennung aromawirksamer Moleküle durch Adsorption. Patent DE19835542.4, 1998.

71. H Pedersen, CK Chin, A Dutta. Yield improvement in plant cell cultures by in situ extraction. In: TJ Fu et al., eds. Plant Cell and Tissue Culture for the Production of Food Ingredients. New York: Kluwer Academic/Plenum Publishers, 1999, pp 129–138.
72. L Tescione, P Asplund, WR Curtis. Reactor design for root culture: oxygen mass transfer limitations. In: TJ Fu et al., eds. Plant Cell and Tissue Culture for the Production of Food Ingredients. New York: Kluwer Academic/Plenum Publishers, 1999, pp 139–156.
73. BJ Cass, F Schade, CW Robinson, JE Thompson, RL Legge. Production of tomato flavor volatiles from a crude enzyme preparation using a hollow-fiber reactor. Biotechnol Bioeng 67:372–377, 2000.
74. G Matheis. Biotechnologische Erzeugung von Aromastoffen und Aromastoffgemischen: Eine Literaturübersicht. Fluess Obst 9:518–524, 1998.
75. P van Berge. Flavors into the 21st Century. Perfum Flavor 23:1–12, 1998.

15

Fine Chemicals from Plants

Michael Keil
Boehringer Ingelheim Pharma KG, Ingelheim, Germany

I. INTRODUCTION

Plants have been used in traditional medicine for a long time. About 13,000 plant species have been used as drugs throughout the world, and approximately 25% of our current materia medica is derived from plants in form of teas, extracts, or pure substances (1). Accordingly, plant secondary metabolites are an important source of various fine chemicals (phytochemicals) that are used directly or as intermediates for the production of pharmaceuticals. Additional applications are in the cosmetic industry and as food or drink additives.

The plant kingdom provides an enormous portfolio of secondary metabolites of which currently about 100,000 compounds from plants are known and 4000 substances newly discovered every year (2). The largest proportion of these compounds consists of terpenoids (more than 30%), followed by alkaloids (approximately 20%) (3). The focus of this chapter is predominantly on alkaloids for pharmaceutical applications.

In recent years, there has been a resurgence of interest in the discovery of new compounds from plants with the aim of finding novel activities against a variety of illnesses. New technical developments in high-throughput screening techniques now allow the screening of substances at a rate not possible previously. In addition, the biological effects of minimal concentrations of secondary metabolites can now be measured due to the much improved sensitivity of the screening systems. There are still many plant species that have not been examined with respect to their biological activities, and there is an expectation that tropical plants in particular will provide compounds with novel activities. Such compounds may be used as lead structures for the development of new drugs by chemical synthesis or, alternatively, may be used as drugs directly or as intermediates for chemical modification following extraction from the plant.

II. PRODUCTION OF PHYTOCHEMICALS

The commercial production of plant secondary metabolites as fine chemicals can be done by either (1) total chemical synthesis, (2) extraction and purification from plant material, or (3) partial chemical synthesis following extraction of biosynthetic precursors from plant material. The decision on which route is taken is made on economical grounds and/or depends on the availability of the plant material. Chemical synthesis of plant-derived compounds is worthwhile only if few synthesis steps are needed and the source chemicals are available at low cost. Molecules of simple chemical structure with few steric centers thus lend themselves to total synthesis. Examples of such compounds are the piperidine alkaloids lobelin and arecoline (Fig. 1). In contrast, complex structures such as the cyclic diterpenoid taxol or alkaloids such as the tropane alkaloid scopolamine or the indole alkaloid camptothecin (Fig. 2) are either extracted from plant material or partially synthesized from extracted biosynthetic precursors. For example, taxol and its analogue taxotere are largely produced by acylation of their biosynthetic precursor 10-deacetylbaccatin III extracted from the needles of *Taxus* (4). The anticancer therapeutic topotecan is obtained semisynthetically from the indol alkaloid camptothecin extracted from the leaves of *Camptotheca acuminata* (5).

Several sources of plant material may be used for the purpose of secondary metabolite extraction. The simplest route is the extraction from plant material that has been harvested from wild plant resources. However, wild plant resources may be limited and hence may not permit sustainable production of phytochemicals. In addition, overexploitation of wild plant resources is undesirable from an environmental point of view (3).

An alternative preferred by producers of phytochemicals is the cultivation of medicinal plants by conventional farming. Conventional farming permits the sustainable production of plant material in the amount required for phytochemical production, provides independence of commercial plant material suppliers, and allows continuous improvement of production levels and economics by breeding and selection of superior genotypes. Disadvan-

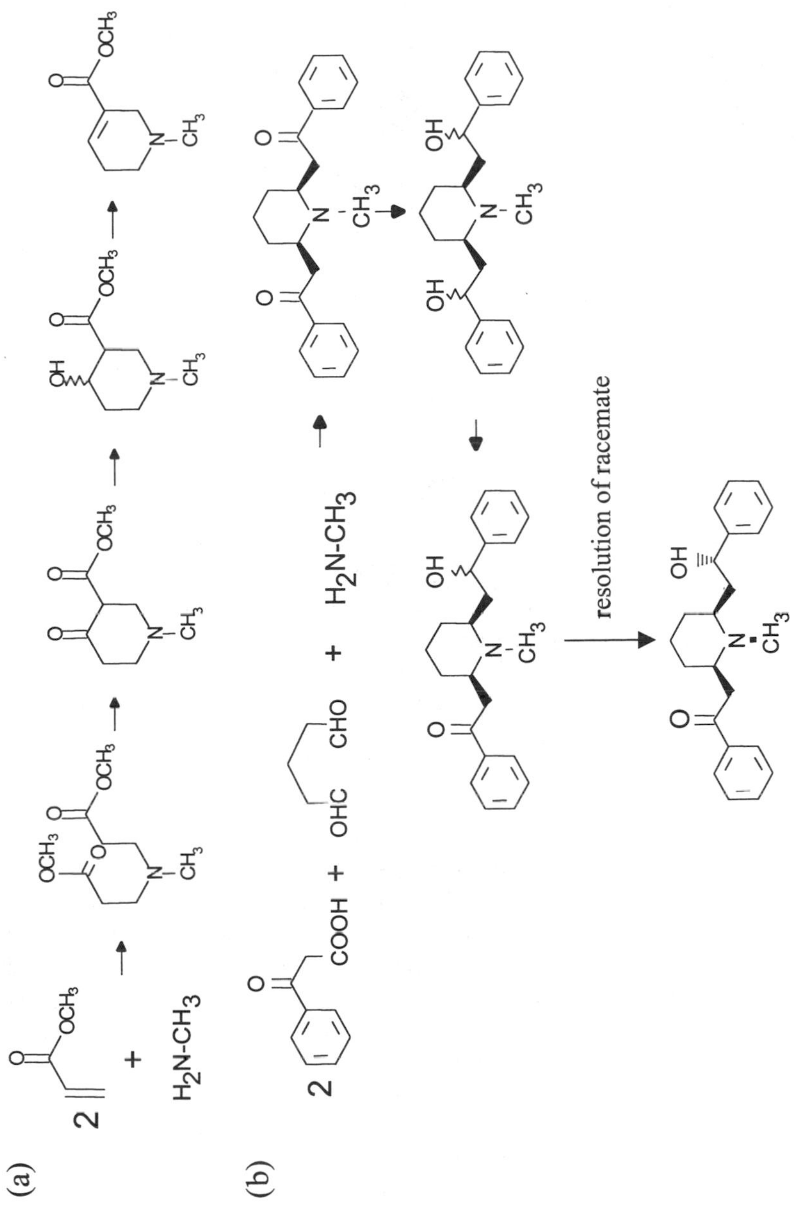

FIGURE 1 Chemical synthesis of the piperidine alkaloids lobelin (a) and arecolin (b).

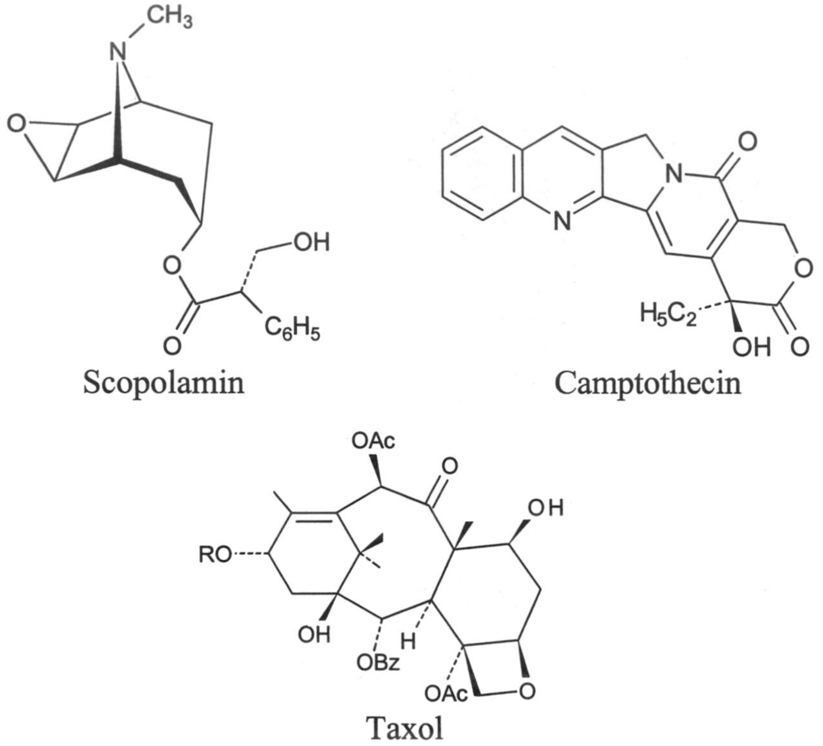

FIGURE 2   Chemical structures of scopolamine, camptothecin, and taxol.

tages are the investments and the long lead times required for the establish-ment of plantations and for responding to changing market demands as well as environmental risks due to adverse weather conditions, pests, and dis-eases. Also, climatic and soil requirements of the plants to be cultivated have to be taken into account. Examples of medicinal plants that are grown on farms for the purpose of phytochemical production are *Pilocarpus microphyllus* (pilocarpine), *Digitalis lanata* (digitalis cardiac glycosides), *Papaver somniferum* (codeine and morphine), and *Duboisia* interspecific hybrids (hyoscyamine and scopolamine).

A great deal of progress has been made in the cultivation of plant cells under controlled conditions in bioreactors that can be operated at virtually any geographic location. The low productivity of plant cell cultures regard-ing secondary metabolites that has often been observed in many cases can be improved significantly by strain selection and elicitation (3,6). As an alternative to plant cell cultures, the use of organ cultures such as fast-growing hairy root cultures obtained after transformation using *Agrobacte-rium rhizogenes* has been proposed (7,8): the main location of secondary metabolite biosynthesis is often in the roots, which maintain secondary me-tabolite production in culture and are genetically stable for long periods of time, in contrast to what has been observed in many plant cell cultures (9). However, the cultivation of organized structures such as hairy roots on a large scale in bioreactors is inherently more difficult than for cell cultures and, thus, has been demonstrated up to the 500 L scale only (10). In contrast, plant cells have been successfully cultivated up to the 60,000 L scale (3). As the costs of large-scale production of phytochemicals using plant cell cultures are still prohibitively high, to date there have been only few ex-amples of their commercial application (2).

III. METABOLIC ENGINEERING FOR IMPROVEMENT OF PRODUCTIVITY

With the advent of molecular biology and the possibility of genetically trans-forming plant cells and obtaining genetically modified (GM) or transgenic plants therefrom, there has been increased interest in improving the produc-tivity and/or quality not only of arable crops but also—to a lesser extent—of medicinal plant species. In addition, molecular biology provides the pos-sibility to transfer single or multiple genes encoding biosynthetic enzymes into microorganisms with the purpose of secondary metabolite production.

The general strategies for the biotechnological exploitation of alkaloid biosynthetic genes have been summarized by Kutchan (11), but they also apply for other secondary metabolites (Fig. 3). Plant secondary metabolite genes can be functionally expressed in microorganisms to produce either single biotransformation steps or short biosynthetic pathways. This approach requires that sufficient quantities of the secondary metabolite precursors needed for biotransformation are available at low cost. Promising results have been achieved in the laboratory of Verpoorte (2), who succeeded in the functional expression of the biosynthetic genes for strictosidine synthase and strictosidine glucosidase in transgenic yeast. After addition of tryptamin to the transgenic yeast cultures growing in a medium containing a secologanin-rich extract of snowberries, strictosidine and cathenamine, both precursors of the potent anticancer alkaloids vinblastine and vincristine, were obtained in high yield (12). However, this approach is difficult or impossible if bio-synthetic enzymes are involved that require plant-specific glycosylation for their activity. In this case, metabolic engineering will be restricted to plants or plant cells. As shown in Fig. 3, overexpression of single biosynthetic genes in plants or plant cells may yield enhanced amounts of the desired secondary metabolites within a pathway or even novel secondary metabolites

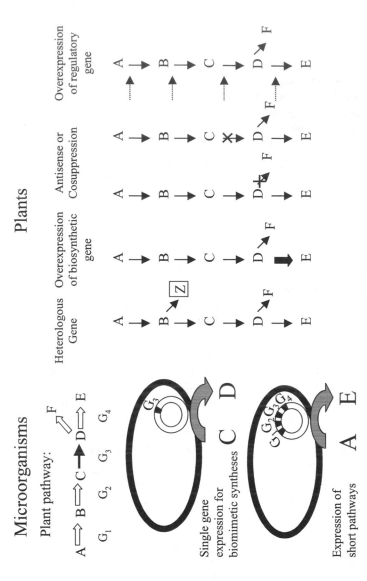

FIGURE 3 Potential biotechnological exploitation of secondary metabolite biosynthetic genes. Plant secondary metabolite genes can be functionally expressed in microorganisms to produce either single biotransformation steps or short biosynthetic pathways. Likewise, using overexpression or antisense or cosuppression technologies, medicinal plants can be tailored to produce important pharmaceutical secondary metabolites by introducing side pathways, eliminating side pathways, or accumulating biosynthetic intermediates. Furthermore, regulatory genes acting on one or more genes within a biosynthetic pathway may be used to up-regulate secondary metabolite biosynthesis and thus enhance product yield. (Adapted from Ref. 11; copyrighted by the American Society of Plant Physiologists and reprinted with permission.)

not normally produced by the plants. Inhibition of the gene activity of biosynthetic enzymes by antisense or cosuppression technology may be utilized to knock out pathway side branches or catabolism of a particular secondary metabolite, thus enhancing product yield. Finally, overexpression of regulatory genes of biosynthetic pathways containing several rate-limiting steps might circumvent the necessitity to overexpress several pathway en-zymes in order to increase the content of the desired secondary metabolite.

The elucidation of the biosynthetic pathways of a number of plant secondary metabolites has advanced to great extent. One of the best known examples is the pathway for the biosynthesis of the benzophenanthridine alkaloid sanguinarine (1). The pathways of a number of other alkaloids (1–3,13,14) and secondary metabolites of pharmaceutical interest, such as the anticancer diterpenoid taxol (4), have been characterized at least partially. However, the biosyntheses of some alkaloids of pharmaceutical importance are still largely unknown. For example, not much is known about the biosynthesis of the jaborandi alkaloids, specifically pilocarpine (15,16). Also, the biosynthetic steps leading from the intermediate strictos-amide to the anticancer indole alkaloid camptothecin still remain rather speculative (17,18).

The sanguinarine biosynthetic pathway is also a good example of the possible complexity of secondary metabolite biosynthesis: 15 enzymatic steps are required for the formation of sanguinarine from the amino acid precursor L-tyrosine. The biosynthesis of bisindole terpenoid indole alkaloids involves over 20 enzymatic steps, which take place in least three subcellular compartments (19). In addition to the potentially large number of biosyn-thetic steps that are involved in secondary metabolite biosynthesis, com-plexity can arise from biosynthetic networks (2) as opposed to linear path-ways. Furthermore, although many enzymes of secondary metabolite biosynthesis are highly substrate specific (13), there are a number of ex-amples in which enzymes can utilize several substrates that may even belong to completely different metabolic pathways (20,21). Metabolic engineering of such enzymes thus may have unwanted pleiotropic effects.

Following the purification and characterization of secondary metabolite biosynthetic enzymes, the corresponding genes can be cloned using reverse genetics. Genes of a large number of secondary metabolite biosynthetic en-zymes have been cloned successfully (Table 1); an extensive list of cloned genes involved in isoprenoid biosynthesis can be found in Ref. 22. Although the full elucidation of a pathway is not an absolute requirement for the purpose of gene cloning and metabolic engineering, thorough knowledge of the regulation and rate-limiting enzymatic steps in the whole pathway is advantageous for the selection of enzymatic steps to be engineered. Without this knowledge, overexpression of biosynthetic enzymes may not lead to

TABLE 1 Examples of Cloned Secondary Metabolite Biosynthetic Genes

Gene[a]	Biosynthetic pathway	Origin	Species in which gene has been studied using transgenic plants (P) or cell/callus (C) or organ (O) cultures[b]	References
Berberine bridge enzyme	Benzyl isoquinoline alkaloids	*Eschscholtzia californica*		56
Berbamunine synthase (CYP80)	Bisbenzyl isoquinoline alkaloids	*Berberis stolonifera*		57
Tropinone reductase II	Calystegins	*Datura stramonium*		58
Tryptophan decarboxylase	Indole alkaloids	*Catharanthus roseus*	*Catharanthus roseus* (C), *Peganum harmala* (O), *Cinchona officinalis* (O)	24,29,30,59–61
Strictosidine synthase	Indole alkaloids	*Catharanthus roseus*, *Rauwolfia serpentina*	Tobacco (P), *C. roseus* (C), *Tabernae montana* (P), *Chinchona officinalis* (O)	29,30,33,35,62
Desacetoxyvindoline-4-hydroxylase	Indole alkaloids	*Catharanthus roseus*		63
ORCA2	Indole alkaloids	*Catharanthus roseus*	*C. roseus* (C)	26
Acetyl-CoA:deacetyl vindoline-4-O-acetyltransferase	Indole alkaloids	*Catharanthus roseus*		64
O-Methyltransferases	Isoquinoline alkaloids Phenylpropanoids	*Thalictrum tuberosum*		20
Tyrosine decarboxylase	Isoquinoline alkaloids	Parsley		65
Cytochrome P450 reductase	Isoquinoline alkaloids	*Eschscholtzia californica*		66
Norcoclaurine-6-O-methyltransferase	Isoquinoline alkaloids			67
(S)-3-Hydroxy-N-methylcoclaurine-4-O-methyltransferase	Isoquinoline alkaloids	*Berberis koetineana*		14

Enzyme	Product	Species	Reference
Codeinone reductase	Morphine	*Papaver somniferum*	68
Homospermidine synthase	Pyrrolizidine alkaloids	*Senecio vernalis*	69
Taxadien synthase	Taxol	*Taxus brevifolia*	70
Taxadien transacetylase	Taxol	*Taxus cuspidata*	71
10-Deacetylbaccatin III-10-*O*-acetyltransferase	Taxol	*Taxus cuspidata*	71
Deoxy-xylulosephosphate synthase	Terpenes	*Mentha piperita*	72–74
Deoxy-xylulosephosphate reductoisomerase	Terpenes	*Mentha piperita* *Arabidopsis thaliana*	74–76
Isopentenyl mono-phosphate kinase	Terpenes	*Mentha piperita*	77
NADPH:cytochrome P450 reductase	Terpenes (coenzyme of geranyl-10-hydroxylase)	*Escherichia coli* *Catharanthus roseus*	19
(S)-N-Methylcoclaurine 3′-hydroxylase (CYP80B1)	Tetrahydrobenzyl isoquinoline alkaloids	*Eschscholtzia californica*	13
HMG-CoA-reductase	Triterpenes Sesquiterpenes	*Camptotheca acuminata*	78,79
Hyoscyamine-6β-hydroxylase	Tropane alkaloids	*Hyoscyamus niger*	40,42,48
		Atropa belladonna (P) *Nicotiana tabacum* (P) *Duboisia* (O) *Hyoscyamus muticus* (O) *Nicotiana rustica* (O)	85
Ornithine decarboxylase	Tropane alkaloids Nicotine	Yeast	80
Arginine decarboxylase	Tropane alkaloids Nicotine	*Escherichia coli* Tomato *Avena sativa*	81–83
Putrescine methyltransferase	Tropane alkaloids Nicotine	*Nicotiana tabacum*	84
Tripinone reductase I	Tropane alkaloids	*Datura stramonium*	58

[a] The list of genes shown is not meant to be comprehensive.
[b] Only plant species are listed; some of the genes shown have also been studied in heterologous systems such as *Escherichia coli* or insect cells.

enhanced production of the secondary metabolite of interest. For example, overexpression of the enzymes phenylalanine lyase (PAL) and cinnamic acid 4-hydroxylase (C4H) in transgenic tobacco plants does not lead to increased production of lignin (23). This suggests the presence of downstream flux control points in the lignin biosynthetic pathway. Similarly, the overexpression of the tryptophan decarboxylase gene in crown gall calli of *Catharanthus roseus* did not increase terpenoid alkaloid production despite increase tryptamine levels in the transgenic tissues (24), again indicating the presence of additional rate-limiting steps downstream in the pathway. In this particular instance it is now known that the secoirodoid secologanin that together with tryptamine is required for the formation of strictosidine is rate limiting (24).

The ever increasing number of genes of known function in DNA sequence databases now provides a fast gene cloning alternative to the reverse genetics approach mentioned before. Using reverse transcription–polymerase chain reaction (RT-PCR) or complementary DNA (cDNA) libraries from tissues or cell-organ cultures specifically producing the secondary metabolite of interest, direct cloning of biosynthetic enzyme genes is possible using DNA primers targeted against consensus sequences within the genes of particular enzyme classes. Using consensus sequences detected within the genes encoding various plant *O*-methyltransferases, direct cloning of novel methyltransferases common to isoquinoline alkaloid and phenylpropanoid biosynthesis has been achieved by Frick and Kutchan (20). Results suggest that this approach should also be successful in the cloning of novel acetyltransferase genes and genes encoding cytochrome P-450 enzymes, cytochrome P-450 reductases, covalently flavinated oxidases, and 2-oxoglutarate–dependent dioxygenases involved in secondary metabolite biosynthesis (13,25).

The utilization of regulatory genes for the up-regulation of alkaloid biosynthetic pathways in transgenic plants or plant cell cultures would be an elegant alternative to the overexpression of individual biosynthetic genes. The transcription factor ORCA2 involved in the jasmonic acid– and elicitor-dependent regulation of the strictosidine synthase gene in *Catharanthus roseus* has been cloned (26). In *C. roseus*, the expression of terpene indole alkaloid biosynthetic genes is coordinately regulated in response to elicitor and jasmonates (26). Hence, overexpression of the ORCA2 gene in transgenic plants or plant cells could potentially increase terpene indole alkaloid content. Of particular interest would be the effect on the alkaloids vinblastine and vincristine used in cancer treatment. In addition, it would be interesting to know whether ORCA2 is also involved in the regulation of the biosynthetic genes leading from strictosidine to the indole alkaloid camptothecin, which possesses potent anticancer activity.

IV. TRANSFORMATION AND REGENERATION OF MEDICINAL PLANTS

Plant cells of virtually any plant species can now be transformed by one or more of the transformation technologies currently available (27). Of particular importance are the particle bombardment and *Agrobacterium*-mediated transformation technologies. At present, genetic transformation studies have been conducted with approximately 200 plant species. Of these, about 70 species belong to medicinal plants (28). Metabolic engineering of medicinal plant cells with the aim of overproducing secondary metabolites of interest is regarded as a promising strategy to make the biotechnological production of phytochemicals in bioreactors economically feasible (29). However, despite stable integration of the introduced genes, gene expression and hence secondary metabolite production are not necessarily maintained in the longer term, as has been shown for transgenic cell and hairy root lines of *C. roseus* containing the genes encoding tryptophan decarboxylase and strictosidine synthase (29,30). In some cases, secondary metabolite production in transgenic cell lines declines over time despite continued expression of the introduced gene(s) at a high level (2). This seems to be associated with the well-known instability of plant cell cultures with respect to secondary metabolite formation that has been reported for numerous other cases (6).

Of the various methods used for the transformation of medicinal plants, *Agrobacterium rhizogenes*–mediated gene transfer has been preferred by many researchers (Ref. 31 and references therein); many medicinal plant species are amenable to transformation by *A. rhizogenes*, yielding hairy roots that are typically more genetically and physiologically stable than cell culture lines (9). This system has been used frequently to cointroduce with the *Agrobacterium* genes additional genes of interest (7,31). As mentioned earlier, the use of hairy roots for the biotechnological production of phytochemicals is currently not feasible on a large scale. However, for a number of plant species, regeneration of shoots from hairy roots has been observed, either spontaneously or after induction using phytohormones (7,31). The shoots root easily and can be transferred to soil. Potentially, such hairy root–derived plants could be used in conventional farming for the production of phytochemicals. However, hairy root–derived plants usually show varying degrees of morphological abnormalities such as stunted growth, reduced apical dominance, abnormal flower production, and wrinkled leaves. These symptoms are known as hairy root syndrome (32). The agronomical performance of such plants in comparison with conventional plants therefore remains to be tested (see also later).

As with arable crops, successful commercial application of genetic modification technology will depend on naked DNA transformation methods or the use of nononcogenic *Agrobacterium* strains, which permit regeneration of phenotypically normal plants (27). Here resides the main problem in the generation of genetically modified medicinal plants: many medicinal plants species are woody and of tropical origin. The development of regeneration protocols has been tedious or impossible in many cases, which is reflected by the preferred use of *A. rhizogenes*–mediated transformation methods, as mentioned before. In some cases, transgenic tobacco plants have been used as a model system to study the expression of secondary metabolite biosynthetic genes (33,34). The obvious disadvantage is that the specific metabolic precursors are not normally present in a heterologous system. For research purposes, this problem may be alleviated to some extent by precursor feeding (34), but this obviously is no solution for production purposes. A better alternative is the use of regenerable plant species that produce secondary metabolites the same as or similar to those of the target species that cannot be regenerated. This strategy has been used by Lopes Cardoso et al. (35), who studied the expression of the strictosidine synthase gene in the Apocynaceous plant *Tabernaemontana pandacaqui* instead of *Catharanthus roseus* of the same family. However, despite an efficient regeneration system for *T. pandacaqui*, the authors could obtain only one transgenic plant, the data for which certainly cannot be generalized. Thus, to date the best example of metabolic engineering of a medicinal plant remains the heterologous expression of the hyoscyamine-6β-hydroxylase gene of *Hyoscyamus niger* in transgenic *Atropa belladonna* plants (36). In this study, the authors found significantly increased scopolamine levels in leaves of the transgenic *A. belladonna* plants, which normally do not contain appreciable amounts of this alkaloid in the leaves.

V. APPLICATION OF BIOTECHNOLOGY TO *DUBOISIA* FOR SCOPOLAMINE PRODUCTION

Boehringer Ingelheim is the major worldwide supplier of the tropane alkaloids hyoscyamine and scopolamine. Scopolamine, which is of higher value than hyoscyamine, is used as a parasymphatolytic and for the production of spasmolytic pharmaceuticals. Previously, the alkaloids were extracted from *Datura* and *Duboisia* plants, both belonging to the Solanaceae family. Because of the low content of scopolamine in *Datura*, extraction of scopolamine from this plant was discontinued in the early 1990s. There are three known species of *Duboisia*, *D. hopwoodii*, *D. myoporoides*, and *D. leichhardtii*, all of which are indigenous to Australia (37,38). In a small area in Western Australia, interspecific hybrids between *D. myoporoides* and *D. leichhardtii* occur naturally. These hybrids were found to have particularly high levels of scopolamine, with an average content of approximately 1% of leaf dry weight. Commercial cultivation of *Duboisia* hybrids began during the 1940s, the main area of cultivation being the South Burnett Region in Queensland, Australia (37). The *Duboisia* hybrids are cultivated in clonal plantations. Figure 4 shows a typical example of a *Duboisia* plantation plant.

FIGURE 4   Photograph of a plantation-grown *Duboisia* hybrid clone in Australia. For harvesting, the plants are usually cut approximately 40 cm above ground. The stem will resprout and the resulting new growth can be harvested again 8 to 12 months later.

A. Conventional Breeding

To improve tropane alkaloid yield, Boehringer Ingelheim started an in-house *Duboisia* breeding program at Ingelheim during the mid-1970s, based on controlled crosses between superior *Duboisia* hybrid parents and recurrent selection on an annual basis. During the time period between 1985 and 1991, the resulting improvement in scopolamine content translated into a 40% increase in total scopolamine production from virtually the same annual amount of *Duboisia* leaf extracted. Additional selection criteria have been tropane alkaloid composition, general growth, leaf density, rooting ability of cuttings, and nematode and insect tolerance. However, insect and nematode damage remains a major problem that severely restricts the productive life of plantation-grown *Duboisia* plants.

B. Tissue Culture Applications

Micropropagation of *Duboisia* species is possible using nodal explants or shoot regeneration from various tissue explants (39–41) (Fig. 5). Micropropagation techniques can be used for mass propagation of plants and for

FIGURE 5 Photographs of *Duboisia* tissue culture plants obtained (a) from nodal stem sections or (b) via callus formation and subsequent regeneration from leaf explants.

plant transport purposes, particularly where local quarantine regulations restrict the import of conventional plants. However, vegetative propagation of *Duboisia* clones using conventional cuttings is relatively straightforward and is done routinely to provide the plants needed for the ongoing planting program on Boehringer Ingelheim's farms. Hence, mass propagation of *Duboisia* clones in vitro is not a cost-effective alternative.

In collaboration with the University of Barcelona in Spain, we also examined the potential of *Duboisia* hairy root cultures for biotechnological production of scopolamine. Hairy roots were obtained after leaf disc infection with the *Agrobacterium rhizogenes* strain A4. Selected hairy root lines were able to produce up to 1.8 mg/L/day in a 4-L airlift bioreactor (42). This productivity of the bioreactor process using hairy roots is far too low to compete with the conventional farming approach. In addition, upscaling of the process to an 80-L bioreactor was found to be difficult. As a consequence, we did not pursue this approach further.

C. Agronomical Performance of Hairy Root–Derived Plants

As with other hairy root systems, *Duboisia* hairy root cultures may regenerate shoots spontaneously (see also earlier). Regenerated plants were transferred to soil and grown under greenhouse conditions. The plants displayed the typical symptoms of the hairy root syndrome to varying extents. We wanted to know whether these plants could outperform conventional *Duboisia* plants and conducted a field trial to analyze growth and scopolamine content of these plants (Fig. 6). Only the plants displaying the strongest hairy root syndrome symptoms had a significantly higher scopolamine content than control plants of the same clone. However, because of their stunted growth, overall productivity was strongly reduced compared with the control plants: the control plants produced almost twice as much scopolamine as the best hairy root plants (Table 2) (43). Thus, our data suggest that hairy root–derived plants do not represent a viable alternative to the cultivation of conventional plants.

D. Genetic Transformation

Genetic transformation of *Duboisia* so far has been reported only for *Agrobacterium rhizogenes*–mediated tranformation methodologies (44). As mentioned before, the disadvantage of this system is that regenerated plants typically display the symptoms of the hairy root syndrome. To supplement our conventional breeding program, we therefore started to develop genetic transformation protocols for *Duboisia* clones based on *Agrobacterium tumefaciens*–mediated transformation of leaf discs. The transformation effi-

FIGURE 6 Field performance of *Duboisia* plants regenerated from hairy roots. (a) Overview of field trial site. Hairy root–derived plants with minor hairy root syndrome symptoms can be seen in the foreground; plants with strong symptoms are located in the background. (b) Photograph of a hairy root–derived *Duboisia* plant with strong hairy root syndrome symptoms (43).

ciency was evaluated using different *A. tumefaciens* strains in combination with a binary vector system harboring an intron-containing β-glucuronidase (GUS) gene (45) and the neomycinphosphotransferase II (*NPTII*) gene as a selectable marker. Using our shoot regeneration protocol for leaf discs following *Agrobacterium* transformation, we were able to obtain transgenic *Duboisia* plants constitutively expressing the GUS gene (Fig. 7). In contrast to the plants regenerated from hairy roots, no morphological difference could be detected between control plants and the plants obtained after *A. tumefaciens*–mediated transformation, as expected.

Of particular interest to the commercial application of genetic modification technology to *Duboisia* is the gene encoding hyoscyamine-6β-hydroxylase (H6H). Based on the published DNA sequence data (46), we cloned the gene from *Hyoscyamus niger* root culture RNA. Transgenic plants expressing the gene under the control of the cauliflower mosaic virus (CaMV) 35S promotor will be tested with respect to its effect on the levels of hyoscyamine, 6β-hydroxy-hyoscyamine, and scopolamine. As *Duboisia* contains high amounts of scopolamine naturally, it will be interesting to see whether overexpression of the h6h gene will lead to a further improvement of the scopolamine content, similar to what has been observed in *Atropa belladonna* plants (36) and *Hyoscyamus muticus* hairy roots (85).

TABLE 2 Size and Scopolamine Content of Field-Tested Plants
Regenerated from *A. rhizogenes*–Transformed Hairy Roots of *Duboisia* (43)

Plant line[a]	Number of plants tested	Average size (cm)	Average scopolamine content (% DW)	Total scopolamine yield (g)
Control	3	253.33 ± 5.77	0.94 ± 0.21	1.50
1	4	130.00 ± 4.76	0.79 ± 0.08	ND
7	4	149.50 ± 5.20	0.73 ± 0.09	ND
10A[b]	3	75.33 ± 10.50	0.84 ± 0.15	ND
10B[b]	3	93.33 ± 16.07	0.84 ± 0.11	ND
15	4	167.50 ± 11.73	0.84 ± 0.12	ND
22A[b]	4	42.50 ± 2.89	1.31 ± 0.06	0.54
22B[b]	4	53.75 ± 4.79	1.16 ± 0.14	0.89
26A[b]	4	53.75 ± 11.09	1.22 ± 0.07	0.81
26B[b]	3	119.67 ± 25.70	0.63 ± 0.03	ND
200	4	167.25 ± 8.62	0.78 ± 0.06	0.83

[a]All lines including the control are derived from a single *Duboisia* clone.
[b]A and B denominate independent regenerants from a single hairy root line.

E. Other Traits of Interest

Apart from modifiying secondary metabolite composition and content, the improvement of agronomic traits such as herbicide resistance, pest and disease resistance, and frost tolerance in *Duboisia* through genetic modification is also of interest. The most widely used systems for herbicide tolerance are the Roundup Ready and Liberty Link technologies, providing tolerance to the herbicides glyphosate and glufosinate, respectively (47–49). *Bacillus thuringiensis* endotoxin genes have been used successfully for the improvement of resistance to various insect pests in transgenic crops (47). There are also promising developments in the field of the genetic modification of nematode tolerance (48). Nematodes are very difficult to control chemically, and the nematocides currently available have highly undesirable environmental characteristics (51). Natural plant nematode tolerance genes such as the *Mi* gene from tomato or the *HS1* gene from sugar beet have been identified and cloned. Other approaches rely on the use of proteinase inhibitor genes such as the oryzacystatin gene from rice or the BARNASE/BARSTAR gene system (50). The improvement of frost tolerance via genetic modification is still in its infancy. The detection and subsequent isolation of plant "antifreeze"

FIGURE 7   (a) Photograph showing two transgenic *Duboisia* hybrid plants in the greenhouse. The plants have been transformed with the *E. coli GUS* gene and the *NPTII* selectable marker gene. No phenotypical difference could be observed in comparison with greenhouse plants of the same clone of the same age. (b) Results of an enzymatic assay demonstrating the activity of the *GUS* gene in leaves of the two transgenic plants shown in (a). Note the dark (blue) appearance of the leaf pieces (middle and bottom), indicating GUS activity, in contrast to the pale appearance of control leaves (top).

proteins (52), however, may open up a new route to the improvement of frost tolerance in transgenic plants.

F. Molecular Markers

Molecular markers can be used for genotype identification (genetic fingerprinting), estimation of the genetic diversity of natural populations or breeding stock, and marker-assisted selection of agronomic traits (53). We have established the random amplified polymorphic DNA (RAPD) marker technology (54,55) for application in *Duboisia* (Fig. 8). Using this technology, we are now able to discriminate and identify our production clones unambigously. This ability may serve as a quality control tool during clonal propagation and as a deterrent against theft of our proprietary elite clones. De-

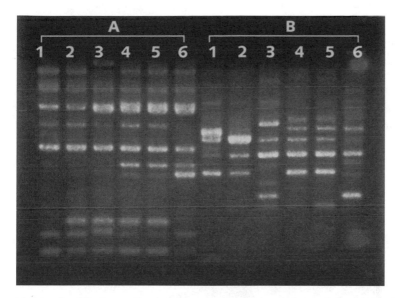

FIGURE 8 Photograph showing genetic fingerprints of five *Duboisia* hybrid clones obtained using the random amplified polymorphic DNA (RAPD) technique. Two random sequence 10mer oligonucleotides denoted A and B were used to obtain the fingerprints shown. Lanes 4 and 5 represent samples from two plants of the same clone; no difference in the DNA profiles can be observed, as expected. All other samples can be readily distinguished by their DNA profiles.

cisions on parent selection in the breeding program will be facilitated using the genetic data obtained by RAPD analysis.

VI. CONCLUSIONS

Advances in the elucidation of secondary metabolite biosynthetic pathways and in the isolation of corresponding biosynthetic genes by reverse genetics or direct cloning make metabolic engineering of these pathways increasingly possible. Whether overexpression of rate-limiting enzymes will lead to enhanced yield of the desired secondary metabolites is a highly debated issue. All theoretical debate will not resolve this issue, and it will need to be addressed experimentally. If successful, metabolic engineering may make biotechnological production processes for phytochemicals using cell or organ cultures economically attractive. This is particularly true if functional expression of biosynthetic gene arrays can be achieved in microorganisms.

The application of metabolic engineering to whole plants will depend on the successful development of methods that allow the regeneration of phenotypically normal plants. In addition, for transgenic medicinal plants to be grown in the field, the regulatory issues regarding the release of genetically modified organisms into the environment and public acceptance need to be taken into account. Public criticism regarding transgenic plants is directed largely against the introduction of such plants or parts therefrom into the human food chain but also against perceived environmental damage by transgenic plants. As medicinal plants are usually highly toxic and thus not eaten by man, the former point should not be an issue; however, the latter point will certainly need to be addressed in a responsible and sensible manner. Apart from genetic modification of medicinal plants, other modern biotechnological approaches such as plant cell and tissue culture and molecular markers also have important applications in the field of medicinal plants. Conventional methods of farming, plant propagation, and breeding will, however, remain the first method of choice in the cultivation and improvement of medicinal plants for the production of phytochemicals.

REFERENCES

1. TM Kutchan. Heterologous expression of alkaloid biosynthetic genes—a review. Gene 179:73–81, 1996.
2. R Verpoorte, R Van Der Heijden, HJG Ten Hoopen, J Memelink. Metabolic engineering of plant secondary metabolite pathways for the production of fine chemicals. Biotechol Lett 21:467–479, 1999.
3. R Verpoorte, R Van Der Heijden, J Schripsema, JHC Hoge, HJG Ten Hoopen. Plant cell biotechnology for the production of alkaloids: present status and prospects. J Nat Prod 56:186–207, 1993.
4. M Hezari, R Croteau. Taxol biosynthesis: an update. Planta Med 63:291–295, 1997.
5. A Bedeschi, I Candiani, C Geroni, L Capolongo. Water-soluble camptothecin derivatives. Drugs Fut 22:1259–1266, 1997.
6. MM Yeoman, CL Yeoman. Manipulating secondary metabolism in cultured plant cells. New Phytol 134:553–569, 1996.
7. PM Doran. Hairy Roots. Culture and Applications. Amsterdam: Harwood Academic Publishers, 1997.
8. JV Shanks, J Morgan. Plant 'hairy root' culture. Curr Opin Biotechnol 10:151–155, 1999.
9. H Wysokinska, A Chmiel. Transformed root cultures for biotechnology. Acta Biotechnol 17:131–159, 1997.
10. PDG Wilson. The pilot-scale cultivation of transformed roots. In: PM Doran, ed. Hairy Roots. Culture and Applications. Amsterdam: Harwood Academic Publishers, 1997, pp 179–190.

11. TM Kutchan. Alkaloid biosynthesis—the basis for metabolic engineering of medicinal plants. Plant Cell 7:1059–1070, 1995.

12. A Geerlings. Strictosidine β-D-glucosidase, an enzyme in the biosynthesis of pharmaceutically important indole alkaloids. PhD dissertation, University of Leiden, The Netherlands, 1999.

13. W-M Chou, TM Kutchan. Enzymatic oxidations in the biosynthesis of complex alkaloids. Plant J 15:289–300, 1998.

14. T Hashimoto, Y Yamada. Alkaloid biogenesis: molecular aspects. Annu Rev Plant Physiol Plant Mol Biol 45:257–285, 1994.

15. E Brochmann-Hanssen, MA Nunes, CK Olah. On the biosynthesis of pilocarpine. Planta Med 28:1–5, 1975.

16. AA Al-Badr, HY Aboul-Enein. Pilocarpine. Anal Profiles Drug Subst 12:385–432, 1983.

17. N Aimi, M Nishimura, A Miwa, H Hoshino, S Sakai, J Haginiwa. Pumiloside and deoxypumiloside; plausible intermediates of camptothecin biosynthesis. Tetrahedron Lett 30:4991–4994, 1989.

18. BK Carte, C DeBrosse, D Eggleston, M Hemling, M Mentzer, B Poehland, N Troupe, JW Westley. Isolation and characterization of a presumed biosynthetic precursor of camptothecin from extracts of *Camptotheca acuminata*. Tetrahedron 46:2747–2760, 1990.

19. AH Meijer, MIL Cardoso, JT Voskuilen, A De Waal, R Verpoorte, JH Hoge. Isolation and characterization of a cDNA clone from *Catharanthus roseus* encoding NADPH:cytochrome P-450 reductase, an enzyme essential for reactions catalysed by cytochrome P-450 monooxygenases in plants. Plant J 4:47–60, 1993.

20. S Frick, TM Kutchan. Molecular cloning and functional expression of O-methyltransferases common to isoquinoline alkaloid and phenylpropanoid biosynthesis. Plant J 17:329–339, 1999.

21. A Finsterbusch, P Lindemann, R Grimm, C Eckerskorn, M Luckner. Δ^5-3β-Hydroxysteroid dehydrogenase from *Digitalis lanata* Ehrh.—a multifunctional enzyme in steroid metabolism? Planta 209:478–486, 1999.

22. PA Scolnik, GE Bartley. A table of some cloned plant genes involved in isoprenoid biosynthesis. Plant Mol Biol Rep 14:305–319, 1996.

23. JW Blount, KL Korth, SA Masoud, S Rasmussen, C Lamb, RA Dixon. Altering expression of cinnamic acid 4-hydroxylase in transgenic plants provides evidence for a feedback loop at the entry point into the phenylpropanoid pathway. Plant Physiol 122:107–116, 2000.

24. OJM Goddijn, EJM Pennings, P Van Der Helm, RA Schilperoort, R Verpoorte, JHC Hoge. Overexpression of a tryptophan decarboxylase cDNA in *Catharanthus roseus* crown gall calluses results in increased tryptamine levels but not in increased terpenoid indole alkaloid production. Transgenic Res 4:315–323, 1995.

25. RA Dixon. Plant natural products: the molecular genetic basis of biosynthetic diversity. Curr Opin Biotechnol 10:192–197, 1999.

26. FLH Menke, A Champion, JW Kijne, J Memelink. A novel jasmonate- and elicitor-responsive element in the periwinkle secondary metabolite biosynthetic gene *Str* interacts with a jasmonate- and elicitor-inducible AP2-domain transcription factor, ORCA2. EMBO J 18:4455–4463, 1999.

27. RG Birch. Plant transformation: problems and strategies for practical application. Annu Rev Plant Physiol Plant Mol Biol 48:297–326, 1997.
28. YPS Bajaj, K Ishimaru. Genetic transformation of medicinal plants. In: YPS Bajaj, ed. Biotechnology in Agriculture and Forestry 45: Transgenic Medicinal Plants. Berlin: Springer, 1999, pp 1–29.
29. C Canel, MI Lopes-Cardoso, S Whitmer, L Van Der Fits, G Pasquali, R Van Der Heijden, JHC Hoge, R Verpoorte. Effects of over-expression of strictosidine synthase and tryptophan decarboxylase on alkaloid production by cell cultures of *Catharanthus roseus*. Planta 205:414–419, 1998.
30. A Geerlings, D Hallard, A Martinez Caballero, I Lopes Cardoso, R van der Heijden, R Verpoorte. Alkaloid production by a *Cinchona officinalis* 'Ledgeriana' hairy root culture containing constitutive expression constructs of tryptophan decarboxylase and strictosidine synthase cDNAs from *Catharanthus roseus*. Plant Cell Rep 19:191–196, 1999.
31. YPS Bajaj, ed. Biotechnology in Agriculture and Forestry 45: Transgenic Medicinal Plants. Berlin: Springer, 1999.
32. MC Christey. Transgenic crop plants using *Agrobacterium rhizogenes*–mediated transformation. In: PM Doran, ed. Hairy Roots—Culture and Applications. Amsterdam: Harwood Academic Publishers, 1997, pp 99–112.
33. TD McKnight, DR Bergey, RJ Burnett, CL Nessler. Expression of enzymatically active and correctly targeted strictosidine synthase in transgenic tobacco plants. Planta 185:148–152, 1991.
34. D-J Yun, T Hashimoto, Y Yamada. Transgenic tobacco plants with two consecutive oxidation reactions catalyzed by hyoscyamine 6β-hydroxylase. Biosci Biotechnol Biochem 57:502–503, 1993.
35. MI Lopes Cardoso, AH Meijer, JHC Hoge. *Agrobacterium*-mediated transformation of the terpenoid indole alkaloid–producing plant species *Tabernaemontana pandacaqui*. Plant Cell Rep 17:150–154, 1997.
36. D-J Yun, T Hashimoto, Y Yamada. Metabolic engineering of medicinal plants: transgenic *Atropa belladonna* with an improved alkaloid composition. Proc Natl Acad Sci USA 89:11799–11803, 1992.
37. J Pearr. Corked up: clinical hyoscine poisoning with alkaloids of the native corkwood, *Duboisia*. Med J Aust 2:422–423, 1981.
38. HA Berens. A survey of duboisias. Chem Druggist 6:593–597, 1953.
39. GD Lin, WJ Griffin. Organogenesis and a general procedure for plant regeneration from callus culture of a commercial *Duboisia* hybrid (*D. leichhardtii* × *D. myoporoides*). Plant Cell Rep 11:207–210, 1992.
40. AK Kukreja, AK Mathur. Tissue culture studies in *Duboisia myoporoides*; plant regeneration and clonal propagation by stem node cultures. Planta Med 2:93–96, 1985.
41. O Luanratana. Micropropagation of *Duboisia* species. In: YPS Bajaj, ed. Biotechnology in Agriculture and Forestry, Vol 40, High-Tech and Micropropagation VI. Berlin: Springer, 1997, pp 313–331.
42. CR Celma. Aplicacion de la tecnologia de las raices transformadas para la produccion en biorreactor de escopolamina. PhD dissertation, University of Barcelona, Barcelona, Spain, 1998.

43. CR Celma, J Palazón, RM Cusidó, T Piñol, M Keil. Reduced scopolamine yield in field-grown *Duboisia* plants regenerated from hairy roots. Planta Med 67:249–253, 2001.

44. T Muranaka, Y Kitamura, T Ikenaga. Genetic transformation of *Duboisia* species. In: YPS Bajaj, ed. Biotechnology in Agriculture and Forestry 45: Transgenic Medicinal Plants. Berlin: Springer, 1999, pp 117–122.

45. G Vancanneyt, R Schmidt, A O'Connor-Sanchez, L Willmitzer, M Rocha-Sosa. Construction of an intron-containing marker gene: splicing of the intron in transgenic plants and ist use in monitoring early events in *Agrobacterium*-mediated plant transformation. Mol Gen Genet 220:245–250, 1990.

46. J Matsuda, S Okabe, T Hashimoto, Y Yamada. Molecular cloning of hyoscyamine-6β-hydroxylase, a 2-oxoglutarate–dependent dioxygenase, from cultured roots of *Hyoscyamus niger*. J Biol Chem 266:9460–9464, 1991.

47. H Daniell. The next generation of genetically engineered crops for herbicide and insect resistance: containment of gene pollution and resistant insects. AgBiotechNet 1:1–7, 1999.

48. MK Saroha, P Sridhar, VS Malik. Glyphosate-tolerant crops: genes and enzymes. J Plant Biochem Biotechnol 7:65–72, 1998.

49. U Wiktorska. A new approach to weed management—the Liberty Link system. Ann Warsaw Agric Univ Agric 32:65–69, 1998.

50. C Jung, U Wyss. New approaches to control plant parasitic nematodes. Appl Microbiol Biotechnol 51:439–446, 1999.

51. IJ Thomason. Challenges facing nematology: environmental risks with nematocides and the need for new approaches. In: JA Veech, DW Dickson, eds. Vistas on Nematology. Hyattsville, MD Society of Nematologists: 1987, pp 469–476.

52. K Meyer, M Keil, MJ Naldrett. A leucine-rich repeat protein of carrot that exhibits antifreeze activity. FEBS Lett 447:171–178, 1999.

53. SP Joshi, PK Ranjekar, VS Gupta. Molecular markers in plant genome analysis. Curr Sci 77:230–240, 1999.

54. JGK Williams, AR Kubelik, KJ Livak, JA Rafalski, SV Tingey. DNA polymorphisms amplified by arbitrary primers are useful as genetic markers. Nucleic Acids Res 18:6531–6535, 1990.

55. SV Tingey, JP del Tufo. Genetic analysis with random amplified polymorphic DNA markers. Plant Physiol 101:349–352, 1993.

56. H Dittrich, TM Kutchan. Molecular cloning, expression, and induction of berberine bridge enzyme, an enzyme essential to the formation of benzophenanthridine alkaloids in the response of plants to pathogenic attack. Proc Natl Acad Sci USA 88:9969–9973, 1991.

57. PFX Kraus, TM Kutchan. Molecular cloning and heterologous expression of a cDNA encoding berbamunine synthase, a C-O phenol-coupling cytochrome P-450 from the higher plant *Berberis stolonifera*. Proc Natl Acad Sci USA 92: 2071–2075, 1995.

58. K Nakajima, T Hashimoto, Y Yamada. Two tropinone reductases with different stereospecificities are short-chain dehydrogenases evolved from a common ancestor. Proc Natl Acad Sci USA 90:9591–9595, 1993.

59. C Poulsen, OJM Goddijn, JHC Hoge, R Verpoorte. Anthranilate synthase and chorismate mutase activities in transgenic tobacco plants overexpressing tryptophan decarboxylase from *Catharanthus roseus*. Transgenic Res 3:43–49, 1994.

60. DD Songstad, V DeLuca, N Brisson, WGW Kurz, CL Nessler. High levels of tryptamine accumulation in transgenic tobacco expressing tryptophan decarboxylase. Plant Physiol 94:1410–1413, 1990.

61. J Berlin. Genetic transformation of *Peganum harmala*. In: YPS Bajaj, ed. Biotechnology in Agriculture and Forestry 45: Transgenic Medicinal Plants. Berlin: Springer, 1999, pp 202–214.

62. TM Kutchan, N Hampp, F Lottspeich, K Beyreuther, MH Zenk. The cDNA clone for strictosidine synthase from *Rauvolfia serpentina*: DNA sequence determination and expression in *Escherichia coli*. FEBS Lett 237:40–44, 1988.

63. F Vazquez-Flota, E De Carolis, A-M Alarco, V De Luca. Molecular cloning and characterization of desacetoxyvindoline 4-hydroxylase, a 2-oxoglutarate dependent dioxygenase involved in the biosynthesis of vindoline in *Catharanthus roseus* (L.) G. Don. Plant Mol Biol 34:935–948, 1997.

64. B St-Pierre, P Laflamme, A-M Alarco, V De Luca. The terminal *O*-acetyltransferase involved in vindoline biosynthesis defines a new class of proteins responsible for coenzyme A–dependent acyl transfer. Plant J 14:703–713, 1998.

65. P Kawalleck, H Keller, K Hahlbrock, D Schell, IE Somssich. A pathogen-responsive gene of parsley encodes tyrosine decarboxylase. J Biol Chem 268:2189–2194, 1993.

66. A Rosco, HH Pauli, W Priesner, TM Kutchan. Cloning and heterologous expression of cytochrome P-450 reductases from the Papaveraceae. Arch Biochem Biophys 348:369–377, 1997.

67. Mitsui-Petrochem. Norcoclaurine 6-*O*-methyltransferase and said enzyme gene–nucleic acid encoding norcoclaurine-6-*O*-methyltransferase, expressed in transgenic plant secondary metabolite. Patent application JP 11178-577, 1997.

68. B Unterlinner, R Lenz, TM Kutchan. Molecular cloning and functional expression of codeinone reductase: the penultimate enzyme in morphine biosynthesis in the opium poppy *Papaver somniferum*. Plant J 18:465–475, 1999.

69. D Ober, T Hartmann. Homospermidine synthase, the first pathway-specific enzyme of pyrrolizidine alkaloid biosynthesis, evolved from deoxyhypusine synthase. Proc Natl Acad Sci USA 96:14777–14782, 1999.

70. MR Wildung, R Croteau. A cDNA clone for taxadien synthase, the diterpene cyclase that catalyzes the committed step of taxol biosynthesis. J Biol Chem 271:9201–9204, 1996.

71. K Walker, R Croteau. Molecular cloning of a 10-deacetylbaccatin III-*O*-acetyl transferase cDNA from *Taxus* and functional expression in *Escherichia coli*. Proc Natl Acad Sci USA 97:583–587, 2000.

72. GA Sprenger, U Schörken, T Wiegert, S Grolle, AA De Graaf, SV Taylor, TP Begley, S Bringer-Meyer, H Sahm. Identification of a thiamin-dependent synthase in *Escherichia coli* required for the formation of the 1-deoxy-D-xylulose 5-phosphate precursor to isoprenoids, thiamin, and pyridoxol. Proc Natl Acad Sci USA 94:12857–12862, 1997.

73. L Lois, N Campos, SR Putra, K Danielsen, M Rohmer, A Boronat. Cloning and characterization of a gene from *Escherichia coli* encoding a transketolase-like enzyme that catalyzes the synthesis of D-1-deoxyxylulose 5-phosphate, a common precursor for isoprenoid, thiamin, and pyridoxol biosynthesis. Proc Natl Acad Sci USA 95:2105–2110, 1998.

74. B Lange, R Croteau. Isoprenoid biosynthesis via a mevalonate-independent pathway in plants: cloning and heterologous expression of 1-deoxy-D-xylulose-5-phosphate reductoisomerase from peppermint. Arch Biochem Biophys 365: 170–174, 1999.

75. J Schwender, C Müller, J Zeidler, HK Lichtenthaler. Cloning and heterologous expression of a cDNA encoding 1-deoxy-D-xylulose-5-phosphate reducto-isomerase of *Arabidopsis thaliana*. FEBS Lett 455:140–144, 1999.

76. S Takahashi, T Kuzuyama, H Watanabe, H Seto. A 1-deoxy-D-xylulose 5-phosphate reductoisomerase catalyzing the formation of 2-C-methyl-D-erythri-tol 4-phosphate in an alternative nonmevalonate pathway for terpenoid biosyn-thesis. Proc Natl Acad Sci USA 95:9879–9884, 1998.

77. BM Lange, R Croteau. Isopentenyl diphosphate biosynthesis via a mevalonate-independent pathway: isopentenyl monophosphate kinase catalyzes the termi-nal enzymatic step. Proc Natl Acad Sci USA 96:13714–13719, 1999.

78. RJ Burnett, IE Maldonado-Mendoza, TD McKnight, CL Nessler. Expression of a 3-hydroxy-3-methylglutaryl coenzyme A reductase gene from *Campto-theca acuminata* is differentially regulated by wounding and methyl jasmonate. Plant Physiol 103:41–48, 1993.

79. IE Maldonado-Mendoza, RJ Burnett, CL Nessler. Nucleotide sequence of a cDNA encoding a 3-hydroxy-3-methylglutaryl coenzyme A reductase from *Ca-tharanthus roseus*. Plant Physiol 100:1613–1614, 1992.

80. JD Hamill, RJ Robins, AJ Parr, DM Evans, JM Furze, MJC Rhodes. Over-expressing a yeast ornithine decarboxylase gene in transgenic roots of *Nicoti-ana rustica* can lead to enhanced nicotine accumulation. Plant Mol Biol 15: 27–38, 1990.

81. E Bell, RL Malmberg. Analysis of a cDNA encoding arginine decarboxylase from oat reveals similarity to the *Escherichia coli* arginine decarboxylase and evidence of protein processing. Mol Gen Genet 224:431–436, 1990.

82. AJ Fleming, T Mandel, I Roth, C Kuhlemeier. The patterns of gene expression in the tomato shoot apical meristem. Plant Cell 5:297–309, 1993.

83. RC Moore, SM Boyle. Nucleotide sequence and analysisi of the *speA* gene encoding biosynthetic arginine decarboxylase in *Escherichia coli*. J Bacteriol 172:4631–4640, 1990.

84. N Hibi, S Higashiguchi, T Hashimoto, Y Yamada. Gene expression in tobacco low-nicotine mutants. Plant Cell 6:732–735, 1994.

85. K Jouhikainen, L Lindgren, T Jokelainen, R Hiltunen, TH Teeri, K-M Oksman-Caldentey. Enhancement of scopolamine production in *Hyoscyamus muticus* L. hairy root cultures by genetic engineering. Planta 208:545–551, 1999.

16

Genetic Engineering of the Plant Cell Factory for Secondary Metabolite Production: Indole Alkaloid Production in *Catharanthus roseus* as a Model

Serap Whitmer, Robert van der Heijden, and Robert Verpoorte
Leiden/Amsterdam Center of Drug Research, Leiden, The Netherlands

I. INTRODUCTION

Metabolism is the closely coordinated series of enzyme-mediated reactions in a living organism such as a plant (1). This complex biochemical network consists of metabolic pathways in which compounds that have direct importance for the vital functions of an organism are synthesized and utilized. This collection of processes is defined as primary metabolism and the compounds involved are primary metabolites, e.g., sugars, amino acids, fatty acids, nucleotides, and the polymers derived from them (polysaccharides, proteins, lipids, DNA, RNA, etc.). Furthermore, in all plants more specialized biochemical pathways exist, known as secondary metabolism, in which a wide range of so-called secondary metabolites are produced, e.g., alkaloids, anthocyanins, flavonoids, quinones, lignans, steroids, and terpenoids (2). Secondary metabolites play a role in the interaction of the plant with its environment, such as attraction of pollinating insects via color and scent and protection against predation by herbivores and insects or against infection by microorganisms (2–4). These secondary metabolite pathways are often restricted to an individual species or genus and might be activated only during particular stages of growth and development or during periods of stress caused by, e.g., wounding, attack by microorganisms, or limitation of nutrients. Many intermediates are utilized in both primary and secondary metabolism. Therefore, the dividing line between primary and secondary metabolism is not always distinct. These overlapping roles of intermediates cause a close interaction between primary and secondary metabolism. Therefore, the regulatory mechanisms in plants that provide control at the cellular, genetic, protein, and molecular levels of secondary metabolism are the subject of considerable research.

The major interest in plant secondary metabolites is due to their functions for the plant, which are often connected with their impressive biological activities (5) and are the basis for commercial applications such as pharmaceuticals, insecticides, flavors, colorants, nutrition, spices, or fragrances (2,4,6). Because of various difficulties involved in growing plants in the field or collecting them in the wild (geographic location, seasonal variation, pests and diseases, risk of species extinction, political instability of the country where the plantations are, etc.), researchers are focusing on plant cell and tissue cultures as a production system as this system can be used under controlled conditions for the production of secondary metabolites of commercial interest. It is widely recognized that cultured plant cells represent a potential source of commercially valuable plant metabolites (6–15).

Some successes in production of secondary metabolites by plant cell cultures have been achieved, e.g., shikonin production by cultures of *Lithospermum erythrorhizon* Siebold & Zucc. (Boraginaceae) (16), berberine production by cultures of *Coptis japonica* (L.) Salisb. (Ranunculaceae) (17,18), and ginsenoside production by cultures of *Panax ginseng* C.A. Mey. (Araliaceae) (19). Taxol is the latest accomplishment with an increase of 10- to 20-fold in productivity compared with the average *Taxus* (Taxaceae) cultures (20). However, there are some major obstacles to the production of secondary metabolites by plant cell cultures: the spectrum of compounds produced is quite different from that in the plant, the yield of the product(s) for economically viable production is too low, and they suffer from heterogeneity and/or instability during long-term culture (21). As in many cases productivity is too low for commercialization, various strategies to improve productivity are being followed:

Cell line improvement: screening and selection for high-producing cell lines

Optimization of growth and production media (phytohormones, nutrients, precursors, antimetabolites)

Optimization of culture conditions (inoculum size, agitation, temperature, light, pH, shaker speed, immobilization, permeabilization, two-phase systems, two-stage systems, elicitation)

Cultures of differentiated cells (in vitro shoot or root cultures, hairy roots)

Metabolic engineering

Although the production of differentiated cells has resulted in improved productivity of the desired compounds (e.g., for tropane alkaloids), for large-scale industrial production the requirement for very special bioreactors is a major economic constraint. Therefore, there is great interest in the last mentioned approach, metabolic engineering, the more so as it has the potential for application to plant cell cultures, microorganisms, and the plant. Also, it has the possibility of application to isolation of enzymes. Further applications of metabolic engineering of secondary metabolism are new compounds with biological activity, new flower or food colors, new taste or fragrance of food, improved nutritional effect of food, decreased levels of undesired compounds in food and fodder, and improved resistance against pests and diseases (6). For increasing the production of desired compounds, different strategies can be considered: increasing the carbon flux through a biosynthetic pathway by affecting, e.g., rate-limiting enzymes, feedback inhibition, and competitive pathways; increasing the number of producing cells; and decreasing catabolism. Besides the previous efforts using random mutation and selection, current genetic engineering offers the tool for a focused approach to improve secondary metabolite production along these lines (Fig. 1) (22).

II. *CATHARANTHUS ROSEUS* AS A SOURCE OF TERPENOID INDOLE ALKALOID PRODUCTION

Catharanthus roseus plant cell cultures are a potential commercial source for the production of some pharmaceutically important alkaloids. *Catharanthus roseus* (L.) G. Don is a tropical plant belonging to the family Apocynaceae and produces a wide range of terpenoid indole alkaloids (TIAs). More than 130 alkaloids have been isolated from different parts of the plant and many of them show biological activities, such as the antineoplastic vinblastine and vincristine, the antihypertensive ajmalicine, and the sedative serpentine (23–25). Cell suspension cultures of *C. roseus* do not represent an exception for secondary metabolite production regarding the obstacles mentioned before. They produce a much simpler alkaloid profile, the yield is less than that from intact plants (26–29), and after prolonged subculture they lose the ability to produce alkaloids at high levels even after extensive

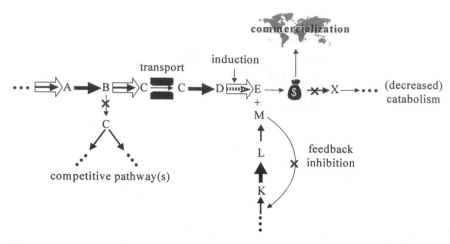

FIGURE 1   Scheme of the targets for genetic engineering of secondary metabolic pathway. Dark arrows, thickness related to flux; open arrows, increased activity; crosses through arrows, blocked steps, ₰, desired product.

screening and selection procedures (30–32). Therefore, the regulation of the biosynthesis of TIAs in cell cultures of *C. roseus* is being extensively studied in our laboratory in a collaborative project, the BSDL–project group Plant Cell Biotechnology, with the Institute of Molecular Plant Sciences Leiden, and the Department of Biochemical Engineering of the Delft University of Technology.

The TIA pathway is a complex metabolic network (Fig. 2). The biosynthetic precursors for TIA biosynthesis are provided by the shikimate pathway and the terpenoid pathway (33–36). It was proved that the terpenoid moiety of the TIAs, secologanin, is not derived from the mevalonate pathway but instead from the MEP* pathway using a cell suspension culture of *C. roseus* (37). Because in the tryptophan branch of the shikimate pathway tryptophan decarboxylase (TDC, EC 4.1.1.28) operates at the interface between primary and secondary metabolism, it has been demonstrated to be a site for regulatory control of TIA biosynthesis (Fig. 3). TDC catalyzes the conversion of L-tryptophan to tryptamine, the indole moiety of the TIA. Strictosidine synthase (STR) catalyzes the condensation of tryptamine with secologanin by a reaction of the Pictet-Spengler type into 3α(*S*)-strictosi-

---

*In the 4th European Symposium on Plant Isoprenoids (April 21–23, 1999, Barcelona) it was agreed to use the term MEP (2-*C*-methyl-D-erythritol 4-phosphate) or Rohmer pathway.

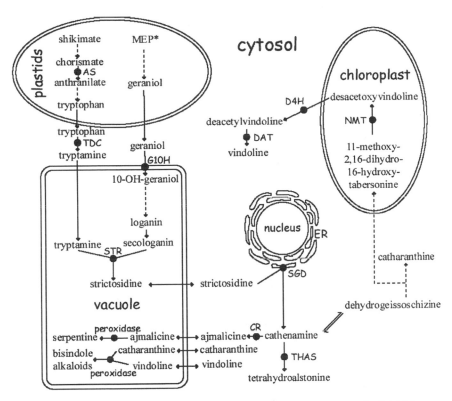

FIGURE 2 Subcellular compartmentation of terpenoid indole alkaloid biosynthesis in *Catharanthus roseus*. AS, anthranilate synthase; CR, NADPH:oath enamine reductase; DAT, deacetylvindoline 17-*O*-acetyltransferase; ER, endoplasmic reticulum; G10H, geraniol 10-hydroxylase; GAP, glyceraldehyde-3-phosphate; D4H, desacetoxyvindoline-4-hydroxylase; MEP; 2-*C*-methyl-D-erythritol 4-phosphate; NMT, *S*-adenosyl-L-methionine methoxy-2,16-dihydro-16-hydroxytabersonine-*N*-methyltransferase; SGD, strictosidine β-glucosidase; STR, strictosidine synthase; TDC, tryptophan decarboxylase; THAS, NADPH: tetrahydroalstonine reductase. Solid arrows represent a single enzymatic step and dashed arrows represent multiple enzyme steps.

dine, the pivotal intermediate of the TIAs (Fig. 3). The properties of TDC and STR from *C. roseus* are listed in Tables 1 and 2, respectively.

A number of enzymes of the TIA biosynthetic pathway have been characterized and the encoding genes have been cloned. This has enabled studies of the overexpression of these genes in various plants and organisms. There are some valuable examples of what metabolic engineering can

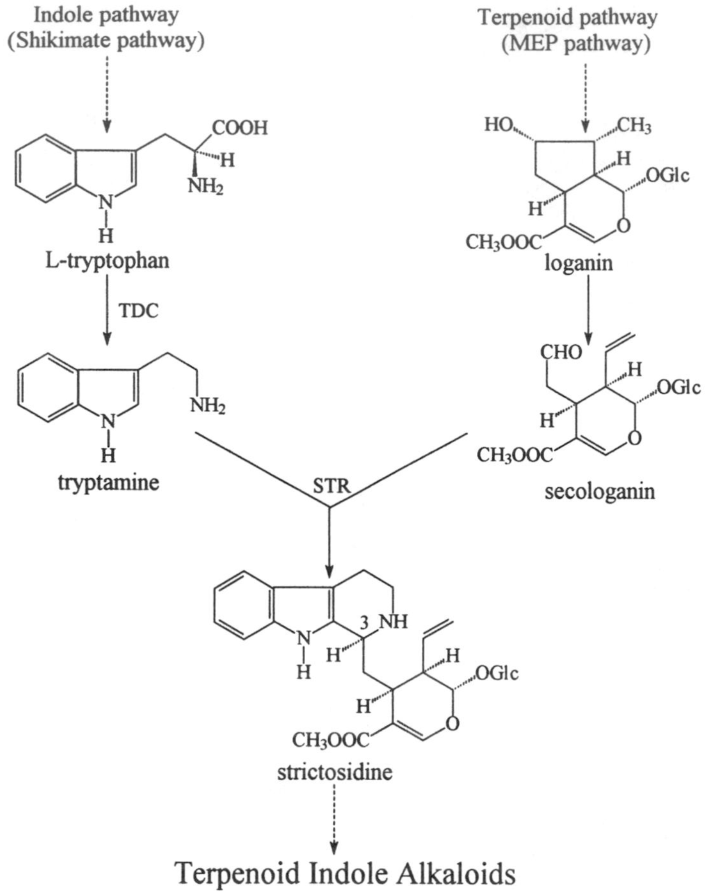

FIGURE 3 Biosynthesis of terpenoid indole alkaloids in *Catharanthus roseus*. STR, strictosidine synthase; TDC, tryptophan decarboxylase.

achieve in secondary metabolism using some of these genes. The seeds of canola (*Brassica napus*) are used as animal feed, but the presence of indole glucosinolates makes the crop less palatable. After the transformation of *B. napus* with the *Tdc* gene from *C. roseus* (Table 3), the mature seeds of the transgenic plants contain reduced levels of indole glucosinolates, which increases the economic value of the plant (38). Another example is the functional expression of the *Str* gene from *Rauwolfia serpentina* in bacteria, yeast, and insect cells, which can be used as a tool for production of large amounts of STR (Table 4) (39–43). Another good example is transgenic

TABLE 1 Properties of the Enzyme Tryptophan Decarboxylase (TDC, EC 4.1.1.28) from *Catharanthus roseus*

Localization	Cytosol (47,48)
Solubility	Soluble (49)
Purification	Apparent homogeneity (49)
Isomers	One (49)
In vivo activity	In vivo equilibrium between an active form and a monomeric form that is susceptible to irreversible inactivation. When the equilibrium is shifted to the monomeric form of TDC, the protein could be degraded in the ubiquitin proteolytic pathway (50).
Relative M_r on PAGE	115 ± 3 kDa (49)
Subunits	Two (54 ± 1 Da) (49), (54 and 56 kDa) (51)
Cofactors	Pyridoxal phosphate (0.5–1 mol) (49), pyrroloquinolinequinone (52)
pl	5.9 (49)
pH optimum	8.5 (49)
Stability (half-life time)	For purified TDC at 0°C, pH 7.5 about 4 days, for partially purified TDC 7 days (49)
	For crude enzyme at $-15°C$ about 7 days (52)
Cooperative effects	
L-Tryptophan	$R_s = 80$, $n_H = 1.06$
Substrate specificity	
Accepted	4-Methyltryptophan, 5-fluorotryptophan (53), L-tryptophan ($K_m = 7.5 \times 10^5$ M), L-5-hydroxytryptophan ($K_m = 1.3 \times 10^3$ M) (49)
Not accepted	p-Fluorophenylalanine, aminoethylcysteine, 5-methyltryptophan (53), L-phenylalanine ($K_m = 8$), L-tyrosine ($K_m = 8$), L-dopa ($K_m = 8$) (49).
Inhibitors	Tryptamine (competitive, K_i 3.1×10^{-4} M), D-tryptophan (noncompetitive), carbon dioxide (noninhibitor) (49)

yeast cultures containing the *Str* and/or *Sgd* (strictosidine glucosidase) genes that can be grown on the juice of *Symphoricarpus* berries, which contains both sugar and secologanin, and are fed tryptamine (Table 4). The system produces large amounts of mostly strictosidine (2 g L^{-1}) and cathenamine (44). However, overexpression of the *Tdc* or *Str* genes does not always lead to the desired results. For example, expression in *Tabernaemontana pandacaqui* did not affect production of TIAs (45) (Table 4), and in transgenic *Peganum harmala* hairy roots overexpressing the *Tdc* gene the production

TABLE 2 Properties of the Enzyme Strictosidine Synthase (STR, EC 4.3.3.2) from *Catharanthus roseus*

Localization	Cytosol (47,54)[a] Vacuole (48,55)[a]											
Solubility	Soluble (56)											
Purification	Partial without isomer recognition (57,58) Apparent homogeneity with recognition of isomers (56,59)[b]											
Relative M_r												
Gel filtration	31 ± 10% kDa (56); 34 kDa (57); 35–37 kDa (59)											
SDS-PAGE	41.5 ± 5% kDa (56); 35 kDa (59)											
Cofactors	Not required (56)											
Isomers	7 (56)[b]				6 (59)[b]							
	I	II	III	IV	A1 (I[b])	A2 (II[b])	B (II[b])	C1 (III[b])	C2 (III[b])	D (IV[b])	average of A1–D
pI	4.8	4.6	4.5	4.3							
pH optimum	6.7	6.7	6.7	6.7							6.8
50% thermal inactivation (°C)	47	48	46	48							
Tryptamine											
K_m (mM[c], μM)	0.9[c]	6.6[c]	1.9[c]	2.2[c]	8.7 ± 0.6	8.2 ± 0.9	9.2 ± 0.3	9.3 ± 1.1	8.4 ± 2.6	9.4 ± 1.1	8.9[d] ± 0.5
V_{max} (pkat[c], nkat)	28.9[c]	93.6[c]	48.0[c]	46.3[c]	215 ± 3	238 ± 14	312 ± 3	303 ± 11	153 ± 15	180 ± 2	234[d] ± 64
Substrate inhibition (mM[c], μM)	>0.83[c]	>1.5[c]	>0.83[c]	>1.0[c]	260 ± 34	442 ± 145	261 ± 14	325 ± 33	248 ± 56	316 ± 33	309[d] ± 73
Secologanin n_H	1.3	2.0	2.2	1.5							

Tryptamine	
K_m (mM)	0.9 immobilized STR (56); 2.3 soluble STR (57); 0.83 soluble STR (58); 8.9[d] soluble STR (59)
V_{max} (pkat)	18.9 immobilized STR (56); 130 soluble STR (57); 234[d] soluble STR (59)
Secologanin	
K_m (mM)	2.1 immobilized STR (56); 3.4 soluble STR (57)
V_{max} (pkat/mg)	9.1 immobilized STR (56); 55 soluble STR (57)
Stability (half-life time)	For crude protein extract (0–70% ammonium sulfate precipitation) 5 hours (60)
	For immobilized STR at 37°C 68 days (60)
	For partly purified soluble STR at 37°C 5 hours (60)
	For partly purified soluble STR at −20°C >2 months (61)
	For purified STR at −20°C up to 1 year (61)
Substrate specificity	
Indoles[e]	Tryptamine (100%), 7-methyltryptamine (15%), 6-hydroxytryptamine (9%), 7-fluorotryptamine (8%), 5-fluorotryptamine (8%), 5-hydroxytryptamine (4%) [57]
Monoterpenoids[e]	Secologanin (K_m = 3.4 × 10⁻³ M), 2′-O-methylsecologanin, 3′-O-methylsecologanin (K_m = 8 × 10⁻³ M (57) and K_m = 3.6 mM, V_{max} = 14.3) (56), dihydrosecologanin (K_m = 3.5 × 10⁻³) (57)
Indoles[f]	5-Methyltryptamine, 5-metoxytryptamine, 5,6-dihydroxytryptamine, 5,7-dihydroxytryptamine, N-methyltryptamine, DL-α-methyltryptamine, α-ethyltryptamine, L-tryptophan, D-tryptophan, phenylethylamine, tyramine, dopamine, homovertrylamine (57)
Monoterpenoids[f]	Secologanic acid, iridotrial, tarennoside (57)
Inhibitors	p-Chloromercuribenzoate at 1 mM (73% inhibition), no inhibition Ca²⁺, Co²⁺, Cu²⁺, Fe²⁺, Mn²⁺, and Zn²⁺ (57)
	Weak product inhibition, no substrate inhibition (59)
	No inhibition by ajmalicine, vindoline, and catharanthine (58)

[a] STR is demonstrated to be located in vacuole; for discussion see Ref. 48.
[b] Identity as speculated by De Waal et al. (59) for isomers 4 of 7 (I–IV) of Pfitzner and Zenk (56) and the purified isoenzymes.
[c] Values are expressed in mM or pkat.
[d] Average values from six isomers isolated by De Waal et al. (59).
[e] Accepted as substrate.
[f] Not accepted as substrate.

TABLE 3 Studies with (Over)expression of the Tryptophan Decarboxylase (*Tdc*) Gene

Tdc (51)

Origin of the clone	*Catharanthus roseus* (L.) G. Don (Apocynaceae)
Promoter	*lac*
Expressed in	*Escherichia coli*
Results	No attempt was made to quantify the level of the TDC activity in *E. coli*.

Tdc (62)

Origin of the clone	*C. roseus* (L.) G. Don (Apocynaceae) (as in Ref. 51)
Method	*Agrobacterium*-mediated transformation using *A. tumefaciens*
Promoter	Cauliflower mosaic virus (CaMV) 35S
Expressed in	*Nicotiana tabacum* 'Xanthi' (Solanaceae)
Obtained as	Transgenic plants
Results	Leaves of transformed plants had 4 to 45 times more TDC activity than the controls.
	Increase in tryptamine accumulation: >1 mg g^{-1} FW plant with the highest TDC activity, a 260-fold increase over control
	The growth and development of the transformed plants appeared normal.

Tdc (63)

Origin of the clone	*C. roseus* (L.) G. Don (Apocynaceae) (as in Ref. 51)
Method	*Agrobacterium*-mediated transformation using *A. tumefaciens*
Promoter	CaMV 35S
Expressed in	*N. tabacum* 'Xanthi' (Solanaceae)
Obtained as	Transgenic plants
Results	TDC activity in the transformed tobacco was 45 times greater than in the controls and resulted in a >200-fold increase in the free tryptamine and >30-fold in free tyramine pools.

Tdc (46)

Origin of the clone	*C. roseus* (L.) G. Don (Apocynaceae) (as in Ref. 64)
Method	*Agrobacterium*-mediated transformation (65)
Promoter	CaMV 35S
Expressed in	*Peganum harmala* L. (Zygophyllaceae)
Obtained as	Cell suspension and hairy root cultures
Results	Despite increased TDC activity, tryptamine level did not increase due to its conversion into serotonin. Consequently, serotonin level increased, but no increase in desired harman-type alkaloids was observed.

Tdc (66)

Origin of the clone	*C. roseus* (L.) G. Don (Apocynaceae) (as in Ref. 37)
Method	*Agrobacterium*-mediated transformation using *A. tumefaciens*
Promoter	CaMV 35S
Expressed in	*N. tabacum* 'Petit Havana' SR1 (Solanaceae)
Obtained as	Plants
Results	The plants with highest TDC activity contained 19 pkat/mg protein. Specific activity of AS and CM was not affected by expression of TDC. Despite increased tryptamine accumulation, the phenotype of the transgenic plants was normal.

Tdc (38)

Origin of the clone	*C. roseus* (L.) G. Don (Apocynaceae) (as in Ref. 51)
Method	*Agrobacterium*-mediated transformation using *A. tumefaciens*
Promoter	CaMV 35S
Expressed in	*Brassica napus* 'Westar' (Brassicaceae)
Obtained as	Transgenic plants
Results	TDC activity in transgenic plants varied and was 4 to 45 times greater than in contrtols Transgenic plants accumulated tryptamine while correspondingly lower levels of tryptophan-derived indole glucosinolates were produced in all plant parts compared with controls. Indole glucosinolate content of mature seeds from transgenic plants was only 3% of that found in nontransformed seeds.

Tdc (67)

Origin of the clone	*C. roseus* (L.) G. Don (Apocynaceae)
Method	*Agrobacterium*-mediated transformation using *A. tumefaciens*
Promoter	CaMV 35S
Expressed in	*C. roseus* 'Morning Mist' (Apocynaceae)
Observed as	Crown-gall calli
Results	An increase in TDC protein level, TDC activity, and tryptamine content but no significant increase in TIA production were observed. In *Tdc*-antisense–containing calli decreased TDC activity was measured.

TABLE 3 Continued

Tdc (68)

Origin of the clone	*C. roseus* (L.) G. Don (Apocynaceae) (as in Ref. 51)
Method	*Agrobacterium*-mediated transformation using *A. tumefaciens*
Promoter	CaMV 35S (as in Ref. 38)
Expressed in	*Solanum tuberosum* 'Désirée' (Solanaceae)
Observed as	Transgenic plants
Results	Phenylpropanoid pathway can be down-regulated by the control of substrate supply created by the introduction of a single gene outside the pathway.
	The redirection of tryptophan to tryptamine resulted in a decrease in the levels of tryptophan, phenylalanine, and phenylalanine-derived phenolic compounds in transgenic tubers.
	Chlorogenic acid was 2- to 3-fold lower in transgenic tubers.
	Transgenic tuber became more susceptible to infection with *Phytophthora infestans* due to alterations in the substrate levels in the shikimate pathway and phenylpropanoid pathways.

Tdc (69)

Origin of the clone	*C. roseus* (L.) G. Don (Apocynaceae) (as in Ref. 51)
Method	*Agrobacterium*-mediated transformation using *A. tumefaciens* (as in Ref. 62)
Promoter	CAMV 35S
Expressed in	*N. tabacum* SR1 (Solanaceae)
Observed as	Transgenic plants

of the desired harman-type alkaloids did not occur due to the conversion of tryptamine to serotonin (Table 3) (46). The results of (over)expressing the enzymes TDC, STR, and TDC and STR in *C. roseus* and other species are shown in more detail in Tables 3, 4, and 5, respectively.

A. Aspects of TIA Formation by Transgenic Cell Lines of *C. roseus* Overexpressing *Tdc* and *Str*

The highest ajmalicine production ever achieved in our laboratories was 77 mg L^{-1}, obtained after 16 days in a bioreactor under optimized conditions with a wild-type cell line of *C. roseus* (79). Such productivity is too low for commercialization; therefore, extensive studies were made to characterize genetically modified *C. roseus* cell cultures, in particular with regard to TIA formation, to investigate the effect of overexpression of the *Tdc* and *Str* genes on TIA formation and to follow their stability during long-term subculture. The available knowledge on wild-type cell cultures of *C. roseus* was the basis for these further studies with transgenic cell lines.

Transgenic cell lines were obtained from *C. roseus* seedling leaf cells by means of *Agrobacterium tumefaciens*–mediated transformation (Table 6). These cells overexpress the genes *Str* (S lines) and/or *Tdc* (T lines). The enzymes STR and TDC catalyze key steps in the biosynthesis of TIAs (Figs. 2 and 3). All our experiments were carried out using a two-stage production system in which biomass and product formation occur sequentially. Sufficient biomass was first obtained in growth medium (GM), and the cells were then inoculated into production medium (PM). Our PM does not contain phosphate, nitrate, or phytohormones, constituents that promote culture growth. Very little cell division therefore occurs in PM; most biomass increase is the result of starch biosynthesis (80).

As shown in Table 6, after the initiation of the transgenic calli lines we obtained 10 transgenic cell lines with six different T-DNA constructs (81). These transgenic cell cultures showed wide phenotypic diversity for TIA accumulation: non–TIA-accumulating and TIA-accumulating cell lines. After the initiation of transgenic cell suspension cultures from these calli lines, several cell lines gradually lost their viability after a number of subcultures. The surviving productive cell lines S10, S1 (S lines, transgenic for STR), and T22 (T line, transgenic for TDC) were chosen for further studies of TIA accumulation. Both S and T lines showed higher STR and TDC activities, respectively, than the wild-type levels. The TIA accumulation profiles of the productive cell lines were as follows: line S10 produced ajmalicine, serpentine, catharanthine, and tabersonine and lines S1 and T22 produced strictosidine, ajmalicine, serpentine, and catharanthine. Most notably, the viable transgenic cell lines showed many years of transgene stability.

TABLE 4 Studies with (Over)expression of the Strictosidine Synthase (*Str*) Gene

Str (39–42)

Origin of the clone	*Rauvolfia serpentina* (L.) Kurz (Apocynaceae)
Method	Baculovirus-based expression (42)
Promoter	Baculovirus polyhedrin (*polh*) (42)
Expressed in	*E. coli* (39,40), *Saccharomyces cerevisiae* (41), *Spodoptera frugiperda* (fall arm worm) (42)
Results	The clone has been identified (39).
	Str has been expressed in an enzymatically active form in *E. coli*. Addition of equimolar (15 mM) concentrations of secologanin and tryptamine to the bacteria in either exponential or stationary phase resulted in the quantitative conversion of the precursors to strictosidine.
	Enzyme was in a soluble form within the bacterium but strictosidine accumulated in the medium (40).
	Expression of *Str* in *S. frugiperda* resulted in the overproduction of STR in a catalytically active form (max. 4 mg L^{-1}) (42).

Str (55)

Origin of the clone	*C. roseus* (L.) G. Don (Apocynaceae) (as in Ref. 70)
Method	*Agrobacterium*-mediated transformation using *A. tumefaciens*
Promoter	CAMV 35S
Expressed in	*N. tabacum* 'Xanthi' (Solanaceae)
Observed as	Transgenic plants
Results	The transgenic plants had 3 to 22 times greater STR activity than *C. roseus* plants.
	One difference in expression of STR between tobacco and *C. roseus* was the presence of a lower molecular weight in tobacco. Two distinct forms of the enzyme were produced in transgenic plants but only a single form was made in *C. roseus* plants. The second form of the protein is not a result of overexpression in tobacco but may represent differences in protein processing between tobacco and *C. roseus*.

Str (71)

Origin of the clone	*C. roseus* (L.) G. Don (Apocynaceae) (as in Ref. 70)
Method	Expression cassette polymerase chain reaction
Promoter	*tac*
Expressed in	*E. coli*
Results	Induction of the *Str* gene resulted in overexpression of the STR, which accumulated mainly as insoluble inclusion bodies.
	Denaturation and refolding of the insoluble protein resulted in the ability to purify up to 6 mg of active enzyme from a single liter of cell culture.
	The recombinant STR activity was about 30 nkat/mg.

Str (45)

Origin of the clone
: *C. roseus* (L.) G. Don (Apocynaceae) (as in Ref. 72),

Method
: *Agrobacterium*-mediated transformation using *A. tumefaciens*

Promoter
: CAMV 35S

Expressed in
: *Tabernaemontana pandacaqui* Poir. (Apocynaceae)

Observed as
: Transgenic plants

Str (43)

Origin of the clone
: *Rauvolfia serpentina* Poir. (Apocynaceae) (as in Ref. 39)

Expressed in
: 22 bacterial strains (*Aeromonas* sp., *Brochothrix thermosphacta, B. campestris, Bacillus lincheniformis, Cellulomonas uda, Citrobacter freundii, Enterobacter aerogenes, E. cloacae, Escherichia coli, E. coli* T, *Klebsiella oxytoca, K. pneumoniae, K. terrigena, Listeria grayi, L. innocus, Proteus mirabilis, P. vulgaris, Salmonella typhimurium, Serratia marcescens, Staphylococcus aureus, S. epidermidis, S. carnosus*)

Results
: Strictosidine was converted by the bacterial strains into vallesiachotamine and isovallesiachotamine.

Str and *Sgd* (73)

Origin of the clone
: *C. roseus* (L.) G. Don (Apocynaceae)

Method
: Lithium acetate procedure

Expressed in
: *Saccharomyces cerevisiae* (yeast)

Results
: With secologanin and tryptamine feeding, transgenic yeast that was introduced with the genes *Str* and *Sgd* (strictosidine glucosidase) produced 2 g L^{-1} strictosidine. Strictosidine is excreted into the medium and only small amounts of cathenamine were found. Growing the transgenic yeast on the sap obtained by pressing *Symphoricarpus* berries, which contain both sugar and secologanin, and feeding tryptamine, the system was able to produce large amounts of strictosidine (44).

TABLE 5 Studies with (Over)expression of the *Tdc* and *Str* Genes

Tdc and *Str* (74)	
Origin of the clone	*C. roseus* (L.) G. Don (Apocynaceae) (45)
Method	*Agrobacterium*-mediated transformation using *A. rhizogenes*
Promoter	CaMV 35S
Expressed in	*Nicotiana tabacum* 'Petit Havana' SR1 (Solanaceae), *Lonicera xylosterum* L. (Caprifoliaceae), *Weigela* 'Styriaca' (Diervillaceae), *C. pusillus* 'Murray') G. Don (Apocynaceae), *Cinchona officinalis* 'Ledgeriana' (Rubiaceae)
Observed as	Transgenic hairy root cultures
Results	In tobacco TDC and STR activities and tryptamine were detected.
	High levels of secologanin (up to 1.25 mg g⁻¹ FW) was produced by *L. xylosterum* cultures.
	The *Weigela* cultures after about 2 months turned brown and stopped growing.
	Different TDC and STR activities between two hairy root cultures of *C. pusillus* were observed.
	TDC and STR activities up to 4.1 and 450 pkat mg⁻¹ of protein, respectively, were recorded. High levels of tryptamine and strictosidine, up to 1.2 and 1.95 mg g⁻¹ DW, respectively, were measured.

Tdc and *Str* (75)	
Origin of the clone	*C. roseus* (L.) G. Don (Apocynaceae)
Method	*Agrobacterium*-mediated transformation using *A. tumefaciens*
Promoter	CaMV 35S
Expressed in	*Nicotiana tabacum* 'Petit Havana' SR1 (Solanaceae)
Observed as	Transgenic cell suspension cultures
Results	Transgenic tobacco cells showed relatively constant TDC and 2- to 6-fold greater STR activities compared with the non-transformed tobacco cell in which the enzyme activities were not detectable.
	The transgenic culture accumulated tryptamine and produced strictosidine upon feeding of secologanin.

Tdc and *Str* (76)	
Origin of the clone	*C. roseus* (L.) G. Don (Apocynaceae)
Method	*Agrobacterium*-mediated transformation using *A. rhizogenes*
Promoter	CaMV 35S
Expressed in	*Cinchona officinalis* 'Ledgeriana' (Rubiaceae)
Observed as	Transgenic hairy root culture
Results	The products of TDC and STR, tryptamine and strictosidine were found in high amounts, 1.2 and 1.9 mg m⁻¹ DW. However, no increase in desired quinoline alkaloids was observed compared with the amounts of 0.05 mg g⁻¹ FW reported by Hamill et al. (77). One year later they lost the capacity to accumulate alkaloids.

Tdc and *Str* (78)

Origin of the clone	*C. roseus* (L.) G. Don (Apocynaceae) (*Tdc* as in Ref. 51; *Str* as in Ref. 70)
Method	Particle bombardment
Promoter	CaMV 35S
Expressed in	*Nicotiana tabacum* 'Samsun' NN
Observed as	Transgenic plants
Results	In 26% of the 150 independent transgenic plants, both the plant *Tdc* and *Str1* transgenes were silenced, 41% demonstrated a preferential silencing of either transgene, 33% of them expressing both transgenes. Seven-week-old seedlings showed 24- and 110-fold variation in levels of TDC and STR1 activities, respectively.

Tdc and *Str* (Hilliou et al., personal communication)

Origin of the clone	*C. roseus* (L.) G. Don (Apocynaceae)
Method	Particle bombardment
Promoter	CaMV 35S
Expressed in	*C. roseus* (L.) G. Don (Apocynaceae)
Observed as	Transgenic calli cultures
Results	Despite an increase in TDC and STR activities observed compared with the nontransformed cultures, TIA accumulation did not increase.

TABLE 6   Gene Constructs and Derived Cultures

| T-DNA construct[a] | Number of established callus lines | | | Established cell cultures |
	Total	Antibiotic resistant	GUS positive	
<npt-gus>-Str>-Tdc1>	38	1	1	ST1
<hpt-<Tdc2-<Str-gus>	146	8	5	None
<npt-<gus-<Str	45	2	2	S1
<hpt-<gus<Str	330	22	19	S4, S6, S7, S10, S11
<npt-gus>-Tdc1>	47	2	2	T11
<hpt-gus>-Tdc1>	288	11	9	T21
<hpt-<Tdc2-<gus	50	2	2	T22

[a]The antibiotic resistance gene is always placed near the left border of the T-DNA; <, > indicate the direction of transcription. In addition, two sets of three lines transformed with either of the unmodified binary vectors were established.

The transgenic cell lines were initiated from seedling leaf cells by means of *A. tumefaciens*. Therefore, these cell lines might arise from one or just a few transformed cells, which carries the risk that effects of the transformation are related to the site of the insertion of the foreign genes. The S lines seemed more productive than the T lines after the initiation of the transgenic cultures. Overexpression of the enzyme TDC increased the tryptamine production in most of the T lines compared with the wild type. When S lines had detectable levels of TDC activities, they were more productive, improving the flux through the pathway before the terpenoid pathway became a limiting factor for TIA formation during long-term subculture.

B.   Effect of Culture Conditions on the Productivity of the Transgenic Cell Lines

Accumulation of TIA was strongly influenced by culture conditions, such as the hormonal composition of the medium and the availability of precursors. Alkaloid accumulation by highly productive transgenic lines showed significant instability in the first year of their maintenance. One of the most productive transgenic cell lines, S10, was used as a model to study the effect of the presence of the synthetic auxins naphthalene acetic acid (NAA) and 2,4-dichlorophenoxy acetic acid (2,4-D) in the culture medium on the accumulation of TIAs. It was found that the presence of NAA during the production phase led to lower levels of alkaloid accumulation, whereas the presence of 2,4-D in the growth medium significantly reduced culture aggregation and measurably repressed the TIA biosynthesis due to reduced capacity to supply biosynthetic precursors (Fig. 4). At the enzyme level, the expression of the reporter gene β-glucuronidase (*Gus*) and the *Str* gene was not affected by auxins under the conditions tested (80). As can be seen in Fig. 5, the productivity of these cell lines increased in the sequence GM < PM < PMa < PMb.

C. Precursor Feeding Experiments: Identification of Bottlenecks and Determination of the Capacity of the Transgenic Cell Lines for TIA Accumulation

As overexpression of the genes *Tdc* and *Str* did not result in a much increased TIA level during long-term subculture, the total capacity of the pathway was assessed by feeding the indole and iridoid precursors. The uptake and utilization rates of the fed precursors differed among cell lines (Table 7). Lines S10, S1, and T22 and the wild-type cell line CRPM were fed in PM with various concentrations and combinations of the indole precursors L-tryptophan and tryptamine and the iridoid precursors loganin and secologanin. First, the optimal conditions for feeding were determined. It was found that addition of precursor(s) in the early culture and to the PM was more efficient than addition to the GM. From feeding a single precursor, it was learned that addition of the indole precursors such as L-tryptophan and tryptamine to the culture medium did not increase TIA production but addition of the iridoids such as loganin and secologanin did increase alkaloid production. This shows that the major bottleneck for the biosynthesis of TIAs was in the terpenoid pathway for both the transgenic lines and the wild-type cell line CRPM (Fig. 3).

Interestingly, feeding with loganin, which first need to be converted to secologanin (Fig. 3), was more efficient than feeding with secologanin. It is thought that exogenous secologanin is trapped in an inactive pool in a different compartment (82); e.g., secologanin can react with free amino groups of proteins. Feeding tryptamine in combination with loganin was also more efficient than the L-tryptophan/loganin combination. This suggests that even in the *Tdc*-overexpressing cell line the conversion of L-tryptophan to tryptamine remains a rate-limiting step for TIA formation. This can be overcome by addition of tryptamine to the culture medium, which results in an increase of TIA accumulation (Figs. 6 and 7) (83). Despite this limitation in the biosynthetic rate of tryptamine, lines S1 and T22 could supply endogenous tryptamine as much as the 0.8 mM and 1.6 mM loganin-fed cultures, respectively (Fig. 7). In line S10 after feeding, no accumulation of tryptamine is observed; instead, it is rapidly converted via strictosidine into TIAs. In lines S1 and T22 strictosidine was accumulated. Considering possible reg-

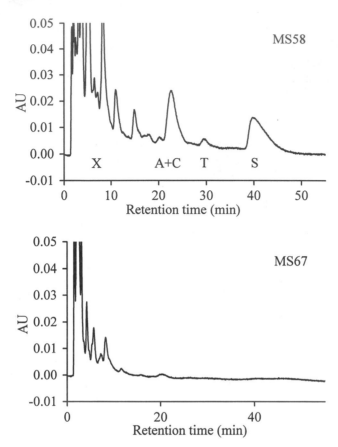

FIGURE 4   HPLC chromatogram showing the effect of growing S10 cells in MS58 [per liter: Murashige and Skoog salts (89), 0.1 g *myo*-inositol, 0.4 mg thiamine, 2 mg NAA, 0.2 mg kinetin, and 30 g sucrose] and MS67 (per liter: MS salts, 0.1 g *myo*-inositol, 1 mg thiamine, 0.5 mg pyridoxine, 0.5 mg niacin, 2 mg glycine, 1 mg 2,4-D, 0.2 mg kinetin, and 30 g sucrose) on alkaloid accumulation. Absorbance at 280 nm by compounds eluding from μBondapak Phenyl column (Waters Chromatography, Etten-Leur, The Netherlands) at a flow rate of 2 mL min^{-1}. The identity of the compounds was established by photodiode array detection of their absorption spectra. X, unidentified non-alkaloid compounds; A, ajmalicine; C; catharanthine; S, serpentine; T, tabersonine.

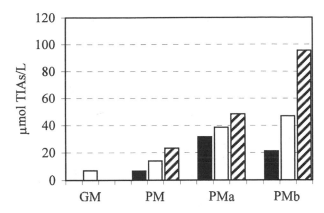

FIGURE 5 Terpenoid indole alkaloid accumulation by transgenic cell suspension cultures of S10 (black bars), S1 (dashed bars) (both transgenic for *Str*), and T22 (white bars) (transgenic for *Tdc*) in different media [GM (89); PM (81); PMa (90); PMb: PMb differ from PMa with 80 g L^{-1} sugar content instead of 50 g L^{-1}]. Duplicate cultures were pooled upon harvesting on day 14 and analyzed as single samples.

ulation aspects of the feeding, it can be concluded that loganin feeding seems to increase production of TIAs by mobilizing tryptamine. This can be either by inducing tryptamine production or by channeling tryptamine away from a competitive pathway into TIA biosynthesis. On the other hand, addition of L-tryptophan or tryptamine does not seem to induce either loganin or secologanin availability.

Under optimized experimental conditions, cultures of line S10 accumulated 350 μmol L^{-1}, line S1 about 600 μmol L^{-1}, and line T22 1200 μmol L^{-1} of TIAs while the wild-type cell line accumulated about 580 μmol L^{-1} (manuscript in preparation). The profile of TIAs was cell line specific, and the time courses of accumulation of individual TIAs differed. The accumulation of large amounts of TIAs by the non–TIA-accumulating wild-type cell line CRPM after feeding tryptamine and loganin (Table 7) suggests that the cells have an enormous capacity to produce TIAs. In all experiments, catabolism of the alkaloids was observed. Very little is known about the products formed in catabolism.

D. Stability of the Transgenic Cell Lines over a Period of 30 Months of Subculture

The transgenic cell lines were not selected for high productivity, but they were maintained in medium containing antibiotic, either hygromycin or kan-

TABLE 7 Summary of TIA Accumulation by Some Cell Lines in Production Medium (PM)

	Fed and harvest (day)	Type of feeding[a]	TIAs[b]		TDC[c]	STR[c]
			Fed	Nonfed		
S10[d]	F0H3	0.5L	183	76	ND	ND
	F11H14	0.5 L	212	194	ND	ND
	F11H14	0.5 LT	352	194	ND	ND
S1[d]	F0H2	0.8–6.4 L	~250	50	*	*
	F0H2	1.6 LT	~700	50	5	229
	F0H2	1 L	250	38	5	114
	F0H14	1 L	600	110	6	381
	F0H2	1 LT	471	38	22	129
	F0H14	1 LT	515	110	11	344
T22[d]	F0H2	1.6–6.4 L	~510	10	*	*
	F0H2	6.4 LT	1100	10	90	89
	F0H2	1 L	250	—	38	71
	F0H14	1 L	48	2	57	141
	F0H2	1 LT	300	—	57	49
	F0H14	1 LT	39	2	117	81
	F0-10 H2	0.4 LT	290	—	8	45
	F0-10 H8	0.4 LT	1250	19	7	35
CRPM[e]	F0H2	1 LT	114	—	ND	ND
	F0H14	1 LT	587	—	ND	ND

[a]Feedings consisted of addition of loganin (L) and loganin/tryptamine (LT), and the concentration is presented in mM.
[b]Concentrations are expressed in μmol L^{-1}.
[c]TDC and STR activities are expressed in pkat mg^{-1} of soluble protein (ND, not determined; —, not detected or too small to quantify; *, enzyme activities will be reported in manuscripts that are in preparation).
[d,e]Wild-type cell line and transgenic cell lines, respectively.

amycin, for transgene selection and gave GUS positive results. The STR and TDC activities showed fluctuations, although they remained significantly higher than in the wild-type line. However, production of TIAs decreased from several mg L^{-1} to an undetectable level during long-term subculture. The decrease in the production of TIAs is thought to be the result of selection of fast-growing cell types, which in terms of secondary metabolism are not active or have new bottlenecks occurring in the secondary metabolite biosynthetic pathways (manuscript in preparation).

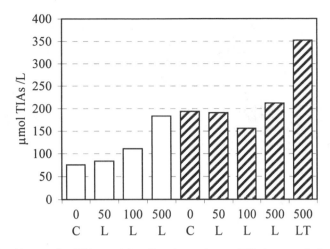

FIGURE 6   Effect of feeding loganin on TIA accumulation by line S10 (transgenic for *Str*) under conditions of low tryptamine availability. White bars represent culture treated on day 0 and sampled on day 3; hatched bars represent culture treated on day 11 and harvested on day 14. The concentration of exogenously added precursors is expressed in μM. C indicates control; L and LT indicate that loganin and both loganin and tryptamine were added, respectively. Triplicate cultures were pooled upon harvesting and analyzed as single samples.

E.   Conclusions and Prospects

Biosynthesis of TIAs involves different cellular compartments (Fig. 2) and even different cells for some of the final steps in biosynthesis of the dimers (84), although it has been shown for anthocyanins in *C. roseus* cell cultures that the level of accumulation in the culture is determined by the percentage of producing cells (85,86). It is also not known whether only a small part of the cells is producing alkaloids. Apparently, overexpression of *Str* in combination with feeding tryptamine and loganin did not lead to increased alkaloid production compared with feeding of wild-type cells (Table 7). This might be explained by having some physiological barriers for alkaloid production. The availability of tryptamine in producing cells might be mentioned in this connection. Maybe overexpression of *Tdc* overcomes the problem of availability of this precursor in the producing cells as it results in a much higher alkaloid production capacity after feeding precursors. Transgenic cell lines with increased TIA accumulation lost this trait after prolonged subculture. Apparently, instability still appears to be a problem as it is for nontransgenic cells. Constitutively increased enzyme activity in the

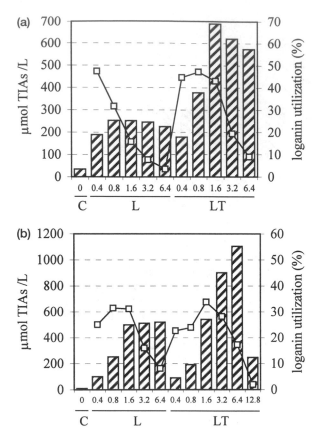

FIGURE 7 Utilization of various concentrations of loganin (L) and loganin/tryptamine (LT) by lines S1 (transgenic for *Str*, (a) and T22 (transgenic for *Tdc*, (b). Feeding consisted of addition of L or LT in a series of concentrations from 0.4 mM up to 6.4 mM at inoculation. Right-hand *y*-axis represents percentages of loganin (□) utilized in TIAs. Duplicate cultures were pooled upon harvesting on day 2 and analyzed as single samples (C, control).

TIA pathway, in this case TDC or STR, was not sufficient to keep elevated TIA levels. Although TDC is known to be a strongly regulated enzyme, which has a transient maximum in activity shortly after subculture on a fresh medium, TDC seems to not be a limiting factor in these cells. Also, STR is not a limiting factor. The following points might be considered in future studies, concerning possible physiological or further biochemical barriers for TIA accumulation in *C. roseus* cell cultures:

The cellular or subcellular localization of enzymes, precursors, and TIAs

Factors involved in transport of the precursors for TIAs

Effect of competitive pathway(s) utilizing TIA precursors

Catabolism of TIAs

Factors that control the carbon flux into the pathway

As all the factors are connected with each other, further studies on the regulatory genes and the overexpression of such genes seem a promising approach. Our studies showed that there is a short-term increase in the TIA levels with overexpression of *Tdc* and *Str* genes; however, in a complex pathway such as the TIA pathway, overexpression of one or two enzymes seemed to be sufficient only when accompanied by a favorable epigenetic environment. This is in contrast to the overexpression of chalcone isomerase in tomatoes, which resulted in a 70-fold increase of flavanoids (87). An encouraging fact is that expression of the transgenes seems to be stable over long periods despite the fact that in cell suspension cultures there is a strong selection for fast-growing cells.

III. THE FUTURE

Since about 6000 BC, when the Sumerians and Babylonians in Mesopotamia produced alcoholic beverages, cheese, and other products, biotechnology has been part of our life and it will only increase in importance. Now, we are at a stage where plant and plant cell culture technology, which has taken big steps since Haberlandt's predictions in 1902 (88) about cultivation of plant tissue, is being combined with genetic engineering. Plants such as potatoes, alfalfa sprouts, and tomatoes can be genetically altered to produce the antigens for hepatitis B and cholera. The next step will be to introduce proteins involved in the biosynthesis of pharmaceuticals such as vinblastine, vincristine, or taxol in the appropriate host to produce the compounds mentioned on an industrial scale with plant cell cultures or microorganisms in bioreactors or by transgenic plants. To accomplish this goal, biosynthetic pathways need to be completely identified on gene, enzyme, and product levels. In particular, identification of factors involved in the regulation of the biosynthetic pathways seems an interesting approach that may lead to regulatory genes controlling the flux through a pathway. This will make possible the production of known or novel fine chemicals by plant cell cultures.

REFERENCES

1. J Mann. Secondary Metabolism. 2nd ed. Oxford: Oxford Science Publications, 1987.

2. JD Phillipson. Plants as sources of valuable products. In: BV Charlwood, MJC Rhodes, eds. Secondary Products from Plant Tissue Culture. Oxford: Clarendon Press, 1990, pp 1–21.
3. T Ersek, Z Kiraly. Phytoalexins: warding-off compounds in plants. Physiol Plant 68:343–346, 1986.
4. M Wink. Plant breeding: importance of plant secondary metabolites for protection against pathogens and herbivores. Theor Appl Genet 75:225–233, 1988.
5. J Stöckigt, P Obitz, H Falkenhagen, R Lutterbach, S Endreß. Natural products and enzymes from plant cell cultures. Plant Cell Tissue Organ Cult 43:97–109, 1995.
6. R Verpoorte, R van der Heijden, HJG ten Hoopen, J Memelink. Metabolic engineering of plant secondary metabolite pathways for the production of fine chemicals. Biotechnol Lett 21:467–479, 1999.
7. F DiCosmo, M Misawa. Plant cell and tissue culture: alternatives for metabolite production. Biotechnol Adv 3:425–453, 1995.
8. J Berlin. Secondary products from plant cell cultures. In: HJ Rehm, G Reed, eds. Biotechnology: A Comprehensive Treatise. Vol 4. Weinheim: Verlag Chemie, Verlagsgesellschaft, 1986, pp 630–658.
9. RM Buitelaar, J Tramper. Strategies to improve the production of secondary metabolites with plant cell cultures: a literature review. J Biotechnol 23:111–141, 1992.
10. AK Lipsky. Problems of optimization of plant cell culture process. J Biotechnol 26:83–97, 1992.
11. GF Payne, V Bringi, CL Prince, ML Schuler. Plant Cell and Tissue Culture in Liquid Systems. Munich: Hanser Publisher, 1992.
12. PM Doran. Designs of bioreactors for plant cells and organs. Adv Biochem Eng Biotechnol 48:113–168, 1993.
13. WW Su. Bioprocessing technology for plant cell suspension cultures. Appl Biochem Biotechnol 2:189–230, 1995.
14. AW Alfermann, M Petersen. Natural product formation by plant cell biology. Plant Cell Tissue Organ Cult 43:199–205, 1995.
15. R Verpoorte, R van der Heijden, W van Gulik, HJG ten Hoopen. Plant biotechnology for the production of alkaloids: present status and prospects. In: A Brossi, ed. The Alkaloids. Vol 40. London: Academic Press, 1991, pp 2–187.
16. M Tabata, H Mizukami, N Hiraoka, M Konoshima. Pigment formation in callus cultures of *Lithospermum erythrorhizon*. Phytochemistry 13:927–993, 1974.
17. Y Fujita, M Tabata, A Nishi, Y Tamada. New medium and production of secondary compounds with two-staged culture method. In: A Fujiwara, ed. Plant Tissue Culture. Tokyo: Maruzen, 1982, pp 399–400.
18. Y Fujita. Industrial production of shikonin and berberine. In: G Bock, J Marsh, eds. Application of Plant Cell and Tissue Culture. New York: Wiley, 1988, pp 228–238.
19. K Ushiyama. Large scale cultivation of ginseng. In: A Komamine, A Misawa, F DiCosmo, eds. Plant Cell Cultures in Japan. Tokyo: CMC, 1991, pp 92–98.
20. M Jaziri, A Zhiri, Y-W Guo, J-P Dupont, K Shimomura, H Hamada, M Vanhaelen, J Homes. *Taxus* sp. cell and organ cultures as alternative sources for

taxoids production: a literature survey. Plant Cell Tissue Organ Cult 46:59–75, 1996.

21. S Ohta, R Verpoorte. Some accounts of variation (heterogeneity and/or instability) in secondary metabolites production by plant cell cultures. Annu Rep Nat Sci Home Econ 32:9–23, 1992.

22. R Verpoorte, R van der Heijden, J Memelink. Plant biotechnology and the production of alkaloids: prospects of metabolic engineering. In: GA Cordell, ed. The Alkaloids. Vol 50. San Diego: Academic Press, 1998, pp 453–508.

23. GH Svoboda, DA Blake. The phytochemistry and pharmacology of *Catharanthus roseus* (L.) G. Don. In: WI Taylor, NR Farnsworth, eds. The *Catharanthus* Alkaloids. New York: Marcel Dekker, 1975, pp 45–83.

24. GA Cordell. The botanical, chemical, biosynthetic and pharmacological aspects of *Catharanthus roseus* (L.) G. Don (Apocynaceae). In: WS Woo, BH Han, eds. Recent Advances in Natural Product Research. Seoul: Seoul National University, The Netherlands Press, 1980, pp 65–72.

25. O van Tellingen, JHM Sips, JH Beijnen, A Bult, WJ Nooijen. Pharmacology, bioanalysis and pharmacokinetics of the vinca alkaloids and semi-synthetic derivatives. Anticancer Res 12:1699–1716, 1992.

26. R van der Heijden, R Verpoorte, HJG ten Hoopen. Cell and tissue cultures of *Catharanthus roseus* (L.) G. Don: a literature survey. Plant Cell Tissue Organ Cult 18:231–280, 1989.

27. M Lounasmaa, J Galambos. Indole alkaloid production *Catharanthus roseus* cell suspension cultures. In: W Hertz, H Grisebach, GW Kirby, CH Tamm, eds. Progress in the Chemistry of the Natural Products. Vol 55. New York: Springer-Verlag, 1989, pp 89–115.

28. PRH Moreno, R van der Heijden, R Verpoorte. Cell and tissue cultures of *C. roseus*: a literature survey II. Updating from 1988 to 1993. Plant Cell Tissue Organ Cult 42.1–25, 1995.

29. R Verpoorte, R van der Heijden, PRH Moreno. Biosynthesis of terpenoid indole alkaloids in *Catharanthus roseus* cells. In: GA Cordell, ed. The Alkaloids. Vol 49. San Diego: Academic Press, 1997, pp 221–299.

30. B Deus-Neumann, MH Zenk. Instability of indole alkaloid production in *Catharanthus roseus* cell suspension cultures. Planta Med 50:427–431, 1984.

31. P Morris. Long term stability of alkaloid production in cell suspension cultures of *Catharanthus roseus*. In: P Morris, AH Scragg, A Stafford, MW Fowler, eds. Secondary Metabolism in Plant Cell Cultures. Cambridge: Cambridge University Press, 1986, pp 257–262.

32. P Morris, K Rudge, R Cresswell, MW Fowler. Regulation of product synthesis in cell cultures of *Catharanthus roseus*. V. Long-term maintenance of cells on a production medium. Plant Cell Tissue Organ Cult 17:79–90, 1989.

33. R Bentley. The shikimate pathway—a metabolic tree with many branches. In: GD Fasman, ed. Critical Reviews in Biochemistry and Molecular Biology. Vol 25. Boca Raton, FL: CRC Press, 1990, pp 307–384.

34. C Poulsen, R Verpooorte. Roles of chorismate mutase, isochorismate synthase and antrhranilate in plants. Phytochemistry 30:377–386, 1991.

35. BK Singh, DL Siehl, JA Connelly. Shikimate pathway: why does it mean so much to so many? In: BJ Miflin, ed. Oxford Surveys of Plant Molecular Biology. Vol 7. New York: Oxford University, Press, 1991, pp 143–185.

36. M Rohmer. The discovery of a mevalonate-independent pathway for isoprenoid biosynthesis in bacteria, algae and higher plants. Nat Prod Rep 16:565–574, 1999.

37. A Contin, R van der Heijden, AWM Lefeber, R Verpoorte. The iridoid glucoside secologanin is derived from the novel triose phosphate/pyruvate pathway in a *Catharanthus roseus* cell culture. FEBS Lett 434:413–416, 1998.

38. S Chavadej, N Brisson, JN McNeil, V De Luca. Redirection of tryptophan leads to production of low indole glucosinolate canola. Proc Natl Acad Sci USA 91:2166–2170, 1994.

39. TM Kutchan, N Hampp, F Lottspeich, K Beyreuther, MH Zenk. The cDNA clone for strictosidine synthase from *Rauvolfia serpentina*. DNA sequence determination and expression in *Escherichia coli*. FEBS Lett 237:40–44, 1988.

40. TM Kutchan. Expression of enzymatically active cloned strictosidine synthase from the higher plant *Rauwolfia serpentina* in *Escherichia coli*. FEBS Lett 257:127–130, 1989.

41. TM Kutchan, H Dittrich, D Bracher, MH Zenk. Enzymology and molecular biology of alkaloid biosynthesis. Tetrahedron 47:5945–5954, 1991.

42. TM Kutchan, A Bock, H Dittrich. Heterologous expression of the plant proteins strictosidine synthase and berberine bridge enzyme in insect cell culture. Phytochemistry 35:353–360, 1994.

43. Z Shen, W Eisenreich, TM Kutchan. Bacterial biotransformation of $3\alpha(S)$-strictosidine to the monoterpenoid indole alkaloid vallesiachotamine. Phytochemistry 48:292–296, 1998.

44. A Geerlings, R Verpoorte, R van der Heijden, J Memelink. Transgenic yeast producing indole alkaloids. European patent 98204116.2, 1998.

45. MI Lopes-Cardoso, MH Meijer, JHC Hoge. *Agrobacterium*-mediated transformation of the terpenoid indole alkaloid producing plant species *Tabernaemontana pandacaqui*. Plant Cell Rep 17:150–154, 1997.

46. J Berlin, C Rügenhagen, P Dietze, LF Fecker, OJM Goddijn, JHC Hoge. Increased production of serotonin by suspension and root cultures of *Peganum harmala* transformed with a tryptophan decarboxylase cDNA clone from *Catharanthus roseus*. Transgen Res 2:336–344, 1993.

47. V De Luca, AJ Cutler. Subcellular localization of enzymes involved in indole alkaloid biosynthesis in *Catharanthus roseus*. Plant Physiol 85:1099–1102, 1987.

48. LH Stevens, TJM Blom, R Verpoorte. Subcellular localization of tryptophan decarboxylase, strictosidine synthase and strictosidine glucosidase in cell suspension cultures of *Catharanthus roseus* and *Tabernaemontana divaricata*. Plant Cell Rep 12:573–576, 1993.

49. W Nóe, C Mollenschot, J Berlin. Tryptophan decarboxylase from *Catharanthus roseus* cell suspension cultures: purification, molecular and kinetic data on the homogeneous protein. Plant Mol Biol 3:281–288, 1984.

50. JA Fernandez, V De Luca. Ubiquitin-mediated degradation of tryptophan decarboxylase from *Catharanthus roseus*. Phytochemistry 5:1123–1128, 1994.

51. V De Luca, C Marineau, N Brisson. Molecular cloning and analysis of cDNA encoding a plant tryptophan decarboxylase: comparison with animal DOPA decarboxylase. Proc Natl Acad Sci USA 86:2582–2586, 1989.

52. EJ Pennings, I Hegger, R van der Heijden, JA Duine, R Verpoorte. Assay of tryptophan decarboxylase from *Catharanthus roseus* plant cell cultures by high performance liquid chromatography. Anal Biochem 165:133–136, 1987.

53. F Sasse, M Buchholz, J Berlin. Selection of cell lines of *Catharanthus roseus* with increased tryptophan decarboxylase activity. Z Naturforsch 38c:916–922, 1983.

54. B Deus-Neumann, MH Zenk. A highly selectable alkaloid uptake system in vacuoles of higher plants. Planta 162:250–260, 1984.

55. TD McKnight, DR Bergey, RJ Burnet, CL Nessler. Expression of enzymatically active and correctly targeted strictosidine synthase in transgenic tobacco plants. Planta 185:148–152, 1991.

56. U Pfitzner, MH Zenk. Homogeneous strictosidine synthase isoenzymes from cell suspension cultures of *Catharanthus roseus*. Planta Med 55:525–530, 1989.

57. JF Treimer, MH Zenk. Purification and properties of strictosidine synthase, the key enzyme in indole alkaloid formation. Eur J Biochem 101:225–233, 1979.

58. H Mizukami, H Nördlov, S Lee, AI Scott. Purification and properties of strictosidine synthase (an enzyme condensing tryptamine and secologanin) from *Catharanthus roseus* cultured cells. Biochemistry 18:3760–3763, 1979.

59. A De Waal, AH Meijer, R Verpoorte. Strictosidine synthase from *Catharanthus roseus*: purification and characterization of multiple forms. Biochem J 306: 571–580, 1995.

60. U Pfitzer, MH Zenk. Immobilization of strictosidine synthase from *Catharanthus roseus* cell cultures and preparative synthesis of strictosidine. J Med Plant Res 46:10–14, 1982.

61. EJM Pennings, BW Gron, JA Duine, R Verpoorte. Tryptophan decarboxylase from *Catharanthus roseus* as a pyridoxo-quinoprotein. FEBS Lett 255:97–100, 1989.

62. DD Songstad, V De Luca, N Brisson, WGW Kurz, CL Nessler. High levels of tryptamine accumulation in transgenic tobacco expressing tryptophan decarboxylase. Plant Physiol 94:1410–1413, 1990.

63. DD Songstad, WGW Kurz, CL Nessler. Tyramine accumulation in *Nicotiana tabacum* transformed with a chimeric tryptophan decarboxylase gene. Phytochemistry 30:3245–3246, 1991.

64. OJM Goddijn, RJ De Kam, A Zanetti, RA Schilperoort, JHC Hoge. Auxin rapidly down-regulates transcription of the tryptophan decarboxylase gene from *Catharanthus roseus*. Plant Mol Biol 18:113–120, 1992.

65. J Berlin, IN Kuzovkina, C Rügenhagen, LF Fecker, U Commandeur, V Wray. Hairy root cultures of *Peganum harmala*—II. Characterization of cell lines and effect of culture conditions on the accumulation of β-carboline alkaloids and serotonin. Z Naturforsch 47c:222–230, 1992.

66. C Poulsen, OMJ Goddijn, JHC Hoge, R Verpooorte. Anthranilate synthase and chorismate mutase activities in transgenic tobacco plants overexpressing tryptophan decarboxylase from *Catharanthus roseus*. Transgen Res 3:43–49, 1994.
67. OJM Goddijn, EJM Pennings, P van der Helm, R Verpoorte, JHC Hoge. Overexpression of a tryptophan decarboxylase cDNA in *Catharanthus roseus* crown gall calluses results in increased tryptamine levels but not in increased terpenoid indole alkaloid production. Transgen Res 4:315–323, 1995.
68. K Yao, V De Luca, N Brisson. Creation of a metabolic sink for tryptophan alters the phenylpropanoid pathway and the susceptibility of potato to *Phytophthora infestans*. Plant Cell 7:1787–1799, 1995.
69. JC Thomas, DA Adams, CL Nessler, JK Brown, HJ Bohnert. Tryptophan decarboxylase, tryptamine, and reproduction of the whitefly. Plant Physiol 109: 717–720, 1995.
70. TD McKnight, CA Roessner, R Devagupta, AI Scott, CL Nessler. Nucleotide sequence of a cDNA encoding the vacuolar protein strictosidine synthase from *Catharanthus roseus*. Nucleic Acids Res 18:4939, 1990.
71. CA Roessner, R Devagupta, M Hasan, HJ Williams, IA Scott. Purification of an indole biosynthetic enzyme, strictosidine synthase, from a recombinant strain of *Escherichia coli*. Protein Express Purif 3:295–300, 1992.
72. G Pasquali, OJM Goddijn, A De Waal, R Verpoorte, RA Schilperoort, JHC Hoge, J Memelink. Coordinated regulation of two indole alkaloid biosynthetic genes from *Catharanthus roseus* by auxin and elicitors. Plant Mol Biol 18: 1121–1131, 1992.
73. A Geerlings, FJ Redondo, A Contin, J Memelink, R van der Heijden and R Verpoorte. Transgenic yeast produces indole alkaloids upon feeding with secologanin and tryptamine. Appl Microbiol Biotechnol, 56:420–424, 2001.
74. D Hallard, A Geerlings, R van der Heijden, MI Lopes-Cardoso, JH Hoge, R Verpoorte. Metabolic engineering of terpenoid indole and quinoline alkaloid biosynthesis in hairy root cultures. In: PM Doran, ed. Hairy Roots Culture and Applications. Harwood Academic Publisher, Amsterdam, 1996, pp 43–49.
75. D Hallard, R van der Heijden, R Verpoorte, MI Lopes Cardoso, G Pasquali, J Memelink, JHC Hoge. Suspension cultured transgenic cells of *Nicotiana tabacum* expressing tryptophan decarboxylase and strictosidine synthase cDNAs from *Catharanthus roseus* produce strictosidine upon secologanin feeding. Plant Cell Rep 17:50–54, 1997.
76. A Geerlings, D Hallard, A Martinez Caballero, I Lopez Cardoso, R van der Heijden, R Verpoorte. Alkaloid production by a *Cinchona officinalis* 'Ledgeriana' hairy root culture containing constitutive expression constructs of tryptophan decarboxylase and strictosidine synthase cDNAs from *Catharanthus roseus*. Plant Cell Rep 19:191–196, 1999.
77. JD Hamill, RJ Robins, MJC Rhodes. Alkaloid production by transformed root cultures of *Cinchona ledgeriana*. Planta Med 55:354–357, 1989.
78. MJ Leech, K May, D Hallard, R Verpoorte, V De Luca, P Christou. Expression of two consecutive genes of a secondary metabolic pathway in transgenic tobacco: molecular diversity influences levels of expression and product accumulation. Plant Mol Biol 38:765–774, 1998.

79. JE Schlatmann, A-M Nuutila, WM van Gulik, HJG ten Hoopen, R Verpoorte, JJ Heijnen. Scale-up of ajmalicine production by plant cell cultures of *Catharanthus roseus*. Biotechnol Bioeng 41:253–262, 1993.

80. S Whitmer, R Verpoorte, C Canel. Influence of auxins on alkaloid accumulation by a transgenic cell line of *Catharanthus roseus*. Plant Cell Tissue Organ Cult 53:135–141, 1998.

81. C Canel, MI Lopes Cardoso, S Whitmer, L van der Fits, G Pasquali, R van der Heijden, JHC Hoge, R Verpoorte. Effects of over-expression of strictosidine synthase and tryptophan decarboxylase on alkaloid production by cell cultures of *Catharanthus roseus*. Planta 205:414–419, 1998.

82. F Naudascher, P Doireau, A Guillot, C Viel, M Thiersault. Time-course studies on the use of secologanin by *Catharanthus roseus* cells cultured in vitro. J Plant Physiol 134:608–612, 1989.

83. S Whitmer, C Canel, D Hallard, C Gonçalves, R Verpoorte. Influence of precursor availability on alkaloid accumulation by transgenic cell line of *Catharanthus roseus*. Plant Physiol 116:853–857, 1998.

84. B St-Pierre, FA Vazquez-Flota, V De Luca. Multicellular compartmentation of *Catharanthus roseus* alkaloid biosynthesis predicts intercellular translocation of a pathway intermediate. Plant Cell 11:887–900, 1999.

85. RD Hall, MM Yeoman. Factors determining anthocyanin yield in cell cultures of *Catharanthus roseus* (L.) G. Don. New Phytol 103:33–43, 1986.

86. RD Hall, MM Yeoman. Intracellular and intercultural heterogeneity in secondary metabolite accumulation in cultures of *Catharanthus roseus* following cell line selection. J Exp Bot 38:1391–1398, 1987.

87. M Verhoeyen, S Muir, G Collin, A Bovy, R de Vos. Increasing flavonoid levels in tomatoes by means of metabolic engineering. Abstract 10th Symposium ALW-discussion group 'Secondary metabolism in plant and plant cell,' Amsterdam, February 11, 2000.

88. G Haberlandt. Culturversuche mit isolierten Pflanzenzellen. Sitzungsberi Math Naturwiss Kl Kais Akad Wiss Wien 111:69–92, 1902.

89. T Murashige, F Skoog. A revised medium for rapid growth and bioassays with tobacco tissue cultures. Plant Physiol 15:473–497, 1962.

90. MH Zenk, H El-Shagi, H Arens, J Stöckingt, EW Weiler, B Deus. Formation of alkaloids serpentine and ajmalicine in cell suspension cultures of *Catharanthus roseus*. In: W Barz, MH Zenk, eds. Plant Tissue Culture and Its Biotechnological Application. New York: Springer Verlag, 1977, pp 27–43.

17

Transgenic Plants for Production of Immunotherapeutic Agents

James W. Larrick, Lloyd Yu, Sudhir Jaiswal, and Keith Wycoff

Planet Biotechnology, Inc. Hayward, and Palo Alto Institute of Molecular Medicine, Mountain View, California, U.S.A.

Since the first reports 20 years ago, genetically engineered plants with improved traits, pest and herbicide resistance, etc., have produced significant agricultural revenues. In contrast, the development of plants as bioreactors to produce transgenic proteins for pharmaceutical use is in its infancy. Numerous immunotherapeutic proteins, antibodies, and vaccines have been produced; however, a limited number have made their way into clinical trials. The most advanced product in human clinical trials is a secretory immunoglobulin A (IgA) antibody composed of four polypeptide chains that inhibits the binding to teeth of *Streptococcus mutans*, the major oral pathogen. This chapter will summarize recent work demonstrating the potential of plants to synthesize and assemble complex proteins suitable for human therapeutic use.

I. PLANT BIOREACTORS FOR IMMUNOTHERAPEUTIC PROTEINS

The first transgenic plants were reported in 1983 (Fraley et al., 1983; Zambryskyi et al., 1983). Since then, many recombinant proteins have been

expressed in several important agronomic species of plants including tobacco, corn, tomato, potato, banana, alfalfa (Austin, 1994), and canola (summarized in Kusnadi et al., 1997, 1998; Hood and Jilka, 1999.). Recent work
suggests that plants will be a facile and economic bioreactor for large-scale
production of industrial and pharmaceutical recombinant proteins (Kusnadi
et al., 1997; Austin et al., 1994; Whitelam et al., 1994). Genetically engineered (transgenic) plants have several advantages as sources of proteins
compared with human or animal fluids or tissues, recombinant microbes,
transfected animal cell lines, or transgenic animals. These include

> Efficiency of the transformation technology and speed of scale-up
> Correct assembly of multimeric antibodies (unlike bacteria)
> Increased safety, as plants do not serve as hosts for human pathogens,
>   such as human immunodeficiency virus (HIV), prions, and hepatitis
>   viruses
> Production of raw material on an agricultural scale at low cost
> Reduced capitalization costs relative to fermentation methods

Depending upon the promoters used, transgenic proteins will be se-
questered throughout the plant or in specific parts of the plants (e.g., seeds)
or specific organelles within a given plant cell. Much of the early work utilized
strong generic promoters, such as the cauliflower mosaic virus 35S promoter
giving widespread protein expression in green biomass. For example, De
Wilde et al. (1998) showed that in *Arabidopsis*, antibody or Fab fragments
bearing conventional leader sequences accumulate at the sites where water
passes on its radial pathway toward and within the vascular bundle. A large
proportion of these proteins are transported into the apoplast of *A. thaliana*,
possibly by the water flow in the transpiration stream. In contrast, numerous
laboratories have shown transgenic protein accumulation in seeds of tobacco
(Fiedler and Conrad, 1995) corn (Russell, 1999; Hood and Jilka, 1999) soybean (Zeitlin et al., 1998) or barley (Horvath et al., 2000).

Table 1 presents a listing of expression levels of several major immunotherapeutic proteins in transgenic plants. The most widely studied proteins have been antigens for use in oral vaccines and antibodies for passive
immunotherapy. Plants are attractive expression vehicles for each of these
applications.

II. ISSUES REGARDING TRANSGENIC PLANT
EXPRESSION OF IMMUNOTHERAPEUTIC PROTEINS

A. Production Costs

Cost benefits of plant production are enormous. For example, Kusnadi et al.
(1997) calculated the cost of producing a recombinant protein in various

TABLE 1 Immunotherapeutic Proteins Synthesized in Transgenic Plants

Recombinant proteins	Plants	Protein production levels	Reference
sIgA Anti–*S. mutans*	Tobacco	200–500 (μg/g)	Ma J, et al. (1995, 1998)
			Ma J, et al. (1998)
HBsAg	Tobacco	0.01% TSP	Mason et al. (1992)
			Thanavala et al. (1995)
Norwalk capsid protein	Tobacco	0.23% TSP	Mason et al. (1996)
Norwalk capsid protein	Potato	10–20 (μg/g)	Mason et al. (1996)
E. coli	Potato	3–4 (μg/g)	Tacket et al. (1998)
LT-B			Haq et al. (1995)
			Mason et al. (1998)
Cholera toxin	Potato	30 (μg/g)	Arakawa et al. (1997)
CT-B			Arakawa et al. (1998)
Mouse GAD67	Potato	150 (μg/g)	Ma S-W, et al. (1997)

TSP—total soluble protein

agricultural crops (see Fig. 1). The cost estimate was based on the commodity price of the crop, the fraction of total protein in the crop, and the not unreasonable assumption that the recombinant protein accumulated to 10% of the total plant protein. Although crops with more protein content (e.g., soybeans 40% versus potatoes 2%) are more cost-effective (see Fig. 1), these costs are 10- to 50-fold less than those for protein produced at a high level in *Escherichia coli* (20% total protein). Depending upon the use of the protein and the requirements for purification for in vivo pharmaceutical use, purification costs will obviously augment final product costs; however, at the hundred kilogram to metric ton level, plant-produced proteins will provide obvious savings.

B. Glycosylation of Transgenic Plant Proteins; Plant Antibody Glycosylation

Cabanes-Macheteau et al. (1999) reported the first detailed analysis of the glycosylation of a functional mammalian glycoprotein expressed in a transgenic plant. The structures of the N-linked glycans attached to the heavy chains of the monoclonal antibody Guy's 13 produced in transgenic tobacco plants (plantibody Guy's 13) were identified and compared with those found in the corresponding IgG1 of murine origin. As in mouse antibodies, both N-glycosylation sites located on the heavy chain of the plantibody Guy's 13 are N-glycosylated. However, the number of Guy's 13 glycoforms is higher in the plant than in the mammalian expressed antibodies. In addition to high-mannose-type N-glycans, 60% of the oligosaccharides N-linked to the plantibody have beta(1,2)-xylose and alpha(1,3)-fucose residues linked to the core Man3GlcNAc2.

These oligosaccharide linkages, not found on mammalian N-linked glycans, are potentially immunogenic and raise the possibility that plantibodies containing N-linked glycans may have limited scope as parenterals or even when applied topically or orally, particularly in patients with severe food allergies. Food allergans bearing beta(1,2)-xylose and alpha(1,3)-fucose have been linked to specific IgE in serum and to biological activity, i.e., histamine release, in allergic patients (Fotisch et al., 1999 and Garcia-Casado et al., 1996). In contrast, the mere presence of serum IgE against cross-reactive carbohydrate determinants, which include the beta(1,2)-xylose and alpha(1,3)-fucose linkages to the core Man3GlcNAc2 of plant glycans, has been shown to be a poor predicter of clinical allergy (van Ree and Aalberse, 1999; Mari et al., 1999; van der Veen et al., 1997). These results preclude generalities about the potential toxicity of plantibody glycans in humans.

When the mouse plantibody (mouse amino acid sequence and plant glycans) was used to immunize mice, there was no, or only a minimally

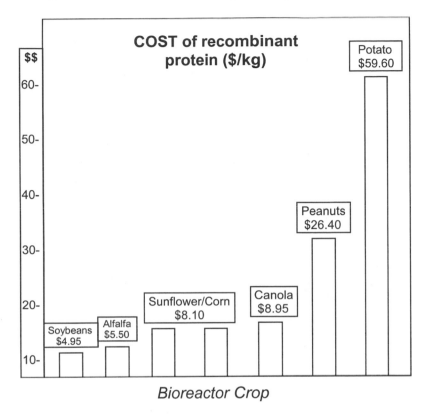

FIGURE 1 Relative cost of protein production in various agricultural crops. (Adapted from Kusnadi et al., 1997.)

detectable, serum immune response (Chargelegue, personal communication, 1999). Thus, in the mouse at least, a "self" primary protein structure, decorated with plant N-linked glycans, can be nonimmunogenic. This plantibody has been applied topically in the mouth of humans with no detection of human antimouse antibodies (Ma et al., 1998).

Humanization of plantibody primary structures may go a long way to obviate the immunogenic potential of plant glycans, and further genetic engineering may, if necessary, alter the glycan structures themselves. Most dramatically, aglycosyl antibodies can be created by altering the peptide recognition sequence for N-linked glycosylation (asn-X-Ser/Thr). High-mannose glycosylation, which does not contain the core-linked xylose and fucose residues, may be favored by the addition of a C-terminal KDEL sequence (Wandelt et al., 1992) and the subsequent targeting of plantibodies to the proximal endoplasmic reticulum. Alternatively, the specific fucosyl (Leiter et al., 1999) or xylosyl transferases, which operate in the trans-Golgi, may be targeted for silencing.

C. Gene Silencing: A Potential Problem of Plant Expression

Levels of transgenic protein accumulation reported in the literature (Table 1) generally represent the highest levels found in primary transformants. It is becoming increasingly clear that stability of expression is just as important as absolute levels of expression. Many laboratories have noted the common occurrence of transgene silencing in plants, in which transgenes can become inactivated either during development or in subsequent generations (Baulcombe, 1999; Depicker and Montagu, 1997; Matzke and Matzke, 1998; Meyer and Saedler, 1996; Stam et al., 1997; Vaucheret et al., 1998). Two types of gene silencing have been observed in plants: transcriptional gene silencing (TGS), in which promoters become silenced and transcription is reduced, and posttranscriptional gene silencing (PTGS), in which transcription continues unabated but transcripts are rapidly degraded. The exact mechanisms of both types of gene silencing are still being discovered, but there are common features to both. For instance, the presence of homologous sequences is frequently involved. If repeats are present within transgenes, or if transgene sequences are homologous to endogenous genes, silencing is more frequent. Methylation of DNA may be involved. In TGS, the promoters of silenced genes are hypermethylated. In PTGS, it is the coding regions that tend to be methylated. de Neve et al. (1999) found both types of silencing in a careful examination of five different homozygous transgenic *Arabidopsis* lines expressing either IgG or Fab.

Silencing seems to be positively correlated with level of expression and with copy number of the transgene. That is, transgenes present in multiple copies or expressed from strong constitutive promoters are more likely to be silenced (Elmayan and Vaucheret, 1996; Jorgensen et al., 1996). This suggests some possibilities for reducing the occurrence of silencing. Silencing may be less of a problem with tissue-specific or weak constitutive promoters, although this may be antithetical to the desire for high levels of expression. The introduction of multiple copies can be reduced by the use of *Agrobacterium*, which tends to result in fewer transgene copies than biolistic transformation. Flanking transgenes with matrix attachment region (MAR) sequences has been shown to reduce the occurrence of gene silencing (Uelker et al., 1999; Vain et al., 1999). This probably works by preventing formation of antisense transcripts by transcription from flanking endogenous genes into the transgene. Viral suppppressors of silencing have been used quite effectively to prevent or reverse PTGS (Anandalakshmi et al., 1998; Kasschau and Carrington, 1998). When plants that are transgenic for a viral protein known as helper component protease (HC-Pro) are crossed with plants in which a transgene is silenced, expression is restored in progeny containing both transgenes. It should also be possible to use HC-Pro−expressing plants as the starting material for transformation with a gene of interest.

Finally, a number of mutants have been isolated in *Arabidopsis* that lack either TGS or PTGS (Elmayan et al., 1998; Furner et al., 1998; Scheid et al., 1998). Some of these mutants appear to be normal in every respect except the inability to silence transgenes. It may eventually be possible to isolate similar mutations in plants that are more practical for the production of transgenic proteins.

D. Purification and Process Development

The potential for cost reduction when using genetically engineered green plants as bioreactors, instead of conventional pharmaceutical factories engineered with concrete and steel, is a powerful argument for the use of transgenic plants to produce recombinant proteins. The commercial-scale production of proteins from transgenic plants generally requires one to envisage the growth and processing of tons of biomass to achieve the economy of scale that would fully exploit the inputs of sunlight, soil, water, and fertilizer. The processing of large amounts of biomass also anticipates large numbers of patients or consumers, large amounts of purified protein needed for each patient or consumer, lower than anticipated levels of expression, and losses during processing.

The realities of commercial-scale production benefit from a demand for simplicity early on in the development of a large-scale process. Bench-scale or laboratory-scale procedures often employ the pampering conditions that are necessary for a proof of concept but are too complicated and expensive for large-scale efforts. For instance, because of issues regarding toxicity and expense, it is preferred not to use protease inhibitors beyond the bench scale. Reagents such as ammonium sulfate are frowned upon because of the disposal-associated costs, as are organic solvents because of their toxicity and flammability. Efforts to avoid the purchase and maintenance of centrifuges will eventually be gratefully acknowledged by maintenance personnel. In addition, and more fundamentally, it is important to realize that a commercial process will not be executed by rocket scientists, each with extensive postdoctoral training, but by more ordinary and, while no less dedicated, almost certainly less well-trained and educated people (i.e., they cost less). Thus, simplicity in the number, as well as the type, of components in a process is paramount to reduce not only costs but also errors.

Minimalist approaches to large-scale process development will often be rewarded even though they may appear simplistic. We found that grinding transgenic tobacco in water alone gave up to 70% of the expected immunoglobulin compared with tobacco ground in a buffer, poised at a specified pH, containing six components in addition to water; each component had been chosen for a particular, biochemically sound, reason. Yet this was evidence which suggested that perhaps not all of the components were vitally important, and we now use a two-component buffer at a specified pH.

Commercial-scale production of recombinant proteins from plants will also benefit from the technology and equipment commonly used in the food and beverage industry. Grain mills and coleslaw slicers will doubtless be useful off-the-shelf machinery for the initial processing of seeds and leafy tissue. The beverage industry has always been concerned with the clarification of juices and their phenolic content (van Sumere et al. 1975) and is a source of knowledge and equipment, both new and used. The chef's trick of delaying the browning of cut fruit by the addition of lemon-acidulated water (Bombauer and Becker, 1975) has led us to formulate buffers to inhibit the "tanning" of proteins in tobacco extracts.

Two constituents of stem and leaf extracts that require special consideration are membranes and cell walls. The green color of stem and leaf extracts indicates the presence of chlorophyll and the suspension of thylakoid membranes. These and other membranes must be removed to allow filterability at or below a pore size of 0.45 μm. This level of filterability helps to ensure, but does not guarantee, the good behavior of the extract during subsequent chromatography and serves to remove bacterial sources of contamination. The cellulose-containing debris, which forms a large part of the insoluble portion of extracts, can be used as an endogenous "filter aid" during the initial clarification to remove these membranous components.

Phenolics are a major concern when extracting most stems and leaves, and one's efforts are rewarded by their early removal. They may interact with proteins and other extract components via hydrophobic interactions, salt bridges, hydrogen bonding, and by additional reactions to nucleophilic centers (Gegenheimer, 1990; van Sumere, et al. 1975; Loomis, 1974). These interactions can dramatically and irreversibly alter the properties of proteins. Fortunately, the majority of released phenolics are generally small in size, as well as water soluble, and may be removed by tangential-flow ultrafiltration/diafiltration, which also serves to concentrate the considerably larger proteins of interest. In addition, other incompatible, water-soluble secondary metabolites, such as neonicotine (anabasine) and nicotine from tobacco, may be removed by ultrafiltration/diafiltration.

After such clarification and concentration steps, recombinant proteins can be assumed to behave independently and with regard to the peculiarities of their own biochemistry. In other words, they are now ready for chromatography. Residual phenolics may still dictate the degree of cleanup necessary before chromatography on an expensive protein A or protein G affinity column is allowed; these are economic considerations common to any purification process.

One additional note of caution: because the processed plant or seed was probably grown in dirt, or some similarly unclean stratum, one may anticipate the presence of a diverse bioburden. Contamination of product with endotoxins and mycotoxins can be minimized by rapid processing and early filtration, but process development must also always recognize the necessity to eliminate compounds as well as the necessity to purify, concentrate, and stabilize a product of interest.

III. ANTIBODIES FROM PLANTS: PLANTIBODIES

Although antibodies were first expressed in plants in the mid-1980s (Steiger, Düring) by two German graduate students (Düring et al., 1990), the first report was published in 1989 (Hiatt et al., 1989). Since then, a diverse group of "plantibody" types and forms have been prepared (see Table 2). Originally, foreign antibody genes were introduced into plant cells by nonpathogenic strains of the natural plant pathogen *Agrobacterium tumefaciens* (Horsch et al., 1985) and regeneration in tissue culture resulted in the recovery of stable transgenic plants. Although this initial work to generate multichain proteins required crossing of plants expressing each chain, more recent studies have shown that multiple chains can be introduced via a single biolistic transformation event (Sanford, 1988; Wycoff et al., unpublished data), greatly reducing the time to final assembled plantibody.

IV. SIgA: A NOVEL ANTIBODY ISOTYPE

This laboratory has focused on the production of secretory IgA (SIgA) plantibodies (Ma et al., 1995) (see Fig. 2). At the present time, plants offer the only large-scale, commercially viable system for production of this unique form of antibody. SIgA is the most abundant antibody class produced by the body (>60% of total immunoglobulin). SIgA is secreted onto mucosal surfaces to provide local protection from toxins and pathogens (see Fig. 3). The SIgA is composed of four different protein chains: heavy and light immunoglobulin chains that form the antigen-binding hypervariable region, a J chain that dimerizes two IgA molecules (SIgA has four antigen binding sites), and a secretory component that is derived from the mucosal epithelial

TABLE 2 Antibody-Derived Molecules Produced in Transgenic Plants

Mab form (no. of chains)	Antigen	Plant species—comment	Reference
Single domain (dAb) (1)	Substance P (neuropeptide)	Nicotiana	Benvenuto et al. (1991)
Single chain Fv (1)	Phytochrome	Nicotiana	Firek et al. (1993), Owen et al. (1992)
Single chain Fv (1)	Artichoke mottled crinkle virus coat	Nicotiana—viral protection	Tavladoraki et al. (1993)
Single chain Fv (1)	Abscisic acid	Nicotiana—wilty phenotype	Artsaenko et al. (1995); Phillips et al. (1997)
Single chain Fv (1)	Root-knot nematode	Nicotiana	Rosso et al. (1996)
Single chain Fv (1)	Beet necrotic yellow vein virus	Nicotiana benthamiana	Fecker et al. (1996)
scFv	Several	Nicotiana, KDEL augments expression	Fiedler et al. (1997); Schouten et al. (1996, 1997); Bruyns et al. (1996)
Fab; IgG (k) (2)	Human creatine kinase	Nicotiana Arabidopsis	de Neve et al. (1993)
IgG (k) (2)	Transition-state analogue	Nicotiana	Hiatt et al. (1989); Hein et al. (1991)
IgG (k) (2)	Fungal cutinase	Nicotiana	van Engelen et al. (1994)
IgG (k) and IgG/A hybrids (2)	Streptococcus mutans adhesin	Nicotiana	Ma et al. (1994)
Secretory IgA/G (4)	S. mutans adhesin	Nicotiana	Ma et al. (1995)
IgM/([]) (2)	NP (4-hydroxy 3-nitrophenylacetyl) (Hapten)	Nicotiana	Düring et al. (1990)

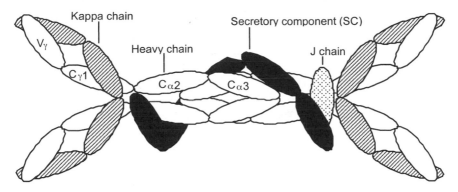

Figure 2 The structure of secretory IgA.

cells. Dimeric IgA containing J chain derived from submucosal B cells binds to the epithelial cell polyimmunoglobulin receptor (PIG^R) that transports the IgA to the mucosal surface. Binding triggers transcytosis to the mucosal surface, where a protease releases a portion of the PIG^R called secretory component conveniently used to bind the SIgA. The secretory component protects the dimeric IgA from proteases and denaturation on the mucosal surface. Previously it was not possible to obtain therapeutic quantities of this class of immunoglobulin. The recent availability of large amounts of secretory IgA plantibodies opens up a number of novel therapeutic opportunities for disorders of the mucosal immune system. These include therapies for intestinal pathogens such as hepatitis viruses, *Helicobacter pylori*, and *enterotoxigenic E. coli*, and cholera; respiratory pathogens such as rhinovirus and influenza; and genitourinary sexually transmitted diseases (e.g., herpes simplex virus) and contraception.

To date, three immunotherapeutic products produced in plants have entered the clinic. These products, listed in Table 3, include two antibodies and an oral vaccine. Clinical studies of the anti-EPCAM plantibody (co-developed by NeoRx and Monsanto) were discontinued due to significant gastrointestinal side effects. The anti–*Streptococcus mutans* antibody is currently in phase II trials.

V. CLINICAL STUDIES OF CaroRx™, AN ANTI–*STREPTOCOCCUS MUTANS* SIgA TO PREVENT DENTAL CARIES

The most clinically advanced SIgA plantibody, called CaroRx, recognizes and inhibits the binding of the major oral pathogen, *Streptococcus mutans*, to teeth. In preliminary work, a series of in vivo passive immunization ex-

SIgA Transcytosis

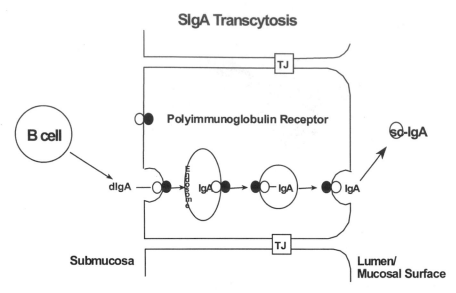

FIGURE 3   SIgA is produced by two cell types. Submucosal lymphoid tissue produces dimeric, IgA, which diffuses to the serosal side of the mucosal epithelial cells, where it binds to the polyimmunoglobulin receptor. Secretory component (SC) is proteolytically derived from the polyimmunoglobulin receptor.

periments was carried out in 84 human subjects using murine anti–*S. mutans* antibodies (Ma et al., 1987, 1989, 1990; Lehner et al., 1975, 1985). Topical application of anti–*S. mutans* antigen SA I/II monoclonal antibodies (MAbs) prevented colonization of artificially implanted exogenous strains of *S. mutans* as well as natural recolonization by indigenous *S. mutans*. In these studies the pathogenic *S. mutans* was replaced by endogenous flora (see Fig. 4).

The presence of the complement-activating and phagocyte-binding sites on the Fc fragment of the MAb was not essential for activity because the F(ab')$_2$ portion of the MAb was as protective as the intact IgG; however, the Fab fragment failed to prevent recolonization of *S. mutans*. Prevention of recolonization was specifically restricted to *S. mutans*, as the proportion of other organisms, such as *S. sanguis,* did not change significantly. The surprising feature of these experiments was that protection from recolonization by *S. mutans* lasted up to 2 years (J. Ma, personal communication), although MAb was applied for only 3 weeks and functional MAb was detected on the teeth for only 3 days following the final application of MAb. All studies indicated that this form of immunotherapy appears to be safe

TABLE 3   Plant-Produced Human Therapeutics (~1998)

Protein	Target	Organization	Stage
SIgA	Caries	PLANET Biotech	Phase II
IgG	Cancer	NeoRx/Monsanto	Phase I/II (cancelled due to unexpected gastrointestinal toxicity) (unpublished)
E. coli LT	E. coli diarrhea	Boyce Thompson	Phase I/II

and well tolerated. The long-term protection could therefore not be accounted for by persistence of MAb on the teeth but may be due to a shift in the microbial balance in which other bacteria occupy the ecological niche vacated by S. mutans, resulting in resistance to recolonization by S. mutans (Fig. 4).

The antigen-binding V regions of the best murine Mab identified by Ma and Lehner, Guy's 13, has been used to create an SIgA plantibody produced in tobacco-designated CaroRx (details in Ma et al., 1995, 1998). Levels of production of CaroRx in tobacco are up to 0.5 mg/g fresh weight. Future plans call for production of CaroRx in corn and other cereal grains.

CaroRx has been produced and purified from tobacco under GMP conditions for clinical testing in the United Kingdom and the United States. CaroRx was engineered with an additional IgG CH2 domain to facilitate purification of the antibody by protein G affinity chromatography. A Poros™ protein G affinity purification was used to obtain >95% pure CaroRx from green plant tissue.

Clinical evaluation of CaroRx in a pilot phase II trial has been completed at Guy's Hospital (Ma et al., 1998). In this trial a functional comparison was made between CaroRx and the parent IgG monoclonal antibody Guy's 13. BIACORE analysis revealed that the affinity of the antibodies for purified S. mutans SA I/II was similar (KD = $0.5-1.3 \times 10^{-9}$ M); however, CaroRx had fourfold higher avidity (functional affinity), a not unexpected result given the tetravalent binding of the SIgA.

Using an experimental design similar to that used to demonstrate activity of the parent Mab, CaroRx gave specific protection against colonization by oral streptococci for over 4 months (details in Ma et al., 1998). In addition to this therapeutic end point, pharmacokinetic studies showed that in the human oral cavity, CaroRx survived for >3 days versus 1 day for the

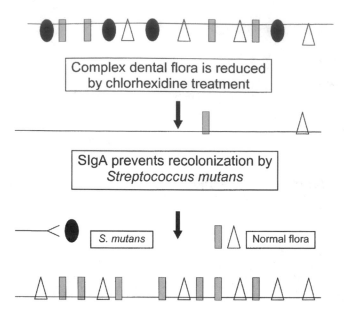

FIGURE 4 Proposed mechanism of CaroRx: the concept of ecological niche. After all flora have been reduced using oral antiseptic, chlorhexidine, a short treatment course with CaroRx provides an opportunity for normal flora to replace *S. mutans*.

IgG antibody and multiple serum antibody samples were negative for human antimouse (HAMA) or antirabbit antibodies. There was no evidence of local or systemic toxicity of the topically applied plantibody.

These initial clinical studies demonstrate that topically applied anti–*S. mutans* SIgA plantibody (CaroRx) is safe (no HAMA, no local or systemic toxicity) and prevents colonization by *S. mutans*, the major cause of human dental caries (Ma et al., 1998). Planet Biotechnology, Inc., has submitted an IND (investigational new drug application) to the U.S. Food and Drug Administration (FDA) and phase I/II confirmatory clinical trials began at the School of Dentistry at the University of California in San Francisco in Autumn 1998.

VI. SUMMARY

Plants offer a cost-effective bioreactor to produce antibodies of diverse types. Recent studies demonstrate that secretory IgA, the predominant antibody isotype of the mucosal immune system, can be made in large quantities in plants. CaroRx, the lead SIgA antibody being developed by Planet Biotechnology, Inc., has demonstrated activity in pilot phase II trials versus *S. mutans*, the major pathogen contributing to development of dental caries. Numerous other SIgA plantibodies are in preclinical development.

REFERENCES

1. R Anandalakshmi, GJ Pruss, X Ge, R Marathe, AC Mallory, TH Smith, VB Vance. 1998. A viral suppressor of gene silencing in plants. Proc Natl Acad Sci USA 95:13079–13084.
2. T Arakawa, DK Chong, JL Merritt, WH Langridge. 1997. Expression of cholera toxin B subunit oligomers in transgenic potato plants. Transgenic Res 6: 403–413.
3. T Arakawa, DK Chong, WH Langridge. 1998. Efficacy of a food plant–based oral cholera toxin B subunit vaccine. Nat Biotechnol 16:292–297.
4. O Artsaenko, M Peisker, U zur Nieden, U Fiedler, EW Weiler, K Müntz, U Conrad. 1995. Expression of a single-chain Fv antibody against abscisic acid creates a wilty phenotype in transgenic tobacco. Plant J 8:745–750.
5. S Austin, ET Bingham, RG Koegel, DE Mathews, MN Shahan, RJ Straub, RR Burgess. 1994. An overview of a feasibility study for the production of industrial enzymes in transgenic alfalfa. Ann NY Acad Sci 721:235–244.
6. DC Baulcombe. 1999. RNA makes RNA makes no protein. Curr Biol 9:R599–R601.
7. TJ Baum, A Hiatt, WA Parrott, LH Pratt, RS Hussey. 1996. Expression in tobacco of a functional monoclonal antibody specific to stylet secretions of the root-knot nematode. Mol Plant Microbe Interact 9:382–387.
8. SY Bednarek, NV Raikhel. 1992. Intracellular trafficking of secretory proteins. Plant Mol Biol 20:133–150.
9. E Benvenuto, RJ Ordas, R Tavazza, G Ancora, S Biocca, A Cattaneo, P Galeffi. 1991. "Phytoantibodies"; a general vector for the expression of immunoglobulin domains in transgenic plants. Plant Mol Biol 17:865–874.
10. IS Bombauer, MR Becker. 1975. Joy of Cooking. 38th ed. Indianapolis: Bobbs-Merril, p 520.
11. A-M Bruyns, G De Jaeger, M De Neve, C De Wilde, M Van Montagu, A Depicker. 1996. Bacterial and plant-produced scFv proteins have similar antigen-binding properties. FEBS Lett 386:5–10.
12. M Cabanes-Macheteau, AC Fitchette-Laine, C Loutelier-Bourhis, C Lange, ND Vine, JK Ma, P Lerouge, L Faye. 1999. *N*-Glycosylation of a mouse IgG expressed in transgenic tobacco plants. Glycobiology 9:365–372.
13. W Cockburn, MRL Owen. 1994. Antibody production in transgenic plants. Transactions 22:940–944.
14. U Conrad, U Fiedler. 1994. Expression of engineered antibodies in plant cells. Plant Mol Biol 26:1023–1030.
15. U Conrad, U Fiedler. 1998. Compartment-specific accumulation of recombinant immunoglobulins in plant cells: an essential tool for antibody production

and immunomodulation of physiological functions and pathogen activity. Plant Mol Biol 38:101–109.

16. G de Jaeger, E Buys, D Eeckhout, C De Wilde, A Jacobs, J Kapila, G Angenon, M Van Montagu, T Gerats, A Depicker. 1999. High level accumulation of single-chain variable fragments in the cytosol of transgenic petunia hybrida. Eur J Biochem 259:426–434.

17. J Denecke, J Botterman, R Deblaere. 1990. Protein secretion in plant cells occur via a default pathway. Plant Cell 2:51–59.

18. M de Neve, S De Buck, C De Wilde, H Van Houdt, I Strobbe, A Jacobs, M Van Montagu, A Depicker. 1999. Gene silencing results in instability of antibody production in transgenic plants. Mol Gen Genet 260:582–592.

19. M de Neve, M De Loose, A Jacobs, H Van Houdt, B Kaluza, U Weidle, A Depicker. 1993. Assembly of an antibody and its derived antibody fragment in *Nicotiana* and *Arabidopsis*. Transgen Res 2:227–237.

20. A Depicker, MV Montagu. 1997. Post-transcriptional gene silencing in plants. Curr Opin Cell Biol 9:373–382.

21. C de Wilde, M De Neve, R De Rycke, AM Bruyns, G De Jaeger, M Van Montagu, A Depicker, G Engler. 1996. Intact antigen-binding MAK33 antibody and Fab fragment accumulate in intercellular spaces of *Arabidopsis thaliana*. Plant Sci 114:233–241.

22. C de Wilde, RTB De Rycke, M De Neve, M Van Montagu, G Engler, A Depicker. 1998. Accumulation pattern of IgG antibodies and Fab fragments in transgenic *Arabidopsis thaliana* plants. Plant Cell Physiol 39:639–646.

23. K Düring, S Hippe, F Kreuzaler, J Schell. 1990. Synthesis and self-assembly of a functional monoclonal antibody in transgenic *Nicotiana tabacum*. Plant Mol Biol 15:281–293.

24. T Elmayan, S Balzergue, F Beon, V Bourdon, J Daubremet, Y Guenet, P Mourrain, JC Palauqui, S Vernhettes, T Vialle, K Wostrikoff, H Vaucheret. 1998. *Arabidopsis* mutants impaired in cosuppression. Plant Cell 10:1747–1758.

25. T Elmayan, H Vaucheret. 1996. Expression of single copies of a strongly expressed 35S transgene can be silenced post-transcriptionally. Plant J Cell Mol Biol 9:787–797.

26. LF Fecker, A Kaufmann, U Commandeur, J Commandeur, R Koenig, W Burgermeister. 1996. Expression of single-chain antibody fragments (scFv) specific for beet necrotic yellow vein virus coat protein or 25 kDa protein in *Escherichia coli* and *Nicotiana benthamiana*. Plant Mol Biol 32:979–986.

27. LF Fecker, R Koenig, C Obermeier. 1997. *Nicotiana benthamiana* plants expressing beet necrotic yellow vein virus (BNYVV) coat protein–specific scFv are partially protected against the establishment of the virus in the early stages of infection and its pathogenic effects in the late stages of infection. Arch Virol 142:1857–1863.

28. U Fiedler, U Conrad. 1995. High-level production and long-term storage of engineered antibodies in transgenic tobacco seeds. Bio/Technology 13:1090–1093.

29. U Fiedler, J Phillips, O Artsaenko, U Conrad. 1997. Optimization of scFv antibody production in transgenic plants. Immunotechnology 3:205–216.
30. S Firek, J Draper, MRL Owen, A Gandecha, B Cockburn, GC Whitelam. 1993. Secretion of a functional single-chain Fv protein in transgenic tobacco plants and cell suspension cultures. Plant Mol Biol 23:861–870.
31. R Fischer, J Drossard, YC Liao, S Schillberg. 1998. Characterisation and applications of plant-derived recombinant antibodies. In: C Cunningham, AJR Porter, eds. Methods in Biotechnology. Vol 3: Recombinant Proteins in Plants: Production and Isolation of Clinically Useful Compounds. Totowa, NJ: Humana Press.
32. R Fischer, Y-C Liao, J Drossard. 1999. Affinity purification of a TMV-specific recombinant full-size antibody from a transgenic tobacco suspension culture. J Immunol Methods 226:1–10.
33. R Fischer, Y-C Liao, K Hoffman, S Schilberg, N Emans. 1999. Molecular farming of recombinant antibodies in plants. Biol Chem 380:825–839.
34. R Fischer, D Schumann, S Zimmerman, J Drossard, M Sack, S Schillberg. 1999. Expression and characterization of bispecific single chain Fv fragments produced in transgenic plants. Eur J Biochem 262:810–816.
35. K Fotisch, F Altmann, D Haustein, S Vieths. 1999. Involvement of carbohydrate epitopes in the IgE response of celery-allergic patients. Int Arch Allergy Immunol 120:30–42.
36. RT Fraley, SG Rogers, RB Horsch, PR Sanders, JS Flick, SP Adams, ML Bittner, LA Brand, CL Fink, JS Fry, GR Galluppi, SB Goldberg, NL Hoffmann, SC Woo. 1983. Expression of bacterial genes in plant cells. Proc Natl Acad Sci. USA 80:4803–4807.
37. R Franconi, P Roggero, P Pirazzi, FJ Arias, A Desiderio, O Bitti, D Pashkoulov, B Mattei, L Bracci, V Masenga, RG Milne, E Benvenuto. 1999. Functional expression in bacteria and plants of an scFv antibody fragment against topoviruses. Immunotechnology 4:189–201.
38. E Franken, U Teuschel, R Hain. 1997. Recombinant proteins from transgenic plants. Curr Opin Biotechnol 8:411–416.
39. IJ Furner, MA Sheikh, CE Collett. 1998. Gene silencing and homology-dependent gene silencing in *Arabidopsis*: genetic modifiers and DNA methylation. Genetics 149:651–662.
40. G Garcia-Casado, R Sanchez-Monge, MJ Chrispeels, A Armentia, G Salcedo, L Gomez. 1996. Role of complex asparagine-linked glycans in the allergenicity of plant glycoproteins. Glycobiology 4:471–477.
41. P Gegenheimer. 1990. Preparation of extracts from plants. Methods Enzymol 182:174–193.
42. WW Gibbs. 1997. Plantibodies. Human antibodies produced by field crops enter clinical trials. Sci Am 277(5):44.
43. OJM Goddin, J Pen. 1995. Plants as bioreactors. Trends Biotechnol 13:379–387.
44. TA Haq, HS Mason, JD Clements, CJ Arntzen. 1995. Oral immunization with a recombinant bacterial antigen produced in transgenic plants. Science 268:714–716.

45. M Hein, Y Tang, DA McLeod, KD Janda, AC Hiatt. 1991. Evaluation of immunoglobulins from plant cells. Biotechnol Prog 7:455–461.
46. A Hiatt. 1990. Antibodies produced in plants. *Nature* 344:469–470.
47. A Hiatt. 1991. Monoclonal antibodies, hybridoma technology and heterologous production systems. Curr Opin Immunol 3:229–232.
48. A Hiatt, R Cafferkey, K Bowdish. 1989. Production of antibodies in transgenic plants. Nature 342:76–78.
49. A Hiatt, M Hein. 1994. Structure, function and uses of antibodies from transgenic plants and animals. In: M Rosenberg, M Gordon, eds. Handbook of Experimental Pharmacology. New York: Springer Verlag.
50. A Hiatt, J Ma. 1993. Characterization and applications of antibodies produced in plants. Int Rev Immunol 10:139–152.
51. A Hiatt, JK-C Ma. 1992. Monoclonal antibody engineering in plants. FEBS Lett 307:71–75.
52. A Hiatt, Y Tang, W Weiser, MB Hein. 1992. Assembly of antibodies and mutagenized variants in transgenic plants and plant cell cultures. Genet Eng 14:49–64.
53. EE Hood, JM Jilka. 1999. Plant-based production of xenogenic proteins. Curr Opin Biotechnol 10:382–386.
54. RB Horsch, JE Fry, NL Hoffmann, D Eichholtz, SG Rogers, RT Fraley. 1985. A simple and general method for transferring genes into plants. Science 227:1229–1231.
55. H Horvath H, J Huang, O Wong, E Kohl, T Okita, LG Kannangara, D von Wettstein. 2000. The production of recombinant proteins in transgenic barley grains. Proc Natl Acad Sci USA 97:1914–1919.
56. RA Jorgensen, PD Cluster, J English, Q Que, CA Napoli. 1996. Chalcone synthase cosuppression phenotypes in petunia flowers: comparison of sense vs. antisense constructs and single-copy vs. complex T-DNA sequences. Plant Mol Biol 31:957–973.
57. J Kapusta, A Modelska, M Figlerowicz, T Pniewski, M Letellier, O Lisowa, V Yusibov, H Koprowski, A Plucienniczak, AB Legocki. 1999. A plant-derived edible vaccine against hepatitis B virus. FASEB J 13:1796–1799.
58. KD Kasschau, JC Carrington. 1998. A counterdefensive strategy of plant viruses: suppression of posttranscriptional gene silencing. Cell 95:461–470.
59. H Khoudi, S Laberge, JM Ferullo, R Bazin, A Darveau, Y Castonguay, G Allard, R Lemieux, LP Vezina. 1999. Production of a diagnostic monoclonal antibody in perennial alfalfa plants. Biotechnol Bioeng 20:135–143.
60. AR Kusnadi, ZL Nikolov, JA Howard. 1997. Production of recombinant proteins in transgenic plants: practical considerations. Biotechnol Bioeng 56:473–484.
61. AR Kusnadi, RL Evangelista, EE Hood, JA Howard, ZL Nikolov. 1998. Processing of transgenic corn seed and its effect on the recovery of recombinant beta-glucuronidase. Biotechnol Bioeng 60:44–52.
62. AR Kusnadi, EE Hood, DR Witcher, JA Howard, ZL Nikolov. 1998. Production and purification of two recombinant proteins from transgenic corn. Biotechnol Prog 14:149–155.

63. JW Larrick, L Yu, J Chen, S Jaiswal, K Wycoff. 1998. Production of antibodies in transgenic plants. Res Immunol 149:603–608.

64. F Le Gall, JM Bove, M Garnier. 1998. Engineering of a single-chain variable-fragment (scFv) antibody specific for the stolbur phytoplasma (Mollicute) and its expression in *Escherichia coli* and tobacco plants. Appl Environ Microbiol 64:4566–4572.

65. T Lehner, J Caldwell, R Smith. 1985. Local passive immunization by monoclonal antibodies against streptococcal antigen I/II in the prevention of dental caries. Infect Immun 50:796.

66. T Lehner, SJ Challaombe, J Caldwell. 1975. Immunological and bacteriological basis for vaccination against dental caries in rhesus monkeys. Nature 254: 517.

67. H Leiter, J Mucha, E Staudacher, R Grimm, J Glossl, F Altmann. 1999. Purification, cDNA cloning, and expression of GDP-L-Fuc:Asn-linked GlcNAcalpha1,3-fucosyltransferase from mung beans. J Biol Chem 274: 21830–21839.

68. WD Loomis. 1974. Overcoming problems of phenolics and quinones in the isolation of plant enzymes and organelles. Methods Enzymol 31:528–544.

69. M Longstaff, CA Newell, B Boonstra, G Strachan, D Learmonth, WJ Harris, AJ Porter, WD Hamilton. 1998. Expression and characterisation of single-chain antibody fragments produced in transgenic plants against the organic herbicides atrazine and paraquat. Biochim Biophys Acta 1381:147–160.

70. JK-C Ma, M Hunjan, R Smith, T Lehner. 1989. Specificity of monoclonal antibodies in local passive immunization against *Streptococcus mutans*. Clin Exp Immunol 77:331–337.

71. JK-C Ma, T Lehner. 1990. Prevention of colonization of *Streptococcus mutans* by topical application of monoclonal antibodies in human subjects. Arch Oral Biol 35:115S 122S.

72. JK-C Ma, R Smith, T Lehner. 1987. Use of monoclonal antibodies in local passive immunization to prevent colonization of human teeth by *Streptococcus mutans*. Infect Immun 55:1274–1278.

73. SW Ma, DL Zhao, ZQ Yin, R Mukherjee, B Singh, HY Qin, CR Stiller, AM Jevnikar. 1997. Transgenic plants expressing autoantigens fed to mice to induce oral immune tolerance. Nat Med 3:793–796.

74. J Ma, B Hikmat, K Wycoff, N Vine, D Chargelegue, L Yu, M Hein, T Lehner. 1998. Characterization of a recombinant plant monoclonal secretory antibody and preventive immunotherapy in humans. Nat. Med 4:601–606.

75. JK Ma, MB Hein. 1995. Plant antibodies for immunotherapy. Plant Physiol 109:341–346.

76. JK Ma, A Hiatt, M Hein, ND Vine, F Wang, P Stabila, C van Dolleweerd, K Mostov, T Lehner. 1995. Generation and assembly of secretory antibodies in plants. Science 268:716–719.

77. JK-C Ma, M Hein. 1995. Immunotherapeutic potential of antibodies produced in plants. TIBTECH 13:522–527.

78. JK-C Ma, MB Hein. 1996. Antibody production and engineering in plants. Ann NY Acad Sci 792:72–81.

79. JK-C Ma, T Lehner, P Stabilia, CI Fux, A Hiatt. 1994. Assembly of monoclonal antibodies with IgG1 and IgA heavy chain domains in transgenic tobacco plants. Eur J Immunol 24:131–138.

80. JK Ma, ND Vine. 1999. Plant expression systems for the production of vaccines. Curr Top Microbiol Immunol 236:275–292.

81. NS Magnuson, PM Linzmaier, JW Gao, R Reeves, G An, JM Lee. 1996. Enhanced recovery of a secreted mammalian protein from suspension culture of genetically modified tobacco cells. Protein Express Purif 7:220–228.

82. A Mari, P Iacovacci, C Afferni, B Barletta, R Tinghino, G Di Felice, et al. 1999. Specific IgE to cross-reactive carbohydrate determinants strongly affects the in vitro diagnosis of allergic diseases. J Allergy Clin Immunol 103: 1005–1011.

83. HS Mason, JM Ball, JJ Shi, X Jiang, MK Estes, CJ Arntzen. 1996. Expression of Norwalk virus capsid protein in transgenic tobacco and its oral immunogenicity in mice. Proc Natl Acad Sci USA 93:5335–5340.

84. HS Mason, DM Lam, CJ Arntzen. 1992. Expression of hepatitis B surface antigen in transgenic plants. Proc Natl Acad Sci USA 89:11745–11749.

85. HS Mason, TA Haq, JD Clements, CJ Arntzen. 1998. Edible vaccine protects mice against *Escherichia coli* heat-labile enterotoxin (LT): potatoes expressing a synthetic LT-B gene. Vaccine 16:1336–1343.

86. MA Matzke, AJ Matzke. 1998. Epigenetic silencing of plant transgenes as a consequence of diverse cellular defense responses. Cell Mol Life Sci 54:94–103.

87. P Meyer, H Saedler. 1996. Homology-dependent gene silencing in plants. Annu Rev Plant Physiol Mol Biol 47:23–48.

88. AA McCormick, MH Kumagai, K Hanley, TH Turpen, I Hakim, LK Grill, D Tuse, S Levy, R Levy. 1999. Rapid production of specific vaccines for lymphoma by expression of the tumor-derived single-chain Fv epitopes in tobacco plants. Proc Natl Acad Sci USA 96:703–708.

89. AS Moffat. 1998. Toting up the early harvest of transgenic plants. Science 282:2176–2178.

90. M Owen, A Gandecha, B Cockburn, G Whitelam. 1992. Synthesis of a functional anti-phytochrome single-chain Fv protein in transgenic tobacco. Bio/Technology 10:790–794.

91. J Phillips, O Artsaenko, U Fiedler, C Horstmann, HP Mock, K Muntz, U Conrad. 1997. Seed-specific immunomodulation of abscisic acid activity induces a developmental switch. EMBO J 16:4489–4496.

92. M-N Rosso, A Schouten, J Roosien, T Borst-Vrenssen, RS Hussey, FJ Gommers, J Bakker, A Schots, P Abad. 1996. Expression and functional characterization of a single chain FV antibody directed against secretions involved in plant nematode infection process. Biochem Biophys Res Commun 220: 255–263.

93. DA Russell. 1999. Feasibility of antibody production in plants for human therapeutic use. Curr Top Microbiol Immunol 236:119–137.

94. JC Sanford. 1988. The biolistic process. Trends Biotechnol 6:299–302.

95. OM Scheid, K Afsar, J Paszkowski. 1998. Release of epigenetic gene silencing by trans-acting mutations in *Arabidopsis*. Proc Natl Acad Sci USA 95: 632–637.
96. S Schillberg, S Zimmerman, K Findlay, R Fischer. 1999. Generation of TMV resistant tobacco by plasma membrane display of single chain Fv fragments. Submitted.
97. S Schillberg, S Zimmermann, A Voss, R Fischer. 1999. Apoplastic and cytosolic expression of full-size antibodies and antibody fragments in *Nicotiana tabacum*. Transgenic Res 8:255–263.
98. A Schouten, J Roosien, JM de Boer, A Wilmink, MN Rosso, D Bosch, WJ Stiekema, FJ Gommers, J Bakker, A Schots. 1997. Improving scFv antibody expression levels in the plant cytosol. FEBS Lett 415:235–241.
99. A Schouten, J Roosien, FA van Engelen, GAM de Jong, AWM Borst-Vrenssen, JF Zilverentant, D Bosch, WJ Steidema, FJ Gommers, A Schots, J Bakker. 1996. The C-terminal KDEL sequence increases the expression level of a single-chain antibody designed to be targeted to both the cytosol and the secretory pathway in transgenic tobacco. Plant Mol Biol 30:781–793.
100. N Shimada, Y Suzuki, M Nakajima, U Conrad, N Murofushi, I Yamaguchi. 1999. Expression of a functional single-chain antibody against $GA_{24/19}$ in transgenic tobacco. Biosci Biotechnol Biochem 63:779–783.
101. L Smolenska, IM Roberts, D Learmonth, AJ Porter, WJ Harris, TM Wilson, S Santa Cruz. 1998. Production of a functional single chain antibody attached to the surface of a plant virus. FEBS Lett 441:379–382.
102. M Stam, JNM Mol, JM Kooter. 1997. The silence of genes in transgenic plants. Ann Bot 79:3–12.
103. G Strachan, SD Grant, D Learmonth, M Longstaff, AJ Porter, WJ Harris. 1998. Binding characteristics of anti-atrazine monoclonal antibodies and their fragments synthesised in bacteria and plants. Biosens Bioelectron 13:665–673.
104. CO Tacket, HS Mason, G Losonsky, JD Clements, MM Levine, CJ Arntzen. 1998. Immunogenicity in humans of a recombinant bacterial antigen delivered in a transgenic potato. Nat Med 4:607–609.
105. P Tavladoraki, E Benvenuto, S Trinca, D Demartinis, A Cattaneo, P Galeffi. 1993. Transgenic plants expressing a functional single-chain Fv antibody are specifically protected from virus attack. Nature 366:469–472.
106. P Tavladoraki, A Girotti, M Donini, FJ Arias, C Mancini, V Morea, R Chiaraluce, V Consalvi, E Benvenuto. 1999. A single-chain antibody fragment is functionally expressed in the cytoplasm of both *Escherichia coli* and transgenic plants. Eur J Biochem 262:617–624.
107. Y Thanavala, YF Yang, P Lyons, HS Mason, C Arntzen. 1995. Immunogenicity of transgenic plant–derived hepatitis B surface antigen. Proc Natl Acad Sci USA 92:3358–3361.
108. B Uelker, GC Allen, WF Thompson, S Spiker, AK Weisinger. 1999. A tobacco matrix attachment region reduces the loss of transgene expression in the progeny of transgenic tobacco plants. Plant J 18:253–263.

109. P Vain, B Worland, A Kohli, JW Snape, P Christou, GC Allen, WF Thompson. 1999. Matrix attachment regions increase transgene expression levels and stability in transgenic rice plants and their progeny. Plant J 18:233–242.

110. MJ van der Veen, R van Ree, RC Aalberse, J Akkerdaas, SJ Koppelman, HM Jansen, JS van der Zee. 1997. Poor biologic activity of cross-reactive IgE directed to carbohydrate determinants of glycoproteins. J Allergy Clin Immunol 100:327–334.

111. FA van Engelen, A Schouten, JW Molthoff, J Roosien, J Salinas, WG Dirkse, A Schots, J Bakker, FJ Gommers, MA Jongsma, D Bosch, WJ Steikema. 1994. Coordinate expression of antibody subunit genes yields high levels of functional antibodies in roots of transgenic tobacco. Plant Mol Biol 26:1701–1710.

112. R van Ree, RC Aalberse. 1999. Specific IgE without clinical allergy. J Allergy Clin Immunol 103:1000–1001.

113. CF van Sumere, J Albrecht, A Dedonder, H de Pooter, I Pé. 1975. Plant proteins and phenolics. In: JB Harborne, CF van Sumere, eds. The Chemistry and Biochemistry of Plant Proteins. London: Academic Press, pp 211–264.

114. C Vaquero, M Sack, J Chandler, J Drossard, F Schuster, M Monecke, S Schillberg, R Fischer. 1999. Transient expression of a tumor-specific single-chain fragment and a chimeric antibody in tobacco leaves. Proc Natl Acad Sci USA 96:11128–11133.

115. H Vaucheret, C Beclin, T Elmayan, F Feuerbach, C Godon, JB Morel, P Mourrain, JC Palauqui, S Vernhettes. 1998. Transgene-induced gene silencing in plants. Plant J 16:651–659.

116. T Verch, V Yusibov, H Koprowski. 1998. Expression and assembly of a full-length monoclonal antibody in plants using a plant virus vector. J Immunol Methods 220:69–75.

117. A Voss, M Niersbach, R Hain, HJ Hirsch, YC Liao, F Kreuzaler, R Fischer. 1995. Reduced virus infectivity in *N. tabacum* secreting a TMV-specific full-size antibody. In: Molecular Breeding: New Strategies in Plant Improvement. Boston: Kluwer Academic Publishers, pp 39–50.

118. CI Wandelt, MRI Khan, S Craig, HE Schroeder, D Spencer, TJV Higgins. 1992. Vicilin with carboxy-terminal KDEL is retained in the endoplasmic reticulum and accumulates to high levels in leaves of transgenic plants. Plant J 2:181–192.

119. GC Whitelam, W Cockburn, MRL Owen. 1994. Antibody production in transgenic plants. Biochem Soc Trans 22:940–943.

120. P Zambryski, et al. 1983. Ti plasmid vector for the introduction of DNA into plant cells without alteration of their normal regeneration capacity. EMBO J 2:2143–2150.

121. L Zeitlin, SS Olmsted, TR Moench, MS Co, BJ Martinell, VM Paradkar, DR Russell, C Queen, RA Cone, KJ Whaley. 1998. A humanized monoclonal antibody produced in transgenic plants for immunoprotection of the vagina against genital herpes. Nat Biotechnol 16:1361–1364.

122. S Zimmerman, S Schillberg, YC Liao. 1998. Intracelluar expression of TMV-specific single-chain Fv fragments leads to improved virus resistance in *Nicotiana tabacum*. Mol Breed 4:369–379.

18

Signal Transduction Elements

Dierk Scheel

Department of Stress and Developmental Biology,
Institute of Plant Biochemistry, Halle (Saale), Germany

I. INTRODUCTION

Optimal growth and differentiation of plants require coordinated regulation
of cellular and intercellular processes and their continuous adaptation to the
variable environment. Efficient mechanisms evolved during plant evolution
for sensing the environment, for translation of these data into biological
information, for transfer of this information within the organism, and for
initiation of appropriate reactions. These processes of biological signal transduction comprise the perception of endogenous and environmental signals,
the generation of endogenous cellular and systemic signals, and their transmission to the appropriate response targets. Receptor proteins of the plasma
membrane or the cytoplasm specifically bind and thereby recognize the signals and either alone or in concert with other proteins initiate cellular signaling processes. Intracellular mediators form the basis of complex cellular
signaling networks that are responsible for signal transmission, integration,
and evaluation and response activation. Although plants lack an equivalent
of the circulating bloodstream, the responses can include the production and
secretion of systemic signals that are transported throughout the plant and
are recognized by receptors of target cells.

In order to allow optimal adaptation to the environment, the signals
perceived by individual receptors need to be integrated and, most important,

evaluated according to their importance. Therefore, linear signal transduction pathways are the exception, if they exist at all. Rather, complex signaling networks with points of signal convergence and divergence support cross talk between the signaling pathways initiated by individual input signals. Thereby, an overall assessment is possible that guarantees an appropriate response to the general environmental situation. This complexity of interconnected signaling networks and the limited knowledge of molecular details of plant signal transduction mechanisms have so far not supported the employment of genes encoding signal transduction components in plant biotechnology. However, some promising results have been reported, primarily using the end points of signal pathways, receptors and transcription factors, respectively. These will be summarized in this chapter in the form of a speculative outlook on future possibilities to modulate developmental processes in plants.

II. HORMONE LEVELS

Alterations in the level of individual or in the balance of different hormones usually result in pleiotropic phenotypes of little interest for application. However, if these changes were spatially and temporally tightly regulated, meaningful effects might be obtained.

Seed dormancy delays germination despite favorable conditions and, thereby, allows seed survival in soil for long periods of time. Although an advantage for weeds, this is undesirable for crop plants. The plant hormone abscisic acid (ABA) is synthesized during seed development and is involved in the induction of primary seed dormancy (1,2). ABA biosynthesis or sensitivity mutants displayed reduced seed dormancy (3,4). Consequently, constitutive expression of the ABA biosynthetic gene *ABA2*, encoding zeaxanthin epoxidase, in *Nicotiana plumpaginifolia* resulted in increased ABA levels and delayed germination, whereas antisense suppression of this gene resulted in reduction of ABA seed levels and rapid germination (5).

Constitutive expression of the *AtGA2ox1* gene from *Arabidopsis thaliana* encoding the key regulatory gene for gibberellin (GA) biosynthesis, GA 20-oxidase, in hybrid aspen resulted in increased levels of several gibberellins in internodes and leaves (6). The transgenic trees showed an improved growth rate, produced larger biomass, and had more numerous and longer xylem fibers than wild-type plants. As one of only a few negative effects, poor root initiation was observed when the transgenic plants were transferred to soil.

Methyl jasmonate (MeJA) is a naturally occurring volatile derivative of the plant hormone jasmonic acid (JA) that also stimulates many typical plant responses to JA when applied exogenously (7). The plant enzyme *S*-

adenosyl-L-methionine:jasmonic acid methyltransferase (JMT) catalyzes the formation of MeJA from JA (8). The gene is differentially expressed in *A. thaliana* during development and with environmental stimuli. Transgenic *A. thaliana* plants constitutively expressing the *JMT* gene from the same plant show increased MeJA but unaltered JA levels. Although visually not distinguishable from wild-type plants, the transgenic plants display elevated transcript levels of JA-responsive genes without any stimulus. Furthermore, their degree of resistance against *Botrytis cinerea* was found to be significantly increased when compared with untransformed *A. thaliana* plants.

The signal molecule salicylic acid (SA) plays a central role in pathogen defense of plants (9) but is also involved in other regulatory processes, such as cell growth, stomatal closure, flower induction, and heat production (10–12). Transformation of tobacco plants with bacterial genes encoding the enzymes isochorismate synthase (ICS) and isochorismate pyruvate lyase (IPL), under control of a constitutive promoter, resulted in strongly increased levels of SA and SA glucoside in healthy plants when the enzymes were targeted to the chloroplast (13). The transgenic plants displayed a normal phenotype but constitutively expressed pathogenesis-related genes and showed elevated resistance against tobacco mosaic virus and the fungal pathogen *Oidium lycopersicon*.

These few examples of rather crude modulation of hormone levels demonstrate the large potential of such an approach for molecular engineering, particularly if more sophisticated regulatory tools become available.

III. RECEPTOR-LIGAND INTERACTIONS

The idea of employing ligand-receptor pairs in transgenic plants has been pursued most intensively with the goal of generating plants with durable disease resistance, although no commercial product has yet been generated by this approach (14). Plants successfully resist the attack of most potential pathogens in their environment through an efficient nonself recognition system that is similar to the innate immunity system of vertebrates and insects (15). Elicitors, originating from the pathogen or released from the plant cell wall during pathogen attack, are specifically recognized by receptors of the plant plasma membrane and thereby initiate a multicomponent defense response (15,16). Although several such elicitors have been purified and shown to bind specifically to binding sites of the plasma membrane (17), only two genes encoding components of the corresponding plant receptors have been cloned so far (18–20). One encodes the 75-kDa plasma membrane–associated binding protein of the hepta-β-glucan elicitor from *Phytophthora sojae* that occurs in various Fabaceae (18,19). Transgenic tomato plants expressing the hepta-β-glucan–binding protein from soybean display high-affinity binding sites for this elicitor (19). It is unknown, however, whether these plants display increased resistance toward pathogens harboring the hepta-β-glucan in their cell walls. A gene encoding a 129-kDa receptor-like kinase apparently involved in the recognition of bacterial flagellin was isolated from *A. thaliana* (20). Point mutations in different regions of this gene resulted in loss of ligand recognition and elicitor responsiveness (21). Although receptor function of the corresponding plasma membrane protein needs to be demonstrated, it represents an essential component in binding of a 22-amino-acid fragment of flagellin and may play an important role in the recognition of bacterial pathogens by plants (20–22). Overexpression of such receptors involved in nonhost recognition has the potential to increase basal pathogen resistance in many different crop plants.

In addition to the basal nonhost resistance, plants have developed a complex host defense system relying on receptor-mediated recognition of pathogen avirulence (*Avr*) gene products by plant resistance (*R*) gene products (23). The existence of corresponding pairs of *Avr* and *R* genes results in an incompatible interaction between the pathogen and its host plant; i.e., disease development is efficiently stopped by the rapid activation of a multicomponent plant defense response. Although evidence for direct ligand/receptor-type interaction of *Avr* and *R* gene products is lacking in most cases, both may be components of larger signal perception complexes, whose formation is required for an incompatible interaction (24). The *R* genes introduced into crop plants by conventional breeding techniques did function in the expected way, but, with a few exceptions, the resulting resistance was found to lack durability in the field (25). Pathogens could rapidly break *R* gene–mediated resistance because conventional breeding allowed only the generation of *R* gene monocultures (23). The availability of a growing number of different cloned *R* genes directed against specific pathogen races would allow generating transgenic *R* gene polycultures of crop plants. Using a population of a given crop plant species consisting of individual plants expressing specific *R* gene patterns would significantly reduce the speed of adaptation of the pathogen and thereby result in an overall reduction of disease development. Interestingly, increasing the level of a specific *R* gene product by overexpression can activate defense responses in the absence of pathogens and thereby result in broad resistance of the transgenic plants as shown for the *Pto* gene in tomato (26,27).

A different strategy has been suggested by de Wit involving the coexpression of pairs of *Avr* and *R* genes in one plant (28). This approach requires constitutive expression of a plant *R* gene and tightly regulated coexpression of a pathogen-derived *Avr* gene under control of a broadly pathogen-responsive promoter. Transformation of tomato plants carrying the *Cf9* resistance gene against *Cladosporium fulvum* with the *Avr9* gene from this fungal pathogen under control of a pathogen-responsive promoter rendered the transgenic plants resistant to a broad spectrum of pathogens (14,29). Although the application of this approach appears to be limited by the lack of functional expression of *Cf9* and *Avr9* genes in certain crop species, different pairs of matching *R* and *Avr* genes may function in different crop species (30–32).

Interestingly, this strategy to generate broad-spectrum disease resistance (28) appears not to be limited to *R/Avr* gene–mediated plant-pathogen recognition. All plant pathogenic *Phytophthora* species analyzed so far secrete elicitins, homologous proteinaceous elicitors (33) that induce an efficient resistance response in tobacco and a few other plant species (34). Tobacco plasma membranes harbor high-affinity binding sites for elicitins that appear to function as elicitin receptors (35,36). *Phytophthora* species pathogenic on tobacco, such as the causal agent of black shank disease, *Phytophthora parasitica* var. *nicotiana*, do not produce elicitins and thereby escape recognition (37). Expression of the elicitin cryptogein from *Phytophthora cryptogea* under control of a strictly pathogen-responsive promoter in tobacco resulted in broad-spectrum resistance of the transgenic plants against *P. parasitica* var. *nicotiana*, *Thielaviopsis basicola*, *Erysiphe cichoracearum*, and *Botrytis cinerea* (38).

IV. GTP-BINDING PROTEINS

In mammals, heterotrimeric G protein complexes link a large number of heptahelical transmembrane G protein–coupled receptors (GPCRs) to cellular signaling networks and thereby regulate a multitude of cellular processes (39). The importance of this signaling mechanism is reflected by the presence of more than 1000 different GPCR-encoding genes and several $G\alpha$, $G\beta$ and $G\gamma$ isoform genes in mammalian genomes (40). Although in plants the existence of only a limited number of GPCRs, $G\beta$ and $G\gamma$, and only a single $G\alpha$ is predicted from the *A. thaliana* genome sequence (41), heterotrimeric G proteins appear to play a central role in plant hormone signaling (40,42).

Knockout mutants of the only $G\alpha$ isoform of *A. thaliana* (*GPA1*) have reduced cell division in their aerial tissues (43) and lack ABA inhibition of guard cell inward K^+ channels as well as pH-independent ABA activation of anion channels (44). Consequently, stomatal opening in these knockout mutants was found to be insensitive to ABA inhibition and, therefore, the water loss rate was greater in mutant than in wild-type plants (44). Because inducible overexpression of *GPA1* in *Arabidopsis* conferred inducible cell division to transgenic tissue, targeted modulation of the status of $G\alpha$ in stoma cells may provide a new tool to control water balance (43,44).

A large group of small GTP-binding proteins in plants are involved in a broad spectrum of cellular signaling processes (41,42). In rice, three genes (*OsRac1–3*) were identified encoding proteins with 60% identity to human Rac proteins (45). Expression of a constitutively active derivative of *OsRac1* in rice resulted in enhanced production of reactive oxygen species (ROS) and phytoalexins, stimulation of programmed cell death, and increased resistance against bacterial and fungal pathogens (45,46). Loss-of-function experiments with a dominant-negative *OsRac1* derivative showed the expected suppression of elicitor-stimulated ROS production and pathogen-induced programmed cell death in transgenic rice (46). These findings suggest that processes regulated via small GTP-binding proteins can be modulated in transgenic plants and cause modification of distinct physiological processes.

V. CALCIUM SIGNALING

Release of Ca^{2+} from internal stores and influx from the apoplastic space resulting in transient increases or oscillation of cytosolic Ca^{2+} levels are involved in many signaling networks in plants (15,47,48). The duration, intensity, and spatial distribution of Ca^{2+} transients appear to encode signal-response specificity. Direct modulation of cytosolic Ca^{2+} levels might therefore be a difficult task. However, downstream targets of Ca^{2+} regulation may represent future tools for engineering plants with improved stress tolerance.

Two calcium-dependent protein kinases (CDPKs) from tobacco (NtCDPK2 and 3) have been shown to be involved in defense and hypo-osmotic stress signaling (49,50). Both CDPKs are phosphorylated and thereby activated in a Ca^{2+}-dependent manner and trigger an oxidative burst and programmed cell death (49). Ectopic expression of a related CDPK from *A. thaliana* in tomato protoplasts stimulated NADPH oxidase activity and an oxidative burst (51). Tightly regulated expression of constitutively active CDPK derivatives might offer a tool to engineer plants with enhanced flooding and disease resistance.

Calmodulin is a well-known target of calcium in multiple signal transduction pathways (47). Transgenic tobacco cells expressing a dominant-acting calmodulin derivative responded with a stronger oxidative burst and enhanced NADPH levels to treatments with various elicitors, infection with avirulent bacteria, and osmotic and mechanical stress compared with wild-type cells (52). Two specific calmodulin isoforms, SCaM4 and 5, and their transcripts were found to accumulate in cultured soybean cells upon elicitor treatment (53). These two calmodulin isoforms are most divergent from other isoforms described from plants and animals (54). Furthermore, their ability to activate calmodulin-dependent enzymes in vitro differed greatly from that of the other calmodulin isoforms of soybean. Transgenic tobacco plants constitutively expressing SCaM4 or SCaM5 exhibited spontaneous lesion formation, constitutive expression of defense-related genes, and increased resistance to virulent viral, bacterial, and oomycete pathogens (53). These findings suggest that specific calmodulin isoforms represent promising tools for engineering stress-tolerant plants.

VI. MITOGEN-ACTIVATED PROTEIN KINASES

Mitogen-activated protein kinase (MAPK) cascades are universal signaling modules of eukaryotic cells that fulfill essential regulatory functions primarily in transduction of extracellular signals to cellular and nuclear responses (55). Each cascade consists of a module of at least three protein kinases (56). The most downstream kinase, MAPK, is activated by dual phosphorylation of a typical threonine-X-tyrosine motif catalyzed by a MAPK kinase (MAPKK). MAPKKs are themselves activated by serine/threonine phosphorylation by a MAPKK kinase (MAPKKK). The activation mechanism of plant MAPKKKs and their linkage to the corresponding receptor are not known. Also, no substrate of a plant MAPK has been identified. According to sequence similarities, the *Arabidopsis* genome contains up to 23 MAPKs, 9 MAPKKs, and at least 25 MAPKKKs (41,55). Therefore, plants have more MAPKs than any other eukaryotic organism. The presence of more MAPKs than MAPKKs furthermore shows that MAPKKs cannot phosphorylate only a single MAPK. Plant MAPK cascades have been found to be involved in hormonal responses, cell cycle regulation, abiotic stress signaling, and pathogen defense, where they exhibit positive as well as negative regulatory functions (55,56). Phenotypes of recently identified MAPK cascade mutants (57,58) and transgenic plants expressing constitutively active MAPKKK (59) or MAPKK derivatives (55,60–62) suggest that it might be possible to employ these signaling elements in order to generate transgenic plants with improved stress tolerance.

The *mpk4* mutant of *A. thaliana* lacks a functional MAPK, AtMPK4 (57). Besides being a severe dwarf, this mutant contains high levels of salicylate, is jasmonate insensitive, and displays enhanced resistance to bacterial and oomycete pathogens. Because active AtMPK4 is required to complement the *mpk4* mutant, this MAPK apparently suppresses salicylate accumulation in wild-type plants and thereby negatively regulates activation of salicylate-dependent defense responses. Although complete lack of AtMPK4 results in this pleiotropic dwarf phenotype, tightly regulated inactivation of this MAPK would possibly increase disease resistance without deleterious effects for the corresponding plant. Another MAPK cascade mutant of *A. thaliana*, *edr1*, carries a mutation in a putative MAPKKK (58). This mutant has a normal phenotype and does not constitutively activate defense responses. However, defense responses are more rapidly activated upon pathogen attack in *edr1* than in the wild-type plant, conferring enhanced resistance against powdery mildew to the mutant. These results suggest that EDR1 negatively regulates the activation of defense reactions and that its inactivation results in enhanced disease resistance.

MAPK cascades also positively regulate disease resistance (15,63). In tobacco, the MAPKs SIPK and WIPK are rapidly activated upon treatment with pathogen-derived elicitors and infection with avirulent pathogens (63). Transient expression of a constitutively active derivative of the tobacco MAPKK, NtMek2, in tobacco leaves resulted in activation of endogenous SIPK and WIPK and furthermore stimulated typical defense reactions, such as programmed cell death and activation of defense-related genes. Transient inducible expression of SIPK in the same experimental system led to increased protein kinase amounts, enhanced enzyme activity caused by phosphorylation, and activation of multiple defense reactions, demonstrating that SIPK alone is sufficient to stimulate the defense response (61). Transgenic *Arabidopsis* plants expressing constitutively active derivatives of the MAPKKs, AtMEK4 and 5, under control of an inducible promoter displayed programmed cell death preceded by activation of endogenous MAPKs and ROS production (62). The *Arabidopsis* orthologue of SIPK, AtMPK6, was also found to be activated by the flagellin-derived oligopeptide elicitor flg22 (64). Expression of components of the corresponding MAPK cascade confers pathogen resistance to leaves of the transgenic *Arabidopsis* plants (55). Together, these findings suggest that spatially and temporally regulated modulation of specific MAPK cascades in transgenic plants can be employed to engineer disease resistance.

Reactive oxygen species such as superoxide anion radical ($O_2^{\cdot-}$), hydroxyl radical ($^{\cdot}OH$), and hydrogen peroxide (H_2O_2) are generated by plant cells in response to diverse abiotic and biotic stresses (65). Although the sensing mechanisms of ROS by plant cells are unknown, it has recently been demonstrated that H_2O_2 activates MAPK cascades in *A. thaliana* (59,66). Specifically, the MAPKKK ANP1 and the two MAPKs AtMPK3 and 6 were found to be activated by H_2O_2 treatment (59). Interestingly, transient expression of a constitutively active version of ANP1 in *Arabidopsis* protoplasts initiated activation of AtMPK3 and 6 in the absence of H_2O_2. Most important, when a constitutively active derivative of the tobacco orthologue of ANP1, NPK1, was expressed in transgenic tobacco, these phenotypically normal plants exhibited enhanced tolerance to various stress conditions, such as freezing, heat shock and salt stress. Thus, modulation of specific MAPKKK activities clearly has the potential to engineer broad stress tolerance in plants.

VII. TRANSCRIPTION FACTORS

Transcription factors represent the final signaling elements that determine the gene expression pattern of a cell. Analysis of the *A. thaliana* genome using a consistent conservative threshold revealed more than 1700 putative transcription factors (41). Among the 29 classes of transcription factors, 16 appear to be unique to plants. Functional analysis by knockout techniques was found to be hampered by the frequent occurrence of functionally redundant genes. In contrast, overexpression of transcription factors in several cases indicated functional links and possible usefulness in molecular engineering.

The tomato *Pti5* gene encodes a pathogen-responsive transcription factor of the ethylene response element–binding protein (AP2/EREBP) class that interacts with the *R* gene product Pto (67). The *Pti5* gene is believed to be specifically involved in pathogen defense because its transcript accumulates in response to infection but not to abiotic stresses or hormone treatments (68). Overexpression of Pti5 in a tomato cultivar lacking the corresponding *R* gene, *Pto*, conferred enhanced resistance against the virulent bacterial pathogen *Pseudomonas syringae* pv. *tomato* to the normal looking transgenic plants (69). Interestingly, the plants did not constitutively express defense-related genes but their activation was drastically accelerated upon infection in comparison with wild-type plants. Overexpression of the *Pti4* gene, also encoding a transcription factor interacting with Pto, in *A. thaliana* resulted only in slightly enhanced resistance to bacterial pathogens (69).

The AP2/EREBP-type transcription factor gene, *Tsi1*, was identified as a salt-responsive transcript in tobacco (70). Transgenic tobacco plants overexpressing *Tsi1* exhibited increased salt tolerance. In addition, these plants showed constitutively elevated transcript levels of defense-related genes and enhanced resistance to the virulent bacterial pathogen, *Pseudomonas syringae* pv. *tabaci*.

The *NPR1* (also designated *NIM1* and *SAI1*) gene of *A. thaliana* is an important regulator of inducible pathogen defense (71–75). It encodes a protein with ankyrin repeats and homology to the human transcription factor IκB (74,76). Despite these homologies and the requirement of its nuclear localization for function (77), it probably does not act directly as transcription factor but acts upon differential interaction with bZIP transcription factors of the TGA family (78–80). Overexpression of *NPR1* renders *Arabidopsis* plants resistant against virulent bacterial and oomycete pathogens without altering their normal phenotype (81). Most interestingly, ectopic expression of the *Arabidopsis NPR1* gene in rice also results in increased resistance of the transgenic plants against the virulent bacterial rice pathogen *Xanthomonas oryzae* pv. *oryzae* (82). The transgenic plants displayed a normal phenotype like the wild-type plants. These results make *NPR1* an excellent candidate tool to engineer broad-spectrum disease resistance in monocotyledonous and dicotyledonous plants.

Overexpression of transcription factors has also been described as a useful approach in engineering abiotic stress tolerance. The dehydration response of *A. thaliana* is regulated in a complex manner via at least four independent pathways, two of which are ABA dependent and two ABA independent (83). The dehydration-responsive *cis*-element (DRE) is essential for regulation of dehydration- and cold-responsive gene expression via one of the ABA-independent signaling pathways (84). The *Arabidopsis* transcription factors, DREB1A and DREB2A, bind to the DRE element and thereby activate transcription of the corresponding gene (85). Constitutive overexpression of DREB1A in *Arabidopsis* plants caused dwarfed phenotypes, expression of stress response genes under nonstressed conditions, and improved tolerance to drought, salt stress, and cold temperatures. However, expression of DREB1A under control of the stress-responsive promoter, *rd29A* (86), did not significantly affect growth of transgenic *Arabidopsis* plants but resulted in even greater drought, salt, and freezing tolerance than constitutive expression (87).

The *Arabidopsis* transcriptional activator CBF1 (C-repeat/DRE binding factor 1) binds to the C-repeat/DRE *cis*-acting element that regulates cold- and drought-responsive gene expression (84,88). Constitutive overexpression of *CBF1* in *A. thaliana* activated cold-responsive genes at normal temperatures and enhanced the freezing tolerance of the transgenic plants without obvious effects on their growth and development (89).

A cold-responsive zinc finger protein from soybean, SCOF-1, appears not to act directly as a transcription factor but rather associates with a soybean G-box binding bZIP transcription factor, SGBF-1, in the nucleus and thereby dramatically enhances its binding affinity to the ABA responsive *cis*-acting element (ABRE) of cold-responsive genes (90). Tobacco plants constitutively overexpressing SCOF-1 had normal phenotypes and did not exhibit altered cold sensitivity. However, cold-stressed SCOF-1–transgenic plants recovered significantly faster than wild-type plants under normal growth conditions. SCOF-1–expressing *Arabidopsis* plants displayed constitutive activation of cold-responsive genes and enhanced freezing tolerance.

The synthesis of heat stress proteins (HSPs) protects plants against damage by high temperatures (91). Heat stress transcription factors (HSFs) regulate the heat-induced transcription of *HSP* genes. Constitutive overexpression of *HSF3* but not *HSF4* stimulated HSP synthesis in *Arabidopsis* plants without altering their normal phenotype (92). Most important, the *HSF3*-transgenic plants displayed increased basal thermotolerance.

The *Alfin1* gene, originally cloned as a salt-responsive gene from alfalfa (93), was found to encode a root-specific zinc finger–type transcription factor (94). Constitutive expression of the *Alfin1* gene in transgenic alfalfa led to increased root growth under normal and saline conditions and enhanced salt tolerance without having adverse effects on plant shoot growth (95,96).

Surprisingly, transcriptional regulators of complex developmental processes, such as flower development, may also be employed in this rather simple approach for genetic improvement of crop species. The *Arabidopsis* genes *LEAFY* (*LFY*) and *APETALA1* (*AP1*) function as meristem identity genes (97) and promote flower initiation when constitutively expressed in *Arabidopsis* (98,99). Constitutive expression of either one of these genes in transgenic citrus resulted in the production of fertile flowers and fruits in the first year and thereby in a significant reduction of the generation time of the transgenic trees (100).

These few examples of molecular engineering via transcriptional regulators nicely demonstrate the enormous potential of this strategy for crop plant improvement.

VIII. CONCLUSIONS AND OUTLOOK

Plants apparently employ signal transduction elements similar to those of animals (47). However, most of them are present in larger numbers (41) and they are often linked in a different and variable way (47). Most if not all plant signaling pathways are organized in networks with points of convergence and divergence allowing cross talk between pathways and evaluation of the importance of simultaneously perceived signals. Significant progress has been made during the last decade in the analyses of all levels of signal transduction. Array and proteomic technologies together with the knowledge of total genome sequences will further complete our understanding of complex signaling processes and allow better targeted approaches than those described. However, simple overexpression studies with model systems have already demonstrated that diverse signaling elements can be employed successfully in molecular engineering of plants. On the other hand, these experiments have also shown that modulation of a specific signaling pathway often interferes with other not related pathways and thereby negatively affects physiological processes. In order to fine-tune expression of transgenes, spatially and temporally tightly regulated promoters are required. The increasing number of functionally characterized *cis*-acting elements might soon allow the construction of artificial promoters that better fulfill this purpose.

In summary, molecular engineering of signal transduction pathways has the potential to modulate all processes of plant growth and development including their responsiveness to biotic and abiotic stresses. In order to use this potential, a detailed understanding of the plant signaling components and their linkages within complex networks is required.

REFERENCES

1.   CD Rock, RS Quatrano. The role of hormones during seed development. In: PJ Davies, ed. Plant Hormones. Dordrecht, The Netherlands: Kluwer Academic Publishers, 1995, pp 671–697.
2.   HWM Hilhorst, CM Karssen. Seed dormancy and germination: the role of abscisic acid and gibberellins and the importance of hormone mutants. Plant Growth Regul 11:225–238, 1992.
3.   DR McCarty. Genetic control and integration of maturation and germination pathways in seed development. Annu Rev Plant Physiol Plant Mol Biol 46: 71–73, 1995.
4.   S Merlot, J Giraudat. Genetic analysis of abscisic acid signal transduction. Plant Physiol 114:751–757, 1997.
5.   A Frey, C Audran, E Marin, B Sotta, A Marion-Poll. Engineering seed dormancy by the modification of zeaxanthin epoxidase gene expression. Plant Mol Biol 39:1267–1274, 1999.
6.   ME Eriksson, M Israelsson, O Olsson, T Moritz. Increased gibberellin biosynthesis in transgenic trees promotes growth, biomas production and xylem fiber length. Nat Biotechnol 18:784–788, 2000.
7.   C Wasternack, B Parthier. Jasmonate-signalled plant gene expression. Trends Plant Sci 2:302–307, 1997.
8.   HS Seo, JT Song, J-J Cheong, Y-H Lee, Y-W Lee, I Hwang, JS Lee, YD Choi. Jasmonic acid carboxyl methyltransferase: a key enzyme for jasmonate-regulated plant responses. Proc Natl Acad Sci USA 98:4788–4793, 2001.
9.   JA Ryals, MG Willits, A Molina, H Steiner, MD Hunt. Systemic acquired resistance. Plant Cell 8:1809–1819, 1996.
10.   I Raskin. Salicylate, a new plant hormone. Plant Physiol 99:799–803, 1992.
11.   I Raskin. Role of salicylic acid in plants. Annu Rev Plant Physiol Plant Mol Biol 43:439–463, 1992.
12.   H Vanacker, H Lu, DN Rate, JT Greenberg. A role for salicylic acid and NPR1 in regulating cell growth in *Arabidopsis*. Plant J 28:209–216, 2001.
13.   MC Verberne, R Verpoorte, JF Bol, J Mercado-Blanco, HJM Linthorst. Overproduction of salicylic acid in plants by bacterial transgenes enhances pathogen resistance. Nat Biotechnol 18:779–783, 2000.
14.   MH Stuiver, JHHV Custers. Engineering disease resistance in plants. Nature 411:865–868, 2001.
15.   T Nürnberger, D Scheel. Signal transmission in the plant immune response. Trends Plant Sci 6:372–379, 2001.

16. D Scheel. Resistance response physiology and signal transduction. Curr Opin Plant Biol 1:305–310, 1998.

17. T Nürnberger. Signal perception in plant pathogen defense. Cell Mol Life Sci 55:167–182, 1999.

18. N Umemoto, M Kakitani, A Iwamatsu, M Yoshikawa, N Yamaoka, I Ishida. The structure and function of a soybean β-glucan-elicitor-binding protein. Proc Natl Acad Sci USA 94:1029–1034, 1997.

19. A Mithöfer, J Fliegmann, G Neuhaus-Url, H Schwarz, J Ebel. The hepta-β-glucoside elicitor-binding proteins from legumes represent a putative receptor family. Biol Chem 381:705–713, 2000.

20. L Gómez-Gómez, T Boller. FLS2: an LRR receptor-like kinase involved in the perception of the bacterial elicitor flagellin in *Arabidopsis*. Mol Cell 5: 1003–1011, 2000.

21. L Gómez-Gómez, Z Bauer, T Boller. Both the extracellular leucin-rich repeat domain and the kinase activity of FLS2 are required for flagellin binding and signalling in *Arabidopsis*. Plant Cell 13:1155–1163, 2001.

22. G Felix, JD Duran, S Volko, T Boller. Plants have a sensitive perception system for the most conserved domain of bacterial flagellin. Plant J 18:265–276, 1999.

23. JL Dangl, JDG Jones. Plant pathogens and integrated defence responses to infection. Nature 14:826–833, 2001.

24. EA van der Biezen, JDG Jones. Plant disease resistance proteins and the gene-for-gene concept. Trends Biochem Sci 23:454–456, 1998.

25. D Pink, I Puddephat. Deployment of disease resistance genes by plant transformation—a "mix and match" approach. Trends Plant Sci 4:71–75, 1999.

26. X Tang, M Xie, YJ Kim, J Zhou, DF Klessig, GB Martin. Overexpression of *Pto* activates defense responses and confers broad resistance. Plant Cell 11: 15–30, 1999.

27. CM Tobias, GE Oldroyd, JH Chang, BJ Staskawicz. Plants expressing the *Pto* disease resistance gene confer resistance to recombinant PVX containing the avirulence gene *AvrPto*. Plant J 17:41–50, 1999.

28. PJGM De Wit. Molecular characterization of gene-for-gene systems in plant-fungus interactions and the application of avirulence genes in control of plant pathogens. Annu Rev Phytopathol 30:391–418, 1992.

29. LS Melchers, MH Stuiver. Novel genes for disease-resistance breeding. Curr Opin Plant Biol 3:147–152, 2000.

30. KE Hammond-Kosack, S Tang, K Harrison, JDG Jones. The tomato *Cf-9* disease resistance gene functions in tobacco and potato to confer responsiveness to the fungal avirulence gene product Avr9. Plant Cell 10:1251–1266, 1998.

31. RAL Van der Hoorn, F Laurent, R Roth, PJGM De Wit. Agroinfiltration is a versatile tool that facilitates comparative analyses of *Avr9/Cf-9*–induced and *Avr4/Cf-4*–induced necrosis. Mol Plant Microbe Interact 13:439–446, 2000.

32. C Hennin, M Höfte, E Diederichsen. Functional expression of *Cf9* and *Avr9* genes in *Brassica napus* induces enhanced resistance to *Leptosphaeria maculans*. Mol Plant Microbe Interact 14:1075–1085, 2001.

33. P Ricci. Induction of the hypersensitive response and systemic acquired re-
     sistance by fungal proteins: the case of elicitins. In: G Stacey, NT Keen, eds.
     Plant-Microbe Interactions. New York: Chapman & Hall, 1997, pp 53–75.
34. S Kamoun, KM Klucher, MD Coffey, BM Tyler. A gene encoding a host-
     specific elicitor protein of *Phytophthora parasitica*. Mol Plant Microbe Inter-
     act 6:573–581, 1993.
35. D Wendehenne, MN Binet, JP Blein, P Ricci, A Pugin. Evidence for specific,
     high-affinity binding sites for a proteinaceous elicitor in tobacco plasma mem-
     brane. FEBS Lett 374:203–207, 1995.
36. S Bourque, MN Binet, M Ponchet, A Pugin, A Lebrun-Garcia. Characteriza-
     tion of the cryptogein binding sites on plant plasma membranes. J Biol Chem
     274:34699–34705, 1999.
37. P Ricci, P Bonnet, JC Huet, M Sallantin, F Beauvais-Cante, M Bruneteau, V
     Billard, G Michel, JC Pernollet. Structure and activity of proteins from path-
     ogenic fungi *Phytophthora* eliciting necrosis and acquired resistance in to-
     bacco. Eur J Biochem 183:555–563, 1989.
38. H Keller, N Pamboukdjian, M Ponchet, A Poupet, R Delon, J-L Verrier. Path-
     ogen-induced elicitin production in transgenic tobacco generates a hypersen-
     sitive response and nonspecific disease resistance. Plant Cell 11:223–235,
     1999.
39. J Gutkind. The pathways connecting G protein–coupled receptors to the nu-
     cleus through divergent mitogen-activated protein kinase cascades. J Biol
     Chem 273:1839–1842, 1998.
40. BE Ellis, GP Miles. One for all? Science 292:2022–2023, 2001.
41. The Arabidopsis Genome Initiative. Analysis of the genome sequence of the
     flowering plant *Arabidopsis thaliana*. Nature 408:796–815, 2000.
42. F Bischoff, A Molendijk, CSV Rajendrakumar, K Palme. GTP-binding pro-
     teins in plants. Cell Mol Life Sci 55:233–256, 1999.
43. H Ullah, J-G Chen, JC Young, K-H Im, MR Sussman, AM Jones. Modulation
     of cell proliferation by heterotrimeric G protein in *Arabidopsis*. Science 292:
     2066–2069, 2001.
44. X-Q Wang, H Ullah, AM Jones, SM Assmann. G protein regulation of ion
     channels and abscisic acid signaling in *Arabidopsis* guard cells. Science 292:
     2070–2072, 2001.
45. T Kawasaki, K Henmi, E Ono, S Hatakeyama, M Iwano, H Satoh, K Shi-
     mamoto. The small GTP-binding protein Rac is a regulator of cell death in
     plants. Proc Natl Acad Sci USA 96:10922–10926, 1999.
46. E Ono, HL Wong, T Kawasaki, M Hasegawa, O Kodama, K Shimamoto.
     Essential role of the small GTPase Rac in disease resistance of rice. Proc Natl
     Acad Sci USA 98:759–764, 2001.
47. S Gilroy, A Trewavas. Signal processing and transduction in plant cells: the
     end of the beginning? Nat Rev Mol Cell Biol 2:307–314, 2001.
48. JI Schroeder, KJ M., GJ Allen. Guard cell abscisic acid signalling and engi-
     neering drought hardiness in plants. Nature 410:327–330, 2001.
49. T Romeis, A Ludwig, R Martin, JDG Jones. Calcium-dependent protein ki-

nases play an essential role in a plant defence response. EMBO J 20:1–12, 2001.
50. T Romeis, P Piedras, JDG Jones. Resistance gene–dependent activation of a calcium-dependent protein kinase in the plant defense response. Plant Cell 12:803–815, 2000.
51. T Xing, X-J Wang, K Malik, BL Miki. Ectopic expression of an *Arabidopsis* calmodulin-like domain protein kinase enhanced NADPH oxidase activity and oxidative burst in tomato protoplasts. Mol Plant Microbe Interact 14:1261–1264, 2001.
52. SA Harding, S-H Oh, DM Roberts. Transgenic tobacco expressing a foreign calmodulin gene shows an enhanced production of active oxygen species. EMBO J 16:1137–1144, 1997.
53. WD Heo, SH Lee, MC Kim, JC Kim, WS Chung, HJ Chun, KJ Lee, CY Park, HC Park, JY Choi, MJ Cho. Involvement of specific calmodulin isoforms in salicylic acid–independent activation of plant disease resistance responses. Proc Natl Acad Sci USA 96:766–771, 1999.
54. SH Lee, JC Kim, MS Lee, WD Heo, HY Seo, HW Yoon, JC Hong, SY Lee, JD Bahk, I Hwang, MJ Cho. Identification of a novel divergent calmodulin isoform from soybean which has differential ability to activate calmodulin-dependent enzymes. J Biol Chem 270:21806–21812, 1995.
55. G Tena, T Asai, W-L Chiu, J Sheen. Plant mitogen-activated protein kinase signaling cascades. Curr Opin Plant Biol 4:392–400, 2001.
56. I Meskiene, H Hirt. MAP kinase pathways: molecular plug-and-play chips for the cell. Plant Mol Biol 42:791–806, 2000.
57. M Petersen, P Brodersen, H Naested, E Andreasson, U Lindhart, B Johansen, HB Nielsen, M Lacy, MJ Austin, JE Parker, SB Sharma, DF Klessig, R Martienssen, O Mattson, AB Jensen, J Mundy. *Arabidopsis* MAP kinase 4 negatively regulates systemic acquired resistance. Cell 103:1111–1120, 2000.
58. CA Frye, D Tang, RW Innes. Negative regulation of defense responses in plants by a conserved MAPKK kinase. Proc Natl Acad Sci USA 98:373–378, 2001.
59. Y Kovtun, W-L Chiu, G Tena, J Sheen. Functional analysis of oxidative stress–activated mitogen-activated protein kinase cascade in plants. Proc Natl Acad Sci USA 97:2940–2945, 2000.
60. K-Y Yang, Y Liu, S Zhang. Activation of a mitogen-activated protein kinase pathway is involved in disease resistance in tobacco. Proc Natl Acad Sci USA 98:741–746, 2001.
61. S Zhang, Y Liu. Activation of salicylic acid–induced protein kinase, a mitogen-activated protein kinase, induces multiple defense responses in tobacco. Plant Cell 13:1877–1889, 2001.
62. D Ren, H Yang, S Zhang. Cell death mediated by mitogen-activated protein kinase is associated with hydrogen peroxide production in *Arabidopsis*. J Biol Chem 277:559–565, 2002.
63. S Zhang, DF Klessig. MAPK cascades in plant defense signaling. Trends Plant Sci 6:520–527, 2001.

64. TS Nühse, SC Peck, H Hirt, T Boller. Microbial elicitors induce activation and dual phosphorylation of the *Arabidopsis thaliana* MAPK 6. J Biol Chem 275:7521–7526, 2000.

65. D Scheel. Oxidative burst and the role of reactive oxygen species in plant-pathogen interactions. In: D Inzé, M van Montagu, eds. Oxidative Stress in Plants. London: Taylor & Francis, 2002, pp 137–153.

66. T Yuasa, K Ichimura, T Mizoguchi, K Shinozaki. Oxidative stress activates ATMPK6, an *Arabidopsis* homologue of MAP kinase. Plant Cell Physiol 42: 1012–1016, 2001.

67. J Zhou, X Tang, GB Martin. The Pto kinase conferring resistance to tomato bacterial speck disease interacts with proteins that bind a *cis*-element of pathogenesis-related genes. EMBO J 16:3207–3218, 1997.

68. VK Thara, YQ Gu, X Tang, GB Martin, J-M Zhou. *Pseudomonas syringae* pv. *tomato* induces the expression of tomato *EREBP*-like genes *Pti4* and *Pti5* independent of ethylene, salicylate and jasmonate. Plant J 20:475–483, 1999.

69. P He, RF Warren, T Zhao, L Shan, L Zhu, X Tang, J-M Zhou. Overexpression of *Pti5* in tomato potentiates pathogen-induced defense gene expression and enhances disease resistance to *Pseudomonas syringae* pv. *tomato*. Mol Plant Microbe Interact 14:1453–1457, 2001.

70. JM Park, C-J Park, S-B Lee, B-K Ham, R Shin, K-H Paek. Overexpression of the tobacco *Tsi* gene encoding an EREBP/AP2-type transcription factor enhances resistance against pathogen attack and osmotic stress in tobacco. Plant Cell 13:1035–1046, 2001.

71. H Cao, SA Bowling, AS Gordon, X Dong. Characterization of an *Arabidopsis* mutant that is nonresponsive to inducers of systemic acquired resistance. Plant Cell 6:1583–1592, 1994.

72. TP Delaney, L Friedrich, JA Ryals. *Arabidopsis* signal transduction mutant defective in chemically and biologically induced disease resistance. Proc Natl Acad Sci USA 92:6602–6606, 1995.

73. J Glazebrook, EE Rogers, FM Ausubel. Isolation of *Arabidopsis* mutants with enhanced disease susceptibility by direct screening. Genetics 143:973–982, 1996.

74. J Ryals, K Weymann, K Lawton, L Friedrich, D Ellis, HY Steiner, J Johnson, TP Delaney, T Jesse, P Vos, S Uknes. The *Arabidopsis* NIM1 protein shows homology to the mammalian transcrition factor inhibitor I kappa B. Plant Cell 9:425–439, 1997.

75. J Shah, F Tsui, DF Klessig. Characterization of a salicylic acid–insensitive mutant (*sai1*) of *Arabidopsis thaliana*, identified in a selective screen utilizing the SA-inducible expression of the *tms2* gene. Mol Plant Microbe Interact 10: 69–78, 1997.

76. H Cao, J Glazebrook, J Clarke, S Volko, X Dong. The *Arabidopsis NPR1* gene that controls systemic acquired resistance encodes a novel protein containing ankyrin repeats. Cell 88:57–63, 1997.

77. M Kinkema, W Fan, X Dong. Nuclear localization of NPR1 is required for activation of PR gene expression. Plant Cell 12:2339–2350, 2000.

78. C Despres, C DeLong, S Glaze, E Liu, PR Fobert. The *Arabidopsis* npr1/ nim1 protein enhances the DNA binding activity of a subgroup of the TGA family of bZIP transcription factors. Plant Cell 12:279–290, 2000.
79. Y Zhang, W Fan, M Kinkema, X Li, X Dong. Interaction of NPR1 with basic leucine zipper protein transcription factors that bind sequences required for salicylic acid induction of the *PR-1* gene. Proc Natl Acad Sci USA 96:6523–6528, 1999.
80. J-M Zhou, Y Trifa, H Silva, D Pontier, E Lam, J Shah, DF Klessig. NPR1 differentially interacts with members of the TGA/OBF family of transcription factors that bind an element of the *PR1* gene required for induction by salicylic acid. Mol Plant Microbe Interact 13:191–202, 2000.
81. H Cao, X Li, X Dong. Generation of broad-spectrum disease resistance by overexpression of an essential regulatory gene in systemic acquired resistance. Proc Natl Acad Sci USA 95:6531–6536, 1998.
82. M-S Chern, HA Fitzgerald, RC Yadav, PE Canlas, X Dong, PC Ronald. Evidence for a disease-resistance pathway in rice similar to the *NPR1*-mediated signaling pathway in *Arabidopsis*. Plant J 27:101–113, 2001.
83. K Shinozaki, K Yamaguchi-Shinozaki. Gene expression and signal transduction in water-stress response. Plant Physiol 115:327–334, 1997.
84. K Yamaguchi-Shinozaki, K Shinozaki. A novel *cis*-acting element in an *Arabidopsis* gene is involved in responsiveness to drought, low temperature, or high-salt stress. Plant Cell 6:251–264, 1994.
85. Q Liu, M Kasuga, Y Sakuma, H Abe, S Miura, K Yamaguchi-Shinozaki, K Shinozaki. Two transcription factors, DREB1 and DREB2, with an EREBP/AP2 DNA binding domain seperate two cellular signal transduction pathways in drought- and low-temperature-responsive gene expression, respectively, in *Arabidopsis*. Plant Cell 10:1391–1406, 1998.
86. K Yamaguchi-Shinozaki, K Shinozaki. Characterization of the expression of a desiccation-responsive *rd29* gene of *Arabidopsis thaliana* and analysis of its promoter in transgenic plants. Mol Gen Genet 236:331–340, 1993.
87. M Kasuga, Q Liu, S Miura, K Yamaguchi-Shinozaki, K Shinozaki. Improving plant drought, salt, and freezing tolerance by gene transfer of a single stress-inducible transcription factor. Nat Biotechnol 17:287–291, 1999.
88. EJ Stockinger, SJ Gilmour, MF Thomashow. *Arabidopsis thaliana CBF1* encodes an AP2 domain–containing transcriptional activator that binds to the C-repeat/DRE, a *cis*-acting DNA regulatory element that stimulates transcription in response to low temperature and water deficit. Proc Natl Acad Sci USA 94:1035–1040, 1997.
89. KR Jaglo-Ottosen, SJ Gilmour, DG Zarka, O Schabenberger, MF Thomashow. *Arabidopsis CBF1* overexpression induces *COR* genes and enhances freezing tolerance. Science 280:104–106, 1998.
90. JC Kim, SH Lee, YH Cheong, C-M Yoo, SI Lee, HJ Chun, D-J Yun, JC Hong, SY Lee, CO Lim, MJ Cho. A novel cold-inducible zinc finger protein from soybean, SCOF-1, enhances cold tolerance in transgenic plants. Plant J 25:247–259, 2001.

91. L Nover, KD Scharf. Heat stress proteins and transcription factors. Cell Mol Life Sci 53:80–103, 1997.
92. R Prandl, K Hinderhofer, G Eggers-Schumacher, F Schöffl. HSF3, a new heat shock factor from *Arabidopsis thaliana*, derepresses the heat shock response and confers thermotolerance when overexpressed in transgenic plants. Mol Gen Genet 258:269–278, 1998.
93. I Winicov. cDNA encoding putative zinc finger motifs from salt-tolerant alfalfa (*Medicago sativa* L.) cells. Plant Physiol 102:681–682, 1993.
94. DR Bastola, VV Pethe, I Winicov. Alfin1, a novel zinc-finger protein in alfalfa roots that binds to promoter elements in the salt-inducible *MsPRP2* gene. Plant Mol Biol 38:1123–1135, 1998.
95. I Winicov, DR Bastola. Transgenic overexpression of the transcription factor *Alfin1* enhances expression of the endogenous *MsPRP2* gene in alfalfa and improves salinity tolerance of the plants. Plant Physiol 120:473–480, 1999.
96. I Winicov. Alfin1 transcription factor overexpression enhances plant root growth under normal and saline conditions and improves salt tolerance in alfalfa. Planta 210:416–422, 2000.
97. M Yanofsky. Floral meristems to floral organs: genes controlling early events in *Arabidopsis* flower development. Annu Rev Plant Physiol Plant Mol Biol 46:167–188, 1995.
98. D Weigel, OA Nilsson. A developmental switch for flower initiation in diverse plants. Nature 377:495–500, 1995.
99. MA Mandel, M Yanofsky. A gene triggering flower formation in *Arabidopsis*. Nature 377:522–524, 1995.
100. L Peña, M Martín-Trillo, J Juárez, JA Pina, L Navarro, JM Martínez-Zapater. Constitutive expression of *Arabidopsis LEAFY* or *Apetala1* genes in citrus reduces their generation time. Nat Biotechnol 19:263–267, 2001.

19

The Plant Cell Wall—Structural Aspects and Biotechnological Developments

Bruno M. Moerschbacher
Department of Plant Biochemistry and Biotechnology,
Westphalian Wilhelm's University Munich, Munich, Germany

I. INTRODUCTION

Plant cell walls are made up of the world's most abundant and most durable organic materials. Small wonder that they have been used by man since the dawn of time. Cell wall–derived materials played crucial roles in the cultural evolution of man, and they continue to be integrated yet often unappreciated ingredients of modern everyday life. Examples are wood as an energy source, used as construction material, or made into pulp and paper; cotton, linen, hemp, ramie, and sisal woven into tissues and/or made into strings and ropes; alginic acid and pectins used as gelling materials; and dietary fibers required for healthy nutrition. Cell walls are essential in material sciences as well as in food and feed technology.

Cell walls evolved to fulfill a range of important tasks in plants, both structural and functional, explaining their versatility when exploited by man. It is clear, however, that plant cell walls never evolved to fulfill the roles we use them for, so that there is a great potential for optimizing cell wall components or architecture for our purposes. This optimization can be done and has been done using traditional breeding procedures leading to materials such as long-staple cotton varieties. Modern methods of genetic engineering will now open up new opportunities that we are only beginning to realize.

Cellulose, the stress-bearing fiber component of plant cell walls, is estimated to be produced at a rate of 10^{10} tons per year, and the existing mass of approximately 10^{11} tons far exceeds that of any other molecule of organic origin. The title "second most abundant organic molecule on earth" is given either to chitin, the fiber-forming component of fungal cell walls, or to lignin, the load-bearing matrix component of many secondary plant cell walls. The different hemicelluloses found in plant cell walls together form about as much biomass as cellulose, and pectins make up another sizable portion of organic plant material. Even structural proteins such as extensin, which are only a minor component of plant cell walls, have been considered to be among the most abundant proteins on earth.

In spite of the enormous importance of cell walls and of our ever increasing knowledge of the components that make up plant cell walls, surprisingly little is known about the biosynthetic machinery responsible for polymerizing the cell wall constituents, and even less is known about their integration and assembly into the architecture of the complex organelle that is a cell wall. Almost terra incognita is the turnover and modification of plant cell wall components during plant cell division, growth, differentiation, senescence, and death. Although the paucity of information available is a serious drawback in genetically optimizing plant cell walls for biotechnological uses, it offers a large and promising field for future studies of both fundamental and applied relevance.

A decade ago, the situation appeared to be far more advanced for lignin biosynthesis. The enzymes necessary to produce the three different monolignols and to polymerize them into the three-dimensional network of the lignin polymer all appeared to be known, and most of the corresponding genes had been cloned from different plant species. Consequently, the genetic engineering of designer lignin seemed an easy task, and studies were initiated to overexpress or silence these genes and to analyze the lignin of the transgenic plants. To our surprise, however, these molecular genetic experiments revealed that the process of monolignol synthesis is far more complex than previously thought—and today we realize that we know less about lignin biosynthesis than we thought we knew. Genetic engineering of lignin remains a challenging and promising task. This is the subject of Chap. 20 of this book, and lignin will therefore not be considered in more detail here.

Drawing up possible biotechnological uses of plant cell walls first requires an understanding of the roles cell walls play in the life of a plant. These roles have driven the evolution of the components and architecture of the cell wall, and we will have to understand which components fulfill the different roles and how they do their job cooperatively in the complex cell wall system. From this knowledge we can define strategies to improve plant performance, and we can deduce possible ex planta uses for the different cell wall components. If optimization is to be achieved by molecular genetic engineering, we have to characterize the enzymatic machinery that builds, assembles, and modifies the cell wall components. The genes coding for these enzymes will eventually be the target for manipulation of plant cell wall components or architecture with the goals of improving plant performance and of obtaining higher quality products specifically optimized for uses in material sciences or food and feed technology.

This chapter aims at giving a—necessarily nonexhaustive—overview of the many roles cell walls fulfill in planta and the many possible ex planta uses for plant cell walls (Fig. 1). It will try to highlight both the enormous opportunities and challenges for the genetic engineering of plant cell walls. Necessarily, this treatise will have to be somewhat cursory, and the reader will, whenever possible, be referred to recent more focused and more detailed reviews.

II. FUNCTIONS OF PLANT CELL WALLS IN PLANTA

Cell walls have long been considered as dead extracellular material possessing solely static structural roles, namely to counteract the turgor pressure and thus give strength and form to plant cells, tissues, organs, and organisms. This view has changed dramatically with the insight modern analytical techniques and molecular probes have allowed into the fine structure and apparent highly complex spatial and temporal regulation of cell wall components and architecture (1). It was concluded that the cell wall must have important functional roles, e.g., in cell-cell communication, transport of metabolites, differentiation, and development. To indicate this new appreciation of the cell wall as a dynamic and regulatory extracellular organelle, some authors prefer to call it an extracellular matrix. The growing realization that this matrix forms an integral and essential part of the plant cell may eventually even lead us to abandon the term extracellular in favor of, e.g., pericellular.

In spite of these considerations, plant cell walls do play important static structural roles. First, they have to bear the turgor pressure, which results in enormous tangentially oriented stress forces in the cell wall plane (2). This stress is taken up by the fibrillar component in the cell wall, namely the cellulose microfibrils, which are interconnected and held in place by hemicelluloses. Second, the cell wall has to take up the pressure exerted on a plant cell by the surrounding tissue. This pressure is taken up by the matrix in which the fibrils are embedded, made up of polysaccharides containing uronic acids, namely pectins and glucuronoarabinoxylans, or—more precisely—by the water molecules held in the cell walls by these negatively

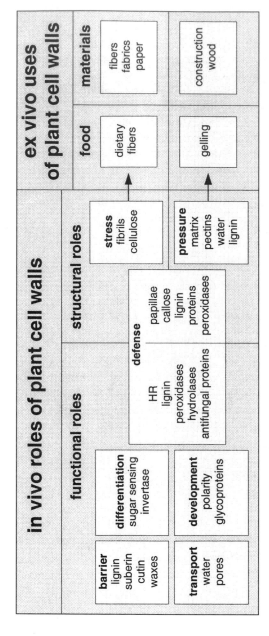

FIGURE 1 In vivo roles and ex vivo uses of plant cell walls and their components. Cell walls fulfill many roles—both structural and functional—in the life of a plant, and different components are responsible for the different roles. Biotechnology may aim at improving these in vivo roles according to human interest, e.g., increasing resistance of crop plants against pathogens (HR, hypersensitive reaction). Ex vivo, different cell wall components can be used according to their physical properties and, again, biotechnology may aim at improving these properties.

charged polymers. In extreme cases, this water is replaced by the incorporation of lignin, a heavily cross-linked three-dimensional polyphenolic network able to withstand very high pressures.

As the hydrophobic lignin replaces the water in the cell wall, lignification also renders cell walls impermeable to water. Other ways to achieve the same goal—impermeability to water—are the incrustation with suberin or the deposition of cutin along with cuticular and epicuticular waxes. These layers form essential barriers to the free transport of water and solutes, such as nutrients and assimilates. Suberin in the Casparian strand of the endodermis effectively seals the central cylinder of the root from the surrounding cortex tissue, allowing control over the uptake of nutrients and water from the soil. Lignin in the tracheary elements of the xylem allows water and solute transport from the roots to the aboveground plant parts by stabilizing the vessel cell walls against the strong internal negative pressure of the transpiration stream and by preventing losses to the surrounding tissue. Cutin and the waxes protect the aboveground plant organs from desiccation by restricting the transpiration to the stomata with their regulated diffusion resistance.

Cell walls restrict and dictate cell size and form (3). Consequently, growth is possible only when the cell walls partially yield to turgor pressure. In effect, plant growth is growth of the plant cell walls. Similarly, the shape of a plant is formed by the shape of the plant cell walls. The processes taking place in the cell wall during growth have been the subject of intense research over many decades (2). Although we are still far from understanding this highly complex and tightly regulated process, it appears today that the hemicelluloses interconnecting cellulose fibrils play a crucial role in this process. Enzymes loosening and retightening these connections, such as expansins and xyloglucanendotransglycosylases, are critically involved in cell wall growth (4–6). Shape of a plant cell, on the other hand, appears to be dictated by the orientation of the cellulose fibrils that is most likely brought about by the movement of the cellulose synthase complex in the plasma membrane, which may or may not be guided and driven by the cytoskeleton (7,8).

In contrast to the preceding morphological roles of plant cell walls, our understanding of the functional roles of cell walls in the physiology of a plant still remains rather patchy. It is clear that cell walls both separate and connect neighboring cells in a tissue, and it can be anticipated that information and metabolites travel from cell to cell both symplastically, i.e., via plasmodesmata, and apoplastically, i.e., across the cell wall. One example of apoplastic transport of metabolites is sucrose on its way from source mesophyll cells to the phloem and from the phloem to sink cells in the root or fruit tissues (9). Sucrose in the cell wall may be hydrolyzed into its constituent hexoses by apoplastic invertase (10), and the glucose produced can act as a local signal in sugar sensing influencing the metabolic state of the plant cells (11,12). Another example of information-bearing molecules traveling in the apoplast is auxin, which is thought to be moved through the plant by polar secretion into the cell wall and uptake from the cell wall by adjacent cells (13).

Although the water present in the cell wall provides the space for intercellular travel of molecules, transport in the cell wall is not unrestricted. We have already seen that special elaborations of the cell wall, such as suberin in the Casparian band, can completely stop flow of water and solutes. In addition, the pore size in the cell wall determines the diffusion rate of larger molecules, such as proteins. The size exclusion limit of plant cell walls has been estimated for globular proteins at between 10 and 30 (and sometimes up to 100) kDa (14). Different types of cross-links between pectins appear to be responsible for the porosity of the plant cell walls (15). Phenolic cross-links possibly involving tyrosine residues of cell wall (glyco)proteins have been postulated to restrict porosity further (16). It remains to be seen whether plant cells are able to change the pore size of their surrounding cell walls actively in reaction to internal or external stimuli such as pathogen attack (17).

Immunological approaches have shown that the wall surrounding a plant cell is far from being a uniform structure. Owing to the availability of a set of two complementary antibodies, the distribution of methyl esterified versus non–methyl esterified pectins has been analyzed in many plant tissues, and distinct distributions both over the thickness and in the plane of plant cell walls have been found (18). Even more spectacular are the distributions of certain proteins that form minor constituents of the cell walls. These epitopes are sometimes restricted to a very small number of cells and their presence may indicate or dictate determination of cell fate, e.g., to become metaxylem cells (19,20). Also, cell polarity may be determined by markers immobilized in the cell wall, as evidenced by a unique localized cell wall domain stabilizing the spatial orientation of the fucus zygote (21). The role of cell walls in the development of plant cells, tissues, organs, and organisms is likely to be a most deserving field for future research.

Cell walls are also of paramount importance in the defense of plants against pests and pathogens and in their tolerance to stress (22). Forming the outer shell of plant cells and, consequently, of total plants, cell walls—and most notably outer epidermal cell walls—are the first barrier put up by the plant against external biotic and abiotic threats. Cell walls are the single most important protection of plants against the myriads of microorganisms that otherwise would be potential plant pathogens. Whereas successful pathogens must have evolved means to degrade and/or penetrate standard plant cell walls, plants have counteracted this attack by a multitude of pre- or postinfection cell wall modifications, such as callose deposition to form a periplasmic papilla, production of active oxygen species in an odixative burst for the peroxidative cross-linking of phenolic acids and proteins, or deposition of suberin or lignin, which as a true polymer is extremely difficult to degrade enzymatically. Active oxygen species produced in the cell wall have also been supposed to act as second messengers inducing the hypersensitive reaction (23), a form of programmed cell death involved in plant disease defense that in some cases may be brought about by the intracellular performance of a process usually confined to the cell wall, namely radical coupling of monolignols to the lignin polymer (24).

The cuticle including the epicuticular wax layers does not only protect plants from desiccation. The microroughness of the plant surface brought about by the wax crystals creates the *Lotus* effect ensuring that dust particles and fungal spores are easily washed away by runoff raindrops (25). The extreme hydrophobicity of the epicuticular wax prevents the formation of a continuous water layer so that motile bacteria cannot easily reach stomates (26). Wax protrusions may protectively cover stomatal openings, effectively reducing stomatal transpiration and preventing recognition of stomates by fungal pathogens searching for easy ingress points (27). Cell walls form the arena where potential microbial pathogens deploy their pathogenicity factors, often a plethora of plant cell wall hydrolyzing enzymes, and where host plant cells mount their defenses, e.g., in the form of pathogenesis-related (pr-) proteins such as inhibitors of the microbial hydrolases or hydrolytic enzymes attacking the microbial cell walls (28).

Not all of the microbes interacting with a plant are potential foes. Both the rhizosphere and the phylloplane are complex ecosystems colonized by a multitude of mutualistic or commensal microbes. The interactions of these microorganisms with the plant are surface interactions and, as such, interactions of the cell walls of both partners, at least initially. However, little is known about the molecular details of these interactions, except for arbuscular mycorrhiza and nitrogen-fixing rhizobia interacting with plant root cells. The fungi and bacteria involved in these interactions form symbiosome structures within the host cells, similar to the haustorial complexes of biotrophic pathogenic fungi such as the rusts and mildews (29). Cell wall penetration must be accomplished by these symbiotic microorganisms in a manner preventing the elicitation of the active resistance mechanisms just described for the defense against pathogenic microorganisms. Mutualists may achieve this by actively suppressing plant defense responses such as callose deposition (30) or by inducing cell wall autolytic processes within the plant cell (31) reminiscent of phragmosome formation during cell division (32). Similarly, successful biotrophic pathogens must continuously suppress resistance reactions by the penetrated host cells, and they may achieve this by generating plant cell wall fragments acting as endogenous suppressors of disease resistance elicitation (33).

Cell walls do not only form the stage of interaction between plant cells and microbial symbionts and pathogens. Cell walls of the male pollen tube and the female pistil cells are also the site where self-recognition in the self-incompatibility system prevents self-pollination (29). This system has been elucidated on the molecular level in *Brassica*, where a receptor kinase in the plasma membrane of the stigma epidermal cells (34) cooperates with a stigma cell wall glycoprotein (35) in the recognition of a cysteine-rich pollen cell wall protein (36). The outcome of this molecular interaction is most likely a multiple response certainly including cell wall modifications that eventually arrest further pollen tube growth (37).

The examples described here are recent corroborations of the much older oligosaccharin concept postulating oligosaccharide fragments from plant cell walls as hormone-like signal molecules influencing plant metabolism and development (38). Intermediate-sized products of fungal endo-polygalacturonase digestion of plant pectins act as endogenous elicitors of active defense reactions in plants. Small products of pectin digestion act as endogenous suppressors of elicitor-induced plant resistance reactions. The biological activity of the endogenous suppressors that are beneficial to the fungal plant pathogens but not to the host plant suggests that these may mimic as yet unknown endogenous plant signals involved in determining differentiation events in plant cells. Indeed, oligomeric fragments of pectin and xyloglucan have been implicated as tissue hormones regulating plant growth and development. Similarly, the nodulation factors of rhizobia initiating meristem formation may mimic endogenous plant signal molecules. It has been speculated that the diverse family of plant arabinogalactan proteins may form a cell wall–located reservoir of such signal molecules involved in local differentiation and development.

III. COMPONENTS AND ARCHITECTURE OF PLANT CELL WALLS

In reviewing the structural and functional roles of plant cell walls, we have already named most of their constituents, and we have indicated which of the different components contribute to the specific functions. In this section, we will briefly summarize our current knowledge of the structure of these components and our understanding of how they interact to assemble into the complex and dynamic three-dimensional pericellular matrix. Over the past decades, a number of ever more refined models of plant cell wall architecture have been proposed, mostly based on detailed investigations of the primary cell walls of suspension-cultured cells (2,39–42). This prototype cell wall of dicot plants is opposed by the somewhat different cell wall of some— the commelinoid—monocot plants, most notably the grasses and cereals (43). The general architectural plan of both types of cell walls, however, is identical: stress-bearing fibrils are embedded in a pressure-bearing matrix (Fig. 2).

In both types of plant cell walls, the fibrillar component is represented by cellulose microfibrils, accounting for about one third of the cell wall dry

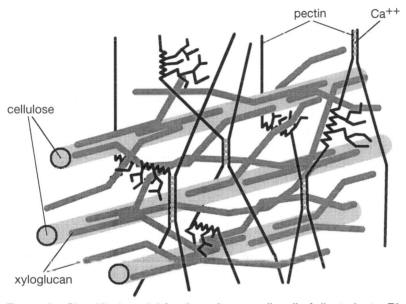

FIGURE 2 Simplified model for the primary cell wall of dicot plants. Plant cell walls are composite materials consisting of stress-bearing fibers made of cellulose embedded in a pressure-bearing matrix consisting of hemicelluloses and pectins. Parallel linear cellulose molecules form the crystalline core of the microfibrils. In a typical primary cell wall of dicot plants, these are thought to be covered by hemicellulosic xyloglucan molecules. Individual xyloglucan molecules may span the interfibrillar space, their ends hydrogen bonded to the surface of different cellulose microfibrils, thus forming a cellulose-xyloglucan net. Hydrophilic, galacturonic acid–containing pectin molecules consisting of linear "smooth" regions and branched "hairy" regions fill in the meshes of this net. The smooth homogalacturonan regions can cross-link via Ca^{2+} bridges, thus forming a second, independent pectin-Ca^{2+} net. A third, independent net of cross-linked glycoproteins may exist in the primary cell wall, possibly involved in defining the precise spacing of the polysaccharide components.

weight. In the dicot cell wall, each cellulose microfibril is thought to be completely wrapped in an envelope of xyloglucan chains mediating the surface interaction between the fibrillar component and the surrounding matrix (44). Xyloglucan molecules are thought to span the interfibrillar space, their ends hydrogen bonding to different cellulose microfibrils. The cellulose and xyloglucan molecules thus form a three-dimensional scaffolding network in the dicot cell wall. In commelinoid monocot cell walls, xyloglucan is mostly replaced by glucuronoarabinoxylan chains, which are similarly hydrogen bonded to several cellulose microfibrils, thus leading to a cellulose-xylan network (45).

The cellulose-xyloglucan network of the dicot cell wall is embedded in the pectic matrix, which makes up about one third of the dry weight of the cell wall (46). The pectins are believed to form a second, independent network in the cell wall that may not be covalently linked to the cellulose-xyloglucan network. The intermolecular cross-links knitting pectic polysaccharides together are most likely Ca^{2+} bridges between nonesterified stretches of homogalacturonan (47), hydrophobic interactions between methyl esterified stretches of homogalacturonan, and, possibly, boron diesters between rhamnogalacturonan II stretches (48). Commelinoid monocot cell walls contain the same pectic polymers as dicot walls but at much lower contents (below 10%). Their role as a water-binding matrix component is substituted by the major hemicellulose, glucuronoarabinoxylan, which alone can account for about half of the dry weight of the commelinoid monocot wall (49). It can be expected that the same types of interpectic cross-links exist in the commelinoid monocot cell wall as discussed for the dicot wall. Many of the arabinose residues of the glucuronoarabinoxylan carry ferulic acid or p-coumaric acid esters, and these may dimerize oxidatively to form diphenolic acid cross-links possibly substituting for missing interpectic links (50). Thus, a pectin and/or a glucuronoarabinoxylan–phenolic acid network may exist in commelinoid monocot cell walls.

Both types of cell wall contain structural glycoproteins—roughly 10–20% of the dry weight of the dicot wall compared with about 2–10% in the commelinoid monocot wall—which may form a third independent network in the primary plant cell wall (51). The major protein component of dicot walls appears to be the hydroxyproline-rich glycoprotein (HRGP) extensin, but proline-rich proteins (PRPs) and glycine-rich proteins (GRPs) may be involved as well. The only cross-links proposed so far to hold the protein network together are intermolecular isodityrosine bridges, but their existence still awaits experimental support.

The classical cell wall models depicting primary plant cell walls as a largely covalently cross-linked system of the constituent polysaccharides and proteins have been extended and improved to yield a composite cell wall model with a more flexible and dynamic architecture (2). According to our current view, primary cell walls of both dicot and monocot plants are built of the three independent interwoven networks already described—the cellulose-glycan network, the pectin network, and the proteinaceous network —held together by noncovalent links. Primary cell walls are capable of enormous rates of largely planar, two-dimensional growth, believed to be restricted by the cellulose-glycan network. In dicot cell walls, controlled activity of expansins and xyloglucanendotransglycosylases is supposed to allow controlled yielding of this network, so that the orientation of the cellulose microfibrils determines the direction of cell elongation. In contrast, some of the xyloglucan-poor monocot cell walls may, during phases of elongation growth, produce an additional polysaccharide component—a mixed-linkage β-1,3-β-1,4-glucan—that may hydrogen bond to cellulose and may transiently replace the xylan, which appears to be turned over rapidly during growth (52,53).

In contrast to the artificial situation of plant cells growing in liquid suspension where all cells are permanently in a similar state of differentiation, preferentially rapid spherical growth, cells of many different types, and in many different states of differentiation, interact with each other to form a highly complex tissue in the intact plant. Cultured plant cells are surrounded only by a primary cell wall. In a plant tissue, all cells still in the process of growing and even many fully differentiated cells, such as leaf mesophyll cells, also contain only a primary and no secondary cell wall. However, many specialized plant cells produce highly sophisticated secondary cell walls, e.g., collenchyma and sclerenchyma cells, tracheids, and pollen cells. Moreover, all the different cells that make up a tissue are "glued" together by the middle lamella, a structure that may not be present in cultured cells, at least if they grow as a very fine suspension of individual cells. Furthermore, it appears that the primary wall of a cell in a tissue is much less uniform than that of a cell grown in suspension culture.

Immunocytological electron microscopic studies using antibodies to different cell wall components revealed microdomains in the seemingly uniform primary cell wall exhibiting striking differences in their molecular composition (54). Antibodies cross-reacting with non-methyl esterified epitopes of pectin stained the middle lamella, especially in cell corners, but also lined the inner surface of the primary cell walls along the plasma membrane. Antibodies cross-reacting with highly methyl esterified pectic components, in contrast, evenly bound across the whole width of the primary cell wall. Cell wall angles where three or more different cells meet are distinct from cell wall stretches in contact with a single neighboring cell, and these again differ from areas that face the intercellular space. The mature primary cell wall of about 80 nm thickness consists of only about three layers of cellulose microfibrils (55). Each layer, thus, resides in a different nanoenvironment, the outer layer facing the pectic middle lamella, the middle layer surrounded by the other two, and the inner layer facing the plasma membrane or, later, the secondary cell wall layers. Virtually nothing is known about the inter-action of the primary cell wall with these neighboring layers. These novel micro- and nanoscopic approaches will certainly shed new light on our un-derstanding of the plant cell wall, but they can be expected to pose more questions initially than they will answer—e.g., concerning assembly and regulation of this enormous complexity outside the cytoplasm.

To make things even more complex, many cells that fulfill special functions within a plant tissue exhibit striking and often highly localized modifications of the standard primary plant cell wall described so far. Epi-dermal cells have to play multiple roles in growth restriction and stress protection, necessitating modifications of the outer primary cell walls. These are thicker than the other epidermal cell walls, and the cutin monomers and cuticular and epicuticular waxes produced in epidermal cells are transported to the outer surface of this outer periclinal wall to reach the plant surface, where they are laid down in layers and where the cutin is polymerized in situ. One of the most prominent examples of secondary cell walls is certainly the intricate elaboration of cell wall thickenings in tracheary elements of the xylem, where additional cellulose fibers and embedding matrix polymers are laid down only on certain parts of the cell walls (56). Eventually, lignin is incorporated into the walls of these cells. The polymerization of the mono-lignols in muro leads to a growing lignin polymer infiltrating the primary cell wall, displacing the water in the process. Eventually, when the poly-merization process is complete, the lignin polymer serves to cement in place all the other components of the cell wall. The wall is then no longer per-meable to water, and the mature tracheid dies. The ensuing loss of turgor pressure would result in a shrinking of the dead cell and most likely a collapse due to the negative pressure (tension) of the transpiration stream, were it not for the lignin impregnating and stiffening the cell wall of the tracheid.

Clearly, the wall of an individual cell is a highly differentiated and coordinated organelle with domains differing in structure and—most likely —also in function. In addition, we can expect a highly dynamic temporal differentiation to be superimposed on this complex spatial differentiation. Cells are born in meristematic zones where cell division occurs, they un-dergo elongation and differentiation, they may eventually senesce, and they will finally die. The wall of a cell most likely changes continuously during these processes, but today little is known of what these changes are and what they may entail.

IV. BIOSYNTHESIS AND ASSEMBLY OF PLANT CELL WALLS

The daunting complexity of the many constituents that make up plant cell walls is rivaled by an equally impressive complexity of genes potentially involved in their synthesis, as revealed by the recent completion of the *Arabidopsis* genome sequencing project. However, little is known about the actual biochemistry of cell wall component biosynthesis, and virtually nothing is known about the processes involved in the assembly of these components into the three-dimensional complex plant cell wall (57). In the plant cell cycle, new cell walls are typically produced during mitosis when the internuclear anaphase spindle is transformed into the phragmoplast (58,59). Within this cytoskeletal apparatus, the new cell plate is assembled from cell wall material synthesized and delivered by the fusing Golgi-derived vesicles. No reliable information is available on the biochemical nature of the cell wall material initially laid down in the outward-expanding cell plate. Circumstantial evidence suggests that callose, hemicelluloses, and pectins are involved, and the finding of a membrane-bound β-1,4-glucanase in the cell plate (60) may even point to cellulose already being synthesized during the birth of the new cell wall.

For decades, the search for cellulose synthase has resembled the quest for the Holy Grail of plant biochemistry (61). The rosette structure seen in the electron microscope at the end of newly deposited cellulose microfibrils has long been suspected to harbor cellulose synthase, but biochemical evidence was and still is lacking. The identification of the first plant gene with sequence similarity to bacterial cellulose synthase only a few years ago (62) has been followed by a deluge of related sequences forming the cellulose synthase superfamily (63–65). At least 12 presumed cellulose synthase genes and around 30 members of the six classes of cellulose synthase–like genes have been identified in the *Arabidopsis* genome (65). They all belong to the class of multipass transmembrane proteins with a large, presumably cytoplasmic catalytic domain.

An antibody raised against a recombinant cellulose synthase confirmed the presence of this enzyme in the rosette structure (66), which is now thought to be composed of six complexes, each with six catalytic subunits (67). It is still unknown whether one or two catalytic sites are involved in chain elongation and whether the chain grows by the addition of one or two glucose units at a time (68). It is generally assumed, though, that the rosette structure allows transmembrane synthesis of an entire cellulose microfibril containing a few dozen parallel cellulose chains through the plasma membrane into the cell wall. The apoplastic self-assembly of the fibril, thus, may be intimately coupled to the biosynthesis of the cellulose molecules. The surface of the nascent microfibril might be immediately covered by hemi-celluloses present on the outside of the plasma membrane, ensuring the individual identity of the nascent fibril by preventing its fusion with con-comitantly synthesized neighboring microfibrils.

The different cellulose synthase genes appear to be expressed in dif-ferent cell types, depending on their state of differentiation and development (69). Different isoenyzmes appear to be involved in the synthesis of cellulose microfibrils for primary and secondary cell walls that differ both in the length of the individual cellulose chains and in the number of chains in individual microfibrils (70). Interestingly, the cellulose synthases may op-erate in pairs in both cases, possibly forming dimers in the rosette structure (69). A mutant unable to assemble rosette structures but unimpaired in cel-lulose biosynthesis forms noncrystalline cellulose aggregates instead of crys-talline microfibrils (71).

Unlike the plasma membrane localization of cellulose synthesis, non-cellulosic polysaccharides are generally believed to be synthesized in the Golgi apparatus (72). This is in agreement with delivery of hemicelluloses and pectins and/or their synthesizing enzymes with Golgi vesicles to the cell plate during mitosis. As for cellulose synthase, the glycosyltransferases in-volved have long resisted biochemical approaches of purification and char-acterization. The first two glycosyltransferases involved in hemicellulose side chain attachment and elongation have only recently been isolated and characterized (73,74). Both enzymes are membrane anchored by a single transmembrane domain near the N-terminus, with the catalytic domain most likely oriented toward the interior of the Golgi vesicles (74–76). This raises the interesting question of how the sugar-nucleotide substrates enter the ves-icles (77).

No biochemical data are yet available on the biosynthesis of the hemi-cellulose backbones, but the multitude of genes encoding cellulose syn-thase–like glycosyltransferases leaves plenty of room for enzymes involved in the generation of linear β-glycosidically linked glycans (63,78). Presum-ably, these cellulose synthase–like polymerases are located in the Golgi vesicle membranes. Their orientation, however, would lead to polysaccharide synthesis in the cytoplasm, as these enzymes are not believed to channel their products through the membrane. The linear glucan and xylan polymers would have to be imported into the Golgi lumen, where the glycosyltrans-ferases just described could attach and elongate side chains to build xylo-glucans and glucuronoarabinoxylans (79). The distribution of the different side chains along the linear backbone of the hemicelluloses is thought to control their roles in the overall architecture of the cell wall. However, noth-ing is known about the regulation of this distribution.

Possibly even less is known about pectin biosynthesis (72). Having a backbone of α-glycosidically linked galacturonic acids, interspersed or not with α-glycosidically linked rhamnose residues, these cannot be synthesized by any of the cellulose synthase—like family-2 glycosyltransferases that produce β-linked glycans from α-linked UDP-sugar donors (80). Homogalacturonan biosynthesis has been achieved in vitro at very low rates (81,82), and pectin methyltransferase activity was associated with the Golgi preparations used (83). Consequently, the pectin is presumed to be delivered into the cell wall in a highly methyl esterified state. Apoplastic pectin methyl esterases are then believed to convert these immature pectin polymers to the diverse partially methyl esterified pectins found in the different domains of mature plant cell walls (84). The presence of up to a dozen isoenzymes of pectin methyl esterase may be related to the many roles pectins have been suggested to play in the physiology of plant tissues.

Nothing is known concerning the biosynthesis of the linear backbone of rhamnogalacturonan I consisting of the α-glycosidically linked repeat disaccharide galacturonic acid $\rightarrow$ rhamnose. The rhamnogalacturonan I backbone appears to be contiguous with the backbone of homogalacturonan (85), posing the interesting question of a possible switch in enzyme activity during biosynthesis. Moreover, those homogalacturonan regions close to rhamnogalacturonan appear to be acetylated but not methylesterified (86)— another unresolved riddle of regulation. Microsomal membranes have been shown in vitro to catalyze the attachment and elongation of galactan side chains to the rhamnogalacturonan I backbone (87), but no activities have been shown to account for arabinan and arabinogalactan side chains. Perhaps the most complex and most puzzling plant cell wall polysaccharide is rhamnogalacturonan II, possessing a linear homogalacturonan backbone decorated with an apparently strictly controlled sequence of four highly complex side chains (48). Unsurprisingly, our knowlegde of its biosynthesis is nil.

Much easier to understand is the biosynthesis of cell wall proteins. These are conventionally polymerized on ribosomes of the rough endoplasmic reticulum and are delivered to the cell wall via vesicle trafficking through the Golgi apparatus (51). Nonstructural cell wall proteins, such as invertase or antimicrobial hydrolytic enzymes, are typically glycoproteins, and the typical plant oligosaccharide side chains are transferred to the polypeptides via the usual dolichol pathway. Secretion into the cell wall appears to be the default pathway for proteins entering the secretory pathway; no further sorting signals appear to be required (88). The situation is less clear for the structural cell wall proteins. Often, these do not carry the typical oligomeric N-linked glycosidic side chains mentioned earlier. Instead or in addition, they may carry O-glycosidically linked oligomeric or polymeric glycan moieties. Moreover, hydroxyproline-rich glycoproteins are characterized by an additional posttranslational modification, the hydroxylation of some but not all proline residues. Again, these modifications appear to be carried out in the endoplasmic reticulum and in the Golgi apparatus (89).

Golgi vesicles contain large amounts of plant proteoglycan—members of the large family of arabinogalactan proteins (AGPs) (90). Their relatively high concentration in the vesicles is in contrast to the low abundance of these proteoglycans in the cell wall. AGPs may, therefore, be interesting candidates for chaperone-like assembly assistants preventing premature self-assembly of cell wall matrix material in the Golgi vesicles and/or ensuring correct assembly of the cell wall components.

Although our knowledge of the synthesis of most cell wall polymers is sketchy at best, we seem to know a little more about the biosynthesis of the monomeric precursors of these polymers. All nucleotide sugars required for polysaccharide biosynthesis are produced from UDP- and GDP-glucose, by the action of epimerases, dehydratases, oxidoreductases, and carboxylases—or they are generated in salvage pathways from cell wall degradation products by the action of kinases and nucleotide pyrophosphorylases (91,92). The nucleotide sugars are synthesized in the cytosol, as are the phenylpropenol monomers of lignin. In contrast, cutin and wax monomers are synthesized in the endoplasmic reticulum, and the hydrophobic, water-insoluble components are most likely transported through the outer epidermal cell wall to the surface of the plant via lipid transfer proteins (93). The only unusual amino acid found in structural plant cell wall proteins is hydroxyproline, but hydroxylation occurs posttranslationally so that no specific amino acid biosyntheses are involved in cell wall generation.

V. BIOTECHNOLOGICAL APPROACHES TO OPTIMIZE CELL WALL PERFORMANCE IN PLANTA

The oldest known fossils are cells with a wall. Cell walls had over 3.5 billion years of time to evolve, so that they can be expected to be fairly optimized for the many roles they play in a plant's life. It may appear totally presumptuous and vain to try to further optimize this admirable pericellular organelle. However, as with all biological structures and systems, the cell wall components and architecture are optimized to fulfill diverse and sometimes contradictory roles concomitantly, so that the result represents an optimal compromise. Consequently, the outcome varies depending on the environmental situation of the species, individual plant, and individual cell—giving rise to the described complexity and diversity of plant cell walls. There is, hence, some room for gentle adjustments of the balance between the different optimization goals so as to better adapt crop plants to the man-made environment of agriculture (or of a fermenter) and to better suit human needs.

However, today our possibilities are greatly limited by the paucity of information available about the biosynthesis, turnover, and modification of the components and by our lack of information on the assembly and regulation of the complex plant cell wall. This relative ignorance of the targets far outweighs the limitations posed by the still less than optimal genetic engineering tools for plant transformation. For the time being, our chances of success are most promising if the aspect to be optimized can clearly be ascribed to a single cell wall component. Ideally, this component should be a nonstructural protein that is synthesized, modified, and delivered to the cell wall by the conventional secretory pathway.

Today's agriculture is characterized by vast homogeneous fields of genetically uniform crop plants. The high-yield cultivars grown are bred for fast growth and maximum yield. Often, the contents of secondary plant metabolites have been reduced in favor of product quality. Typically, disease resistance has not been a primary breeding goal, and the inherent resistance traits of the wild progenitors have largely been lost in our crop species. There is a clear need to breed old and new resistance traits back into modern crop cultivars, and one way to achieve this is by genetic engineering. We have seen that cell walls play crucial roles in plant defense against microbial pathogens so that here are promising targets. The most easily handled are cell wall–located pathogenesis-related (pr-) proteins (94). These include antimicrobial defensins, thionines, proteases, peroxidases, β-1,3-glucanases, and chitinases. Transgenic plants overexpressing a single pr-protein usually do not exhibit significantly increased disease resistance, but combinations, e.g., of fungal cell wall hydrolyzing β-1,3-glucanases and chitinases, are more efficient (95). The hyphal cell walls of some phytopathogenic fungi growing endophytically in the host plant tissues have been shown to contain chitosan rather than chitin, so that chitosanases may be more effective than chitinases in protecting plants from disease (96). Systematic approaches to analyze cooperative, synergistic actions of the different classes of pr-proteins are currently under way.

A number of further strategies toward transgene-mediated disease resistance are currently being pursued, and most of them involve the plant cell wall. The most efficient resistance reaction of many plants against a variety of pathogens is the hypersensitive response, which is often induced as a result of the molecular interaction of the direct or indirect products of plant resistance genes and microbial avirulence genes (24). The former are believed to act as plasma membrane–bound receptors for the latter, which are therefore termed elicitors. Expression and cell wall secretion of a microbial avirulence gene under the control of a strictly pathogen-induced promoter in a plant carrying the corresponding resistance gene has been shown to lead to pathogen-induced triggering of an artificial hypersensitive response (97). Similarly, the tightly regulated expression of glucose oxidase can lead to the generation of an artificial oxidative burst (98), which in turn activates a hypersensitive-like reaction. As this reaction involves the activation of a cell death program, tight regulation of the process—both in time and in space, and under a broad regime of environmental conditions—is required for this approach.

Of the incredible biodiversity of plant species, only a very few have been domesticated to provide the mainstay of the human food supply. Valuable traits from other plant species that cannot be inbred into our crop species can now be introduced by genetic engineering. Examples of important traits are the production of pharmacologically active secondary plant metabolites and mechanisms of stress tolerance, such as salt tolerance and drought tolerance. The main problem with these complex traits is that they are brought about by the consorted action of multiple genes and are therefore difficult to transfer (99). We have seen that the outer epidermal cell walls are crucial for drought tolerance, mainly by virtue of the cuticle. The classical *eceriferum* mutants of barley implicated around 100 genes as involved in the biosynthesis and assembly of the cuticle (100). Currently, tagged *eceriferum* mutants of barley and *Arabidopsis* are being selected to shed light on the molecular machinery involved. Eventually, we will have to transfer and broaden the knowledge gained from these model plants to include, e.g., drought-resistant desert plants. Only then can we hope to learn on a molecular basis how these plants cope with their hostile environment, and we may be able to profit from this insight by generating drought-resistant crop plants that may flourish in the growing semiarid areas. Similar approaches are being undertaken for other stress factors. Cell walls can be anticipated to be prominently involved in many of these stress resistance mechanisms.

Comparing modern crop plants with their wild relatives reveals the striking changes in morphology that sometimes thousands of years of breeding have generated and that are witness to the enormous flexibility in plant organization. Today's maize cultivars producing corn cobs with up to 24 rows of perhaps 50 kernels each were bred from wild progenitors having cobs the size of a pencil eraser. Sugar beets with a diameter of up to 20 cm and a sugar content of up to 20% were bred from small beets with only trace amounts of sucrose within less than 150 years. Plant morphology is dictated by the cell walls: walls shape the forms of plant cells and organs; walls restrict the growth of plant cells and organs; walls are responsible for the strength of plant cells and organs; and cell wall enzymes and oligosaccharin wall fragments are involved in the development of different types of plant cells and organs. In this case, our basic handicap for biotechnological manipulations is the sparse knowledge we have of how cell walls influence these different aspects and the realization that these are multifactorial and multigenic traits. However, first results have been obtained, and if these have not yet led to commercially useful crop plants, they do provide proof of principle for the general approach.

The changes forced upon the crop plants by conventional breeding all involved redistribution of photoassimilates toward the sink organs of human interest: shorter stems in favor of more grains, less aboveground growth in favor of more and thicker tubers, etc. A century ago, the harvest index of wheat straw/grain was 70/30; it is 40/60 today. The regulation of source-sink ratios is achieved by modifications in the relative phloem transport rates from the different source tissues toward the different sink tissues of a plant (9). Both phloem loading and phloem unloading, which are thought to be the main regulatory steps in phloem transport in most plants, involve an apoplastic step. Cell wall–located invertases are likely to be involved in the process of sugar sensing, which determines both source and sink activities. Overexpressing a fungal invertase in the apoplast of potato leaves or tubers led to dramatic phenotypes, including modified root-to-shoot ratios, changed number and weight of tubers, and altered morphology of leaves and tubers (101). This is another example of genetic engineering of plants via manipulation of cell walls where the target is a nonstructural cell wall protein.

Cell walls are essential for food texture and fruit juice turbidity, and modifications in cell wall composition and architecture are among the most prominent changes during fruit maturation. Interfering with these changes by genetic engineering promises to increase shelf life and to improve the manufacturing processes in food technology. Perhaps the best known example of "novel food" is the Flavr-Savr tomato expressing an antisense gene for endopolygalacturonase (102). Decreasing the activity of this cell wall enzyme prevents maceration of the tomato tissue so that red-ripe fruits can be transported over long distances and stored for prolonged times. However, the processes of fruit maturation are still far from being understood, and cause-consequence relationships are mostly unknown. The partial breakdown of cell wall components during maturation is most likely an integral part of a sequence of events, and disturbing one link of the chain will have unpredictable consequences for all downstream events. Flavr-Savr tomatoes have been reported to look but not to taste like red-ripe tomatoes.

VI. BIOTECHNOLOGICAL APPROACHES TO OPTIMIZE CELL WALL COMPONENTS FOR USES EX PLANTA

Plant cell walls are the most abundant renewable resource available to man, and they have accompanied mankind from the beginning. Wood, entirely consisting of plant cell walls, is used in the form of timber for construction purposes and as a raw material to produce pulp and paper. Cell wall fibers have provided man with strings to be woven into tissues and drilled into ropes, and they are increasingly manufactured into insulation materials. Plant fibers are also healthy ingredients of man's diet and may actually have antitumor activities, and other cell wall components are used as gelling materials in food technology. Many other uses of plant cell walls have been proposed even though they may not yet have been verified. Optimizing plant cell wall components for these ex planta uses appears a much more easily achievable goal in the short range, as cell walls are not naturally optimized for these applications. However, as attempts at genetically engineering a reduced lignin content in wood for the pulp and paper industries have amply proved, surprises can result from this strategy, too.

Genetic engineering of lignin is extensively described in Chap. 20 and will, therefore, not be treated in detail here. But we should reflect briefly on the lessons to be learned from these approaches as they relate to the general subject of biotechnology of plant cell walls. Most of the initial experiments aimed at reducing the lignin content by expressing antisense genes for lignin biosynthetic enzymes, with the goal of improving rumen digestibility of forage grasses and easing the delignification step during paper production (103). Invariably, lignin content was almost unchanged in these transgenic plants, but lignin composition often changed to some extent. Clearly, the cells that survived the switching off of specific steps in lignin biosynthesis did so by making use of the inherent flexibility in cell wall architecture. If some monolignol was no longer available due to a block in its biosynthesis, other, often unusual, monomers were incorporated into the lignin of the transgenics, and the resulting altered lignins were apparently able to take over at least partially the roles of normal lignin. Similar flexibility can be expected in the case of other cell wall components as well, as exemplified by suspension-cultured cells growing in the presence of cellulose biosynthesis inhibitors (104) or fucosyltransferase-lacking mutants of *Arabidopsis* (105). Consequently, the results of even very pinpointed genetic engineering experiments can hardly be predicted with certainty.

Although we will not ponder further on genetic engineering of lignin biosynthesis per se, it is important to realize that lignin always forms part of complex cell walls, and the composite material is not uniform but able to react to environmental influences with compositional and, most likely, architectural changes. If woody plants are misaligned from the vertical axis, so-called reaction wood is formed to realign the stem (106). Xylem cells involved in the transformation undergo massive changes in their cell wall composition. In gymnosperms, compression wood is formed in the areas of the stem that are subject to increased pressure, and in angiosperms, tension wood is formed in the areas of the stem subjected to increased stress. Compression wood is increased in lignin content and decreased in cellulose content, while tension wood contains more cellulose and less lignin (107). It might be conceivable, then, to generate wood that is specifically adapted to certain construction needs. Such a manipulation would have to consider both the polysaccharides and the polyphenolics present in the cell wall.

Fibers of plant cell walls form the basis of millennia-old techniques such as warping and weaving. Yarns, strings, ropes, fabrics, and tissues were and still are to some extent produced from plant cell wall–based fibers. Increasingly, natural fibers are also used in the manufacturing of insulation materials for construction purposes. Cotton is the most widely used plant fiber today. Conventional breeding has already produced cotton varieties with cellulose fibrils of different lengths and colors. When we understand the roles of the different cellulose synthase isoenzymes in producing cellulose microfibrils of different lengths and diameters, we might be able to generate an even greater variety of cotton-based raw materials. Flax, hemp, sisal, and ramie yield fibers used on a large scale, and the material properties of these fibers differ from those of cotton fibers owing to the different cell wall polysaccharides accompanying the cellulose microfibrils in the cell walls of these plants (108). Breeding and biotechnology with these fiber plants are far less advanced than with cotton, so that there is a great potential for future improvements. Many other plants such as palm trees and agava plants are used locally for fiber production. These plants are typically integrated in a complex system of human uses not restricted to fiber production, so that there is less room for optimization in a single direction.

Pectins are widely used as gelling agents in the food industries. However, only pectins of a very few plant species have been shown to be well suited for the purpose, as other pectins available in large quantities at competitive prices proved inadequate. Large volumes of pectin-rich waste pulp are produced by the potato starch industries, but the pectin present is less suitable for the food industries partly due to its high content of rhamnogalacturonan side chains, particularly neutral galactans, and its low degree of methyl esterification. Expressing a fungal galactanase in potato tubers led to the formation of galactan-poor pectins with increased solubility characteristics, but the transgenic plants showed no obvious altered phenotype (109). It remains to be seen whether this recombinant pectin is better suited for industrial uses and whether the genetic engineering approach is superior to conventional postharvest enzymic treatments. But clearly, expressing hydrolases in the cell walls is successful and interesting because it is presumably a very versatile means of manipulating plant cell wall composition and architecture in planta for uses ex planta.

Many other biotechnological uses for plant cell walls and their components have been described or proposed. In terms of market size, medicinal applications are among the most attractive. Cell wall components—in the form of nondigestible oligosaccharides (110) or as dietary fibers (111)—are increasingly suspected to be responsible for some of the pronounced health effects of vegetables and fruits. Dietary fibers, and particularly pectins, have been implicated in protection against cancer, coronary heart disease, and vascular diseases (112). No causal relationships have yet been established, but functional foods are already on the horizon. Cell wall components may also be useful in environmental protection. For instance, the cation chelating properties of pectins have been used for heavy metal decontamination of wastewaters (113).

Increased knowledge of biosynthesis and assembly of cell wall constituents may eventually allow us to draw up totally new designer polysaccharides not known from any natural source. As an example, let us consider again the biosynthesis of cellulose and other β-glycosidically linked hemicellulose backbones. All of these linear polysaccharides consist of disaccharide repeat units. The two glycosyl units involved are most likely attached to the growing polysaccharide chain by two independent catalytic sites— present in either one or two cellulose synthase or cellulose synthase–like proteins (68). If this concept proves to be right, and when different catalytic domains with specificities for different nucleotide sugar substrates become known and available (114,115), domain swapping may lead to the creation of novel polysaccharides. Even though these may not properly and cooperatively function in a primary cell wall, it may still be possible to produce them, e.g., in cotton fibers—and it may thus become possible to tailor novel designer polysaccharides for specific ex planta uses.

VII. PROSPECTS FOR THE FUTURE OF PLANT CELL WALL BIOTECHNOLOGY

We have seen that the major limitation for biotechnology with plant cell walls today is our limited knowledge about the biosynthesis of the cell wall components and their assembly into the complex pericellular matrix. The genomic approaches initiated recently promise rapid progress in identifying the genes potentially involved in these processes, at least for some model plants such as *Arabidopsis*, maize, rice, barley, poplar, and pine (116–120). Proteomic approaches will complement these studies, and they will help in identifying the enzymes involved in the biosynthesis and assembly of specific cell wall domains elaborated under certain sets of environmental conditions. As seen with the cellulose synthase genes and, even more clearly, with the six classes of cellulose synthase–like genes, extensive functional genomic experiments will be needed to assign specific functions to the genes identified, and these will keep us occupied for years to come. Functional genomics will allow us to analyze which enzymes are involved in biosynthesis and assembly of complex cell walls, but they will not shed light on how these enzymes do their jobs. Methods for functional proteomics may have to be developed to investigate the molecular interactions between proteins (e.g., Ref. 121), polysaccharides, and the other cell wall constituents to understand how the cell wall is assembled and how it functions. This knowlegde, of course, is a prerequisite for targeted genetic engineering of plant cell walls for biotechnological uses.

Functional proteomics will have to make use of the newly developed technical possibilities offered by the nanosciences. Nanoanalytical tools allow the visualization and the determination of surface properties of individual molecules (122). They also permit direct measurement of intermolecular forces holding together the noncovalent networks that make up plant cell walls. Interestingly, nanosciences not only offer analytical tools but also are currently developing manipulating tools with which to handle and influence individual molecules. These tools can be increasingly used in vivo, and this will allow us to learn about the discrete roles that individual molecules play in the complex fabric of the cell wall. Eventually, cell walls may serve as ideal models for complex, versatile, and very stable composite materials, and the molecular machinery for assembly or the principles of self-assembly of the different components to build the complex cell wall may well serve as a model for biomimetic processes to be used in the construction of nanomachines or nanomaterials using nonnatural or semisynthetic molecules.

It should be clear from the preceding discussion that we are only beginning to see the surface of the incredible biodiversity inherent in plant cell walls—as it is in almost every aspect of living systems. We have seen the flexibility and dynamics in cell wall architecture of a single plant, and we have pointed to the differences in cell wall composition and architecture between dicot plants and the commelinoid monocot plants. However, this treatise has focused on the cell walls of the angiosperms while those of the gymnosperms have not been mentioned. But to be comprehensive, we would even have to include cell walls of the nonflowering ferns and mosses and the green, red, and brown algae. Fungal cells are also surrounded by cell walls that function according to the same principle as plant cell walls— stress-bearing fibers in a pressure-bearing matrix—but this is realized using quite different polysaccharides and proteins, and the diversity within the fungal kingdom is enormous. Furthermore, almost all prokaryotic cells possess cell walls, but these are built in a completely different way, such as the murein sacculus, which is one giant covalently linked molecule. We know little about most of these cell walls—and this is an exciting prospect for future researchers who are not intimidated by the enormous complexity of the pericellular matrices.

Clearly, plant cell walls are an incredibly rich renewable resource for sustainable biotechnologies. However, we still understand little of the cause-and-effect relationships in most of the physiological processes involving plant cell wall components. Manipulating single components of such a complex structure has almost invariably yielded results that were different from what was expected. We should seize the opportunities molecular genetic tools offer, but we should take our time to analyze the effects of genetic engineering of cell wall components, assembly, and architecture. There should be no haste to introduce transgenic plants and their products to the environment and to the market prematurely. Only well-planned, well-done, and well-analyzed transgenic plants with clear environmental and consumer benefits will stand a chance to overcome reasonable and irrational fears of the consumers.

ACKNOWLEDGMENTS

I thank Markus Pauly and Steven Lenhert for critically reading this review.

REFERENCES

1. NC Carpita, M McCann. The cell wall. In: BB Buchanan, W Gruissem, RL Jones, eds. Biochemistry and Molecular Biology of Plants. Rockville, MD: American Society of Plant Physiologists, 2000, pp 52–108.
2. NC Carpita, DM Gibeaut. Structural models of primary-cell walls in flowering plants—consistency of molecular-structure with the physical-properties of the walls during growth. Plant J 3:1–30, 1993.
3. PB Green, RO Erickson, PA Richmond. On the physical basis of cell morphogenesis. Ann NY Acad Sci 175:712–731, 1970.
4. DJ Cosgrove. Enzymes and other agents that enhance cell wall extensibility. Annu Rev Plant Physiol Plant Mol Biol 50:391–417, 1999.
5. HT Cho, DJ Cosgrove. Altered expression of expansin modulates leaf growth and pedicel abscission in *Arabidopsis thaliana*. Proc Natl Acad Sci USA 97: 9783–9788, 2000.
6. K Vissenberg, IM Martinez-Vilchez, JP Verbelen, JG Miller, SC Fry. In vivo colocalization of xyloglucan endotransglycosylase activity and its donor substrate in the elongation zone of *Arabidopsis* roots. Plant Cell 12:1229–1237, 2000.
7. TH Giddings, LA Staehelin. The cell wall: microtubule-mediated control of microfibril deposition—a re-examination of the hypothesis. In: CW Lloyd, ed. The Cytoskeletal Basis of Plant Growth and Form. London: Academic Press, 1991, pp 85–99.

8. GO Wasteneys. The cytoskeleton and growth polarity. Curr Opin Plant Biol 3:503–511, 2000.
9. E Zamski, A Schaffer, eds. Photoassimilate Distribution in Plants and Crops. Source-Sink Relationships. New York: Marcel Dekker, 1996.
10. A Sturm. Invertases. Primary structures, functions, and roles in plant development and sucrose partitioning. Plant Physiol 121:1–7, 1999.
11. KE Koch. Carbohydrate-modulated gene expression in plants. Annu Rev Plant Physiol Plant Mol Biol 47:509–540, 1996.
12. FB Abeles, WL Hershberger, LJ Dunn. Hormonal regulation, and intracellular localization of a 33-kD cationic peroxidase in excised cucumber cotyledons. Plant Physiol 89:664–668, 1989.
13. L Gälweiler, C Guan, A Müller, E Wisman, K Mendgen, A Yephremov, K Palme. Regulation of polar auxin transport by AtPIN1 in *Arabidopsis* vascular tissue. Science 282:2226–2229, 1998.
14. N Carpita, D Sabularse, D Montezinos, DP Delmer. Determination of the pore size of cell walls of living plant cells. Science 205:1144–1147, 1979.
15. C Fleischer, R Titel, R Ehwald. The boron requirement and cell wall properties of growing and stationary suspension-cultured *Chenopodium rubrum album* L. cells. Plant Physiol 117:1401–1410, 1998.
16. G Wallace, SC Fry. Phenolic components of the plant cell wall. Int Rev Cytol 151:229–267, 1994.
17. A Levine, R Tenhaken, R Dixon, C Lamb. H_2O_2 from the oxidative burst orchestrates the plant hypersensitive disease resistance responses. Cell 79: 583–593, 1994.
18. JP Knox, PJ Linstead, J King, C Cooper, K Roberts. Pectin esterification is spatially regulated both within cell walls and between developing tissues of root apices. Plant 181:512–521, 1990.
19. M Smallwood, A Deven, N Donovan, SJ Neill, J Peart, K Roberts, JP Knox. Localization of cell wall proteins in relation to the developmental anatomy of the carrot root apex. Plant J 5:237–246, 1994.
20. JP Knox, S Day, K Roberts. A set of cell surface glycoprotein forms an early marker of cell position, but not cell type, in the root apical meristem of *Daucus carota* L. Development 106:47–56, 2001.
21. KD Belanger, RS Quatrano. Polarity: the role of localized secretion. Curr Opin Plant Biol 3:67–72, 2000.
22. B Moerschbacher, K Mendgen. Structural aspects of defense. In: AJ Slusarenko, RSS Fraser, LC van Loon, eds. Mechanisms of Resistance to Plant Diseases. Dordrecht: Kluwer Academic Publishers, 2000, pp 231–277.
23. R Mittler, EH Herr, BL Orvar, W van Camp, H Willekes, CD Inz, BE Ellis. Transgenic tobacco plants with reduced capability to detoxify reactive oxygen intermediates are hyperresponsive to pathogen infection. Proc Natl Acad Sci USA 96:14165–14170, 1999.
24. BM Moerschbacher, HJ Reisener. The hypersensitive resistance reaction. In: H Hartleb, R Heitefuss, HH Hoppe, eds. Resistance of Crop Plants against Fungi. Jena: Gustav Fischer Verlag, 1997, pp 126–158.

25. C Neinhuis, W Barthlott. Characterization and distribution of water-repellent, self-cleaning plant surfaces. Ann Bot 79:667–677, 1997.
26. J Huang. Ultrastructure of bacterial penetration in plants. Annu Rev Phytopathol 24:141–157, 1986.
27. D Rubiales, RE Niks. Avoidance of rust infection by some genotypes of *Hordeum chilense* due to their relative inability to induce the formation of appressoria. Physiol Mol Plant Pathol 49:89–101, 1996.
28. E Stahl, JG Bishop. Plant-pathogen arms races at the molecular level. Curr Opin Plant Biol 3:299–304, 2000.
29. M Parniske. Intracellular accommodation of microbes by plants: a common developmental program for symbiosis and disease? Curr Opin Plant Biol 3: 320–328, 2000.
30. B Ahlborn, D Werner. Inhibition of 1,3-β-glucan synthase from *Glycine max* and *Pisum sativum* by exopolysaccharides of *Bradyrhizobium japonicum* and *Rhizobium leguminosarum*. Physiol Mol Plant Pathol 39:299–307, 1991.
31. JA Munoz, C Coronado, J Perez-Hormaeche, A Kondorosi, P Ratet, AJ Palomares. *MSPG3*, a *Medicago sativa* polygalacturonase gene expressed during the alfalfa *Rhizobium meliloti* interaction. Proc Natl Acad Sci USA 95:9687–9692, 1998.
32. AAN Van Brussel, R Bakhuizen, PC Van Spronsen, HP Spaink, T Tak, BJJ Lugtenberg, JW Kijne. Induction of preinfection thread structures in the leguminous host plant by mitogenic lipooligosaccharides of *Rhizobium*. Science 257:70–72, 1992.
33. B Moerschbacher, M Mierau, B Graeßner, U Noll, AJ Mort. Small oligomers of galacturonic acid are endogenous suppressors of disease resistance reactions in wheat leaves. J Exp Bot 50:605–612, 1999.
34. JC Stein, R Dixit, ME Nasralla, JB Nasralla. SRK, the stigma specific S locus receptor kinase of *Brassica*, is targeted to the plasma membrane in transgenic tobacco. Plant Cell 8:429–445, 1996.
35. K Kandasamy, D Paolillo, C Faraday, JB Nasralla, ME Nasralla. The S locus specific glycoproteins of *Brassica* accumulate in the cell wall of developing stigma papillae. Dev Biol 134:462–472, 1989.
36. CR Schopfer, ME Nasralla, JB Nasralla. The male determinant of self-incompatibility in *Brassica*. Science 286:1697–1700, 1999.
37. HG Dickinson. Dry stigmas, water and self-incompatibility. Sex Plant Reprod 8:1–10, 1995.
38. A Darvill, C Augur, C Bergmann, RW Carlson, JJ Cheong, S Eberhard, MG Hahn, VM Lo, V Marfa, B Meyer, D Mohnen, MA O'Neill, MD Spiro, H van Halbeek, WS York, P Albersheim. Oligosaccharins—oligosaccharides that regulate growth, development and defence responses in plants. Glycobiology 2:181–198, 1992.
39. DTA Lamport. The primary cell wall. In: RA Young, RM Rowell, eds. Cellulose: Structure, Modification and Hydrolysis. New York: Wiley, 1986, pp 77–90.
40. M McNeil, AG Darvill, SC Fry. Structure and function of the primary cell walls of plants. Annu Rev Biochem 53:625–663, 1984.

41. K Keegstra, KW Talmadge, WD Bauer, P Albersheim. Structure of plant cell walls. III. A model of the walls of suspension-cultured sycamore cells based on their interconnections of the macromolecular components. Plant Physiol 51:188–197, 1973.

42. K Iiyama, TBT Lam, BA Stone. Covalent cross-links in the cell wall. Plant Physiol 104:315–320, 1994.

43. NC Carpita. Structure and biogenesis of the cell walls of grasses. Annu Rev Plant Physiol Plant Mol Biol 47:445–476, 1996.

44. S Levy, G MacLachlan, LA Staehelin. Xyloglucan sidechains modulate binding to cellulose during in vitro binding assays as predicted by conformational dynamics simulations. Plant J 11:373–386, 1997.

45. NC Carpita. Fractionation of hemicelluloses from maize cell walls with increasing concentrations of alkali. Phytochemistry 23:1089–1093, 1984.

46. J Visser, AGJ Voragen. Pectins and pectinases. Progress in Biotechnology. Vol 14. Amsterdam: Elsevier Science, 1996.

47. I Braccini, RP Brasso, S Perez. Conformational and configurational features of acidic polysaccharides and their interactions with calcium ions: a molecular modeling investigation. Carbohydr Res 317:119–130, 1999.

48. M O'Neill, D Warrenfeltz, K Kates, P Pellerin, T Doco, AG Darvill, P Albersheim. Rhamnogalacturonan-II, a pectic polysaccharide in the walls of growing plant cell, forms a dimer that is covalently cross-linked by a borate ester. J Biol Chem 271:22923–22930, 1996.

49. KCB Wilkie. The hemicelluloses of grasses and cereals. Adv Carbohydr Chem Biochem 36:215–264, 1979.

50. J Ralph, S Quideau, JH Grabber, RD Hatfield. Identification and synthesis of new ferulic acid dehydrodimers present in grass cell walls. J Chem Soc Perkin Trans 1:3485–3498, 1994.

51. GI Cassab. Plant cell wall proteins. Annu Rev Plant Physiol Plant Mol Biol 49:281–309, 1998.

52. MT Herrera, I Zarra. Changes in the hemicellulosic polysaccharides of *Avena* coleoptile cell walls during growth of intact seedlings. Can J Bot 66:949–954, 1988.

53. DG Luttenegger, DJ Nevins. Transient nature of a (1-3),(1-4)-β-D-glucan in *Zea mays* coleoptile cell walls. Plant Physiol 77:175–178, 1985.

54. MS Bush, MC McCann. Pectic epitopes are differentially distributed in the cell walls of potato (*Solanum tuberosum*) tubers. Physiol Plant 107:201–213, 1999.

55. MC McCann, K Roberts. Architecture of the primary cell wall. In: CW Lloyd, ed. The Cytoskeletal Basis of Plant Growth and Form. London: Academic Press, 1992, pp 109–129.

56. K Roberts, MC McCann. Xylogenesis: the birth of a corpse. Curr Opin Plant Biol 3:517–522, 2000.

57. DM Gibeaut, NC Carpita. Biosynthesis of plant cell wall polysaccharides. FASEB J 8:904–915, 1994.

58. M Otegui, LA Staehelin. Cytokinesis in flowering plants: more than one way to divide a cell. Curr Opin Plant Biol 3:493–502, 2000.

59. AL Samuels, TH Giddings, LA Staehelin. Cytokinesis in tobacco BY-2 and root tip cells: a new model of cell plate formation in higher plants. J Cell Biol 130:1–13, 1995.

60. J Zuo, QW Niu, N Nishizawa, Y Wu, B Kost, N-H Chua. KORRIGAN, an *Arabidopsis* endo-1,4-β-glucanase, localizes to the cell plate by polarized targeting and is essential for cytokinesis. Plant Cell 12:1137–1152, 2000.

61. DP Delmer. Cellulose biosynthesis: exciting times for a difficult field of study. Annu Rev Plant Physiol Plant Mol Biol 50:245–276, 1999.

62. J Pear, Y Kawagoe, W Schreckengost, DP Delmer, D Stalker. Higher plants contain homologs of the bacterial *CelA* genes encoding the catalytic sub-unit of the cellulose synthase. Proc Natl Acad Sci USA 93:12637–12642, 1996.

63. IM Saxena, RM Brown. Cellulose synthases and related enzymes. Curr Opin Plant Biol 3:523–531, 2000.

64. S Cutler, C Somerville. Cellulose synthesis: cloning in silico. Curr Biol 7: R108–R111, 1997.

65. T Richmond, C Somerville. The cellulose synthase superfamily. Plant Physiol 124:495–498, 2000.

66. S Kimura, W Laosinchai, T Itoh, X Cui, C Linder, RM Brown Jr. Immunogold labeling of rosette terminal cellulose-synthesizing complexes in the vascular plant *Vigna angularis*. Plant Cell 11:2075–2085, 1999.

67. RMJ Brown, IM Saxena. Cellulose biosynthesis: a model for understanding the assembly of biopolymers. Plant Physiol Biochem 38:57–67, 2000.

68. IM Saxena, RMJ Brown, M Fevre, RA Geremia, B Henrissat. Multidomain architecture of ß-glycosyl transferases: implications for mechanism of action. J Bacteriol 177:1419–1424, 1995.

69. N Holland, D Holland, T Helentjaris, K Dhugga, B Xoconostle-Cazare, DP Delmer. A comparative analysis of the plant cellulose synthase (*CesA*) gene family. Plant Physiol 123:1313–1324, 2000.

70. NG Taylor, WR Scheible, S Cutler, CR Somerville, SR Turner. The irregular xylem3 locus of *Arabidopsis* encodes a cellulose synthase required for secondary cell wall synthesis. Plant Cell 11:769–779, 1999.

71. T Arioli, L Peng, AS Betzner, J Brun, W Wittke, W Herth, C Camilleri. Molecular analysis of cellulose biosynthesis in *Arabidopsis*. Science 279:717–720, 1998.

72. D Mohnen. Biosynthesis of pectins and galactomannans. In: BM Pinto ed. Comprehensive Natural Products Chemistry. Vol. 3: Carbohydrates and Their Derivatives Including Tannins, Cellulose and Related Lignins. Amsterdam: Elsevier, 1999, pp 497–527.

73. RM Perrin, AE deRocher, M Bar-Peled, W Zeng, L Norambuena, A Orellana, NV Raikhel, K Keegstra. Xyloglucan fucosyltransferase, an enzyme involved in plant cell wall biosynthesis. Science 284:1976–1979, 1999.

74. ME Edwards, CA Dickson, S Chengappa, C Sidebottom, MJ Gidley, JSG Reid. Molecular characterization of a membrane-bound galactosyltransferase of plant cell wall matrix polysaccharide biosynthesis. Plant J 19:691–697, 1999.

75. A Faik, M Bar-Peled, AE deRocher, WQ Zeng, RM Perrin, C Wilkerson, NV Raikhel, K Keegstra. Biochemical characterization and molecular cloning of an α-1,2-fucosyltransferase that catalyzes the last step of cell wall xyloglucan biosynthesis in pea. J Biol Chem 275:15082–15089, 2000.

76. JC Paulson, KJ Colley. Glycosyltransferases—structure, localization, and control of cell type-specific glycosylation. J Biol Chem 264:17615–17618, 1989.

77. C Wulff, L Norambuena, A Orellana. GDP-fucose uptake into the Golgi apparatus during xyloglucan biosynthesis requires the activity of a transporter-like protein other than the UDP-glucose transporter. Plant Physiol 122:867–877, 2000.

78. C Somervill, S Cutler. Use of genes encoding xylan synthase to modify plant cell wall composition. International patent application, 1998.

79. DM Gibeaut. Nucleotide sugars and glycosyltransferases for synthesis of cell wall matrix polysaccharides. Plant Physiol Biochem 38:69–80, 2000.

80. YA Campbell, GJ Davies, V Bulone, B Henrissat. A classification of nucleotide-diphospho-sugar glycosyl transferases based upon amino-acid similarities. Biochem J 326:929–939, 1997.

81. HV Scheller, RL Doong, BL Ridley, D Mohnen. Pectin biosynthesis: a solubilized alpha 1,4-galacturonosyltransferase from tobacco catalyzes the transfer of galacturonic acid from UDP-galacturonic acid onto the non-reducing end of homogalacturonan. Planta 207:512–517, 1999.

82. RL Doong, D Mohnen. Solubilization and characterization of a galacturonosyltransferase that synthesizes the pectic polysaccharide homogalacturonan. Plant J 13:363–374, 1998.

83. F Goubet, D Mohnen. Subcellular localization and topology of homogalacturonan methyltransferase in suspension-cultured Nicotiana tabacum cells. Planta 209:112–117, 1999.

84. JTDM Gaffe, AK Handa. Pectin methylesterase isoforms in tomato (Lycopersicon esculentum) tissues. Effects of expression of a pectin methylesterase antisense gene. Plant Physiol 105:199–203, 1994.

85. M McNeill, A Darvill, P Albersheim. Structure of plant cell walls X. Rhamnogalacturonan I, a structurally complex pectic polysaccharide in the walls of suspension-cultured sycamore cells. Plant Physiol 66:1128–1134, 1980.

86. P Komalavilas, AJ Mort. The acetylation at O-3 of galacturonic acid in the rhamnose-rich portion of pectins. Carbohydr Res 189:261–272, 1989.

87. N Geshi, B Jorgensen, HV Scheller, P Ulvskov. In vitro biosynthesis of 1,4-β-galactan attached to rhamnogalacturonan I. Planta 210:622–629, 2000.

88. JL Hadlington, J Denecke. Sorting of soluble proteins in the secretory pathway of plants. Curr Opin Plant Biol 3:461–468, 2000.

89. LA Staehelin, I Moore. The plant Golgi apparatus: structure, functional organization and trafficking mechanisms. Annu Rev Plant Physiol Plant Mol Biol 46:261–288, 1995.

90. M Vicre, A Jauneau, JP Knock, A Driouich. Immunolocalization of beta-(-1) and beta-(1-6)-D-galactan epitopes in the cell wall and Golgi stacks of developing flax foot tissues. Protoplasma 203:26–34, 1998.

91. CP Bonin, WD Reiter. A bifunctional epimerase-reductase acts downstream of the *MUR1* gene product and completes the de novo synthesis of GDP-fucose in *Arabidopsis*. Plant J 21:445–454, 2000.

92. B Seitz, C Kos, M Wurm, R Tenhaken. Matrix polysaccharide in *Arabidopsis* cell walls are synthesized by alternate pathways with organ-specific expression patterns. Plant J 21:537–546, 2000.

93. JC Kader. Lipid-transfer proteins in plants. Annu Rev Plant Physiol Plant Mol Biol 47:627–654, 1996.

94. CA De, WJ Van, P Dhaese, FG Loontiens. The carbohydrate-binding specificity of the lectin from *Vicia faba*. Arch Int Physiol Biochim 84:150–151, 1976.

95. LS Melchers, MH Stuiver. Novel genes for disease-resistance breeding. Curr Opin Plant Biol 3:147–152, 2000.

96. NE El Gueddari, B Moerschbacher. A rust fungus turns chitin into chitosan upon plant tissue colonization to evade recognition by the host. Adv Chitin Sci 4:588–592, 1999.

97. G Strittmatter, J Janssens, C Opsomer, J Botterman. Inhibition of fungal disease development in plants by engineering controlled cell death. Bio/Technology 13:1085–1089, 1995.

98. G Wu, BJ Shortt, EB Lawrence, KC Fitzsimmons, DM Shak. Disease resistance conferred by expression of a gene encoding H_2O_2-generating glucose oxidase in transgenic potato plants. Plant Cell 7:1357–1368, 1995.

99. JC Cushman, HJ Bohnert. Genomic approaches to plant stress tolerance. Curr Opin Plant Biol 3:117–124, 2000.

100. D Post-Beittenmiller. Biochemistry and molecular biology of wax production in plants. Annu Rev Plant Physiol Plant Mol Biol 47:405–430, 1996.

101. U Sonnewald, MR Hajirezaei, J Kossmann, A Heyer, RN Trethewey, L Willmitzer. Increased potato tuber size resulting from apoplastic expression of a yeast invertase. Nat Biotechnol 15:794–797, 1997.

102. G Tucker, J Zhang. Expression of polygalacturonase and pectinesterase in normal and transgenic tomatoes. I. In: J Visser, AGJ Voragen, eds. Pectins and Pectinases. Amsterdam: Elsevier, 1996, pp 347–368.

103. J Grima-Pettenati, D Goffner. Lignin genetic engineering revisited. Plant Sci 145:51–65, 1999.

104. E Shedletzky, M Shmuel, T Trainin, S Kalman, D Delmer. Cell wall structure in cells adapted to growth on the cellulose-synthesis inhibitor 2,6-dichlorobenzonitrile. Plant Physiol 100:120–130, 1992.

105. E Zablackis, WS York, M Pauly, S Hantus, WD Reiter, CCS Chapple, P Albersheim, A Darvill. Substitution of L-fucose by L-galactose in cell walls of *Arabidopsis* mur1. Science 272:1808–1810, 1996.

106. KV Sarkanen, CH Ludwig. Definition and nomenclature. In: KV Sarkanen, CH Ludwig, eds. Lignins: Occurrence, Formation, Structure, and Reactions. New York: Wiley, 1971, pp 1–18.

107. LG Wu, SP Joshi, VL Chiang. A xylem-specific cellulose synthase gene from aspen (*Populus tremuloides*) is responsive to mechanical stress. Plant J 22:495–502, 2000.

108. TA Gorshkova, SE Wyatt, VV Salnikov, DM Gibeaut, MR Ibragimov, VV Lozovaya, NC Carpita. Cell-wall polysaccharides of developing flax plants. Plant Physiol 110:721–729, 1996.

109. S Oxenboll Sorensen, M Pauly, M Bush, Michael Skjot, MC McCann, B Borkhardt, P Ulvskov. Pectin engineering: modification of potato pectin by in vivo expression of an endo-1,4-β-D-galactanase. Proc Natl Acad Sci USA 97:7639–7644, 2000.

110. AGJ Voragen. Technological aspects of functional food-related carbohydrates. Trends Food Sci Technol 9:328–335, 1998.

111. M Lahaye, B Kaeffer. Seaweed dietary fibres: structure, physicochemical and biological properties relevant to intestinal physiology. Sci Aliments 17:563–584, 1997.

112. H Yamada. Contribution of pectins in health care. In: J Visser, AGJ Voragen, eds. Pectins and Pectinases. Amsterdam: Elsevier, 1996, pp 173–190.

113. MAV Axelos, C Garnier, CMGC Renard, JF Thibault. Heavy metals binding by pectins: selectivity, quantification and characterization. In: J Visser, AGJ Voragen, eds. Pectins and Pectinases. Amsterdam: Elsevier, 1996, pp 535–540.

114. W Jing, PL DeAngelis. Dissection of the two transferase activities of the *Pasteurella multocida* hyaluronan synthase: two active sites exist in one polypeptide. Glycobiology 10:883–889, 2000.

115. PL DeAngelis. Hyaluronan synthases: fascinating glycosyltransferases from vertebrates, bacterial pathogens, and algal viruses. Cell Mol Life Sci 56:670–682, 1999.

116. WD Reiter, C Chapple, CR Somerville. Mutants of *Arabidopsis thaliana* with altered cell wall polysaccharide composition. Plant J 12:335–345, 1997.

117. I Allona, M Quinn, E Shoop, K Swope, SS Cyr, J Carlis, J Riedl, E Retzel, MM Campbell, R Sederoff, RW Whetten. Analysis of xylem formation in pine by cDNA sequencing. Proc Natl Acad Sci USA 95:9693–9698, 1998.

118. F Sterky, S Regan, J Karlson, M Hertzberg, A Rohde, A Holmberg, B Amini, R Bhalerao, M Larsson, R Villaroel. Gene discovery in the wood-forming tissues of poplar: analysis of 5692 expressed sequence tags. Proc Natl Acad Sci USA 95:13330–13335, 1998.

119. M Fagard, H Höfte, S Vernhette. Cell wall mutants. Plant Physiol Biochem 38:15–25, 2000.

120. W-D Reiter. The molecular analysis of cell wall components. Trends Plant Sci 3:27–32, 1998.

121. AJ Walhout, M Vidal. A genetic strategy to eliminate self-activator baits prior to high-throughput two-hybrid screens. Genome Res 9:1128–1134, 1999.

122. A Janshoff, M Neitzert, Y Oberdörfer, H Fuchs. Force spectroscopy of molecular systems—single molecule spectroscopy of polymers and biomolecules. Angew Chem Int Ed 39:3212–3237, 2000.

20

Lignin Genetic Engineering: A Way to Better Understand Lignification beyond Applied Objectives

Alain-M. Boudet

Institute of Plant Biotechnology, UMR CNRS/UPS 5546, Castanet-Tolosan, France

I. INTRODUCTION

Over the last 35 years, the production of paper has more than tripled. To be competitive, to reduce the uncontrolled exploitation of wild forests, and to limit energy and environmental problems, the forest products industry must develop genetically improved woody materials. Until now, genetic improvement of trees was achieved using conventional breeding methods, which are particularly difficult to exploit for woody species because of long generation times and evaluation of specific characters at the adult stage. Among the various facets of plant biotechnology, genetic engineering offers new opportunities and attractive perspectives for improving woody species. The first transgenic tree was obtained in 1987. Since that time, the development of genetic transformation techniques for several tree species and progress in the characterization of several important genes in woody plants have opened the door to wide exploitation of genetic engineering in forest management.

Genomic and genetic transformations are entering the field of plant biology at an extremely rapid pace, and their impact should enhance understanding of plant function and lead to the improvement of crops and forest products. This may correspond to increases in productivity but also to optimization of the chemical composition to provide new products better suited to nutritional, pharmaceutical, or industrial purposes.

As the composition of wood is very important for the pulp industry, lignin genetic engineering is a very active area of research that has been stimulated in recent years by the characterization of important genes controlling lignification. A significant number of transformed plants exhibiting qualitative and quantitative changes in their lignins have already been obtained (1–3) and, in a limited number of cases, preliminary data have demonstrated the industrial usefulness of the resulting products.

Pulp and paper production requires the use of costly, energy-consuming, and often polluting treatments to separate lignins from cellulose, and it is clear that genetically engineered trees that produce modified lignins (in lower amounts or that can be more easily processed) would lead to substantial industrial and environmental benefits.

The successful extension to clonal forestry of the efficient strategies defined on model systems has at least four prerequisites: convenient germ plasm, large-scale vegetative propagation, convenient transformation techniques, and long-term in-field confirmation of the effects observed in model plants. Investigations are being actively conducted along these different lines and, in parallel, investigations are conducted on easy-to-transform model plants in order to define the best ways to engineer new cell wall compositions.

This chapter will critically highlight recent results in this area, suggest new strategies for the future, and emphasize the long-term benefits that can be expected from manipulating lignin metabolic pathways.

II. RECENT ACHIEVEMENTS IN THE FIELD OF LIGNIN GENETIC ENGINEERING

Among the polymers that contribute to the plant cell wall, little has been done with genetic manipulation of cellulose and hemicelluloses because of a lack of convenient regulatory genes. In contrast, over the last decade, it has become clear that genetic engineering can tackle lignin composition and content. This is due to a reasonable knowledge of the biosynthetic pathways (and corresponding enzymes) leading to lignins even though recent results have revealed unexpected new pathways or new regulatory circuits. Figure 1 summarizes our present knowledge of lignin biosynthetic pathways.

It is clear that the synthesis of lignins from phenylalanine proceeds through three different sets of reactions.

1. The common phenylpropanoid pathway, which provides general intermediates (cinnamoyl-CoAs) for different phenolic syntheses
2. The monolignol specific pathway, which gives rise to the three building units of the lignin polymer, the coumaryl, coniferyl, and sinapyl alcohols
3. The polymerization of monolignols, for which peroxidases and laccases are candidates

Most of the genes encoding the enzymes along these pathways have now been characterized with the exception of the hydroxylase converting coumaroyl–coenzyme A (CoA) into caffeoyl-CoA. These genes have been exploited through antisense or cosuppression strategies in order to reduce the activity of the corresponding enzymes and alter the functioning of the lignin pathway.

The modifications of lignin profiles that have been envisaged until now have dealt either with reduction in lignin content or with changes in the lignin monomeric composition because this parameter is important for lignin extractability. However, it soon appeared that experiments aiming to decrease the lignin content indirectly altered the chemical composition of transgenic lignins. In the following sections significant results obtained through the down-regulation of different target genes will be highlighted irrespective of the initial objective.

A. Down-regulation of *O*-Methyltransferases

Higher plants contain at least three classes of *O*-methyltransferases that can control the degree of methylation of the monolignols. Genetic engineering experiments have confirmed that caffeic acid 3-*O*-methyltransferases (COMTs) give rise to sinapic acid from 5-hydroxyferulic acid and that caffeoyl-CoA 3-*O*-methyltransferases (CCoAOMTs) lead to feruloyl-CoA. In addition, other methyltransferases can probably convert hydroxyconiferaldehyde and hydroxyconiferyl alcohol into sinapaldehyde and sinapyl alcohol, respectively.

The down-regulation of COMT both in tobacco and in poplar (4–6) through the antisense strategy did not decrease the lignin content, whereas the monomeric composition was dramatically modified. Transgenic lignin has a decreased S content while the amounts of G units remained relatively unchanged. In addition, these transformed lines accumulated a novel 5-hydroxyguaiacyl unit in their lignins. Chemical and pulping experiments revealed that these lignins contain, as could be predicted, more condensed bonds (C—C) and are more difficult to extract from the cell wall (higher kappa number) than the control lignins. These transformed plants are not particularly of interest to the pulp industry but may be of interest if the

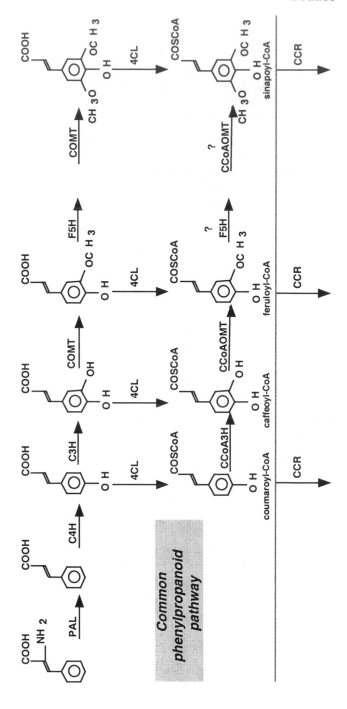

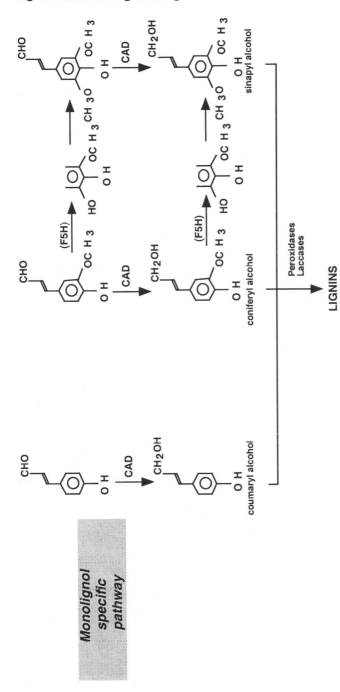

FIGURE 1 The lignin biosynthetic pathway. PAL, phenylalanine ammonia-lyase; C4H, cinnamate-4-hydroxylase; C3H, 4-coumarate-3-hydroxylase; COMT, caffeic acid 3-O-methyltransferase; CCoAOMT, caffeoyl-CoA 3-O-methyltransferase; F5H, ferulate-5-hydroxylase; 4CL, hydroxycinnamate-CoA-ligase; CCR, cinnamoyl-CoA-reductase; CAD, cinnamyl alcohol dehydrogenase. F5H may hydroxylate coniferyl aldehyde and coniferyl alcohol with better efficiency than ferulic acid.

wood is used for fuel because the calorific value of condensed lignins should be improved.

Several groups have envisaged down-regulating another methyltransferase (CCoAOMT) considered to be particularly involved in the synthesis of coniferyl alcohol (7) (M. Legrand, personal communication; W. Boerjan, personal communication). CCoAOMT has been shown to be encoded by multigene families in different plants including poplar (8) and tobacco (9), and specific members of these multigene families have been used in antisense or cosuppression experiments.

Reduction of CCoAOMT alone in transgenic tobacco plants resulted in a decreased lignin content but a simultaneous reduction in CCoAOMT and COMT activities induced a further reduction in lignin content, confirming that both enzymes are indeed involved in methylation reactions in lignin biosynthesis (7). Surprisingly, although the transgenic plants showed a 40 to 60% reduction in lignin content, they appeared to grow normally under greenhouse conditions even though they exhibited a deformation of vessel elements. These results have been partly confirmed by M. Legrand's group in Strasbourg (France) (personal communication) on the same material. However, in addition to a reduction in lignin content, CCoAOMT down-regulation induced a decreased growth rate and a dramatic disorganization of vascular tissues (reduction of xylem thickness, reduction of vessel diameter). These discrepancies concerning the growth of the plants have not yet been explained. W. Boerjan et al. (personal communication) obtained poplar lines cosuppressed for CCoAOMT that exhibit a slight reduction in lignin content. It thus appears that CCoAOMT alone or in combination with COMT is an interesting target for lignin genetic engineering, particularly if the developmental effects observed for some transgenic lines can be reduced. From a qualitative point of view, G units were preferentially reduced in the β-O-4-linked monomer fraction of these transgenic plants, leading to an interesting increase in the S/G ratio of the lignins.

B. Down-regulation of 4-Coumarate CoA Ligase (4CL)

This enzyme governs a key step of the common phenylpropanoid pathway leading to hydroxycinnamoyl CoAs. A down-regulation of its activity should lead to pleiotropic effects, such as those observed when phenylalanine ammonia-lyase (PAL) and cinnamate-4-hydroxylase (C4H) are down-regulated, if the targeted enzymes (isoforms) are not strictly dedicated to monolignol biosynthesis. However, interesting results have been obtained in *Arabidopsis*, tobacco, and poplar 4CL down-regulated plants, suggesting that this enzymatic step is an interesting candidate for lignin genetic engineering (10–12). In both *Arabidopsis* and tobacco, a nearly total block in 4CL activity led to only a modest reduction in total lignin content and the plants were morphologically normal. In contrast, transgenic aspen exhibited a substantial reduction in lignin quantity and a 15% relative increase in cellulose content (13). In addition, tree growth was substantially enhanced and phenolic profiles of the upper leaves and shoot apex were significantly altered in transgenic lines (14). Even though their rationale is not understood at the moment, these secondary effects on plant growth and cellulose content are particularly interesting in the context of the pulp industry and these investigations merit further attention.

C. Cinnamoyl CoA Reductase (CCR) Down-regulation

As the first committed enzyme in monolignol biosynthesis, CCR channels phenylpropanoid metabolites into the biosynthesis of lignins. Significant down-regulation of CCR activity in tobacco was obtained by ectopic expression of the homologous tobacco antisense gene (15). The CCR down-regulated tobacco plants with a moderate decrease of CCR activity exhibited only a very slight reduction in lignin content and had a normal phenotype, whereas the most severely inhibited transformant showed a 50% reduction in lignin content and an abnormal, heritable phenotype (reduced growth, abnormal leaf morphology). In these plants, the hypolignified xylem vessels were unable to withstand the compressive forces generated during transpiration and tended to collapse inward. In these transformants, the yield of thioacidolysis products was reduced, reflecting a lower proportion of β-O-4 bonds in lignin. However, this uncondensed lignin fraction was enriched in S units. A substantial increase in G units containing free phenolic groups was also observed (C. Lapierre, personal communication). In addition, an increase of cell wall–linked phenolics released by mild alkaline hydrolysis occurred, the main enrichment concerning ferulic and sinapic acids and acetosyringone. It was suggested that the incorporation of ferulic acid into the cell wall was responsible for the brown-orange coloration observed as a consequence of CCR silencing (15). Additional studies using ^{13}C nuclear magnetic resonance (NMR) revealed the presence of ferulate tyramine in cell walls of the CCR antisense line (16). These different phenolic compounds could be an integral part of the lignin polymer because, at least in grasses, Jacquet et al. (17) and Ralph et al. (18) have shown that ferulate esters can be converted to phenoxy radicals that copolymerize with lignin polymers. It is clear that CCR silencing induces a marked reorientation of phenolic metabolism, resulting in a new partitioning of phenolic precursors into lignins and other related phenolic sinks.

Simulated pulping experiments performed on CCR down-regulated tobacco plants obtained in our laboratory or in collaboration with Zeneca (19) revealed a significant decrease in kappa number with little modification of other parameters such as cellulose yield or cellulose degree of polymerization (DP).

Taken together, these data suggest that CCR controls the entry of carbon flux into the lignin pathway. However, the resulting effects—reduction in lignin, increase in cell wall–associated phenolics and in soluble phenolics, potential decrease in other monolignol-derived compounds such as lignans and/or dehydroconiferyl alcohol glucosides—may have an impact on plant development. Whatever the molecular mechanisms involved, it is clear that these results illustrate the unexpected effects that can be associated with a dramatic decrease in lignin content (and associated responses). A compromise should thus be attained between a moderate reduction in lignin content and normal development.

D. Cinnamyl Alcohol Dehydrogenase (CAD) Down-regulation

CAD down-regulation was one of the most impressive successes of lignin genetic engineering research a few years ago (20). Since that time, the results initially obtained on tobacco have been confirmed by independent reports of CAD suppression in tobacco and have been extended to poplar (6,21–23). The plants with a strong reduction of CAD activity have normal development and a novel red color of the xylem at certain developmental stages. Their lignin content is not (tobacco) or slightly (poplar) altered but their lignins are enriched in coniferyl and sinapaldehydes, the aldehyde substrates of CAD. The newly formed lignins, which have been characterized in detail by NMR (16), are more easily extracted from antisense plant samples by sodium hydroxide or thioglycolic acid, suggesting that the structural changes in the antisense lignin make it more extractable.

Kraft pulping experiments have shown more extensive delignification of low-CAD tobacco and poplar plants when compared with control plants, demonstrating the benefit of this genetic manipulation for pulp and paper making (6), and field trials of these CAD down-regulated poplars are currently being performed in France and in the United Kingdom. A confirmation of the effects of CAD down-regulation has been indirectly obtained from the study of a CAD-deficient *Pinus taeda* mutant (24) that accumulates aldehyde components and substantial levels of dehydroconiferyl alcohol in its lignins. Here again, the mutant plant stands normally and has a normal phenotype (except for the red coloration of xylem) and the resulting lignins are more easily extracted. These is no doubt that these CAD-deficient plants are particularly promising for the pulp industry.

E. Double Transformants

In order to combine the advantages of the down-regulation of individual genes and to yield new and useful lignins, several groups have envisaged producing double and triple transformants by crossing existing plants with single antisense genes or using transgenes containing multiple genes under the control of a single promoter.

Our recent results on CCR/CAD down-regulated plants illustrate the potential interest of this approach. By crossing homozygous tobacco lines down-regulated for CCR and for CAD, we have obtained hybrid lines down-regulated for both enzymes. If the hybrid exhibits intermediate values between the two parents for some characteristics, its lignin content is surprisingly lower than for the CCR down-regulated homozygous line. However, in contrast to this parent line, its size and morphology are not affected. This hybrid line with only 40% of residual lignin was processed in simulated kraft pulping experiments by the Centre Technique du Papier, Grenoble (France) (M. Petit-Conil, personal communication), and a 35% decrease in kappa number, a 24% increase in yield, and a 13% increase in cellulose DP were observed. These interesting results, which have been confirmed several times on tobacco plants obtained in culture-room conditions, should now be extended to other species grown in natural conditions. They suggest that the CCR and CAD transgenes work in synergy in a still undefined way to give rise to new lignin profiles without altering plant morphology.

III. POTENTIAL SIDE EFFECTS AND COMPENSATION MECHANISMS ASSOCIATED WITH LIGNIN MODIFICATIONS

One envisaged drawback related to lignin manipulation in plants could be the potential attenuation of defense mechanisms against pathogens. Indeed, lignification is considered one of the adaptive plant responses to pathogen attacks (25). Down-regulating lignin synthesis could reduce the ability of plants to protect themselves against pests. Such a hypothesis has not yet been substantiated using the transgenic plants already obtained, but the use of chemical inhibitors of lignification resulted in decreasing the hypersensitive response in wheat plants challenged by *Puccinia graminis* (26). In order to minimize the drop in these defense mechanisms, which usually occur at the surface of the tissues, the use of specific promoters to drive transgene expression has been envisaged. Ideally, such a promoter should be xylem specific, avoiding a decrease in stress lignins produced on the plant surfaces.

Another important aspect of lignin manipulation is the potential disturbance of plant development. Reduction or modification of the lignin content of plant cell walls may theoretically exert pleiotropic effects on plant functions through changes in the strength of plant organs, sap conduction through the xylem, or permeability of cell wall barriers (caspary band). Changes in plant development have been observed for strongly depressed CCR tobacco lines even though it has not been clearly demonstrated whether the resulting phenotype (reduced growth) is due to changes in cell wall properties or to changes of soluble phenolic profiles. Such a reduction in size has also been demonstrated for CCoAOMT down-regulated tobacco plants. Surprisingly, CCR/CAD double transformants exhibit a strong decrease in lignin content but no changes in morphology (at least in culture room conditions). This last observation suggests that the decrease in lignin content per se is not directly responsible for the developmental changes observed but that other unknown metabolic changes are probably involved.

It is worthwhile mentioning that in our hands the same tobacco transgenic line (CCR down-regulated) that exhibited normal growth in culture room conditions was developmentally altered when grown in field conditions. It is possible that the outside habitat provides more adverse conditions (e.g., water stress, mechanical stress) and that the transgenic line adapts itself less easily to this natural context.

It is clear that for applied purposes the transformed plants should behave as wild-type plants in terms of biomass production and agronomic performance. This has been verified for 4-year-old CAD down-regulated poplars, which appear to be promising candidates for immediate application. The same evaluation procedures at the field level should be performed in the long term for each new transformed line.

From the functional point of view, plants with altered lignin synthesis seem to adopt compensatory strategies for maintaining the integrity of their cell walls. This is clearly illustrated in the case of CAD down-regulated plants that utilize cinnamaldehydes instead of cinnamyl alcohols as building blocks of their transgenic lignins. CCR down-regulated tobacco plants partially compensate for the lack of lignin in their cell walls by a striking increase in tyramine ferulate (16). Additional reinforcement of the cell wall may also be provided in these transgenic lines by a very significant increase in wall-associated phenolic compounds (15).

The question is still open whether these new biochemical phenolic decorations of the cell wall or of the cytosolic fraction have a positive or a negative impact on the pulping process. A negative example is provided by *Eucalyptus camaldulensis* CAD down-regulated plants that overproduce tannins, interfering with the bleaching step during the pulping process (T. Ona, personal communication). In contrast, if the compensation strategy could concern useful components of the cell wall, the benefits would be obvious. This situation is illustrated by the exciting results obtained by Chiang's group (14) for 4CL down-regulated aspen, for which the down-regulation of lignin was accompanied by an increase in cellulose leading to a nearly doubled cellulose/lignin ratio. If the required defense, water transfer, and other properties survive, as they appear to in these particularly vigorous plants, the potential for improving plant utilization is enormous.

IV. FUTURE TARGETS FOR ENGINEERING NEW LIGNINS

In addition to relatively well-defined strategies using, for example, CAD down-regulation, which can be of immediate utility, and other potential interesting target genes such as CCoAOMT, 4CL, or CCR whose interest has to be confirmed, other approaches and more diversified plant material will be exploited in the future for lignin genetic engineering.

Some of the prerequisites for efficient targeted genetic modification of lignin profiles are

A new cell wall composition more adapted to specific agroindustrial purposes

A lack of secondary effects potentially inducing a reduction in plant productivity or in the efficiency of the pulping process

Easy adaptation to plant species of high economic interest

Different strategies aiming to fulfill the first two criteria are to be envisaged in the future. Among them are the use of new target genes downstream in the lignin pathway such as laccase/peroxidase genes potentially involved in the polymerization step of lignins and/or UDPG glucosyltransferase and coniferin β-glucosidase genes potentially involved in the transport, storage, and mobilization of monolignol precursors. The exploitation of these new targets should reduce the pleiotropic effects due to a shortage in the production of monolignols. However, it is difficult to predict whether the down-regulation of these genes would avoid, through sequential feedback mechanisms, the accumulation of intermediates upstream in the pathway.

Other strategies that have been particularly successful in the case of flavonoid metabolism manipulation deal with the use of transcription factor genes regulating a whole pathway. Genes encoding transcription factors of the MYB family are good candidates because in many of the promoters of the genes encoding enzymes of the phenylpropanoid and lignin branch pathway, binding sites for these proteic factors have been characterized. In addition, L. Tamagnone et al. (27) have shown that ectopic expression of MYB genes in tobacco induced a reduction in lignin content. Interestingly, A. Kawaoka and H. Ebinuma (28) have identified a new transcription factor that controls gene expression in the lignin biosynthesis pathway (Nt lim 1).

Its deduced amino acid sequence is similar to that of the LIM protein family that contains a zinc finger motif. Transgenic tobacco plants with antisense Nt lim 1 showed low expression levels of phenylalanine ammonia-lyase, 4-coumarate CoA ligase, and cinnamyl alcohol dehydrogenase and a decrease in lignin content. This simultaneous down-regulation of several enzymes is particularly interesting because it can prevent the unwanted accumulation of phenolic intermediates along the phenylpropanoid and lignin pathways and facilitate the reorientation of carbon flux into primary metabolism.

The production of multiple transformants down-regulated for several genes is also interesting to fine-tune lignin profiles. This strategy is elegantly exploited by C. Halpin et al. (29) through the use of multiple genes under the control of a single promoter, and in our hands it has already suggested the potential interest of CCR/CAD double transformants.

In addition to down-regulation of specific genes, the overexpression or the combined down-regulation and overexpression of key genes along the pathway may be envisaged. That is particularly illustrated by the overexpression of an angiosperm gene *F5H* in gymnosperms to increase their syringyl unit content and to make the lignin more readily extractable. Several groups are initiating research projects on this topic.

Whatever the gene or the combination of genes used, the limitation of transgene expression to specific tissues or even to specific types of cells within the lignified tissues may be an important way to avoid undesirable effects. Until now, promoters of lignification genes such as *CCR* or *CAD* were, at least in our hands, less efficient than the 35S CAMV promoter in driving *CAD* and *CCR* antisense expression. However, K. Meyer et al. (30) have shown that the C4H promoter was more efficient than the 35S CAMV promoter in driving the expression of F5H in *Arabidopsis*. These results show the potential exploitation of specific promoters for limiting the expression of the transgenes. More sophisticated approaches may be further envisaged through the use of promoters expressed in specific parts of the xylem such as fibers or vessels (W. Boerjan, personal communication) in order to alter fiber composition, for example, without disturbing the structure and function of the vessels.

V. A SOCIOECONOMIC PERSPECTIVE

World pulp production reached 198.2 million tons in 1997, and the European Union contributed 33.1 million tons to this total. It is expected that the success in lignin content reduction, in lignin solubility increase, or in lignin structure modification will have a very important impact on the pulp and paper industry. There are a number of projects in progress around the world aiming to document the effect on trees of the modifications of lignification gene expression. One of the most advanced, partially funded by the Euro-pean Union (EU), began in 1989 (OPLIGE) and gave rise to field tests of genetically engineered poplars that are under way in the United Kingdom and France. Shell is developing an experiment with CAD down-regulated *Eucalyptus* trees in Uruguay. The trees were planted at the end of 1997 and have another year to go before they will be mature enough for felling and testing. Advanced Technologies Cambridge (ATC) in collaboration with our laboratory is also involved in the transformation of *Eucalyptus camaldulensis* with different antisense constructs containing lignification genes. Other programs within the EU-supported project TIMBER aim to introduce such genes into spruce or into *E. globulus*, an important species for southern Europe. Additional target species for which genetic transformation appears to be realistic are larch and the *Pinus* species *Pinus radiata* and *Pinus taeda*, for which efficient protocols using biolistic transformation of embryogenic tissues have been designed.

One initiative that attracted a great deal of media attention was the $60 million joint venture between Fletcher Challenge Forest, International Paper, Westvaco, and Monsanto announced in April 1999. Genesis Research and Development, Auckland, New Zealand is providing the genomic research. The experiment will concentrate initially on eucalyptus and poplar species, radiata pine, loblolly pine, and sweet gum. The genetic improvements aimed for include herbicide tolerance, higher growth rates, and improved fiber quality in order to meet the world's wood and fiber needs without increasing pressure on native forest. In addition to the European investments illustrated by the EU OPLIGE and TIMBER projects, this new initiative demonstrates the worldwide interest in these new biotechnologies applied to trees.

In the future, it would be crucial to combine the availability of convenient genes and the ability to transform important tree species easily. Protection of intellectual property is an important aspect of these new strategies, and several companies are launching research programs in order to characterize new interesting genes to avoid sublicensing procedures.

The European groups have already filed various patents protecting the exploitation of different key genes (*COMT*, *CCR*, *CAD*) and are well advanced in the transfer of the technology to species of economic interest. In addition to previous studies on the poplar hybrid INRA 717 B4 *Populus tremula* × *P. alba*, easy to transform but of limited economic interest, significant results have been obtained on the economically important poplar OGY cultivar (*Populus deltoides* × *P. nigra*).

One of the main concerns about genetically modified trees is their acceptance in public opinion. The fact that these transformed plants are nonfood products designed for industrial purposes should facilitate public acceptability. However, the recent destruction of genetically modified poplar trees at the Jealott's Hill experimental station of ZENECA near Bracknell Berkshire indicates the strong negative reactions of anti-GMO (genetically modified organisms) activists. One of the main potential problems in addition to the unknown long-term effects is the risk of cross-pollination and the possible effect on biodiversity. In the ZENECA field trial, the trees were female and not able to produce pollen. But several approaches are being exploited worldwide for genetically engineering sterility at the same time as introducing the constructs of interest (31) in order to avoid the release of pollen from transgenic lines.

Risk assessment for transgenic trees presents special challenges because of the space and time scale involved but also because of the complex interplay of ecological, genetic, agronomic, and social factors. Altering the lignin profiles of plants should not provide a competitive advantage to the transformed lines but could limit their resistance to different pests. A careful examination of these new transgenic products is then absolutely necessary in order to evaluate the potential negative impacts, if any, that could outweigh the already identified advantages for the competitiveness of the pulp industry. The process of genetically modifying trees is very much in its infancy, and the public perception of the new products is a critical aspect of its wide development.

VI. CONCLUDING REMARKS AND PROSPECTS

It is clear that the lignin content and composition of crops and woody species can be manipulated through genetic engineering. In addition to the application-orientated objectives that have triggered most of the research in this area, the impressive amount of data that have been accumulated in the last few years has generated new fundamental concepts about lignin biosynthesis pathways and lignin chemical flexibility. As with other metabolic engineering experiments, the results obtained have also revealed pleiotropic and unexpected effects beyond the targeted pathway. Some of these could be explained through cross-talks between genes because it has been recently shown by G. Pinçon et al. (32) that COMT gene down-regulation has an impact on the expression of other genes of the lignification pathway. Lignin genetic engineering is also a way to explore the subtle regulations occurring during the lignification process. The analysis of our CCR × CAD tobacco double transformants has shown that in addition to a strong reduction in lignin content, the simultaneous down-regulation of these two activities has a significant and specific impact on the nature of lignins synthesized and on the preferential cell localization of the residual lignins. In collaboration with K. Ruel and A. Yoshinaga (33), it was demonstrated that the transgenic lignins were enriched in noncondensed units and that the reduction in lignin content was more important in the fibers than in the vessels.

The new transformed plants may also help better identify the chemical characteristics of lignocelluloses, which are beneficial in wood processing. Indeed, comparisons performed on wild-type and transgenic lines represent a new exploratory approach to reevaluate the criteria important for the efficiency of the pulping/bleaching process (in addition to the classical S/G ratio). Such chemical studies have been initiated by C. Lapierre (personal communication) using a range of sophisticated methods. For example, lignins from CCR and CAD down-regulated plants are enriched in free phenolic groups, which improve their extractability.

At the present stage of progress, and in addition to the evaluation of the stability of the traits in the long term, which has already been discussed, many tests still have to be performed on the existing transgenic material in order to evaluate its industrial usefulness. Simulated pulping experiments have been performed by the CTP (Centre Technique du Papier, Grenoble, France) on tobacco stems and poplar branches and trunks of different transformants with decreased COMT, CCoAOMT, CCR, and CAD activities (6) showing several interesting modifications in terms of kappa number, cellulose yield, and cellulose DP.

The most interesting lignin modifications should now be integrated in the pulping/bleaching process taking into account the constraints and limitations associated with each technological stage. Breeders and chemists involved in the pulp industry are usually looking for genetically engineered trees with the lowest lignin content possible but enough lignin (or the equivalent!) to enable them to withstand environmental hardships (winds, heavy rains, etc.). At the same time, they want to process trees with lignin that would be easier to dissolve, thus requiring less chemicals, less heat, and less retention time in the digester. The resultant pulp should have equal or improved quality (attributes desired in the pulp and in the end-use product, paper) when compared with current bleached kraft market pulp.

From the same biological resource, the quality of the pulp can be improved by modifying intermediate steps in the process and particularly in the cooking and the bleaching stages. However, this improvement is usually associated with a decrease in yield and an increase in the amount of chemicals and energy required. Thus, it is clear that most of the time, a compromise has to be found between opposing objectives. An optimized resource should increase the yield and the quality of the pulp and/or decrease the quantity of energy and chemicals used in the transformation process.

Taking into account these potential limitations, the existing technology should now be rapidly extended to different woody species of economic interest in light of the progress observed in tree genetic transformation [see, for example, D. Clapham et al. (34) and M. Maralejo et al. (35)]. New avenues are also open to optimize and to extend the first results. They include, as partially described, the fine tuning of the lignin content or composition through the use of a combination of genes and the exploitation of transcription factors.

Finally, progress in the characterization of the genes involved in the biosynthesis of the polysaccharides of the cell wall (36) opens the way to targeted genetic manipulation of cellulose and hemicelluloses in plants. A combined modification of polysaccharides and lignin profiles can also be envisaged.

The resulting transformants should increase the natural variability of cell wall components in plants. They should then contribute directly or through integration in classical breeding programs to the improvement of the resource used for pulp and paper making.

ACKNOWLEDGMENT

We would like to thank the European Community (DGXII) for its support of the OPLIGE "ECLAIR programme (Contract n° AGRE-CT92-0021)" and TIMBER "FAIR programme (Contract n° FAIR-CT95-0424) (DGXII)" projects. The collaboration of Christine Guidice in the preparation of this chapter is acknowledged.

Noted added in proof: After the writing of this chapter several new results have been obtained concerning the lignin biosynthetic pathway: the gene encoding 4-coumarate 3-hydroxylase has been characterized: Schoch et al. J Biol Chem 276(29):36566–36574, 2001. A new gene encoding a hydroxycinnamyl dehydrogenase converting specifically sinapaldehyde into synapyl alcohol (the SAD gene) has been discovered: Li et al. Plant Cell 13:1–20, 2001. Finally several arguments support the idea that free sinapic acid is not a precursor of monolignols: Harding et al. Plant Physiol 128(2): 428–438, 2002.

REFERENCES

1. A-M Boudet, J Grima-Pettenati. Lignin genetic engineering. Mol Breed 2:25–39, 1996.
2. M Baucher, B Monties, M Montagu, W Boerjan. Biosynthesis and genetic engineering of lignin. Crit Rev Plant Sci 17:125–197, 1998.
3. J Grima-Pettenati, D Goffner. Lignin genetic engineering revisited. Plant Sci 145:51–65, 1999.
4. R Atanassova, N Favet, F Martz, B Chabbert, M-T Tollier, B Monties, B Fritig, M Legrand. Altered lignin composition in transgenic tobacco expressing *O*-

methyltransferase sequences in sense and antisense orientation. Plant J 8:465–477, 1995.

5. J Van Doorsselaere, M Baucher, E Chognot, B Chabbert, M-T Tollier, M Petit-Conil, J-C Leplé, G Pilate, D Cornu, B Monties, M Van Montagu, D Inzé, W Boerjan, L Jouanin. A novel lignin in poplar trees with a reduced caffeic acid/5-hydroxyferulic acid O-methyltransferase activity. Plant J 8:855–864, 1995.

6. C Lapierre, B Pollet, M Petit-Conil, G Toval, J Romero, G Pilate, J-C Leplé, W Boerjan, V Ferret, V De Nadai, L Jouanin. Structural alterations of lignins in transgenic poplars with depressed cinnamyl alcohol dehydrogenase or caffeic acid O-methyltransferase activity have an opposite impact on the efficiency of industrial kraft pulping. Plant Physiol 119:153–163, 1999.

7. R Zhong, WH Morrison, J Negrel, Z-H Ye. Dual methylation pathways in lignin biosynthesis, Plant Cell 10:2033–2046, 1998.

8. C. Chen, H Meyermans, M Van Montagu, W Boerjan. Cell-specific expression of chimeric CCoAOMT genes in transgenic poplar. Proceedings of 8th International Cell Wall Meeting, Norwich, 1998.

9. F Martz, S Maury, G Pinçon, M Legrand. cDNA cloning, substrate specificity and expression study of tobacco caffeoyl-CoA 3-O-methyltransferase, a lignin biosynthetic enzyme. Plant Mol Biol 36:427–437, 1998.

10. D Lee, K Meyer, C Chapple, CJ Douglas. Antisense suppression of 4-coumarate: coenzyme 1 ligase activity in *Arabidopsis* leads to altered lignin subunit composition. Plant Cell 9:1985–1998, 1997.

11. S Kajita, Y Katayama, S Omori. Alterations in the biosynthesis of lignin in transgenic plants with chimeric genes for 4-coumarate:CoA ligase. Plant Cell Physiol 37:957–965, 1996.

12. S Kajita, S Hishiyama, Y Tomimura, Y Katayama, S Omori. Structural characterization of modified lignin in transgenic tobacco plants in which the activity of 4-coumarate:coenzyme A ligase is depressed. Plant Physiol 114:871 879, 1997.

13. J Hu, J Lung, JL Popko, SA Harding, A Kawaoka, Y-Y Kao, S Hideki, DD Stokke, PL Rinaldi, C-J Tsai, VL Chiang. Reduced lignin quality, increased cellulose content, and enhanced growth in transgenic aspen trees. The 8th Meeting of the Conifer Biotechnology working group, Rutgers University, 1998.

14. C-J Stai, SA Harding, W-J Hu, J Lung, JL Popko, VL Chiang. Enhanced growth and cellulose accumulation in lignin reduced transgenic aspen with down-regulated 4-coumarate-CoA ligase. Proceedings of joint meeting of The International Wood Biotechnology Symposium "Forest Biotechnology 99" (IUFRO Working party 2.04-06, Molecular genetics of trees), Oxford, 1999, Seminar 28.

15. J Piquemal, C Lapierre, K Myton, A O'Connell, W Schuch, J Grima-Pettenati, A-M Boudet. Down-regulation of cinnamoyl CoA reductase induces significant changes of lignin profiles in transgenic tobacco plants. Plant J 13:71–83, 1998.

16. J Ralph, RD Hatfield, J Piquemal, N Yahiaoui, M Pean, C Lapierre, A-M Boudet. NMR characterization of altered lignins extracted from tobacco plants down-regulated for lignification enzymes cinnamyl-alcohol dehydrogenase and cinnamoyl-CoA reductase. Proc Natl Acad Sci USA 95:12803–12808, 1998.

17. G Jacquet, B Pollet, C Lapierre, F Mhamdi, C Rolando. New ether-linked

ferulic acid–coniferyl alcohol dimers identified in grass straws. J Agric Food Chem 43:2746–2751, 1995.

18. J Ralph, JH Grabber, RD Hatfield. Lignin-ferulate crosslinks in grasses: active incorporation of ferulate polysaccharide esters into ryegrass lignins. Carbohydr Res 275:167–178, 1995.

19. A O'Connell, K Holt, J Piquemal, J Grima-Pettenati, A Boudet, C Lapierre, M Petit-Conil, W Schuch, C Halpin. Improved paper pulp from plants with suppressed cinnamoyl-CoA reductase or cinnamyl alcohol dehydrogenase. Transgenic Research, 2002 (in press).

20. C Halpin, ME Knight, CA Foxon, MM Campbell, A-M Boudet, JJ Boon, B Chabbert, MT Tollier, W Schuch. Manipulation of lignin quality by down regulation of cinnamyl alcohol dehydrogenase. Plant J 6:339–350, 1994.

21. N Yahiaoui, C Marque, K Myton, J Negrel, A-M Boudet. Impact of different levels of cinnamyl dehydrogenase down regulation on lignins of transgenic tobacco plants. Planta 204:8–15, 1998.

22. N Yahiaoui, C Marque, H Corbière, A-M Boudet. Comparative efficiency of different constructs for down regulation of tobacco cinnamyl alcohol dehydrogenase. Phytochemistry 49:295–306, 1998.

23. T Hibino, K Takabe, T Kawazu, D Shibata, T Higuchi. Increase of cinnamaldehyde groups in lignin of transgenic tobacco plants carrying an antisense gene for cinnamyl alcohol dehydrogenase. Biosci Biotechnol Biochem 59:929–931, 1995.

24. RR Sederoff, JJ Mac Cay, J Ralph, RD Hatfield. Unexpected variation lignin. Curr Opin Plant Biol 2:145–152, 1999.

25. MH Walter. Regulation of lignification in defense. In: T Boller, F Meins, eds. Plant Gene Research. Genes Involved in Plant Defense. Vienna: Springer, 1992, pp 327–352.

26. B Moershbacher, U Noll, L Gorrichon, HJ Reisener. Specific Inhibition of lignification breaks hypersensitive resistance of wheat to stem rust. Plant Physiol 93:465–470, 1990.

27. L Tamagnone, A Merida, A Parr, S MacKay, FA Culinanez-Marcia, K Roberts, C Martin. The AmMYB308 and AmMYB330 transcription factors from *Anthirrhinum* regulate phenylpropanoid and lignin biosynthesis in transgenic tobacco. Plant Cell 10:135–154, 1998.

28. A Kawaoka, H Ebinuma. Cloning and characterization of transcription factor Ntlim1 involved in lignin biosynthesis. Proceedings of 10th International Symposium on Wood and Pulping Chemistry, Yokohama, 1999, Vol 1 pp 510–515.

29. C Halpin, A Barakate, J Abbott, B Askari, A O'Connell, W Schuch, MD Ryan. Enabling technologies for the complex manipulation of lignin biosynthesis. Proceedings of joint meeting of The International Wood Biotechnology Symposium "Forest Biotechnology 99" (IUFRO Working party 2.04-06, Molecular genetics of trees), Oxford, 1999, Seminar 23.

30. K Meyer, AM Shirley, JC Cusumano, D Bell-Lelong, C Chapple. Lignin monomer composition is determined by the expression of a cytochrome P450–dependent monooxygenase in *Arabidopsis*. Proc Natl Acad Sci USA 95:6619–6623, 1998.

31. J Lemmetyinen, M Lännenpää, A Elo, I Porali, P Hyttinen, K Keinonen-Mettälä, T Sopanen. Towards non-flowering birches. Proceedings of joint meeting of The International Wood Biotechnology Symposium "Forest Biotechnology 99" (IUFRO Working party 2.04-06, Molecular genetics of trees), Oxford, 1999, Seminar 44.

32. G Pinçon, M Chabannes, M Lapierre, B Pollet, K Ruel, J-P Joseleau, AM Boudet, M Legrand. Simultaneous down-regulation of caffeic/5-OH ferulic acid-*O*-methyltransferase I (COMT I) and cinnamoyl-CoA reductase (CCR) in the progeny from a cross between tobacco lines homozygous for each transgene: consequences for plant development and lignin synthesis. Plant Physiology 126:145–155, 2001.

33. M Chabannes, K Ruel, A Yoshinaga, B Chabbert, A Jauneau, J-P Joseleau, AM Boudet. In situ analysis of lignins from tobacco transgenic lines down-regulated for several enzymes of monolignol synthesis reveals a differential impact of transformation on the spatial patterns of lignin deposition at the cellular and subcellular levels. Plant Journal 28(3):271–282, 2001.

34. D Clapham, RJ Newton, S Sen, S Von Arnold. Transformation of *Picea* species. In: M Jain, SC Minocha, eds. Molecular Biology of Woody Plants. Boston: Kluwer Academic Publishers 2:102–118, 2000.

35. M Moralejo, F Rochange, A-M Boudet, C Teulières. Generation of transgenic *Eucalyptus globulus* plantlets through *Agrobacterium tumefaciens* mediated transformation. Aust J Plant Physiol 25:207–212, 1998.

36. WD Reiter. The molecular analysis of cell wall components. Trends Plant Sci 3(1):27–32, 1998.

21

Transgenic Plants Expressing Tolerance Toward Oxidative Stress

Frank Van Breusegem and Dirk Inzé
Flanders Interuniversity Institute for Biotechnology, Ghent University, Ghent, Belgium

A variety of environmental stresses (such as chilling, ozone, high light, drought, and heat) can severely damage crop plants with consequent high yield losses. A common factor in all these adverse conditions is the occurrence of oxidative stress. Oxidative stress can be defined as the enhanced accumulation of active oxygen species (AOS) within several subcellular compartments of the plant. The AOS can react very rapidly with DNA, lipids, and proteins, with cellular damage as a result. Under normal growth conditions, AOS are efficiently scavenged by both enzymatic and nonenzymatic detoxification mechanisms. Nevertheless, during prolonged stress conditions, this defense system becomes saturated and cellular damage is inevitable. The key players in the defense system are superoxide dismutases, ascorbate peroxidase, and catalases. These antioxidant enzymes directly eliminate AOS. This chapter gives an overview of transgenic plants with modulated antioxidant enzyme levels (focusing on superoxide dismutases, ascorbate peroxidase, and catalase) that are produced to test the potential use of antioxidant enzymes in improving stress tolerance during adverse environmental conditions.

I. OXIDATIVE STRESS

A. Foe . . .

The AOS are generated as side products of regular cellular metabolism, but most of the AOS are formed by dysfunction of enzymes or electron transport systems as a result of a stress situation. In the cell, AOS are mainly produced in the organelles with high oxidizing metabolic activities or with intense electron flows. Together with the electron transport chain of mitochondria, chloroplasts are hence the main sites of AOS production in plants. During the reduction steps to H_2O, several AOS can be formed. Incomplete reduction of O_2 or, alternatively, acceptance of either electrons or excess energy by O_2 leads to the formation of superoxide radicals ($O_2^{\cdot-}$) and hydrogen peroxide (H_2O_2). Other sources of AOS production are the microbodies. In the glyoxisomes, H_2O_2 is produced during fatty acid degradation in the glyoxylate cycle. In the peroxisomes, photorespiration is responsible for H_2O_2 production (1).

Superoxide is a moderately reactive, short-lived AOS with a half-life of approximately 2–4 μsec. The $O_2^{\cdot-}$ cannot cross biological membranes and is readily dismutated to H_2O_2. Alternatively, $O_2^{\cdot-}$ can reduce quinones and transition metal complexes of Fe^{3+} and Cu^{2+}, thus affecting the activity of metal-containing enzymes. On the contrary, hydroperoxyl radicals ($HO_2^{\cdot}$) that are formed from $O_2^{\cdot-}$ by protonation in aqueous solutions can cross the biological membranes and subtract hydrogen atoms from polyunsaturated fatty acids and lipid hydroperoxides, thus initiating lipid autooxidation. H_2O_2 is moderately reactive and a relatively long-lived molecule (1 msec) that can diffuse some distances from its site of production (2). It may inactivate enzymes by oxidizing their thiol groups. Enzymes of the Calvin cycle as well as copper/zinc (Cu/Zn) and iron (Fe) superoxide dismutases (SODs) are also inactivated by H_2O_2. Although neither $O_2^{\cdot-}$ nor H_2O_2 seems particularly harmful at physiological concentrations, their toxicity in vivo is enhanced by a metal ion–dependent conversion into hydroxyl radicals ($OH^{\cdot}$), one of the most reactive species known in chemistry: $O_2^{\cdot-} + H_2O_2 \rightarrow O_2 + 2OH^{\cdot}$ (Haber-Weiss reaction). Hydroxyl radicals can react indiscriminately to cause lipid peroxidation, the denaturation of proteins, and the mutation of DNA (3,4).

Fortunately, plants have the capacity to cope with these aggressive agents by eliminating them with an AOS-scavenging system (see later). Under moderate stress conditions, the produced radicals can be efficiently scavenged. During periods of more severe stress, however, these scavenging systems become saturated with the increased rate of radical production. The presence of excessive AOS results, for instance, in damage to the photosynthetic apparatus, bleaching of leaves by oxidation of the pigments resulting in severe yield losses. The AOS are excessively produced under a wide variety of stresses (such as drought, heat, salt, pollution, chilling, and ozone stress). These, at first sight unrelated, stresses all have in common the electron leakage from electron transport chains in chloroplasts and mitochondria (1).

B.   . . . And Friend

In certain situations, the plant uses AOS in a beneficial way. The AOS play an important role in the induction of protection mechanisms during biotic and abiotic stresses. The best known example is their role in the activation of resistance responses during incompatible plant-pathogen interactions. Upon infection, an NADPH oxidase of the plasma membrane is activated, producing superoxide radicals (5). Superoxide radicals are, through spontaneous dismutation or SOD activity, converted into H_2O_2. The defensive properties of H_2O_2 are situated on several levels.

1.   High levels of H_2O_2 are toxic for both pathogen and plant cells. Killing the plant cells surrounding the infection site inhibits spreading of a (biotrophic) pathogen (6).
2. H_2O_2 can serve as a substrate in peroxidative cross-linking reactions of lignin precursors (7) and induces cross-linking of cell wall proteins (8). A reinforced plant cell wall slows down the spreading of the pathogen and makes new infections more difficult.
3.   Because H_2O_2 is relatively stable and diffusible through membranes (in contrast to superoxide), it is a good candidate to act as a signal molecule during stress responses. H_2O_2 induces several pathogen defense genes (coding for pathogenesis-related proteins and phytoalexin production) and a programmed cell death pathway (9,10).

These different effects of H_2O_2 are thought to be regulated not only through the level and timing of H_2O_2 induction but also through interactions with other potential signals, such as salicylic acid, superoxide, and nitric oxide (11–13). The role of H_2O_2 as a molecular signal for the induction of gene expression may not be limited to plant-pathogen interactions. Studies of maize hypocotyls and potato and mustard seedlings have shown that H_2O_2 mediates subsequent cold and heat tolerance, respectively (14–17). In barley, H_2O_2 treatment broke dormancy of seeds (18). Hence, AOS production is vital to plant development and in defense against pathogens, but it needs to be tightly controlled in order to avoid cellular damage.

II. OXIDATIVE STRESS DEFENSE MECHANISMS IN PLANTS

To limit cellular damage caused by excessive AOS levels, plants have evolved a broad variety of nonenzymatic and enzymatic protection mechanisms that efficiently scavenge AOS (19). The best-known nonenzymatic antioxidants are ascorbate, glutathione, α-tocopherol, and carotenoids. They are present in relatively high concentrations within plant cells. For a detailed overview on these components, the reader is referred to Alscher and Hess (20).

Because hydroxyl radicals are too reactive to be directly controlled, aerobic organisms prefer to eliminate the less reactive precursor forms, such as superoxide and H_2O_2, and hence prevent the formation of hydroxyl radicals. Superoxide dismutases scavenge superoxide radicals, whereas catalases and peroxidases remove H_2O_2. Catalases consume the bulk of H_2O_2. Ascorbate peroxidases remove H_2O_2 that is not accessible for catalase, because of their higher affinity and their diverse subcellular locations. Other enzymes that are involved in the removal of AOS are monodehydroascorbate reductase, dehydroascorbate reductase, glutathione reductase, and glutathione peroxidase. However, these enzymes will not be described in detail in this chapter. For an overview of these enzymes the reader is referred to Noctor and Foyer (21).

A. Superoxide Dismutases

Superoxide dismutases (SOD; superoxide:superoxide oxidoreductase; EC 1.15.1.1) can be considered key enzymes of the antioxidative stress defense mechanism (22). They directly determine the cellular concentrations of $O_2^{\cdot-}$ and H_2O_2 because they dismutate $O_2^{\cdot-}$ into O_2 and H_2O_2.

$$2H^+ + 2O_2^{\cdot-} \xrightarrow{\text{SOD}} H_2O_2 + O_2$$

Apart from a few exceptions, SODs are present in all aerobic organisms and in all subcellular compartments that have to deal with AOS. They are classified according to their metal cofactor as isozymes containing Cu/Zn, Fe, and manganese (Mn). The FeSOD and MnSOD proteins are structurally similar (23), whereas the Cu/ZnSOD family is structurally unrelated (24). The SODs catalyze the disproportionation of superoxide through an oxidation–reduction cycle of the prosthetic Cu, Mn, or Fe cofactor. Experimentally, the three SOD types can easily be distinguished via in situ gel staining (25). Incubating the gels with KCN or H_2O_2 allows discrimination between the different classes. The Cu/ZnSOD is characterized as being sensitive to both H_2O_2 and KCN; FeSOD is sensitive only to H_2O_2, whereas MnSOD is resistant to both inhibitors. However, an FeSOD from rice was shown to be resistant to both inhibitors (26). Besides their differential sensitivity toward inhibitors, these enzymes also have a distinctive subcellular distribution. The amount and relative abundance of SOD isozymes vary within each organism. Developmental control and environmental stresses that generate AOS (ultraviolet, ozone, air pollutants, low temperatures, salt stress, drought, heat shock, pathogen infections, etc.) can induce plant SOD activities (22). The MnSOD is found in the mitochondria of all eukaryotic cells, including plants; Cu/ZnSODs are found in the cytosol, peroxisomes, and chloroplasts of higher plants. Multiple Cu/ZnSOD isoforms are also present in the extracellular fluids of Scots pine (27,28). Until now, FeSODs were found only in prokaryotes and in the chloroplasts of plants.

B. Ascorbate Peroxidase

Peroxidases are ubiquitous enzymes found in plants. Besides the peroxidases, whose oxidation products play mainly physiological roles (lignification, cross-linking of cell wall matrices), a second class is also part of the AOS defense system. Ascorbate peroxidases (APXs; EC 1.11.1.11) destroy harmful H_2O_2 via the ascorbate–glutathione pathway in chloroplasts and cytosol of plants, algae, and some cyanobacteria (21). Also, APX activity has been identified in insects and purified from bovine eye tissue (29,30). The ascorbate-glutathione pathway provides protection against oxidative stress by a series of coupled redox reactions, particularly in photosynthetic tissues (21) but also in mitochondria and peroxisomes (31). The APX eliminates H_2O_2 by using ascorbate as an electron donor in an oxidation–reduction reaction. Ascorbate is then oxidized to monodehydroascorbate (MDHA).

$$2 \text{ ascorbate} + H_2O_2 \xrightarrow{\text{APX}} 2 \text{ MDHA} + 2H_2O$$

Ascorbate is regenerated from MDHA in the chloroplast membrane by ferredoxin (32) or in the stroma by MDHA reductase at the expense of NADPH. The MDHA can also spontaneously dissociate into ascorbate and dehydroascorbate (DHA), which is rereduced by DHA reductase (DHAR). The DHAR uses reduced glutathione as an electron donor. Oxidized glutathione is then recycled by NADPH-consuming glutathione reductase (GR). The APX is directly involved in the defense response against oxidative stress. In pea and radish, APX activities are induced by Fe excess and salt and drought stresses (33–35).

Winter-acclimated pine needles contain up to 65-fold more APX than summer needles (36), whereas in maize APX (together with SOD) is constitutively higher in a chilling-resistant than in a sensitive line (37). After anoxic stress, APX activity rises in wheat roots and rice seedlings (38,39).

Also in wheat, correct temporal expression of APX is an important factor for efficient seed germination. APX activity increases during germination in parallel with the rise of ascorbate levels (40).

Based on the available sequence data, seven different APXs are distinguished in plants: two soluble cytosolic forms, three cytosol membrane-bound types including a glyoxisome-bound form, one chloroplastic stromal, and one thylakoid membrane–bound APX (41). The various isoforms differ in several molecular and enzymatic properties, such as molecular weight, electron donor specificity, lability in the absence of ascorbate, pH optimum, and ascorbate and H_2O_2 affinity (42). In general, the chloroplastic isoforms are very specific for ascorbate as electron donor, whereas the cytosolic APX can also oxidize pyrogallol (43).

C. Catalases

Plants, unlike animals, have multiple forms of catalase ($H_2O_2:H_2O_2$ oxido-reductase; EC 1.11.1.6) that are mainly found in peroxisomes and glyoxisomes. Catalase activity was also found in the mitochondria of maize (44,45). Catalases directly consume H_2O_2 or oxidize substrates (R), such as methanol, ethanol, formaldehyde, and formic acid.

$$2H_2O_2 \xrightarrow{\text{catalase}} 2H_2O + O_2$$

$$H_2O_2 + RH_2 \xrightarrow{\text{catalase}} 2H_2O + R$$

There are some striking similarities in the organization of the catalase gene family in different species. Our laboratory showed that catalases can be divided into three classes according to their expression (46). The transition from glyoxisomes to leaf peroxisomes during seedling development is associated with the disappearance of class III catalases and the induction of class I catalases. In maize, however, both class I and class III are expressed in seeds, indicating that the class I catalase has a dual function in maize. Class I is most prominent in photosynthetic tissues, where they are involved in the removal of photorespiratory H_2O_2. Class II catalases are highly expressed in vascular tissues, where they might play a role in lignification, but their exact biological role remains unknown. As mentioned before, class III is abundant only in seeds and young seedlings and its activity is linked to the removal of excessive H_2O_2 that is produced during fatty acid degradation in the glyoxylate cycle in the glyoxisomes. Because catalase isozymes are rapidly induced by ultraviolet B (UV-B), ozone, and also chilling, they may play a direct role in stress protection (47,48).

III. TRANSGENIC PLANTS WITH MODIFIED ANTIOXIDANT ENZYME LEVELS

Because of the involvement of AOS in a wide variety of environmental stresses, antioxidant enzymes are interesting molecular targets for the production of new plant varieties that can cope with these stresses. Several antioxidative stress enzymes have been genetically engineered into plants to assess their potential capacity for enhancing oxidative stress tolerance. The beneficial effects observed in some of these transgenic plants can lead to interesting agronomic applications. The following section gives an overview of the state of the art concerning transgenic plants with modified levels of antioxidant enzyme levels.

A. Transgenic Plants with Elevated Superoxide Dismutase Levels

The first report on transgenic plants overproducing SOD was, however, not promising. Transgenic tobacco plants with 30- to 50-fold increased SOD activity levels due to the production of a petunia chloroplastic Cu/ZnSOD were not more resistant against methyl viologen (MV) (49). The tolerance toward oxidative stress is easily tested in vitro with MV (also known as paraquat), which is a light-activated herbicide. In the light, MV becomes an electron acceptor from photosystem I (PSI), subsequently reducing dioxygen to $O_2^{\cdot-}$. In this way, the herbicide strongly enhances the formation of superoxide radicals and is a fairly good mimic of the superoxide-forming process that occurs in vivo in illuminated chloroplasts. The MV also accepts electrons from the respiratory electron transport chain in the mitochondria and forms superoxide radicals in the dark too. Transgenic tomato plants with two- to fourfold increased SOD levels by producing the same chloroplastic Cu/ZnSOD were not better protected against photoinhibitory conditions known to increase oxidative stress (high light, low temperatures, and low CO_2 concentrations).

Pitcher et al. (50) showed that the same transgenic tobacco plants were not protected against ozone stress. Although the activity levels of enzymes involved in H_2O_2 scavenging (APX, GR, etc.) were not measured in both cases, the lack of induced tolerance is plausibly explained by the inability of the plants to cope with the elevated levels of H_2O_2 that were produced by the enhanced dismutation of superoxide radicals. Transgenic potato plants overproducing tomato cytosolic or chloroplastic Cu/ZnSOD lacked a chlorotic and wilting phenotype that was seen in wild-type plants after dipping their shoots in 100 μM MV. In culture medium with 10 μM MV, root cultures from the same transgenic potatoes grew at rates similar to those in control medium (51). In contrast to the very high overproduction of SOD reported in the transgenic tobacco plants (49), these transgenic potato plants had only a modest increase in Cu/ZnSOD activity. In this way, too high accumulation of H_2O_2 is avoided, preventing promotion of hydroxyl formation by H_2O_2 and $O_2^{\cdot-}$ interaction. A correlation between MV tolerance and chilling tolerance was found in transgenic tobacco plants that overproduce a chloroplast-localized Cu/ZnSOD from pea at twofold higher levels than the endogenous FeSOD (52,53). In agreement, moderate cytosolic overproduction of a pea Cu/ZnSOD in transgenic tobacco confers partial resistance to ozone-induced foliar necrosis (54). Up to sixfold enhancement of SOD activity led to 30 to 50% less visible foliar damage. The beneficial effect of the transgene could, however, be detected only in older leaves (leaves 3 to 5). In the youngest leaves (5 and 6), no difference between transgenic and wild-type plants could be observed. Because Cu/ZnSODs are sensitive to H_2O_2, it is also possible that some of the introduced Cu/ZnSOD activity is inhibited by its own end product. In this way, MnSOD, which is insensitive to H_2O_2, was introduced as a better candidate for engineering oxidative stress tolerance.

The MnSOD from *Nicotiana plumbaginifolia* has indeed been more consistently reported to confer resistance to oxidative stress. Transgenic tobacco plants overproducing the nuclear-encoded mitochondrial MnSOD in either mitochondria or chloroplasts (by replacing the mitochondrial transit sequence with a chloroplast transit peptide) were less sensitive to MV. Overproduction of MnSOD in the mitochondria rendered the plants more tolerant to MV in dark but not in light incubations with MV. By contrast, MnSOD overproduction in the chloroplasts made the plants more tolerant to MV in both dark and light incubations (55). The SOD overproduction in the chloroplasts also led to the induction of other antioxidant enzymes (FeSOD, APX, DHAR peroxidase) and higher glutathione and ascorbate levels (56). Enhanced SOD activity in the mitochondria had only a minor effect on ozone tolerance in transgenic tobacco. However, overproduction of SOD in the chloroplasts resulted in a three- to fourfold reduction of visible ozone injury (Fig. 1) (57). In addition to the fact that MnSOD is resistant against H_2O_2, the subcellular location may probably play an important role in determining whether positive or negative effects arise from SOD overproduction. Targeting SODs to organelles with a higher risk of AOS production seems more efficient than cytosolic overproduction. Because during most abiotic stresses the chloroplast is the main site of AOS production, it is considered the target organelle for protection against oxidative stress.

Because of their original presence in plant chloroplasts, FeSOD proteins are theoretically better candidates for overproduction in the chloroplasts. Under conditions in which H_2O_2 accumulation is not the limiting factor, the biochemical properties of the FeSOD proteins should be better

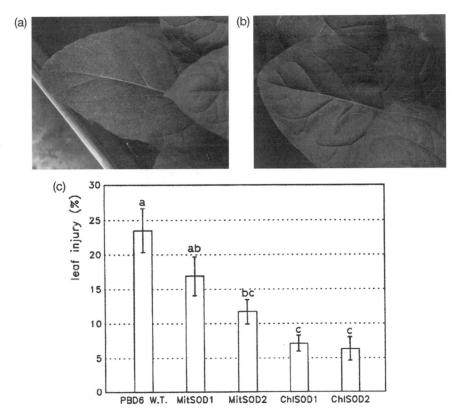

FIGURE 1 Foliar injury of leaf 6 from wild-type (a) and transgenic tobacco overproducing superoxide dismutase (b) after 7 days of ozone exposure. (c) Leaf injury in transgenic versus wild-type tobacco plants after 1 week of ozone fumigation. The values given show the mean ± standard error of the mean for wild-type plants (PBD6 W.T.), transgenic plants overproducing SOD targeted to the mitochondria (MitSOD1 and MitSOD2), and transgenic plants overproducing SOD targeted to the chloroplasts (ChlSOD1 and ChlSOD2). Means with the same letter are not significantly ($P < 0.05$) different. (©Macmillan Publishers, reprinted with permission.)

adapted to function optimally under the specific conditions within the chloroplast. This characteristic was indeed shown by overproduction in transgenic tobacco of the mature FeSOD from *Arabidopsis thaliana* coupled to a chloroplast transit peptide (58). Transgenic FeSOD protected both the plasmalemma and PSII against superoxide generated during illumination of leaf discs impregnated with MV. Overproduction of a mitochondrial MnSOD in the chloroplasts protected only the plasmalemma, but not PSII, against MV. This difference was attributed to the higher membrane affinity of FeSOD, allowing this enzyme to scavenge superoxide radicals at the site of their formation, i.e., near PSI (58).

The potential of SOD overproduction in stress engineering strategies has been reported for three crops, cotton, alfalfa, and maize. Transgenic cotton plants that produce chloroplast-localized MnSOD are more tolerant against chilling-induced oxidative stress (59). Transgenic alfalfa plants with twice the amount of total SOD-overproducing *N. plumbaginifolia* MnSOD in both the mitochondria and the chloroplast were tested for increased tolerance against oxidative stress by incubating leaves with the herbicide acifluorfen. Acifluorfen is a photobleaching, *p*-nitrodiphenyl ether herbicide that promotes accumulation of the chlorophyll precursor protoporphyrin; in the light it generates singlet oxygen that causes peroxidation in the tonoplast, plasmalemma, and chloroplast envelope. Increased resistance against acifluorfen correlated with increased freezing stress tolerance. One transgenic plant, propagated by cuttings, had increased regrowth capacities compared with nontransgenic plants when subjected to sublethal freezing temperatures from $-8°C$ to $-16°C$ (60). When viability was measured by electrolyte leakage or by tetrazolium staining, differences between transgenic and wild-type plants were minor or nonexistent, despite differences in winter survival among the plants (61). Two other experiments with two transgenic alfalfa plants producing MnSOD indicated the value of engineering oxidative stress tolerance to improve environmental stress tolerance in crop plants. For the first time, transgenic plants overproducing antioxidant enzymes were also tested in a natural field environment. The transgenic plants tended to show reduced injury from water deficit stress as determined by chlorophyll fluorescence, electrolyte leakage, and regrowth from crowns. Over a 3-year field trial, vigor and survival of transgenic plants were significantly improved (62).

It should be noted that the preceding experiments were done by using only a few transgenic plants obtained by transformation of an alfalfa cultivar with poor agronomic performance and originally not well adapted to winter conditions. McKersie et al. (62) expanded their original observations by transforming two elite alfalfa plants that are adapted to the field environment, and they examined many more independent transgenics under both laboratory and field conditions. This detailed assessment confirmed the earlier observation that transgenics had a greater survival and yield (total shoot dry matter production) in the field.

Increased winter survival was also correlated with increased SOD levels in transgenic alfalfa plants overproducing FeSOD but not with any beneficial effects on photosynthesis, growth, or oxidative stress tolerance in leaves. This correlation can be explained by the increased superoxide-scavenging capacity in the root, thereby enhancing recovery from primary freezing injury (63). Another more attractive explanation relates to enhanced levels of H_2O_2 produced by SOD overproduction. In addition to its potential toxicity, H_2O_2 is well recognized to act as a signal molecule in several defense responses (10,64,65). Altered H_2O_2 levels might induce an acclimation process that induces a general defense response to tolerate better the adverse environmental conditions during winter survival (63).

We evaluated transgenic maize plants that overproduce an *N. plumbaginifolia* MnSOD or an *A. thaliana* FeSOD complementary DNA (cDNA) fused to a chloroplast transit peptide from pea under control of the cauliflower mosaic virus 35S (CaMV35S) promoter. The recombinant MnSOD was correctly targeted to chloroplasts and its enzymatic activity could be distinguished on SOD activity gels. One transgenic line showed enhanced tolerance to MV. The growth characteristics of transgenic maize lines were followed during growth at ambient (22–25°C) and chilling temperatures (15–17°C) in growth cabinets. Although the transgenic lines in all experiments had a growth advantage compared with the wild-type lines, no statistically significant increase in growth could be observed (66).

To extrapolate the in vitro oxidative stress tolerance to improved growth effects during environmental stress conditions, such as chilling, we are currently evaluating the initiated field trials at different locations in Europe to scale up the experiments and to check the behavior of the transgenic plants under natural environmental stress conditions. Immunolocalization experiments revealed that in transgenic maize the recombinant MnSOD is mainly (although not exclusively) located in the chloroplasts of the bundle sheath cells (67). The low levels of MnSOD in the mesophyll cells could be attributed to a different expression capacity of the CaMV35S promoter, but posttranscriptional or posttranslational regulation of the recombinant MnSOD cannot be excluded. Wilson et al. (68) have also shown the vascular-specific activity of β-glucuronidase (GUS) in leaves of maize plants transformed with a CaMV35S-GUS fusion.

Because endogenous SOD and APX activities are restricted to the bundle sheath cells, whereas GR and DHAR activities could be detected only in the mesophyll tissue of maize, a clear partitioning of the AOS defense system between the mesophyll and bundle sheath cells must be present (69). This differential localization correlates with the need for NADPH of the respective enzymes. Because NADPH is limited in the bundle sheath cells, GR and DHAR activities are rate limited in this compartment. This differential distribution of antioxidants is, of course, crucial for maize plants to deal with oxidative stress. The overproduction of AOS scavengers within each cell type could restore this natural imbalance and, hence, confer a higher tolerance to maize plants against chilling-associated oxidative stress. The exclusive presence of the transgenic MnSOD in the bundle sheath cells probably influenced the expected protective effect of MnSOD in the transgenic maize plants during chilling stress.

Transgenic maize lines overproducing an A. thaliana FeSOD in the chloroplasts also suffered less from paraquat damage than controls as indicated by decreased membrane leakiness and by higher photosynthetic activity. In contrast to the MnSOD lines, the transgenic FeSOD maize plants also exhibited a significantly increased growth rate at low temperatures (as estimated from fresh weight and summed leaf length determinations) (70). These and previous results in tobacco suggest that FeSOD is a better candidate enzyme to protect plants against oxidative stress. The reason could be the difference in suborganellar location between the overproduced MnSOD and FeSOD. Because of their chloroplastic and mitochondrial origin, respectively, FeSOD and MnSOD might have different properties. Van Camp et al. (58) showed that in tobacco the transgenic FeSOD is at least partially bound to the chloroplast membrane, whereas transgenic MnSOD behaves more like a stromal enzyme. This differential subcellular location might provoke different protective effects against oxygen radicals. Transgenic FeSOD might be able to bind electrostatically to the chloroplast membrane in the vicinity of the site of radical production, resulting in increased protective properties.

B. Transgenic Plants with Modulated Ascorbate Peroxidase or Catalase Levels

Overproduction of an A. thaliana APX in tobacco chloroplasts provided almost complete protection of the PSII reaction center against aminotriazole. Aminotriazole inhibits catalase and in this way provokes accumulation of H_2O_2. Tolerance to MV was slightly better, but tolerance to eosin (a singlet oxygen generator) or chilling-induced photoinhibition was not enhanced (L. Slooten, personal communication). Transgenic tobacco plants overproducing an A. thaliana peroxisomal APX were protected against aminotriazole but not against MV (causing mainly AOS formation in the chloroplast) (71). These data show that H_2O_2 formed in the peroxisomes can diffuse and affect photosynthetic activities in the chloroplast. Protecting plant cells against oxidative stress during photorespiratory conditions can hence be done by scavenging H_2O_2 at the place of production in the peroxisomes or by providing the chloroplast with extra H_2O_2-scavenging capacity. In an ozone-sensitive transgenic tobacco, a 10-fold increase in chloroplastic APX activity was not effective against cellular injury caused by ozone stress (72). Overproduction in tobacco chloroplasts of the other major H_2O_2 scavenger, catalase, led to tolerance against MV and drought stress conditions (73,74).

By combining superoxide scavengers and H_2O_2 scavengers, even better results can be envisaged, the most spectacular results having been reported in insects. In *Drosophila melanogaster*, the overproduction of SOD and catalase resulted in a delay of aging and greater longevity, whereas overproduction of either of the two enzymes alone had only minor effects (75,76). Transgenic tobacco overproducing an *Escherichia coli* GR or a rice Cu/ ZnSOD were fivefold more resistant against ion leakage caused by MV treatments. Crossings between both transgenic lines were highly tolerant to MV concentration of 50 μM, whereas control plants were sensitive to MV concentrations as low as 1 μM (77).

Underproduction of antioxidative stress enzymes often increases sensitivity to the experienced stress. Tobacco plants with decreased APX or GR activity, obtained by antisense technology, are more susceptible to ozone stress and MV, respectively (78,79). Transgenic plants deficient in catalase (class I) can be grown only under low light conditions (<100 μmol m^{-2} sec^{-1} photosynthetic photon fluence rate). When exposed to higher light intensities, these plants developed white necrotic lesions on the leaves after 1 to 2 days. Lesion formation was induced by photorespiration because damage was prevented under elevated CO_2. Stress analysis revealed that Cat1-deficient plants were more sensitive to paraquat, salt, and ozone stress, indicating that Cat1 is a key component of several stress defenses (10,65,80,81).

IV. PERSPECTIVES

The production of crop plants that can cope with adverse environmental conditions is a very important research objective within the agroindustry. Improved production rates of crops during stress situations, such as drought and chilling, or resistance against pathogen attacks will certainly improve the life quality of the ever-growing world population in the next century. The rationale for the production of stress-tolerant crop lines is to reduce the yearly losses due to adverse environmental conditions or to expand the growing range to currently less favorable regions.

The strategy that was surveyed in this chapter is the study (and eventual modification) of the antioxidants in different crop species. Besides the production and evaluation of transgenic lines, we should also focus on a more detailed characterization of the plant AOS-scavenging machinery. From this basic research, valuable information will be gained that can be applied to future engineering strategies. In maize, SOD activity is apparently different between bundle sheath and mesophyll cells, which could be a main cause of the chilling susceptibility of maize (70). Specific enhancement of SOD activity in the mesophyll cells could lead to chilling-tolerant maize lines. The isolation of novel isoforms of AOS enzymes in plants will also provide better insight into its defense mechanisms. The molecular characterization of several chloroplastic APXs in *A. thaliana* clearly shows that the AOS-scavenging enzyme families in plants are larger than previously thought (41). With the help of genome-wide transcript profiling methods, such as cDNA-amplified fragment length polymorphism and microarray technology, or through the outcome of several genome and cDNA sequence initiatives, new isoforms of the different AOS-scavenging enzyme families will certainly be discovered.

There is evidence for a dual role for H_2O_2 (66). At higher concentrations it can promote cell death, but at lower levels it serves as a signal molecule to induce stress defense responses that can eventually lead to acclimation. Controlled modulation of H_2O_2 levels is, hence, an interesting second route to identify and improve the oxidative stress signal transduction pathway(s), leading to broader and more sustainable resistance. As described before, such a situation in which altered H_2O_2 levels induce stress tolerance might already exist in SOD-overproducing plants. Together with the identification of novel antioxidant isozymes, a better characterization of signal transduction pathways involved in oxidative stress tolerance will certainly open new opportunities for the engineering of stress tolerance in plants.

ACKNOWLEDGMENTS

The authors thank Martine De Cock for help in preparing the manuscript. FVB is indebted to the Vlaams Instituut voor de Bevordering van het Wetenschappelijk-Technologisch Onderzoek in de Industrie for a postdoctoral fellowship.

REFERENCES

1. CH Foyer, PM Mullineaux. Causes of Photooxidative Stress and Amelioration of Defense Systems in Plants. Boca Raton, FL: CRC Press, 1994.
2. A Levine, R Tenhaken, R Dixon, C Lamb. H_2O_2 from the oxidative burst orchestrates the plant hypersensitive disease resistance response. Cell 79:583–593, 1994.
3. AP Breen, JA Murphy. Reactions of oxyl radicals with DNA. Free Radic Biol Med 18:1033–1077, 1995.
4. M Desimone, A Henke, E Wagner. Oxidative stress induces partial degradation of the large subunit of ribulose-1,5-bisphosphate carboxylase/oxygenase in isolated chloroplasts of barley. Plant Physiol 111:789–796, 1996.
5. R Desikan, JT Hancock, MJ Coffey, NJ Neill. Generation of active oxygen in elicited cells of *Arabidopsis thaliana* is mediated by a NADPH oxidase-like enzyme. FEBS Lett 382:213–217, 1996.

6. A Levine, RI Pennell, ME Alvarez, R Palmer, C. Lamb. Calcium-mediated apoptosis in a plant hypersensitive disease resistance response. Curr Biol 6: 427–437, 1996.

7. RA Dixon, CJ Lamb, S Masoud, VJ Sewalt, NL Paiva. Metabolic engineering: prospects for crop improvement through the genetic manipulation of phenyl-propanoid biosynthesis and defense responses—a review. Gene 179:61–71, 1996.

8. LF Brisson, R Tenhaken, C. Lamb. Function of oxidative cross-linking of cell wall structural proteins in plant disease resistance. Plant Cell 6:1703–1712, 1994.

9. R Desikan, A Reynolds, JT Hancock, SJ Neill. Harpin and hydrogen peroxide both initiate programmed cell death but have differential effects on defence gene expression in *Arabidopsis* suspension cultures. Biochem J 330:115–120, 1998.

10. S Chamnongpol, H Willekens, W Moeder, C Langebartels, H Sandermann Jr, M Van Montagu, D Inzé, W Van Camp. Defense activation and enhanced pathogen tolerance induced by H_2O_2 in transgenic plants. Proc Natl Acad Sci USA 95:5818–5823, 1998.

11. ME Alvarez, RI Pennell, P-J Meijer, A Ishikawa, RA Dixon, C Lamb. Reactive oxygen intermediates mediate a systemic signal network in the establishment of plant immunity. Cell 92:773–784, 1998.

12. M Delledonne, Y Xia, RA Dixon, C. Lamb. Nitric oxide functions as a signal in plant disease resistance. Nature 394:585–588, 1998.

13. W Van Camp, M Van Montagu, D Inzé. H_2O_2 and NO: redox signals in disease resistance. Trends Plant Sci 3:330–334, 1998.

14. TK Prasad, MD Anderson, BA Martin, CR Stewart. Evidence for chilling-induced oxidative stress in maize seedlings and a regulatory role for hydrogen peroxide. Plant Cell 6:65–74, 1994.

15. TK Prasad, MD Anderson, CR Stewart. Acclimation, hydrogen peroxide, and abscisic acid protect mitochondria against irreversible chilling injury in maize seedlings. Plant Physiol 105:619–627, 1994.

16. JF Dat, H Lopez-Delgado, CH Foyer, IM Scott. Parallel changes in H_2O_2 and catalase during thermotolerance induced by salicylic acid or heat acclimation in mustard seedlings. Plant Physiol 116:1351–1357, 1998.

17. H Lopez-Delgado, JF Dat, CH Foyer, IM Scott. Induction of thermotolerance in potato microplants by acetylsalicylic acid and H_2O_2. J Exp Bot 49:713–720, 1998.

18. O Fontaine, C Huault, N Pavis, J-P Billard. Dormancy breakage of *Hordeum vulgare* seeds: effects of hydrogen peroxide and scarification on glutathione level and glutathione reductase activity. Plant Physiol Biochem 32:677–683, 1994.

19. D Inzé, M Van Montagu. Oxidative stress in plants. Curr Opin Biotechnol 6: 153–158, 1995.

20. RG Alscher, JL Hess. Antioxidants in Higher Plants. Boca Raton, FL: CRC Press, 1993.

21. G Noctor, CH Foyer. Ascorbate and glutathione: Keeping active oxygen under control. Annu Rev Plant Physiol Plant Mol Biol 49:249–279, 1998.

22. W Van Camp, M Van Montagu, D Inzé. Superoxide dismutases. In: CH Foyer, PM Mullineaux, eds. Causes of Photooxidative Stress and Amelioration of Defense Systems in Plants. Boca Raton, FL: CRC Press, 1994, pp 317–341.

23. WC Stallings, KA Pattridge, RK Strong, ML Ludwig. Manganese and iron superoxide dismutases are structural homologs. J Biol Chem 259:10695–10699, 1984.

24. C Bowler, W Van Camp, M Van Montagu, D Inzé. Superoxide dismutase in plants. CRC Crit Rev Plant Sci 13:199–218, 1994.

25. CO Beauchamp, I Fridovich. Superoxide dismutase: Improved assays and an assay applicable to acrylamide gels. Anal Biochem 44:276–287, 1971.

26. H Kaminaka, S Morita, M Tokumoto, H Yokoyama, T Masumura, K Tanaka. Molecular cloning and characterization of a cDNA for an iron-superoxide dismutase in rice (*Oryza sativa* L.). Biosci Biotechnol Biochem 63:302–308, 1999.

27. S Streller, G. Wingsle. *Pinus sylvestris* L. needles contain extracellular CuZn superoxide dismutase. Planta 192:195–201, 1994.

28. H Schinkel, S Streller, G Wingsle. Multiple forms of extracellular superoxide dismutase in needles, stem tissues and seedlings of Scots pine. J Exp Bot 49: 931–936, 1998.

29. MC Mathews, CB Summers, GW Felton. Ascorbate peroxidase: A novel antioxidant enzyme in insects. Arch Insect Biochem Physiol 34:57–68, 1997.

30. N Wada, S Kinoshita, M Matsuo, K Amako, C Miyake, K Asada. Purification and molecular properties of ascorbate peroxidase from bovine eye. Biochem Biophys Res Commun 242:256–261, 1998.

31. A Jiménez, JA Hernández, LA del Río, F Sevilla. Evidence for the presence of the ascorbate–glutathione cycle in mitochondria and peroxisomes of pea leaves. Plant Physiol 114:275–284, 1997.

32. C Miyake, K Asada. Ferredoxin-dependent photoreduction of the monodehydroascorbate radical in spinach thylakoids. Plant Cell Physiol 35:539–549, 1994.

33. G Vansuyt, F Lopez, D Inzé, JF Briat, P Fourcroy. Iron triggers a rapid induction of ascorbate peroxidase gene expression in *Brassica napus*. FEBS Lett 410:195–200, 1997.

34. JA Hernández, E Olmos, FJ Corpas, F Sevilla, LA Del Río. Salt-induced oxidative stress in chloroplasts of pea plants. Plant Sci 105:151–167, 1995.

35. F Lopez, G Vansuyt, F Casse-Delbart, P Fourcroy. Ascorbate peroxidase activity, not the mRNA level, is enhanced in salt-stressed *Raphanus sativus* plants. Physiol Plant 97:13–20, 1996.

36. G Ievinsh, A Valcina, D Ozola. Induction of ascorbate peroxidase activity in stressed pine (*Pinus sylvestris* L.) needles: A putative role for ethylene. Plant Sci 112:167–173, 1995.

37. DM Hodges, CJ Andrews, DA Johnson, RI Hamilton. Antioxidant compound responses to chilling stress in differentially sensitive inbred maize lines. Physiol Plant 98:685–692, 1996.

38. T Ushimaru, Y Maki, S Sano, K Koshiba, K Asada, H Tsuji. Induction of enzymes involved in the ascorbate-dependent antioxidative system, namely, ascorbate peroxidase, monodehydroascorbate reductase and dehydroascorbate reductase, after exposure to air of rice (*Oryza sativa*) seedlings germinated under water. Plant Cell Physiol 38:541–549, 1997.

39. S Biemelt, U Keetman, G Albrecht. Re-aeration following hypoxia or anoxia leads to activation of the antioxidative defense system in roots of wheat seedlings. Plant Physiol 116:651–658, 1998.

40. L De Gara, MC de Pinto, O Arrigoni. Ascorbate synthesis and ascorbate peroxidase activity during the early stage of wheat germination. Physiol Plant 100:894–900, 1997.

41. HM Jespersen, IVH Kjærsgård, L Østergaard, KG Welinder. From sequence analysis of three novel ascorbate peroxidases from *Arabidopsis thaliana* to structure, function and evolution of seven types of ascorbate peroxidase. Biochem J 326:305–310, 1997.

42. GP Creissen, EA Edwards, PM Mullineaux. Glutathione reductase and ascorbate peroxidase. In: CH Foyer, PM Mullineaux, eds. Causes of Photooxidative Stress and Amelioration of Defense Systems in Plants. Boca Raton, FL: CRC Press, 1994, pp 343–364.

43. T Koshiba. Cytosolic ascorbate peroxidase in seedlings and leaves of maize (*Zea mays*). Plant Cell Physiol 34:713–721, 1993.

44. H Willekens, R Villarroel, M Van Montagu, D Inzé, W Van Camp. Molecular identification of catalases from *Nicotiana plumbaginifolia* (L.). FEBS Lett 352: 79–83, 1994.

45. LQ Guan, JG Scandalios. Molecular evolution of maize catalases and their relationship to other eukaryotic and prokaryotic catalases. J Mol Evol 42:570–579, 1996.

46. H Willekens, D Inzé, M Van Montagu, W Van Camp. Catalases in plants. Mol Breed 1:207–228, 1995.

47. H Willekens, W Van Camp, M Van Montagu, D Inzé, H Sandermann Jr, C Langebartels. Ozone, sulfur dioxide, and ultraviolet B have similar effects on mRNA accumulation of antioxidant genes in *Nicotiana plumbaginifolia* (L.). Plant Physiol 106:1007–1014, 1994.

48. C-K Auh, JG Scandalios. Spatial and temporal responses of the maize catalases to low temperature. Physiol Plant 101:149–156, 1997.

49. JM Tepperman, P Dunsmuir. Transformed plants with elevated levels of chloroplastic SOD are not more resistant to superoxide toxicity. Plant Mol Biol 14: 501–511, 1990.

50. LH Pitcher, E Brennan, A Hurley, P Dunsmuir, JM Tepperman, BA Zilinskas. Overproduction of petunia chloroplastic copper/zinc superoxide dismutase does not confer ozone tolerance in transgenic tobacco. Plant Physiol 97:452–455, 1991.

51. A Perl, R Perl-Treves, S Galili, D Aviv, E Shalgi, S Malkin, E Galun. Enhanced oxidative-stress defense in transgenic potato expressing tomato Cu,Zn superoxide dismutases. Theor Appl Genet 85:568–576, 1993.

52. A Sen Gupta, JL Heinen, AS Holaday, JJ Burke, RD Allen. Increased resistance
     to oxidative stress in transgenic plants that overexpress chloroplastic Cu/Zn
     superoxide dismutase. Proc Natl Acad Sci USA 90:1629–1633, 1993.
53. A Sen Gupta, RP Webb, AS Holaday, RD Allen. Overexpression of superoxide
     dismutase protects plants from oxidative stress. Plant Physiol 103:1067–1073,
     1993.
54. LH Pitcher, BA Zilinskas. Overexpression of copper/zinc superoxide dismutase
     in the cytosol of transgenic tobacco confers partial resistance to ozone-induced
     foliar necrosis. Plant Physiol 110:583–588, 1996.
55. C Bowler, L Slooten, S Vandenbranden, R De Rycke, J Botterman, C Sybesma,
     M Van Montagu, D Inzé. Manganese superoxide dismutase can reduce cellular
     damage mediated by oxygen radicals in transgenic plants. EMBO J 10:1723–
     1732, 1991.
56. L Slooten, K Capiau, W Van Camp, M Van Montagu, C Sybesma, D Inzé.
     Factors affecting the enhancement of oxidative stress tolerance in transgenic
     tobacco overexpressing manganese superoxide dismutase in the chloroplasts.
     Plant Physiol 107:737–750, 1995.
57. W Van Camp, H Willekens, C Bowler, M Van Montagu, D Inzé, P Reupold-
     Popp, H Sandermann Jr, C Langebartels. Elevated levels of superoxide dis-
     mutase protect transgenic plants against ozone damage. Bio/Technology 12:
     165–168, 1994.
58. W Van Camp, K Capiau, M Van Montagu, D Inzé, L Slooten. Enhancement
     of oxidative stress tolerance in transgenic tobacco plants overexpressing Fe-
     superoxide dismutase in chloroplasts. Plant Physiol 112:1703–1714, 1996.
59. RD Allen. Dissection of oxidative stress tolerance using transgenic plants. Plant
     Physiol 107:1049–1054, 1995.
60. BD McKersie, Y Chen, M de Beus, SR Bowley, C Bowler, D Inzé, K
     D'Halluin, J Botterman. Superoxide dismutase enhances tolerance of freezing
     stress in transgenic alfalfa (Medicago sativa L.). Plant Physiol 103:1155–1163,
     1993.
61. BD McKersie, SR Bowley, KS Jones. Winter survival of transgenic alfalfa
     overexpressing superoxide dismutase. Plant Physiol 119:839–847, 1999.
62. BD McKersie, SR Bowley, E Harjanto, O Leprince. Water-deficit tolerance and
     field performance of transgenic alfalfa overexpressing superoxide dismutase.
     Plant Physiol 111:1177–1181, 1996.
63. BD McKersie, J Murnaghan, KS Jones, SR Bowley. Iron-superoxide dismutase
     expression in transgenic alfalfa increases winter survival without a detectable
     increase in photosynthetic oxidative stress tolerance. Plant Physiol 122:1427–
     1437, 2000.
64. TK Prasad. Mechanisms of chilling-induced oxidative stress injury and toler-
     ance in developing maize seedlings: Changes in antioxidant system, oxidation
     of proteins and lipids, and protease activities. Plant J 10:1017–1026, 1996.
65. J Dat, F Van Breusegem, S Vandenabeele, E Vranová, M Van Montagu, D
     Inzé. Active oxygen species and catalase during plant stress responses. Cell
     Mol Life Sci 57:779–795, 2000.

66. F Van Breusegem, L Slooten, J-M Stassart, J Botterman, T Moens, M Van Montagu, D Inzé. Effects of overexpression of tobacco MnSOD in maize chloroplasts on foliar tolerance to cold and oxidative stress. J Exp Bot 50:71–78, 1999.

67. F Van Breusegem, S Kushnir, L Slooten, G Bauw, J Botterman, M Van Montagu, D Inzé. Processing of a chimeric protein in chloroplasts is different in transgenic maize and tobacco plants. Plant Mol Biol 38:491–496, 1998.

68. HM Wilson, WP Bullock, JM Dunwell, JR Ellis, B Frame, J Register III, JA Thompson. Maize. In: K Wang, A Herrera-Estrella, M Van Montagu, eds. Transformation of Plants and Soil Microorganisms. Plant and Microbial Biotechnology Research Series, Vol 3. Cambridge, UK: Cambridge University Press, 1995, pp 65–80.

69. AG Doulis, N Debian, AH Kingston-Smith, CH Foyer. Differential localization of antioxidants in maize leaves. Plant Physiol 114:1031–1037, 1997.

70. F Van Breusegem, L Slooten, J-M Stassart, T Moens, J Botterman, M Van Montagu, D Inzé. Overproduction of *Arabidopsis thaliana* FeSOD confers oxidative stress tolerance to transgenic maize. Plant Cell Physiol 40:515–523, 1999.

71. J Wang, H Zhang, RD Allen. Overexpression of an *Arabidopsis* peroxisomal ascorbate peroxidase gene in tobacco increases protection against oxidative stress. Plant Cell Physiol 40:725–732, 1999.

72. G Torsethaugen, LH Pitcher, BA Zilinskas, EJ Pell. Overproduction of ascorbate peroxidase in the tobacco chloroplast does not provide protection against ozone. Plant Physiol 114:529–537, 1997.

73. T Shikanai, T Takeda, H Yamauchi, S Sano, K-I Tomizawa, A Yokota, S Shigeoka. Inhibition of ascorbate peroxidase under oxidative stress in tobacco having bacterial catalase in chloroplasts. FEBS Lett 428:47–51, 1998.

74. Y Miyagawa, M Tamoi, S Shigeoka. Evaluation of the defense system of chloroplast to photooxidative stress caused by paraquat using transgenic tobacco plants expressing catalase from *Escherichia coli*. Plant Cell Physiol 41:311–320.

75. WC Orr, RS Sohal. Extension of life-span by overexpression of superoxide dismutase and catalase in *Drosophila melanogaster*. Science 263:1128–1130, 1994.

76. RS Sohal, A Agarwal, S Agarwal, WC Orr. Simultaneous overexpression of copper- and zinc-containing superoxide dismutase and catalase retards age-related oxidative damage and increases metabolic potential in *Drosophila melanogaster*. J Biol Chem 270:15671–15674, 1995.

77. M Aono, H Saji, A Sakamoto, K Tanaka, N Kondo, K Tanaka. Paraquat tolerance of transgenic *Nicotiana tabacum* with enhanced activities of glutathione reductase and superoxide dismutase. Plant Cell Physiol 36:1687–1691, 1995.

78. M Aono, H Saji, K Fujiyama, M Sugita, N Kondo, K Tanaka. Decrease in activity of glutathione reductase enhances paraquat sensitivity in transgenic *Nicotiana tabacum*. Plant Physiol 107:645–648, 1995.

79. BL Örvar, BE Ellis. Transgenic tobacco plants expressing antisense RNA for

cytosolic ascorbate peroxidase show increased susceptibility to ozone injury. Plant J 11:1297–1305, 1997.

80.   S Chamnongpol, H Willekens, C Langebartels, M Van Montagu, D Inzé, W Van Camp. Transgenic tobacco with a reduced catalase activity develops necrotic lesions and induces pathogenesis-related expression under high light. Plant J 10:491–503, 1996.

81.   H Willekens, S Chamnongpol, M Davey, M Schraudner, C Langebartels, M Van Montagu, D Inzé, W Van Camp. Catalase is a sink for H_2O_2 and is indispensable for stress defence in C_3 plants. EMBO J 16:4806–4816, 1997.

22

Transgenic Plants with Increased Resistance and Tolerance against Viral Pathogens

Holger Jeske

Department of Molecular Biology and Plant Virology, University of Stuttgart, Stuttgart, Germany

I. INTRODUCTION

Compared with fungi and bacteria, plant viruses are the third important pathogens on cultured plants (1,2). They rely completely on host cells for replication and gene expression, and that is why numerous attempts to cure infected plants by chemical treatment have been unsuccessful. Therefore, plant virologists have focused their research on quarantine measures (diagnosis, establishment of virus-free cultures), control of virus vectors (insects, fungi, and nematodes), and resistance breeding (3). Genetic traits of wild and cultivated plant species introduced by classical crossings have become the most sustainable instruments to protect plants from viruses. These tools are now being augmented by gene technological engineering. Molecular biology has identified and characterized known resistance genes and created new resistance traits. Classical genetics and molecular biology have developed fruitful interactions and rely on each other for long-lived defense strategies.

In contrast to animals, plants do not have a real immune system to combat invading viruses. Whereas animals encounter viruses from the outside of the cell, plants receive viruses by wounding or direct injection into the cytoplasm. As a rule, plant viruses stay within symplastic tissues during further spread through plants without the need to cross membrane borders. Exceptionally, those plant viruses have evolved mechanisms to bud out through membranes that multiply in their insect vectors. Natural as well as artificial resistance mechanisms have to acknowlege these peculiarities of plant viruses.

A. Definition of Terms

"Resistance" and "tolerance" have specific meanings in plant virology in contrast to other areas of plant pathology. Unfortunately, the terms are often intermingled in the literature. It is, therefore, necessary to define the terms more precisely, at least to understand this review (Table 1), following the principles of Matthews' classical textbook on plant virology (3). In spite of

TABLE 1 Definition of Terms Following the Principles of Matthews' Classical Textbook on Plant Virology (3)

1. *Nonhosts* (immune plants)	Viruses do not replicate in protoplasts, cells, or plants. After inoculation they might be disassembled but do not multiply, e.g., due to lack of essential host factors. (The term "immune" is frequently used in a general meaning and might be misunderstood; plants do not have a real immune system and the failure to multiply in nonhosts does not rely on an equivalent mechanism.)
2. *Hosts* (infectible plants)	Viruses can infect and replicate in protoplasts.
2.1. *Resistant* plants	Viruses remain confined to the primary infected cell(s). Further spread is suppressed either by lack of compatible host transport factors or by defense reactions (e.g., hypersensitive response).
2.2. *Susceptible* plants	Viruses replicate and spread systemically through the plant.
2.2.1. *Sensitive* plants	Plants react with more or less severe symptoms.
2.2.2. *Tolerant* plants	Viruses do not induce obvious symptoms.
2.2.3. *Recovery*	Some organs recover from infection either because they become accidentally virus free or because they develop a somaclonal resistance.

widespread misuse of terms in the literature, for practical applications it is crucial to assign the genetic traits accurately. In case of uncertainty, "protection" might be used instead of resistance, leaving open whether viral multiplication, spread, or symptoms are ameliorated.

II. DEFENSE RESPONSES OF PLANTS

Following inoculation, plants activate a broad spectrum of responses, either general ones after several types of stress, pathogens and wounding, or specific ones that require the recognition of a particular pathogen. Both reaction types may be linked or act separately and determine whether the virus is *virulent* (the host susceptible) or *avirulent* (the host resistant). The resulting interaction is mostly the outcome of a race between virus multiplication and the velocity of defense.

A. General Responses

As with other pathogens (see Chap. 23), viruses encounter an oxidative burst early (4), followed by the transcriptional activation of pathogenesis-related (PR) proteins (5), accumulation of phytoalexins (6), and a systemic signal mediated by salicylic acid, leading to systemically acquired resistance (SAR) in noninoculated tissues or organs (7). No evidence is available that PR proteins interact directly with plant viruses, and it is believed that the broad defense reaction is rationalized by the evolutionary experience that a virus seldom comes alone in nature (8). A further general response has been attributed to ribosome-inactivating proteins (RIPs), which accumulate in certain plants to high levels in an inactive pre-form that is rendered active during wounding to destroy the translational machinery (9–12). Several components of the general response system have been manipulated in order to obtain protection against viruses but so far with only limited success (13,14).

B. Virus-Specific Responses

As Flor (15) elaborated for fungi, genetic traits of a pathogen coevolve with the respective host resistance genes in a pairwise manner. This gene-for-gene concept also holds true for plant viruses. Gene products of a virus are recognized by a host and lead to a defense response. The viral genes are then said to be avirulence (*avr*) genes, although they fulfill different functions in the context of the viral life cycle. Almost all viral genes may function as *avr* genes by triggering a host response (16–37) (Table 2). Breaking resistance is most frequently caused by mutations in *avr* genes, and during classical breeding the main task has been to cope with this challenge by

TABLE 2 Examples of the Relation between Host Resistance (*R* Genes) and Virus (*avr* Genes)

Host species	*R* gene	Virus	*avr* gene	Ref.
Arabidopsis thaliana	RCY1	Cucumber mosaic	Coat protein	16
A. thaliana	HRT	Turnip crinkle	Coat protein	17
Capsicum chinense	L3	Pepper mild mottle	Coat protein	18
Lycopersicon esculentum	Tm2	Tobacco mosaic L	ORF(30kDa) (movement protein)	19
L. esculentum	Tm2^2	Tomato mosaic	ORF(30kDa) (movement protein)	20
L. esculentum	Tm-1	Tobacco mosaic	Helicase-like domain in ORF 126/183 (replicase)	21,22
Nicandra physaloides	?	Potato A	6K2 and VPg (protein linked to the 5' end of genomic RNA)	23
Nicotiana clevelandii, *N. edwardsonii*	ccd1	Cauliflower mosaic	Gene VI (viroplasma protein, transactivator of translation)	24–26
N. glutinosa	N	Tobacco mosaic	Helicase-like domain in ORF (126/183kDa) (replicase)	27,28
N. sylvestris	N'	Tobacco mosaic	Coat protein	29–31
N. tabacum Samsun nn	?	Tomato aspermy	TAV2b (suppressor of PTGS)	32
Phaseolus vulgaris	?	Bean golden mosaic	BV1 (movement protein)	33
Solanum acaule	Rx	Potato X	Coat protein	34
Solanum stoloniferum	Ry	Potato Y	NIa proteinase (function?)	35
Vigna unguiculata	Cry	Cucumber mosaic	ORF 2a (replicase)	36,37

introducing new resistance genes into crop plants, essentially by crossing wild predecessors and cultivated plants (3).

The *avr* gene products elicit host responses on different levels. One example, *extreme resistance* (38), inhibits viral multiplication in the inoculated plant cells and in protoplasts. The fact that this genetic trait is terminologically considered resistance and not immunity is merely based on the observation that other strains of the same virus multiply readily in single cells.

A second and most common example, the *hypersensitive reaction* (HR) (38), allows multiplication in the inoculated cells and limited spread in the tissues, but then infected parts die and further infection is impaired. It has been questioned whether resistance is solely due to this "suicide for survival" (8,36,39), and at least in some cases resistance can be uncoupled from cell death (26,33).

Extreme resistance and hypersensitive reactions are currently the best known barriers for viral infections; others are being investigated to elucidate additional mechanisms in the near future (40,41). In particular, *apoptosis*, the programmed cell death of single cells thereby preventing necrosis of tissues, might be an interesting candidate. This mechanism has been intensively studied in animals (42) and might also be responsible for resistance traits in plants (43–45). Most virus-specific host responses rely on the general signal transduction cascades of eukaryotic cells, but for plants signal perception between virus and host takes place mostly inside the cell (in the symplast).

C. Recovery

The word "symptom" originates from the Greek συμπτωμα, which means "chance, transient peculiarity" (46). Correspondingly, symptoms may disappear during the infection process. Some shoots may escape from infection just by chance; others may develop virus-specific resistance that does not allow reinfection by the same virus, but unrelated viruses are propagated without problems. This recovery phenomenon gave rise to intensive research and led to the discovery of virus-induced gene silencing, discussed later (47,48).

III. HOST RESISTANCE GENES

Numerous genes conferring resistance to viruses have been characterized by classical genetics and localized to certain chromosomes (38). The two best analyzed examples will be reported here in greater detail to exemplify current experimental strategies to transmit virus resistance from wild to domestic plants. One of the most fascinating outcomes of this research was the discovery that resistance genes directed against pathogens as diverse as fungi, bacteria, and viruses produce effector proteins with similar molecular architectures (49–51) (Fig. 1).

A. *N* Gene

The *N* gene is the classical example of virus resistance associated with a hypersensitive reaction. It is specifically directed against *tobacco mosaic*

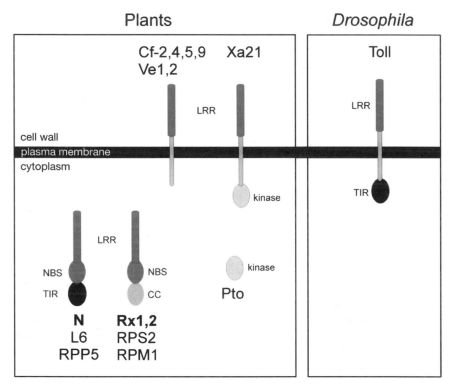

FIGURE 1   Similarity of plant resistance gene products and the signal receptor Toll of *Drosophila*. The proteins are composed of modular domains: leucine-rich repeats (LRRs), kinase, Toll/interleukin-1 receptor (TIR), nucleotide-binding site (NBS), and coiled coil (CC) domain. The following resistance genes are compared: against *Cladosporium fulvum* (Cf), *Verticillium dahliae* (Ve), *Xanthomonas oryzae* (Xa), *Pseudomonas syringae* (RPS, RPM, Pto), *Melamspora lini* (L6), *Peronospora parasitica* (RPP), tobacco mosaic virus (N), and potato virus X (Rx).

virus (TMV). After its discovery in *Nicotiana glutinosa*, it was introgressed into the Samsun cultivar of *N. tabacum* (52). The "N" refers to necrosis on foliage leaves a few days post inoculation (dpi). The development of local necrotic lesions restricts systemic spread of TMV in a temperature-dependent manner. Above 28°C, HR is suppressed and TMV can infect further tissues. Decreasing the temperature then again induces systemic necrosis in large areas, as far as TMV has spread. Exploiting this peculiar feature of the dominant *N* gene, it was possible to clone and sequence a virus resistance gene for the first time by a combination of transposon tagging and positional cloning (53).

An isolated genomic clone harboring the *N* gene was transferred to the susceptible cultivar *N. tabacum* cv. Petite Havana SR1 (53) and to *Lycopersicon esculentum* (54), proving that this clone was necessary and sufficient to establish TMV-specific resistance.

The *N* gene encodes a protein in which one of the domains shows high similarity to the *Drosophila* Toll gene and the interleukin-1 receptor of mammals (named the TIR domain), indicating that it might participate in the general signal transduction pathway of eukaryotic cells (53) (Fig. 1). It contains further domains with a nucleotide binding site (NBS) and with leucine-rich repeats (LRRs). Both signatures are commonly found in animal receptor proteins (42) that perceive and transmit signals from outside the cell at the plasma membrane. In the case of the *N* gene product, however, no transmembrane anchor is present. Therefore, it is suggested that the *N* gene protein recognizes TMV inside the cell.

The *N* gene is composed of five exons and four introns, which are alternatively spliced (53,55–57). Interestingly, a larger transcript is constitutively expressed before and after TMV inoculation. The alternative, shorter transcript is induced during infection. To provide resistance, both transcripts have to be expressed, leading to the assumption that two versions of N proteins cooperate in signal transduction. Consequently, it is necessary to transfer the genomic clone rather than a complementary DNA (cDNA) clone to new species to obtain resistance.

Using such transgenic plants as indicators, the avirulence gene of TMV, the helicase domain of the 126-kDa and 183-kDa protein, could be defined very precisely (28).

B. *Rx* Gene

The *Rx* gene of *Solanum acaule* is directed against *potato virus X* (PVX). It confers extreme resistance, which works at the single-cell level. The corresponding *avr* gene is the coat protein gene of PVX (34). For the molecular analysis, the *Rx* locus was defined by classical and bacterial artificial chromosome (BAC) mapping. Using cobombardment of BAC clones with reporter constructs of virulent or avirulent modified PVX strains, the *Rx* gene could be identified and transferred to susceptible potato and a heterologous plant species (*N. benthamiana*) (58). The two transformed plant species exhibited extreme resistance against *avr* PVX but not against a resistance-breaking strain. These results proved that the isolated gene was necessary and sufficient for the strain-specific resistance against PVX.

The *Rx* gene with three exons and two introns encodes a protein of 937 amino acid residues (107.5 kDa), which revealed surprising similarities with HR-inducing receptor proteins of the NBS-LRR class (58) (Fig. 1). Although an HR is not observed during normal infection of an *avr* PVX on an *Rx* plant, reinvestigation of this relation showed that if the coat protein of *avr* PVX was ectopically expressed under the control of 35S CaMV promoter in an *Rx* gene–containing plant, HR was obviously induced (58). Moreover, the avr coat protein of PVX was expressed from a TMV vector in NN tobacco also containing the *Rx* gene. Extreme resistance was found in this interaction but no *N*-mediated HR, suggesting that *Rx* is epistatic to HR (58).

The *Rx* prevents the multiplication of PVX inside the cell but has no influence on the transport of the virus. Double-grafting experiments showed that *avr* PVX moved from an infected wild-type rootstock through an *Rx* gene–containing scion to a second wild-type scion, inducing symptoms only in the latter (58). These experiments indicate that the signal-inducing interaction between the *avr* and *R* gene does not occur in all cells of a plant.

In summary, currently available evidence suggests that a continuum between extreme resistance, the micro-hypersensitive response, and the fully developed local or systemic HR exists in plants. The question is whether the defense reaction is triggered early or late in the infection process.

C. *Avr* Genes as Tools for Resistance Breeding

The discovery that even proteins responsible for extreme resistance can induce HR under certain conditions allowed the development of a versatile tool for the identification of unknown resistance genes as far as the viral *avr* genes of interest are delimited (59). Using an *Agrobacterium*-mediated transient expression assay, further *Rx* loci were cloned and sequenced (59). The same assay was used to answer a reciprocal question, to identify an *avr* gene if the resistance gene is known (35). Ry provides extreme resistance in potato against *potato virus Y* (PVY). However, because no resistance-breaking strains of PVY are known, it was unclear which viral gene is the elicitor. *Agrobacterium*-mediated transient expression assays assigned the NIa proteinase domain to the *avr* gene and provided evidence for the necessity of an intact catalytic center for HR induction (35). Although it cannot be completely ruled out that the structure of the center is responsible for virus recognition, the experimental results promote the intriguing hypothesis that the proteolytic function of this domain is essential.

IV. PATHOGEN-DERIVED RESISTANCE

A. Classical Cross-Protection

In the early days of plant virology, a phenomenon was observed and called "cross-protection" (60): Preinoculating a plant with a mild virus can protect it from a subsequently inoculated severe and related virus. Classically, such experiments were used to determine the relatedness of two viruses. Although this phenomenon formally resembles that of immunization, it differs significantly in that the effect is obtained only if the mild virus continues to multiply. This circumstance implies that viruses are disseminated continuously when this strategy is used. Because the term "mild" or "severe" does not define the virus proper but refers to a relationship between pathogen and host, the classical approach contains the possible impact of creating new diseases on other plants. In consequence, the application of such techniques has remained limited, although prominent success has been obtained against *Citrus tristeza virus* (61) and *Papaya ringspot virus* (62). The classical approaches using cross-protection and their respective risks have been reviewed in detail (63,64).

A variation on this theme is the use of satellite viruses or satellite RNA as protective agents (65–68). These small entities need a helper virus for multiplication, and they gained special interest because they modulate symptom expression. Dependent on the combination of helper virus, satellite, and host plant, symptoms may be increased or ameliorated. As with cross-protection, the effect of satellites depends on their multiplication and, in principle, involves the same risks (68,69). Nevertheless, the approach has been utilized with some success using preinoculation of a mild combination of helper virus and satellite (70–75).

Envisioning the advantages as well as the safety impacts of this strategy, it was consequently a challenge to scrutinize whether the useful properties of a virus could be dissected from the harmful ones. Pertinent questions were (1) whether parts of the viral genomes are sufficient for the protection and (2) whether viral protein or RNA is responsible for the effect. Starting in the 1980s, molecular biology has answered these questions unequivocally and opened new, unprecedented areas of resistance that have also revolutionized our understanding of basic gene regulation in eukaryotes on the level of epigenetics. During this period it became obvious that cross-protection can be obtained on a variety of routes including viral proteins and RNA. In summary, cross-protection came out to be a more general term.

B. Protein-Mediated Protection

Several viral proteins can mediate cross-protection. Some provide protection only to the closest relatives, others to a broader community of related viruses. Among the strategies using expressed viral proteins, two have been predominantly proved successful: expression of functional coat proteins or of defective proteins from genes with dominant negative mutations.

1. Coat Proteins

One early idea to explain cross-protection focused on the assembly of virus particles. For TMV it has been shown that disassembly of invading particles starts by the interaction of virions and ribosomes (76). This cotranslational disassembly needs a partial free 5' end of the viral RNA for first recognition events, which is normally obtained by removal of five to seven coat protein molecules according to the thermodynamic equilibrium between viral particles and components inside the cytoplasm. It should, therefore, be possible to shift the equilibrium to the formation of stable particles by increasing the amount of coat protein within the cells. The elevated level of coat protein could be produced by a preinoculated virus or, more conveniently, by ectopic expression of the coat protein in transgenic plants. In fact, this concept proved to be successful not only for TMV as the first virus (77) but also for a vast variety of viruses (78–85).

In the light of the additionally found RNA-mediated protection (see later), it was questioned whether the effector in the case of coat protein expression was protein or RNA. The role of coat protein, however, has been confirmed by several lines of evidence (reviewed in Ref. 80).

1. The coat protein–mediated resistance can be overcome by inoculating naked viral RNA.
2. Mutations in the coat protein genes that affect assembly also abolish protection.
3. The protection occurs only for viruses with proteins that might coassemble to mixed-coat virions (phenotypic mixing).
4. Virus particles are stable for hours in protoplasts transgenic for coat proteins, whereas they are disassembled within a few seconds in wild-type protoplasts.

All these data have led to the conclusion that although RNA-mediated effects might overlap, particular cases are based on the presence of the correct coat protein. These cases are usually recognized by a correlation between the amount of expressed coat protein and the degree of protection.

With one exception (86), this strategy was reliable for RNA- but not for DNA-containing viruses. Stimulated by plant virologists' work, it was also applied to animals (87).

2. Dominant Negative Mutations

A second idea has been raised by the perception that several viral proteins are multifunctional and are therefore composed of multiple domains. Such domains must cooperate, e.g., for RNA binding and RNA polymerization or for RNA binding and plasmodesmata gating. It was predicted that if one of the functions of the domains was inactivated and the mutated protein still formed complexes with the RNA, the mutation might be dominant over the wild-type gene (dominant negative mutations) (88,89). Conceivably, most work concentrated on replicase-mediated protection because it was anticipated that the earliest defect during viral multiplication should be the most useful one. Nevertheless, every viral protein might be prone to such an effective modification. The role of nonstructural viral proteins in protecting plants has been extensively reviewed (80–85,90–93), and it has been shown that RNA- as well as DNA-containing viruses may be impaired by this strategy.

In addition to the interaction of viral proteins with each other, in certain cases the binding of a viral with an unrelated protein was useful for protection. Potyviruses need viral proteinases for the processing of their large preprotein (94). This function can be blocked in transgenic plants by ectopic expression of a cysteine proteinase inhibitor (95), providing broad resistance against distinct potyviruses.

C. RNA-Mediated Protection

During early work on protein-mediated cross-protection, some transgenic constructs behaved unexpectedly. While expressing low or no proteins, they nevertheless provided some degree of protection (78,96). In these cases no correlation between protein concentration and protective effect was observed, and sometimes, in control experiments, an untranslatable messenger RNA (mRNA) serendipitously appeared to be the effector molecule. This initial evidence founded the hypothesis that RNA by itself might mediate protection.

1. Antisense and Sense RNAs

An alternative idea finally led to a similar conclusion. To block translation of viral RNA by means of hybridization, antisense RNA was expressed in various plants (97–102). Surprisingly, in some cases not only antisense but also sense RNA, from a control construct, provided protective effects (103).

The inhibition of viruses by sense constructs resembled a simultaneously discovered phenomenon perceived by plant molecular biologists and has been termed "cosuppression" (104) (see Chap. 26): If several copies of the same gene or of the same promoter were artificially integrated into plant chromosomes, their expression was not increased but abolished.

Antisense as well as sense RNA strategies have, subsequently, proved to be useful in certain cases (85) but, even more important, have opened a new field of research that is now called "silencing" (see later), emphasizing the role of RNA in protection against viruses.

D. Interfering Replicons

During natural virus infections replicons may accumulate that are derived from the master virus but have a smaller size (105) or are predominantly unrelated to the helper virus (satellites) (68). Moreover, some replicons appeared to be recombination products of a helper virus nucleic acid and unrelated sequences (106). These additional replicons might enhance the effect of the virus, ameliorate the symptoms, or behave neutrally (68). When they interfere with virus multiplication or symptom development, they are called defective interfering (DI) nucleic acids.

1. Defective Interfering (DI) Nucleic Acids

DI nucleic acids have been observed for RNA- and DNA-containing viruses (105,107). DI-DNA molecules of geminiviruses have attracted special interest (Fig. 2). Geminiviruses encapsidate single-stranded circular DNA and replicate in nuclei. Over the past three decades they raised worldwide devastating epidemics, predominantly in tropical and subtropical countries (108,109) but also in the United States (*beet curly top virus*, BCTV) (110) and around the Mediterranean sea (*tomato yellow leaf curl virus*, TYLCV) (111). Their genomes consist of either one component (genera *Mastre-*, *Curto-*, and *Topocuvirus*) or two components (most of the genus *Begomovirus*), which are called DNA A and DNA B (112). DI-DNA has been investigated in more detail for BCTV, with a set of half and smaller size DNA circles from a single genomic component (113) (Fig. 2), and for *African cassava mosaic begomovirus*, which accumulates smaller molecules derived from the DNA B component (114,115). All these DI molecules harbor the origin of replication and parts of predominantly the left halves of the genomic components. They usually do not contain intact open reading frames with the exception of BCTV-DI, possessing the small ORF C4 influencing symptom expression (116).

Integrating tandem copies of DI-DNA into plant chromosomes does not disturb the development of plants, and DI-DNA is not replicated because

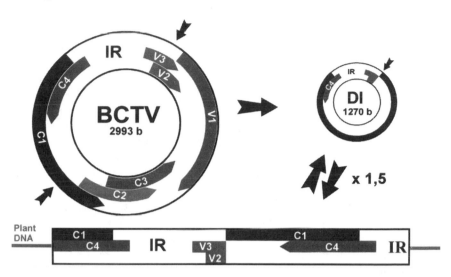

FIGURE 2 Defective-interfering (DI) DNA of beet curly top geminivirus (BCTV) to protect plants from virus infection. Open reading frames are named according to their complementary (C) or viral (V) orientation. IR assigns the intergenic region harboring viral promoters and the origin of replication. Bitmers (×1.5) were integrated into plant chromosomes, which are transreplicated upon infection with the cognate virus (BCTV).

the viral replication initiator protein (AC1 or C1) is lacking. Upon challenge by the cognate virus, DI-DNA replicates and symptoms are ameliorated, leading to a recovery after longer time periods of infection (114,115,117). During this process, DI-DNA accumulates and full-length viral DNA is reduced compared with infection of control plants. It has been suggested that down-regulation of viral multiplication is caused by the competition of DI-DNA with viral DNA for replication complexes, but it is not completely excluded that intact proteins (C4), defective proteins (ΔAC1, ΔBC1) as dominant-negative effectors, or RNA-mediated processes participate in the protection (105). In this context it is interesting to note that smaller DI-DNA may replicate to even higher levels, but such an increase did not lead to enhanced inhibition (118). DI-DNA is packed into half-size particles (119) and the amount of DI molecules is elevated after serial passages to further plants, rendering this strategy especially useful for field applications.

A limitation results from the necessity to recognize a specific origin of replication by the compatible replication-associated (Rep) protein of the cognate virus to induce DI-DNA multiplication. Therefore, DI-DNA can act only against closely related viruses that are able to transreplicate each other.

2. Satellites

Satellites are small, a few hundred nucleotides in size, RNA (satRNA) or DNA (satDNA) molecules that are transreplicated by a helper virus (67,68). Some of them, encoding their own coat protein for packaging their RNA, are called satellite viruses. The other satellites might also contain small open reading frames (ORFs) of unknown function.

Satellites modulate symptom expression of their helper virus, either increasing or decreasing severity. Their particular effects are governed by the triangular interrelationship between virus strain, satellite, and host genotype. Expression of putative proteins from ORFs of satRNAs has been prevented by site-directed mutagenesis without abolishing symptom modulation (65). Therefore, it is believed that the secondary structure determines the effect rather than the coding capacity of the RNA.

It was shown in model plants that satRNA expressed from transgenes can confer some protection against the helper virus effects (70,71). For field application this approach was frequently questioned, especially because of the variability of possible effects upon coinfection with other viruses or because satRNA might be transferred to other nontarget hosts (65,68,69). Nevertheless, this strategy using either classical coinoculation techniques or transgenic means has been successfully applied in Asia (72–75), resulting in a considerable reduction of yield losses.

V. GENE SILENCING

Silencing of genes is a long known process providing programmed differential gene expression during the individual development of all organisms. It can occur at different levels, repressing genes in prokaryotic operons, combining sets of transcription factors in eukaryotic cells, or condensing chromatin structures. In addition to these regulatory means, it was already perceived in classical genetics that certain genes can be suppressed (phenocopied) and that such suppression can be inherited (paramutation, imprinting) (120,121). It was proposed that this type of silencing might rely, at least for some examples, on the interaction of homologous sequences (ectopic pairing of homologous chromatin) followed by the inactivation of the particular gene, perhaps through methylation. However, whether methylation, which is regularly seen in such genes, is the cause or the result of silencing remains to be shown.

From a historical point of view, one of the most fascinating chapters of molecular biology was the perception during the past 10 years of how results of completely unrelated areas of research have converged to throw light on the puzzle of silencing mechanisms. Phenomena called "cosuppression" for transgenic plants, "virus-induced gene silencing" for plant viruses, "RNA interference" for *Drosophila* and *Caenorhabditis*, and "quelling" for *Neurospora* turned out to rely on common molecular pathways (122–133).

Summarizing the evidence, homology-dependent gene silencing (HDGS) may occur on two different levels: (1) transcriptional gene silencing (TGS) based on promoter inactivation, methylation, and chromatin remodeling and (2) posttranscriptional gene silencing (PTGS) caused by sequence-specific destruction of transcripts. The removal of sequence-specific RNA in PTGS was shown to be based on the presence of homologous interfering double-stranded RNA (dsRNA). Some evidence suggests that a feedback loop of regulation from PTGS to TGS exists (134,135).

Originally, these mechanisms were thought to control the copy number of genes to protect the genome from too many mobile genetic elements and viruses. Currently, however, the possibility arises that similar mechanisms are also involved in the programmed differential gene activation during ontogeny (136).

A. Transcriptional Gene Silencing (TGS)

The TGS process (122,135,137) is an unpredictable one in which multiple copies of a gene or a promoter influence each other, leading to the inactivation of a particular gene. In experiments it is recognized by a lack of steady-state transcripts and the absence of the particular transcript in nuclear run-on assays. Frequently, it is associated with increased methylation of the promoter and the gene. TGS may be present in the whole plant or in sectors of various organs. The pattern of TGS may be inherited by the next generation (imprinting) and reverted to an active state of transcription under certain conditions (138).

B. Posttranscriptional Gene Silencing (PTGS)

The PTGS process relies on the multiplicity of genes; their inactivation is also unpredictable, as for TGS, but the effective homology must reside within the transcript (123,129–132,139–141). Experimentally, it is detected by reduced steady-state amounts of RNA as in TGS but the presence of nuclear run-on transcripts. Moreover, the appearance of small RNA molecules (21–25 nt) of both polarities is diagnostic for PTGS. PTGS is not present throughout the whole life cycle of a plant but appears late during development in individual plants of silencing lines (104).

C. Virus-Induced Gene Silencing (VIGS)

The first evidence for VIGS came from the observation of recovery from symptoms after inoculating a potyvirus on transgenic plants containing a potyviral gene (47). The recovered shoots remained protected from a second infection with the same virus but were susceptible to others. The interpretation of this observation became a heuristic idea, namely that effector molecules, presumably complementary RNA, had been generated to signal virus-specific inhibition (123,124). In a similar approach it was shown that viroids can also trigger gene silencing, underscoring that the effector molecule might be RNA (134). Interestingly, these peculiar experiments with viroids led to inactivation of its homologous transgene, associated with methylation of the DNA copy. In other cases VIGS did not change the transcription but induced PTGS, which was best shown for PVX, a virus that replicates exclusively in the cytoplasm. Transgenes, e.g., glucuronidase (GUS), were silenced by PTGS if the plant was infected by a chimeric PVX containing the GUS gene, too. Conversely and surprisingly, a plant transgenic for GUS became resistant to this hybrid PVX (141).

D. Systemic Acquired Gene Silencing (SAS)

Once established at some site, the silencing signal may spread throughout the plant, which is most convincingly demonstrated by grafting experiments (142–145). A silenced rootstock can transfer the signal to an unsilenced scion. This effect is maintained as long as the homologous gene is present in the recipient plant.

Which molecules trigger SAS is still unclear. It has been suggested that the gene-specific small RNAs participate in the signal transduction, but recent grafting experiments have questioned this idea (145). Alternatively, longer dsRNA might induce SAS, but evidence for this assumption is still lacking.

E. Mechanistic Aspects of Posttranscriptional Gene Silencing

The mechanism of PTGS has been elucidated in different organisms, converging to a fundamental concept during the last few years (126–133) (Fig. 3). The trigger for PTGS is dsRNA (blunt end or with a few protruding $3'$-nucleotides) homologous to a target RNA. Several genes are involved in this process in plants, animals, and fungi coding for (1) a ribonuclease (RNase) III-type enzyme cutting dsRNA, (2) an RNA-dependent RNA polymerase (RdRp), (3) a helicase, and (4) a protein with homology to a eukaryotic initiation factor (eIF2C) (146–148). These factors interact in (trans-)gene silencing, whereas for viruses a viral RdRp may replace or modulate the analogous host enzyme.

One key enzyme is now called DICER (148) because it chops small pieces of about 22 bp (siRNA, for short interfering RNA) proceeding from the ends of dsRNA with a type III–like RNase activity. The protein contains two RNase III domains, a putative helicase domain, and a PAZ domain, which is possibly responsible for protein-protein interaction with other factors (148).

According to the currently discussed model (148–150) (Fig. 3), siRNA remains bound to DICER, is melted by a helicase, and subsequently guides the sequence-specific nuclease to homologous RNA of (+) or (−) orientation. In this reaction, target RNA is cleaved in the middle of the recognized 22-nt sequence (149), leading to a periodicity in the digested RNA similar to that of siRNA but with a phase shift. The advantage of the DICER reaction lies in the amplification of the signal because several siRNAs are generated from a single dsRNA molecule, each one ready to attack several new target RNAs.

An even more efficient amplification of the signal for destruction would be obtained if target ssRNA, loaded with siRNA as a primer, was complemented by an RdRp to produce new dsRNA to be recognized by DICER again. However, such a feedback cycle has yet to be shown experimentally. In any case, it would explain the functionality of host or viral RdRp genes in PTGS as inferred from genetic analysis. The demonstrated role of an inducible host RdRp in plant antiviral defense might be a further hint in the same direction (151).

F. Suppressors of PTGS

The PTGS process is thought to be a defense response against foreign genes, especially of viral origin. Therefore, it is not surprising that viruses have developed ways to combat such strategies. Consequently, certain viral genes were discovered to suppress PTGS (145,152–159). Two of them, HC-Pro (helper component-proteinase) of potyviruses and ORF 2b of cucumoviruses, have been intensively investigated in this function. Although PVX was initially thought to lack such a capability, its p25 protein (part of the triple gene block with movement functions) inhibits PTGS under certain conditions (156). These three examples, among several others (157,159), also represent suppression of PTGS at different steps in the course of the process. HC-Pro is able to switch off existing PTGS and leads to a decrease of siRNA levels (145). In contrast, CMV 2b has no influence on established PTGS but prevents its initiation (152), and PVX p25 abolishes the mobile signal of PTGS (156). Whenever suppression of PTGS has occurred, it promotes not

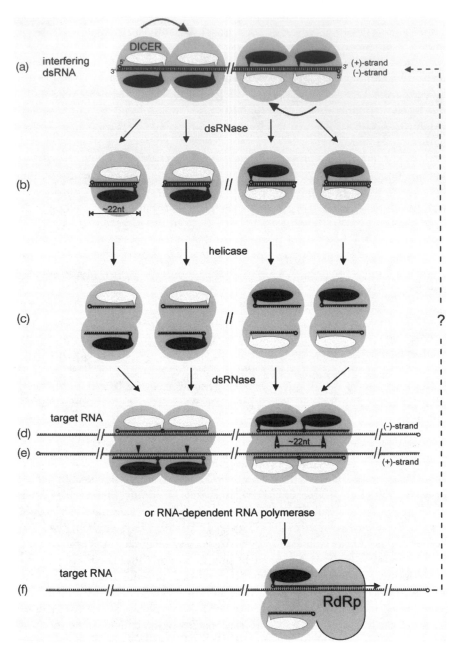

FIGURE 3   Current model of posttranscriptional silencing. (a) A dsRNA-specific RNase (DICER) binds to interfering dsRNA at both ends and digests it consecutively in both directions, resulting in bits of small dsRNA (~22 nt in length)

only parental but also unrelated viruses in multiplication and spread. Such synergistic effects were observed upon coinfection of two viruses (160) as well as upon single-virus inoculation on transgenic plants harboring a viral PTGS-suppressor gene (161).

VI. SUSTAINABILITY CONCERNS

Resistance breeding has always been a race between host response and viral evolution. Molecular techniques provide the breeder with new tools to accelerate transmission or selection of interesting host genes. Nevertheless, creating a new cultivar that is resistant to a particular virus and otherwise true to type for the market is still tedious work. It is therefore necessary for agriculture and horticulture to choose sustainable strategies to be applied in practice.

Most of the work discussed here has been performed with a few model plant species and laboratory strains of viruses. The current challenge is to introduce resistance traits into agronomically relevant cultivars that have to encounter a variety of viral strains or viral species. For a growing number of virus problems, this aim has already been reached (162); for very important virus epidemics, e.g., geminiviruses (108), it is still a task. To reduce the number of setbacks during this long-lasting process, a series of risk concerns have been evaluated by aimed experiments as well as field surveys. Most of them have been covered by a very comprehensive and detailed review (85), and therefore it is possible to summarize the most important topics here.

A. Tolerance

Contrary to real resistance, a tolerant variety, whether it is created by classical or molecular means, will accumulate virus populations that are as vir-

---

attached to DICER (b). After strand separation by a helicase activity (c), the nucleoprotein complex can scan target RNAs of both polarities for complementarity (d or e) and cut the complementary sequence in the middle. In each complex only one binding site is used for cutting. Hypothetically, some complexes of DICER, siRNA, and target RNA might serve as templates for RNA-dependent RNA polymerases (RdRp) to recycle interfering dsRNA (f). The proteins are symbolic drawings. It has yet to be shown how many proteins cooperate for the respective functions. Equally, the interaction of DICER complexes with target RNA (d or e) symbolizes only the principal possibilities of action, which may not occur simultaneously.

ulent as before on related plant species lacking a tolerance-inducing gene. In geographically isolated field conditions such an approach might be helpful if no alternative is seen, but in the majority of agronomically relevant cases it might promote new epidemics of the same virus and/or accelerate the evolution of new virulent virus strains.

B. Resistance Breakage

In general, resistance based on multiple genes is less easily broken than a monogenically based resistance. Monogenic exceptions to this rule that were extremely durable (e.g., for the N gene) are obvious. Again, the problem is similar irrespective of classical or molecular breeding. Pathogen-derived protein-mediated approaches tend to have a smaller spectrum of protective effects, whereas RNA-based strategies sometimes provide broader efficiency (96). In every case, there is no a priori indication of which trait will be broken soon and which will be long-lived. The probability has to be tested through case-by-case experiments and field trials.

C. Transcapsidation

Transcapsidation of two discrete but related viruses has been observed in classical field situations, where it may create an epidemic problem if changes in vector transmissibility are associated (3). Molecular studies have defined conditions for and frequencies of transcapsidation (163). In the worst case, transcapsidation will create new virions that can be transmitted, e.g., from a transgenic crop plant to a next neighbor plant, but such a virion will not be harmful to a larger agricultural area because it is lost in the recipient plant. Nevertheless, a careful resistance strategy should aim to avoid any pathogen-derived sequence in the protective gene that contributes to vector transmission.

D. Recombination

Recombination between plant viruses is a well-established fact now (164). Some viruses use template switch as an obligatory step of replication (pararetroviruses) (3) and are therefore prone to recombination. In other virus families, recombination is an accidental phenomenon. Although it occurs at a low frequency, it is worth considering because of the huge numbers of viruses in plants. Mostly, recombinational effects are overlooked because wild-type viruses overgrow the recombinants. Consequently, evidence for recombination in field isolates of RNA plant viruses is limited. In contrast to most RNA viruses, the DNA-containing geminiviruses show multiple footprints of recombination in their sequences (165), which has been associated with the evolution of resistance-breaking strains (166,167). Recently, a possible explanation for this tendency was found in that geminiviruses may replicate using a recombination-dependent pathway in addition to the classical rolling-circle replication (168).

In summary, accepting that recombination occurs with different frequencies but always frequently enough to create new variants, the real critical question is whether new viral capabilities are selected that provide invasion of new plants or vectors. This concern is equally relevant for natural coinfection of different viruses and for viruses infecting a cultivar containing viral transgenes because invading viruses have been shown to take up transgene sequences in different virus families (169–171). In conclusion, viral sequences coding for vector transmissibility, tissue tropism, host range, or symptom expression should be identified and omitted from transgenic constructs.

E. Synergism

Suppression of silencing, as discussed earlier, is also a lesson for risk concerns using pathogen-derived resistance. It is necessary but not sufficient to analyze the manipulated trait by challenging with the target virus. In the field situation, unrelated viruses might be promoted rather than inhibited by the viral gene product. The history of the discovery of PVX suppressor capabilities has shown that functions of these types are not always obvious but have to be elucidated by intelligent means (156). In conclusion, resistance traits to be released in the field have to be challenged in the respective crop plants by relevant unrelated viruses in addition to the usually tested cognate target virus. Moreover, because of the uncertainties of such tests, surveillance monitoring has to be carried out.

VII. PERSPECTIVES

Pathogen-derived resistance (PDR) of the first generation had its merits as a step toward engineering virus resistance and, more important, in investigating the interaction of virus and host. However, it also has its limits and risks. Now, the second generation of PDR in the form of gene silencing is superior in abandoning the necessity for functional viral genes, providing the possibility to silence the viral silencing suppressors, and extending the reservoir of effector molecules that might be directed against a broad spectrum of viruses. Possibly, this approach will gain the same role for plants as immunization for animals. In addition to pathogen-derived resistance, the overwhelming progress in deciphering host responses and virus-specific recognition will change the main fields of research on resistance. Identifying receptor proteins and other components of the signal transduction pathway will lead to a profound understanding of host defense and open the possibility to transfer individual components of the resistance network to the crop plants of choice.

ACKNOWLEDGMENTS

I am very grateful to Dr. Christina Wege and Prof. Dr. R. Ghosh for critically reading the manuscript and for helpful discussions.

REFERENCES

1. GN Agrios. Plant Pathology. San Diego: Academic Press, 1997.
2. R Hull, JW Davies. Approaches to nonconventional control of plant virus diseases. Crit Rev Plant Sci 11:17–33, 1992.
3. R Hull. Matthew's Plant Virology. 4th ed. San Diego: Academic Press, 2001.
4. C Lamb, RA Dixon. The oxidative burst in plant disease resistance. Annu Rev Plant Physiol Plant Mol Biol 48:251–275, 1997.
5. DJ Bowles. Defense-related proteins in higher plants. Annu Rev Biochem 9: 873–907, 1990.
6. J Kuc. Phytoalexins: stress metabolism and disease resistance in plants. Annu Rev Phytopathol 33:275–297, 1995.
7. L Sticher, B Mauch-Mani, JP Metraux. Systemic acquired resistance. Annu Rev Phytopathol 35:235–270, 1997.
8. AM Murphy, S Chivasa, DP Singh, JP Carr. Salicylic acid–induced resistance to viruses and other pathogens: a parting of the ways? Trends Plant Sci 4: 155–160, 1999.
9. F Stirpe, L Barbieri, MG Battelli, M Soria, DA Lappi. Ribosome-inactivating proteins from plants: present status and future prospects. Biotechnology 10: 405–410, 1992.
10. MR Hartley, JA Chaddock, MS Bonness. The structure and function of ribosome-inactivating proteins. Trends Plant Sci 1:254–260, 1996.
11. JD Irvin, FM Uckun. Pokeweed antiviral protein: ribosome inactivation and therapeutic applications. Pharmacol Ther 55:279–302, 1992.
12. S Taylor, A Massiah, G Lomonossoff, LM Roberts, JM Lord, M Hartley. Correlation between the activities of five ribosome-inactivating proteins in depurination of tobacco ribosomes and inhibition of tobacco mosaic virus infection. Plant J 5:827–835, 1994.
13. JK Lodge, WK Kaniewski, NE Tumer. Broad-spectrum virus resistance in transgenic plants expressing pokeweed antiviral protein. Proc Natl Acad Sci USA 90:7089–7093, 1993.
14. Y Hong, K Saunders, MR Hartley, J Stanley. Resistance to geminivirus infection by virus-induced expression of dianthin in transgenic plants. Virology 220:119–127, 1996.

15. HH Flor. Current status of the gene-for-gene concept. Annu Rev Phytopathol 9:275–296, 1971.

16. H Takahashi, M Suzuki, K Natsuaki, T Shigyo, K Hino, T Teraoka, D Hosokawa, Y Ehara. Mapping the virus and host genes involved in the resistance response in cucumber mosaic virus–infected *Arabidopsis thaliana*. Plant Cell Physiol 42:340–347, 2001.

17. MB Cooley, S Pathirana, HJ Wu, P Kachroo, DF Klessig. Members of the *Arabidopsis* HRT/RPP8 family of resistance genes confer resistance to both viral and oomycete pathogens. Plant Cell 12:663–676, 2000.

18. A Berzal-Herranz, A de la Cruz, F Tenllado, JR Diaz-Ruiz, L Lopez, AI Sanz, C Vaquero, MT Serra, I Garcia-Luque. The *Capsicum* L3 gene–mediated resistance against the tobamoviruses is elicited by the coat protein. Virology 209:498–505, 1995.

19. T Meshi, R Motoyoshi, T Maeda, S Yoshiwoka, H Watanabe, Y Okada. Mutations in the tobacco mosaic virus 30-kD protein gene overcome Tm-2 resistance in tomato. Plant Cell 1:515–522, 1989.

20. H Weber, S Schultze, AJ Pfitzner. Two amino acid substitutions in the tomato mosaic virus 30-kilodalton movement protein confer the ability to overcome the *Tm-2(2)* resistance gene in the tomato. J Virol 67:6432–6438, 1993.

21. T Meshi, F Motoyoshi, A Adachi, Y Watanabe, N Takamatsu, Y Okada. Two concomitant base substitutions in the putative replicase genes of tobacco mosaic virus confer the ability to overcome the effects of a tomato resistance gene, *Tm-1*. EMBO J 7:1575–1581, 1988.

22. KW Buck. Replication of tobacco mosaic virus RNA. Philos Trans R Soc Lond Ser B 354:613–627, 1999.

23. ML Rajamaki, JPT Valkonen. The 6K2 protein and the VPg of potato virus A are determinants of systemic infection in *Nicandra physaloides*. Mol Plant Microbe Interact 12:1074–1081, 1999.

24. L Kiraly, AB Cole, JE Bourque, JE Schoelz. Systemic cell death is elicited by the interaction of a single gene in *Nicotiana clevelandii* and gene VI of cauliflower mosaic virus. Mol Plant Microbe Interact 12:919–925, 1999.

25. K Palanichelvam, AB Cole, M Shababi, JE Schoelz. Agroinfiltration of cauliflower mosaic virus gene VI elicits hypersensitive response in *Nicotiana* species. Mol Plant Microbe Interact 13:1275–1279, 2000.

26. AB Cole, L Kiraly, K Ross, JE Schoelz. Uncoupling resistance from cell death in the hypersensitive response of *Nicotiana* species to cauliflower mosaic virus infection. Mol Plant Microbe Interact 14:31–41, 2001.

27. HS Padgett, Y Watanabe, RN Beachy. Identification of the TMV replicase sequence that activates the N gene–mediated hypersensitive response. Mol Plant Microbe Interact 10:709–715, 1997.

28. FL Erickson, S Holzberg, A Calderon-Urrea, V Handley, M Axtell, C Corr, B Baker. The helicase domain of the TMV replicase proteins induces the N-mediated defence response in tobacco. Plant J 18:67–75, 1999.

29. T Saito, T Meshi, N Takamatsu, Y Okada. Coat protein gene sequence of tobacco mosaic virus encodes a host response determinant. Proc Natl Acad Sci USA 84:6074–6077, 1987.

30. DA Knorr, WO Dawson. A point mutation in the tobacco mosaic virus capsid protein gene induces hypersensitivity in *Nicotiana sylvestris*. Proc Natl Acad Sci USA 85:170–174, 1988.

31. ZF Taraporewala, JN Culver. Identification of an elicitor active site within the three-dimensional structure of the tobacco mosaic tobamovirus coat protein. Plant Cell 8:169–178, 1996.

32. HW Li, AP Lucy, HS Guo, WX Li, LH Ji, SM Wong, SW Ding. Strong host resistance targeted against a viral suppressor of the plant gene silencing defence mechanism. EMBO J 18:2683–2691, 1999.

33. ER Garrido-Ramirez, MR Sudarshana, WJ Lucas, RL Gilbertson. Bean dwarf mosaic virus BV1 protein is a determinant of the hypersensitive response and avirulence in *Phaseolus vulgaris*. Mol Plant Microbe Interact 13:1184–1194, 2000.

34. A Bendahmane, BA Köhm, C Dedi, DC Baulcombe. The coat protein of potato virus X is a strain-specific elicitor of Rx1-mediated virus resistance in potato. Plant J 8:933–941, 1995.

35. P Mestre, G Brigneti, DC Baulcombe. An Ry-mediated resistance response in potato requires the intact active site of the NIa proteinase from potato virus Y. Plant J 23:653–661, 2000.

36. CH Kim, P Palukaitis. The plant defense response to cucumber mosaic virus in cowpea is elicited by the viral polymerase gene and affects virus accumulation in single cells. EMBO J 16:4060–4068, 1997.

37. A Karasawa, I Okada, K Akashi, Y Chida, S Hase, Y Nakazawa-Nasu, A Ito, Y Ehara. One amino acid change in cucumber mosaic virus RNA polymerase determines virulent/avirulent phenotypes on cowpea. Phytopathology 89: 1186–1192, 1999.

38. KE Hammond-Kosack, JDG Jones. Plant disease resistance genes. Annu Rev Plant Physiol Plant Mol Biol 48:575–607, 1997.

39. H Takusari, T Takahashi. Studies on viral pathogenesis in host plants. IX. Effect of citrinin on the formation of necrotic lesions and virus localisation in the leaves of "Samsun NN" tobacco plants after tobacco mosaic virus infection. Phytopathol Z 96:324–329, 1979.

40. ST Chisholm, SK Mahajan, SA Whitham, ML Yamamoto, JC Carrington. Cloning of the *Arabidopsis RTM1* gene, which controls restriction of long-distance movement of tobacco etch virus. Proc Natl Acad Sci USA 97:489–494, 2000.

41. SA Whitham, RJ Anderberg, ST Chisholm, JC Carrington. Arabidopsis *RTM2* gene is necessary for specific restriction of tobacco etch virus and encodes an unusual small heat shock–like protein. Plant Cell 12:569–582, 2000.

42. L Aravind, VM Dixit, EV Koonin. The domains of death: evolution of the apoptosis machinery. Trends Biochem Sci 24:47–53, 1999.

43. DG Gilchrist. Programmed cell death in plant disease: the purpose and promise of cellular suicide. Annu Rev Phytopathol 36:393–414, 1998.

44. JT Greenberg. Programmed cell death in plant-pathogen interactions. Annu Rev Plant Physiol Plant Mol Biol 48:525–545, 1997.

45. EA van der Biezen, JDG Jones. The NB-ARC domain: a novel signalling motif shared by plant resistance gene products and regulators of cell death in animals. Curr Biol 8:R226–R227, 1998.
46. H Jeske. Pathogens and symbionts as growth modulators. In: P Westhoff, H Jeske, K Kloppstech, G Link, eds. Molecular Plant Development—From Gene to Plant. Oxford: Oxford University Press, 1998, pp 220–255.
47. JA Lindbo, L Silva-Rosales, WM Proebsting, WG Dougherty. Induction of a highly specific antiviral state in transgenic plants: implications for regulation of gene expression and virus resistance. Plant Cell 5:1749–1759, 1993.
48. JA Lindbo, WP Fitzmaurice, G della-Cioppa. Virus-mediated reprogramming of gene expression in plants. Curr Opin Plant Biol 4:181–185, 2001.
49. JL Dangl. Pièce de résistance: novel classes of plant disease resistance genes. Cell 80:363–366, 1995.
50. BJ Staskawicz, MB Mudgett, JL Dangl, JE Galan. Common and contrasting themes of plant and animal diseases. Science 292:2285–2289, 2001.
51. EA van der Biezen, JDG Jones. Plant disease-resistance proteins and the gene-for-gene concept. Trends Biochem Sci 23:454–456, 1998.
52. FO Holmes. Inheritance of resistance to tobacco-mosaic disease in tobacco. Phytopathology 28:553–561, 1938.
53. S Whitham, SP Dinesh-Kumar, D Choi, R Hehl, C Corr, B Baker. The product of the tobacco mosaic virus resistance gene N: similarity to toll and the interleukin-1 receptor. Cell 78:1101–1115, 1994.
54. S Whitham, S McCormick, B Baker. The N gene of tobacco confers resistance to tobacco mosaic virus in transgenic tomato. Proc Natl Acad Sci USA 93: 8776–8781, 1996.
55. FL Erickson, SP Dinesh-Kumar, S Holzberg, CV Ustach, M Dutton, V Handley, C Corr, BJ Baker. Interactions between tobacco mosaic virus and the tobacco N gene. Philos Trans R Soc Lond Ser B 354:653–658, 1999.
56. SP Dinesh-Kumar, WH Tham, BJ Baker. Structure-function analysis of the tobacco mosaic virus resistance gene N. Proc Natl Acad Sci USA 97:14789–14794, 2000.
57. SP Dinesh-Kumar, BJ Baker. Alternatively spliced N resistance gene transcripts: their possible role in tobacco mosaic virus resistance. Proc Natl Acad Sci USA 97:1908–1913, 2000.
58. A Bendahmane, K Kanyuka, DC Baulcombe. The Rx gene from potato controls separate virus resistance and cell death responses. Plant Cell 11:781–791, 1999.
59. A Bendahmane, M Querci, K Kanyuka, D Baulcombe. *Agrobacterium* transient expression system as a tool for the isolation of disease resistance genes: application to the Rx2 locus in potato. Plant J 21:73–81, 2000.
60. HH McKinney. Mosaic diseases in the Canary Islands, West Africa, and Gibraltar. J Agric Res 39:557–578, 1929.
61. AS Costa, GW Müller. Tristeza control by cross protection. Plant Dis 64: 538–541, 1980.
62. SD Yeh, D Gonsalves. Evaluation of induced mutants of papaya ringspot virus for control by cross protection. Phytopathology 74:1086–1091, 1984.

63. GA de Zoeten. Risk assessment: do we let history repeat itself? Phytopathology 81:585–586, 1991.

64. RW Fulton. Practices and precautions in the use of cross protection for plant virus disease control. Annu Rev Phytopathol 24:67–81, 1986.

65. CW Collmer, SH Howell. Role of satellite RNA in the expression of symptoms caused by plant viruses. Annu Rev Phytopathol 30:419–442, 1992.

66. JA Dodds. Satellite tobacco mosaic virus. Annu Rev Phytopathol 36:295–310, 1998.

67. RIB Francki. Plant virus satellites. Annu Rev Microbiol 39:151–174, 1985.

68. MJ Roossinck, D Sleat, P Palukaitis. Satellite RNAs of plant viruses: structures and biological effects. Microbiol Rev 56:265–279, 1992.

69. P Palukaitis, MJ Roossinck. Spontaneous change of a benign satellite RNA of cucumber mosaic virus to a pathogenic variant. Nat Biotechnol 14:1264–1267, 1996.

70. DC Baulcombe, GR Saunders, MW Bevan, MA Mayo, BD Harrison. Expression of biologically active viral satellite RNA from the nuclear genome of transformed plants. Nature 321:446–449, 1986.

71. BD Harrison, MA Mayo, DC Baulcombe. Virus resistance in transgenic plants that express cucumber mosaic virus satellite RNA. Nature 328:799–805, 1987.

72. P Tien, G Wu. Satellite RNA for the biocontrol of plant disease. Adv Virus Res 39:321–339, 1991.

73. P Tien, X Zhang, B Qiu, B Qin, G Wu. Satellite RNA for the control of plant diseases caused by cucumber mosaic virus. Ann Appl Biol 111:143–152, 1987.

74. Y Yie, ZX Wu, SY Wang, SZ Zhao, TQ Zhang, GY Yao, P Tien. Rapid production and field testing of homozygous transgenic tobacco lines with virus resistance conferred by expression of satellite RNA and coat protein of cucumber mosaic virus. Transgen Res 4:256–263, 1995.

75. H Sayama, T Sato, M Kominato, T Natsuaki, JM Kaper. Field testing of a satellite-containing attenuated strain of cucumber mosaic virus for tomato protection in Japan. Phytopathology 83:405–410, 1993.

76. TMA Wilson. Cotranslational disassembly of tobacco mosaic virus in vitro. Virology 137:255–265, 1984.

77. P Abel, RS Nelson, B De, N Hoffmann, SG Rogers, RT Fraley, RN Beachy. Delay of disease development in transgenic plants that express the tobacco mosaic virus coat protein gene. Science 232:738–743, 1986.

78. JH Fitchen, RN Beachy. Genetically engineered protection against viruses in transgenic plants. Annu Rev Microbiol 47:739–763, 1993.

79. GP Lomonossoff. Pathogen-derived resistance to plant viruses. Annu Rev Phytopathol 33:323–343, 1995.

80. M Bendahmane, RN Beachy. Control of tobamovirus infections via pathogen-derived resistance. Adv Virus Res 53:369–387, 1999.

81. D Gonsalves. Control of papaya ringspot virus in papaya: a case study. Annu Rev Phytopathol 36:415–437, 1998.

82. R Grumet. Development of virus resistant plants via genetic engineering. Plant Breed Res Rev 12:47–79, 1994.

83. R Michelmore. Molecular approaches to manipulation of disease resistance genes. Annu Rev Phytopathol 33:393–427, 1995.

84. TMA Wilson. Strategies to protect crop plants against viruses: pathogen-derived resistance blossoms. Proc Natl Acad Sci USA 90:3134–3141, 1993.

85. J Hammond, H Lecoq, B Raccah. Epidemiological risks from mixed virus infections and transgenic plants expressing viral genes. Adv Virus Res 54: 189–314, 1999.

86. T Kunik, R Salomon, D Zamir, N Navot, M Zeidan, I Michelson, Y Gafni, H Czosnek. Transgenic tomato plants expressing the tomato yellow leaf curl capsid protein are resistant to the virus. Biotechnology 12:500–504, 1994.

87. JE Clements, RJ Wall, O Narayan, D Hauer, R Schoborg, D Sheffer, A Powell, LM Carruth, MC Zink, CE Rexroad. Development of transgenic sheep that express the visna virus envelope gene. Virology 200:370–380, 1994.

88. CJ Braun, CL Hemenway. Expression of amino-terminal portions or full-length viral replicase genes in transgenic plants confers resistance to potato virus X infection. Plant Cell 4:735–744, 1992.

89. M Longstaff, G Brigneti, F Boccard, S Chapman, D Baulcombe. Extreme resistance to potato virus X infection in plants expressing a modified component of the putative viral replicase. EMBO J 12:379–386, 1993.

90. JP Carr, A Gal-On, P Palukaitis, M Zaitlin. Replicase-mediated resistance to cucumber mosaic virus in transgenic plants involves suppression of both virus replication in the inoculated leaves and long-distance movement. Virology 199:439 447, 1994.

91. JP Carr, M Zaitlin. Resistance in transgenic tobacco plants expressing a nonstructural gene sequence of tobacco mosaic virus is a consequence of markedly reduced virus replication. Mol Plant Microbe Interact 4:579–585, 1991.

92. KH Hellwald, P Palukaitis. Viral RNA as a potential target for two independent mechanisms of replicase-mediated resistance against cucumber mosaic virus. Cell 83:937–946, 1995.

93. Y Hong, J Stanley. Virus resistance in *Nicotiana benthamiana* conferred by African cassava mosaic virus replication-associated protein (AC1) transgene. Mol Plant Microbe Interact 9:219–225, 1996.

94. WG Dougherty, BL Semler. Expression of virus-encoded proteinases: functional and structural similarities with cellular enzymes. Microbiol Rev 57: 781–822, 1993.

95. R Gutierrez-Campos, JA Torres-Acosta, LJ Saucedo-Arias, MA Gomez-Lim. The use of cysteine proteinase inhibitors to engineer resistance against potyviruses in transgenic tobacco plants. Nat Biotechnol 17:1223–1226, 1999.

96. P De Haan, JJL Gielen, M Prins, IG Wijkamp, A vanSchepen. Characterization of RNA-mediated resistance to tomato spotted wilt virus in transgenic tobacco plants. Biotechnology 10:1133–1137, 1992

97. M Bendahmane, B Gronenborn. Engineering resistance against tomato yellow leaf curl virus (TYLCV) using antisense RNA. Plant Mol Biol 33:351–357, 1997.

98. M Cuozzo, KM O'Connell, W Kaniewski, RX Fang, NH Chua, NE Tumer.
     Viral protection in transgenic tobacco plants expressing the cucumber mosaic
     virus coat protein or its antisense RNA. Biotechnology 6:549–557, 1988.
99. AG Day, ER Bejarano, KW Buck, M Burrell, CP Lichtenstein. Expression of
     an antisense viral gene in transgenic tobacco confers resistance to the DNA
     virus tomato golden mosaic virus. Proc Natl Acad Sci USA 88:6721–6725,
     1991.
100. C Hemenway, RX Fang, WK Kaniewski, NH Chua, NE Tumer. Analysis of
     the mechanism of protection in transgenic plants expressing the potato virus
     X coat protein or its antisense RNA. EMBO J 7:1273–1280, 1988.
101. PA Powell, DM Stark, PR Sanders, RN Beachy. Protection against tobacco
     mosaic virus in transgenic plants that express tobacco virus antisense RNA.
     Proc Natl Acad Sci USA 86:6949–6952, 1989.
102. MA Rezaian, KGM Skene, JG Ellis. Anti-sense RNAs of cucumber mosaic
     virus in transgenic plants assessed for control of the virus. Plant Mol Biol 11:
     463–471, 1988.
103. R Chasan. Making sense (suppression) of viral RNA–mediated resistance.
     Plant Cell 6:1329–1331, 1994.
104. F Meins. RNA degradation and models for post-transcriptional gene silencing.
     Plant Mol Biol 43:261–273, 2000.
105. T Frischmuth, J Stanley. Strategies for the control of geminivirus disease.
     Semin Virol 4:329–337, 1993.
106. XH Li, LA Heaton, JT Morris, A Simon. Turnip crinkle virus defective in-
     terfering RNAs intensify viral symptoms and are generated de novo. Proc
     Natl Acad Sci USA 86:9173–9177, 1989.
107. KBG Scholthof, HB Scholthof, AO Jackson. The effect of defective interfer-
     ing RNAs on the accumulation of tomato bushy stunt virus proteins and
     implications for disease attenuation. Virology 211:324–328, 1995.
108. A Moffat. Geminiviruses emerge as serious crop threat. Science 286:1835,
     1999.
109. ER Rybicki, G Pietersen. Plant virus disease problems in the developing
     world. Adv Virus Res 53:127–175, 1999.
110. AC Magyarosy, JE Duffus. The occurrence of highly virulent strains of the
     beet curly top virus in California. Plant Dis Rep 61:248–251, 1997.
111. H Czosnek, H Laterrot. A worldwide survey of tomato yellow leaf curl vi-
     ruses. Arch Virol 142:1391–1406, 1997.
112. EP Rybicki, RW Briddon, JK Brown, CM Fauquet, DP Maxwell, BD Harri-
     son, PG Markham, DM Bisaro, D Robinson, J Stanley. Family Geminiviridae.
     In: MHV van Regenmortel, CM Fauquet, DHL Bishop, eds. Virus Taxonomy
     —Classification and Nomenclature of Viruses. San Diego: Academic Press,
     2000, pp 285–297.
113. T Frischmuth, J Stanley. Characterization of beet curly top virus subgenomic
     DNA localizes sequences required for replication. Virology 189:808–811,
     1992.
114. T Frischmuth, J Stanley. African cassava mosaic virus DI-DNA interferes with
     the replication of both genomic components. Virology 183:808–811, 1991.

115. J Stanley, T Frischmuth, S Ellwood. Defective viral DNA ameliorates symptoms of geminivirus infection in transgenic plants. Proc Natl Acad Sci USA 87:6291–6295, 1990.
116. J Stanley, JR Latham. A symptom variant of beet curly top geminivirus produced by mutation of open reading frame C4. Virology 190:506–509, 1992.
117. T Frischmuth, J Stanley. Beet curly top virus symptom amelioration in *Nicotiana benthamiana* transformed with a naturally-occurring viral subgenomic DNA. Virology 200:826–830, 1994.
118. T Frischmuth, M Engel, H Jeske. Beet curly top virus DI DNA–mediated resistance is linked to its size. Mol Breed 3:213–217, 1997.
119. T Frischmuth, M Ringel, C Kocher. The size of encapsidated single-stranded DNA determines the multiplicity of African cassava mosaic virus particles. J Gen Virol 82:673–676, 2001.
120. MA Matzke, AJM Matzke, WB Eggleston. Paramutation and transgene silencing: a common response to invasive DNA? Trends Plant Sci 1:382–388, 1996.
121. M Matzke, AJM Matzke. Genomic imprinting in plants: parental effects and trans-inactivation phenomena. Annu Rev Plant Physiol Plant Mol Biol 44:53–76, 1993.
122. P Meyer, H Saedler. Homology-dependent gene silencing in plants. Annu Rev Plant Physiol Plant Mol Biol 47:23–48, 1996.
123. FG Ratcliff, SA MacFarlane, DC Baulcombe. Gene silencing without DNA: RNA-mediated cross-protection between viruses. Plant Cell 11:1207–1215, 1999.
124. F Ratcliff, BD Harrison, DC Baulcombe. A similarity between viral defense and gene silencing in plants. Science 276:1558–1560, 1997.
125. SN Covey, NS Al-Kaff, A Lángara, DS Turner. Plants combat infection by gene silencing. Nature 385:781–782, 1997.
126. SM Hammond, E Bernstein, D Beach, GJ Hannon. An RNA-directed nuclease mediates post-transcriptional gene silencing in *Drosophila* cells. Nature 404: 293–296, 2000.
127. A Fire, S Xu, MK Montgomery, SA Kostas, SE Driver, CC Mello. Potent and specific genetic interference by double-stranded RNA in *Caenorhabditis elegans*. Nature 39:806–811, 1998.
128. W Reik, A Murrell. Silence across the border. Nature 405:408–409, 2000.
129. T Sijen, JM Kooter. Post-transcriptional gene-silencing: RNAs on the attack or on the defense? Bioessays 22:520–531, 2000.
130. DR Smyth. Gene silencing: plants and viruses fight it out. Curr Biol 9:R100–R102, 1999.
131. V Vance, H Vaucheret. RNA silencing in plants—defense and counterdefense. Science 292:2277–2280, 2001.
132. PM Waterhouse, NA Smith, MB Wang. Virus resistance and gene silencing: killing the messenger. Trends Plant Sci 4:452–457, 1999.
133. PM Waterhouse, MB Wang, T Lough. Gene silencing as an adaptive defence against viruses. Nature 411:834–842, 2001.

134. M Wassenegger, S Heimes, L Riedel, HL Sänger. RNA-directed de novo methylation of genomic sequences in plants. Cell 76:567–576, 1994.
135. MF Mette, W Aufsatz, J van der Winden, MA Matzke, AJM Matzke. Transcriptional silencing and promoter methylation triggered by double-stranded RNA. EMBO J 19:5194–5201, 2000.
136. V Ambros. Dicing up RNAs. Science 293:811–813, 2001.
137. M Fagard, H Vaucheret. (Trans)gene silencing in plants: how many mechanisms? Annu Rev Plant Physiol Plant Mol Biol 51:167–194, 2000.
138. O Mittelsten Scheid, K Afsar, J Paszkowski. Release of epigenetic gene silencing by trans-acting mutations in *Arabidopsis*. Proc Natl Acad Sci USA 95:632–637, 1998.
139. M Stam, R de Bruin, R van Blokland, RAL van der Hoorn, JNM Mol, JM Kooter. Distinct features of post-transcriptional gene silencing by antisense transgenes in single copy and inverted T-DNA repeat loci. Plant J 21:27–42, 2000.
140. DC Baulcombe, JJ English. Ectopic pairing of homologous DNA and post-transcriptional gene silencing in transgenic plants. Curr Biol 7:173–180, 1996.
141. JJ English, E Mueller, DC Baulcombe. Suppression of virus accumulation in transgenic plants exhibiting silencing of nuclear genes. Plant Cell 8:179–188, 1996.
142. JC Palauqui, S Balzergue. Activation of systemic acquired silencing by localised introduction of DNA. Curr Biol 9:59–66, 1999.
143. JC Palauqui, T Elmayan, JM Pollien, H Vaucheret. Systemic acquired silencing: transgene-specific post-transcriptional silencing is transmitted by grafting from silenced stocks to non-silenced scions. EMBO J 16:4738–4745, 1997.
144. S Sonoda, M Nishiguchi. Graft transmission of post-transcriptional gene silencing: target specificity for RNA degradation is transmissible between silenced and non-silenced plants, but not between silenced plants. Plant J 21: 1–8, 2000.
145. AC Mallory, L Ely, TH Smith, R Marathe, R Anandalakshmi, M Fagard, H Vaucheret, G Pruss, L Bowman, VB Vance. HC-pro suppression of transgene silencing eliminates the small RNAs but not transgene methylation or the mobile signal. Plant Cell 13:571–583, 2001.
146. T Dalmay, A Hamilton, S Rudd, S Angell, DC Baulcombe. An RNA-dependent RNA polymerase gene in *Arabidopsis* is required for posttranscriptional gene silencing mediated by a transgene but not by a virus. Cell 101:543–553, 2000.
147. P Mourrain, C Béclin, T Elmayan, F Feuerbach, C Godon, JB Morel, D Jouette, AM Lacombe, S Nikic, N Picault, K Rémoué, M Sanial, TA Vo, H Vaucheret. *Arabidopsis SGS2* and *SGS3* genes are required for posttranscriptional gene silencing and natural virus resistance. Cell 101:533–542, 2000.
148. E Bernstein, AA Caudy, SM Hammond, GJ Hannon. Role for a bidentate ribonuclease in the initiation step of RNA interference. Nature 409:363–366, 2001.

149. SM Elbashir, J Harborth, W Lendeckel, A Yalcin, K Weber, T Tuschl. Duplexes of 21-nucleotide RNAs mediate RNA interference in cultured mammalian cells. Nature 411:494–498, 2001.
150. PD Zamore, T Tuschl, PA Sharp, DP Bartel. RNAi: double-stranded RNA directs the ATP-dependent cleavage of mRNA at 21 to 23 nucleotide intervals. Cell 101:25–33, 2000.
151. Z Xie, B Fan, C Chen, Z Chen. An important role of an inducible RNA-dependent RNA polymerase in plant antiviral defense. Proc Natl Acad Sci USA 98:6516–6521, 2001.
152. G Brigneti, O Voinnet, WX Li, LH Ji, SW Ding, DC Baulcombe. Viral pathogenicity determinants are suppressors of transgene silencing in *Nicotiana benthamiana*. EMBO J 17:6739–6746, 1998.
153. LH Ji, SW Ding. The suppressor of transgene RNA silencing encoded by cucumber mosaic virus interferes with salicylic acid–mediated virus resistance. Mol Plant Microbe Interact 14:715–724, 2001.
154. LK Johansen, JC Carrington. Silencing on the spot. Induction and suppression of RNA silencing in the *Agrobacterium*-mediated transient expression system. Plant Physiol 126:930–938, 2001.
155. AP Lucy, HS Guo, WX Li, SW Ding. Suppression of post-transcriptional gene silencing by a plant viral protein localized in the nucleus. EMBO J 19:1672–1680, 2000.
156. O Voinnet, C Lederer, DC Baulcombe. A viral movement protein prevents spread of the gene silencing signal in *Nicotiana benthamiana*. Cell 103:157–167, 2000.
157. O Voinnet, YM Pinto, DC Baulcombe. Suppression of gene silencing: a general strategy used by diverse DNA and RNA viruses of plants. Proc Natl Acad Sci USA 96:14147–14152, 1999.
158. CN Mayers, P Palukaitis, JP Carr. Subcellular distribution analysis of the cucumber mosaic virus 2b protein. J Gen Virol 81:219–226, 2000.
159. F Di Serio, H Schöb, A Iglesias, C Tarina, E Bouldoires, F Meins. Sense- and antisense-mediated gene silencing in tobacco is inhibited by the same viral suppressors and is associated with accumulation of small RNAs. Proc Natl Acad Sci USA 98:6506–6510, 2001.
160. G Pruss, X Ge, XM Shi, JC Carrington, VB Vance. Plant viral synergism: the potyviral genome encodes a broad-range pathogenicity enhancer that transactivates replication of heterologous viruses. Plant Cell 9:859–868, 1997.
161. VB Vance, PH Berger, JC Carrington, AG Hunt, XM Shi. 5' proximal potyviral sequences mediate potato virus X/potyviral synergistic disease in transgenic tobacco. Virology 206:583–590, 1995.
162. OECD database of field trials. http://www.olis.oecd.org/biotrack.nsf
163. M Varrelmann, E Maiss. Mutations in the coat protein gene of plum pox virus suppress particle assembly, heterologous encapsidation and complementation in transgenic plants of *Nicotiana benthamiana*. J Gen Virol 81:567–576, 2000.
164. MMC Lai. RNA recombination in animal and plant viruses. Microbiol Rev 56:61–79, 1992.

165. M Padidam, S Sawyer, CM Fauquet. Possible emergence of new geminiviruses by frequent recombination. Virology 285:218–225, 1999.

166. X Zhou, Y Liu, L Calvert, C Munoz, GW Otim-Nape, DJ Robinson, BD Harrison. Evidence that DNA-A of a geminivirus associated with severe cassava mosaic disease in Uganda has arisen by interspecific recombination. J Gen Virol 78:2101–2111, 1997.

167. XP Zhou, YL Liu, DJ Robinson, BD Harrison. Four DNA-A variants among Pakistani isolates of cotton leaf curl virus and their affinities to DNA-A of geminivirus isolates from okra. J Gen Virol 79:915–923, 1998.

168. H Jeske, M Lütgemeier, W Preiß. DNA forms indicate rolling circle and recombination-dependent replication of Abutilon mosaic virus. EMBO J 20: 6158–6167, 2001.

169. T Frischmuth, J Stanley. Recombination between viral DNA and the transgenic coat protein gene of African casava mosaic geminivirus. J Gen Virol 79:1265–1271, 1998.

170. A Greene, RF Allison. Deletions in the 3′ untranslated region of cowpea chlorotic mottle virus transgene reduce recovery of recombinant viruses in transgenic plants. Virology 225:231–234, 1996.

171. L Kiraly, JE Bourque, JE Schoelz. Temporal and spatial appearance of recombinant viruses formed between cauliflower mosaic virus (CaMV) and CaMV sequences present in transgenic Nicotiana bigelovii. Mol Plant Microbe Interact 11:309–316, 1998.

23

Transgenic Plants with Enhanced Tolerance against Microbial Pathogens

Raimund Tenhaken

Department of Plant Physiology, University of Kaiserslautern, Kaiserslautern, Germany

Disease resistance in crop plants is a major challenge in plant breeding. Within industrialized agriculture, the need for efficient self-protection of cultivated plants is likely to increase in the future. To achieve this goal, diverse strategies will be necessary and compete. Conventional breeding has made great progress in incorporating natural defense genes, but the limitations of this method are also obvious. The progress in plant molecular biology now allows the generation of transgenic plants, thereby exploiting the mechanisms nature has developed to control and to limit the infection of microbial pathogens on plants. An additional benefit of transgenic plants is the possibility to validate the usefulness of endogenous plant genes in defense strategies. Changes in the expression or composition of a desired gene may still be achieved by conventional breeding assisted by the results from transgenic plants.

The possibility to transform all major crop plants—although sometimes tedious and labor intensive—opens a new opportunity for novel enhanced plant tolerance toward microbial pathogens. Over the last few decades many of the natural plant defense strategies were thoroughly investigated, and as a result numerous mechanisms are known by now. For example, the concept of pathogenesis-related proteins (PR-proteins) has revealed an inducible defense system that is thought to have antimicrobial properties. The constitutive expression of a single or a few genes of this group of defense genes was started in the late 1980s but with limited success. This will be covered in the first section of this chapter. On the other hand, new strategies for enhanced pathogen tolerance try to use the plants' signaling network for defense activation. As a consequence, a whole battery of diverse defense reactions is triggered (variable in different plant species), which seems to be more powerful than the overexpression of single antimicrobial proteins. These approaches acting on complex defense networks will be described in the second part.

I. SINGLE-GENE DEFENSE MECHANISMS

A. Pathogenesis-Related Proteins

About 40 years ago Ross (1,2) performed experiments with tobacco Xanthi nc plants, which after viral inoculation exhibit a hypersensitive reaction in which a limited number of cells die and form a lesion. Plants that were inoculated with the tobacco mosaic virus showed much smaller lesions after a challenge infection 7 days later than newly inoculated leaves. This phenomenon is now well known as systemic acquired resistance (SAR). Searching for the mechanisms underlying SAR, PR-proteins were discovered (3). Subsequently some of them were biochemically identified as hydrolases of fungal cell walls, namely β-1,3-glucanase (4,5) and chitinase (6). However, the biochemical properties and enzymatic function of other PR-proteins such as the well-characterized PR-1 remain puzzling (7).

Transgenic tobacco plants overexpressing a chitinase gene from tobacco showed enhanced tolerance against fungal infection by *Cercospora nicotianae* (8) indicating that high levels of plant hydrolyzing enzymes are a suitable strategy to increase pathogen tolerance. By using a combination of a β-1,3-glucanase and a chitinase gene, Broglie et al. (9) demonstrated the synergistic effect of both enzymes, yielding a higher degree of resistance compared with the expression of each gene alone. A similar conclusion was drawn from a study in tobacco, in which a gene for a basic chitinase from rice and a gene for an acidic glucanase from alfalfa were coexpressed after appropriate crossing of individual transgenic plants (10). The different gene combinations were compared, as was the level of expression of these hydrolytic enzymes in homozygous versus heterozygous plants. Based on the reduction of lesion size after *C. nicotianae* infection, it was shown that the combination of glucanase and chitinase expression even at a moderate level is more beneficial than the expression of either gene alone at a much higher level (10). The concept of the constitutive expression of hydrolytic enzymes was later extended from tobacco to crop plants. The expression of chitinase in oilseed rape (*Brassica napus*) to obtain enhanced tolerance was tested in field trials after inoculation with three different fungal pathogens (11). Although the overall protection of the transgenic plants in the field trials seems to be smaller than previously shown in experiments in a greenhouse, the enhanced tolerance in chitinase-expressing rape was effective against several fungi under natural field conditions. Many other plant species were transformed with plant glucanases or chitinases, and most research groups were able to find at least one fungus that is sensitive to these hydrolytic enzymes (12–14).

The reports on transgenic plants overexpressiong hydrolytic enzymes imply that this strategy might already be sufficient to combat most fungal pathogens. However, this is obviously not the case. Quite often fungi were chosen for pathogenicity assays that are known to be sensitive to chitinases and/or glucanases, *Rhizoctonia solani* and *Trichoderma viride* being examples. In contrast, well-recognized plant pathogens are often not significantly inhibited in their pathogenicity (15). Thus, plant hydrolases are useful to limit the spread of some fungal pathogens and are likely to increase the basal tolerance of plants against fungal infections. Whether the increased tolerance of transgenic plants with elevated levels of hydrolytic enzymes is the result of the direct inhibition of the fungal (tip) growth needs to be carefully analyzed in future work. Enzymatic cleavage of fungal cell walls liberating chitin or glucan oligomers will also activate diverse plant defense responses as these carbohydrate oligomers are potent elicitors in almost every plant (16,17). The transcriptional activation of genes involved in lignin precursor formation or antimicrobial phytoalexins by elicitors is well known (18,19). The genes for PR-proteins of unknown biochemical function such as PR1 or PR5 were also constitutively expressed in tobacco plants. The PR1a-overexpressing plants have a higher degree of resistance against a limited number of fungal pathogens from oomycetes (20). However, no enhanced tolerance against other pathogenic fungi of tobacco or tobacco mosaic virus was achieved. It was later shown that PR1-proteins from tobacco and tomato inhibit the spore germination of *Phytophthora infestans* and reduce the lesion size of diseased tomato leafs (21). Notably, the PR1a isoform, used in the transgenic lines (20), was very inefficient and showed only 10% of the biological activity of the most potent antifungal isoform PR1g (21).

The overexpression of a rice thaumatin-like gene of the PR5 family in rice plants gave enhanced tolerance against *Rhizoctonia solani*, the agent causing sheath blight disease. The infected leaf area was reduced to one fifth in the best lines, indicating that the thaumatin-like gene is highly effective against *R. solani* infection (22). Expression of a similar osmotin-like gene in potato also enhanced tolerance against the late blight disease caused by *Phytophthora infestans* (23).

One interesting point about PR-proteins in cereals is worth mentioning. Wheat, for instance, expresses hydrolytic enzymes (β-1,3-glucance, chitinase) after pathogen infection. In contrast to this situation in *Arabidopsis* and tobacco, the status of a systemic acquired resistance in wheat is not correlated with the constitutive expression of these PR-genes (24). This raises the question of whether hydrolytic enzymes are an important part of the plant defense system at all. Alternatively, other as yet unidentified genes (or mechanisms) are important players in plant tolerance to pathogens. Support for the latter view comes from experiments with *Arabidopsis* DNA microarrays in which many genes are transcriptionally induced in resistant plants (25). The fact that 413 genes (out of ~7000 genes representing 25% of the genome) were reported to show a consistently higher expression level in SAR or plant resistance indicates that there will be more than a thousand genes that are significantly induced in local resistance or SAR. This incredibly high number draws a far more complex picture of plant resistance than previously thought. As well as being frustrating for researchers trying to dissect single-gene function in SAR, the high number of SAR-related genes is a huge challenge and opportunity for the plant molecular biologist to create novel transgenic lines with enhanced tolerance against a variety of microbial pathogens.

B. Defense Peptides

Plant and animals have developed an efficient mechanism to combat pathogens by using small antimicrobial peptides collectively termed defensins. These peptides are now divided into several families on the basis of sequence homology and structural properties. They are usually relatively small (<60 amino acids) and in the case of plant defensins contain several cysteine residues that form one or more stable disulfide bridges. Defense peptides exhibit a broad spectrum of antimicrobial activity against bacteria, fungi, and even enveloped viruses. In animal cells even parasites and tumor cells are inhibited (26). Most of the known defensin genes are from insects showing mainly antibacterial activity (27,28). The large diversity of these peptides (more than 500 varieties are known) provides a huge reservoir of genes that can be expressed in plants to enhance tolerance against pathogens, an approach that has just started to be exploited by plant molecular biologists.

A plant defensin from radish (RsAFLP2) was expressed at high levels in tobacco plants. Subsequent infection with the fungal pathogen *Alternaria longipes* revealed efficient protection of the transgenic plants resulting in more than 80% reduction of lesion sizes (29).

Lipid transfer proteins were initially described as shuttle proteins, involved in the transfer of lipids between organelles (30). Later, the antimicrobial activity of these peptides was discovered (31,32). Transgenic *Arabidopsis* plants overexpressing the lipid transfer protein LTP2 from barley showed a strong reduction in disease symptoms after infection with *Pseudomonas syringae* pv. *tomato* (33).

Thionins are often found in quite high amounts in the endosperm of cereals and other plants. They are toxic to plant pathogenic fungi but only at relatively high concentrations. Bohlmann et al. (34) showed inhibition of the barley pathogen *Drechslera teres* at concentrations of 500 μM. Nevertheless, transgenic *Arabidopsis* plants expressing high thionin levels were shown to be more resistant to *Fusarium oxysporum* (35) and *Plasmodiophora brassicae* (36).

The overexpression of plant defense peptide genes to achieve higher tolerance against a broad range of pathogens was only partly successful. A few pathogens (mainly bacteria) were restricted in their growth, but many other pathogens were not. This problem was overcome by the expression of chimeric defense peptides in potato plants. The synthetic peptides consist of two domains derived from cecropin (from the giant silk moth *Hyalaphora cecropia*) and melittin (the major component of bee venom) (37). The N-terminus of the chimeric peptide had to be modified in order to be tolerated as nontoxic by the plant cells. The transgenic potato plants were almost totally resistant against *Erwinia carotovora* even over a very long period of time (e.g., 6 months of tuber storage). This high degree of resistance was also observed after infection of potato plants with different fungi including *Phytophthora cactorum* and *Fusarium solani* (37). As mentioned before, some defense peptides are also toxic to animals, implying the need for great care in the use of these defense molecules in edible plants. The transgenic potatoes containing the chimeric cecropin-melittin peptide were fed to mice for several weeks without any notable change in animal behavior or body weight. The large number of known potent defense peptides from insects combined with molecular biology tools will make it possible to exploit these natural defense mechanisms on a broad basis.

C. Ribosome-Inactivating Proteins

Ribosome-inactivating proteins (RIPs) are widely found in plants. They exhibit a specific RNA-*N*-glycosidase activity that selectively cleaves off an adenine residue from a conserved site of the 28S rRNA. This prevents binding of the elongation factor 2 and consequently leads to an arrest in protein biosynthesis (38). RIPs do not inactivate the ribosomes of their own species but inactivate those of distantly related species. Expression of the barley

seed RIP in tobacco under the control of a wound-inducible promoter resulted in enhanced tolerance against fungal infections with *Rhizoctonia solani* (39). However, attempts to express the barley RIP in wheat plants were
unsuccessful, indicating that the constitutive expression of this protein is
toxic for wheat to allow regeneration of transgenic lines (14). RIPs are
primarily antiviral proteins, and overexpression of RIPs often leads to resistance of the transgenic plant against a broad spectrum of plant viruses. However, an additional antifungal activity of RIP was observed when overexpressing pokeweed antiviral protein and mutant forms, which exhibit no
N-glycosidase activity typical of the RIP function (40). The advantage of
the mutant RIPs is their low toxicity compared with the original protein.

D. Plants with Elevated Levels of Antimicrobial Secondary Compounds

The efficient protection of many wild-type plants against microbial pathogens is thought to be mediated at least in part by toxic secondary metabolites
from plants (41,42). Often these compounds require very complex biosynthesis, and many of the genes involved in their formation are not known or
characterized. These limitations usually make the formation of secondary
metabolites with complex biosynthesis in transgenic plants very difficult.
Therefore, today's strategies are based on the expression of a single gene
(or a few genes) to equip a plant with a novel secondary metabolite.

A successful example is the expression of a stilbene synthase gene
from grapevine (*Vitis vinifera*) in tobacco and crop plants under the control
of its native pathogen-inducible promoter. The enzyme requires only one *p*-
coumaroyl-coenzyme A (CoA) (a lignin precursor) and three malonyl-CoA
to form one molecule of the stilbene resveratrol, which has antimicrobial
properties in plants. The novel phytoalexin accumulated after *Botrytis
cinerea* infection to low millimolar levels within a few days, resulting in a
reduction of diseased leaf area of roughly two thirds (43). The same gene
was later expressed in tomato (44), rice (45), and other crop plants. Again,
a strong increase in basal pathogen tolerance was obtained. One drawback,
however, is the observation that tomato plants producing resveratrol showed
an increase in resistance against *Phytophthora infestans* (~50% reduction
in diseased leaf area) but not against *Botrytis cinerea* (44), which was efficiently restricted on resveratrol-producing tobacco plants. In retrospect, it is
amazing to see that the grapevine stilbene synthase promoter is pathogen
inducible in so many plants and that this inducibility is essential as the
constitutive expression of the stilbene synthase gene results in detrimental
effects such as male sterility in tobacco (46).

II. MULTIGENE DEFENSE MECHANISMS

A. Elevation of Endogenous Levels of Salicylic Acid

In the first section, strategies were applied that are based on the expression of a single gene (or a few genes) and a single target mechanism. From epidemiological studies it is evident that such strategies have a high chance of being overcome by pathogens, similar to resistance against chemical pesticides. From this point of view, more complex defense strategies should be beneficial for enhanced long-term pathogen tolerance. One obvious strategy is to make use of the plant's own defense system, for instance, by lowering the threshold level above which a plant mounts an efficient set of defense reactions to combat microbial pathogens. Verbene et al. (47) described the expression of two bacterial genes in plant chloroplasts that lead to salicylic acid biosynthesis from chorismate. Transgenic tobacco plants have high levels ($\sim$100 μM) of salicylic acid glucoside (the plant's vacuolar storage form) and are resistant against the fungus *Oidium lycopersicon*. Similarly, lesions after infection with tobacco mosaic virus were much smaller than those in wild-type plants, resembling the establishment of systemic acquired resistance (SAR). The constant production of salicylic acid in the transgenic plants turns on the immune system of plants. Although this strategy was so far tested only in tobacco, it seems to be very promising for many other plants.

B. Constitutive Systemic Acquired Resistance

Many laboratories have developed mutagenesis-based screens to look for plants with a defect in the plant immune system, simplified set equivalent to SAR. A single mutant was isolated independently by three groups. The gene is synonymously called *npr* (48) (nonexpresser of PR-genes), *nim* (49) (non-immunity), or *sai* (50) (salicylic acid insensitive). Although the biochemical function of the NPR protein is not well understood, it seems to interact with transcription factors required for PR-gene expression in a salicylic acid–dependent manner (51). One obvious experiment after the isolation of the *NPR* gene was to increase the expression level of this gene in transgenic plants. Luckily, the transgenic *Arabidopsis* lines showed strongly enhanced tolerance against infection by the biotrophic fungus *Peronospora parasitica*. This was achieved by a threefold increase in the protein level of NPR, which turned out to be sufficient to activate the complex plant defense system constitutively, resulting in SAR and resistance against *P. parasitica* (52).

C. Hydrogen Peroxide

A surprisingly small and reactive molecule plays a key role in plant defense: hydrogen peroxide (H_2O_2). Previously, H_2O_2 was regarded as an unavoidable by-product of respiration and photosynthesis for which the plant cell has no use, and thus it is rapidly detoxified by means of, for instance, catalase or ascorbate peroxidase. Over the past decade a totally different picture of the function of H_2O_2 and other reactive oxygen species has emerged from many studies (53,54). For instance, H_2O_2 drives the cross-linking of plant cell wall structural proteins. The local toughening of the cell wall is a barrier for penetrating fungi and is beneficial for the plant to mount a local defense to stop the ingression of the fungus. Cross-linked cell walls were shown to be much more resistant towards microbial cell wall cleaving hydrolyses (55,56). Along the same line, the expression of a wheat germin protein in wheat leaves results in enhanced resistance against *Blumeria graminis* f.sp. *tritici*, the causal agent of wheat powdery mildew (57). It was shown that germin is oxidatively cross-linked in wheat cell walls at sites of attempted fungal penetration. From a biochemical point of view, it is also clear that other compounds can be polymerized in H_2O_2-dependent processes. Many other phenolic secondary compounds are candidates for mixed polyphenols beside the lignin precursors (58). Thus, it is conceivable that part of the enhanced pathogen tolerance of stilbene-producing plants is mediated by the supply of the phenolic compound resveratrol for polyphenol formation rather than the direct antimicrobial activity of this phytoalexin alone.

The plant-pathogen interaction that results in programmed cell death of the hypersensitive reaction (HR) exhibits an extra oxidative burst several hours after contact between plant and microbe. This phenomenon is widespread and therefore seems to be important in plant resistance. The H_2O_2 from the oxidative burst was shown to be a signal that diffuses locally around the site of the infection and thereby transcriptionally induces genes in neighboring cells (59). Furthermore, H_2O_2 pulses (<10 minutes) induce cell death in soybean cell cultures several hours later. Although the concentrations used for the H_2O_2 pulses were relatively high (2–5 mM), a similar result can be obtained by supplying a constant H_2O_2 concentration of 10–50 μM for several hours using the enzyme glucose oxidase (R. Tenhaken and C. Rübel, unpublished).

Following these ideas, transgenic plants were generated in which a catalase gene was largely suppressed by an antisense strategy (60,61). As a consequence, the steady-state levels of H_2O_2 are higher and these plants are more sensitive to microbial pathogen attack, resulting in enhanced tolerance. These observations resemble findings in transgenic potato plants that express a glucose oxidase gene from *Aspergillus* in their cell wall (62,63). The slightly elevated levels of H_2O_2 in the potato plants have a strong impact on pathogen resistance.

Why and how a small increase in H_2O_2 causes the greatly increased pathogen tolerance remain a mystery waiting to be solved in future studies.

As a final remark, the catalase antisense plants are more resistant to pathogens but they are also more sensitive to oxidative stress caused by photosynthesis under high light conditions. One should be cautious about the likely reduction in cold tolerance and drought stress, two unfavorable environmental conditions that put oxidative stress on the plants. It seems that the concentration of H_2O_2 has to remain within a certain margin in order to avoid detrimental side effects on the plant's tolerance to unusual environmental conditions.

D. Cell Death as a Trigger of Plant Resistance

1. Mutants with Spontaneous Cell Death

The most potent plant defense against microbial pathogens is the hypersensitive reaction, a form of programmed cell death in plants. The principle of the HR is based on the early recognition of a pathogen, which then actively triggers a particular cell death program in the attacked cell (64). Although it is easily conceivable that an attack of a biotrophic fungus can be stopped efficiently by killing the plant cell in contact with the pathogen, the molecular basis for the general success of the HR to stop pathogen ingression remains to be solved in detail (65). The scientific problem comes down to the question of what else is turned on by the programmed cell death of the HR that is not activated by cell death per se (for instance, wounding).

In general, an HR in one leaf will establish an immune response (SAR) in the whole plant that allows a more efficient HR in the next infection event, as mentioned before for the tobacco mosaic virus–inoculated tobacco plants (2). A lesson from these and other studies is that the plant's endogenous cell death program is a valuable tool for plant resistance.

Plant breeders have used programmed cell death in plants for decades to achieve resistance against pathogens. The barley *Mlo* gene is probably the best understood example (66). In rare cases, a mutation in the *Mlo* locus causes programmed cell death of single cells in leaves, a phenomenon termed lesion mimic. The important thing about lesion mimic is the discovery that this cell death causes plant responses very similar to those that HR of a few cells would cause after a primary infection. Thus, lesion mimic is a physiological equivalent of an HR. The *Mlo* gene was identified by positional cloning and predicted to be a membrane protein that is a negative regulator (repressor) of cell death (67). About 10 years ago, several groups started to look for lesion mimic mutants in *Arabidopsis* (68,69). Many mutants have been identified, and at least for some of them the molecular mechanism is known. In contrast to the hidden cell death in *Mlo* barley plants, the *Arabidopsis* lesion mimics often show drastic phenotypes including dwarfism (70).

2.   Transgenic Plants with Induced Limited Cell Death

Meanwhile, several different approaches have been taken to induce cell death in plants. Often the cell death behaves like the cell death in the hypersensitive reaction after pathogen contact and subsequently triggers a whole battery of plant defenses, usually including elevated levels of salicylic acid (SA), induction of PR-genes, and immunity by the SAR process. One example is the bacterial ribonuclease barnase (71), which is a very potent enzyme that kills eukaryotic cells when present at a few molecules per cell. This enzyme is effectively inhibited by the small protein barstar. Using a pathogen-inducible promoter of a glutathione-*S*-transferase gene from potato, Strittmatter et al. (72) expressed the barnase gene in potato plants. To avoid killing of the whole plant by the leaky promoter, it was necessary to coexpress the barstar inhibitor. Potato lines in which the strength of the pathogen-inducible promoter led to a surplus of free active barnase showed pathogen-dependent cell death after *Phytophthora infestans* inoculation. Thus, a totally artificial cell death process was able to enhance plant tolerance toward fungal infection. The limited success of the study is most likely due to a nonoptimal choice of the promoters used, but nevertheless it demonstrates impressively the power of artificial cell death–inducing systems.

Another cell death–inducing molecule is the fungal protein cryptogein, secreted by *Phytophthora cryptogea*. It is one of many similar proteins of *Phytophthora* species, which are collectively termed elicitins. Expression of the gene encoding cryptogein under the control of the pathogen-inducible *hsr203J* promoter in tobacco resulted in enhanced tolerance of transgenic lines against *Phytophthora parasitica*, *Thielaviopsis basicola*, and *Erysiphe cichoracearum* infection (73). The mode of action seems to be the induction of a cell death program by the cryptogein protein, which then activates the general plant defense machinery. It is, however, currently anticipated that elicitins cause cell death only in tobacco species, thus limiting the use of this system in crop plants.

A third example of induced cell death was published by Tang et al. (74). The researchers constitutively expressed the *pto*-kinase gene in tomato, which was identified some years ago and shown to be required for resistance against the bacterial pathogen *Pseudomonas syringae* pv. *tomato*. The constitutive expression of *pto*-kinase is already sufficient to induce programmed cell death in a limited number of cells in tomato leaves, subsequently activating the SAR response in tomato. These transgenic plants exhibit broadspectrum pathogen tolerance as expected from the activated SAR response.

Although the latter examples of engineered cell death are specific for particular plant species, a novel more generally applicable system for the induction of artificial cell death emerges from many studies of plant resistance and microbial avirulence genes. As predicted from the early studies of Flor (1947), the interaction of a plant resistance gene product with a microbial avirulence gene product is the basis for the programmed cell death in the HR. Whereas the first *avr* genes were cloned from phytopathogenic bacteria in the early 1980s, the first plant resistance genes (*R* genes) were identified in 1994. The rigorous test of whether the *avr* gene product by itself is sufficient to induce an HR was impressively answered and confirmed by studies in which the *avr* gene was transiently expressed in plant cells.

Plant *R* genes seem to work functionally at least in closely related species. For instance, the tomato *Cf-9* resistance gene was transformed into tobacco plants and the resulting transgenic lines still responded to the corresponding avr9 peptide from the tomato pathogenic fungus *Cladosporium fulvum* (75). When this tobacco line was crossed with a transgenic plant expressing the *avr9* avirulence gene, the siblings showed a whole-plant HR and died at the seedling stage, thus confirming the concept. The expression of both *avr* and *R* genes in a strictly controlled pathogen-dependent manner is one of the most promising strategies for engineered cell death and subsequent activation of the plant's immune system. A great variety of suitable promoter elements for such experiments is currently under development. The use of two different pathogen-inducible promoters for the resistance and the avirulence gene will help to minimize the detrimental effects of unintentional cell death.

Thus, the regulation of cell death is an important issue and understanding it is necessary for further exploitation of programmed cell death as a novel mechanism for enhanced pathogen tolerance.

III. SALICYLIC ACID FUNCTION IN PROGRAMMED CELL DEATH

A central player in many but not all forms of programmed cell death in plants is salicylic acid (76). This was first demonstrated in transgenic tobacco and *Arabidopsis* plants expressing a salicylate hydroxylase gene from *Pseudomonas putida* called *nahG* (77). These plants are still capable of synthesizing SA, but as soon as SA starts to accumulate it is converted into catechol by the enzyme salicylate hydroxylase. *Arabidopsis* plants of the Col-O ecotype, which as wild-type plants are genetically resistant to *Peronospora parasitica* infection, were totally colonized by the fungus in *nahG*-expressing plants (76). This conversion of an incompatible to a compatible interaction could be reverted by spraying with high concentrations of salicylic acid or the synthetic analogue INA (2,6-dichloro-isonicotinic acid), indicating that SA is a key control molecule in plant resistance. Some of the dwarf lesion mimic mutants of *Arabidopsis* mentioned before could also be reverted to wild-type–like plants when crossed with *nahG* plants (78). Thus, the question arises of how SA controls programmed cell death in plants.

The discovery tour begins with a binding protein for SA, characterized from tobacco, that was shown to be a catalase (79), pointing to increased levels of H_2O_2 as the driving force of programmed cell death. This view was later extended by the finding that the other major H_2O_2-scavenging enzyme, ascorbate peroxidase, can also be inhibited in tobacco by SA (80). This model is attractive at first glance as H_2O_2 is an important signaling molecule in plant defense, as outlined earlier (53). However, many other groups have tested the phenomenon in various plant species, and none of their studies has shown a significant inhibition of catalase or ascorbate peroxidase at physiologically relevant concentrations (<100 μM). For example, in soybean SA (100 μM) neither inhibits both enzymes in vitro nor interferes with the capacity of soybean cell cultures to metabolize H_2O_2, which reflects the in vivo situation (81).

In a model system for programmed cell death in plants, we investigate the HR in soybean cell cultures (cv. Williams 82) triggered by the avrA protein of the bacterial pathogen *Pseudomonas syringae* pv. *glycinea*. The HR occurs according to the gene-for-gene hypothesis requiring the *Rps2* resistance gene in soybean and the *avrA* avirulence gene in *Pseudomonas*. The advantage of the cell culture system is its flexibility and the possibility to manipulate the system easily using inhibitors or activators. The HR re-quires SA (~50 μM), which can be added to the cell culture medium to complete the cell death program. This phenotype in soybean resembles the studies with *nahG Arabidopsis* plants, in which the HR is also dependent on SA. The addition of SA to soybean cell cultures strongly accelerates the HR, finally leading to faster cell death (82). By using the vital stain Evans Blue, we showed loss of membrane control as the final sign of cell death in the HR after about 8 hours in the SA-accelerated HR compared with about 14 hours in control inoculated cells without SA (similar to the time range of the HR in plants).

Using this acceleration for a screenable test system, we identified sev-eral chemicals that can substitute for SA (83). To our surprise, the chemicals were diverse in structure and thus made it initially difficult to come up with a model to explain their biochemical function. We have collectively termed these chemicals FASs (functional analogues of SA). A rigorous check of the literature revealed an emerging function of FAS compounds in animal cells. Almost all of them are ligands of a transcription factor system in humans, the peroxisome proliferator–activated receptors (PPARs). The PPARs are a subfamily within the superfamily of nuclear hormone receptors, which are posttranslationally activated by binding their specific ligands (84). One ther-apeutic aspect of PPARs in humans is their relevance in type II diabetes, which can be treated by the uptake of synthetic PPAR ligands (such as troglitazone) on a daily basis (85). This drug can also be used in soybean to complement and accelerate the HR, triggered by *Pseudomonas* bacteria, and by this criterion is an FAS compound. The outlined model predicts that genes in plants are transcriptionally induced by SA or FAS compounds. We have described a novel putative lectin-encoding gene from soybean that is induced by both groups of compounds as predicted (83). Using differential display and subtractive suppression hybridization, we have identified a larger set of genes, which are transcriptionally induced by salicylic acid and FAS compounds (C Anstätt, N Ausländer, A Ludwig, G Schwerdtfeger, M Hansen, R Tenhaken, unpublished results). This set probably includes novel genes, which are required for the execution of programmed cell death in a salicylic acid–dependent manner.

Chemicals such as INA or the benzothiadiazole Bion® are thought to mimic SA and are thus able to activate the signaling process leading to SAR (86). This raises the question of whether FAS chemicals are similar to SAR inducers in their mode of action. *Arabidopsis* and tobacco treated with SA or Bion show a strong induction of the PR-1 messenger RNA (mRNA) (Fig. 1). In contrast, the FAS chemical flufenamate was unable to induce the same response in tobacco, pointing to different signal transduction pathways. In the model plant tobacco, several SA-inducible genes were described besides the classical PR genes, for instance, in a study by Horvath and Chua (87,88). Whether theses genes are novel PR genes in the sense that they are induced in plant-pathogen interactions remains to be analyzed. We have chosen the *C14-1b* (88) gene as a probe to test whether FAS compounds will induce genes in tobacco as well as in soybean. The sensitive induction of the tobacco *C14-1b* gene by flufenamate, which occurs at a much lower concentration compared with SA, is an example (Fig. 2). As also shown in Fig. 2, another tobacco gene, *G8-1* (88), is strongly induced by salicylic acid but almost not induced by the FAS compound flufenamate. These observations underline the different signal transduction pathways and mode of actions in which salicylic acid is involved in plants. The data can be summarized in a new model in which SA controls plant resistance by at least two independent processes. In the first place, high concentrations of SA are needed to execute the plant cell death program, and the predicted mode of action is the transcriptional activation of genes (which mostly remains to be identified in future studies). This function of SA can be mimicked by FAS compounds as shown for soybean. The second major function of SA is the participation in the establishment of the plant immunity (SAR). This function of SA can be potently mimicked by INA or Bion.

Our current goal is to identify an SA/FAS-regulated transcription factor that is involved in programmed cell death in plants. By functional analysis

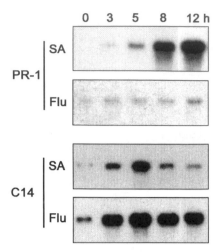

FIGURE 1 RNA blot with salicylic acid–regulated genes from tobacco. A BY-2 tobacco cell culture was treated with 250 μM salicylic acid (SA) or 25 μM flufenamate (Flu) and samples were taken at designated time points. The upper set of samples was hybridized with a tobacco *PR-1a* gene, showing the beginning of the induction of the *PR-1* gene by salicylic acid whereas flufenamate is inactive in inducing the *PR-1* gene. *C14-1b* is a tobacco gene of unknown function (88) that is only weakly induced by salicylic acid but strongly induced by flufenamate.

of the promoter, we identified a *cis* element in an FAS-responsive gene that is also specifically bound by nuclear protein extracts in gel retardation assays (C. Anstätt and R. Tenhaken, unpublished results). Controlled activation or repression of this factor will probably be a valuable tool to modify programmed cell death in plants. We anticipate a novel mode of enhanced plant tolerance toward microbial pathogens by engineering the execution of a natural cell death program and making use of the powerful natural broad-spectrum resistance that plants have developed during evolutionary history.

IV. CONCLUSIONS AND FUTURE PERSPECTIVES

The dream of transgenic plants resistant to viruses, bacteria, fungi, and insects is still (and is likely to be forever) a fantasy. What has been achieved in the last 15 years in creating transgenic plants with enhanced tolerance against pathogens is, however, very remarkable. The rapid success took advantage of decades of biochemical work of plant pathologists and has now opened a wide field of approaches for novel resistance principles in plants

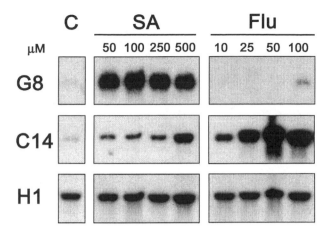

FIGURE 2 RNA blot with a dose-response experiment on salicylic acid–regulated genes in tobacco. A tobacco cell line (BY-2) was treated with different concentrations of salicylic acid (SA) or flufenamate (Flu) and collected for RNA isolation 5 hours later. The blot was hybridized with two tobacco genes (88) of unknown function, which exhibit different induction patterns. Whereas salicylic acid strongly induces the *G8-1* gene, flufenamate is largely inactive in this response. In contrast, the *C14-1b* gene shows a reversed gene induction pattern with weak induction by salicylic acid and a strong increase in mRNA caused by flufenamate. The equal loading was verified by rehybridization of the blot with a gene for the tobacco histone 1 (H1).

as a defense against microbial pathogens. Clearly, from today's point of view, the approaches that activate the host's defense machinery will be successful. Artificial cell death triggered under the strict control of pathogen-inducible promoters is my personal favorite. Although we can already use the plant's innate immune system, we basically understand very little of how it works. Detailed knowledge of single or network-like mechanisms will be a prerequisite for a next generation of engineered plants that specifically activate certain parts of the complex immune response required in particular for individual pathogens.

REFERENCES

1. AF Ross. Localized acquired resistance to plant virus infection in hypersensitive hosts. Virology 14:329–339, 1961.
2. AF Ross. Systemic acquired resistance induced by localized virus infections in plants. Virology 14:340–358, 1961.

3. LC Van Loon, A van Kammen. Polyacrylamide disc electrophoresis of the soluble leaf proteins from *Nicotiana tabacum* var "Samsun" and "Samsun NN." II Changes in protein constitution after infection with tobacco mosaic virus. Virology 40:199–211, 1970.

4. A Schlumbaum, F Mauch, U Vögeli, T Boller. Plant chitinases are potent inhibitors of fungal growth. Nature 324:365–367, 1986.

5. S Kauffmann, M Legrand, P Geoffroy, B Fritig. Biological function of "pathogenesis-related" proteins: four PR proteins of tobacco have 1,3-β-glucanase activity. EMBO J 6:3209–3212, 1987.

6. M Legrand, S Kauffmann, P Geoffroy, B Fritig. Biological function of pathogenesis-related proteins: four tobacco pathogenesis-related proteins are chitinases. Proc Natl Acad Sci USA 84:6750–6754, 1987.

7. LC Van Loon, EA Van Strien. The families of pathogenesis-related proteins, their activities, and comparative analysis of PR-1 type proteins. Physiol Mol Plant Pathol 55:85–97, 1999.

8. JM Neuhaus, P Ahl-Goy, U Hinz, S Flores, F Meins Jr. High level expression of a tobacco chitinase gene in *Nicotiana sylvestris*. Plant Mol Biol 16:141–151, 1991.

9. K Broglie, I Chet, M Holliday, R Cressman, P Biddle, S Knowlton, CJ Mauvais, R Broglie. Transgenic plants with enhanced resistance to the fungal pathogen *Rhizoctonia solani*. Science 254:1194–1197, 1991.

10. Q Zhu, EA Maher, S Massoud, RA Dixon, CJ Lamb. Enhanced protection against fungal attack by constitutive co-expression of chitinase and glucanase genes in transgenic tobacco. Biotechnology 12:807–812, 1994.

11. R Grison, B Grezes-Besset, M Schneider, N Lucante, L Olsen, JJ Leguay, A Toppan. Field tolerance to fungal pathogens of *Brassica napus* constitutively expressing a chimeric chitinase gene. Nat Biotechnol 14:643–646, 1996.

12. H Vierheilig, M Alt, JM Neuhaus, T Boller, A Wiemken. Colonization of transgenic *Nicotiana sylvestris* plants, expressing different forms of *Nicotiana tabacum* chitinase, by the root pathogen *Rhizoctonia solani* and by the mycorrhizal symbiont *Glomus mossae*. Mol Plant Microbe Interact 6:261–264, 1993.

13. W Lin, CS Anuratha, K Datta, I Potrykus, S Muthukrishnan, SK Datta. Genetic engineering of rice for resistance to sheath blight. Biotechnology 13:686–691, 1995.

14. M Bliffeld, J Mundy, I Potrykus, J Fütterer. Genetic engineering of wheat for increased resistance to powdery mildew disease. Theor Appl Genet 98:1079–1086, 1999.

15. E Kombrink, IE Somssich. Pathogenesis-related proteins and plant defense. In: GC Carroll, P Tudzynski, eds. The Mycota V. Part A: Plant Relationships. Berlin: Springer, 1997, pp 107–128.

16. T Boller. Chemoperception of microbial signals in plant cells. Annu Rev Plant Physiol Plant Mol Biol 46:189–214, 1995.

17. J Ebel. Oligoglucoside elicitor-mediated activation of plant defense. Bioessays 20:569–576, 1998.

18. RA Dixon, CJ Lamb. Molecular communication in interactions between plants and microbial pathogens. Annu Rev Plant Physiol Plant Mol Biol 41:339–367, 1990.

19. RA Dixon, MJ Harrison, CJ Lamb. Early events in the activation of plant defense responses. Annu Rev Phytopathol 32:479–501, 1994.

20. D Alexander, RM Goodman, M Gut-Rella, C Glascock, K Weymann, L Friedrich, D Maddox, P Ahl-Goy, T Luntz, E Ward, J Ryals. Increased tolerance to two oomycete pathogens in transgenic tobacco expressing pathogenesis-related protein 1a. Proc Natl Acad Sci USA 90:7327–7331, 1993.

21. T Niderman, I Genetet, T Bruyère, R Gees, A Stintzi, M Legrand, B Fritig, E Mösinger. Pathogenesis-related PR-1 proteins are antifungal. Plant Physiol 108: 17–27, 1995.

22. K Datta, R Velazhahan, N Oliva, I Ona, T Mew, GS Khush, S Muthukrishnan, SK Datta. Over-expression of the cloned rice thaumatin-like protein (PR-5) gene in transgenic rice plants enhances environmental friendly resistance to *Rhizoctonia solani* causing sheath blight disease. Theor Appl Genet 98:1138–1145, 1999.

23. B Zhu, THH Chen, PH Li. Analysis of late-blight disease resistance and freezing tolerance in transgenic potato plants expressing sense and antisense genes for an osmotin-like protein. Planta 198:70–77, 1996.

24. J Görlach, S Volrath, G Knauf-Beiter, G Hengy, U Beckhove, KH Kogel, M Oostendorp, T Staub, E Ward, H Kessmann, J Ryals. Benzothiadiazole, a novel class of inducers of systemic acquired resistance, activates gene expression and disease resistance in wheat. Plant Cell 8:629–643, 1996.

25. K Malek, A Levine, T Eulgem, A Morgan, J Schmid, KA Lawton, JL Dangl, RA Dietrich. The transcriptome of *Arabidopsis thaliana* during systemic acquired resistance. Nat Genet 26:403–410, 2000.

26. RE Hancock, R Lehrer. Cationic peptides: a new source of antibiotics. Trends Biotechnol 16:82–88, 1998.

27. M Meister, C Hetru, JA Hoffmann. The antimicrobial host defense of *Drosophila*. Curr Top Microbiol Immunol 248:17–36, 2000.

28. JA Hoffmann, FC Kafatos, CA Janeway Jr, RAB Ezekowitz. Phylogenetic perspectives in innate immunity. Science 284:1313–1318, 1999.

29. FRG Terras, K Eggermont, V Kovaleva, NV Raikhel, RW Osborn, A Kester, SB Rees, S Torrekens, F van Leuven, J Vanderleyden, BPA Cammue, WF Broekaert. Small cysteine-rich antifungal proteins from radish: their role in host defense. Plant Cell 7:573–588, 1995.

30. JC Kader. Lipid-transfer proteins in plants. Annu Rev Plant Physiol Plant Mol Biol 47:627–654, 1996.

31. FRG Terras, HME Schoofs, MFC De Bolle, F Van Leuven, SB Rees, J Vanderleyden, BPA Cammue, WF Broekart. Analysis of two novel classes of antifungal proteins from radish (*Raphanus sativus* L.) seeds. J Biol Chem 267: 15301–15309, 1992.

32. A Molina, F García-Olmedo. Developmental and pathogen-induced expression of three barley genes encoding lipid transfer proteins. Plant J 4:983–991, 1993.

33. A Molina, F García-Olmedo. Enhanced tolerance to bacterial pathogens caused by the transgenic expression of barley lipid transfer protein LTP2. Plant J 12: 669–675, 1997.

34. H Bohlmann, S Clausen, S Behnke, H Giese, C Hiller, U Reimann-Philipp, G Schrader, V Barkholt, K Apel. Leaf-specific thionins of barley—a novel class of cell wall proteins toxic to plant pathogenic fungi and possibly involved in the defense mechanism of plants. EMBO J 7:1559–1565, 1988.

35. P Epple, K Apel, H Bohlmann. Overexpression of an endogenous thionin enhances resistance of *Arabidopsis* against *Fusarium oxysporum.* Plant Cell 9: 509–520, 1997.

36. S Holtorf, J Ludwig-Müller, K Apel, H Bohlmann. High level expression of a viscotoxin in *Arabidopsis thaliana* gives enhanced resistance against *Plasmodiophora brassicae.* Plant Mol Biol 36:673–680, 1998.

37. M Osusky, G Zhou, L Osuska, RE Hancock, WW Kay, S Misra. Transgenic plants expressing cationic peptide chimeras exhibit broad-spectrum resistance to phytopathogens. Nat Biotechnol 18:1162–1166, 2000.

38. F Stirpe, L Barbieri, MG Batelli, M Soria, DA Lappi. Ribosome inactivating proteins from plants: present status and future prospects. Biotechnology 10: 405–412, 1992.

39. J Logemann, G Jach, H Tommerup, J Mundy, J Schell. Expression of a barley ribosome-inactivating protein leads to increased fungal protection in transgenic tobacco plants. Biotechnology 10:305–308, 1992.

40. O Zoubenko, F Uckun, Y Hur, I Chet, N Tumer. Plant resistance to fungal infection by nontoxic pokeweed antiviral protein mutants. Nat Biotechnol 15: 992–996, 1997.

41. RL Nicholson, R Hammerschmidt. Phenolic compounds and their role in disease resistance. Annu Rev Phytopathol 30:369–389, 1992.

42. JP Morrissey, AE Osbourn. Fungal resistance to plant antibiotics as a mechanism of pathogenesis. Microbiol Mol Biol Rev 63:708–724, 1999.

43. R Hain, HJ Reif, E Krause, R Langebartels, H Kindl, B Vornam, W Wiese, E Schmelzer, PH Schreier, RH Stöcker, K Stenzel. Disease resistance results from foreign phytoalexin expression in a novel plant. Nature 361:153–156, 1993.

44. J Thomzik, K Stenzel, R Stöcker, PH Schreier, R Hain, D Stahl. Synthesis of a grapevine phytoalexin in transgenic tomatoes (*Lycopersicon esculentum* Mill.) conditions resistance against *Phytophthora infestans.* Physiol Mol Plant Pathol 51:265–278, 1997.

45. P Stark-Lorenzen, B Nelke, G Hännsler, H Mühlbach, JE Thomzik. Transfer of a grapevine stilbene synthase gene to rice (*Oryza sativa* L.). Plant Cell Rep 16:668–673, 1997.

46. R Fischer, I Budde, R Hain. Stilbene synthase gene expression causes changes in flower colour and male sterility in tobacco. Plant J 11:489–498, 1997.

47. MC Verberne, R Verpoorte, JF Bol, J Mercado-Blanco, HJM Linthorst. Overproduction of salicylic acid in plants by bacterial transgenes enhances pathogen resistance. Nat Biotechnol 18:779–783, 2000.

48. H Cao, J Glazebrook, JD Clarke, S Volko, X Dong. The *Arabidopsis NPR1* gene that controls systemic acquired resistance encodes a novel protein containing ankyrin repeats. Cell 88:57–63, 1997.

49. JA Ryals, K Weymann, K Lawton, L Friedrich, D Ellis, HY Steiner, J Johnson, TP Delaney, T Jesse, P Vos, S Uknes. The *Arabidopsis Nim1* protein shows homology to the mammalian transcription factor inhibitor I-κB. Plant Cell 9: 425–439, 1997.

50. J Shaw, F Tsui, DF Klessig. Characterization of a salicylic acid insensitive mutant (*sai1*) of *Arabidopsis thaliana*, identified in a selective screen utilizing the SA-inducible expression of the *tms2* gene. Mol Plant Microbe Interact 10: 69–78, 1997.

51. Y Zhang, W Fan, M Kinkema X Li, X Dong. Interaction of NPR1 with basic leucine zipper protein transcription factors that bind sequences required for salicylic acid induction of the *PR-1* gene. Proc Natl Acad Sci USA 96:6523–6528, 1999.

52. H Cao, X Li, X Dong. Generation of broad-spectrum disease resistance by overexpression of an essential regulatory gene in systemic acquired resistance. Proc Natl Acad Sci USA 95:6531–6536, 1998.

53. R Tenhaken, A Levine, LF Brisson, RA Dixon, C Lamb. Function of the oxidative burst in hypersensitive disease resistance. Proc Natl Acad Sci USA 92: 4158–4163, 1995.

54. C Lamb, RA Dixon. The oxidative burst in plant disease resistance. Annu Rev Plant Physiol Plant Mol Biol 48:251 275, 1997.

55. LF Brisson, R Tenhaken, CJ Lamb. Function of oxidative cross-linking of cell wall structural proteins in plant disease resistance. Plant Cell 6:1703–1712, 1994.

56. O Otte, W Barz. The elicitor-induced oxidative burst in cultured chickpea cells drives the rapid insolubilization of two cell wall structural proteins. Planta 200: 238–246, 1996.

57. P Schweizer, A Christoffel, R Dudler. Transient expression of members of the germin-like gene family in epidermal cells of wheat confers disease resistance. Plant J 20:540–552, 1999.

58. R Vogelsang, E Berger, T Hagedorn, U Mühlenbeck, R Tenhaken, W Barz. Characterization of metabolic changes involved in hypersensitive-like browning reactions of chickpea (*Cicer arietinum* L.) cell cultures following challenge by *Ascochyta rabiei* culture filtrate. Physiol Mol Plant Pathol 44:141–155, 1994.

59. A Levine, R Tenhaken, RA Dixon, CJ Lamb. H_2O_2 from the oxidative burst orchestrates the plant hypersensitive disease resistance response. Cell 79:583–593, 1994.

60. S Chamnongpol, H Willekens, W Moeder, C Langebartels, H Jr Sandermann, M Van Montagu, D Inze, W Van Camp. Defense activation and enhanced pathogen tolerance induced by H_2O_2 in transgenic tobacco. Proc Natl Acad Sci USA 95:5818–5823, 1998.

61. H Du, DF Klessig. Role for salicylic acid in the activation of defense responses in catalase-deficient transgenic tobacco. Mol Plant Microbe Interact 10:922–925, 1997.

62. G Wu, BJ Shortt, EB Lawrence, EB Levine, KC Fitzsimmons, DM Shah. Disease resistance conferred by expression of a gene encoding H_2O_2 generating glucose oxidase in transgenic potato plants. Plant Cell 7:1357–1368, 1995.

63. G Wu, BJ Shortt, EB Lawrence, J Leon, KC Fitzsimmons, EB Levine, I Raskin, DM Shah. Activation of host defense mechanisms by elevated production of H_2O_2 in transgenic plants. Plant Physiol 115:427–435, 1997.

64. JL Dangl, RA Dietrich, MH Richberg. Death don't have no mercy: cell death programs in plant-microbe interactions. Plant Cell 8:1793–1807, 1996.

65. C Richael, D Gilchrist. The hypersensitive response: a case of hold or fold? Physiol Mol Plant Pathol 55:5–12, 1999.

66. M Wolter, K Hollrichter, F Salamini, P Schulze-Lefert. The *mlo* resistance alleles to powdery mildew infection in barley trigger a developmentally controlled defense mimic phenotype. Mol Gen Genet 239:122–128, 1993.

67. R Büschges, K Hollricher, R Panstruga, G Simons, M Wolter, A Frijters, R van Daelen, T van der Lee, P Diergaarde, J Groenendijk, S Töpsch, P Vos, F Salamini, P Schulze-Lefert. The barley *Mlo* gene: a novel control element of plant pathogen resistance. Cell 88:695–705, 1997.

68. RA Dietrich, TP Delaney, SJ Uknes, ER Ward, JA Ryals, JL Dangl. *Arabidopsis* mutants simulating disease resistance response. Cell 77:565–577, 1994.

69. JT Greenberg, A Guo, DF Klessig, FM Ausubel. Programmed cell death in plants: a pathogen-triggered response activated coordinately with multiple defense functions. Cell 77:551–563, 1994.

70. JT Greenberg. Programmed cell death in plant-pathogen interactions. Annu Rev Plant Physiol Plant Mol Biol 48:525–545, 1997.

71. RW Hartley. Barnase and barstar: two small proteins to fold and fit together. Trend Biochem Sci 14:450–454, 1988.

72. G Strittmatter, J Janssens, C Opsomer, J Botterman. Inhibition of fungal disease development in plants by engineering controlled cell death. Biotechnology 13: 1085–1089, 1995.

73. H Keller, N Pamboukdjian, M Ponchet, A Poupet, R Delon, JL Verrier, D Roby, P Ricci. Pathogen-induced elicitin production in transgenic tobacco generates a hypersensitive response and nonspecific disease resistance. Plant Cell 11: 223–235, 1999.

74. X Tang, M Xie, YJ Kim, J Zhou, DF Klessig GB Martin. Overexpression of *Pto* activates defense responses and confers broad resistance. Plant Cell 11: 15–29, 1999.

75. KE Hammond-Kosack, S Tang, K Harrison, JDG Jones. The tomato *Cf-9* disease resistance gene functions in tobacco and potato to confer responsiveness to the fungal avirulence gene product Avr9. Plant Cell 10:1251–1266, 1998.

76. TP Delaney, S Uknes, B Vernooij, L Friedrich, K Weymann, D Negrotto, T Gaffney, M Gut-Rella, H Kessmann, E Ward, JA Ryals. A central role of salicylic acid in plant disease resistance. Science 266:1247–1250, 1994.

77. L Friedrich, B Vernooij, T Gaffney, A Morse, J Ryals. Characterization of tobacco plants expressing a bacterial salicylate hydroxylase gene. Plant Mol Biol 29:959–968, 1995.

78. K Weymann, M Hunt, S Uknes, U Neuenschwander, K Lawton, HY Steiner, J Ryals. Suppression and restoration of lesion formation in *Arabidopsis lsd* mutants. Plant Cell 7:2013–2022, 1995.
79. Z Chen, H Silva, DF Klessig. Active oxygen species in the induction of plant systemic acquired resistance by salicylic acid. Science 262:1883–1886, 1993.
80. J Durner, DF Klessig. Inhibition of ascorbate peroxidase by salicylic acid and 2,6-dichloroisonicotinic acid, two inducers of plant defense responses. Proc Natl Acad Sci USA 92:11312–11316, 1995.
81. R Tenhaken, C Rübel. Salicylic acid is needed in hypersensitive cell death in soybean but does not act as a catalase inhibitor. Plant Physiol 115:291–298, 1997.
82. A Ludwig, R Tenhaken. Defence gene expression in soybean is linked to the status of the cell death program. Plant Mol Biol 44:209–218, 2000.
83. R Tenhaken, C Anstätt, A Ludwig, K Seehaus. WY-14,643 and other agonists of the peroxisome proliferator-activated receptor reveal a new mode of action for salicylic acid in soybean disease resistance. Planta 212:888–895, 2001.
84. K Schoonjans, B Staels, J Auwerx. The peroxisome proliferator activated receptors (PPARs) and their effects on lipid metabolism and adipocyte differentiation. Biochim Biophys Acta 1302:93–109, 1996.
85. SA Kliewer, TM Willson. The nuclear receptor PPARγ—bigger than fat. Curr Opin Genet Dev 8:576–581, 1998.
86. KA Lawton, L Friedrich, M Hunt, K Weymann, T Delaney, H Kessmann, T Staub, J Ryals. Benzothiadiazole induces disease resistance in *Arabidopsis* by activation of the systemic acquired resistance signal transduction pathway. Plant J 10:71–82, 1996.
87. DM Horvath, NH Chua. Identification of an immediate-early salicylic acid–inducible tobacco gene and characterization of induction by other compounds. Plant Mol Biol 31:1061–1072, 1996.
88. DM Horvath, DJ Huang, NH Chua. Four classes of salicylate-induced tobacco genes. Mol Plant Microbe Interact 11:895–905, 1998.

24

Transgenic Crop Plants with Increased Tolerance to Insect Pests

Danny J. Llewellyn and Thomas J. V. Higgins
CSIRO Plant Industry, Canberra, Australia

I. INTRODUCTION

Control of insect pests represents one of the major input costs of world agricultural production and consumes billions of dollars annually, predominantly in chemical pesticides and lost production. Insect control with pesticides is, however, becoming more problematic in both developed and third world countries. Farmers and consumers are increasingly aware of the environmental and ecological impacts of excessive pesticide use. As the target insects are repeatedly exposed to the same pesticide chemistries, they often develop heritable resistance to those pesticides and require the use of higher and higher doses or more toxic insecticide blends to achieve economic levels of pest control. Plant breeders, seed companies, and farmers are now turning to biotechnological solutions to these important pest problems in agriculture because of the environmental and economic advantages of deploying inbuilt insect control strategies as opposed to external applications. Over the past 10 years, transgenic crops protected against insects have been generated in some of the important broadacre crops, including cotton and corn, as well as in horticultural crops such as potato, and have been released commercially (1).

II. *BACILLUS THURINGIENSIS CRY* GENES AS A SOURCE OF INSECTICIDAL GENES FOR CROPS

The bacterium *Bacillus thuringiensis* has had a long history of use as a sprayable microbial biopesticide (2). It has been used predominantly in horticulture and to a smaller extent in broadacre crops, especially those grown organically. During sporulation, the bacterium produces a series of insecticidal proteins assembled into a parasporal crystal. A mixture of the dried spores and crystals is reconstituted with water and sprayed on plants. On ingestion by the target insects, the crystals are solubilized and activated by digestive proteases to form highly potent and specific toxins that bind to particular gut receptors. The toxins are thought to aggregate and form ion-permeable pores that lead to gut dysfunction, lysis of gut epithelial cells, and the eventual death of the insect (see review in Ref. 3). Individual crystals in the bacterium may be a complex mixture of toxins each with its own range of insect specificities. The individual proteins within the crystals are encoded by both plasmid and chromosomal *Cry* genes that were identified as early targets for incorporation into crop plants to protect them from insects. There are now hundreds of identified *Cry* genes, and there is a standard nomenclature to describe them based on sequence similarities (see http://epunix.biols.susx.ac.uk/Home/Neil_Crickmore/Bt/ and indicated links).

Initial attempts to express *Cry* genes in transgenic plants were not particularly successful, and this was attributed to the bacterial origins of these genes and especially their high AT content, which resulted in low levels of insecticidal protein. Resynthesis of the *Cry* genes removing cryptic plant polyadenylation signals, putative messenger RNA (mRNA) destabilization signals, improving the codon usage bias, and increasing the GC content were strategies that increased expression to a useful level (0.2–0.3% of total soluble protein) in transgenic plants (4). Modified and unmodified, full-length and truncated *Cry* genes have so far been expressed in an array of plant species (from trees such as poplar, larch, and eucalypt; cereals like wheat, maize, and rice; legumes like chickpeas, soybeans, and peanuts; vegetables such as potato, tomato, cabbage, broccoli, and sweet potato; and fruits like apples and strawberries) (see references in Refs. 5–9), but intellectual property ownership has restricted commercialization to a few high-value agricultural species, such as cotton and corn and to a lesser extent potato.

III. COMMERCIALIZED *BACILLUS THURINGIENSIS* (Bt) CROPS

A. Bt-Corn

The United States is the largest corn-producing country in the world. One of the significant pests of U.S. and Canadian corn is the European corn borer

(ECB) (*Ostrinia nubilalis*), a lepidopteran insect, whose larvae are difficult to control as they bore into the corn stalk, where they are protected from applied pesticides. Only a few percent of the total acreage of corn is sprayed for control of ECB, so economic losses from this insect can be high. Bt-corn varieties targeted at ECB control were first released in the United States in 1996 and were so rapidly adopted by farmers that over a quarter of all the corn sown by 1999 was Bt-corn. Five different types of Bt-corn are currently registered by the U.S. Environmental Protection Agency (EPA), each with either different *Cry* genes (*Cry1Ab*, *Cry1Ac*, or *Cry9C*) or different levels or patterns of expression. The Monsanto and Syngenta Yieldgard corns express truncated codon-modified *Cry1Ab* genes derived from *B. thuringiensis* subsp. *kurstaki* (10,11). The Syngenta corn line has been backcrossed into both field and sweet corn varieties and, as with the Monsanto line, the *Cry1Ab* gene is expressed throughout the plant, including the silks and kernels, but is particularly high in leaves. Bt-corns called KnockOut (Novartis) or NatureGard (Mycogen) also express *Cry1Ab* but use a novel two-gene strategy to target expression to ECB-susceptible tissues. One *Cry1Ab* is controlled by the corn phosphoenolpyruvate carboxylase (PEPC) gene (12) directing expression in green photosynthetically active tissues. A pollen-specific promoter from the maize calcium-dependent protein kinase (CDPK) gene (13) controls the second. The combination of PEPC and pollen promoters provides high *Cry1Ab* gene expression in leaves and pollen, where it is highly effective in controlling European corn borer. Changes in business structures have resulted in the phaseout of KnockOut and NatureGard corns, and their registration will not be renewed. All existing stocks of these events can now be used only until the 2003 corn-growing season.

A Dekalb Bt-corn called Bt-Xtra contains three genes, the *Cry1Ac* gene from *B. thuringiensis* subsp. *kurstaki*, the *bar* gene from *Streptomyces hygroscopicus*, and the potato proteinase inhibitor *pinII*. While proteinase inhibitor genes expressed at high levels can inhibit insect digestive proteases and confer insecticidal properties to transgenic plants, the *pinII* gene in Bt-Xtra corn was truncated during integration into the corn genome and no protein expression is detectable. Cry1Ac is very similar to Cry1Ab but has a slightly different insecticidal activity spectrum. Since the purchase of Dekalb by Monsanto, the Bt-Xtra corn has been phased out in favor of Yieldgard. Finally, StarLink is a Bt-corn expressing a completely different type of *Cry* gene, the *Cry9C* from *B. thuringiensis* subsp. *tolworthi*. The Cry9C protein is active against ECB and some other lepidopteran pests of corn, including black cutworm (*Agrotis ipsilon*), but not the corn earworm (*Helicoverpa zea*). StarLink corn was registered only for animal feed and nonfood industrial use as there were some questions raised at registration concerning its potential allergenicity in humans. Subsequently, *Cry9C* was detected in corn chips and other corn food products, in violation of the original animal feed use registration. StarLink was voluntarily withdrawn from registration in 2000, and there is now an extensive program in place to track and remove remnants of StarLink corn from the human food chain. At the time of its withdrawal, it had been sown on less than 1% of the corn area in the United States. Thus, only two of the five original transgenic Bt-corn lines will be available commercially in the next few years.

Bt-corn varieties have been shown to be very effectively protected against ECB, but levels of infestation vary from year to year and region to region. It has been estimated that in 10 of the last 13 years growers would have received an economic benefit had they been growing Bt-corn when ECB pest pressures were high (14). Since their release, modest reductions (1.5%) in pesticide usage have been realized (15), but given that insecticides are rarely effective for ECB control and hence seldom used, this is still significant. Indirect benefits are also likely as the reduced ECB damage to Bt-corn is also likely to reduce sites for entry of pathogens, particularly those that produce mycotoxins that are health hazards to humans and farm animals eating the corn. The *Cry1*-expressing Bt-corn varieties have clearly allowed farmers to better control a very difficult agricultural pest (ECB), for which effective and affordable pest control options were unavailable. Yield losses from this insect have reached as much as 300 million bushels of corn in a single year. Such losses could be virtually eliminated using Bt-corn.

New additions to the current suite of Bt-corn varieties are likely in the next few years; the first, a Cry1F-expressing corn (lepidopteran active), received conditional registration in 2001 (16). Of particular interest will be the first Bt-corn targeted at beetle pests, especially the corn rootworm (*Diabrotica* sp.). Corn rootworm is really a complex of three or four different species of *Diabrotica* that are serious pests in the United States, with estimated costs of over U.S. $1 billion in chemical control and lost production. The larvae of these beetles attack the roots, and so control is primarily with soil insecticides, crop rotations, and foliar insecticide applications to kill the adult beetles and prevent further egg laying. Monsanto have developed a new transgenic corn (MaxGard) expressing the beetle-active *Cry3Bb* gene that appears to be very effective in controlling the larval stages of corn rootworm (17). The gene construct is believed to be a codon-modified synthetic *Cry3Bb* with a leader region from the wheat chlorophyll *a/b* binding protein gene, an intron from the rice actin gene, and a terminator from the wheat *hsp17* gene (18). Monsanto applied for full registration of a number of events of its MaxGard corn in late 1999 (19). The petition for nonregulated status was withdrawn in 2001, but Monsanto continues to do large-scale trials all over the U.S. corn belt. The transgenic corn will have to compete with other nontransgenic corn rootworm-resistant varieties (20). These nontransgenic varieties may be more acceptable to consumers who are concerned about GM corn products.

B. Bt-Cotton

Cotton is a major world crop grown primarily for its fiber but also for its oil and seed meal. Cotton is a demanding crop requiring significant inputs of both water and agricultural chemicals (insecticides, herbicides, fungicides, defoliants, and fertilizers). High inputs and high returns have made cotton one of the leading crops for the application of biotechnology, and by 1999–2000 about 12% of the world cotton area was sown with genetically modified varieties. In 2000 over 70% of the U.S. crop was sown with insect- and/or herbicide-tolerant GM cotton varieties: about 60% were herbicide-tolerant varieties and 40% were Bt varieties (about three quarters of the latter were stacked with a herbicide tolerance trait). The Bt varieties (Bollgard) were first released in the United States in 1996 primarily for the control of *Heliothis virescens* (tobacco budworm, TBW) but also for other pests such as *Helicoverpa zea* (cotton bollworm, CBW, or corn earworm) and *Pectinophora gossypiella* (pink bollworm). Bollgard cotton varieties contain a single insertion of a codon-modified hybrid *Cry1Ab/Cry1Ac* gene from *B. thuringiensis* subsp. *kurstaki* (21). Three lines of Bt-cotton, 531, 757, and 1076, were initially deregulated, but only one, 531 (Bollgard), was released commercially in the United States. Initial sowings of Bollgard cotton in 1996 were about 1 million hectares (about 12% of the U.S. cotton area). This was a difficult year for a commercial launch because there were unusually high levels of CBW rather than the more usual cotton pest, TBW (CBW is less susceptible to *Cry1Ab/c* than TBW). Many thousands of hectares of Bt-cotton in the U.S. southeast became infested with CBW and had to be sprayed with pesticides. This poor performance was blamed on the increased corn area that year (which increased populations of CBW) and the hot dry weather that appeared to decrease expression of the *Cry1Ab/c* gene, particularly later in the season (22). Subsequent years proved less difficult for growers, and the performance of Bt varieties has improved as growers have adjusted their adoption and management of Bt-cotton to suit their particular pest problems and economics.

The benefits of Bt-cotton varieties in the United States have been difficult to quantify despite extensive surveys since their introduction. Attributing effects on yield, pesticide usage, and profits due to Bt is a statistical challenge as many other factors can influence yield and pesticide usage even on a single farm. However, a number of analyses (23) have indicated that there are at least three benefits to growers of Bt-cotton. First, there have been statistically significant reductions in yield losses due to the target pests in many areas (24). Pesticide usage has also declined in many areas, but again statistical analyses suggest that this decrease is small and mainly for the pesticides other than major organophosphates and synthetic pyrethrins (23). These increases in yields and decreases in pesticide usage translate to higher net returns to growers who adopt Bt-cotton technology.

Australia saw the first release of Bt-cotton in 1996, as INGARD varieties (Bollgard was already registered as a proprietary name for a different technology), on a relatively small area (30,000 ha or about 10% of total cotton area). In Australia, the main insect pests of cotton, as in the United States, are also the larvae of heliothine moths, in this case *Helicoverpa armigera* and *H. punctigera*. Infestations are normally extreme by comparison with that in the United States, and it is not unusual to spray 12–14 times per season to control these pests (compared with 2–3 in the United States). The total crop in 2000 occupied about 500,000 ha (about 180,000 of it INGARD) and insect control costs on conventional cotton often exceed AUS$100 million annually, for a crop worth about AUS$1.6 billion. Both *H. armigera* and *H. punctigera* are naturally less susceptible (by about 10-fold) to the Cry1Ab/c protein expressed in INGARD (or Bollgard) cotton and are much more abundant in cotton-producing areas than the similar pests in the United States. Early research trials indicated that although INGARD cotton controlled *H. armigera* well (Fig. 1), particularly during the early part of the season, efficacy declined late in the season when fruit was being set (25).

Annual surveys of the usage of pesticides and economic returns to growers using Bt-cotton have been carried out in Australia since 1996. Because INGARD cotton can be only a proportion of the cotton on any one farm, detailed paired comparisons between the performance of INGARD and corresponding conventional varieties were possible (26). Pest control costs represent a significant proportion of the annual variable costs of cotton production in Australia, so the impact of INGARD on pesticide use appears to be much more apparent than in the United States. In all 4 years there was a significant reduction in overall use of pesticides (across all pesticide groups) and particularly those used against the *Helicoverpa* caterpillars. In 1999–2000, for example, there was a 40% decrease in pesticides targeted at all pests (down from 10.3 sprays in conventional cotton to 6.2 in Bt-cotton) or 47% reduction in pesticides targeted specifically at *Helicoverpa* species (from 9.7 down to 5.1 sprays on Bt-cotton) (Fig. 2). There was a 75% reduction in endosulfan use (a pesticide under threat in Australia because of its high toxicity to fish and the very low tolerance set for endosulfan contamination in rivers) and a 43% reduction in pyrethrins. This should

FIGURE 1   INGARD cotton provides good early season control of Australia's two main caterpillar pests, *Helicoverpa armigera* and *H. punctigera*. Research plots at the Australian Cotton Research Institute of a conventional variety on the left and its INGARD equivalent on the right. The fields were not sprayed at all for insect pests, and the conventional variety failed to yield any harvestable cotton as most flower buds and bolls were devoured by caterpillars. (Photo courtesy of Cheryl Mares, CSIRO.)

reduce the pressure on synthetic pyrethroids that have selected a high incidence of resistance among cotton pests.

The main concern to growers has been the high variability of lepidopteran control in Bt-cotton crops. The number of effective days of *Helicoverpa* control (time from planting to first foliar spray of insecticide) varies from as little as 20 days up to 140 days with considerable differences in the performances of different transgenic varieties (26). Reduced control has been correlated with a decrease in both Cry1Ab/c protein and mRNA (27,28) and possibly increases in secondary metabolites that decrease the efficacy of Bt-toxins (29). Most of the benefit of Bt-cotton is realized in the first half of the season, when little insecticide now has to be used to control both major lepidopteran pest species. Careful crop monitoring and spraying when pests reach their economic thresholds (new thresholds were established specifically for INGARD varieties) has stabilized the performance of INGARD cotton. There is a strong tendency to sow INGARD cotton in environmentally sensitive areas (fields close to waterways and towns as Australia has very strong pollution control laws in relation to pesticide contamination of

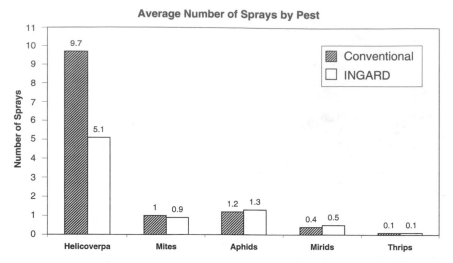

FIGURE 2 INGARD cotton under commercial production required significantly fewer pesticide sprays for caterpillar pests than similar conventional varieties in the 1999–2000 season. An area of 191,000 ha of cotton was surveyed, of which 28% was INGARD cotton. Pesticide applications to INGARD were 47% less than on conventional fields, and there were no major differences in pesticides applied for other pests. (Adapted from Ref. 26.)

rivers etc.) to capture the significant environmental benefits of reduced pesticide applications. The economic benefit to the grower has been variable, in line with the variability in successful insect control. Of the paired comparisons carried out, 54% recorded some economic benefit and the rest were about break-even.

The Bt technology clearly has benefits, but some limitations have emerged in Australia (and presumably in Asia, where *H. armigera* is a major pest of cotton). In response to the poorer efficacy of Bt-cotton in Australia (and to enhanced *H. zea* control and better resistance management in the United States), a double gene product is being developed that contains two different Bt-toxin genes, the *Cry1Ab/c* gene (in INGARD and Bollgard) and a *Cry2Ab* gene also from *Bacillus thuringiensis*. These two gene cottons (Bollgard II) are still in the breeding stage, but initial field studies have indicated much better *H. armigera* control.

Bt-cotton is now approved for food use in Argentina, Australia, Canada, China, Japan, Mexico, and South Africa, as well as the United States. Bt-cotton was grown on about 0.5 million hectares in mainland China in 2000—much of it Bollgard, but also some Bt-cotton developed in China, e.g., (30)—with outcomes very similar to that observed in Australia (31), where the lepidopteran pests are similar. A number of Asian, European, and Latin American countries have experimental crops of both Bt-cotton and Bt-corn for precommercial evaluation.

C. Bt-Potatoes

In 1996 about 1% of the total U.S. potato acreage was planted to Bt-potatoes, and this rose to about 3% in 2000. The target pest for Bt-potatoes was the Colorado potato beetle (CPB) (*Leptinotarsa decemlineata*), whose adults and larvae defoliate potatoes, reducing tuber yields to the extent that potato production has had to cease in some areas. Although nonchemical methods of control are available, none are as cost-effective or practical as chemical pesticides. CPB easily develops resistance to chemical insecticides, and many populations are now resistant to a broad array of chemical types. Microbial biopesticides containing Cry3A Bt-toxins are effective in killing the young larvae but, in practice, are not very effective against the most damaging life stages, the third and fourth instar larvae and adults. Expressing the *Cry3A* gene in the plant has considerable advantages in targeting the most vulnerable stage of the insect's life cycle, neonates, but also, if the expression levels are high enough, in controlling larger larvae and adults. Several different transgenic cultivars of potato (sold as NewLeaf varieties) express the *Cry3A* gene from *B. thuringiensis* subsp. *tenebrionis* (32). Because potato is a vegetatively propagated plant, with low or no fertility, each cultivar or variety must be separately transformed, whereas with Bt-corn or Bt-cotton a single event can be backcrossed into several different elite cultivars or hybrids.

NewLeaf potatoes expressing Cry3A proteins were the first to be released, and control of CPB was very effective together with over 40% reduction in insecticide applications to the NewLeaf variety (15). In most potato-growing areas, aphids are also a significant pest, not only for the damage they can do to the plant but also because they transmit harmful plant viruses such as potato leaf roll virus (PLRV) and potato virus Y (PVY). The introduction of NewLeafPlus and NewLeafY potato varieties that have a built-in transgene protection against PLRV and PVY, respectively, as well as the Bt gene, might have made an even bigger impact on the use of Bt-potatoes in areas where these diseases are prevalent. However, large buyers of potato products have refused to buy GM potatoes and processors have followed suit, reducing the market for NewLeaf varieties despite good evidence that they provide considerable economic benefit to the farmer. After the poor economic performance of its NewLeaf potatoes, Monsanto appears to be reluctantly withdrawing completely from the transgenic potato market to concentrate on its other more successful biotech crops (33).

D.   Resistance Management and Bt Crops

One of the key concerns with deploying single *Cry* genes into crop plants
is that the increased exposure will select for resistance in the target insects.
This could destroy the potential for long-term control using this technology
and affect other users of microbial pesticides. Unfortunately, field resistance
to microbial Bt formulations has already occurred in both the Indian meal
worm (*Plodia interpunctella*) in stored grains and the diamondback moth
(*Plutella xylostella*) in a number of geographically isolated insect populations attacking vegetable crops (34). Laboratory-selected resistance has also
been observed in a number of species that attack Bt-crops, including *H.
virescens* (35) and more recently *H. armigera* (36) and *P. gossypiella* (37).
These relatively isolated occurrences of resistant insects have, however, provided models for how to manage the development of resistance to Bt-crops
in an agricultural setting (38).

The provision of non-Bt plants or refugia nearby to generate sufficient
numbers of susceptible adults is the most commonly used resistance management strategy. This ensures that any rare homozygous resistant insects
selected on the Bt crop will find only fully susceptible mates, thereby con-
tinually diluting the resistance genes in the target insects. For Bt-cotton, the
refugia can be unsprayed or sprayed cotton or other crops that are hosts for
the pest species and generate adult insects at the same time as those that
might emerge on Bt-cotton. The area and location of these refugia relative
to the Bt crop are critical and in Australia, at least, have been determined
empirically for Bt-cotton from the numbers of pupae of *H. armigera* generated by crops treated in different ways. The requirements for refugia and
other resistance management procedures are legislated into the registration
label for INGARD cotton. The IPM approach adopted in Australia has been
summarized (39). At present, refugia requirements are easily achieved as the
total Bt area for any one farm can only be 30% of the total cotton planted.
This area limit is likely to be lifted only when cotton with two Bt genes
becomes available around 2003 or 2004. Similar refuges for Bt-cotton and
Bt-corn have been strongly recommended by U.S. regulators in response to
scientific concerns that the technology was at risk from overuse. The effectiveness of refuge strategies is dependent on many factors including the
biology and population dynamics of the pest species, the frequency and
mechanism of resistance, the placement and management of the crop and
neighboring crops, and the expression level of the transgenes.

E.   Nontarget Effects of Bt Crops

A second concern with Bt crops, or any other insecticidal transgenic plant,
is that when released on a large scale they may have unintended, or even

unnoticed, effects on other organisms in agricultural or neighboring ecosystems. Because Bt-corn is thought to exude Bt-toxins into the soil (40), where they can persist for relatively long periods bound to soil particles (41), and Bt-containing biomass is regularly returned to the soil during agricultural production, impacts on soil insects and microbes must also be considered. Two instances of possible nontarget effects have been described. First, Bt-corn has been reported to have unexpected effects on a beneficial insect, the green lacewings that feed on the European corn borer (ECB) (42). Green lacewings fed ECB that had eaten Bt-corn had a higher death rate and delayed development compared with the controls, and this was attributed to the poor nutritional quality of ECB larvae exposed to the Bt toxin. This study has not been extended to an agricultural setting.

In a second case of possible nontarget impacts of insecticidal crops, Bt-corn was believed to accelerate the demise of the endangered monarch butterfly (*Danaus plexippus*) in the United States (43). Monarch caterpillars feed on milkweed plants that often grow near cornfields. In a laboratory simulation, milkweed leaves were dusted with Bt-corn pollen and these leaves were fed to monarch larvae. After 4 days, only 56% of the larvae survived relative to larvae fed on plants dusted with non-Bt pollen or no pollen at all. Surviving larvae were also smaller than the controls. This laboratory study raised questions about the validity of the regulatory assessment of the environmental impacts of transgenic plants and prompted new field studies (summarized in Ref. 44) that did not confirm the laboratory results. The field study showed that Bt-corn would not have any adverse effects on nontarget lepidoptera such as monarch butterflies and demonstrated that the potential exposure of larvae to Bt-corn pollen was, in fact, very low. The risks to monarch butterflies from Bt crops must be considered small compared with the risks posed by habitat destruction and must be weighed against the effects of chemical pesticides on all environments adjacent to cropping areas. This incident highlights the need for field-based confirmation of any laboratory study that makes a claim for benefits or risks associated with any new biotech product.

IV. OTHER SOURCES OF INSECTICIDAL GENES FOR CROP PROTECTION

Bacillus thuringiensis and its close relatives have proved to be a valuable source of insecticidal genes. Nevertheless, after the screening of many thousands of isolates, new types of insecticidal Bt genes have been difficult to find. The commercial and environmental successes of Bt-crops have provided the impetus for a widespread search for other potent insecticidal genes from microbes, insects, plants, and even animals. These new insecticidal genes could be used instead of Cry-toxins or used to augment Cry-toxins to strengthen resistance management programs for the current generation of transgenic crops that rely on a single insecticidal activity. The main requirements for these new genes are that the insecticidal agents must be orally active and preferably encoded by relatively few genes as transformation technologies can still handle only a couple of genes at a time. Novel insecticidal genes have been found using mass screening or the rational approach of seeking specific inhibitors of an insect's digestive or neural physiology. Several potential new technologies have emerged (reviewed in Ref. 9), but none of these has yet had the commercial impact of the *Cry* genes.

A. New Oral Toxins

Different *Cry* genes have been isolated, covering a spectrum of insecticidal activities, but many important pests still cannot be controlled by Cry proteins. Most mass screening efforts to identify novel *Bacillus* strains have concentrated on the identification of new crystal proteins and hence have been carried out with sporulating cultures. *B. thuringiensis* isolates with high activity against corn rootworms (*Diabrotica* sp.), for example, have been found only recently, but a novel screening approach has been used to discover a rootworm-active insecticidal complex produced during the vegetative growth phase of *B. cereus* (45). One non–crystal-forming isolate, AB78, contained a proteinaceous insecticidal activity produced during log phase growth that was absent at sporulation. This was traced to the presence of two proteins designated Vip1A and Vip2A (vegetative insecticidal protein) that, together, appear to have activity against corn rootworms. The corresponding *Vip* genes were cloned and were found to be widespread in *Bacillus* species (46). Vip2 has been crystallized and has been shown to be an ADP-ribosylating toxin (47). During the vegetative phase screening, a strain of *B. thuringiensis*, AB88, was also identified. This strain had high activity against many lepidopteran pests but no activity against coleopterans, and from this strain a third type of vegetative insecticidal protein, Vip3A, has been isolated. *Vip3* has no homology to the *Cry* genes or any other protein in the sequence databases and thus represents a totally new class of insecticidal genes. Vip3 expressed in transgenic tobacco achieved partial protection against *Spodoptera litura* (48), but no extensive analyses of transgenic plants expressing this or other *Vip* genes have yet been published. The existence of the *Vip* genes will encourage a new round of screening of the existing *Bacillus* strain collections.

The venom of many insect predators such as spiders and scorpions contains potent neurotoxins, many of which are insect specific. These have been used to enhance the speed of kill of insect viruses (e.g., Ref. 49) but seem unlikely candidates for expression in plants because of the food safety concerns, real or imagined, that they would arouse in consumers (50).

B. Inhibitors of Digestion

The insect gut is the obvious target for orally acting insecticidal agents to be expressed in plants (e.g., Ref. 50). The membranes lining the gut appear to be the target for the Cry toxins and probably the Vips, through their pore-forming or lytic activities. The insect's digestive biochemistry is another target. Disruption should reduce nutrient intake, resulting in mortality or at least in slowing the growth of larvae to such an extent that they are subject to a greater degree of parasitism and predation in the field. Different components of the digestive system can be targeted depending on whether the food source is rich in protein or carbohydrate. Plants already appear to have adopted this as a defense strategy, as many accumulate high levels of proteinaceous inhibitors of digestive enzymes in their storage organs, such as seeds, fruits, tubers, and corms. A number of plants contain wound-inducible inhibitors of insect digestive enzymes that are switched on during herbivory.

1. Protease Inhibitors

To utilize the proteins present in their diets, insects rely predominantly on the serine proteases trypsin, chymotrypsin, and elastase. Trypsin is the dominant protease type in lepidopteran larvae. Plants accumulate high levels of trypsin inhibitors in their seed. These are thought to act as insect antifeedants as well as nitrogenous seed reserves. A number of genes for trypsin-chymotrypsin inhibitors have been cloned and expressed in plants (reviewed in Refs. 51 and 52). These transgenic plants demonstrated antifeedant activity against a number of lepidopteran larvae. Larval growth rates were reduced, and in some cases mortality was markedly increased. Levels of the inhibitors had to be over 1% of leaf soluble protein in order to afford any protection. Similar results have been reported in transgenic rice, wheat, tobacco, lucerne, and poplar plants with a variety of serine and cysteine protease inhibitors from both plants and insects (reviewed in Ref. 7).

For example, a gene for a trypsin-chymotrypsin inhibitor from the giant taro plant, *Alocasia mycrorrhiza* (53), was expressed in transgenic tobacco. This inhibitor retarded the growth of *Helicoverpa armigera*, but the larvae rapidly altered their digestive physiology to compensate for the presence of the inhibitor in their diet. Although very active against mammalian chymotrypsin, the giant taro inhibitor turned out to be active only against *H. armigera* trypsin; its chymotrypsin inhibitor domain proved to be inactive in the insect. Insects fed on transgenic plants containing the giant taro inhibitor (or artificial diets containing purified inhibitor) adapted by elevating the levels of chymotrypsin and elastase in the midgut to counter the reduction in trypsin activity (54). Similar responses have been noted for other protease inhibitors, although the mechanisms of adaptation can differ (55–57). This process of adaptation may be responsible for the failure of any transgenic plants containing protease inhibitors to reach the marketplace, although there are a couple of reports of field tolerance, e.g., to stem borers in rice expressing the cowpea trypsin inhibitor genes (58). It will be important to include multiple protease inhibitors with differing specificities in order to develop robust insect tolerance in plants. For example, the stigma-specific protease inhibitor from *Nicotiana alata* (ornamental tobacco) is produced as a single polyprotein that is subsequently processed into six different trypsin and chymotrypsin inhibitors (59,60) and the gene confers tolerance to *H. punctigera* and *H. armigera* when expressed in transgenic plants (61,62).

2. Alpha Amylase Inhibitors

Some insects have diets rich in starch and utilize the enzyme α-amylase to digest the starch to simple sugars. α-Amylase inhibitors (αAIs) are produced in the seeds of many plants and may confer some insect tolerance (63). Seeds of many domesticated crops have been selected for reduced levels of their inhibitors of digestion, as they are antinutritional components for human or animal consumption. This has made some grain crops more vulnerable to attack, particularly in storage. Field peas are widely grown across southern Australia for both human and animal feed, but the green pods are prone to attack by the bruchid beetle, pea weevil (*Bruchus pisorum*). Some legumes are highly tolerant to attack by beetles and produce high levels of αAI, among other antinutritional compounds in their seeds. A gene for αAI from the common bean *Phaseolus vulgaris* was linked to a strong seed-specific promoter and used to generate transgenic field peas expressing levels of αAI as high as those in bean seeds (64). The inhibitor was stably expressed through to at least the T5 generation and the pea seeds were resistant to damage by pea weevil (Fig. 3), both in the glasshouse and in the field (65,66), as well as to a number of pests of stored grain (cowpea weevil, *Callosobruchus maculatus* and azuki bean weevil, *C. chinensis*) (64). The seeds were shown to have no detectable antinutritional effects in animal feeding trials (67). The same construct in transgenic azuki beans confers resistance to three stored grain pests, *C. chinensis*, *C. maculatus*, and *C. analis*, but not to the South American bruchid *Zabrotes subfasciatus* (68). This technology therefore holds promise for protecting grain legumes against certain coleopteran pests.

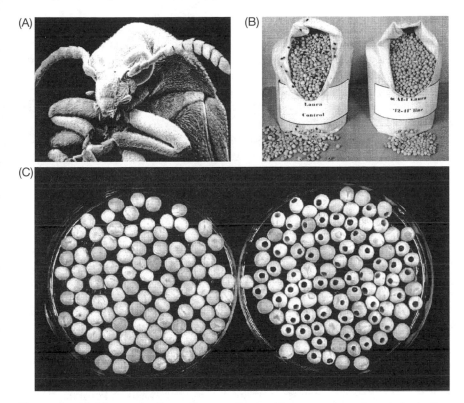

FIGURE 3 Field peas in Australia are attacked by a number of pests including the pea weevil, *Bruchus pisorum* (A), which also affects the stored grain. Larvae burrow into the seed and consume much of the internal tissues. They complete their development in the seed, emerging as adults and leaving a large exit hole. This reduces the nutritional value of the seed for animals and the economic value of stored grain. (B) Transgenic peas (cultivar "Laura") expressing the bean α-amylase gene (right) prevent larval development in the seed and very few adults emerge from seed grown in an infested field and stored for a few months compared with the control nontransformed variety (left) that is almost completely infested with larvae and emerged adults. (C) Closeup of the stored grain showing the extensive damage to the conventional variety and limited damage to the transformed "Laura" (left).

3. Lectins and Assorted Insecticidal Proteins

Few insecticidal proteins have been found that are active against sap-sucking insects such as plant hoppers, leafhoppers, and aphids, yet these insects represent a significant component of the pest problem in many crops. Frequently, they are vectors for serious viral diseases devastating world agriculture. Sugar binding proteins, or lectins, have been purified from many plants, and a number of lectins have been reported (69) to have toxic effects on sap-sucking insects that derive most of their energy reserves from the sugars being transported in the phloem. The mannose binding lectin from the bulb of snowdrop (*Galanthus nivalis*) is toxic to the peach-potato aphid (*Myzus persicae*), the glasshouse potato aphid (*Aulacorthum solani*), and the rice brown planthopper (*Nilaparvata lugens*). When the snowdrop gene is expressed in transgenic tobacco from either a constitutive or a phloem-specific promoter, the lectin has been shown to be ingested by the aphids, as it can be detected in their honeydew. The gene confers some tolerance to aphids in whole plant bioassays (70). Wang and Guo (71) have expressed both the gene for the Cry1Ab Bt toxin and the gene for snowdrop lectin in transgenic tobacco and have generated plants showing high insecticidal activity to both *H. armigera* and *M. persicae*, suggesting that it may be possible to stack genes active against both lepidopteran and sucking pests.

When three genes (*Cry1Ac*, *Cry2A*, and the gene for snowdrop lectin) were stacked in transgenic rice, protection against three important pests (a leaf folder, a stem borer, and a plant hopper) was achieved (72). Similarly, a pea lectin in transgenic tobacco was thought to have an additive effect with protease inhibitors in conferring tolerance to lepidopteran larvae, again highlighting the importance of multiple components of the defensive system (73). Snowdrop lectin in transgenic plants was also reported to have antinutritional effects on some lepidopterans (74), as was wheat germ agglutinin and jacalin (reviewed in Ref. 75). Many of these lectins are generally toxic, and experiments with transgenic potatoes containing the snowdrop lectin have been implicated in downstream effects on predators of the target sucking insect pests (76). Subsequent analyses, however, indicate that this is due not to the toxicity of the lectin to the insect predator but to the lowered nutritional value of the affected aphids (77). No effects were reported for the development of a wasp parasitoid fed on *Lacanobia oleracea* larvae raised on GNA-expressing potatoes, and there was a significant reduction in plant damage, at least in the glasshouse, using the combination of transgenic plant and parasitic wasp (78). However, as yet, no lectin-expressing transgenic plant has reached the marketplace. The biotin binding proteins avidin and streptavidin have also been shown to have strong insecticidal activity and have been expressed at high levels in the kernels of maize (79). Although originally produced for the commercial production of these proteins, the plants show good levels of resistance to a number of stored grain pests of corn (80). A thorough analysis of the human food safety issues associated with avidin and streptavidin corn will be necessary before the grain can enter the food chain.

Cholesterol oxidase (EC 1.1.3.6) from a *Streptomyces* species is a potent oral inhibitor of the cotton boll weevil (*Anthonomonas grandis grandis* Boheman) and has reduced activity against a number of lepidopteran species. It is thought to act through the oxidation of cholesterol in the membranes of the midgut brush border, and low concentrations cause mortality in larvae and reduced fertility in adults. The reason for the specificity of the protein is unclear as both susceptible and tolerant insect species have similar cholesterol levels. Gut pH may be involved, as the pH optimum of the enzyme does not favor its activity in the high pH found in the midguts of lepidopteran larvae (81). The gene for cholesterol oxidase has been cloned and expressed as enzymatically active protein in plant cells (82). The specificity of cholesterol oxidase makes it an attractive target for expression in transgenic cotton to enhance the current conventional approaches to boll weevil control. However, the success of the current boll weevil eradication program in the United States, which relies on conventional technology, may be a contributing factor in the lack of commercial incentive to progress the cholesterol oxidase gene as an insect control option in cotton. Other enzymes, chitinase and lipoxygenase, have also been reported to confer some insecticidal activity, but this has not been demonstrated in transgenic plants (referenced in Ref. 7).

C.   Secondary Metabolites

Plants produce an abundance of secondary chemicals that serve as a defense against insects. In some cases, they act as olfactory or oral cues to both beneficial and pest species. Some of the defensive chemicals are being evaluated with the aim of manipulating the genes for their biosynthesis in transgenic plants to enhance the host plant resistance or for the biological production of commercial pesticides (83,84). Thomas et al. (85) created transgenic tobacco plants expressing the *Catharanthus roseus* gene for tryptophan decarboxylase (TDC), which converts tryptophan to the insecticidal indole alkaloid tryptamine. Sweet potato whiteflies (*Bemisia tabaci*) fed on these plants showed a dramatic decrease in fertility. However, others have reported undesirable plant phenotypes as well as elevated tryptamine levels (86).

Alterations in metabolite profiles brought about by overexpressing key branch point enzymes, such as TDC, can have both desirable and undesirable pleiotrophic effects, as has been reported in transgenic potatoes and canola (87,88). In potato, overexpression of TDC resulted in an altered balance of key substrates in the shikimate and phenypropanoid pathways, leading to reduced levels of phenolics and other defense compounds. These plants were more susceptible to fungal pathogens. In canola, reductions in the available tryptophan pool resulted in reductions in the levels of tryptophan-derived indole glucosinolates as well as increased tryptamine. Complex regulation of secondary metabolic pathways may also have allowed Smigocki et al. (89,90) to inadvertently elevate secondary metabolites in transgenic tobacco plants expressing the cytokinin-producing *ipt* gene from *Agrobacterium tumefaciens* driven by a wound-inducible promoter, either through the activation of cytokinin-regulated biosynthetic genes or through changes in pools of core metabolites shared between secondary metabolism and cytokinin biosynthesis. Their plants showed a considerably enhanced tolerance to tobacco hornworm (*Manduca sexta*) and the green peach aphid (*Myzus persicae*).

Transgenic plants with altered disease tolerance were generated by Hain et al. (91) when they introduced into tobacco two genes from grapevine encoding the enzyme stilbene synthase that converts 4-coumaroyl CoA and malonyl CoA into the toxic phytoalexin resveratrol. Plants that produced resveratrol constitutively showed enhanced tolerance to the fungal pathogen *Botrytis cinerea*, but this was associated with altered flower color and male sterility (92), highlighting the importance of regulating the production of toxic defense chemicals. Using the native grapevine genes, however, with their own highly regulated pathogen and wound-inducible promoters has allowed the production of transgenic tomatoes without the deleterious side effects (93). These plants were protected from infection by the fungus *Phytophthora infestans*, but not by *B. cinerea* and *Alternaria solani*, even though these pathogens induced the accumulation of resveratrol.

Before we can routinely engineer pest and pathogen tolerance by manipulating secondary metabolism in plants, we must develop a better understanding of the complex interactions between different metabolic pathways and their regulation.

V. CONCLUSIONS

Insect control through biotechnology is an outstanding example of the use of gene technology to enhance the efficiency and sustainability of production of broadacre crops. It is not without its risks, such as the development of resistance by the target insects and potential nontarget impacts in the environment. There must be a concerted effort to develop new insect tolerance genes to ensure that the existing technologies are not lost by overuse in the short term. However, finding highly potent insecticidal genes that are as effective as the first generation of genes based on delta endotoxins from *Bacillus thuringiensis* is not a simple matter. The new class of vegetative insecticidal genes from *Bacillus* species holds the most promise for the control of difficult lepidopteran and coleopteran pests, but they have yet to be expressed effectively in transgenic plants. Inhibitors of digestion used against lepidopterans and/or sucking pests have received a great deal of publicity but are yet to prove themselves commercially and will be challenged by the great adaptability of insects and the problems of antinutritional activity against humans and animals if they are used in food or fodder crops. They are most likely to be at their best with highly specialized insects that are very host specific, such as some of the stored grain pests, and do not possess well-developed adaptive mechanisms to cope with a wide array of digestive inhibitors in their diets. The area of secondary metabolite manipulation in host plants also holds promise, but a fuller understanding of the biochemical pathways involved and the complex interrelationships between both secondary and primary metabolism in plants is essential. The next decade will be an exciting time for biotechnologists and entomologists as they explore the full-scale commercial production of transgenic insect-protected crops.

REFERENCES

1. M Peferoen. Progress and prospects for field use of Bt genes in crops. Trends Biotechnol 15:173–177, 1997.
2. E Schnepf, N Crickmore, J Van Rie, D Lereclus, J Baum, J Feitelson, DR Zeigler, DH Dean. *Bacillus thuringiensis* and its pesticidal crystal proteins. Microbiol Mol Biol Rev 62:775–806, 1998.
3. SS Gill, EA Cowles, PV Pietrantonio. The mode of action of *Bacillus thuringiensis* endotoxins. Annu Rev Entomol 37:615–636, 1992.
4. FJ Perlak, RL Fuchs, DA Dean, SL McPherson, DA Fishhoff. Modification of the coding sequence enhances plant expression of insect control protein genes. Proc Natl Acad Sci USA 88:3324–3328, 1991.
5. JJ Estruch, NB Carozzi, N Desai, NB Duck, GW Warren, MG Koziel. Transgenic plants: an emerging approach to pest control. Nat Biotechnol 15:137–141, 1997.
6. L Jouanin, M Bonadé-Bottino, C Girard, G Morrot, M Giband. Transgenic plants for insect resistance. Plant Sci 131:1–11, 1998.
7. VA Hilder, D Boulter. Genetic engineering of crop plants for insect resistance—a critical review. Crop Prot 18:177–191, 1999.
8. RA de Maagd, D Bosch, W Stiekema. *Bacillus thuringiensis* toxin-mediated insect resistance in plants. Trends Plant Sci 4:9–13, 1999.
9. HC Sharma, KK Sharma, N Seetharama, R Ortiz. Prospects for using transgenic resistance to insects in crop improvement. Electron J Biotech [on-line], Vol 3 No 2, issue of August 15, 2000. Available at http://www.ejb.org/content/vol3/issue2/full/3/
10. JT Odell, F Nagy, N-H Chua. Identification of DNA sequences required for activity of the cauliflower mosaic virus 35S promoter. Nature 313:810–812, 1985.

11. IK Vasil. Phosphinothricin resistant crops. In: SO Duke, ed. Herbicide Resistant
     Crops. New York: Lewis Publishers, 1996, pp 85–91.
12. RL Hudspeth, JW Grula. Structure and expression of the maize gene encoding
     the phosphoenolpyruvate carboxylase enzyme involved in C4 photosynthesis.
     Plant Mol Biol 12:579–589, 1989.
13. JJ Estruch, S Kadwell, E Merlin, L Crossland. Cloning and characterization of
     a maize pollen-specific calcium-dependent calmodulin-independent protein ki-
     nase. Proc Natl Acad Sci USA 91:8837–8841, 1994.
14. JE Carpenter, LP Gianessi. Agricultural biotechnology: updated benefit esti-
     mates. National Center for Food and Agricultural Policy (NCFAP), 2001, pp
     1–46. Available at http://www.ncfap.org/pup/biotech/updatedbenefits.pdf
15. JE Carpenter. Case studies in benefits and risks of agricultural biotechnology:
     Roundup Ready soybeans and Bt field corn. National Center for Food and
     Agricultural Policy (NCFAP), 2001, pp 1–54. Available at http://www.ncfap.
     org/pup/biotech/benefitsandrisks.pdf
16. W Vogt. New biotech corn gets conditional clearance. Farm Prog, May 31,
     2001.
17. T Nowatzki, J Oleson, J Tollefson, M Rice. Corn rootworm management up-
     date. 2000. Available at http://www.ent.iastate.edu/pest/rootworm/
18. DA Fischoff, RL Fuchs, PB Lavrik, SA McPherson, FJ Perlak. Genetically
     transformed plants with toxicity to Coleopteran pests—obtained using chimeric
     gene containing sequence encoding toxin protein of *Bacillus thuringiensis*. US
     patent 5,763,241, 1998.
19. D Ferber. New corn plant draws fire from GM food opponents. Science 287:
     1370, 2000.
20. B Hibbard, LL Darrah, BD Barry. Combining ability of resistance leads and
     identification of a new resistance source for western corn rootworm (Coleop-
     tera: Chrysomelidae) larvae in corn. Maydica 44:133–139, 1999.
21. FJ Perlak, RW Deaton, TA Armstrong, RL Fuchs, SR Sims, JT Greenplate, DA
     Fischoff. Insect resistant cotton plants. Biotechnology 8:839–943, 1990.
22. JT Greenplate. Quantification of *Bacillus thuringiensis* insect control protein
     Cry1Ac over time in Bollgard cotton fruit and terminals. J Econ Entomol 92:
     1377–1383, 1999.
23. J Fernandez-Cornejo, WD McBride. Genetically engineered crops for pest
     management in U.S. agriculture: farm level effects. Economic Research Ser-
     vice, U.S. Department of Agriculture, Agricultural Economics Report 786,
     2000. Available at http://www.ers.usda.gov/publications/aer786/aer786.pdf
24. R Heimlich, J Fernandez-Cornejo, WD McBride, C Klotz-Ingram, S Jans, N
     Brooks. Genetically engineered crops: has adoption reduced pesticide use? Ec-
     onomic Research Service, U.S. Department of Agriculture. Agricultural Out-
     look/August 2000, 2000, pp 13–17. Available at http://www.ers.usda.gov/
     publications/agoutlook/aug2000/ao273f.pdf
25. GP Fitt, CL Mares, DJ Llewellyn. Field evaluation and potential ecological
     impact of transgenic cottons (*Gossypium hirstum*) in Australia. Biocontrol Sci
     Technol 4:535–548, 1994.

26. B Pyke, ed. The performance of INGARD cotton in Australia during the 1999/ 2000 season. Cotton Research and Development Corporation (CRDC). CRDC Ocacasional Papers: Transgenics. Narrabri, NSW, Australia: CRDC, 2000.

27. H Holt. Season-long monitoring of transgenic cotton plants—development of an assay for the quantification of *Bacillus thuringiensis* insecticidal protein. Proceedings of the 9th Australian Cotton Conference, Broadbeach, Queensland, Australia, 1998, pp 331–335.

28. EJ Finnegan, DJ Llewellyn, GP Fitt. Expression of a Bt transgene in field grown cotton in Australia. In: PJ Larkin, ed. Agricultural Biotechnology: Laboratory, Field and Market. Proceedings 4th Asia-Pacific Conference on Agricultural Biotechnology (Darwin, July 13–16, 1998), 1998, pp 225–227.

29. KM Olsen, JC Daly. Plant-toxin interactions in transgenic Bt-cotton and their effect on mortality of *Helicoverpa armigera* (Lepidoptera: Noctuidae). J Econ Entomol 93:1293–1299, 2000.

30. CM Tang, J Sun, XF Zhu, WZ Guo, TZ Zhang, JL Shen, CF Gao, WJ Zhou, ZX Chen, SD Guo. Inheritance of resistance to *Helicoverpa armigera* of three kinds of transgenic Bt strains available in upland cotton in China. Chin Sci Bull 45:363–367, 2000.

31. TH Zhang, CM Tang. Commercial production of transgenic Bt insect-resistant cotton varieties and the resistance management for bollworm (*Helicoverpa armigera* Hubner). Chin Sci Bull 45:1249–1257, 2000.

32. FJ Perlak, TB Stone, YN Muskopf, LJ Petersen, GB Parker, SA McPherson, J Wyman, S Love, G Reed, D Biever, DA Fischoff. Genetically improved potatoes: protection from damage by Colorado potato beetle. Plant Mol Biol 22: 313–321, 1993.

33. P Reschke. Monsanto pulls plug on NatureMark spuds. Ontario Farmer, March 6, 2001.

34. BE Tabashnik. Evolution of resistance to *Bacillus thuringiensis*. Annu Rev Entomol 39:47–79, 1994.

35. F Gould, A Martinez-Ramirez, A Anderson, J Ferre, FJ Silva, WJ Moar. Broadspectrum resistance to *Bacillus thuringiensis* toxins in *Heliothis virescens*. Proc Natl Acad Sci USA 89:7986–7988, 1992.

36. R Akhurst, B James, L Bird. Resistance to INGARD® cotton by the cotton bollworm, *Helicoverpa armigera*. Proceedings of the 10th Australian Cotton Conference, Broadbeach, Queensland, Australia, 2000.

37. YB Liu, BE Tabashnik, SK Meyer, Y Carriere, AC Bartlett. Genetics of pink bollworm resistance to *Bacillus thuringiensis* toxin Cry1Ac. J Econ Entomol 94:248–252, 2001.

38. R Brousseau, L Massson, D Hegedus. Insecticidal transgenic plants: arc thcy irresistible? AgBiotechNet 1:1–10, 1999. Available at http://www.agbios.com/ articles/abn022brousseau.htm

39. GP Fitt. An Australian approach to IPM in cotton: integrating new technologies to minimise insecticide dependence. Crop Prot 19:793–800, 2000.

40. D Saxena, S Flores, G Stotzky. Insecticidal toxin in root exudates from Bt-corn. Nature 402:480, 1999.

41. H Tapp, G Stotzky. Persistence of the insecticidal toxin from *Bacillus thuringiensis* subsp. *kurstaki* in soil. Soil Biol Biochem 30:471–476, 1998.
42. A Hilbeck, M Baumgartner, PM Fried, F Bigler. Effects of transgenic *Bacillus thuringiensis* corn-fed prey on mortality and development time of immature *Chrysoperla carnea* (Neuroptera: Chrysopidae). Environ Entomol 27:480–487, 1998.
43. JJ Losey, L Raynor, ME Cater. Transgenic pollen harms monarch larvae. Nature 399:214, 1999.
44. FJ Perlak, RL Fuchs, DA Dean, SL McPherson, DA Fischoff, D Powell. Update: potential impacts of pollen from Bt-corn, 1999. http://www.plant. uoguelph.ca/safefood/gmo/bt-corn-update-nov99.htm
45. GW Warren. Vegetative insecticidal proteins: novel proteins for control of corn pests. In: NB Carozzi, MG Koziel, eds. Advances in Insect Control: The Role of Transgenic Plants. New York: Taylor & Francis. 1997, pp 109–121.
46. JJ Estruch, GW Warren, MA Mullins, GJ Nye, JA Craig, MG Koziel. Vip3A, a novel *Bacillus thuringiensis* vegetative insecticidal protein with a wide spectrum of activities against lepidopteran pests. Proc Natl Acad Sci USA 93:5389–5394, 1996.
47. S Han, JA Craig, CD Putnam, NB Carozzi, JA Tainer. Evolution and mechanism from structures of an ADP-ribosylating toxin and NAD complex. Nat Struct Biol 6:932–936, 1999.
48. RK Bhatnagar. ICGEB Activity report 1999. Mammalian biology: insect resistance. International Centre for Genetic Engineering and Biotechnology, New Delhi, 1999. Available at http://www.icgeb.trieste.it/RESEARCH/ND/ Insectresistance99.htm
49. LMD Stewart, M Hirst, ML Ferber, AT Merryweather, PJ Cayley, RD Possee. Construction of an improved baculovirus insecticide containing an insect-specific toxin gene. Nature 352:85–88, 1991.
50. M Robins, M Le Page. Mean and green—does a deadly spider hold the key to eco-pesticides? New Sci 166:5, 2000.
51. AMR Gatehouse, JR Gatehouse. Identifying proteins with insecticidal activity: use of encoding genes to produce insect-resistant transgenic crops. Pestic Sci 52:165–175, 1998.
52. CA Ryan. Protease inhibitors in plants: genes for improving defences against insects and pathogens. Annu Rev Phytopathol 28:425–449, 1990.
53. A Mathews, DJ Llewellyn, Y Wu, ES Dennis. Isolation and characterisation of full-length cDNA clones of the giant taro (*Alocasia macrorrhiza*) trypsin/chymotrypsin inhibitor. Plant Mol Biol 30:1035–1039, 1996.
54. Y Wu, DJ Llewellyn, A Mathews, ES Dennis. Adaptation of *Helicoverpa armigera* (Lepidoptera: Noctuidae) to a proteinase inhibitor expressed in transgenic tobacco. Mol Breed 3:371–380, 1997.
55. RM Broadway. Dietary proteinase inhibitors alter complement of midgut proteases. Arch Insect Biochem Physiol 32:39–53, 1996.
56. M Bonnade-Bottino, J Lerin, B Zaccomer, L Jouanin. Physiological adaptation explains the insensitivity of *Baris coerulescens* to transgenic oilseed rape expressing oryzacystatin I. Insect Biochem Mol Biol 29:131–138, 1999.

57. P Lara, F Ortego, E Gonzales-Hidalgo, P Castanera, P Carbonero, I Diaz. Adaptation of *Spodoptera exigua* to barley trypsin inhibitor BTI/Cme expressed in transgenic tobacco. Transgen Res 9:169–178, 2000.

58. D Xu, Q Xue, D McElroy, Y Mawal, VA Hilder, R Wu. Constitutive expression of a cowpea trypsin inhibitor gene, CpTi, in transgenic rice confers resistance to two major rice insect pests. Mol Breed 2:167–173, 1996.

59. AH Atkinson, RL Heath, RJ Simpson, AE Clarke, MA Anderson. Proteinase inhibitors in *Nicotiana alata* are derived from a precursor protein which is processed into five homologous inhibitors. Plant Cell 5:203–213, 1993.

60. HJ Schirra, MJ Scanlon, MCS Lee, MA Anderson, DJ Craik. The solution structure of C1-T1, a two domain proteinase inhibitor from a circular precursor from *Nicotiana alata*. J Mol Biol 306:69–79, 2001.

61. RL Heath, G McDonald, JT Christeller, M Lee, K Bateman, J West, R Van-Heeswjick, MA Anderson. Proteinase inhibitors from *Nicotiana alata* enhance plant resistance to insect pests. J Insect Physiol 43:833–842, 1997.

62. JA Charity, MA Anderson, DJ Bittisnich, M Whitecross, TJV Higgins. Transgenic tobacco expressing a proteinase inhibitor from *Nicotiana alata* have increased insect resistance. Mol Breed 5:357–365, 1999.

63. M Chrispeels, M de Sa Grossi, TJV Higgins. Genetic engineering with α-amylase inhibitors makes seeds resistant to bruchids. Seed Sci Res 8:257–263,1998.

64. RE Shade, HE Schroeder, JJ Pueyo, LM Tabe, LL Murdock, TJV Higgins, MJ Chrispeels. Transgenic pea seeds expressing the alpha amylase inhibitor of the common bean are resistant to bruchid beetles. Biotechnology 12:793–796, 1994.

65. HE Schroeder, S Gollasch, A Moore, LM Tabe, S Craig, DC Hardie, MJ Chrispeels, D Spencer, TJV Higgins. Bean alpha amylase inhibitor confers resistance to the pea weevil (*Bruchis pisorum*) in transgenic pea (*Pisum sativum* L.). Plant Physiol 107:1233–1239, 1995.

66. RL Morton, HE Schroeder, KS Bateman, MJ Chrispeels, E Armstrong, LM Tabe, TJV Higgins. Bean α-amylase inhibitor 1 in transgenic peas (*Pisum sativum*) provides complete protection from pea weevil (*Bruchus pisorum*) under field conditions. Proc Natl Acad Sci USA 97:3820–3825, 2000.

67. A Pusztai, G Grant, S Bardocz, R Alonso, MJ Chrispeels, HE Schroeder, LM Tabe, TJV Higgins. Expression of the insecticidal bean α-amylase inhibitor transgene has minimal detrimental effect on the nutritional value of peas fed to rats at 30% of the diet. J Nutr 129:1597–1603, 1999.

68. M Ishimoto, T Sato MJ, Chrispeels, K Kitamura. Bruchid resistance of transgenic azuki bean expressing seed alpha amylase inhibitor of common bean. Entomol Exp Appl 79:309–315, 1996.

69. Y Rahbe, N Sauvion, G Febvay, WJ Peumans, AMR Gatehouse. Toxicity of lectins and processing of ingested proteins in the pea aphid *Acyrthosiphon pisum*. Entom Exp Appl 76:143–155, 1995.

70. VA Hilder, KS Powell, AMR Gatehouse, JA Gatehouse, Y Shi, WDO Hamilton, A Merryweather, CA Newell, JC Timans, WJ Peumans, E Damme, D Boulter,

E Van Damme. Expression of snowdrop lectin in transgenic tobacco plants results in added protection against aphids. Transgen Res 4:18–25, 1995.

71. ZB Wang, SD Guo. Expression of two insect-resistant genes CryIA (b&c)/ GNA in transgenic tobacco plants results in added protection against both cotton bollworm and aphids. Chin Sci Bull 44:2051–2058, 1999.

72. SB Maqbool, S Riazuddin, NT Loc, AMR Gatehouse, JA Gatehouse, P Christou. Expression of multiple insecticidal genes confers broad resistance against a range of different rice pests. Mol Breed 7:85–93, 2001.

73. D Boulter, GA Edwards, AMR Gatehouse, JA Gatehouse, VA Hilder. Additive protective effects of different plant derived insect resistance genes in transgenic tobacco plants. Crop Prot 9:351–354, 1990.

74. AMR Gatehouse, GM Davison, CA Newell, A Merryweather, WDO Hamilton, EPJ Burgess, RJC Gilbert, JR Gatehouse. Transgenic potato plants with enhanced resistance to the tomato moth, *Lacanobia oleracea*: growth room trials. Transgen Res 3:49–63, 1997.

75. TH Czapla. Plant lectins as insect control proteins in transgenic plants. In: NB Carozzi, MG Koziel, eds. Advances in Insect Control: The Role of Transgenic Plants. New York: Taylor & Francis. 1997, pp 123–138.

76. ANE Birch, IE Geoghegan, MEN Majerus, JW McNicol, CA Hackett, AMR Gatehouse, JA Gatehouse. Tri-trophic interactions involving pest aphids, predatory 2-spot ladybirds and transgenic potatoes expressing snowdrop lectin for aphid resistance. Mol Breed 5:75–83, 1999.

77. RE Down, L Ford, SD Woodhouse, RJM Raemaekers, B Leitch, JA Gatehouse, AMR Gatehouse. Snowdrop lectin (GNA) has no acute toxic effects on a beneficial insect predator, the 2-spot ladybird (*Adalia bipunctata* L.). J Insect Physiol 46:379–391, 2000.

78. HA Bell, EC Fitches, GC Marris, J Bell, JP Edwards, JA Gatehouse, AMR Gatehouse. Transgenic GNA expressing potato plants augment the beneficial biocontrol of *Lacanobia oleracea* (Lepidoptera; Noctuidae) by the parasitoid *Eulophus pennicornis* (Hymenoptera; Eulohidae). Transgen Res 10:35–42, 2001.

79. EE Hood, DR Witcher, S Maddock, T Meyer, C Baszczynski, M Bailey, P Flynn, J Register, L Marshall, D Bond, E Kulisek, A Kusnadi, R Evangelista, Z Nikolov, C Wooge, RJ Mehigh, R Hernan, WK Kappel, D Ritland, CP Li, JA Howard. Commercial production of avidin from transgenic maize: characterization of transformant, production, processing, extraction and purification. Mol Breed 3:291–306,1997.

80. KJ Kramer, TD Morgan, JE Throne, M Bailey, JA Howard. Transgenic maize expressing avidin is resistant to storage pests. Nat Biotechnol 18:670–674, 2000.

81. Z Shen, DR Corbin, JT Greenplate, RJ Grebenok, DW Galbraith, JP Purcell. Studies on the mode of action of cholesterol oxidase on insect midgut membranes. Arch Insect Biochem Physiol 34:429–442, 1997.

82. DR Corbin, JT Greenplate, EY Wong, JP Purcell. Cloning of an insecticidal cholesterol oxidase and its expression in bacteria and plant protoplasts. Appl Environ Microbiol 60:4239–4244, 1994.

83. S Chilton. Genetic engineering of plant secondary metabolism for insect protection. In: NB Carozzi, MG Koziel, eds. Advances in Insect Control: The Role of Transgenic Plants. New York: Taylor & Francis, 1997, pp 237–269.

84. J George, HP Bais, GA Ravishankar. Biotechnological production of plant-based insecticides. Crit Rev Biotechnol 20:49–77, 2000.

85. JC Thomas, DG Adams, CL Nessler, JK Brown, HJ Bohnert. Tryptophan decarboxylase, tryptamine, and reproduction of the whitefly. Plant Physiol 109: 717–720, 1995.

86. G Guillet, J Poupart, J Basurco, V De Luca. Expression of tyryptophan decarboxylase and tyrosine decarboxylase genes in tobacco results in altered biochemical and physiological phenotypes. Plant Physiol 122:933–944, 2000.

87. KN Yao, VD Luca, N Brisson. Creation of a metabolic sink for tryptophan alters the phenylpropanoid pathway and the susceptibility of potato to *Phytophthora infestans*. Plant Cell 7:1787–1799, 1995.

88. S Chavadej, N Brisson, JN McNeil, VD Luca. Redirection of tryptophan leads to production of low indole glucosinolate canola. Proc Natl Acad Sci USA 91: 2166–2170, 1994.

89. A Smigocki, JW Neal, I McCanna, L Douglass. Cytokinin-mediated insect resistance in *Nicotiana* plants transformed with the *ipt* gene. Plant Mol Biol 23:325–335, 1993.

90. A Smigocki, S Heu, G Buta. Analysis of insecticidal activity in transgenic plants carrying the *ipt* plant growth hormone gene. Acta Physiol Plant 22:295–299, 2000.

91. R Hain, HJ Reif, E Ktause, R Langebartels, H Kindl, B Vornam, W Wiese, E Schmelzer, PH Schreier, RH Stocker, K Stenzel. Disease resistance results from foreign phytoalexin expression in a novel plant. Nature 361:153–156, 1993.

92. R Fischer, I Budde, R Hain. Stilbene synthase gene expression causes changes in flower colour and male sterility in tobacco. Plant J 11:489–498, 1997.

93. JE Thomzik, K Stenzel, R Stocker, PH Schreier, R Hain, DJ Stahl. Synthesis of a grapevine phytoalexin in transgenic tomatoes (*Lycopersicon esculentum* Mill.) conditions resistance against *Phytophthora infestans*. Physiol Mol Plant Pathol 51:265–278, 1997.

25

Transgenic Herbicide-Resistant Crops—Advantages, Drawbacks, and Failsafes

Jonathan Gressel

Department of Plant Sciences, The Weizmann Institute of Science, Rehovot, Israel

I. INTRODUCTION

There has been and is an intimate, bidirectional relationship between molecular biology—a most basic biological science—and weed control, the most base as well as basic agricultural practice. Weed control, as practiced for millennia, is base because it has employed the majority of mankind on this planet since crops were domesticated yet has been the lowest paying mass employment known. Weed control can be one of the most physically taxing tasks, yet typically more women are employed in weeding than men. Without weed control there is virtually no crop to harvest. The advent of mechanization (tractor plowing and cultivating) replaced much of the hand labor in the developed world as well as the developing parts of the third world. Mechanical weed control is fraught with high energy costs, facilitates

International Standards Organization (ISO) accepted common names of herbicides are used throughout. The chemical names can be found in Tomlin (1994). Occasionally, molecular biologists have used nonaccepted common names for herbicides. Their synonyms will be given in parentheses at first use in this chapter.

soil erosion and compaction, and has been largely replaced by chemical weed control using herbicides that can selectively eradicate weeds from crops. As countries industrialize and develop economically, cheap farm labor becomes unavailable, increasing the necessity for cost-effective chemical weed control. Selectivity between a crop and its associated weeds is no mundane trait to find in a chemical; it is based on the immense biodiversity of metabolic pathways abounding in the plant kingdom. Too often there is no selective chemical that can control a particular weed in a particular crop, and the weed infestations understandably worsen based on simple ecological, genetic, and biochemical principles.

No single agricultural practice used extensively in the past to control weeds has remained as useful as initially presumed. After one weed problem is solved, there are other weed species that will fill the ecological vacuum left behind. Thus, either weeds have evolved resistance to most commonly used herbicides or weed species that had never been controlled by the particular herbicide replace the weed species controlled. This is most apparent in monoculture, where weeds closely related to crops are becoming the greatest problems: grass weeds in grain crops, legumes in soybeans, brassicas in oilseed rape, etc. As most selectivities between crop and weed are due to catabolic degradation of the herbicide by the crop, closely related weeds are to be expected to have catabolic pathways similar to those of the crop. This is one major reason that transgenic herbicide-resistant crops (T-HRCs) have become so useful and that biotechnology has been utilized to produce such crops as well as to find new herbicide targets (see the excellent review by Cole and Rodgers 2000). Selectivity is enhanced by inserting exogenous resistance genes into the crops or by selecting natural mutations. This also leads to one major concern about T-HRCs, that the transgene will genetically introgress into related weeds.

These and other issues are reviewed in this chapter. A call for new concepts of using molecular biology for weed control that do not necessarily utilize HRC, with some possible examples for consideration, is made in the article from which this chapter is derived (Gressel, 2000). Indeed, it is highly unfortunate that a World Bank report (Kendall et al., 1997) dealing with how molecular biology can alleviate constraints to world food production has delineated only insects, pathogens, and lack of water as the major constraints where molecular biology can assist. That report ignored weeds, even though the greatest variable input into agriculture is the cost of weed control, whether (fe)manual, mechanical, or chemical, with or without transgenics. And there are many unsolved weed problems lowering yields over vast areas of the world. If the weeds could be controlled, crop yields would increase without additional inputs.

II. CONTRIBUTIONS OF WEED SCIENCE TO BIOCHEMISTRY AND MOLECULAR BIOLOGY

It was already stated that there has been a bidirectional relationship often not realized between weed control and molecular biology and biochemistry. Most believe that the relationship has been unidirectional from biochemistry and molecular biology to the farmer. This is hardly so. The dire need for developing cost-effective chemical weed control systems has led to a vast industrial investment to find and develop selective herbicides and later T-HRCs. Virtually all herbicides marketed are the result of random screening of chemicals to obtain active leads, as the result of huge efforts of procurement and chemical synthesis. Once a lead has been obtained, further syntheses around it are used to find compounds with greater activity and then selectivity. After such compounds have been found and marketed, they opportunistically become widely used tools of the physiologists and biochemists, first to find the site of action and then as "antimetabolites" to further understand and modulate metabolic pathways. Thus the advent of 2,4-dichlorophenoxy acetic acid (2,4-D) assisted in understanding auxin action, atrazine and diuron (DCMU) in understanding photosystem II, paraquat for photosystem I, dinitroanilines in dealing with tubulin assembly into microtubules, dichlobenil for cellulose biosynthesis, etc. Herbicides are the antimetabolites of choice in dealing with key enzymes such as glutamine synthase (glufosinate, phosphinothricin), acetolactate synthase (ALS) (many herbicides), acetyl-coenzymeA (CoA) carboxylase (ACCase) (many herbicides), dihydropteroate synthase (asulam), enolpyruvate-shikimate phosphate (EPSP) synthase (glyphosate), phytoene desaturase (many herbicides). We know far more about the enzymes blocked by herbicides than most other enzymes in plants. The genes for most of these enzymes have been isolated and used in transgenic programs. Such research transcended plant biochemistry and agriculture. For example, it was discovered through comparative genomics that plant and trypanosome β-tubulins were similar to each other and different from mammalian β-tubulin. The dinitroaniline herbicides then proved to be excellent trypanocides (Bell, 1998)

The repetitive (mis)use of single herbicides in monoculture over many years predictably led to the evolution of herbicide-resistant weeds (Gressel and Segel, 1978; Powles and Holtum, 1994). The advent of triazine resistance was crucial to the understanding of the role of the psbA gene product in the photosystem II binding site, leading to innumerable studies of photosynthesis, biophysics, and biochemistry correlated with molecular structure of the gene product. Again, this transcended plant molecular biology and weeds when, based on this agricultural phenomenon, similar mutations were generated in photosynthetic bacteria. The mutant and natural psbA gene products were crystallized and analyzed, leading to new insights into "drug" (ligand) binding and design, as well as a Nobel Prize in Medicine (Diesenhofer and Michel, 1989).

Information from herbicide resistance provided the theoretical underpinning for designing transient drought-resistant plants. Harvey and Harper (1982) first promoted the idea that paraquat resistance can be similar to oxidative stress tolerance. This was later extrapolated to being similar to transient drought tolerance (Malan et al., 1990). This has allowed the development of quick pretests with paraquat to ascertain the level of transient drought tolerance of transgenic plants bearing genes designed to confer oxidative stress resistance.

Genes coding for herbicide resistance developed for agriculture became the selectable markers of choice for generating transgenics, supplanting antibiotic resistance, even when there is no plan for registering the herbicide for use in that crop.

The huge corporate investment in HRC and *Bacillus thuringiensis* (Bt) toxin–containing crops due to perceived market size resulted in the gain of much of our knowledge on promoters, organelle-specific and transit peptides, and more recently organelle transformation. This corporate investment in basic plant molecular biology was greater than the public sector effort, and the spillover was great. Let us remember that it was market-driven transgenic research, and the market is for weed control.

III. NEEDS FOR TRANSGENICS NOT BEING MET

Because much of the basic and applied effort is market driven, the predominant weed problems being solved are those where corporations perceive maximal gain. Thus, the parasitic flowering *Striga* spp. that are claimed to decimate half the maize, sorghum, and legume yields of 100 million farmers in Africa are not considered a market. This is despite the demonstration that genes already available allow selective control of these and similar parasites on transgenic crops (Joel et al., 1995). The genes are not even being made available by the multinationals to public sector researchers for crossing or insertion into African crop varieties (Anonymous, 1999b).

There is a dire need for HR wheat and rice to enable the control of grass weeds that have evolved resistance to most graminicides (grass-killing herbicides) that can be used in these two crops. As these crops are predominantly cultivated with farmer-saved seed, especially in marginal areas, the commercial gain from seed sales does not attract the multinationals. Again, many of the genes that could be used are already available and could be cheaply inserted. The cost of isolating genes and producing constructs that allow normal yields, as well as ascertaining toxicological safety, has already been incurred. Introducing the same constructs in other crops is far less costly, but the corporate reluctance remains.

Much hand labor and mechanical methods are still used for controlling weeds in vegetable crops, even in developed countries, because of a paucity of registered selective herbicides. Such crops are especially adaptable to a transgenic approach using existing genes and herbicides. A large premium could be added to seed price when weed control costs are lowered, but the small market size and minuscule additional herbicide sales have hampered the registration of appropriate herbicides and have not enticed the development of such crops.

IV. THE SUCCESS OF TRANSGENIC HERBICIDE--RESISTANT CROPS

Millions of hectares are being planted with T-HRCs, with insect resistance in second place, with both traits often "stacked" in the same seeds to enhance their value. Farmers perceive a utility in T-HRCs, as they have many alternatives and still repeatedly purchase seed of T-HRCs. The real values of T-HRCs come from instances in which there really are no viable weed control methods (including the situation due to evolved herbicide resistances in weeds) and the fact that such T-HRCs could lead to more sustainable world food production. T-HRCs are the subject of a book (Duke, 1996) as well as in-depth reviews of the molecular aspects (Cole and Rodgers, 2000; Gressel, 2000, 2002).

The easiest way to obtain selectivity among closely related species is to engineer resistance to a general herbicide into the crop. For example, it has already been shown that red rice (*Oryza sativa*) is easily controlled by glufosinate in transgenic rice (Sankula et al., 1997) bearing the *bar* gene that confers resistance to this herbicide. The immediate answer to multiple resistance problems in weeds of wheat in major growing areas is to engineer resistances to inexpensive herbicides (Gressel, 1988, 2002). Neither the chemical nor the biotechnological industries have shown particular interest in generating T-HRCs in wheat and rice. Because too little profit is perceived to come from wheat or rice seed or even from generic herbicides, it may be necessary to have wheat and rice engineered by the public sector. Glufosinate resistance has been engineered into wheat, more as a marker gene than for agronomic utility (Weeks et al., 1993). A request has been made for field use of glufosinate-resistant rice in the United States (Anonymous, 1999a), which has been allowed, despite the fact that it is well known that the gene can introgress into conspecific red rice (Sankula et al., 1998).

BD-HR wheats and rices may be an answer to the major problems of these crops. Insert a gene into wheat or rice conferring resistance to a broad-spectrum herbicide and you can control weeds that evolved resistance in wheat and even closely related grasses, including red, weedy, and other wild rices (Gressel, 1999a, 1999b, 1999c). The transgenes will allow overcoming the problems of resistance that have evolved, especially the problems of cross-resistances (where one evolutionary step conferred resistance to a variety of chemicals) and multiple resistance (where a sequence of evolutionary steps with different selectors conferred resistance to a variety of chemicals). The use of nonplant transgenes may also allow farmers to overcome the natural resistances in weeds closely related to the crop. One can even choose bacterial genes for inexpensive long-established herbicides such as dalapon (Buchanan-Wollaston et al., 1992), modify them for plant codon usage, and transform them into grain crops (Gressel, 1997a). Such genes have the advantage that it is harder, albeit not impossible, for the weeds to mimic bacteria than to mimic plants in evolving resistance.

The problems of interclass cross-resistances and multiple resistances in wheat have necessitated considering generating BD-HR oilseed rape, especially in Australia (Gressel, 1999b). Oilseed rape has become an excellent rotational crop, alternating with wheat in many places where wheat is grown. There are many agronomic advantages to rotating a dicot with a monocot, especially vis-à-vis weeds. It should be far easier to eliminate grass weeds in oilseed rape than in wheat, as there are more selective graminicides available for use in dicot crops. This is usually correct except in Australia, where *Lolium rigidum* has successfully evolved widespread target site and intergroup cross-resistances as well as multiple resistances to all these graminicides (although not in all fields). It is additionally necessary to generate BD-HR oilseed rape (*Brassica napus*) to allow control of closely related *Brassica* weeds including those such as *Raphanus raphanistrum*. Here we can make the case for the need for management strategies with T-HRCs as with any selective herbicide. Atrazine-resistant BD-HR oilseed rape was widely and exclusively used to control *R. raphanistrum* in Australia. It was generated by backcrossing the gene in from a mutant *B. campestris* that evolved in Canada (Souza-Machado, 1982). Inevitably, there are reports that this weed has evolved resistance to triazines (Cheam et al., 2000) as well as resistance to ALS inhibitors (Cheam et al., 1999) and to diflufenican (Cheam et al., 2000). Thus, one must engineer resistance into oilseed rape that allows control of both *Lolium rigidum* and *Raphanus raphanistrum*, with all their resistances. To delay further resistances to the few remaining herbicides, perhaps gene stacking and using herbicide mixtures are necessary (see later).

Only the first generations of BD-HCRs have been released, and considerable improvements are to be expected. There are far too few concrete molecular and biochemical data published about the properties of these crops, and thus there are problems in evaluating their properties to allow suggestions for improvements. Thus, some of what will be said here should be considered speculative. Two types of genes have been used to generate herbicide-resistant crops: (1) those whose gene product detoxifies the herbicide and (2) those for which the herbicide target has been modified such that it no longer binds the herbicide. One could envisage other types such as exclusionary mechanisms or sequestration, but they have yet to be found and thus not utilized. The resulting T-HRCs with each type of resistance are rather different and thus will be discussed separately.

A. Metabolically Resistant T-HRCs

Many crops bearing transgenes coding for highly specific enzymes that metabolically catabolize herbicides have been generated (Cole and Rodgers, 2000). These include, for example, bromoxynil resistance crops bearing a nitrilase (Stalker et al., 1996), glufosinate resistance crops bearing an acetyltransferase (Vasil, 1996), 2,4-D–resisting crops bearing a highly specific soluble cytochrome P-450 monooxygenase (Streber and Willmitzer, 1989) phenmidipham-resisting crops bearing a bacterial gene (Streber et al., 1994), and dalapon-resisting crops bearing a dehalogenase (Buchanan-Wollaston et al., 1992). Of these, only the bromoxynil- and glufosinate-resistant crops have reached commercialization. All of the genes used commercially are of bacterial or actinomycete origin, despite the fact that plants contain genes for herbicide resistance, which is the basis for most natural metabolic selectivities used for 45 years. Nevertheless, plant genes conferring metabolic resistance have not yet been used commercially. There are reports on using nonprokaryotic genes to confer resistance, but none of this work is yet commercial, and whether they confer sufficient resistance is not clear. A rabbit esterase gene conferred resistance to thiazopyr via degradation (Feng et al., 1998). The expression of plant and animal P-450 transgenes conferred phenylurea resistance (Inui et al., 1999; Siminsky et al., 1999). Transgenes encoding maize glutathione transferases increased the level of herbicide resistance (Jepson et al., 1997).

Unlike the target site resistances that are fraught with problems (see later), the crops generated with metabolic resistances seem to be problem free, with little metabolic load conferred by generating the small amount of enzyme needed. The toxicology is simplified because the transgene product typically initiates a cascade of events whereby the herbicide eventually disappears.

There has been an assumption that one cannot use catabolic enzymes to confer resistance to fast-acting herbicides. This view is of doubtful relevance as biotechnologists should be able to perform as well as plants. Inhibitors of protoporphyrinogen IX oxidase (protox), which actually cause the accumulation of the photodynamically toxic product (Böger and Wakabayashi, 1999), induce photodynamic death of plants within 4–6 hours in bright sunlight. Beans are immune to some members of this group, e.g., acifluorfen, because they possess a specific homoglutathione transferase and contain enough homoglutathione to degrade these herbicides stoichiometrically before they can damage the crop (Skipsey et al., 1997). Similarly, strains of *Conyza bonariensis* contain a complex of enzymes capable of detoxifying the reactive oxygen species generated by the photosystem I blocker paraquat and keeping the plants alive until the paraquat is dissipated (Ye and Gressel, 2000).

As almost all herbicides are degraded either in the soil or in some plant species, one should be able to find more genes for catabolic resistance to those herbicides and then rapidly generate herbicide-resistant crops with metabolic resistance rather than with target site resistance.

B. Target Site Resistant Crops

Biotechnology has been used to generate target site HRC crops by two means—transfer of field- or laboratory-generated mutants into crop varieties by genetic means or transgenically.

One must assume a fitness penalty with any target site HRC. This is simply due to the fact that if the resistant mutation was neutral or very nearly neutral there would be naturally resistant populations preexisting in the wild due to Sewall Wright drift. As all the mutations were found at a low frequency, one in a million or less, the resistant traits are not near neutral fitness. This was most obvious with triazine resistance, which was crossed from weedy populations of *Brassica campestris* (=*B. rapa*) that evolved resistance into oilseed rape (*B. napa*) (Souza-Machado, 1982). This *psbA* gly264ser mutation is plastid inherited, allowing researchers to test fitness in the near-isonuclear eighth backcross (Gressel and Ben-Sinai, 1985), and in isonuclear reciprocal hybrids (Beversdorf et al., 1988). There was a consistent 20% productivity loss when the crop was grown alone and a much greater fitness loss when the resistant biotypes were grown in competition with the sensitive biotype. A 20% yield loss does not mean that such crops are without value; this triazine-resistant oilseed rape is widely grown in Australia because some pernicious weeds can be inexpensively controlled. Still, the triazine-resistant varieties will probably eventually be supplanted by transgenics, which do not have the fitness penalty.

Despite a plethora of papers claiming no fitness penalty in acetolactate synthase mutant and transgenic plants (see Saari et al., 1994), elegant experiments by Bergelson et al. (1996) with transgenic *Arabidopsis* did indeed show that such a penalty exists. Previous competitive fitness experiments started with transplants and/or did not measure seed output. The two most competitive phases in a plant's life cycle are establishment, when large populations of seedlings self-thin to a single plant per parent, and fertilization, when thousands of pollen grains compete on stigma and through style. If competition is not measured seed to seed, then the fitness penalty is unknown.

The natural resistance mutations to ALS and ACCase derived from maize tissue cultures or via pollen mutagenesis and then backcrossed into inbreds seem to have produced normally yielding hybrids. There have been more problems with the transgenics than the mutants, as successful as they have been in the marketplace. The magnitude of resistance with transgenics is not as high as it is with the natural mutations, with both ALS and EPSP synthase target site resistances. The first single-site mutant pro101ser EPSP synthase transgenic plants tested were insufficiently resistant to glyphosate for field use. Thus, considerable effort was made to increase the level of resistance and repair the huge shift in K_m toward natural substrates by both adding second site-directed mutations as well as enhanced promoters (Padgette et al., 1996; Lebrun et al., 1997). The necessity to do this was the basis for the claimed invincibility of glyphosate-resistant crops from evolving resistance in weeds; it would take too many simultaneous mutations for resistance to evolve (Bradshaw et al., 1997). This thesis was published after there was already known natural metabolic resistance in legumes (Komossa et al., 1992) as well as natural resistance due to enhanced levels of the target enzyme in *Lotus corniculatus* (Boerboom et al., 1990).

These flaws in the invincibility theory were pointed out by Dyer (1994) and reiterated and updated by Gressel (1996). The only difference that could be found between a recently evolved glyphosate-resistant *Lolium rigidum* population having a sevenfold increase in the I_{50} value and the wild type was the presence of a doubled level of EPSP synthase (Gruys et al., 1999). *Lotus corniculatus* with a doubled level of EPSP synthase also had a sevenfold increase in glyphosate resistance (Boerboom et al., 1990). This is not the only case in which it has been possible to confer herbicide resistance by overexpression of the gene encoding its target site. Glufosinate resistance was achieved in tissue culture by this means with overproduction of glutamine synthase (Deak et al., 1988). Similarly, poplars were transformed for elevated glutamine synthase to enhance nitrogen utilization (Gallardo et al., 1999). The transgenic poplars are more resistant to glufosinate than the wild type (F. M. Cánovas, personal communication). In addition, transgenic tobacco plants with fivefold overexpression of protoporphyrinogen IX oxidase in chloroplasts are resistant to a discriminatory dose of 20 μM acifluorfen, which severely inhibited the wild type (Lermontova and Grimm, 2000). Unfortunately, in the latter case there are no quantitative data on the magnitude of resistance.

The nail in the coffin of invincibility came with the identification of glyphosate-resistant *Eleucine indica* in Malaysia (Lee and Ngim, 2000). This material was resistant to far greater than typical field application rates of glyphosate. It was disclosed that this material bore a target site resistance to glyphosate (Tran et al., 1999). The single mutation in the binding domain of this material was at a position equivalent to the pro101ser in the *Salmonella typhimurium aroA* resistance gene (Tran et al., 1999), the same position that was regarded as insufficiently resistant in trangenic resistant genes (Padgette et al., 1996).

In retrospect, there is good reason to expect that transgenics will be less resistant than target site mutants bearing the same transversion. Recurrent selection of a natural mutant will select for homozygous resistant individuals. This cannot happen with the present generation of transgenics. They bear and express both the native and transgenic enzymes (although the ratios of the two have never been published to the best of my knowledge). Thus, transgenic plants with target site resistance are functionally heterozygous and will remain so despite recurrent selection. Homozygosity can be achieved at present only when totally exogenous genes such as those for metabolic resistance are used, as they do not compete with native genes. When the herbicide is applied, target site HRC must depend on the transgenic derived enzyme while the native is inhibited, perhaps even causing phytotoxic precursors to accumulate.

There have been continuing problems with some glyphosate-resistant crops, suggesting that the critical balance of transgenic and native enzymes is not optimal. Initial varieties of maize, about to be released 4 years ago, were retracted because late-season herbicide application caused pollen sterility. Similar problems of obtaining vegetative (only) glyphosate-resistant cotton plants continue to plague cotton growers. The period of pollen production is well known to be subject to inhibition by minor chemical, metabolic, or environmental perturbations. There have been reports of stem brittleness and cracking of untreated glyphosate-resistant soybeans (Anonymous, 2000), suggesting an overproduction of product leading to increased lignin formation when both the native and transgene-derived enzymes are operative.

One does not know whether such glyphosate-resistant transgenic plants are transiently weakened until the herbicide is dissipated. This period after glyphosate application is analogous to sublethal treatments of nontransgenic plants with glyphosate. When legumes are given sublethal applications of glyphosate, they are unable to produce phytoalexins after elicitation and are far more susceptible to disease (Sharon et al., 1992). A study purporting to ascertain whether phytoalexin biosynthesis was compromised after glyphosate treatment of glyphosate-resistant soybeans (Lee et al., 2000) was artifactual. The authors measured only constitutive levels of glyceollin, without elicitation, and their conclusions, concerning phytoalexins, which are by definition elicited, are without basis.

There are at least two apparent solutions to the problems arising from the functional heterozygote status of transgenic target site resistant plants. One solution is to try to enhance the metabolism of the herbicide, obviating the problem. This may have been done using the *gox* gene coding for an enzyme-degrading glyphosate together with target site resistance (Mannerlöf, 1997). There are no data indicating how quickly the transformants degraded the glyphosate in this case. A better solution might be to begin with a "clean slate." Target site resistant plants should be generated via replacement, i.e., in plants that have had the native gene deleted or functionally irreversibly "knocked out." This can be done by classical deletion mutagenesis with ultraviolet (UV) or gamma irradiation, but this causes multiple lesions and the screening for particular mutants is not easy. Molecular techniques show far more promise as they are more "surgical," i.e., there is less likelihood of multiple deletions along a chromosome, leading to complications from linked deletions. These techniques include T-DNA tagging (Zupan and Zambryski, 1995; Choe and Feldmann, 1998) and transposon mutagenesis (Tissier et al., 1999). There are an estimated 3000 genes that might be target sites of herbicides (Berg et al., 1999), giving mutations that will have to be rescued by the specific gene products of interest as the selectable markers. The most elegant new technique is with specific viral vectors spliced to the gene to be silenced. After cloning into the crop, some "black magic" mechanism virus-induced gene silencing turns off both the viral and the endogenous gene (Baulcombe, 1999). At least one of the virus vectors used is transmitted through seeds and by pollen, so following generations remain with the gene suppressed (Lister and Murant, 1967), but whether the gene is never reactivated is not certain.

After the native gene is suppressed, it should be possible to obtain transgenics with various levels of transgene expression in functional homozygotes and choose those with the optimal expression level for agricultural release. The gene used may have to be very different in sequence from the native gene as viral-induced gene silencing can suppress the native gene.

V. RESISTANCE MANAGEMENT—THE NEED FOR STACKED GENES

Probably >10 million ha of agricultural lands are infested with weeds that are resistant to one or more classes of diverse herbicide chemistry (Heap,

2000). Proactive resistance management is finally being considered; previously, there was a smug assumption that industry would continue to develop many new chemicals. Now there is a realization that T-HRC is almost all there will be that is new for a long while, and measures should be instituted to delay weeds from evolving resistance. One way to delay the evolution of herbicide resistance in weeds is to stack two genes for herbicide resistance and require that the farmers use only a mixture of herbicides. This is useful because it considerably lowers the mutation frequency for resistance in the weed. For example, if the frequency of resistance to one herbicide is 10^{-6} and the other 10^{-8}, the resistance to them stacked together is 10^{-14} on condition that both herbicides are used. The use of stacking is very important where one wishes to preserve the ability of an excellent herbicide such as glyphosate and one has a pernicious weed with a propensity to evolve resistance, e.g., *Lolium rigidum* in Australia.

There are two hazards that must be assessed with glyphosate and other BD-HR wheats: the risk of grass weeds such as *Lolium rigidum* evolving resistance to the herbicide and the risk of introgression of the gene into related weeds. If glyphosate resistance is to be engineered into wheat, it should be stacked with a gene coding for resistance to a second unrelated graminicide. The herbicide mixtures should always be used when such wheat is cultivated. Glyphosate used alone will clearly further engender evolution of glyphosate resistance (Gressel, 1996; Powles et al., 1998; Pratley et al., 1999), and/or bring about a shift in weed spectra toward weeds that have never been controlled by glyphosate (Owen, 1997). The use of stacked BD-HR wheat and herbicide mixtures should delay the resistance of *Lolium* to glyphosate. The risk of introgression of stacked genes into wheat-related weeds will be discussed later.

When stacked resistance genes and herbicide mixtures are used, it is possible that there will be cases of enhanced weed control by the mixture. Still, some mixtures may be contraindicated vis-à-vis resistance management. Simply combining herbicides with different modes of action will not result in delaying resistance if the efficacy and temporal activity characteristics of the mixed herbicides do not match (Roush et al., 1990). Both mixing partners must effectively inhibit the weeds most sensitive to the vulnerable herbicide because the greater selection pressure weed species are the most likely to evolve target site resistance (Gressel and Segel, 1982; Maxwell et al., 1990). Resistance could quickly evolve in a weed species that is naturally resistant to one of the herbicides in a mixture.

The components of the mixture need to have similar persistence or the mixing partner with a low mutation frequency must have longer persistence than the vulnerable one with a high mutation frequency of resistance. Otherwise, there will be a period when only the vulnerable partner is present, and it will select for resistance in the target weed as if there were no mixture at all. This would be the case when a persistent resistance-vulnerable ALS inhibitor is mixed with a short-lived phenoxy herbicide (Wrubel and Gressel, 1994). Unlike crops, which have been selected to germinate uniformly shortly after planting, seeds of many weed species display many flushes of germination during a cropping season (Bewley and Black, 1982). If a resistance-prone weed species has multiple flushes during the season and the vulnerable herbicide has a longer period of activity than the mixing partner does, then the vulnerable herbicide selects for individuals resistant only to it after the mixing partner has dissipated. A mixture that is not well matched for persistence can still be effective if all the weeds germinate over a short period of time and both herbicides outlast the germination.

The ideal mixing partner should have three other properties (Wrubel and Gressel, 1994) in addition to equal persistence:

1. It should have a different target site of action from the vulnerable herbicide.
2. The mixing partner should not be degraded in the same manner as the vulnerable herbicide. For example, if the vulnerable herbicide is degraded in the crop by a glutathione transferase, the mixing partner should have no chemical site that can be attacked by the same enzyme.
3. Another useful attribute in a mixing partner would be to possess negative cross-resistance, i.e., where individuals resistant to the vulnerable herbicide are more susceptible to the mixing partner than the wild type. This would actually reduce the frequency of resistant alleles in the weed population. This strategy was first proposed for herbicides on the basis of laboratory data (Gressel and Segel, 1990), and information on the existence of negative cross-resistance has been published at the whole plant level (Gadamski et al., 2000).

VI. INTRODUCTION OF TRANSGENES FROM CROPS TO WEEDS

We have been warned that T-HRCs can lead to the evolution of "super-weeds" that will inherit the earth (Kling, 1996). The rapid commercial release of such crops has often been without broad-based scientific scrutiny with the most competent experts being involved, and there is little published physiological, biochemical, ecological, genetic, and agronomic data to scrutinize. This leads to a certain degree of skepticism among scientists, which contributes to the public questioning the needs, utility, risks, and values associated with the use of T-HRCs. The severe pressures by detracting groups on policy makers make it politically incorrect to pursue public sector research in this area, which prejudices the ability to perform experiments to obtain accurate information about the risks. These pressures also prevent generating crops needing resistance to herbicides where the agrochemical or seed industry perceives little profit.

Herbicide-resistant rice and wheat can be of great benefit only if used with care to prevent or mitigate gene transfer to related feral and wild rices and related weeds of wheat such as *Aegilops cylindrica*. Whereas the wild and feral rices are widely distributed, *A. cylindrica* has a relatively narrow distribution, although it can be quite pernicious. No solution in agriculture has been forever; farmers have always had to deal with evolution. More sophistication will be needed than is presently being used with transgenic wheat and rice to mitigate introgression delaying such evolution.

Discussions of T-HRCs have rarely dealt with the risks from a weed biology perspective. The main risk stated by the detractors is claimed to be that of the T-HRCs becoming "volunteer" weeds (in following crops) or their introgressing traits into a wild relative, rendering it weedier: the superweeds of the mass media. An attempt at such an assessment based on weed science was made using a defined set of uniform criteria in a decision tree format (Gressel and Rotteveel, 2000). Decision trees, by requiring discrete answers to sequential, stepped questions, lower the bias in arriving at conclusions vis-à-vis the risks deriving from a given hazard.

Conversely, there are wild species that are unlikely ever to become weeds unless they evolve a large number of weedy traits (Keeler et al., 1996). Unfortunately, too many risk studies do not differentiate between weedy relatives of crops and wild relatives (e.g., Sindel, 1997). Risk assessment must be performed on a local or regional basis, as the risks from the same T-HRC will vary greatly from one agricultural ecosystem to another. A second assessment should be done (but has not been done in the past) about the effects of introgression from transgenic crops on weed flora of countries that import bulk unprocessed commodities such as wheat and oilseed rape (Gressel, 1997b).

The risks of introgression have been assessed on a case-by-case basis by regulatory authorities. The Canadians delineated criteria before even having to evaluate plants with novel traits, whether or not transgenic (Anonymous, 1994a), and then specifically evaluated oilseed rape in the context of these criteria (Anonymous, 1994b). In a series of documents they further evaluated oilseed rapes resistant to imidazolinone (Anonymous, 1995a), glyphosate (Anonymous, 1995b), and glufosinate (Anonymous, 1996). The decision process was based on their perception of the risks to the regional agricultural ecosystems in western Canada, without considering other regions that may be importing the crops. Internationally, the OECD (Organization for Economic Cooperation and Development) and UNIDO (United Nations Industrial Development Organization) are developing a series of consensus documents on the biology of various crops (with regard also to related weeds) so that there is a common starting point to evaluate each cropping situation (Anonymous, 1997). Most of the stated hazards of interspecific introgression from T-HRCs are based on unpredictive laboratory experiments, which prove that introgressions could occur. Thus, they show that the hazard exists but give little indication of risk: how quickly such transfers will occur in the field or how fit recipients will be to cope with competition in a multfactorial situation. The time factor is not inconsequential; if resistance introgresses to produce resistant populations more slowly than natural mutational evolution of resistance, what is the significance of introgression?

A. Vertical, Horizontal, and Diagonal Gene Transfer

Two types of gene transfer are widely discussed: (1) vertical, within a species, and (2) horizontal, transfer among unrelated species, usually by asexual means. Biology is not as clear-cut; there can be some sexual transfer between plant species in the same genus and closely related genera. These are typically included in discussions of introgression as horizontal gene transfer. Horizontal gene transfer via plasmids is common in prokaryotic organisms. Because extrapolations are often made from these rare cases of gene transfer among closely related species to "prove" that all horizontal transfers are possible, I suggest terming these special interspecific cases in plants as "diagonal" gene transfer, denoting the gray area where they exist.

Vertical and diagonal transfer possibilities are obvious to any biologists, but horizontal transfers, in eukaryotes, with their potentially disastrous implications for agriculture are not. The possibilities of true horizontal transfers are extrapolated from the intergeneric and interfamilial plasmid-mediated transfer of traits among microorganisms, which have allowed transfer of antibiotic resistance (analogous to herbicide resistance) among unrelated pathogens. The claim continues that because plasmids are often used as vectors in the genetic engineering of crops, interfamilial transfers will become commonplace or at least inevitable. This claim does not stand up to epidemiological experience with organisms such as *Agrobacterium tumefaciens* and *A. rhizogenes*. The plasmids used for laboratory gene transfers from these *Agrobacterium* spp. naturally infect a broad range of dicots, using the plasmid as part of the infection process.

If such interfamilial transfers were to occur via *Agrobacterium*, they would have been seen over the past 50 years with naturally occurring herbicide resistances. There are no known cases where such genes have trans-ferred interfamiliarly from any crop to weed via *Agrobacterium*, despite the great selective advantages that such weeds would have and the ubiquity of *Agrobacterium* in the environment. The more than 10 million hectares of herbicide-resistant weeds that have appeared in the past 30 years can all be traced to evolution resulting from mutant selection, not to plasmid-mediated horizontal gene transfer. In addition, an extensive survey of the GenBank database found few *Agrobacterium* DNA sequence pieces in any of the plant genes, which would have been expected in the millions of years of cohab-itation. This matching task took hundreds of hours of computer time (Rubin and Levy, 1999). Horizontal gene transfer need not be discussed further; diagonal gene transfer is a hazard for which the risks must be estimated in particular cases.

B. Generalizing from Hazards to Risks

Because of the genetic variability of crops and weeds and the chemical variability in herbicides having different effects and modes of action, one cannot make easy generalizations about the risks of introgression of resis-tance. Each case needing a prediction of the risk of introgression must be evaluated on its merits, often after basic biological, genetic, and epidemio-logical studies. More important, other issues must be considered:

1.   What is the benefit to agriculture of introducing resistance into a certain crop, especially in management of evolved herbicide re-sistance in weeds?
2.   What are the possibilities of, and implications of, having herbicide resistance pass from the crop into a weedy or wild species?
3.   What are the possibilities of and implications of having the T-HRC becoming a volunteer resistant weed in agricultural ecosys-tems or become an alien weed in ruderal or more pristine ecosys-tems?

The final decision on risk/benefit ratio is ultimately a balance between science, economics, local benefits, local values, pressure groups, and local politics. The politicians often misuse or ignore science for clearly political decisions (Powell, 1997). Still, there is good reason that the criteria for risk assessment for T-HRCs should be uniform, using universal criteria and pro-cesses of examination.

1. Risks of Introgression of Transgenes to Related Weeds

In most of the world there can be virtually no risk of transgenes carrying herbicide resistance moving from crop to related weeds in such major crops as maize, soybeans, potatoes, and cotton. Outside the centers of origin, there are no related weeds or even remotely interbreeding wild species that could become weeds. Clearly, experimentation would be required in the centers of origin to assay the risk, but these centers represent a minuscule portion of the areas where these crops are grown.

Genes from wheat easily introgress into the genomes of some related wild species, wild species that are related to the progenitors of domestic wheat. There is much information on introgression of genes from hexaploid wheat into *Aegilops cylindrica*, a very problematic weed in the western plain states and the Pacific northwest of the United States (Zemetra et al., 1998; Seefeldt, 1998). This species shares a D genome with hexaploid wheat, and homologous recombination occurs naturally under field conditions. It would theoretically be much harder to obtain transfer resistance from AABB tetraploid (durum) wheat to this species, as durum lacks the D genome. Thus, homeologous recombination (crossing over between related but dissimilar chromosomes) would have to occur when herbicide-resistant durum is used.

Genes for various traits have been transferred from many wild grasses to domestic wheat, especially from the genus *Aegilops*. Many *Aegilops* species are in fact considered to be *Triticum* species (Kimber and Sears, 1987). The only other *Aegilops* sp. besides *A. cylindrica* known to have a wheat-homologous genome is *A. squarrosa* = *Triticum tauschii* (Kimber and Sears, 1987). Only homeologous recombination can occur between hexaploid and durum wheats with the other less-related species. It takes much more than a single gene, even one for herbicide resistance, to turn a wild species into a weed (Baker, 1991). The risks of introgression of transgenes from wheat to these wild species are very low, even though breeders have moved them. Many intergeneric hybrids generated by breeders do not occur in the field; natural alleles have not passed from wheat to wild, related but mainly ruderal species and vice versa. The breeders have to resort to forced crosses and then use techniques such as embryo rescue in tissue culture to save the hybrid embryos that would otherwise abort.

Most oilseed rapes cultivated are *Brassica napus*, a recent tetraploid derived from the CC genome of *B. oleracea* and the AA genome of *B. campestris* = *B. rapa* (U, 1935). Thus, the only weed where homologous recombination can occur is *B. campestris*. Still, homeologous recombination is known, especially in the laboratory, with many related species. Interspecifically this required hand pollination after emasculation of the weed, male sterility or self-incompatibility in the weed, massive amounts of crop pollen, and/or embryo rescue of the rare progeny, which are mostly sterile or runts (Darmency, 1994a; Bing et al., 1996; Landbo et al., 1996; Lefol et al., 1996; Mikkelsen et al., 1996; Scheffler et al., 1995). Initially, the general fear of T-HRCs precluded performing such experiments in the field, especially in

Europe, where the public fears are greatest. This hysteria prevents obtaining field data that might substantiate or more likely allay the fears of introgression. A large-scale field experiment in Australia found no introgression from oilseed rape to the related *Raphanus raphanistrum* (Rieger et al., 1999).

The significance of laboratory introgression studies to the field situation was evaluated by Kareiva et al. (1996). The older epidemiological or apocryphal reports are actually more relevant to risk analysis than many of the artificial laboratory experiments; the older results could indicate that such transfers can occur in the field as well as the time until predominance and the competitive advantage (if any) of such introgressions. Few studies dare to estimate comparatively how long it will take to have resistance introgress and predominate in field weed populations versus how long it would take resistance to evolve by natural selection versus the expected commercial lifetime of the herbicide.

It has already been shown that transgenic glufosinate resistance can be transferred genetically from rice to red rice (Sankula et al., 1998). How easily this will occur in the field is unclear, as rice is predominately self-pollinated, before flower opening. Less is clear about the other rice species. Rice is a species for which there is far too little field information despite considerable information in the breeders' and cytogeneticists' laboratories. Cultivated rice *Oryza sativa* has an AA genome, as does the red and many feral forms of weedy rice that are also *Oryza sativa* (Aggarwal et al., 1997; Aswidinnoor et al., 1995; Brar and Khush, 1997; Khush, 1997). Genes readily move between the cultivated and feral forms despite rice being predominantly self-pollinating (Brar and Khush, 1997; Aswidinnoor et al., 1995; Langevin et al., 1990; Kaushal and Ravi, 1998; Majumder et al., 1997; Mariam et al., 1996). There are many wild and weedy rice species, one of which, *O. rufipogon*, has an AA genome and is considered a major weed of rice (Holm et al., 1997). Another major weedy rice, *O. officinalis*, has a CC genome.

There are many other diploid and tetraploid wild (but not weedy) rices bearing genomes through the alphabet from AA to HHJJ. It is possible to transfer genes from many of these wild and weedy species to rice, which is how the breeders introduce new traits, despite the chromosomal incompatibilities (Brar and Khush, 1997; Khush, 1997). This often requires embryo rescue and/or intermediate crosses through bridge species. The significance of this to field problems is unclear. The taxonomic differentiation among these species is also rather unclear. In many places where weedy rices are even major problems, it is not known whether the weeds are the conspecific red and feral forms or the other species (Cohen et al., 1999; Baltazar and Janiya, 1999). Indeed, studies have shown that the material in genetic resource depositories has been misclassified. Much of this ambiguity will be clarified by modern molecular taxonomic techniques using DNA fingerprinting (Martin et al., 1997; Suh et al., 1997), so that better predictions can be made of risk in different locales.

C. Assaying Introgression in the Field

There are ways to ascertain the rapidity of gene movement without causing lasting damage to agriculture and/or the environment: simply insert a gene conferring resistance to a rarely used herbicide (Gressel, 2000). There would be little consequence of the herbicide becoming unusable due to the resistance disseminating into the wild, as herbicides are not used on wild populations. Another tracing system could be used with acetolactate synthase (ALS) resistance. At present, if a weed becomes resistant to any ALS herbicide, it cannot be known whether it evolved resistance naturally by mutation or was introgressed through cross-pollination with a nontransgenic T-HRC. The highly mutable ALS gene ($\sim 10^{-6}$ natural resistance frequency in populations) quickly evolves naturally in weeds (Saari et al., 1994). Engineering the same ALS allele either with a two-base-change coding difference from the natural resistance allele or with different introns would allow easy differentiation between mutational events and introgression. This would indicate the rate of evolution due to introgression versus the rate of evolution from natural mutation.

Triazine resistance has been found to be maternally inherited in many species, so one might assume it will never introgress from crops to weeds (Daniell, 1999). Maternal inheritance is not absolute; 0.4% pollen transfer of triazine resistance was found with genetic markers in weeds (Darmency, 1994b) and 0.5% for tentoxin resistance in tobacco (Avni and Edelman, 1991). Maternal inheritance of chloroplast-encoded traits is typical, but there are many cases in higher species, including some crops where such traits are biparentally inherited (Corriveau and Coleman, 1988; Reboud and Zeyl, 1994). Corriveau and Coleman (1988) developed a rapid cytological assay to screen for biparental inheritance, which they did not find in the major agronomic crops but did find in sweet potato, chickpea, vetch, alfalfa, common beans, and geranium. In all, they found evidence for biparental inheritance in 14% of the 192 plant species assayed. Their evidence and the genetic evidence they cite would not have found the 0.4% paternal inheritance described by Darmency (1994b). Thus, it will be necessary to ascertain frequencies of paternal inheritance with each crop situation where maternal inheritance is predominant. There is a good chance that more pollen transmission will be found with herbicide resistance, and the estimates will be more accurate. Herbicide resistance is a better selectable marker than those previously available to the geneticists dealing with chloroplast inheritance, and because of the field size experiments, the numbers of plants tested will be much larger than had been possible. Polymerase chain reaction (PCR) amplification of chloroplast primers discriminates between *Brassica napus* and *B. rapa = campestris* in the ~0.6% natural hybrids that occur between the two species (Scott and Wilkinson, 1999). A low number of hybrids were examined (47), and all were indicative of maternal inheritance. It would have been far more interesting if they had used chloroplast DNA encoded triazine resistance, using a semidominant nuclear marker to verify the hybrid nature of random field crosses of both species. The analysis would be far easier and could be confirmed by PCR amplification.

Susceptible weeds growing among T-HRCs will not introgress the resistance genes if the herbicide is used; dead weeds do not have sex, but the genes could introgress into nearby unsprayed weeds. Seed set on emasculated plants from oilseed rape was measured 1.5 km from the pollen source (Timmons, 1996), but can the progeny compete or survive in feral populations without a selector? Without emasculation of the recipient, resistant pollen fertilized 24% of conspecific plants in the immediate vicinity but <0.027% just 10 m away. If T-HRC seeds carry over as volunteers to other crops where related species are serious weeds, gene exchange might happen through this route and long-distance dispersal becomes unnecessary.

Many weed scientists were surprised that the Canadian authorities allowed the field use of ALS inhibitors, glyphosate, glufosinate, (and soon) bromoxynil-engineered transgenic oilseed rapes. The surprise was due to the known introgression of herbicide-resistant genes into weeds, including the problematic *Brassica campestris = B. rapa* (e.g., Mikkelsen et al., 1996; Kerlan et al., 1993; Lefol et al., 1996; Snow et al., 1999). *Brassica campestris* has been domesticated to become various crops in many places (Polish oilseed rape, turnip, Chinese cabbage, and pak choi). The more ancient conspecific wild form as well as the feral forms are pernicious weeds in other areas (Holm et al., 1997). While botanically identical, they have very different morphotypes, with very different phenologies, biologies, and competitiveness. Most of the *B. campestris* crop types are easily controlled in fields, and as easily controlled volunteer weeds they have never left agricultural or ruderal areas. Their botanically but not phenotypically identical weedy twin can be very problematic in agroecosystems and was a predominant weed in grains before the advent of selective herbicides (Holm et al., 1997). It is hard to predict what will become of the feral populations of Polish oilseed rape, which still remain ruderal. They could evolve to become more weedy.

Deleterious weed gene introgression from feral populations is common into both commercial oilseed rapes, lowering yield and oil quality (e.g., McMullan et al., 1994). A major use of herbicide-resistant oilseed rapes is to facilitate control of its wild relatives. The Canadian authorities subjected nontransgenic ALS-resistant oilseed rape derived by mutagenesis to full regulatory scrutiny before release to the market. Their decisions (Anonymous, 1995a, 1995b, 1996) allow unrestricted field cultivation, while noting the likelihood of introgression into *B. campestris* and stating that the worst case would be the loss of the particular herbicide to control such weeds (Anonymous, 1995a). The decision, partly based on release studies (Crawley et al., 1993), stated that introgression would not increase weediness of the crop or related weeds outside agriculture. *B. campestris* is the weediest of species related to oilseed rape (Holm et al., 1997) and the one species with demonstrated field transfer of genes (Mikkelsen et al., 1996), but not quite as readily as initially presumed (Landbo et al., 1996). No formal or informal monitoring system was instituted. This is unfortunate as much has been written on how comparatively easy it is to eradicate small pest populations and how impossible it is after they have reached a critical size (Moody and Mack, 1988; Thill and Mallory-Smith, 1997), yet few learn these lessons.

Instead, a survey has shown widespread introgression of herbicide resistance traits among the various volunteer herbicide-resistant oilseed rape populations in western Canada (Hall et al., 2000). Many volunteer populations already contain all the three released herbicide resistances, to ALS-inhibiting herbicides, glufosinate, and glyphosate resistances. The regulatory system cannot preclude reintroducing triazine-resistant varieties, and possibly could not prevent the use of 2,4-D−resistant varieties if such were generated from known transgenes (Streber and Willmitzer, 1989) because each case is considered on its own merits. There soon could be "volunteer" weed populations of oilseed rape in subsequent crops in rotational cycles that cannot be controlled by any herbicides available in the farmers' arsenal. There is clear reason to consider requiring the use of certified seed for all plantings along with a requirement that the certified seed contains a single herbicide-resistant gene, unless there is a premeditated reason for stacking and a requirement for the use of herbicide mixtures.

The question of how quickly resistance genes will move from *B. napus* to *B. campestris* may be moot, as Polish oilseed rape (*B. campestris*) with various herbicide resistance genes has also been released in western Canada. The crop as a volunteer weed could become a problem, especially if it also becomes multiply resistant via cross-breeding to all four herbicides introduced in various varieties.

VII.   PREVENTING AND MITIGATING INTROGRESSION FROM T-HRCs TO WEEDS

There are various failsafe mechanisms that can be used to prevent or mitigate the risk of introgression, when and if it does occur (Gressel, 1999a). These vary from management practices (weed-free zones around transgenic crops) to techniques that involve breeding or more biotechnology. Those using a transgenic approach are discussed here, but other approaches such as using apomyxis are also conceivable (Gressel, 2000).

A.   Gene Placement Failsafes

The particular placement of a transgene in crop genomes can affect its movement to other varieties and species.

1.   Chromosomal

Wheat and oilseed rape are composed of multiple genomes derived from different wild sources (Kimber and Sears, 1987; U, 1935). In any given locale it is possible that only one of the genomes of the crop is identical to that of a related weed, allowing easy gene transfer. As the D genome of wheat is compatible with the D genome of *Aegilops cylindrica* and the B genome of oilseed rape with many brassica weeds, transgenes easily introgress from them to wild species. One can perform the cytogenetic localization to ensure that the transgene is on the incompatible A or B genomes of wheat or the C genome of oilseed rape and then ascertain whether there is no homeologous introgression, i.e., crossing over between the nonhomologous chromosomes. There is evidence, though, for a considerable extent of homeologous introgression between oilseed rape and *B. campestris* (Mikkelsen et al., 1996), so there is no utility in this type of failsafe in oilseed rape, but there is little information for assessing homeologous wheat introgression into *Aegilops cylindrica*.

2.   Hybrids

A simple failsafe can be found with hybrid crops. If a dominant transgene for herbicide resistance is placed in the male sterile line in close linkage with the male sterility gene, there will be no possibility of introgression in crop production areas. Care will have to be taken only in the seed production areas when the male sterile line is restored. Such areas must be kept free of related weeds, a typical precaution in seed production generally practiced before the advent of transgenics.

3.   Plastome or Chondriome

If the transgene for herbicide resistance is placed on the mitochondrial or plastid genomes, as has been done in tobacco (Daniell, 1999; Khan and

Maliga, 1999), there should be limited possibility of gene flow, due to the maternal inheritance of these genomes. Species such as tobacco that are often claimed to have no paternal inheritance often have about 0.1–0.5% pollen transfer of traits (Avni and Edelman, 1991; Darmency, 1994b). The risk of transgenes being established in related populations is far greater with herbicide-resistant traits where the herbicide exerts selection pressure, than traits such as production of medicinals, with other traits in between. Large-scale experiments should be performed with such crops to ensure that the level of paternal transfer of traits is sufficiently low to justify using this strategy, and the risk analysis must consider the transformed traits.

4. Transient Transgenics

It would be conceivable to insert certain metabolic traits (e.g., catabolic herbicide resistance) on RNA viruses or in endomycorrhizae that are expressed in the plant but are not carried through meiosis into reproductive cells. One would have to transfect the crop every generation, but the transgenes could not spread sexually.

B. Transgenetic Mitigation (TM)

Genetic engineering can be used to mitigate any positive survival traits transgenes may confer. If the herbicide resistance gene engineered into the crop is flanked on either side by a TM gene in a tandem construct, the overall effect would be deleterious to weeds introgressing the construct from a crop (Gressel, 1999a, 2002). This is based on three premises:

1. Tandem constructs of genes act genetically as tightly linked genes and their segregation from each other is exceedingly rare.
2. There are traits that are either neutral or positive for a crop that would be deleterious to a typical or volunteer weed, or to a wild species.
3. Because weeds are strongly competitive among themselves and have large seed outputs, individuals bearing even mildly deleterious traits are quickly eliminated from populations. Even if one of the TM alleles mutates, is deleted, or crosses over, the other flanking TM gene will remain, providing mitigation.

TM traits that could be used are best visualized when observing the differences between crops and weeds. This is best illustrated with two cases: (1) wheat and weedy relatives and (2) oilseed rape (*Brassica napus*) and feral and weedy Polish rape/wild radish *B. campestris* = *B. rapa*, as summarized in the following.

1. Traits for Transgenetic Mitigation

    a. Seed Dormancy. Weed seeds typically have secondary dormancy with seeds from one harvest germinating throughout the following season and over a number of years. This evolutionary trait is considered to be a risk-spreading strategy that maximizes fitness while reducing losses due to sib competition (Hyatt and Evans, 1998; Lundberg et al., 1996). Staggered secondary dormancy prevents all the weeds from being controlled by a single agronomic procedure. However, crops have lost secondary dormancy as a result of domestication. Genetically abolishing secondary dormancy would be neutral to both crops but deleterious to their related weeds. Tillage, crop rotation, and preplant use of herbicides, all standard practices, would control the uniformly germinating TM weed seeds lacking secondary dormancy in rotational crops.

    b. Ripening and Shattering. Weeds disperse their seed over a period of time and much of the ripe seed "shatters" to the ground, ensuring continuity. A proportion of the weed seed is harvested with crop seed, contaminating the crop seed and facilitating weed dispersal to wherever the crop seed will be grown. Weeds have evolved morphological and phenological "mimicries" to the crop seed (Gould, 1991), necessitating continual evolution and refinement of techniques to remove the contaminating weed seed. Recently domesticated crops such as oilseed rape still suffer from shattering (Price et al., 1996). In addition to the loss of yield, the shattering of crop seeds results in their becoming a volunteer weed, especially in oilseed rape (Lutman, 1993). Uniform ripening and antishattering genes would be detrimental to weeds but neutral for crops that ripens uniformly and not shatter and positive for oilseed rape, which still has a shattering problem.

    c. Dwarfing. Crops have been selected for height, to outgrow weeds. Weed evolution kept apace, selecting for taller weeds. The advent of selective herbicides to kill weeds allowed genetic dwarfing of these crops, with more seed harvest and less straw. Various new systems of genetically engineered height reduction are being introduced. These include genes related to hormone production (Azpiroz et al., 1998; Schaller et al., 1998; Peng et al., 1999) as well as to shade avoidance. Shade avoidance is advantageous when competing with other species but not in a weed-free crop stand where only siblings are competing. The overexpression of specific phytochrome genes prevents recognition of shading, and thus the plant remains short (Robson et al., 1996). This is advantageous for a crop and could also be used where the present dwarfing genes prevent obtaining the highest yields. Dwarfing would be disadvantageous for a weed that must compete with the crops; it would be shaded over by the crop.

2. Balancing Primary and TM Traits

If the primary transgenic trait confers an advantage to a weed, how much will TM traits actually mitigate that advantage? Weeds are not only highly competitive with crops, they are competitive with weeds of other species as well as within their own species. Weeds typically produce thousands of seeds in steady-state conditions to replace a single plant, suggesting extreme competition to be the replacement; the selection for the highest competitive fitness is intense.

Other transgenes can be considered for mitigating the risks of integration with other crops. These include the following characteristics:

1. Eliminating seed coat characters that allow weed seeds to pass through animal digestive tracks intact and then be dispersed.
2. Genes promoting partitioning to roots would be advantageous to cultivated root crops such as beets but detrimental to related crops such as annual wild beets.
3. Genes that prevent premature flowering (bolting), i.e., promote biennial growth, would be excellent for carrots, celery, cabbage, lettuce, beets, and related crops but would be highly deleterious to related weeds.
4. Antiflowering genes would prevent introgression of genes from potatoes into their wild Andean relatives and prevent volunteer potatoes arising from true seed.

After a weed introgresses a transgene and then stabilizes (eliminates cytogenetic incompatibilities), the trait will quickly spread through a population, even if it has a marginally positive fitness advantage (Thill and Mallory-Smith, 1997). Conversely, one can balance the disadvantage of TM traits against the advantage of the primary trait. This must be done in both the presence and absence of the herbicide, as herbicide resistance provides an advantage only when the herbicide is used. Indeed, when the herbicide is not present, the transgene resistance trait can be disadvantageous; as demonstrated with an ALS resistance transgene (Bergelson et al., 1996).

Each TM trait should work in a balance with the primary trait, and it might be necessary to have more than one TM trait in a construct to obtain balance. The risks of introgression can be further decreased by combining TM traits with a cytogenetic failsafe, where these are available. Even if one or two TM genes were to confer some unforeseen and unforeseeable advantage in the future, this would be akin to evolving resistance to a herbicide. The first case would be reviewed and a decision could be made whether the situation warrants removing the TM transgenic crop from market to prevent the occurrence of further cases. The only aspect that is predictable is the segregation of tightly linked genes, and that is why flanking the primary gene on both sides could be used. The segregation of a gene would be visible, allowing any such material to be removed from the breeding population.

3. TM Genes Are Available to Mitigate Movement of Resistance

Some possible traits for TM constructs just exist as named genes that are inherited, others are also mapped to positions on various chromosomes, and a few are actually characterized as sequenced genes. Thus, not all TM traits have genes that are immediately available for insertion in tandem constructs. Still, there can be many different ways for a plant to confer a TM trait, and thus more than one gene might be available.

a. Secondary Dormancy. Unfortunately, *Arabidopsis*, the typical source of genes, has already been sufficiently domesticated that it is unlike cruciferous weeds; the laboratory strains no longer have strong secondary dormancy (Van der Schaar et al., 1997). A mutant that is insensitive to abscisic acid and lacks secondary dormancy was found in a wild, undomesticated *Arabidopsis* strain (Steber et al., 1998).

b. Shattering. Physiologically, one way to avoid premature seed shattering is to have uniform ripening. Early maturing seeds of oilseed rape on indeterminate, continuously flowering varieties typically shatter. Determinacy, with its single uniform flush of flowering is one method to prevent shattering, but this often shortens the season, reducing yield. The hormonology of the abscission zone controls whether shattering will occur, and it is possible that if cytokinins are overproduced, shattering will be delayed. As with secondary dormancy, no sequenced genes are yet at hand except for cytokinin overproduction.

c. Dwarfing. Many of the genes used for breeding dwarfism seem to have an unknown function. Still, many genes are known that control height.

d. Gibberellins. Preventing the biosyntheses of gibberellins reduces height (Webb et al., 1998), which is the basis of many chemical dwarfing agents used commercially on wheat. The three enzymes and genes controlling various steps in gibberellin biosyntheses are known and have been cloned (Yamaguchi et al., 1998; Helliwell et al., 1998). *Arabidopsis* mutations bearing mutations in any of them are dwarfed, and the dwarfing is reversible by gibberellin treatment. Overexpression of a gene coding for *ent*-kaurene synthase, causing cosuppression, mimicked the mutant phenotype.

In addition, a GA receptor gene has been isolated that confers gibberellin insensitivity when a truncated form is transformed into grains (*GAI*) and thereby induces dwarfing (Peng et al., 1999).

Some processes such as flower stalk bolting are controlled by specific gibberellins; in radish, GA_1 and GA_4 are responsible (Nishijima et al., 1998). It may be necessary to characterize the genes coding for the monooxygenases and dioxygenases that are responsible for the later steps (Hedden and Kamiya, 1997). Some of these genes have been isolated as well (Kusaba et al., 1998).

e. Brassinosteroids. This group of hormones also causes elongation of stems in many plant species, and their absence results in dwarf plants. A 22 d-hydroxylase cytochrome P-450 has been isolated that controls a series of these steps in brassinosteroid biosynthesis (Choe and Feldmann, 1998) and plants lacking the enzyme are dwarfed (Azpiroz et al., 1998). Plants are also dwarfed when they produce normal levels of these growth regulators but are mutated in the *bri1* gene coding for the receptor (Noguchi et al., 1999). In addition, suppressive overexpression of a sterol C24-methyl transferase in the pathway also causes dwarfing (Schaller et al., 1998).

f. Shade Avoidance. Various forms of the pigment phytochrome interact to detect whether a plant is being shaded (Torii et al., 1998). The engineering of suppressive overexpression constructs of one of these phytochromes led to plants that did not elongate in response to shading (Robson et al., 1996).

VIII. CONCLUDING REMARKS

One can and should envisage many other ways besides T-HRC in which genetic engineering can be used to replace partially or augment herbicides in controlling weeds. Thus, it is unfortunate or deplorable (depending on one's outlook) that because of an antichemical bias among some of the authors of the World Bank report, weeds were not among the constraints to agriculture that could be overcome by biotechnology (Kendall et al., 1997). With weeds not listed, many biotechnologists do not realize the problems facing agriculture and will not be stimulated to support and study the positive aspects and drawbacks of the first generation of T-HRCs and then innovate new solutions. Thus, one can quote Confucius: "If language is incorrect, what is said is not meant. If what is said is not meant, what ought to be done remains undone." If weeds are not considered a constraint, biotechnological solutions will not be found to deal with them.

ACKNOWLEDGMENTS

The author deeply appreciates the questioning comments and suggestions of Dr. David Cole. The author's research on the assessment of gene introgression from crops to weeds is supported by EU-INCO project ERB IC18 CT98 0391, the research on transgenic mitigation by the Levin Fund, and that on *Striga* control on BD-HR maize by the Rockefeller Foundation. J.G. is the Gilbert de Botton Professor of Plant Sciences. This chapter is condensed and modified from a review article (J Gressel, Molecular biology of weed control, *Transgen Res* 9:355–382, 2000) and is reproduced with the kind permission of the copyright holder, Kluwer Academic Publishers, Dordrecht, Netherlands.

REFERENCES

Aggarwal RK, Brar DS, Khush GS. 1997. Two new genomes in the *Oryza* complex identified on the basis of molecular divergence analysis using total genomic DNA hybridization. Mol Gen Genet 254:1–12.

Anonymous. 1994a. Assessment criteria for determining environmental safety of plants with novel traits. Directive Dir 94-08, Plant Products Division, Agriculture and Agri-Food Canada, Nepean Ontario (http://www.cfia-acia.agr.ca/English/food/pbo/dir9408html).

Anonymous. 1994b. The biology of *Brassica napus* L (canola/rapeseed). Directive Dir. 94-09, Plant Products Division, Agriculture and Agri-Food Canada, Nepean Ontario (http://www.cfia-acia.agr.ca/English/food/pbo/dir9409html).

Anonymous. 1995a. Determination of environmental safety of Pioneer Hi-Bred International Inc's imidazolinone-tolerant canola line. Decision Document DD95-03, Plant Products Division, Agriculture and Agri-Food Canada, Nepean Ontario (http://www.cfia-acia.agr.ca/English/food/pbo/dd9503ehtml).

Anonymous. 1995b. Determination of environmental safety of Monsanto Canada Inc's Roundup herbicide tolerant *Brassica napus* canola line GT73. Decision Document 95-02, Plant Products Division, Agriculture and Agri-Food Canada, Nepean Ontario (http://www.cfia-acia.agr.ca/English/food/pbo/dd9502html).

Anonymous. 1996. Determination of environmental safety of AgrEvo Canada Inc's glufosinate-ammonium herbicide tolerant *Brassica napus* canola line. Decision Document 96-11, Plant Products Division, Agriculture and Agri-Food Canada, Nepean Ontario (http://www.cfia-acia.agr.ca/English/food/pbo/dd9611html).

Anonymous. 1997. Consensus document on the biology of *Brassica napus* L (oilseed rape) Series on the harmonization of regulatory oversight in biotechnology No 7. Environmental Directorate, Organization for Economic Cooperation and Development, Paris.

Anonymous. 1999a. AgrEvo USA Co. Receipt of petition for determination of non-regulated status for rice genetically engineered for glufosinate herbicide tolerance. (US) Federal Register 64 (16), Notice 3924.

Anonymous. 1999b. Access issues may determine whether agri-biotech will help the world's poor. Nature 402:341–344.

Anonymous. 2000. Modified soya beans are cracking up in the heat. Trends Plant Sci 5:55.

Aswidinnoor H, Nelson RJ, Gustafson JP. 1995. Genome-specific repetitive DNA probes detect introgression of *Oryza minuta* genome into cultivated rice, *Oryza sativa*. Asia Pacific J Mol Biol 3:215–223.

Avni A, Edelman M. 1991. Direct selection for paternal inheritance of chloroplasts in sexual progeny of *Nicotiana*. Mol Gen Genet 225:273–277.

Azpiroz R, Wu Y, LoCascio JC, Feldmann KA. 1998. An *Arabidopsis* brassinosteroid-dependent mutant is blocked in cell elongation. Plant Cell 10:219–230.

Baker HG. 1991. The continuing evolution of weeds. Econ Bot 45:445–449.

Baltazar AM, Janiya JD. 1999. Weedy rice in the Philippines. In: Wild and Weedy Rices. International Rice Research Institute (IRRI) Limited Proceedings No 2, IRRI, Manilla.

Baulcombe D. 1999. Fast forward genetics based on virus induced gene silencing. Curr Opin Plant Biol 2:109–113.

Bell A. 1998. Microtubule inhibitors as potential antimalarial agents. Parasitology Today 14:234–240.

Berg D, Tietjen K, Wollweberet D, Hain R. 1999. From genes to targets: impact of functional genomics on herbicide discovery. Proceedings of the Brighton Crop Protection Conference, pp 491–500.

Bergelson J, Purrington CB, Palm CJ, Lopez-Guitierrez JC. 1996. Costs of resistance —a test using transgenic *Arabidopsis thaliana*. Proc R Soc Lond Ser B Biol Sci 262:1659–1663.

Beversdorf WD, Hume DJ, Donnelly-Vanderloo JJ. 1988. Agronomic performance of triazine-resistant and susceptible reciprocal canola hybrids. Crop Sci 28:932–934.

Bewley JD, Black M. 1982. Physiology and Biochemistry of Seeds in Relation to Germination. Vol II: Viability, Dormancy and Environmental Control. Berlin; Springer-Verlag.

Bing DB, Downey RK, Rakow GFW. 1996. Assessment of transgene escape from *Brassica rapa (B. campestris)* into *B. nigra* or *Sinapis arvensis*. Plant Breeding 115:1–4.

Boerboom CM, Wyse DL, Somers DA. 1990. Mechanism of glyphosate tolerance in birdsfoot trefoil (*Lotus maizeiculatus*). Weed Sci. 38:463–467.

Böger P, Wakabayashi K, eds. 1999. Peroxidizing Herbicides. Berlin: Springer.

Bradshaw LD, Padgette SR, Kimball SK, Wells BH. 1997. Perspectives on glyphosate resistance. Weed Technol 11:189–198.

Brar DS, Khush GS. 1997. Alien introgression in rice. Plant Mol Biol 35:35–47.

Buchanan-Wollaston V, Snape A, Cannon F. 1992. A plant selectable marker gene based on the detoxification of the herbicide dalapon. Plant Cell Reports 11:627–631.

Cheam A, Lee S, Bowran D, Hashem A. 1999. Control of group B resistant wild radish in wheat. Proceedings of the 12th Australian Weeds Conference, Hobart, p 219.

Cheam A, Lee S, Bowran D, Nicholson D, Hashem A. 2000. Triazine and diflufenican resistance in wild radish (*Raphanus raphanistrum* L) in Australia. Proceedings of the 3rd International Weed Science Congress. Foz do Iguassu, Brazil, p. 303.

Choe S, Feldmann KA. 1998. T-DNA mediated gene tagging. In: Lindsey K, ed. Transgenic Plant Research. Amsterdam: Harwood Academic Press, pp 57–73.

Choe S, Dilkes BP, Fugioka S, Takatsuto S, Sukarai A, Feldmann KA. 1998. The *DWF4* gene of *Arabidopsis* encodes a cytochrome P450 that mediates multiple 22 α-hydroxylation steps in brassinosteroid biosynthesis. Plant Cell 10:231–144.

Cohen MB, Jackson MT, Lu BR, Morin SR, Mortimer AM, Pham JL, Wade LJ. 1999. Predicting the environmental impact of transgene outcrossing in wild and weedy rices in Asia. In: Gene Flow in Agriculture: Relevance for Transgenic Crops. Farnham, UK: British Crop Protection Council, pp 151–157.

Cole DJ, Rodgers MW. 2000. Plant molecular biology for herbicide-tolerant crops and discovery of new herbicide targets. In: Cobb AH, Kirkwood RC, eds. Mechanisms of Herbicide Action. Sheffield: Sheffield Academic Press, pp. 239–278.

Corriveau JP, Coleman AW. 1988. Rapid screening method to detect potential biparental inheritance of plastid DNA and results for over 200 angiosperm species. Am J Bot 75:1443–1458.

Crawley MJ, Hails RS, Rees M, Kohn D, Buxton J. 1993. Ecology of transgenic oilseed rape in natural habitats. Nature 363:620–623.

Daniell H. 1999. Environmentally friendly approaches to genetic engineering. In Vitro Cell Dev Biol Plants 35:361–368.

Darmency H. 1994a. The impact of hybrids between genetically modified crop plants and their related species: introgression and weediness. Mol Ecol 3:37–40.

Darmency H. 1994b. Genetics of herbicide resistance in weeds and crops. In: Powles SB, Holtum JAM, eds. Herbicide Resistance in Plants. Boca-Raton, FL: Lewis, pp 263–297.

Deak M, Donn G, Feher A, Dudits D. 1988. Dominant expression of a gene amplification–related herbicide resistance in *Medicago* cell hybrids. Plant Cell Rep 7: 158–161.

Deisenhofer J, Michel H. 1989. The photosynthetic reaction center from the purple bacterium (*Rhodopseudomonas viridis*). Science 245:1463–1473.

Duke SO, ed. 1996. Herbicide Resistant Crops: Agricultural, Environmental, Economic, Regulatory, and Technical Aspects. Boca Raton, FL: CRC Press.

Dyer WE. 1994. Resistance to glyphosate. In: Powles SB, Holtum JAM, eds. Herbicide Resistance in Plants. Boca-Raton, FL: Lewis, pp 229–241.

Feng PCC, Ruff TG, Rangwala SH, Rao SR. 1998. Engineering plant resistance in thiazopyr herbicide via expression of a novel esterase deactivation enzyme. Pestic Biochem Physiol 59:89–103.

Gadamski G, Ciarka D, Gressel J, Gawronski SW. 2000. Negative cross-resistance in triazine resistant biotypes of *Echinochloa crus-galli* (barnyardgrass) and *Conyza canadensis* (horseweed). Weed Sci 48:176–180.

Gallardo F, Fu J, Cantón FR, Garcia-Gutiérrez A, Cánovas FM, Kirby EG. 1999. Expression of a conifer glutamine synthetase gene in transgenic poplar. Planta 210:19–26.

Gould F. 1991. The evolutionary potential of crop pests. Am Sci 79:496–507.

Gressel J. 1988. Multiple resistances to wheat selective herbicides: new challenges to molecular biology. Oxford Surv Plant Mol Cell Biol 5:195–203.

Gressel J. 1996. Fewer constraints than proclaimed to the evolution of glyphosate-resistant weeds. Resistant Pest Manage 8:2–5.

Gressel J. 1997a. Genetic engineering can either exacerbate or alleviate herbicide resistance. Proceedings of the 50th New Zealand Plant Protection Conference, Christchurch, pp 298–306.

Gressel J. 1997b. Can herbicide resistant oilseed rapes from commodity shipments potentially introgress with local *Brassica* weeds, endangering agriculture in importing countries? Pest Resist Manage Winter, pp 2–5.

Gressel J. 1999a. Tandem constructs; preventing the rise of superweeds. Trends Biotechnol 17:361–366.

Gressel J. 1999b. Needed: new paradigms for weed control. In: Weed Management in the 21st Century: Do We Know Where We Are Going? Proceedings of the 12th Australian Weeds Conference, Hobart, Tasmania, pp 462–486.

Gressel J. 1999c. Herbicide resistant tropical maize and rice: needs and biosafety considerations. Brighton Crop Protection Conference—Weeds, pp 637–645.

Gressel J. 2000. Molecular biology of weed control. Transgen Res 9:355–382.

Gressel J. 2002. Molecular biology of weed control. London: Taylor and Francis.

Gressel J, Ben-Sinai G. 1985. Low intra-specific competitive fitness in a triazine resistant, nearly nuclear-isogenic line of *Brassica napus*. Plant Sci Lett 38:29–32.

Gressel J, Rotteveel AW. 2000. Genetic and ecological risks from biotechnologically-derived herbicide-resistant crops: decision trees for risk assessment. Plant Breed Rev 18:251–303.

Gressel J, Segel LA. 1978. The paucity of genetic adaptive resistance of plants to herbicides: possible biological reasons and implications. J Theor Biol 75:349–371.

Gressel J, Segel LA. 1982. Interrelating factors controlling the rate of appearance of resistance. In: Le Baron II, Gressel J, eds. Herbicide Resistance in Plants. New York: Wiley, pp 325–347.

Gressel J, Segel LA. 1990. Negative cross-resistance; a possible key to atrazine resistance management: a call for whole plant data. Z Naturforsch 45c:470–473.

Gruys KJ, Biest-Taylor NA, Feng PCC, Baerson SR, Rodriguez DJ, You J, Tran M, Feng Y, Kreuger RW, Pratley JE, Urwin NA, Stanton RA. 1999. Resistance of glyphosate in annual ryegrass (*Lolium rigidum*). II. Biochemical and molecular analyses. Weed Sci Soc Am Abstr 39:163.

Hall LM, Huffman J, Topinka K. 2000. Pollen flow between tolerant canola (*Brassica napus*) is the cause of multiple resistant canola volunteers. Weed Sci Soc Am Abstr 40:117.

Harvey BMR, Harper DB. 1982. Tolerance to bipyridylium herbicides. In: LeBaron H, Gressel J, eds. Herbicide Resistance in Plants. New-York: Wiley, pp 215–234.

Heap IM. 2000. International survey of herbicide-resistant weeds and state by state surveys of the US, Weed Smart, Corvallis, OR, http://www weedscience, com

Hedden P, Kamiya Y. 1997. Gibberellin biosynthesis: enzymes, genes and their regulation. Annu Rev Plant Physiol Plant Mol Biol 48:431–460.

Helliwell CA, Sheldon CC, Olive MR, Walker AR, Zeevaart, JAD, Peacock WJ, Dennis ES. 1998. Cloning of the *Arabidopsis ent*-kaurene oxidase gene GA3. Proc Nat Acad Sci USA 95:9019–9024.

Holm L, Doll J, Holm E, Pancho J, Herberger J. 1997. Worlds Weeds: Natural Histories and Distributions. New York: Wiley.

Hyatt LA, Evans AS. 1998. Is decreased germination fraction associated with risk of sibling competition? Oikos 83:29–35.

Inui H, Ueyama Y, Shiota N, Ohkawa Y, Ohkawa H. 1999. Herbicide metabolism and cross-tolerance in transgenic potato plants expressing human CYP1A1. Pestic Biochem Physiol 64:33–46.

Jepson I, Holt DC, Roussel V, Wright SY, Greenland AJ. 1997. Transgenic plant analysis as a tool for the study of maize glutathione transferases. In: Hatzios KK, ed. Regulation of Enzymatic Systems Detoxifying Xenobiotics in Plants. Dordrecht: Kluwer, pp 313–323.

Joel DM, Kleifeld Y, Losner-Goshen D, Herzlinger G, Gressel J. 1995. Transgenic crops against parasites. Nature 374:220–221.

Kareiva P, Parker IM, Pascual M. 1996. Can we use experiments and models in predicting the invasiveness of genetically engineered organisms? Ecology 77: 1670–1675.

Kaushal P, Ravi. 1998. Crossability of wild species of *Oryza* with *O. sativa* cvs PR 106 and Pusa Basmati 1 for transfer of bacterial leaf blight resistance through interspecific hybridization. J Agric Sci 130:423–430.

Keeler KH, Turner CE, Bollick MR. 1996. Movement of crop transgenes into wild plants. In: Duke SO, ed. Herbicide Resistant Crops: Agricultural, Environmental, Economic, Regulatory, and Technical Aspects. Boca Raton, FL: CRC Press, pp 303–330.

Kendall HW, Beachy R, Eisner T, Gould F, Herdt R, Raven PH, Schell JS, Swaminathan MS. 1997. The Bioengineering of Crops. Washington, DC: The World Bank.

Kerlan MC, Chevre AM, Eber F. 1993. Interspecific hybrids between a transgenic rapeseed (*Brassica napus*) and related species: cytogenetical characterization and detection of the transgene. Genome 36:1099–1106.

Khan MS, Maliga P. 1999. Fluorescent antibiotic resistance marker for tracking plastid transformation in higher plants. Nat Biotechnol 17:910–915.

Khush GS. 1997. Origin, dispersal, cultivation and variation of rice. Plant Mol Biol 35:125–134.

Kimber G, Sears ER. 1987. Evolution of the genus *Triticum* and the origin of cultivated wheat. In: Heyne EG, ed. Wheat and Wheat Improvement. Agronomy Monograph No 13. Madison, WI: Agronomy Society of America, pp 154–164.

Kimber G, Feldman M. 1987. Wild wheats, an introduction. Special Report 353, College of Agriculture, Univiversity of Missouri, Columbia.

Kling J. 1996. Could transgenic supercrops one day breed superweeds? Sci 274: 180–181.

Komossa D, Gennity I, Sandermann H Jr. 1992. Plant metabolism of herbicides with C-P bonds: glyphosate. Pestic Biochem Physiol 43:85–94.

Kusaba S, Fukumoto M, Honda C, Yamaguchi I, Sakamoto T, Kano-Murakami Y. 1998. Decreased GA(1) content caused by the overexpression of *OSH1* is accompanied by suppression of GA_{20}^- and oxidase gene expression. Plant Physiol 117: 1179–1184.

Landbo L, Andersen B, Jorgensen RB. 1996. Natural hybridization between oilseed rape and a wild relative: hybrids among seeds from weedy *B. campestris*. Hereditas 125:89–91.

Langevin SA, Clay K, Grace JB. 1990. The incidence and effects of hybridization between cultivated rice and its related weed red rice (*Oryza sativa* L.). Evolution 44:1000–1008.

Lebrun M, Leroux B, Sailland A. 1997. A chimeric gene for the transformation of plants. US patent 5,633,448.

Lee LJ, Ngim J. 2000. A first report of glyphosate resistant (*Eleucine indica*) in Malaysia. Pest Manage Sci 56:336–340.

Lee C, Penner D, Renner K. 2000. The influence of glyphosate on glyceollin production in glyphosate-resistant soybeans. Weed Sci Soc Am Abstr 40:293.

Lefol E, Fleury A, Darmency H. 1996. Gene dispersal from transgenic crops II Hybridization between oilseed rape and the wild hoary mustard. Sex Plant Reprod 9:189–196.

Lermontova I, Grimm B. 2000. Overexpression of plastidic protoporphoyrinogen IX oxidase leads to resistance to the diphenyl-ether herbicide acifluorfen. Plant Physiol 122:75–83.

Lister RM, Murant AF. 1967. Seed transmission of nematode-borne viruses. Ann Appl Biol 59:59–62.

Lundberg S, Nilsson P, Fagerstrom T. 1996. Seed dormancy and frequency dependent selection due to sib competition: the effect of age specific gene expression. J Theor Biol 183:9–17.

Lutman PJW. 1993. The occurrence and persistence of volunteer rapeseed. *Brassica napus*). Aspects Appl Biol 35:29–36.

Majumder ND, Ram T, Sharma AC. 1997. Cytological and morphological variation in hybrid swarms and introgrossed population of interspecific hybrids (*Oryza rufipogon* Griff × *Oryza sativa* L) and its impact on evolution of intermediate types. Euphytica 94:295–302.

Malan C, Greyling MM, Gressel J. 1990. Correlation between antioxidant enzymes, CuZn SOD and glutathione reductase, and environmental and xenobiotic stress tolerance in maize inbreds. Plant Sci 69:157–166.

Mannerlöf M, Tuvesson S, Steen P, Tenning P. 1997. Transgenic sugar beet tolerant to glyphosate. Euphytica 94:83–91.

Mariam AL, Zakri AH, Mahani MC, Normah MN. 1996. Interspecific hybridization of cultivated rice, *Oryza sativa* L with the wild rice, *O. minuta* Presl. Theor Appl Genet 93:664–671.

Martin C, Juliano A, Newbury HJ, Lu BR, Jackson MT, Ford Lloyd BV. 1997. The use of RAPD markers to facilitate the identification of *Oryza* species within a germplasm collection. Genet Resour Crop Evol 44:175–183.

Maxwell BD, Roush ML, Radosevich S. 1990. Predicting the evolution and dynamics of herbicide resistance in weed populations. Weed Technol 4:2–13.

McMullan PM, Daun JK, DeClercq DR. 1994. Effect of wild mustard (*Brassica kaber*) competition on yield and quality of triazine-tolerant and triazine-susceptible canola (*Brassica napus* and *Brassica rapa*). Can J Plant Sci 74:369–374.

Mikkelsen TR, Jensen J, Jorgensen RB. 1996. Inheritance of oilseed rape (*Brassica napus*) RAPD markers in a backcross with progeny with *Brassica campestris*. Theor Appl Genet 92:492–497.

Moody ME, Mack RN. 1988. Controlling the spread of plant invasions: the importance of nascent foci. J Appl Ecol 25:1009–1021.

Nishijima T, Katsura N, Koshioka M, Yamazaki H, Nakayama M, Yamane H, Yamaguchi I, Yokota T, Murofushi N, Takahashi N, Nonaka M. 1998. Effects of gibberellins and gibberellin-biosynthesis inhibitors on stem elongation and flowering of *Raphanus sativus* L. J Japan Soc Hortic Sci 67:325–330.

Noguchi T, Fujioka S, Choe S, Takatsuto S, Yoshida S, Yuan H, Feldmann KA, Tax FE. 1999. Brassinosteroid-insensitive dwarf mutants of *Arabidopsis* accumulate brassinosteroids. Plant Physiol. 121:743–752.

Oard JH, Linscombe SD, Bravermann MP, Jodari F, Blouin DC, Leech M, Kohli A, Vain P, Cooley JC, Christou P. 1996. Development, field evaluation, and agronomic performance of transgenic herbicide resistant rice. Mol Breed 2:359–368.

Owen MDK. 1997. North American development of herbicide resistant crops. Brighton Crop Protection Conference, Weeds, pp 955–963.

Padgette SR, Re DB, Barry GF, Eichholtz DA, Delannay X, Fuchs RL, Kishore GM, Fraley RT. 1996. New weed control opportunities: development of soybeans with a Roundup Ready™ gene. In: Duke SO, ed. Herbicide-Resistant Crops: Agricultural, Economic, Environmental, Regulatory, and Technological Aspects. Boca Raton, FL: CRC Press, pp 53–84.

Peng J, Richards DE, Hartle NM, Murph GP, Devos KM, Flintham JE, Beales J, Fish LJ, Worland AJ, Pelica F, Sudhaka D, Christou P, Snape JW, Gale MD, Harberd NP. 1999. 'Green revolution' genes encode mutant gibberellin response modulators. Nature 400:256–261.

Powell M. 1997. Science in sanitary and phytosanitary dispute resolution. Discussion Paper 97-50, Resources for the Future, Washington, DC.

Powles SB, Holtum JAM, eds. 1994. Herbicide Resistance in Plants. Boca Raton, FL: Lewis.

Powles SB, Loraine-Colwill DF, Dellow JJ, Preston C. 1998. Evolved resistance to glyphosate in rigid ryegrass (*Lolium rigidum*) in Australia. Weed Sci 46:604–607.

Pratley J, Urwin N, Stanton R, Baines P, Broster J, Cullis K, Schafer D, Bohn J, Krueger R. 1999. Resistance to glyphosate in *Lolium rigidum* 1. Bioevaluation. Weed Sci 47:405–411.

Price JS, Hobson RN, Nealle MA, Bruce DM. 1996. Seed losses in commercial harvesting of oilseed rape. J Agric Res 65:183–191.

Reboud X, Zeyl C. 1994. Organelle inheritance in plants. Heredity 72:132–140.

Rieger MA, Preston C, Powles SB. 1999. Risks of gene flow from transgenic herbicide-resistant canola (*Brassica napus*) to weedy relatives in southern Australian cropping systems. Aust J Agric Res 50:115–128.

Robson PRH, McCormac AC, Irvine AS, Smith H. 1996. Genetic engineering of harvest index in tobacco through overexpression of a phytochrome gene. Nat Biotechnol 14:995–998.

Roush ML, Radosevich SR, Maxwell BD. 1990. Future outlook for herbicide-resistance research. Weed Technol 4:208–214.

Rubin E, Levy A. 1999. Database against database: a computational approach to assess the fequency of horizontal gene transfer. The Second Symposium on Plant Genomics, Maagan, Israel, PF 3.

Saari LL, Cotterman JC, Thill DC. 1994. Resistance to acetolactate synthase inhibiting herbicides. In: Powles SB, Holtum JAM, eds. Herbicide Resistance in Plants. Boca Raton, FL: Lewis, pp 83–139.

Sankula S, Braverman MP, Linscombe SD. 1997. Response of *bar*-transformed rice (*Oryza sativa*) transformed with the *bar* gene and red rice to glufosinate application timing. Weed Technol 11:303–307.

Sankula S, Braverman MP, Oard JH. 1998. Genetic analysis of glufosinate resistance in crosses between transformed rice (*Oryza sativa*) and red rice (*Oryza sativa*). Weed Technol 12:209–214.

Schaller H, Bouvier-Navé P, Benveniste P. 1998. Overexpression of an *Arabidopsis* cDNA encoding a sterol-C24-methyltransferase in tobacco modifies the ratio of 24-methyl cholesterol to sitosterol and is associated with growth reduction. Plant Physiol 118:461–469.

Scheffler JA, Parkinson R, Dale PJ. 1995. Evaluating the effectiveness of isolation distances for field plots of oilseed rape (*Brassica napus*) using a herbicide resistant transgene as a selectable marker. Plant Breed 114:317–321.

Scott SE, Wilkinson MJ. 1999. Low probability of chloroplast movement from oilseed rape (*Brassica napus*) into wild *Brassica rapa*. Nat Biotechnol 17:390–392.

Seefeldt SS, Zemetra R, Young FL, Jones SS. 1998. Production of herbicide-resistant jointed goatgrass (*Aegilops cylindrica*) × wheat (*Triticum aestivum*) hybrids in the field by natural hybridization. Weed Sci 46:632–634.

Sharon A, Amsellem Z, Gressel J. 1992. Glyphosate suppression of induced defense responses: increased susceptibility of *Cassia obtusifolia* to a mycoherbicide. Plant Physiol 98:654–659.

Siminsky B, Corbin FT, Ward ER, Fleishmann TJ, Dewey R. 1999. Expression of a soybean cytochrome P_{450} monooxygenase cDNA in yeast and tobacco enhances the metabolism of phenylurea herbicides. Proc Nat Acad Sci USA 96:1750–1755.

Sindel BM. 1997. Outcrossing of transgenes to weedy relatives. In: McLean GD, Waterhouse PM, Evans G, Gibbs MJ, eds. Commercialisation of Transgenic Crops: Risk, Benefit and Trade Considerations. Canberra: Cooperative Research Center for Plant Sciences and Bureau of Resource Science, pp 43–81.

Skipsey M, Andrews CJ, Townson JK, Jepson I, Edwards R. 1997. Substrate and thiol specificity of a stress-inducible glutathione transferase from soybean. FEBS Lett 409:370–374.

Snow AA, Andersen B, Jorgensen RB. 1999. Costs of transgenic herbicide resistance introgressed from *Brassica napus* into weedy *Brassica rapa*. Mol Ecol 8:605–615.

Souza-Machado V. 1982. Inheritance and breeding potential of triazine tolerance and resistance in plants. In: LeBaron HM, Gressel J, eds. Herbicide Resistance in Plants. New York: Wiley, pp 257–273.

Stalker DM, Kiser JA, Baldwin G, Coulombe B, Houck CM. 1996. Cotton weed control using the BXN system. In: Duke SO, ed. Herbicide Resistant Crops: Agricultural, Environmental, Economic, Regulatory, and Technical Aspects. Boca Raton, FL: CRC Press, pp 93–106.

Steber CM, Cooney SE, McCourt P. 1998. Isolation of the GA-response mutant *sly1* as a suppressor of ABI1-1 in *Arabidopsis thaliana*. Genetics 149:509–521.

Streber WR, Willmitzer L. 1989. Transgenic tobacco plants expressing a bacterial detoxifying enzyme are resistant to 2,4-D. Biotechnology 7:811–816.

Streber WR, Kutschka U, Thomas F, Pohlenz HD. 1994. Expression of a bacterial gene in transgenic plants confers resistance to the herbicide phenmedipham. Plant Mol Biol 25:977–987.

Suh HS, Sato YI, Morishima H. 1997. Genetic characterization of weedy rice (*Oryza sativa* L) based on morpho-physiology, isozymes and RAPD markers. Theor Appl Genet 94:316–321.

Thill DC, Mallory-Smith CA. 1997. The nature and consequence of weed spread in cropping systems. Weed Sci 45:337–342.

Timmons AM, Charters YM, Crawford, JW, Burn D, Scott SE, Dubbels SJ, Wilson, NJ, Robertson A, O'Brien ET, Squire GR, Wilkinson MJ. 1996. Risks from transgenic crops. Nature 380:487.

Tissier A, Marillonnet S, Klimyuk V, Patel K, Torres M, Murphy G, Jones J. 1999. Multiple independent suppressor-mutator transposon insertions in *Arabidopsis*: a tool for functional genomics. Plant Cell 11:1841–1852.

Tomlin C, ed. 1994. The Pesticide Manual. Farnham, UK: British Crop Protection Council.

Torii KU, McNellis TW, Deng XW. 1998. Functional dissection of *Arabidopsis* COP1 reveals specific roles of its three structural modules in light control of seedling development. EMBO J 17:5577–5587.

Tran M, Baerson S, Brinker R, Casagrande L, Faletti M, Feng Y, Nemeth M, Reynolds T, Rodriguez D, Schafer D, Stalker D, Taylor N, Teng Y, Dill G. 1999. Characterization of glyphosate resistant *Eleusine indica*. Proceedings 17th Asian-Pacific Weed Science Society Conference, Bangkok, pp 527–536.

U N. 1935. Genome analysis in *Brassica* with special reference to the experimental formation of *B. napus* and the peculiar mode of fertilization. Jpn J Bot 7:389–452.

Van der Schaar W, Blanco CLA, Kloosterziel KM, Jansen RC, Van Ooijen JW, Koornneef M. 1997. QTL analysis of seed dormancy in *Arabidopsis* using recombinant inbred lines and MQM mapping. Heredity 79:190–200.

Vasil IK. 1996. Phosphinothricin resistant crops. In: Duke SO, ed. Herbicide Resistant Crops: Agricultural, Environmental, Economic, Regulatory, and Technical Aspects. Boca Raton, FL: CRC Press, pp 85–92.

Webb SE, Appleford NEJ, Gaskin P, Lenton JR. 1998. Gibberellins in internodes and ears of wheat containing different dwarfing alleles. Phytochemistry 47:671–677.

Weeks JT, Anderson OD, Blechl AE. 1993. Rapid production of multiple independent lines of fertile transgenic wheat (*Triticum aestivum*). Plant Physiol 102:1077–1084.

Wrubel RP, Gressel J. 1994. Are herbicide mixtures useful for delaying the rapid evolution of resistance? A case study. Weed Technol. 8:635–648.

Yamaguchi S, Sun TP, Kawaide H, Kamiya Y. 1998. The GA$_2$ locus of *Arabidopsis thaliana* encodes *ent*-kaurene synthase of gibberellin biosynthesis. Plant Physiology 116:1271–1278.

Ye B, Gressel J. 2000. Transient, oxidant-induced antioxidant transcript and enzyme levels correlate with greater oxidant-resistance in paraquat-resistant *Conyza bonariensis*. Planta 211:50–61

Zemetra RS, Hansen J, Mallory-Smith CA. 1998. Potential for gene transfer between wheat (*Triticum aestivum*) and jointed goatgrass (*Aegilops cylindrica*). Weed Sci 46:313–317.

Zupan JR, Zambryski P. 1995. Transfer of T-DNA to *Agrobacterium* to the plant cell. Plant Physiol 107:1041–1047.

26

Plants and Environmental Stress Adaptation Strategies

Hans J. Bohnert
Department of Plant Biology, University of Illinois,
Urbana, Illinois, U.S.A.

John C. Cushman
Department of Biochemistry, University of Nevada,
Reno, Nevada, U.S.A.

I. INTRODUCTION

Being rooted in place, plants cannot evade attacks by pathogens and predators and they are permanently exposed to their physical environment, with changes in seasonal and diurnal cycles and extreme and stressful climatic variations. Among the abiotic factors, on which we will focus, a number of conditions are meaningful stresses because they can be life threatening and in combination determine plant distribution and the productivity of crop species (1). Prolonged drought, high or fluctuating salinity, and low and freezing temperatures account for most production losses, but flooding, high light, ozone, ion deficiency or imbalance, heavy metals, and soil structure are other factors that threaten plant life.

The foremost abiotic stress is water deficit. Water must supply the needs of both agriculture and a still increasing human population. This competition for water resources poses problems in areas where water is an increasingly precious commodity, most significantly in Australia, countries of the Middle East, North Africa, the west and midwest of the United States, parts of the Indian subcontinent, and central Asia (http://www.undp.org/popln/fao/water.html). Even in areas with typically ample precipitation, transient drought can lead to economic hardship for farmers or inconvenience urban populations. At the same time, predictions of possible global climatic changes seem to indicate that the distribution of rain might become more erratic than in the past. Lack of water prolongs the agricultural growing cycle, increases vulnerability to pathogens, and ultimately results in decreased yield. Also, agricultural practices jeopardize productivity in many irrigated areas because long-term irrigation leads to the buildup of sodium chloride and other salts in the soil, with up to half of the area under irrigation affected (2,3). What is happening in many growing areas with elaborate irrigation schemes is reminiscent of events that led to the decline of ancient civilizations, the former "fertile crescent" from Egypt to Persia being the prime example. How will we provide a stable supply of food, feed, and fiber for a human population that may reach 9 billion people within the next two generations (4)? One approach is to study plant responses to abiotic stresses and to understand why and how some plants can tolerate water deficits while maintaining at least some productivity.

II. DEFINING THE PROBLEM AND SEARCHING FOR SOLUTIONS

How to obtain water is the problem. For continued vegetative growth and the development of reproductive organs under stress, plants must, above all, obtain water. Each of many diverse mechanisms must be subordinate to this essential goal. When stomata close to limit water loss, a series of events adjusts photosynthesis, carbon fixation, and carbohydrate transport, initiating processes to maintain the integrity of the photosynthetic and carbon fixation apparatus (5–7). Objectives must be to obtain more water or to separate sodium from the essential macronutrient, potassium, while being able to take up water or at low temperature assure that water arrives in the leaves and apical meristems.

In the search for solutions, plant stress biology has relied on physiological studies for many years. More recently, progress has resulted from molecular genetic and genetic analyses, which must be correlated with physiological observations of biophysical and biochemical principles outlined by generations of researchers (8). The last 10 years have brought a principal change in our views by incorporating molecular genetics in the analysis. Transcript abundance and stress-induced change in many species and the complete sequence of the *Arabidopsis* genome have made it apparent that many genes for proteins that provide abiotic stress tolerance are common to all plants. In fact, the evolutionary appearance of entirely novel genes and mechanisms seems unlikely considering that plant speciation is a recent evolutionary event (9). Genetic uniformity indicates that tolerance is determined by how effectively, or how fast, the biochemical response machinery can be brought into play. Stress tolerance, then, seems largely determined by stress sensing, signal transduction, and the networking of sensory stimuli rather than by the biochemical hardware (10–12). However, it may be that signaling pathways evolved in an order-, family-, or species-specific fashion, and this may determine tolerance.

At present, we stand at yet another threshold because technologies have emerged that permit the analysis of plant stress responses on the level of entire genomes. Expressed sequence tags (ESTs), microarrays, the clustering of expression patterns, and the mapping of quantitative trait loci (QTLs) are techniques for assessing the contribution of many genes to stress responses, tolerance, and yield under stress (13–17). Conclusions can now be drawn that supersede correlative analyses. Now is the time to examine the physiological principles established in the past and to integrate the results from physiological studies with molecular genetics and genomics studies for the genetic engineering of crop species that tolerate environmental insults.

III. ENVIRONMENTAL STRESS ADAPTATION STRATEGIES

A. How Many Strategies Are There?

It would be impossible to list all reported data sets on stress adaptations, each of which provides a glimpse of how plants maintain growth during stressful times or at least how they survive, but there are now sufficient data to identify probable protective mechanisms in a general sense. In broad categories, one can identify three levels of response: (1) the immediate reactions, (2) adjustments to reach a new equilibrium, and (3) long-term developmental changes. Most is known about the "downstream" reactions, the induction of new pathways or the repression of enzymes and biochemical pathways, which have been established during normal growth and which functioned prior to stress (Fig. 1). Immediate stress reactions, which represent the first line of defense, are mediated by the cellular defense mechanisms that are evolutionarily well conserved. A second category, including sensing and signaling circuits, initiates the immediate response, establishes a new equilibrium for further growth, and prepares for long-term changes. Responding to stress signaling and the concomitantly altered hormonal state are developmental changes that may provide the best answers to long-term stress.

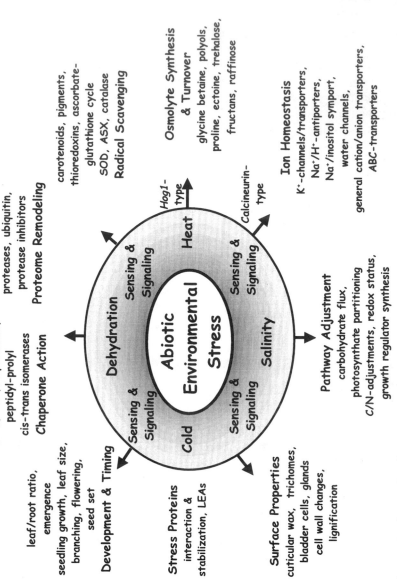

FIGURE 1 Strategies for stress tolerance. The scheme symbolizes sensing and signaling pathways for dehydration, cold, heat, and salinity stress recognition that are largely unknown. A Hog1-like pathway leading to osmolyte synthesis and a calcineurin-type pathway for ion homeostasis, both similar to yeast signal transduction pathways, are indicated. Included are boxes that summarize biochemical, metabolic, molecular, and developmental elements that contribute to stress tolerance.

B. Protection of Downstream Reactions

Several mechanisms are known for which the evidence for a protective role seems clear: (1) scavenging of radical oxygen species (18,19), (2) controlled ion and water uptake (11,20–24), and (3) management of accumulating reducing power through adjustments in carbon-nitrogen allocation and the accumulation of compatible solutes (25–29). Responses on this level of defense are immediate, taken in an enhanced form from the repertoire of daily plant life. They encompass, for example, the down-regulation of the light-harvesting and water-splitting complexes, enhancement of the water-water cycle, increased synthesis and action of radical-scavenging enzymes and nonenzymatic scavenging molecules, the storage of reducing power in a reusable form (e.g., as proline, polyols, or fructans), and the engagement of proton pumping and ion and water transport systems. Also, the stabilization of proteins and membranes through enhanced chaperone action belongs into this category of reactions (30,31). Such mechanisms accomplish ordinary adjustments, although reaction bandwidth, speed, and magnitude seem to delineate species-specific boundaries. Several of these emergency measures have been enhanced in transgenic models with limited, albeit recognizable, success (28,32–34), but prolonged stress requires additional mechanisms. This view then assumes a common set of genes for stress responses, similar or identical for glycophytes, halophytes, and xerophytes, as well as low-temperature tolerant species. The differences between species lie in how, how fast, or how resolutely the response can be initiated, elaborated, or sustained.

1. Osmotic Adjustment

This term was coined to explain functions of accumulating metabolites. As the external osmotic potential decreases, internal accumulations restore the differential necessary for the plant to take up water. Potassium, if available, can serve in this function. Also, amino acids and some amino acid derivatives, sugars, acyclic and cyclic polyols, fructans, and quaternary amino and sulfonium compounds accumulate (28,35–38). Typically, pathways leading to their synthesis are connected to pathways in general metabolism with high flux rates. Examples are the proline biosynthetic pathway (36), glycine betaine synthesis (35,39), sulfonium compounds (40), and the pathway leading to the methylated inositol, D-pinitol (41–43). In the past, it had been assumed that the enzymatic activities that lead to accumulation of one or the other metabolite indicated differences in gene complement between species. This seems not to be the case; rather, different plant species activate different sets of genes (40). The osmotic effect of high accumulation is documented by the ice plant (*Mesembryanthemum crystallinum*), although this may be an extreme case. This plant, a salt accumulator, contains sodium in vacuoles of epidermal bladder cells at a concentration exceeding 1 M and a methylated inositol, D-pinitol, in the cytosol at a concentration that may be as high as 600–800 mM (44–46).

Whether the function of these accumulators is simply or only in mass action is questionable. In experiments with transgenic plants (28,32,47), low to moderate accumulation of mannitol, glycine betaine, D-ononitol, or sorbitol has had a measurable protective effect. Experiments indicate that these osmolytes at low concentration within cells also have a protective function by preventing the formation of hydroxyl radicals (48–50). Another possibility, yet to be rigorously tested, is that transiently accumulating metabolites could defuse reductive power during stress (26,51). As the stress-induced decline in growth increases the ratio of $[NAD(P)H + H^+]$ to $[NAD(P)^+]$, the accumulation of an osmolyte could lead to adjustment of the cellular redox state. Thus, current understanding places less emphasis on the previously assumed function of osmolytes, the stabilization of proteins, protein complexes, or membranes by mass action, or either protection or replacement of the water shell around proteins (52,53).

2. Radical Scavenging Capacity

The production of reactive oxygen species (ROS) is unavoidable in chloroplasts, but ROS are also produced in mitochondria, peroxisomes, and cytosol in all cells. The ROS, including singlet oxygen, superoxide, hydrogen peroxide, and hydroxyl radicals, react with and can damage proteins, membrane lipids, and other cellular components (19,54). The ROS also serve as signaling molecules, for example, in the recognition of attack by pathogens (55). Under water deficit conditions, radical production increases in plants (18,19). ROS toxicity is clearly evident during drought (18), chilling stress (56), and high salinity (50) (see Ref. 28 for a review). Superoxide and H_2O_2 concentrations increase during drought and at low temperature (57). Enhanced production of ROS results in an increase in lipid peroxidation, as documented by a more than fivefold increase of malonaldehyde production in wheat (58). The concentration of nonbound iron increases under drought stress (59), which stimulates the production of hydroxyl radicals in the presence of H_2O_2 in a Fenton reaction. Compared with superoxide and H_2O_2, hydroxyl radicals oxidize a variety of molecules at nearly diffusion-controlled rates. Finally, levels of nonenzymatic radical scavengers, such as ascorbate, carotenoids, flavonoids, sugar polyols, and proline, increase and can complement the existing enzyme-based protection systems, such as superoxide dismutase, ascorbate peroxidase, or catalase. Together, the scavenging enzymes and nonenzymatic antioxidants provide sufficient protection under normal growth conditions and can handle moderate increases of ROS, unless long-term stress exceeds the detoxification capacity (6,18).

Evidence for a protective effect of enhanced ROS scavenging systems has been provided by the overexpression of an enzyme with the combined activities of glutathione S-transferase (GST), and glutathione peroxidase (GPX) (60). With doubling of the GST-GSX activity in transgenic tobacco, the seedlings and plants showed significantly faster growth than the wild type during chilling and salt stress episodes. The increased enzyme activities resulted in higher amounts of oxidized glutathione (GSSG) in the stressed plants, indicating that the oxidized form could provide an increased sink for reducing power. Even more striking is the detection of a gain-of-function mutation, improved growth under salt stress, in *Arabidopsis* (61). The recessive mutation is characterized by constitutively increased activities for ROS-scavenging enzymes. In a different strategy, the transgenic increase of a catalase in tobacco resulted in protection of leaves against paraquat-induced bleaching (62). The experiment indicated as a major stress factor the depletion of chloroplastic ASA affecting ascorbate peroxidase amounts.

Another set of experiments shed light on the relationships between ROS and the accumulation of polyols. When a bacterial gene (*mtlD*) encoding mannitol-1-phosphate dehydrogenase was modified so that the enzyme was expressed in chloroplasts, transgenic tobacco contained 60–100 mM mannitol in the plastids. Using transgenic plants, freshly prepared cells, or a thylakoid in vitro system, the protective effect exerted by mannitol on photosynthesis characteristics could be shown (50,63). The presence of mannitol resulted in increased resistance to oxidative stress generated by methylviologen, and cells exhibited significantly higher CO_2 fixation rates than controls during stress. After impregnation of tissues, isolated cells, or a reconstituted thylakoid system with dimethyl sulfoxide, a hydroxyl radical generator, mannitol-containing cells showed a lower rate of methanesulfinic acid production than the wild type, indicating that mannitol acted specifically as a hydroxyl radical scavenger. Calvin cycle enzymes sustained primary damage rather than components of the light-harvesting and electron transfer systems. Phosphoribulokinase (PRK) and probably other SH enzymes of the Calvin cycle showed sensitivity to hydroxyl radicals, and the activity of PRK was protected by the presence of mannitol (63,64). At present, a possible interpretation is that mannitol reduces the amount of hydroxyl radicals produced in Fenton reactions, possibly by complexing free iron (Fe^{2+}) (G. Bongi, personal communication).

3. Water and Ion Relations

Abiotic stresses affect plant water and ion relations, and both are intricately connected. The most crucial aspects during drought and salinity stress are continued uptake of water and potassium and the exclusion or internal sequestration of sodium ions. Lowering the plant's internal osmotic potential to match or exceed that of the medium is the generally assumed function of potassium and the accumulating osmolytes. In reality, little is known about the relationships between the adaptive responses that initiate solute production and accumulation, water transport (either loss, acquisition, or movement), and ion uptake under water deficit (or sodium uptake or exclusion in high salinity). Loss of turgor following water deficit caused either by lowering of water uptake through roots or continued evapotranspiration through stomata very likely constitutes a signal, possibly transduced through phosphorylation cascades similar to the yeast high osmolarity glycerol (HOG) and calcineurin osmotic and ionic signaling pathways (10,65).

a. *Water Channels.* Such channels (aquaporins) enhance membrane permeability to water in both directions, depending on osmotic pressure differences across a membrane. They are present in all organisms as facilitators for the movement of water and metabolites, such as urea or glycerol (65–67). In plants, two subfamilies can be distinguished: aquaporins targeted to the plasma or vacuolar membrane, termed PIP and TIP, respectively. As in animal cells (68), plant aquaporins seem to be controlled by posttranslational protein modification and cycling through the endomembrane system (69–72). For example, a putative plasma membrane aquaporin from spinach is reversibly phosphorylated at multiple sites in response to changes in calcium and a lowering of the apoplastic water potential (69). The function of water channels in maintaining water balance under osmotic stress is still debated. Their presence increases water movement across a membrane by a certain factor, which is different for each channel, but water movement through lipid membranes alone is still substantial. The significance of these channels seems to lie in determining the flux rate of water and solutes through tissues rather than on their maintaining water relations of individual cells. Important to the stress aspect is that the protein amount and the spatial expression patterns of water channels are regulated. There is evidence that location may vary with changes in physiological state (67). Plasma membrane–located proteins appear in membrane fractions that are cell internal under salt stress in *Mesembryanthemum* (72) (H. J. Bohnert, unpublished data). High salinity leads to the disappearance of tonoplast-localized channels, whereas drought causes the decline of plasma membrane–localized water channel proteins (72) (R. Vera-Estrella, B. Barkla, H. J. Bohnert, unpublished results).

Plant genomes contain more than 23 genes for water (and/or solute) channels (73,74). Studies indicate that water channel transcripts are expressed with developmental and spatial complexity in diverse tissues

(73,75,76). An analysis of more than 3000 transcripts comparing control and salt-stressed maize ESTs documented that water channels are highly abundant in roots and roots contain a greater diversity of different transcripts than leaves with a more than 10-fold decline in transcript number following salinity stress (H. Wang, G. Zepeda, H. J. Bohnert, unpublished results). These dynamic changes indicate that control over water channel distribution or activity is important under stress conditions, but additional data are required to verify this. At present, we know only that transcriptional and post-transcriptional controls exist, which seem to involve synthesis, membrane trafficking, and possibly reversible insertion into or removal from membranes and aquaporin half-life.

     b. Proton-ATPases and Vacuolar Pyrophosphatase. Plasma membrane and vacuolar proton transporters play essential roles in plant salinity stress tolerance by maintaining the transmembrane proton gradient that ensures control over ion fluxes and pH regulation (77–79). Three proteins or protein complexes exist for this purpose: the plasma membrane (H^+)-adenosine triphosphatase (ATPase) (P-ATPase) and two vacuolar transport systems, an (H^+)-ATPase (V-ATPase) and a pyrophosphatase (PP$_i$ase).

     The plant P-ATPase (100 kDa) is represented by a family of more than 10 genes with homology to the yeast PMAs (80,81). As the main proton pump in the plasma membrane, this ATPase shows increased activity accompanying salt stress. Halophytic plants have been shown to increase pump activity under salt stress conditions more drastically than glycophytes (82,83), but we know little about regulatory circuits that lead to increased activity during salt stress. The V-ATPase, a multisubunit complex homologous to organellar, yeast (VMA), and bacterial F_0F_1-ATPases, has already been shown to be important in plant salinity tolerance. Electrophysiological studies revealed increased activity of this ATPase when cells or tissues from stressed plants were analyzed (84). Transcripts for subunits of the V-ATPase are up-regulated following salt shock (85–87). In *Mesembryanthemum*, V-ATPase activity increases severalfold following stress without affecting protein amount, although some of the subunits are expressed in increased abundance in complementary DNA (cDNA) libraries from the stressed plant (88–90). The response is specific for NaCl and is not elicited by mannitol-induced osmotic stress. Little is known about the function of the PP$_i$ase enzyme in stressed plants. PPase activity declined under salt stress in some species but increased in others (91–93). Overexpression of the Na^+/H^+ antiporter from *Arabidopsis* has been shown to confer salt tolerance to the salt-sensitive ena1 mutant of yeast; however, this phenotype required the expression of both a chloride ion transporter and an Na^+/H^+-antiporter (94).

c. Potassium Transporters and Channels. One possible passage for sodium across the plasma membrane is through transport systems for other monovalent cations (11,22–24). Among those, the most significant is the uptake system for potassium, the most abundant cation in the cytosol with important roles in plant nutrition, development, and physiological regulation. Physiological observations indicating a biphasic uptake of K^+ into roots (95) gave rise to the assumption that two uptake entities should be involved, a high-affinity system functioning at μM concentrations of external K^+ and a low-affinity system active in the mM range of potassium. In reality, a complex assortment of different proteins participate in potassium uptake. Analysis of transcripts, proteins, and electrophysiological studies indicated not only cell-, tissue-, and stress-specific expression differences but also regulation that allows some transport systems to function in both high- and low-affinity mode, depending on external potassium concentrations (96,97). In contrast to earlier assumptions, K^+ channels are highly selective against Na^+ (98) and thus play an insignificant role in inadvertent sodium uptake (22).

K$^+$ transporters, first isolated from wheat (HKT1) (99), operate at low external potassium and may mediate entry of sodium in saline soil. The high-affinity HKT1 was indicated as a K^+/Na^+ symporter with high-affinity binding sites for both K^+ and Na^+ (99) (see Ref. 23). Another line of evidence for the involvement of the high-affinity K^+ uptake system in salt tolerance came from the study of salt-sensitive mutants. The *sos1* mutant of *Arabidopsis thaliana* is hypersensitive to Na^+ and Li^+ and unable to grow on low potassium (100). The *Sos1* gene encodes a sodium/proton antiporter (J. K. Zhu, personal communication) different from the proteins detected by other groups (94,101). Other transporters in a different protein family complicate the picture even further; these take up potassium with dual affinity (102,103).

d. Sodium Transport Systems. Sodium is thought to enter plant cells by two different routes: through the apoplast or through the symplast. Multiple cellular uptake mechanisms through the plasma membrane have been documented that display varying amounts of permeability to Na^+ (23). In particular, voltage-insensitive monovalent cation (VIC) channels have been argued to play important roles in allowing the bulk of Na^+ influx (22,24,104). Therefore, genetic manipulation of VIC channels to improve their selectivity against Na^+ influx or alter their expression and regulation represents possible strategies for reducing Na^+ uptake (24).

The existence of an antiport system, connected to transmembrane proton gradients, had been observed before (105). Increased sodium/proton antiport activity during salt stress has been measured in several model systems, tissues, cells, and isolated vacuoles; it parallels the increase in proton-pumping V-ATPase activity (46,90,106,107). It was shown that sodium transport from cytosol to vacuoles is accomplished by a sodium/proton antiporter (34,53,101). These *Arabidopsis* antiporters seem to be associated with vacuolar or prevacuolar membranes, suggesting that sodium would either be deposited into the central vacuole or partitioned into an exocytotic pathway (108). The overexpression of this antiporter, NHE1, in *Arabidopsis* had a remarkably protective effect under salt stress conditions (34). Ultimately, however, the capacity to sequester Na^+ into the vacuole via Na^+/H^+ antiport activity will depend on H^+-ATPase and/or H^+-PPiase proton-pumping capacity. Thus, engineering strategies exploiting Na^+/H^+ antiport overexpression as a means of improving Na^+ compartmentation must also increase proton-pumping activities, as was demonstrated by the expression of a plant vacuolar H^+-PPiase in yeast (94). Furthermore, the inhibition of Na^+/H^+ antiport activity by high salinity can be influenced by the degree of membrane desaturation in membrane lipids. *Synechocystis* mutants deficient in desaturase genes and thus lacking polyunsaturated fatty acids were found to be more sensitive to salinity-mediated inhibition of photosystem II (PSII) activity (109).

Yet another pathway for sodium transport may exist that functions in concert with the antiporters. In *Mesembryanthemum*, the synthesis of *myo*-inositol and its transport through phloem to the roots are correlated with sodium and *myo*-inositol transport from roots to the leaves (43,110). Only in the leaves, sodium accumulates to high amounts in vacuoles and methylated inositols accumulate in the cytosol of mesophyll cells. Stress-induced sodium/*myo*-inositol symporters (111), together with Na^+/H^+-antiporters (34), could account for long-distance transport of sodium. Also, an $Na^+/$ *myo*-inositol symport mechanism provides an attractive hypothesis considering that the stress-enhanced passage of *myo*-inositol through the ice plant vasculature connects leaf photosynthesis to root sodium uptake and its sequestration into vacuoles (110).

e. The Essentiality of Calcium. Increasing calcium improves the salinity tolerance of plants. The effect is mediated through an increase of intracellular calcium, changes in vacuolar pH, and activation of the vacuolar Na^+/H^+-antiporter (112–114). The strict control over calcium concentrations in the cytosol and calcium storage in a number of locations (vacuole, mitochondria, endoplasmic reticulum) demonstrates the crucial role of calcium in plant salinity stress responses. An *Arabidopsis* mutant, *sos3*, with hypersensitivity to NaCl has been characterized (115). The mutant phenotype can be masked by the external addition of calcium, and it reveals the link between calcium and salinity stress tolerance. *Sos3* encodes a subunit of a calcineurin-related signal transduction chain (115) involved in altering $K^+/$ Na^+ selectivity.

C. Ubiquitous Cellular Stress Tolerance Mechanisms— Comparison of Models

A comparison of stress-regulated transcripts detected in model organisms seems to represent a common set of transcripts for the cellular complement of genes for immediate, stress-relieving emergency reactions (11,116). Many up-regulated genes reveal stress-regulated functions found in cyanobacteria and yeasts, which are similar to those regulated in xerophytic, halophytic, or glycophytic species. Figure 2 summarizes these in terms of proteins that

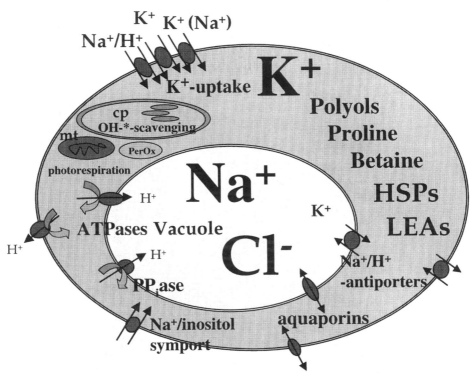

FIGURE 2 Cellular stress tolerance elements. The schematic drawing of a cell includes the vacuole into which sodium is partitioned, a chloroplast, mitochondrium, and peroxisome, which are prone to oxygen radical damage. Included are symbols for H^+-ATPases/PP$_i$ases that maintain transmembrane proton gradients and symbols for potassium uptake, sodium transport, and controlled water flux. Also included are protective proteins (chaperone [heat shock proteins] and late embryogenesis abundant [LEA] proteins) and products (glycine betaine, polyols, proline) of biochemical pathways that aid in osmotic adjustment. (Modified from Ref. 11.)

represent these "downstream" biochemical mechanisms, compiled from work mainly with *Arabidopsis* mutants, yeast, and several halophytic and xerophytic models.

Transgenic tobacco expressing selected genes has frequently been utilized for its usefulness as a biochemical model. Effects have been studied in transgenic plants in which several mechanisms were tested: enhanced osmotic adjustment, the expression of late embryogenesis abundant (LEA) proteins, or proteins leading to radical scavenging through the overexpression of foreign proteins (33). Moderate improvements of tolerance have been documented. In other models, *Arabidopsis thaliana, Medicago sativa,* and *Oryza sativa,* equally successful minor improvements of abiotic stress tolerance have been reported (38,117–121) (for reviews see Refs. 38,120,121).

Saccharomyces cerevisiae is by far the best model for understanding salinity tolerance mechanisms at the cellular level (20,33,122). Yeast is salt tolerant, and salt-sensitive mutants are readily obtained. These mutants allow the identification of important salinity tolerance genes, and complementation allows the identification of homologues from other species as well as providing useful strains for a variety of physiological and transgenic experiments at the cellular level. The yeast genome includes approximately 6000 open reading frames (123). Based on several analyses and the following considerations, approximately 100 ORFs seem to provide the basis for salinity stress tolerance (33). For example, through the deletion of *PBS2,* encoding a mitogen-activated protein (MAP) kinase kinase of the HOG osmotic stress signaling pathway, proteins controlled by this pathway could be documented by their disappearance from two-dimensional gels (124). The authors found 29 proteins affected by this deletion. Assuming that only a small number of the yeast proteins are of sufficiently high abundance to be visible, by extrapolation a number of approximately 100 closely stress-associated genes seems a reasonable estimate. That a larger number should aid in tolerance but not be absolutely necessary is reasonable, and this consideration puts the estimate of stress-affected genes in yeast in the low hundreds.

An analysis by microarray of all yeast ORFs indicated that approximately 300 transcripts, or 5% of the yeast genes, are significantly increased following salt stress and approximately 200 transcripts are down-regulated to a similar degree (J. Yale, H. J. Bohnert, unpublished results). When the cells experienced oxidative stress or heat shock, similar numbers of up-regulated transcripts were observed, although the overlap between the different stresses was only approximately 25%. The processes controlled by the known yeast genes are similar to those found or suggested as essential reactions of stressed plants. The majority of these genes, including those most strongly up-regulated, encode functions in energy metabolism, ion homeostasis, cell defense, chaperone functions, and transport facilitation. Approximately one third of the up-regulated yeast transcripts encode proteins of unknown function. We think that in plants a similar number of genes, maybe up to 10% of the plant genome or roughly 3000 genes, may participate in adaptive stress reactions (116).

IV. STRESS PROTECTION BASED ON SENSING AND SIGNAL TRANSDUCTION

During the past decade, many sensing and signaling components of abiotic stress signaling pathways have emerged. Such pathways play primary roles in reacting to short-term or "emergency" insults brought about by rapid changes in environmental conditions. More important, stress sensing and signaling pathways set into motion "intermediary" responses that lead to important, long-term adaptations for abiotic stress tolerance involving changes in plant development, growth dynamics, vegetative-floral meristem transitions, and seed production and maturation programs. Such long-term adaptations have perhaps the greatest impact for engineering the retention or improvement of agricultural productivity under abiotic stress conditions. Many sensing and signaling pathway components in plants, such as those that participate in MAP kinase and phosphatase cascades, have counterparts in animals and yeast (33,64,125–128). For example, a plant MAP kinase homologue effectively replaced the yeast HOG1 kinase under high osmotic stress conditions (129). Other signaling components are unique to plants, such as calcium-dependent protein kinases (130,131).

A histidine kinase has been identified as an osmosensor in *Arabidopsis* (132), which shares homology with other eukaryotic two-component response regulators. Stress-dependent regulation of an increasing number of elements in several signaling pathways has been reported (133–136). These pathways apparently function at different levels, are interconnected, and respond differentially to a variety of environmental stimuli through a series of signaling molecules including calcium, abscisic acid (ABA) (137), gibberellins (138), ethylene, phosphatidylinositols, cytokinins (38), brassinosteroids (139), or jasmonates (12,134,136,140–145).

Differences in how stresses are perceived and how such information is processed into biochemical responses between, for example, glycophytes and halophytes are likely to account for tolerance phenotypes. Halophytes may have evolved distinct stress recognition or signaling pathways and regulatory controls to confer stress protection. For example, the halophyte *Aster tripolium* possesses an Na^+-sensing system that down-regulates K^+ uptake by guard cells in response to high salt concentrations (146). This results in stomatal closure and prevents excessive Na^+ uptake via the transpiration stream. In contrast, nonhalophytic relatives appear to lack this specialized sensing ability and may actually respond to Na^+ ions by increasing stomatal apertures. Assuming that the *Arabidopsis* genome contains a gene complement that is largely similar to that present in the genome of the halophytic *Mesembryanthemum*, differences in stress perception and signaling may be explained by regulatory changes that evolved as a result of duplication and alterations in genes encoding signal transduction components (e.g., receptors, protein kinases, and phosphatases). Other distinctions may be the result of differences in the numbers and specificity of transcription factors and the *cis*-regulatory elements they recognize.

V. CHANGES IN GROWTH AND MORPHOLOGY UNDER STRESS

The third category of stress preparedness involves long-term adaptations. They are set in motion by signaling pathways that respond to immediate changes following stress and lead to changes in development and growth. The classical explanation of a stress-mediated decline in photosynthesis followed by a retardation of plant growth has been based on metabolic correlations, i.e., assuming altered development as a consequence of impaired metabolism (147). Another interpretation suggests that dehydration and salinity stress initiate a sequence of events in which plants move toward a state of minimal metabolic activity that persists until the stress is relieved (148). In this view, Ca^{2+} and ABA signaling lead to adjustments in proton gradients across membranes and the associated K^+ and water movements generate signals. Yet it seems probable that other, as yet unknown signaling pathways exist that actively slow or stop meristem activity, cell expansion, and growth. Suppression of growth may be an active process rather than a passive response to stress. Compared with normal developmental growth processes, stress-induced developmental changes could be indicators of tolerance mechanisms.

Apart from the cessation or delay of normal growth, stress reactions include a host of morphological changes. Often observed are a shift in the root-to-shoot ratio (149), changes in lignification, epidermis thickening and the development of trichomes, glands or wax additions to the cuticle, altered branching, different leaf shape and size, or stress-enhanced entry into flowering and seed production (14,150–156). Less obvious in their relevance to stress yet equally important will be mutations that alter meristematic activity by either fewer or more cell divisions or by changes in the elongation rate of meristematic cells or mutants in which the control of development by growth regulators (e.g., ABA) is altered.

Considering that mild abiotic stresses are generally accompanied by an acceleration of ontogeny, a mutational or transgenic approach to alter development in crop species can be imagined. Mutants in *Arabidopsis* and other species have been described that show such morphological and developmental changes, but to our knowledge these mutants have not been investigated from the perspective of stress tolerance. Another example, in a different category, is provided by the *Arabidopsis* or corn *cer* mutants, which are deficient in epidermal wax deposition (118,157). The plants desiccate and in extreme cases are not viable. A transgenic approach enhancing wax production has, to our knowledge, not been attempted. Similarly, we can imagine mutants with deeper roots, or roots with larger xylem capacity, or plants with large leaves early in the growing season and small leaves in hotter, dryer times later in the season. The realization of such concepts is not yet possible, but the completed sequence of the *Arabidopsis* genome, saturation mutagenesis in *Arabidopsis* and corn (158,159), and other ongoing high-throughput sequencing projects may soon provide the gene material to consider such work.

VI. STRESS GENOMICS

The analysis of gene expression changes is now possible on a much larger scale. Information on global expression changes in response to salinity, drought, and low temperature will soon be available from the analysis of transcript patterns in EST sequencing projects and microarray analyses (116).

Replacing the gene-by-gene approach for the analysis of mechanisms involved in stress responses, new technologies can now be employed that are based on the ease of DNA manipulation and the feasibility of analyzing large sets of genes facilitated by advanced bioinformatics tools. A first step toward categorizing genetic complexity in relationship to abiotic stresses includes gene discovery by large-scale, partial sequencing of randomly selected complementary DNA (cDNA) clones (ESTs). Such EST collections already exist for a number of plants (http://www.ncbi.nlm.nih.gov/dbEST/dbEST_summary.html), with the number of entries rapidly growing. They represent more than half of the total expected plant gene complement (~28,000) as estimated from the gene content of the entirely sequenced chromosome 2 in *Arabidopsis* (160). These collections are biased toward high to moderate abundance classes from different tissues, organs, or cells; developmental states; and various external stimuli and plant growth regulator treatments. Only a few studies have focused specifically on ESTs from rice or *Arabidopsis* exposed to abiotic stresses; however, these studies included rather small numbers of ESTs (161,162).

The scarcity of ESTs derived from cDNAs of stressed tissues of glycophytes indicates the absence of stress-relevant transcripts from the existing EST collections. An EST sequencing initiative with a stress focus is now under way using more than 40 cDNA libraries for specific cells, tissues, and developmental stages. The libraries were generated with RNA from salt-stressed glycophytes (*Arabidopsis* and *Oryza sativa*) and also *Mesembryanthemum* and *Dunaliella salina*, which serve as halophytic model species (listed at http://www.biochem.arizona.edu/BOHNERT/functgenomics/front2.html). Preliminary data sets are available at http://stress-genomics.org. As part of the gene discovery effort in maize, EST collections are also being established from cDNA libraries prepared from salt-stressed roots and shoots (http://www.zmdb.iastate.edu/zmdb/EST_project.html). Similar projects are under way with a focus on drought tolerance–specific ESTs (163,164). EST collections have also been initiated in the moss *Physcomitrella patens*, a model in which gene knockout is possible. To identify genes associated with desiccation tolerance, ESTs were characterized from moss protonema following ABA treatment. Most ESTs (69%) shared homology with known sequences, and many clones encoded proteins known to be induced during heat shock, cold acclimation, oxidative stress adaptation, and xenobiotic detoxification (165).

A sampling of approximately equal numbers of ESTs from well-watered and salinity-stressed *M. crystallinum* leaves revealed that stressed plants contain approximately 15% more functionally unknown genes than unstressed plants (166). This supports the notion that ESTs related to salinity stress are underrepresented in the nonredundant GenBank database. Sampling differences between unstressed and stressed plants revealed pronounced down-regulation in transcript abundance for components of the photosynthetic apparatus, water channels, or the translation machinery and concomitant up-regulation of constituents involved with proteome restructuring (e.g., proteases, ubiquitinases, chaperonins), osmotic and dehydration stress adaptation, and oxidative stress detoxification. Among the stress-dependently up-regulated *Mesembryanthemum* transcripts many known functions, as discussed before, appeared, but most dramatic is the increase of sequences that have no known function.

In all EST analyses, the precise function of a large proportion of the predicted proteins remains unknown (9) and requires that high-throughput sorting and categorizing of the information be completed before functional assignments can be attempted. One tool is provided by microarray analysis of large numbers of ESTs printed on slides and probed with fluorescence-labeled probes in comparative hybridizations between the stressed and unstressed states. Microarrays are currently the most parsimonious way to gain insights into the complexity of gene function and regulatory control and to approach functional analysis of unknown sequences (13,167). This will be aided by establishing differential expression patterns, by which functionally unidentifiable ESTs are clustered with the expression profiles of known sequences. The kinetics of expression and congruent changes in expression are expected to provide clues about functions (15,168). Large-scale cDNA microarray analyses of salinity stress–responsive gene expression profiles are now under way for *Mesembryanthemum*, rice, and *Arabidopsis* in our laboratories (http://stress-genomics.org). Although the initial analyses include only a small fraction of the entire gene complement, they will provide an important starting point for prioritizing transcripts for further functional analysis. The analyses are also expected to provide functional information about the role of unknown ORFs in cellular stress adaptation. Closer analysis of the data sets already indicated that a number of the up-regulated ORFs in yeast (J. Yale, H. J. Bohnert, unpublished results) have counterparts in *Synechocystis* PCC6803 (R. Burnap, unpublished results), *Aspergillus nidulans* (R. Prade, unpublished data), and *Mesembryanthemum* following salt stress (J. C. Cushman, H. J. Bohnert, unpublished data). Comparisons among these evolutionarily divergent organisms have begun to reveal the gene complement that delineates cell-based tolerance mechanisms.

VII. CONCLUSIONS

Enhancing environmental stress tolerance of crop species through classical breeding, while retaining high productivity, has not been successful. We see three strategies for turning apparent failure into success: enhancing metabolic reactions, by, for example, the transfer of multiple genes; the engineering stress sensory pathways; or the use of information about genome-wide expression changes, either QTL-anchored or microarray-based, for additional breeding attempts using marker-assisted selection strategies (169). All three scenarios make use of formidable technological breakthroughs in genomics, bioinformatics, and molecular genetics.

Engineering of sensing and signaling pathways and engineering of metabolism in crop plants are technically possible. Several prerequisites are in place, among them knowledge about the main targets of stress and genes that provide protection and advanced transformation technologies for the major crops. The ability now exists for the transfer of fragments of DNA, which could include hundreds of genes (170), to an increasing number of plants, including many important crop species (171,172). Removing coding regions by targeted gene disruption through homologous recombination poses challenges in plants but is no longer an insurmountable problem (173,174). However, one limitation still exists. Currently missing is a sufficiently large and complex set of plant promoters with cell-specific, tissue-specific, developmental stage–specific, and stress-inducible patterns of expression. Ongoing EST sequencing projects in conjunction with microarray analysis of stress-induced global gene expression patterns are expected to provide this information. Conducting these analyses using a collection of both halophytic and glycophytic model and crop species can be expected to eliminate the problem in the near future by providing a library of plant promoters with the desired specificities.

Even with the appropriate tools and a sufficient amount of correlative evidence, metabolic engineering remains risky without detailed knowledge about gene product function and regulation (39,175). At present, attempts at engineering salinity tolerance in multicellular organisms such as plants can be compared to following single, random pages of an instruction manual without page numbers. Analyzing yeast has provided individual pages and even some sections, but we still have to guess at how the sections fit together for understanding the cellular salinity stress tolerance manual in plants. For some sections, which cover developmental aspects of stress tolerance, tissue and organ interactions, and metabolic flux within tissues, for example, yeast will be a less suitable guide. Metabolic engineering of crop plants to improve stress tolerance should start with the isologs of some of the identified yeast genes, in both metabolic functions and stress signaling, and a number of the unknown genes for which functions are being established at present. Considerable time and effort will have to be devoted to testing appropriate expression characteristics of these transgenes so that the transgenic plants can become material for traditional breeding programs (176).

ACKNOWLEDGMENTS

We thank Ray Bressan, Mike Hasegawa, and Jian-Kang Zhu for many discussions. Work in our laboratories is or has been, off and on, supported by the U.S. Department of Energy, the National Science Foundation, the U.S. Department of Agriculture, and the Arizona and Oklahoma Agricultural Experiment Stations. Visiting scientists have been supported by CONACyT (Mexico City), Agricultural Research Council (London), Rockefeller Foundation (New York), Smithsonian Institution/Carnegie-Mellon Foundation (New York), Japanese Society for the Promotion of Science (Tokyo), New Energy Development Organization (Tokyo), Max-Kade-Foundation (New York), and Deutsche Forschungsgemeinschaft (Bonn). A genomics approach to finding stress tolerance determinants is funded by the National Science Foundation (DBI-9813330).

REFERENCES

1. JS Boyer. Plant productivity and environment. Science 218:443–448, 1982.
2. D Seckler, U Amarasinghe, D Molden, R da Silva, R Barker. World water demand and supply, 1990 to 2025: scenarios and issues. IWMI Research Report 19. Colombo, Sri Lanka: IWMI, 1998.

3. TJ Flowers, AR Yeo. Breeding for salinity tolerance in crop plants: where next? Aust J Plant Physiol 22:875–884, 1995.
4. GS Khush. Green revolution: preparing for the 21st century. Genome 42:646–655, 1999.
5. SP Long, S Humphries, PG Falkowski. Photoinhibition of photosynthesis in nature. Annu Rev Plant Physiol Plant Mol Biol 45:633–663, 1994.
6. CB Osmond, SC Grace. Perspectives on photoinhibition and photorespiration in the field: quintessential inefficiencies of the light and dark reactions of photosynthesis? J Exp Bot 46:1351–1362, 1995.
7. P Horton, AV Ruban, RG Walters. Regulation of light harvesting in green plants. Annu Rev Plant Physiol Plant Mol Biol 47:655–684, 1996.
8. J Levitt. Responses of Plant to Environmental Stress Chilling, Freezing, and High Temperature Stresses. 2nd ed. Mew York: Academic Press, 1980.
9. C Somerville, S Somerville. Plant functional genomics. Science 28:380–383, 1999.
10. K Shinozaki, K Yamaguchi-Shinozaki. Gene expression and signal transduction in water-stress response. Plant Physiol 115:327–334, 1997.
11. PM Hasegawa, RA Bressan, JK Zhu, HJ Bohnert. Plant cellular and molecular responses to high salinity. Annu Rev Plant Physiol Plant Mol Biol 51:463–499, 2000.
12. T Mizoguchi, K Ichimura, R Yoshida, K Shinozaki. MAP kinase cascades in *Arabidopsis*: their roles in stress and hormone responses. Results Probl Cell Differ 27:29–38, 2000.
13. B Lemieux, A Aharoni, M Schena. Overview of DNA technology. Mol Breed 4:277–289, 1998.
14. SA Quarrie, DA Laurie, J Zhu, C Lebreton, A Semikhodskii, A Steed, H Wisenboer, C Calestani. QTL analysis to study the association between leaf size and abscisic acid accumulation in droughted rice leaves and comparison across cereals. Plant Mol Biol 35:155–165, 1997.
15. MB Eisen, PT Spellman, PO Brown, D Botstein. Cluster analysis and display of genome-wide expression patterns. Proc Natl Acad Sci USA 95:14863–14868, 1998.
16. C Frova, A Caffulli, E Pallavera. Mapping quantitative trait loci for tolerance to abiotic stresses in maize. J Exp Zool 282:164–170, 1999.
17. DM Kehoe, P Villand, S Somerville. DNA microarrays for studies of higher plants and other photosynthetic organisms. Trends Plant Sci 4:38–41, 1999.
18. G Noctor, CH Foyer. Ascorbate and glutathione: keeping active oxygen under control. Annu Rev Plant Physiol Plant Mol Biol 49:249–279, 1998.
19. K Asada. The water-water cycle in chloroplasts: scavenging of active oxygens and dissipation of excess photons. Annu Rev Plant Physiol Plant Mol Biol 50:601–639, 1999.
20. R Serrano. Salt tolerance in plants and microorganisms: toxicity targets and defense responses. Int Rev Cytol 165:1–52, 1996.
21. JK Zhu, PM Hasegawa, RA Bressan. Molecular aspects of osmotic stress in plants. Crit Rev Plant Sci 16:253–277, 1997.

22. A Amtmann, D Sanders. Mechanisms of Na^+ uptake by plant cells. Adv Bot Res 29:75–112, 1999.

23. D Schachtmnan, W Liu. Molecular pieces to the puzzle of the interaction between potassium and sodium uptake in plants. Trends Plant Sci 4:281–286, 1999.

24. PJ White. The molecular mechanism of sodium influx to root cells. Trends Plant Sci 4:245–246, 1999.

25. KE Koch. Carbohydrate-modulated gene expression in plants. Annu Rev Plant Physiol Plant Mol Biol 47:509–540, 1996.

26. R Ansell, K Granath, S Hohmann, JM Thevelein, L Adler. The two isoenzymes for yeast NAD-dependent glycerol 3-phosphate dehydrogenase encoded by *GPD1* and *GPD2* have distinct roles in osmoadaptation and redox regulation. EMBO J 16:2179–2187, 1997.

27. J Sheen. Ca^{2+}-dependent protein kinases and stress signal transduction in plants. Science 274:1900–1902, 1996.

28. HJ Bohnert, H Su, B Shen. Molecular mechanisms of salinity tolerance. In: K Shinozaki, ed. Cold, Drought, Heat, and Salt Stress: Molecular Responses in Higher Plants. Austin, TX: RG Landes, 1999, pp 29–60.

29. T Roitsch. Source-sink regulation by sugar and stress. Curr Opin Plant Biol 2:198–206, 1999.

30. RS Boston, PV Viitanen, E Vierling. Molecular chaperones and protein folding in plants. Plant Mol Biol 32:191–222, 1996.

31. M Schroda, O Vallon, FA Wollman, CF Beck. A chloroplast-targeted heat shock protein 70 contributes to the photoprotection and repair of photosystem II during and after photoinhibition. Plant Cell 11:1165–1178, 1999.

32. RK Jain, G Selvaraj. Molecular genetic improvement of salt tolerance in plants. Biotechnol Annu Rev 3:245–267, 1997.

33. D Nelson, B Shen, HB Bohnert. Salinity tolerance—mechanisms, models, and the metabolic engineering of complex traits. In: J. K. Setlow, ed. Genetic Engineering, Principles and Methods. Vol 20. New York: Plenum Press 1998, pp 153–176.

34. MP Apse, GS Aharon, WA Snedden, E Blumwald. Salt tolerance conferred by overexpression of a vacuolar Na^+/H^+-antiport in *Arabidopsis*. Science 285: 1256–1258, 1999.

35. KF McCue, AD Hanson. Drought and salt tolerance: towards understanding and application. Biotechnology 8:358–362, 1990.

36. AJ Delauney, DPS Verma. Proline biosynthesis and osmoregulation in plants. Plant J 4:215–223, 1993.

37. D Bartels, DE Nelson. Approaches to improve stress tolerance using molecular genetics. Plant Cell Environ 17:659–667, 1994.

38. PD Hare, WA Cress, J van Staden. The involvement of cytokinins in plant responses to environmental stress. Plant Growth Regul 23:79–103, 1997.

39. ML Nuccio, D Rhodes, SD McNeil, AD Hanson. Metabolic engineering of plants for osmotic stress resistance. Curr Opin Plant Biol 2:128–134, 1999.

40. AD Hanson, B Rathinasabapathi, J Rivoal, M Burnet, MO Dillon, DA Gage. Osmoprotective compounds in the Plumbaginaceae: a natural experiments in

metabolic engineering of stress tolerance. Proc Natl Acad Sci USA 91:306–310, 1994.

41. DM Vernon, HJ Bohnert. A novel methyl transferase induced by osmotic stress in the facultative halophyte *Mesembryanthemum crystallinum*. EMBO J 11:2077–2085, 1992.

42. M Ishitani, AL Majumder, A Bornhouser, CB Michalowski, RG Jensen, HJ Bohnert. Coordinate transcriptional induction of *myo*-inositol metabolism during environmental stress. Plant J 9:537–548, 1996.

43. DE Nelson, G Rammesmayer, HJ Bohnert. The regulation of cell-specific inositol metabolism and transport in plant salinity tolerance. Plant Cell 10: 753–764, 1998.

44. P Adams, DE Nelson, S Yamada, W Chmara, RG Jensen, HJ Bohnert, H Griffiths. Growth and development of *Mesembryanthemum crystallinum* (Aizoaceae). New Phytol 138:171–190, 1998.

45. P Adams, JC Thomas, DM Vernon, HJ Bohnert, RG Jensen. Distinct cellular and organismic response to salt stress. Plant Cell Physiol 33:1215–1223, 1992.

46. BJ Barkla, R Vera-Estrella, J Camacho-Emiterio, O Pantoja. Na^+/H^+-antiport in the halophyte *Mesembryanthemum crystallinum* L. is correlated to sites of Na^+ storage. (submitted).

47. A Sakamoto, Alia, N Murata. Metabolic engineering of rice leading to biosynthesis of glycine betaine and tolerance to salt and cold. Plant Mol Biol 38:1011–1019, 1998.

48. B Halliwell, M Grootveld, JMC Gutteridge. Methods for the measurement of hydroxyl radicals in biochemical systems: deoxyribose degradation and aromatic hydroxylation. Methods Biochem Anal 33:59–90, 1988.

49. N Smirnoff, QJ Cumbes. Hydroxyl radical scavenging activity of compatible solutes. Phytochemistry 28:1057–1060, 1989.

50. B Shen, RG Jensen, HJ Bohnert. Increased resistance to oxidative stress in transgenic plants by targeting mannitol biosynthesis to chloroplasts. Plant Physiol 113:1177–1183, 1997.

51. B Shen, S Hohmann, RG Jensen, HJ Bohnert. Roles of sugar alcohols in osmotic stress adaptation. Replacement of glycerol by mannitol and sorbitol in yeast. Plant Physiol 121:45–52, 1999.

52. B Schobert, H Tschesche. Unusual solution properties of proline and its interaction with proteins. Biochim Biophys Acta 541:270–277, 1978.

53. PH Yancey, ME Clark, SC Hand, RD Bowlus, GN Somero. Living with water stress: evolution of osmolyte system. Science 217:1214–1222, 1982.

54. B Halliwell, MC Gutteridge. Role of free radicals and catalytic metal ions in human disease: an overview. Methods Enzymol 186:1–85, 1990.

55. C Lamb, RA Dixon. The oxidative burst in plant disease resistance. Annu Rev Plant Physiol Plant Mol Biol 48:251–275, 1997.

56. RR Wise. Chilling-enhanced photooxidation: the production, action and study of reactive oxygen species produced during chilling in the light. Photosynth Res 45:79–97, 1995.

57. TK Prasad. Mechanisms of chilling-induced oxidative stress injury and tolerance in developing maize seedlings: changes in antioxidant system, oxidation of proteins and lipids, and protease activities. Plant J 10:1017–1026, 1996.

58. AH Price, GAF Handry. Iron-catalyzed oxygen radical formation and its possible contribution to drought damage in nine native grasses and three cereals. Plant Cell Environ 14:477–484, 1991.

59. JF Moran, M Becana, I Iturbe-Ormaetxe, S Frechilla, RV Klucas, P Aparicio-Tejo. Drought induces oxidative stress in pea plants. Planta 194:346–352, 1994.

60. VP Roxas, RK Smith Jr, ER Allen, RD Allen. Overexpression of glutathione S-transferase/glutathione peroxidase enhances the growth of transgenic tobacco seedlings during stress. Nature Biotechnol 15:988–991, 1997.

61. K Tsugane, K Kobayashi, Y Niwa, Y Ohba, K Wada, H Kobayashi. A recessive *Arabidopsis* mutant that grows photoautotrophically under salt stress shows enhanced active oxygen detoxification. Plant Cell 11:1195–1206, 1999.

62. Y Miyagawa, M Tamoi, S Shigeoka. Evaluation of the defense system of chloroplast to photooxidative stress caused by paraquat using transgenic tobacco plants expressing catalase from *Escherichia coli*. Plant Cell Physiol 41: 311–320, 2000.

63. B Shen, RG Jensen, HJ Bohnert. Mannitol protects against oxidation by hydroxyl radicals. Plant Physiol 115:527–532, 1997.

64. H Hirt. MAP kinases in plant signal transduction. Results Prob Cell Differ 27:1–9, 2000.

65. C Maurel. Aquaporins and water permeability of plant membranes. Annu Rev Plant Physiol Plant Mol Biol 48:399–429, 1997.

66. B Yang, AS Verkman. Water and glycerol permeabilities of aquaporins 1–5 and MIP determined quantitatively by expression of epitope-tagged constructs in *Xenopus* oocytes. J Biol Chem 272:16140–16146, 1997.

67. S Tyerman, HJ Bohnert, C Maurel, E Steudle, JAC Smith. Plant aquaporins: their molecular biology, biophysics and significance for plant water relations. J Exp Bot 50:1055–1071, 1999.

68. S Nielsen, CL Chou, D Marples, EI Christensen, BK Kishore, MA Knepper. Vasopressin increases water permeability of kidney collecting duct by inducing translocation of AQP-CD water channels to plasma membranes. Proc Natl Acad Sci USA 92:1013–1017, 1995.

69. I Johansson, M Karlsson, VK Shukla, MJ Chrispeels, C Larsson, P Kjellbom. Water transport activity of the plasma membrane aquaporin PM28A is regulated by phosphorylation. Plant Cell 10:451–459, 1998.

70. BJ Barkla, R Vera-Estrella, H-H Kirch, O Pantoja, HJ Bohnert. Aquaporin localization—how valid are the TIP and PIP labels? Trends Plant Sci 4:86–88, 1999.

71. GY Jauh, TE Phillips, JC Rogers. Tonoplast intrinsic protein isoforms as markers for vacuolar functions. Plant Cell 11:1867–1882, 1999.

72. HH Kirch, R Vera-Estrella, D Golldack, B Barkla, F Quigley, CB Michal-

owski, HJ Bohnert. Cellular localization of MIP proteins during growth and under salt stress in *Mesembryanthemum crystallinum*. Plant Physiol 123:111–124, 2000.

73. A Weig, C Deswarte, MJ Chrispeels. The major intrinsic protein family of *Arabidopsis* has 23 members from three distinct groups with functional aquaporins in each group. Plant Physiol 114:1347–1357, 1997.

74. P Gerbeau, J Guclu, P Ripoche, C Maurel. Aquaporin Nt-TIPa can account for the high permeability of tobacco cell vacuolar membrane to small neutral solutes. Plant J 18:577–587, 1999.

75. F Chaumont, F Barrieu, EM Herman, MJ Chrispeels. Characterization of a maize tonoplast aquaporin expressed in zones of cell division and elongation. Plant Physiol 117:1143–1152, 1998.

76. T Fukuhara, HH Kirch, HJ Bohnert. Expression of *Vp1* and water channel proteins during seed germination. Plant Cell Environ 22:417–424, 1999.

77. TH Stevens, M Forgac. Structure, function and regulation of the vacuolar (H$^+$)-ATPase. Annu Rev Cell Dev Biol 13:779–808, 1997.

78. B Michelet, M Boutry. The plasma membrane H$^+$-ATPase—a highly regulated enzyme with multiple physiological functions. Plant Physiol 108:1–6, 1995.

79. H Sze, XH Li, MG Palmgren. Energization of plant cell membranes by H$^+$-pumping ATPases: regulation and biosynthesis. Plant Cell 11:677–689, 1999.

80. MR Sussman. Molecular analysis of proteins in the plant plasma membrane. Annu Rev Plant Physiol Plant Mol Biol 45:211–234, 1994.

81. ND DeWitt, B Hong, MR Sussman, JF Harper. Targeting of two *Arabidopsis* H$^+$-ATPase isoforms to the plasma membrane. Plant Physiol 112:833–844, 1996.

82. X Niu, RA Bressan, PM Hasegawa, JM Pardo. Ion homeostasis in NaCl stress environments. Plant Physiol 109:735–742, 1995.

83. M Weiss, U Pick. Primary structure and effect of pH on the expression of the plasma membrane H$^+$-ATPase from *Dunaliella acidophila* and *Dunaliella salina*. Plant Physiol 112:1693–1702, 1996.

84. F Ayala, JW O'Leary, KS Schumaker. Increased vacuolar and plasma membrane H$^+$-ATPase activities in *Salicornia bigelovii* Torr. in response to NaCl. J Exp Bot 47:25–32, 1995.

85. MS Tsiantis, DM Bartholomew, JA Smith. Salt regulation of transcript levels for the c subunit of a leaf vacuolar H$^+$-ATPase in the halophyte *Mesembryanthemum crystallinum*. Plant J 9:729–736, 1996.

86. R Löw, B Rockel, M Kirsch, R Ratajczak, B Horntensteiner, E Martinoia, U Luttge, T Rausch. Early salt stress effects on the differential expression of vacuolar H$^+$-ATPase genes in roots and leaves of *M. crystallinum*. Plant Physiol 110:259–265, 1996.

87. A Lehr, M Kirsch, R Viereck, J Schiemann, T Rausch. cDNA and genomic cloning of sugar best V-type H$^+$-ATPase subunit a and c isoforms: evidence for coordinate expression during plant development and coordinate induction in response to high salinity. Plant Mol Biol 39:463–475, 1999.

88. KJ Dietz, B Arbinger. cDNA sequence and expression of subunit E of the vacuolar H^+-ATPase in the inducible Crassulacean acid metabolism plant *Mesembryanthemum crystallinum*. Biochim Biophys Acta 1281:134–138, 1996.

89. B Rockel, U Lüttge, R Ratajczak. Changes in message amount of V-ATPase subunits during salt-stress induced C3-CAM transition in *Mesembryanthemum crystallinum*. Plant Physiol Biochem 36:567–573, 1998.

90. R Vera-Estrella, BJ Barkla, HJ Bohnert, O Pantoja. Salt stress in *Mesembryanthemum crystallinum* L. cell suspensions activates adaptive mechanisms similar to those observed in the whole plant. Planta 207:426–435, 1999.

91. C Bremberger, U Lüttge. Dynamics of tonoplast proton pumps and other tonoplast proteins of *Mesembryanthemum crystallinum* L. during the induction of Crassulacean acid metabolism. Planta 188:575–580, 1992.

92. L Zingarelli, P Anzani, P Lado. Enhanced K^+-stimulated pyrophosphatase activity in NaCl-adapted cells of *Acer pseudoplatanus*. Physiol Plant 91:510–516, 1994.

93. E Fischer-Schliebs, E Ball, E Berndt, E Besemfelder-Butz, ML Binzel, M Drobny, D Muhlenhoff, ML Muller, K Rakowski, R Ratajczak. Differential immunological cross-reactions with antisera against the V-ATPase of *Kalanchoe daigremontiana* reveal structural differences of V-ATPase subunits of different plant species. Biol Chem 378:1131–1139, 1997.

94. RA Gaxiola, R Rao, A Sherman, P Grisafi, SL Alper, GR Fink. The *Arabidopsis thaliana* proton transporters, AtNhx1 and Avp1, can function in cation detoxification in yeast. Proc Natl Acad Sci USA 96:1480–1485, 1999.

95. E Epstein, DW Rains, OE Elzam. Resolution of dual mechanisms of potassium absorption by barley roots. Proc Natl Acad Sci USA 49:684–692, 1963.

96. FJM Maathuis, AM Ichida, D Sanders, et al. Roles of higher plants K^+ channels. Plant Physiol 114:1141–1149, 1997.

97. K Czempinski, N Gaedeke, S Zimmermann, B Mueller-Roeber. Molecular mechanisms and regulation of plant ion channels. J Exp Bot 50:955–766, 1999.

98. A Bertl, JA Anderson, CL Slayman, RF Gaber. Use of *Saccharomyces cerevisiae* for patch-clamp analysis of heterologous membrane proteins: characterization of Kat1, an inward-rectifying K^+ channel from *Arabidopsis thaliana*, and comparison with endogeneous yeast channels and carriers. Proc Natl Acad Sci USA 92:2701–2705, 1995.

99. F Rubio, W Gassman, JI Schroeder. Sodium-driven potassium uptake by the plant potassium transporter HKT1 and mutations conferring salt tolerance. Science 270:1660–1663, 1995.

100. S Wu, L Ding, J Zhu. SOS1, a genetic locus essential for salt tolerance and potassium acquisition. Plant Cell 8:617–627, 1996.

101. JM Pardo, MP Reddy, S Yang, A Maggio, GH Huh, T Matsumoto, MA Coca, M Paino-D'Urzo, H Koiwa, DJ Yun, AA Watad, RA Bressan, PM Hasegawa. Stress signaling through Ca^{2+}/calmodulin-dependent protein phosphatase calcineurin mediates salt adaptation in plants. Proc Natl Acad Sci USA 4:9681–9686, 1998.

102. EJ Kim, JM Kwak, N Uozumi, JI Schroeder. *AKUP1*: an *Arabidopsis* gene encoding high-affinity potassium transport activity. Plant Cell 10:51–62, 1998.

103. HH Fu, S Luan. AtKuP1: a dual-affinity K$^+$ transporter from *Arabidopsis*. Plant Cell 10:63–73, 1998.

104. SD Tyerman, IM Skerrett. Root ion channels and salinity. Sci Hortic 78:175–235, 1999.

105. E Blumwald, RJ Poole. Salt tolerance in suspension-cultures of sugarbeet—induction of Na$^+$/H$^+$ antiport activity at the tonoplast by growth in salt. Plant Physiol 83:884–887 1987.

106. BJ Barkla, O Pantoja. Physiology of ion transport across the tonoplast of higher plants. Annu Rev Plant Physiol Plant Mol Biol 47:159–184, 1996.

107. J Garbarino, FM Dupont. NaCl induces a Na$^+$/H$^+$ antiport in tonoplast vesicles from barley roots. Plant Physiol 86:231–236, 1988.

108. WB Frommer, U Ludewig, D Rentsch. Taking transgenic plants with a pinch of salt. Science 28:1222–1223, 1999.

109. SI Allakhverdiev, Y Nishiyama, I Suzuki, Y Tasaka, N Murata. Genetic engineering of the unsaturation of fatty acids in membrane lipids alters the tolerance of *Synechocystis* to salt stress. Proc Natl Acad Sci USA 96:5862–5867, 1999.

110. DE Nelson, M Koukoumanos, HJ Bohnert. *Myo*-inositol amounts in roots control sodium uptake in a halophyte. Plant Physiol 119:165–172, 1999.

111. S Chauhan, N Forsthoefel, Y Ran, F Quigley, DE Nelson, HJ Bohnert. Na$^+$/myo-inositol symporters and Na$^+$/H$^+$-antiport in mesembryanthemum crystallinum. Plant J 24(4):511, 2000.

112. TD Colmer, TMW Fan, RM Higashi, A Läuchli. Interactions of Ca^{2+} and NaCl stress on the ion relations and intracellular pH of *Sorghum bicolor* root tips—an in vivo ^{31}P-NMR study. J Exp Bot 45:1037–1044, 1994.

113. V Martinez, A Läuchli. Effects of Ca^{2+} on the salt-stress response of barley roots as observed by in-vivo ^{31}P-nuclear magnetic resonance and in-vitro analysis. Planta 190:519–524, 1993.

114. GR Cramer. Sodium-calcium interactions under salinity stress. In: A Läuchli, U Lüttge, eds. Salinity: Environment, Plants, Molecules. Boston: Kluwer, in press.

115. J Liu, JK Zhu. A calcium sensor homolog required for plant salt tolerance. Science 280:1943–1945, 1998.

116. JC Cushman, HJ Bohnert. Genomic approaches to plant stress tolerance. Curr Opin Plant Biol 3:117–124, 2000.

117. BD McKersie, SR Bowley, E Harjanto, et al. Water-deficit tolerance and field performance of transgenic alfalfa over-expressing superoxide dismutase. Plant Physiol 111:1177–1181, 1996.

118. D Xu, X Duan, B Wang, B Hong, T-HD Ho, R Wu. Expression of a late embryogenesis abundant protein gene, *HVA1*, from barley confers tolerance to water deficit and salt stress in transgenic rice. Plant Physiol 110:249–257, 1996.

119. SD McNeil, ML Nuccio, AD Hanson. Betaines and related osmoprotectants.

Targets for metabolic engineering of stress resistance. Plant Physiol 120:945–950, 1999.

120. HJ Bohnert, E Sheveleva. Plant stress adaptations—making metabolism move. Curr Opin Plant Biol 1:267–274, 1998.

121. A Sakamoto, N Murata. Genetic engineering of glycine betaine synthesis in plants; current status and implications for enhancement of stress tolerance. J Exp Bot 51:81–88, 2000.

122. ML Guerinot, D Eide. Zeroing in on zinc uptake in yeast and plants. Curr Opin Plant Biol 2:244–249, 1999.

123. B Dujon. The yeast genome project: what did we learn? Trends Genet 12: 263–270, 1996.

124. N Akhtar, A Blomberg, L Adler. Osmoregulation and protein expression in a pbs2Δ mutant of *Saccharomyces cerevisiae* during adaptation to hypersaline stress. FEBS Lett 403:173–180, 1997.

125. MC Gustin, J Albertyn, M Alexander, K Davenport. MAP kinase pathways in the yeast *Saccharomyces cerevisiae*. Microbiol Mol Biol Rev 62:1264–1294, 1998.

126. I Meskiene, L Bogre, W Glaser, J Balog, M Brandstotter, K Zwerger, G Ammerer, H Hirt. MP2C, a plant protein phosphatase 2C, functions as a negative regulator of mitogen-activated protein kinase pathways in yeast and plants. Proc Natl Acad Sci USA 95:1938–1943, 1998.

127. S Miyazaki, R Koga, HJ Bohnert, T Fukuhara. Tissue- and environmental response–specific expression of 10 PP2C transcripts in *Mesembryanthemum crystallinum*. Mol Gen Genet 261:307–316, 1999.

128. T Munnik, W Ligterink, I Meskiene, O Calderini, J Beyerly, A Musgrave, H Hirt. Distinct one-sensing protein kinase pathways are involved in signaling moderate and severe hyper-osmotic stress. Plant J 20:381–388, 1999.

129. B Popping, T Gibbons, MD Watson. The *Pisum sativum* MAP kinase homologue (PsMAPK) rescues the *Saccharomyces cerevisiae hog1* deletion mutant under conditions of high osmotic stress. Plant Mol Biol 31:355–363, 1996.

130. EM Hrabak. Calcium-dependent protein kinases and their relatives. In: M Kreis, JC Walker, eds. Plant Protein Kinases. Advances in Botanical Sciences. New York: Academic Press, in press.

131. AC Harmon, M Gribskov, JF Harper. CDPK's—a kinase for every Ca^{2+} signal? Trends Plant Sci 5:154–159, 2000.

132. T Urao, B Yakubov, R Satoh, K Yamaguchi-Shinozaki, M Seki, T Hirayama, K Shinozaki. A transmembrane hybrid-type histidine kinase in *Arabidopsis* functions as an osmosensor. Plant Cell 11:1743–1754, 1999.

133. KP Dixon, JR Xu, N Smirnoff, NJ Talbot. Independent signaling pathways regulate cellular turgor during hyperosmotic stress and appressorium-mediated plant infection by *Magnaporthe grisea*. Plant Cell 11:2045–2058, 1999.

134. C Pical, T Westergren, SK Dove, C Larsson, M Sommarin. Salinity and hyperosmotic stress induce rapid increases in phosphatidylinositol 4,5-bisphosphate, diacylglycerol pyrophosphate, and phosphatidylcholine in *Arabidopsis thaliana* cells. J Biol Chem 274:38232–38240, 1999.

135. J Shi, KN Kim, O Ritz, V Albrecht, R Gupta, K Harter, S Luan, J Kudla. Novel protein kinases associated with calcineurin b–like calcium sensors in *Arabidopsis*. Plant Cell 11:2393–2406, 1999.
136. H Knight. Calcium signaling during abiotic stress in plants. Int Rev Cytol 195:269–324, 2000.
137. PK Busk, M Pages. Regulation of abscisic acid–induced transcription. Plant Mol Biol 37:425–435, 1998.
138. NN Vettakkorumakankav, D Falk, P Saxena, RA Fletcher. A crucial role for gibberellins in stress protection of plants. Plant Cell Physiol 40:542–548, 1999.
139. S Dhaubhadel, S Chaudhary, KF Dobinson, P Krishna. Treatment with 24-epibrassinolide, a brassinosteroid, increases the basic thermotolerance of *Brassica napus* and tomato seedlings. Plant Mol Biol 40:333–342, 1999.
140. JR Ecker. The ethylene signal transduction pathway in plants. Science 268: 667–675, 1995.
141. J Giraudat. Abscisic acid signaling. Curr Opin Cell Biol 7:232–238, 1995.
142. P Reymond, EE Farmer. Jasmonate and salicylate as global signals for defense gene expression. Curr Opin Plant Biol 1:404–411, 1998.
143. BK Drobak, RE Dewey, WF Boss. Phosphoinositide kinases and the synthesis of polyphosphoinositides in higher plant cells. Int Rev Cytol 189:95–130, 1999.
144. D Sanders, C Brownlee, JF Harper. Communicating with calcium. Plant Cell 11:691–706, 1999.
145. AJ Trewavas, R Malho. Ca^{2+} signaling in plant cells: the big network! Curr Opin Plant Biol 1:428–433, 1998.
146. AA Very, MF Robinson, TA Mansfield, D Sanders. Guard cell cation channels are involved in Na^+-induced stomatal closure in a halophyte. Plant J 14:509–521, 1998.
147. R Munns. Physiological processes limiting plant-growth in saline soils—some dogmas and hypothesis. Plant Cell Environ 16:15–24, 1993.
148. G Netting. pH, abscisic acid and the integration of metabolism in plants under stressed and non-stressed conditions: cellular responses to stress and their implication for plant water relations. J Exp Bot 51:147–158, 2000.
149. RE Sharp, Y Wu, GS Voetberg, IN Saab, ME LeNoble. Confirmation that abscisic acid accumulation is required for maize primary root elongation at low water potentials. J Exp Bot 45:1743–1751, 1994.
150. J Dorweiler, A Steca, J Kermicle, J Doebley. Teosinte-glume-architecture-1, a genetic-locus controlling a key step in maize evolution. Science 262:233–235, 1993.
151. R Wu, HD Bradshaw, RF Stettler. Molecular genetics of growth and development in *Populus* (Salicaceae). 5. Mapping quantitative trait loci affecting leaf variation. Am J Bot 84:143–153, 1997.
152. C Giauffret, R Bonhomme, M Derieux. Heterosis in maize for biomass production, leaf area establishment, and radiation use efficiency under cool spring conditions. Maydica 42:13–19, 1997.

153. RN Sarma, BS Gill, G Galiba, J Sutka, DA Laurie, JW Snape. Comparative mapping of the wheat chromosome 5A Vrn-A1 region with rice and its relationship to QTL for flowering time. Theor Appl Genet 97:103–109, 1998.

154. S Serce, JP Navazio, AF Goke, JE Staub. Nearly isogenic cucumber genotypes differing in leaf size and plant habit exhibit differential response to water stress. J Am Soc Hortic Sci 124:358–365, 1999.

155. JR Caradus, DR Woodfield. Genetic control of adaptive root characteristics in white clover. Plant Soil 200:63–69, 1998.

156. MAR Mian, R Wells, TE Carter, DA Ashley, HR Boerma. RFLP tagging of QTLs conditioning specific leaf weight and leaf size in soybean Theor Appl Genet 96:354–360, 1998.

157. V Negruk, P Yang, M Subramanian, JP McNevin, B Lemieux. Molecular cloning and characterization of the *CER2* gene of *Arabidopsis thaliana*. Plant J 9:137–145, 1996.

158. V Walbot. Genes, genomes, genomics. What can plant biologists expect from the 1998 National Science Foundation Plant Genome Research Program? Plant Physiol 119:1151–1155, 1999.

159. PJ Krysan, JC Young, MR Sussman. T-DNA as an insertional mutagen in *Arabidopsis*. Plant Cell 11:2283–2290, 1999.

160. X Lin and 36 coauthors. Sequence and analysis of chromosome 2 of the plant *Arabidopsis thaliana*. Nature 402:761-769, 1999.

161. M Umeda, C Hara, Y Matsubayashi, HH Li, Q Liu, F Tadokoro, S Aotsuka, H Uchimiya. Expressed sequence tags from cultured cells of rice (*Oryza sativa* L.) under stressed conditions: analysis of transcripts of genes engaged in ATP-generating pathways. Plant Mol Biol 25:469–478, 1994.

162. KY Pih, HJ Jang, SG Kang, HL Piao, I Hwang. Isolation of molecular markers for salt stress responses in *Arabidopsis thaliana*. Mol Cell 7:567-571, 1997.

163. C Bockel, F Salamini, D Bartels. Isolation and characterization of genes expressed during early events of the dehydration process in the resurrection plant *Craterostigma plantagineum*. J Plant Physiol 152:158–166, 1998.

164. AJ Wood, MJ Oliver. Translational control in plant stress: the formation of messenger ribonucleoprotein particles (mRNPs) in response to desiccation of *Tortula ruralis* gametophytes. Plant J 18:359–370, 1999.

165. J Machuka, S Bashiardes, E Ruben, K Spooner, A Cuming, C Knight, D Cover. Sequence analysis of expressed sequence tags from an ABA-treated cDNA library identifies stress response genes in the moss *Physcomitrella patens*. Plant Cell Physiol 40:378–387, 1999.

166. MA Cushman, M Dennis, D Bufford, I Akselrod, D Landrith, J Maroco, S Kore-eda, JC Cushman. Monitoring salt stress responses by expressed sequence tags in *Mesembryanthemum crystallinum* leaves. Plant Physiol, in preparation.

167. D Baldwin, V Crane, D Rice. A comparison of gel-based, nylon filter and microarray techniques to detect differential RNA expression in plants. Curr Opin Plant Biol 2:96–103, 1999.

168. P Tamayo, D Slonim, J Mesirov, Q Zhu, S Kitareewan, E Dmitrovsky, ES Lander, TR Golub. Interpreting patterns of gene expression with self-orga-

nizing maps: methods and application to hematopoietic differentiation. Proc Natl Acad Sci USA 96:2907–2912, 1999.

169. TJ Flowers, ML Koyama, SA Flowers, C Sudhakar, KP Singh, AR Yeo. QTL: their place in engineering tolerance of rice to salinity. J Exp Bot 51:99–106, 2000.

170. CM Hamilton, A Frary, C Lewis, SD Tanksley. Stable transfer of intact high molecular weight DNA into plant chromosomes. Proc Natl Acad Sci USA 93: 9975–9979, 1996.

171. JD Heath, MI Boulton, DM Raineri, SL Doty, AR Mushegian, TC Charles, JW Davies, EW Nester. Discrete regions of the sensor protein VirA determine the strain-specific ability of *Agrobacterium* to agroinfect maize. Mol Plant Microbe Interact 10:221–227, 1997.

172. T Komari, Y Hiei, Y Saito, N Murai, T Kumashiro. Vectors carrying two separate t-DNAs for cotransformation of higher plants mediated by *Agrobacterium tumefaciens* and segregation of transformants free from selection markers. Plant J 19:165–174, 1996.

173. SA Kempin, SJ Lijergren, LM Block, SD Rounsley, MF Yanofsky. Targeted gene disruption in *Arabidopsis*. Nature 389:802–803, 1997.

174. AC Vergunst, LE Jansen, PJ Hooykaas. Site-specific integration of *Agrobacterium* T-DNA in *Arabidopsis thaliana* mediated by *Cre* recombinase. Nucleic Acids Res 26:2729–2734, 1998.

175. AD Hanson, DA Gage, Y Shachar-Hill. Plant one-carbon metabolism and its engineering. Trends Plant Sci 5:206–213, 2000.

176. B Miflin. Crop improvement in the 21st century. J Exp Bot 51:1–8, 2000.

27

Molecular Mechanisms that Control Plant Tolerance to Heavy Metals and Possible Roles in Manipulating Metal Accumulation

Stephan Clemens
Leibniz Institute of Plant Biochemistry, Halle (Saale), Germany

Sébastien Thomine
Institute of Plant Sciences–CNRS, Gif-Sur-Yvette, France

Julian I. Schroeder
Division of Biology, University of California, San Diego, La Jolla, California, U.S.A.

I. INTRODUCTION

Soils and waters with high levels of toxic metals such as cadmium, arsenic, lead, and mercury are detrimental to human and environmental health. Toxic heavy metals contaminate soils and waters in industrialized nations as well as in developing nations. The four heavy metals arsenic, lead, mercury, and cadmium have been identified as belonging to the five priority most hazardous substances found at toxic Superfund sites in the United States (1). Many human disorders have been attributed to ingestion of heavy metals including learning disabilities in children, dementia, and increased rates of cancer in response to cadmium (Cd) (2,3). Removal of heavy metals from highly contaminated soils and waters is therefore a very costly but necessary process that is currently being pursued at contaminated sites worldwide.

Remediation of soils containing high levels of toxic heavy metals is pursued by physical removal of metals because most of these metals cannot be degraded in the soil, in contrast to many organic contaminants (4). Current practical methods used to decontaminate such sites involve physical excavation of top soils, transport, and reburial elsewhere. However, these cleanup methods are feasible only for small soil areas and are very costly. For example, costs for cleaning 1 ha to a depth of 1 m have been reported between $600,000 and $3,000,000 (5). Furthermore, excavation strategies are not applicable to contaminated waters. Alternatively, metals can be immobilized in soils. This approach can carry the long-term risk of resolubilization because of chemical or biological changes (e.g., acidification) (6). Complementary approaches involving both removal and immobilization will be needed to remediate heavy metal–contaminated sites.

Research and applications indicate that uptake of heavy metals into plants via the root system could provide a cost-effective approach for toxic metal removal and remediation of heavy metal–laden soils and waters. Plant roots have been shown to remediate waters by removal of heavy metals (7). Furthermore, increasing the accumulation of valuable metals could eventually lead to "phytomining," as proposed, for instance, for gold (8).

On the other hand, uptake of toxic heavy metals in crop plants is known to cause health problems (see earlier). Understanding of the mechanisms underlying toxic metal uptake and sequestration could lead to engineering of crops that avoid toxic metal accumulation in edible parts of plants.

Toxic heavy metals are transported across the plasma membrane into plant root cells via physiological metal uptake transporters. However, for plants to accumulate large amounts of toxic metals for bioremediation purposes, many mechanisms and genes need to be identified and modified in plants. Several rate-limiting steps are critical for effective removal of heavy metals from soils. These include making the contaminants biologically accessible in the soil by chelation or external acidification and subsequent transport of metals or complexed metals across the plasma membrane of root cells. Upon uptake of heavy metals into plant cells, intracellular detoxification and transport through plant tissues are required.

Other metals, such as iron and zinc, are essential nutrients. Iron deficiency is the most widely spread micronutrient deficiency worldwide, affecting up to 3.7 billion people, particularly women (9), and zinc deficiency is a significant limiting factor for production and quality of cereals (10). Identification of molecular mechanisms that enhance accumulation of essential nutrients in plants could lead to development of crops that help to address major nutritional problems in humans and could also play an important role in plant nutrition and growth.

Phytoremediation involves the biotechnological exploitation of metal tolerance mechanisms. The same applies for the nutrition-related objectives of minimizing the concentration of toxic metals in edible parts and enriching crops for essential metals. In the present chapter we summarize the molecular understanding that has been gained to date and review the transgenic approaches that have been pursued to increase plant metal tolerance. This includes work that does not exploit plant tolerance mechanisms but instead utilizes bacterial or mammalian genes. The heavy metals mainly considered are cadmium, mercury, copper, zinc, iron, and nickel. Also, even though aluminum, strictly speaking, is not a heavy metal (because the specific weight is under 5 g/cm^3), we include work on Al tolerance because of its agronomic importance (11). Most of the fundamental insight into metal homeostasis in eukaryotes has come from studies involving *Saccharomyces cerevisiae*. For many of the molecular components identified in yeast, plant and mammalian homologues have been found, indicating a high degree of conservation. Thus, the work on *S. cerevisiae* is covered in some detail.

II. TOLERANCE MECHANISMS

The toxic effects of plant exposure to elevated metal concentrations, such as growth inhibition and chlorosis, are well documented. However, as for other organisms, the biochemical understanding of metal toxicity is limited. Class B metals such as Cu(I) and Hg and borderline metals such as Cu(II), Zn, Ni, and Cd (12) can form complexes with nitrogen and sulfur atoms in proteins, thereby potentially inactivating them. The nonessential metals Cd and Pb compete for binding with the essential metals Ca and Zn. Also implicated in symptoms of metal toxicity is the redox activity of some metals (e.g., Cu, Fe) that may result in the formation of reactive oxygen species.

A complex network enables plants to control tightly the intracellular concentrations and distribution of essential heavy metals such as copper and to minimize the cytosolic concentrations of nonessential heavy metals such as cadmium. The interplay of mainly transport and chelation processes that constitutes this network results in a "basic metal tolerance," the distribution, sequestration, and exclusion of potentially toxic heavy metal ions. In addition, a number of plant species show "metal hypertolerance." They can grow on soil that naturally or because of human activities contains heavy metal concentrations that are growth prohibiting to most plants. These species belong to a specialized flora that has colonized Ni-rich serpentine soils or Zn- and Cd-polluted areas (13). Two principal responses to otherwise toxic heavy metal concentrations can be found, metal exclusion and metal hyperaccumulation, with the latter mainly restricted to Ni, Zn, and Se. About 400 different species belonging to a wide range of taxa have been described as hyperaccumulators, about 75% of which are Ni hyperaccumulators (13).

It is a general characteristic of both basic metal tolerance of most organisms and hypertolerance of some specialized plants that most of the mechanisms involved appear to be specific for a certain metal or a small

DNAs (cDNAs) in *S. cerevisiae* (14,28). The analysis of a *Schizosaccharomyces pombe* PCS knockout strain and of *S. cerevisiae* cells expressing PCS genes from different sources showed that the formation of PCs confers Cd tolerance and to a limited degree Cu, arsenate, and Hg tolerance. Based on these yeast studies, there is no indication for a role in, for instance, Zn, Co, or Ni detoxification. In the meantime, PC-mediated As tolerance was confirmed for plants (29,30).

The expression of PCS genes in baker's yeast, which resulted in a more than 15-fold increase in Cd tolerance (14,28), exemplified the potential of PC synthesis to enhance Cd tolerance in organisms that do not normally produce PCs. However, the situation may be different in plants because apparently all plant species express phytochelatin synthases (31). There have been numerous attempts to enhance PC formation and thereby plant metal tolerance by generating plants overproducing the PC precursor glutathione. *Brassica juncea* plants overexpressing the *Escherichia coli gsh1* gene, encoding γ-glutamylcysteine synthetase with a plastid targeting sequence under control of the 35S promoter, display up to fivefold higher enzyme activity accompanied by an enhancement of PC2 and PC3 formation. This resulted in an increase in Cd accumulation and Cd tolerance by about twofold as determined from root growth in the presence of Cd^{2+} (32). Similar effects were generated by the 35S-driven overexpression of an *E. coli* glutathione synthase (encoded by the *gsh2* gene) in the cytosol of *B. juncea* plants (33). Again, the higher activity of a glutathione biosynthetic enzyme led to an increase in PC formation, Cd tolerance, and Cd accumulation. These results are interpreted as an indication that glutathione biosynthesis represents a rate-limiting step for PC synthesis. This view is supported by the basal constitutive expression of both PCS activity and PCS genes (24,27). However, there are indications for an up-regulation of PCS genes under metal stress (14), and the effects of PCS overexpression on Cd tolerance in plants have yet to be reported. Also, a strategy to boost PC synthesis further could be to combine PCS overexpression with the overproduction of glutathione and/or cysteine. These experiments will show whether PCS overexpression in plants—despite the apparent noninvolvement of phytochelatin formation in naturally selected hypertolerance (see later)—can be utilized for a significant enhancement of tolerance toward Cd and potentially other metals.

2. Metallothioneins

Metallothioneins (MTs) are a ubiquitous group of small cysteine-rich proteins (34). The class I MTs of mammalian cells are involved in cytosolic metal ion buffering. In yeast, MTs bind predominantly copper ions, whereas mammalian MTs bind mainly Zn. The copper-inducible *S. cerevisiae* CUP1 contributes to copper detoxification (35). A disruption of cup1 causes Cu hypersensitivity; the overexpression of cup1 enhances Cu tolerance. CUP1 as well as the mammalian MTs can also bind Cd. MT-Cd complexes are formed in *S. pombe* cells expressing the *S. cerevisiae* MT gene (36). For a number of mammalian cell lines it was shown that MT deficiency results in Cd hypersensitivity and that MT overexpression leads to an elevation of Cd tolerance (37). Still, in the light of numerous genetic experiments demonstrating the nonessentiality of MTs under normal growth conditions, the actual physiological function of MTs is considered enigmatic (38).

To date, more than 50 MT-like sequences have been found in a variety of plant species (39). Plant metallothioneins are similar to class II MTs; i.e., they differ considerably from class I MTs in the alignment of Cys residues. MT genes from *Arabidopsis* complement MT-deficient yeast and *Synechococcus* mutants, which are Cu and Zn hypersensitive, respectively (40,41), and they confer Cd tolerance to *S. cerevisiae* when overexpressed (14). It has been shown that MT genes are expressed in various plant tissues, and there are examples of induction by Cu (42). A correlation between MT expression and Cu tolerance has been observed for 10 different *Arabidopsis* ecotypes. However, because genetic data and direct evidence for the binding of metals have been scarce, the role of metallothioneins in plant metal homeostasis is understood to an even lesser extent than in yeast and mammals. It remains a question whether MTs can function as metal chaperones analogous to ATX1 and others (see later).

Metallothioneins from various sources have been constitutively expressed in plants, resulting in elevated Cd tolerance and changes in the Cd distribution between roots and shoots. The first such report dates back to 1987 (43). A Chinese hamster MT II cDNA was inserted into a cauliflower mosaic virus. This recombinant virus was used to express metallothionein in infected *Brassica campestris* tissue, resulting in an increase in Cd-binding capacity and Cd tolerance. Pan et al. (44) introduced a synthetic gene coding for the α-domain of mouse MT I into tobacco. They were able to detect mouse metallothionein in leaves of transgenic plants and found significantly enhanced Cd tolerance (44). Similarly, human metallothionein was expressed in tobacco. Here, the authors analyzed the distribution of Cd in transgenic lines compared with the wild type and found that root-to-shoot translocation was reduced and the shoot content was 60–70% lower (45). More recently, the transformation of *Nicotiana tabacum* with an MT gene from *Nicotiana plumbaginifolia* was reported (46). Again, the 35S-driven expression led to an elevation of Cd tolerance as determined by measuring growth and chlorophyll content. Data on the tolerance of the various MT-overexpressing plants toward other metals have not been reported. From the preceding results concerning MTs in yeast and mammals, one might expect to see some enhancement of Cu tolerance as well.

The MT genes are also being exploited for enriching crops with micronutrients. In combination with genes encoding a phytase and ferritin (see later), an MT gene was introduced into rice to help increase the Fe content and improve Fe absorption in the human body (47).

3. Other Proteinaceous Chelators

In *S. cerevisiae*, Cu^+ ions are taken up by the high-affinity uptake systems CTR1 and CTR3 (48) following reduction of Cu^{2+} by plasma membrane reductases. Once they are inside the cell, Cu^1 ions are bound by metallothioneins and at least three different chelators that serve specialized functions in copper delivery to different enzymes. ATX1, originally identified as a suppressor of oxygen toxicity in sod$^-$ strains, and its human homologue, HAH1, direct Cu to a post-Golgi compartment via Cu-pumping ATPases (49). Yeast two-hybrid experiments provided evidence for direct interaction of ATX1 with the yeast P-type ATPase CCC2. Similarly, HAH1 appears to bind to the Wilson disease adenosine triphosphatase (ATPase) depending on available copper, and the mutations in the transporter associated with the genetic defect impair Cu delivery to the ATPase (50). Other known Cu chelators in *S. cerevisiae* are COX17, involved in Cu delivery to mitochondria for insertion into cytochrome *c* oxidase (51), and LYS7, which functions as a Cu chaperone for the Cu,Zn-superoxide dismutase (52). Both COX17 and LYS7 have human homologues. An ATX1 homologue was identified in *Arabidopsis* that complements an ATX1-deficient yeast strain (53), indicating the presence of similar chelation networks in plants. Furthermore, the *Arabidopsis* RAN1 represents a homologue of the Wilson disease ATPase. It is also suggested to be located in post-Golgi vesicles and to be involved in Cu delivery to the ethylene receptor (54).

Metal tolerance appears to be not directly related to this homeostatic Cu chaperone-chelator network. Yeast cells lacking one of the Cu chelators display a range of phenotypes ranging from oxygen radical susceptibility (lys7$^-$) to respiration deficiency (cox17$^-$). However, no absolute requirement for Cu detoxification was observed. Instead, metallothioneins seem to play the key role as cytosolic Cu buffers in yeast cells (55). The *Arabidopsis* ATX1 homologue is down-regulated under Cu stress. Nonetheless, it remains to be analyzed whether an increase in Cu binding capacity by overexpression of the Cu chelators would result in an enhancement of Cu tolerance.

Other specific cytosolic chelator proteins involved in trafficking of essential metal ions await identification in plants. Known and used biotechnologically is the Fe-binding protein ferritin, which can be found in bacteria, animals, and plants and functions in iron storage. Overexpression of a soybean ferritin gene specifically in rice seeds under the control of a glutelin promoter resulted in transgenic rice plants with a threefold higher seed iron content (56).

4. Organic Acids, Amino Acids

The reactivity of metal ions with S, N, and O renders organic and amino acids potential ligands. A considerable number of studies have been published on variations in the content of citric, oxalic, and malic acid in relation to metal tolerance. Computer modeling has been used to predict metal speciation in different compartments and dependent on pH, nutrient, and metal supply (reviewed in Ref. 39). Unequivocal evidence, however, for a role of organic and amino acids in plant metal tolerance has to date been mainly confined to the histidine response in some Ni hyperaccumulators (15) and the excretion of malate or citrate as a mechanism of Al tolerance (11).

Alyssum lesbiacum, a highly Ni-tolerant Ni hyperaccumulator, contains significantly more Ni in shoots than the nonhyperaccumulating relative *A. montanum* when grown in the presence of Ni. This was found to be correlated with a dramatic increase in histidine concentration of the xylem sap of *A. lesbiacum*, a response that was not observed for *A. montanum* (15). Histidine is an efficient chelator of Ni, and it was found by X-ray absorption spectroscopy that most of the Ni in *A. lesbiacum* is bound to histidine. The essential role of histidine for Ni hypertolerance was confirmed by the finding that spraying of *A. montanum* with histidine resulted in a doubling of biomass production under Ni stress. More recently, however, studies on another Ni hyperaccumulator, *Thlaspi goesingense*, showed that the histidine response appears not to be a general Ni tolerance mechanism. Free histidine concentrations in shoot and xylem sap are no higher in *T. goesingense* than in the nontolerant *T. arvense* (57). Genes encoding histidine biosynthetic enzymes are not up-regulated upon Ni exposure.

Nonetheless, the effect of spraying histidine on *A. montanum* suggests that increasing the histidine content could enhance Ni tolerance, which is hypothesized to be the main determinant for Ni hyperaccumulation (58). It remains to be seen whether the efforts to generate His-overproducing plants by manipulating the histidine biosynthetic pathway will lead to the development of Ni-hyperaccumulating transgenic plants (Krämer, U., Kim, E. J., Schroeder, J. I., unpublished results).

Aluminum is the third most abundant element by percent weight in the outer earth's crust. Low pH in the soil enhances solubilization of bound Al. Because most plants are sensitive to micromolar concentrations of aluminum, Al toxicity is a major growth-inhibiting factor on acid soils (11). Thus, the objective of engineering Al tolerance is to improve crop yields. Exclusion of Al from the root apex, which is the main site of Al uptake, has been established as a tolerance mechanism. In snapbean (59) a correlation between Al-induced release of citrate and Al tolerance was observed, and for near-isogenic wheat lines differing in Al tolerance (Alt1 locus) it was shown that this trait cosegregates with high rates of malate release from the roots (60,61). The protective effect of organic acid release is probably based on the fact that malate and citrate can chelate Al^{3+}, the presumed primary toxic Al species, and thereby reduce uptake of this ion into the symplast.

These findings inspired a transgenic approach aiming at the production of Al-tolerant plants by increasing the citrate content (62). A citrate synthase gene from *Pseudomonas aeruginosa* was overexpressed in tobacco and papaya under 35S control. The protein was targeted to the cytosol. Four tobacco lines with two- to threefold higher citrate synthase activity were analyzed. They contained up to 10-fold more citrate and released up to 4-fold more citrate into the medium. Staining of root tissue with hematoxylin indicated that less Al is taken up by the citrate synthase overexpressors. In growth assays in the presence of Al, these lines showed a slight increase in tolerance. For instance, at 100 μM root growth was inhibited by 20–30% in the transgenic lines compared with about 50% in wild-type plants. At 200 μM the inhibition was 40–60% instead of 85%. The benefits of higher citrate content were more pronounced for papaya plants (62). Transgenics with again two- to threefold higher citrate synthase activity developed roots in medium containing up to 300 μM Al, whereas control plants failed to develop roots in the presence of 50 μM Al.

The nonprotein amino acid nicotianamine, which is synthesized from *S*-adenosylmethionine and is a precursor of the phytosiderophore mugeneic acid, forms stable complexes with Fe(II). It may function as a transport form for Fe(II) and possibly other metal ions (63). The cloning of genes encoding nicotianamine synthases (64,65) now makes it possible to study the role of this ubiquitous compound for metal ion mobility and regulation of metal responses in molecular detail.

A considerable number of metal chelators and chaperones remain to be identified. Little is known about the network governing the intracellular distribution of essential ions. The transport forms of metal ions in the xylem and phloem are not determined in sufficient detail. Furthermore, even the exact function of known chelators is largely unclear. Are metallothioneins involved in Zn and Cu buffering, and do phytochelatins play a role in metal homeostasis? The Cd tolerance that is conferred by these compounds may well represent a product rather than their physiological function. Understanding and engineering the trafficking and translocation of metal ions have great potential for the generation of plants suitable for phytoremediation as well as of crops enriched in micronutrients.

B. Sequestration

Excess heavy metal ions are removed from the cytosol by sequestration. The compartment mainly involved in this process in yeast and plant cells is the vacuole. Mediators of sequestration are transporters in the respective membrane. The actual substrates for most of these transporters are as yet unknown.

1. Roles of ABC-type Transporters in Sequestration

Hmt1 from *S. pombe* was the first gene identified that encodes a protein involved in vacuolar sequestration of heavy metals (66). *Hmt1* was cloned on the basis of the complementation of a Cd-hypersensitive *S. pombe* strain lacking vacuolar sulfide-containing PC-metal complexes. The corresponding protein belongs to the large family of ABC-type transporters. It localizes to the vacuolar membrane and transports PC-Cd complexes and apo-PCs (67). *S. pombe* cells overexpressing *hmt1* accumulate more Cd^{2+} and are more Cd^{2+} tolerant. This indicates the potential of corresponding experiments in plants. However, no transgenic plants expressing *hmt1* have been described yet. Also, the respective ABC transporters that account for the transport of PC-metal complexes in plants are not yet known. The *Arabidopsis* MRPs 1–4 were shown to mediate the vacuolar sequestration of various xenobiotics conjugated to glutathione but lack PC transport activity (68). This led to the speculation that additional factors might be involved in the sequestration of Cd ions as PC complexes in plants (68).

In *S. cerevisiae*, the transporter YCF1 is required for Cd tolerance of yeast cells (69). YCF1-deficient cells are Cd hypersensitive. From transport studies with vacuolar membrane vesicles purified from wild-type and ycf⁻ *S. cerevisiae* cells it is known that YCF1 mediates the Mg-ATP–dependent transport of bis(glutathionato)cadmium into vacuoles (70). YCF1 apparently also represents one of two pathways in *S. cerevisiae* for arsenite detoxification (71). A ycf1-knockout strain is As(III) hypersensitive and deficient in MgATP- and glutathione-dependent vacuolar uptake of ^{73}As(III).

2. Cation Diffusion Facilitators (CDFs)

The cation diffusion facilitator family constitutes another group of proteins involved in metal ion transport. First found in bacteria (72), members of this family have now been described in yeast, animals, and plants (73,74). The *S. cerevisiae* proteins COT1 and ZRC1 confer tolerance of cobalt (COT1) (75) and zinc/cadmium (ZRC1) (76) when overexpressed. The observation that both proteins localize to the vacuolar membrane (77) suggests a role in metal transport into the vacuole. *S. pombe* expresses one ZRC homologue that appears to be involved in Zn accumulation and Zn tolerance as a knockout strain is Zn hypersensitive and accumulates less Zn (Clemens, S., Nies, D. H., unpublished).

Four mammalian zinc transporters (ZnT 1–4) of the CDF family are known. ZnT-1 was cloned as a rat cDNA complementing the zinc sensitivity of a hamster cell line (78). The protein was detected in the plasma membrane and proposed to mediate Zn efflux. Both ZnT-2 and ZnT-3 reside in endomembranes, suggesting a function in zinc transport into the respective compartments: lysosomes (ZnT-2) (79) and synaptic vesicles (ZnT-3) (80). To date, one CDF has been studied in *Arabidopsis*, at least two additional sequences can be found in the genome (Accession numbers AL353032, AC004561). The *Arabidopsis* cDNA ZAT was isolated and, because sequence alignments indicated potential significance for metal transport, introduced into *Arabidopsis* in sense and antisense orientation fused to the 35S promoter (74). The sense lines showed a slight Zn tolerance phenotype when grown alongside control plants on medium containing toxic Zn concentrations. No differences were observed in Cd sensitivity. When Zn accumulation was measured in a hydroponic culture, the sense line contained significantly more Zn in the roots than the control line, whereas shoot contents were similar for both lines. From these results the authors speculate that the ZAT protein might be a vacuolar transporter involved in the sequestration of Zn.

3.   Other Transporters

Direct transport of Cd^{2+} ions could represent a pathway of Cd sequestration in addition to the transport of PC-Cd complexes. In tonoplast-enriched vesicles from oat roots, Salt and Wagner (81) detected a Cd^{2+}/H^+ antiport activity. They demonstrated saturable ΔpH-dependent uptake of Cd^{2+} with an apparent K_M of 5.5 μM. The authors estimated this K_M to lie in the range of cytoplasmic Cd^{2+} concentration for plants growing in Cd-contaminated soil, meaning that this antiport activity could play a role in Cd accumulation in the vacuole. Molecularly, the Cd^{2+}/H^+ antiport might be attributable to the same transporter as the Ca^{2+}/H^+ antiport in the vacuolar membrane. This was suggested by Salt and Wagner and others (82). Ca^{2+}/H^+ antiporters CAX1 and CAX2 have been cloned from *Arabidopsis* on the basis of the complementation of a Ca-hypersensitive *S. cerevisiae* mutant (83). CAX1 was shown to mediate high-affinity Ca^{2+}/H^+ exchange when expressed in yeast, and the K_M of CAX2 for Ca^{2+} is too high to be of any physiological relevance. Instead, the authors mention for CAX2 high-affinity and high-capacity H^+/heavy metal cation antiporter activity. This would make CAX2 another target for engineering higher Cd accumulation in plants. No reports have been published yet on CAX2 overexpressors.

C. Efflux

The pumping of toxic heavy metal ions out of the cell represents the main tolerance mechanism in bacteria. Chromosomally or plasmid-encoded efflux systems have been found in all eubacterial groups studied so far (84). The metals for which specific transporters exist include copper, cadmium, zinc, silver, lead, and arsenite. The respective transporter genes are generally part of metal tolerance operons also containing regulatory genes and genes coding for metal-binding proteins. Most of the bacterial metal transporters belong to the family of CPx-type ATPases, a subclass of the P-type ATPases (85). Well-studied examples are the Cd^{2+}-specific pump CadA from *Staphylococcus aureus* and the Cu pumps CopA and CopB from *Enterococcus hirae* (86,87). Copper-transporting CPx-type ATPases have now also been characterized in eukaryotes. The human Wilson and Menkes disease proteins as well as the *S. cerevisiae* CCC2 protein are involved in transport of Cu ions into a post-Golgi compartment (88–90). Defects in these systems lead to deficiencies in copper-dependent iron uptake systems. In addition to functioning in the intracellular Cu distribution, eukaryotic CPx-type ATPases can mediate Cu tolerance in a fashion similar to that in the bacterial systems. In *Candida albicans* a Cu pump was identified that is essential for Cu tolerance and localizes primarily to the plasma membrane, suggesting an efflux pump activity (91).

The one functionally characterized CPx-type ATPase from plants, RAN1 from *Arabidopsis*, probably serves a function similar to CCC2 and the Wilson and Menkes disease proteins. It is proposed to be involved in supplying Cu to the ethylene receptor (54). However, there are numerous other sequences in the *Arabidopsis* genome containing the signature motifs of CPx-type ATPases, such as the putative metal-binding CxxC elements in the amino terminus. At least two of the protein sequences show homology to the Cd^{2+} pump CadA from *S. aureus*. Thus, there could well exist efflux systems in plants conferring heavy metal tolerance. To date, no such activity has been described physiologically in plants and the functional data have not been published for any of the other putative plant metal pumps. It is conceivable that these pumps, as well as the bacterial pumps, can be exploited for enhancing the tolerance of plants to Cu, Cd, and possibly other metals.

D. Engineering Metal Uptake Systems

We have seen tremendous advances in our molecular understanding of plant metal uptake. By complementation of respective yeast mutants, uptake transporters have been cloned for the micronutrients Fe, Cu, and Zn. Potential pathways for the entry of the nonessential metals Pb and Cd have been identified. The cloning of transporters and the thorough characterization with regard to expression levels, tissue distribution, cellular localization, and biophysical parameters will make it possible to enhance plant metal tolerance or plant metal accumulation by altering the properties of plant metal uptake. Conceivable strategies include the overexpression or the down-regulation of particular transporters as well as the expression of transporters with altered affinities and specificities.

1. The ZIP Family of Transporters

The ZIP family of metal ion transporters (for ZRT-like, IRT-like proteins) (10) comprises proteins from eukaryotic organisms as diverse as trypanosomes and humans. The first member isolated was IRT1 from *Arabidopsis*. It complements the iron uptake deficiency of the *S. cerevisiae fet3fet4* mutant and encodes an Fe^{2+} transporter (92). This activity and the expression in roots, which is inducible by iron limitation, indicate that IRT1 mediates uptake of Fe^{2+} from the soil. Additional experiments in yeast showed that IRT1 has a broad substrate range and also transports Mn^{2+}, Zn^{2+}, and, judging from competition experiments, Cd^{2+} (93). The yeast homologues ZRT1 and ZRT2 were identified on the basis of sequence similarities to IRT1. They were subsequently shown to mediate high-affinity and low-affinity Zn^{2+} uptake, respectively (94,95). A zrt1/zrt2 mutant was then used to clone plant Zn uptake transporters. Three *Arabidopsis* cDNAs were found that complement the zinc uptake deficiency, designated ZIP1-3 (96). They encode transporters with K_M values for Zn^{2+} in the nanomolar range, suggesting a role in plant Zn^{2+} uptake.

Expression levels of ZIP transporters could represent one factor determining metal ion uptake and accumulation. When the Zn hyperaccumulator *Thlaspi caerulescens* and its nonaccumulating relative *T. arvense* were analyzed, it was found that Zn uptake into *T. caerulescens* roots shows a K_M similar to that of the uptake in *T. arvense* but a 4.5-fold higher V_{MAX} (97). Subsequently, a Zn^{2+} transporter was cloned from *T. caerulescens* that complemented the Zn uptake–deficient yeast strain zhy3 (98) and displays kinetic properties matching the data for Zn uptake into *T. caerulescens* roots. According to Northern analysis, this Zn transporter, termed ZNT1, is expressed at very high levels in *T. caerulescens* irrespective of the Zn status whereas in *T. arvense* the ZNT1 homologue mRNA is of low abundance in Zn-sufficient media and up-regulated under Zn-limiting conditions. These observations suggest that by altering the expression levels of uptake systems it is possible to engineer metal accumulation in plants.

2. *Nramp* Metal Transporters

Genes encoding members of the *Nramp* family of integral membrane proteins were identified through very diverse genetic screens (99). *Nramp1* determines sensitivity to mycobacterium infection such as tuberculosis or leprosy and gave its name, natural resistance–associated macrophage protein, to the gene family (100). *Nramp* homologous sequences have now been identified in bacteria, fungi, plants, and animals. It has been shown that the *Nramp* homologues SMF1 in yeast and DCT1/Nramp2 in mammals mediate the uptake of a broad range of metals (101–104). In plants, the ethylene insensitivity gene *EIN2*, which functions in signal transduction, contains an *Nramp* homologous domain but has no demonstrated metal transport function (105). *Arabidopsis* homologues of *Nramp* genes have been cloned and their function as metal transporters has been shown both in the yeast heterologous expression system and in planta (19,106).

Disruption of the *AtNramp3* gene leads to an increase in Cd^{2+} resistance, whereas overexpression of this gene confers increased Cd^{2+} sensitivity in *Arabidopsis* (19). These results point to a role of *AtNramps* in physiological Cd^{2+} transport and Cd^{2+} sensitivity in plants. *AtNramp3*, *AtNramp4*, and to some extent *AtNramp1* complement the phenotype of *fet3fet4*, a yeast mutant deficient in Fe uptake (see earlier). This result indicates that two families of Fe transporters, *IRTs* and *AtNramps* may contribute to Fe homeostasis in plants (19,106). The role of *AtNramps* in Fe transport in planta is supported by both the induction of *AtNramps* upon Fe starvation and the observation that *AtNramp3*-overexpressing plants can accumulate higher levels of Fe upon Cd^{2+} treatment (19), whereas *AtNramp1*-overexpressing plants confer resistance to toxic levels of Fe.

Thus, the broad specificity of *Nramp* transporters may render them good target genes to engineer metal nutrient uptake and transport in plants as well as toxic metal sensitivity and accumulation in plants.

3. The Uptake of Nonessential Metal Ions

There is no known biological function for the highly toxic metals Pb and Cd. Thus, it is assumed that no transporters specific for these metals exist. Instead, Pb^{2+} and Cd^{2+} ions most likely enter the cytosol of plant cells via nonselective cation transporters. Candidate proteins have been described. The screening for wheat cDNAs that complement a K^+ uptake deficiency of *S. cerevisiae* strain CY162 led to the isolation of a low-affinity cation transporter named LCT1 (107). Testing of effects on Cd^{2+} sensitivity of expressing yeast cells resulted in the observation that LCT1 mediates Cd^{2+} uptake (17). Further studies revealed that the physiological substrates for LCT1 are probably Ca^{2+} and perhaps Na^+ (17,107). Other potential pathways include some of the Nramp and ZIP transporters (see earlier). The IRT1-mediated Fe^{2+}, Mn^{2+}, and Zn^{2+} uptake is inhibited by Cd^{2+}. For ZNT1, low-affinity Cd^{2+} uptake activity has been demonstrated directly (98).

A cyclic nucleotide–gated channel from tobacco (NtCBP4) has been described as a first example of a plant transporter mediating Pb^{2+} uptake (18). Originally identified as a calmodulin-binding protein, this channel was found to localize to the plasma membrane. NtCBP4-overexpressing tobacco plants exhibit increased sensitivity toward Pb^{2+}, which correlates with enhanced Pb^{2+} accumulation. Interestingly, NtCBP4 overexpressors at the same time are more Ni^{2+} tolerant. Possible explanations for this effect are, as suggested by the authors, interaction of NtCBP4 with Ni^{2+}, which attenuates uptake, or the suppression of other, more Ni^{2+} selective channels, by NtCBP4 overexpression.

The dual effect described in this study provides another example of the potential of engineered changes in the expression of metal ion uptake systems for enhancing plant metal tolerance or accumulation. This approach will certainly be developed further with a more detailed understanding of the molecular structure of transporter proteins. Elucidation of specificity and affinity determinants, for instance, will allow the expression of mutated proteins that exhibit fewer undesired activities. A more indirect way of potentially minimizing toxic metal uptake is related to iron nutrition. *S. cerevisiae* cells are rendered more metal sensitive by defects in the high-affinity Fe uptake system. This is due to the Fe deficiency–induced expression of low-affinity uptake transporters such as FET3, which are less selective and facilitate entry of other heavy metal ions (77). Similarly, Cd^{2+} uptake into root cells of Fe-deficient pea seedlings is sevenfold higher than into Fe-sufficient pea seedlings (108). Because the kinetic properties are not affected by the iron status, this difference is attributable to the induction of iron-regulated IRT1-like transporters. Thus, enhancing the iron efficiency of plants may lead to a reduction in undesired uptake activities.

E. Biotransformation and Direct Removal

Mercury is one of the most hazardous heavy metal contaminants worldwide and has caused numerous ecological disasters, e.g., the poisoning of Minamata Bay in Japan in the 1950s and 1960s. Toward the goal of developing plants suitable for the phytoremediation of Hg-contaminated sites, genes of the *mer* operon of gram-negative bacteria have been very successfully used. MerA is a mercuric ion reductase catalyzing the production of elemental volatile Hg^0 from Hg^{2+}. Following extensive modification of the coding region for plant-optimized codon usage and insertion of a consensus plant translation signal into the 5' upstream region, the *merA* gene was first introduced into *Arabidopsis* under 35S control. The resulting transgenic lines grew and developed in the presence of Hg^{2+} concentrations (50–100 μM) that wild-type plants cannot tolerate (109). In accordance with the observation that *E. coli* expressing merA are more resistant to Au^{3+} ions, the transgenic *Arabidopsis* lines also grew better than the wild type on Au^{3+}-containing growth medium. When Hg^0 evolution was assayed in several lines, it was found to correlate well with merA mRNA levels and mercury resistance.

In order to evaluate the feasibility of this approach for phytoremediation, *merA* was also expressed in the forest tree yellow poplar (*Liriodendron tulipifera*) (110). Three constructs differing in the extent of modification of the coding region were used to transform proembryonic masses by particle bombardment. Again, *merA* expression led to the generation of Hg^{2+} resistance. Transgenic colonies, embryos, and plantlets were able to grow in the presence of normally toxic $HgCl_2$ concentrations. The most extensively altered construct produced the highest number of Hg^R colonies. The Hg^{2+} resistance correlated with the presence of the MerA protein and Hg^0 release.

The principal form of mercury that biomagnifies, i.e., accumulates in ecosystems, and caused the poisoning of Minamata Bay is methyl mercury (CH_3Hg^+), which arises through bacterial methylation of Hg(II). It is about 100-fold more toxic than Hg(II), probably because it can easily diffuse through biological membranes. The *mer* operon that was identified in bacteria isolated from mercury-contaminated environments consists of *merA*, a few other genes encoding Hg transporters and regulatory factors, and *merB*, which codes for an organomercurial lyase. This enzyme catalyzes the formation of Hg(II) from CH_3Hg^+ via protonolysis of the C—Hg bond. In order to generate plants suitable for phytoremediation of methyl mercury, the *merB* gene was modified and expressed in *Arabidopsis*, analogous to *merA* (111). Transgenic plants transcribing *merB* mRNA and synthesizing MerB protein were tested for growth on phenylmercuric acetate and monomethylmercuric chloride. At concentrations (0.5–2 μM) that are growth prohibiting for wild-type plants, the *merB* expressors germinated and grew well. The accumulation of Hg^{2+}, which has to be assumed for plants containing an organomercurial lyase and growing in the presence of methyl mercury, is apparently tolerable because Hg(II) is far less toxic and phytochelatins are likely to sequester some of it. Nonetheless, plants showing both the lyase and the reductase activity would represent even more efficient tools for the removal of mercury from contaminated sites. Consequently, the group of Richard Meagher crossed *Arabidopsis merA* and *merB* lines (112). The respective plants were found to be up to 10-fold more tolerant to methyl mercury than transgenics expressing *merB* alone, demonstrating the potential of this approach.

Selenium is an essential element that can be toxic when high levels are reached, for instance, in wetlands as a consequence of irrigation. Because selenium is similar to sulfur, selenocysteine can be formed, which, when incorporated into proteins, may lead to dysfunction (113). Two principal pathways for detoxification can be found in plants, chemical reduction and incorporation into organic compounds. The uptake of Se occurs mainly as selenate via high-affinity sulfate transporters. Because the reduction appeared to be the main factor limiting accumulation, the *Arabidopsis* ATP sulfurylase (APS), hypothesized to be the enzyme catalyzing selenate reduction, was overexpressed in *Brassica juncea* (114). The transgenic lines showed twofold higher APS activity and increases in selenate reduction, selenate accumulation, and Se tolerance.

F. Repair Mechanisms

One component of metal tolerance could be the repair of metal-induced damage. Copper is known to cause plasma membrane leakage resulting in K^+ efflux (115). A correlation was found in differentially Cu-sensitive *Arabidopsis* ecotypes between Cu sensitivity and the ability to reverse the K^+ efflux following Cu exposure (116). Also, lipid metabolism genes, possibly involved in membrane repair, are induced in *Arabidopsis* by Cu treatment (Murphy and Taiz, cited in Ref. 117).

Cadmium can denature proteins by interacting with SH groups and by replacing Zn. That the removal of abnormal proteins formed upon Cd exposure is of importance for the basic cellular cadmium tolerance was demonstrated for *S. cerevisiae* (118). Yeast cells with defects in either specific ubiquitin-conjugating enzymes or the proteasome are Cd hypersensitive. Moreover, expression of the ubiquitin system is induced by Cd treatment. This finding could not be confirmed for the *Arabidopsis* homologues (119). Thus, it remains unclear whether the UBCs play a similar role in plants.

III. MECHANISMS OF HYPERTOLERANCE

Species and genotypes, which are part of the heavy metal plant communities established on sites enriched in toxic heavy metals, are necessarily hypertolerant. The molecular mechanisms underlying this trait are little understood. The typical plant response to metal exposure, the formation of phytochelatins (see earlier), appears not to be involved in naturally occurring hypertolerance. *Silene vulgaris* ecotypes differing in Cu, Zn, or Cd tolerance have been investigated thoroughly in this respect. When Cd-tolerant and -sensitive lines were compared, it was found that the sensitive lines form more PCs in their roots upon Cd exposure. Furthermore, synthesis of PC-Cd complexes is more rapid in sensitive lines and the PC composition is the same in sensitive and tolerant lines (120). Equivalent results were obtained for *S. vulgaris* ecotypes with differential Cu and Zn tolerance

(121,122). At Cu concentrations that were affecting tolerant and sensitive plants equally, the phytochelatin accumulation was also equal. The Zn-tolerant *Silene* plants produced less PCs than Zn-sensitive plants after exposure to both the same Zn concentration and equal-effect Zn concentrations. The chain length distribution of PCs in both ecotypes was the same, indicating that there is no difference in PC stability explaining the differential tolerance.

The more rapid PC synthesis in sensitive *Silene* plants is interpreted as an indication of higher Zn concentrations in the cytosol. This difference is due not to a reduction of uptake but to increased compartmentalization of Zn in tolerant plants. There is evidence for enhanced Mg-ATP–energized vacuolar transport of Zn in tolerant ecotypes (123). This 2.5- to 3-fold difference cosegregates with Zn tolerance in crosses between Zn-tolerant and Zn-sensitive plants (124), which clearly shows the importance of vacuolar sequestration for Zn tolerance. Together with the effects of Zn transporter overexpression in *Arabidopsis* (74) (see earlier), these data represent a promising lead for the generation of Zn-hypertolerant plants.

The other example of a clear correlation between metal hypertolerance and a specific response or mechanism is the previously mentioned Ni-induced synthesis of histidine in some Ni hyperaccumulators, which results in efficient chelation of nickel (15).

IV. CONCLUSIONS

Mainly owing to physiological studies on model species and the use of *S. cerevisiae* for the functional expression of plant genes, significant advances have been achieved in the molecular understanding of plant metal tolerance and homeostasis mechanisms. A number of metal transporters involved in uptake and sequestration have been identified, and genes encoding enzymes involved in chelator synthesis have been cloned. The observations that Ni hyperaccumulators synthesize more histidine and Zn-tolerant *Silene vulgaris* lines show enhanced vacuolar Zn uptake provide valuable leads for the elucidation of naturally selected plant metal tolerance. Furthermore, using genes predominantly from bacterial sources, transgenic plants have been generated with enhanced mercury, methyl mercury, aluminum, and cadmium tolerance.

However, we are only beginning to understand the complexity of metal tolerance and accumulation. Many fundamental questions remain to be answered. For instance, we completely lack understanding of root-to-shoot translocation. Different plant species and even different genotypes of the same species vary greatly in the root/shoot ratio of metal accumulation. This ratio is the net result of the interplay of many different factors including uptake into root cells, sequestration in root cells, efflux, long-distance transport, uptake into leaf cells, and storage. Similarly, most of the players in-

requiring apoenzymes by phytochelatin-metal complexes. FEBS Lett 284:66–69, 1991.

26. R Howden, PB Goldsbrough, CR Andersen, CS Cobbett. Cadmium-sensitive, *cad1* mutants of *Arabidopsis thaliana* are phytochelatin deficient. Plant Physiol 107:1059–1066, 1995.

27. SB Ha, AP Smith, R Howden, WM Dietrich, S Bugg, MJ O'Connell, PB Goldsbrough, CS Cobbett. Phytochelatin synthase genes from *Arabidopsis* and the yeast *Schizosaccharomyces pombe*. Plant Cell 11:1153–1163, 1999.

28. O Vatamaniuk, S Mari, Y Lu, P Rea. AtPCS1, a phytochelatin synthase from *Arabidopsis*: isolation and in vitro reconstitution. Proc Natl Acad Sci USA 96:7110–7115, 1999.

29. ME Schmöger, M Oven, E Grill. Detoxification of arsenic by phytochelatins in plants. Plant Physiol 122:793–801, 2000.

30. I Pickering, R Prince, M George, R Smith, G George, DE Salt. Reduction and coordination of arsenic in indian mustard. Plant Physiol 122:1171–1178, 2000.

31. W Gekeler, E Grill, E-L Winnacker, MH Zenk. Survey of the plant kingdom for the ability to bind heavy metals through phytochelatins. Z Naturforsch 44c:361–369, 1989.

32. YL Zhu, EA Pilon-Smits, AS Tarun, SU Weber, L Jouanin, N Terry. Cadmium tolerance and accumulation in indian mustard is enhanced by overexpressing gamma-glutamylcysteine synthetase. Plant Physiol 121:1169–1178, 1999.

33. YL Zhu, EA Pilon-Smits, L Jouanin, N Terry. Overexpression of glutathione synthetase in indian mustard enhances cadmium accumulation and tolerance. Plant Physiol 119:73–80, 1999.

34. DH Hamer. Metallothionein. Annu Rev Biochem 55:913–951, 1986.

35. LT Jensen, W Howard, J Strain, DR Winge, V Culotta. Enhanced effectiveness of copper ion buffering by CUP1 metallothionein compared with CRS5 metallothionein in *Saccharomyces cerevisiae*. J Biol Chem 271:18514–18519, 1996.

36. W Yu, V Santhanagopalan, AK Sewell, LT Jensen, DR Winge. Dominance of metallothionein in metal ion buffering in yeast capable of synthesis of (gamma EC)nG isopeptides. J Biol Chem 269:21010–21015, 1994.

37. BA Masters, EJ Kelly, CJ Quaife, RL Brinster, RD Palmiter. Targeted disruption of metallothionein I and II genes increases sensitivity to cadmium. Proc Natl Acad Sci USA 91:584–588, 1994.

38. RD Palmiter. The elusive function of metallothioneins. Proc Natl Acad Sci USA 95:8428–8430, 1998.

39. WE Rauser. Structure and function of metal chelators produced by plants. Cell Biochem Biophys 31:19–48, 1999.

40. J Zhou, PB Goldsbrough. Functional homologs of fungal metallothionein genes from *Arabidopsis*. Plant Cell 6:875–884, 1994.

41. NJ Robinson, J Wilson, J Turner. Expression of the type 2 metallothionein-like gene *MT2* from *Arabidopsis thaliana* in Zn^{2+}-metallothionein-deficient

Synechococcus PCC 7942: putative role for MT2 in Zn^{2+} metabolism. Plant Mol Biol 30:1169–1179, 1996.

42. A Murphy, L Taiz. Comparison of metallothionein gene expression and non-protein thiols in ten *Arabidopsis* ecotypes. Correlation with copper tolerance. Plant Physiol 109:945–954, 1995.

43. DD Lefebvre, BL Miki, J-F Laliberte. Mammalian metallothionein functions in plants. Biotechnology 5:1053–1056, 1987.

44. A Pan, M Yang, F Tie, L Li, Z Chen, B Ru. Expression of mouse metallothionein-I gene confers cadmium resistance in transgenic tobacco plants. Plant Mol Biol 24:341–351, 1994.

45. T Elmayan, M Tepfer. Synthesis of a bifunctional metallothionein/beta-glucuronidase fusion protein in transgenic tobacco plants as a means of reducing leaf cadmium levels. Plant J 6:433–440, 1994.

46. MC Suh, D Choi, JR Liu. Cadmium resistance in transgenic tobacco plants expressing the *Nicotiana glutinosa* L. metallothionein-like gene. Mol Cells 8: 678–684, 1998.

47. I Potrykus, cited in Ref. 9.

48. A Dancis, DS Yuan, D Haile, C Askwith, D Eide, C Moehle, J Kaplan, RD Klausner. Molecular characterization of a copper transport protein in *S. cerevisiae*: an unexpected role for copper in iron transport. Cell 76:393–402, 1994.

49. RA Pufahl, CP Singer, KL Peariso, SJ Lin, PJ Schmidt, CJ Fahrni, VC Culotta, JE Penner-Hahn, TV O'Halloran. Metal ion chaperone function of the soluble Cu(I) receptor ATX1. Science 278:853–856, 1997.

50. I Hamza, M Schaefer, LW Klomp, JD Gitlin. Interaction of the copper chaperone HAH1 with the Wilson disease protein is essential for copper homeostasis. Proc Natl Acad Sci USA 96:13363–13368, 1999.

51. DM Glerum, A Shtanko, A Tzagoloff. Characterization of *COX17*, a yeast gene involved in copper metabolism and assembly of cytochrome oxidase. J Biol Chem 271:14504–14509, 1996.

52. VC Culotta, LW Klomp, J Strain, RL Casareno, B Krems, JD Gitlin. The copper chaperone for superoxide dismutase. J Biol Chem 272:23469–23472, 1997.

53. E Himelblau, H Mira, SJ Lin, VC Culotta, L Penarrubia, RM Amasino. Identification of a functional homolog of the yeast copper homeostasis gene *ATX1* from *Arabidopsis*. Plant Physiol 117:1227–1234, 1998.

54. T Hirayama, JJ Kieber, N Hirayama, M Kogan, P Guzman, S Nourizadeh, JM Alonso, WP Dailey, A Dancis, JR Ecker. RESPONSIVE-TO-ANTAGONIST1, a Menkes/Wilson disease-related copper transporter, is required for ethylene signaling in *Arabidopsis*. Cell 97:383–393, 1999.

55. J Valentine, E Gralla. Delivering copper inside yeast and human cells. Science 278:817–818, 1997.

56. F Goto, T Yoshihara, N Shigemoto, S Toki, F Takaiwa. Iron fortification of rice seed by the soybean ferritin gene. Nat Biotechnol 17:282–286, 1997.

57. MW Persans, X Yan, JM Patnoe, U Krämer, DE Salt. Molecular dissection

of the role of histidine in nickel hyperaccumulation in *Thlaspi goesingense.* Plant Physiol 121:1117–1126, 1999.

58. U Krämer, RD Smith, WW Wenzel, I Raskin, DE Salt. The role of metal transport and tolerance in nickel hyperaccumulation by *Thlaspi goesingense* Halacsy. Plant Physiol 115:1641–1650, 1997.

59. SC Miyasaka, JG Buta, RK Howell, CD Foy. Mechanism of aluminum tolerance in snapbeans: root exudation of citric acid. Plant Physiol 96:737–743, 1991.

60. E Delhaize, S Craig, CD Beaton, RJ Bennet, VC Jagadish, PJ Randall. Aluminum tolerance in wheat (*Triticum aestivum* L.). I. Uptake and distribution of aluminum in root apices. Plant Physiol 103:685–693, 1993.

61. E Delhaize, PR Ryan, PJ Randall. Aluminum tolerance in wheat (*Triticum aestivum* L.). II. Aluminum-stimulated excretion of malic acid from root apices. Plant Physiol 103:685–693, 1993.

62. JM de la Fuente, V Ramirez-Rodriguez, JL Cabrera-Ponce, L Herrera-Estrella. Aluminum tolerance in transgenic plants by alteration of citrate synthesis. Science 276:1566–1568, 1997.

63. UW Stephan, G Scholz. Nicotianamine: mediator of transport of iron and heavy metals in the phloem? Physiol Plant 88:522–527, 1993.

64. HQ Ling, G Koch, H Bäumlein, MW Ganal. Map-based cloning of chloronerva, a gene involved in iron uptake of higher plants encoding nicotianamine synthase. Proc Natl Acad Sci USA 96:7098–7103, 1999.

65. K Higuchi, K Suzuki, H Nakanishi, H Yamaguchi, NK Nishizawa, S Mori. Cloning of nicotianamine synthase genes, novel genes involved in the biosynthesis of phytosiderophores. Plant Physiol 119:471–480, 1999.

66. DF Ortiz, L Kreppel, DM Speiser, G Scheel, G McDonald, DW Ow. Heavy metal tolerance in the fission yeast requires an ATP-binding cassette-type vacuolar membrane transporter. EMBO J 11:3491–3499, 1992.

67. DF Ortiz, T Ruscitti, KF McCue, DW Ow. Transport of metal-binding peptides by HMT1, a fission yeast ABC-type vacuolar membrane protein. J Biol Chem 270:4721–4728, 1995.

68. PA Rea, ZS Li, YP Lu, YM Drozdowicz, E Martinoia. From vacuolar GS-X pumps to multispecific ABC transporters. Annu Rev Plant Physiol Plant Mol Biol 49:727–760, 1998.

69. MS Szczypka, JA Wemmie, WS Moye-Rowley, DJ Thiele. A yeast metal resistance protein similar to human cystic fibrosis transmembrane conductance regulator (CFTR) and multidrug resistance–associated protein. J Biol Chem 269:22853–22857, 1994.

70. ZS Li, YP Lu, RG Zhen, M Szczypka, DJ Thiele, PA Rea. A new pathway for vacuolar cadmium sequestration in *Saccharomyces cerevisiae*: YCF1-catalyzed transport of bis(glutathionato)cadmium. Proc Natl Acad Sci USA 94: 42–47, 1997.

71. M Ghosh, J Shen, BP Rosen. Pathways of As(III) detoxification in *Saccharomyces cerevisiae*. Proc Natl Acad Sci USA 96:5001–5006, 1999.

72. DH Nies. The cobalt, zinc, and cadmium efflux system CzcABC from *Al-*

caligenes eutrophus functions as a cation-proton antiporter in *Escherichia coli*. J Bacteriol 177:2707–2712, 1995.

73. IT Paulsen, MH Saier Jr. A novel family of ubiquitous heavy metal ion transport proteins. J Membr Biol 156:99–103, 1997.

74. BJ van der Zaal, LW Neuteboom, JE Pinas, AN Chardonnens, H Schat, JA Verkleij, PJ Hooykaas. Overexpression of a novel *Arabidopsis* gene related to putative zinc-transporter genes from animals can lead to enhanced zinc resistance and accumulation. Plant Physiol 119:1047–1056, 1999.

75. DS Conklin, JA McMaster, MR Culbertson, C Kung. *COT1*, a gene involved in cobalt accumulation in *Saccharomyces cerevisiae*. Mol Cell Biol 12:3678–3688, 1992.

76. A Kamizono, M Nishizawa, Y Teranishi, K Murata, A Kimura. Identification of a gene conferring resistance to zinc and cadmium ions in the yeast *Saccharomyces cerevisiae*. Mol Gen Genet 219:161–167, 1989.

77. L Li, J Kaplan. Defects in the yeast high affinity iron transport system result in increased metal sensitivity because of the increased expression of transporters with a broad transition metal specificity. J Biol Chem 273:22181–22187, 1998.

78. RD Palmiter, SD Findley. Cloning and functional characterization of a mammalian zinc transporter that confers resistance to zinc. EMBO J 14:639–649, 1995.

79. RD Palmiter, TB Cole, SD Findley. ZnT-2, a mammalian protein that confers resistance to zinc by facilitating vesicular sequestration. EMBO J 15:1784–1791, 1996.

80. RD Palmiter, TB Cole, CJ Quaife, SD Findley. ZnT-3, a putative transporter of zinc into synaptic vesicles. Proc Natl Acad Sci USA 93:14934–14939, 1998.

81. DE Salt, GJ Wagner. Cadmium transport across tonoplast of vesicles from oat roots. Evidence for a Cd^{2+}/H^+ antiport activity. J Biol Chem 268:12297–12302, 1993.

82. BJ Barkla, O Pantoja. Physiology of ion transport across the tonoplast of higher plants. Annu Rev Plant Physiol Plant Mol Biol 47:159–184, 1994.

83. KD Hirschi, RG Zhen, KW Cunningham, PA Rea, GR Fink. CAX1, an H^+/Ca^{2+} antiporter from *Arabidopsis*. Proc Natl Acad Sci USA 93:8782–8786, 1996.

84. S Silver. Bacterial resistances to toxic metal ions—a review. Gene 179:9–19, 1996.

85. M Solioz, C Vulpe. CPx-type ATPases: a class of P-type ATPases that pump heavy metals. Trends Biochem Sci 21:237–241, 1997.

86. KJ Tsai, KP Yoon, AR Lynn. ATP-dependent cadmium transport by the cadA cadmium resistance determinant in everted membrane vesicles of *Bacillus subtilis*. J Bacteriol 174:116–121, 1992.

87. A Odermatt, H Suter, R Krapf, M Solioz. Primary structure of two P-type ATPases involved in copper homeostasis in *Enterococcus hirae*. J Biol Chem 268:12775–12779, 1993.

88. C Vulpe, B Levinson, S Whitney, S Packman, J Gitschier. Isolation of a candidate gene for Menkes disease and evidence that it encodes a copper-transporting ATPase. Nat Genet 3:7–13, 1993.

89. PC Bull, GR Thomas, JM Rommens, JR Forbes, DW Cox. The Wilson disease gene is a putative copper transporting P-type ATPase similar to the Menkes gene. Nat Genet 5:327–337, 1993.

90. D Fu, TJ Beeler, TM Dunn. Sequence, mapping and disruption of CCC2, a gene that cross-complements the Ca^{2+}-sensitive phenotype of csg1 mutants and encodes a P-type ATPase belonging to the Cu^{2+}-ATPase subfamily. Yeast 11:283–292, 1995.

91. Z Weissman, I Berdicevsky, BZ Cavari, D Kornitzer. The high copper tolerance of *Candida albicans* is mediated by a P-type ATPase. Proc Natl Acad Sci USA 97:3520–3525, 2000.

92. D Eide, M Broderius, J Fett, ML Guerinot. A novel iron-regulated metal transporter from plants identified by functional expression in yeast. Proc Natl Acad Sci USA 93:5624–5628, 1996.

93. YO Korshunova, D Eide, WG Clark, ML Guerinot, HB Pakrasi. The IRT1 protein from *Arabidopsis thaliana* is a metal transporter with a broad substrate range. Plant Mol Biol 40:37–44, 1999.

94. H Zhao, D Eide. The yeast *ZRT1* gene encodes the zinc transporter protein of a high-affinity uptake system induced by zinc limitation. Proc Natl Acad Sci USA 93:2454–2458, 1996.

95. H Zhao, D Eide. The ZRT2 gene encodes the low affinity zinc transporter in *Saccharomyces cerevisiae*. J Biol Chem 271:23203–23210, 1996.

96. N Grotz, T Fox, E Connolly, W Park, ML Guerinot, D Eide. Identification of a family of zinc transporter genes from *Arabidopsis* that respond to zinc deficiency. Proc Natl Acad Sci USA 95:7220–7224, 1998.

97. MM Lasat, AIM Baker, LV Kochian. Physiological characterization of root Zn^{2+} absorption and translocation to shoots in Zn hyperaccumulator and non-accumulator species of *Thlaspi*. Plant Physiol 112:1715–1722, 1996.

98. NS Pence, PB Larsen, SD Ebbs, DL Letham, MM Lasat, DF Garvin, D Eide, LV Kochian. The molecular physiology of heavy metal transport in the Zn/Cd hyperaccumulator *Thlaspi caerulescens*. Proc Natl Acad Sci USA 97: 4956–4960, 2000.

99. M Cellier, G Prive, A Belouchi, T Kwan, V Rodrigues, W Chia, P Gros. Nramp defines a family of membrane proteins. Proc Natl Acad Sci USA 92: 10089–10093, 1995.

100. S Vidal, D Malo, K Vogan, E Skamene, P Gros. Natural resistance to infection with intracellular parasites: isolation of a candidate for Bcg. Cell 73:469–485, 1993.

101. F Supek, L Supekova, H Nelson, N Nelson. A yeast manganese transporter related to the macrophage protein involved in conferring resistance to mycobacteria. Proc Natl Acad Sci USA 93:5105–5110, 1996.

102. H Gunshin, B Mackenzie, UV Berger, Y Gunshin, MF Romero, WF Boron, S Nussberger, JL Gollan, MA Hediger. Cloning and characterization of a mammalian proton-coupled metal-ion transporter. Nature 388:482–488, 1997.

103. X Liu, F Supek, N Nelson, VC Culotta. Negative control of heavy metal uptake by the *Saccharomyces cerevisiae BSD2* gene. J Biol Chem 272:11763–11769, 1997.

104. X Chen, J Peng, A Cohen, H Nelson, N Nelson, MA Hediger. Yeast SMF1 mediates H^+-coupled iron uptake with concomitant uncoupled cation currents. J Biol Chem 274:35089–35094, 1999.

105. J Alonso, T Hirayama, G Roman, S Nourizadeh, J Ecker. EIN2, a bifunctional transducer of ethylene and stress responses in *Arabidopsis*. Science 248: 2148–2152, 1999.

106. C Curie, JM Alonso, M Le Jean, JR Ecker, JF Briat. Involvement of NRAMP1 from *Arabidopsis thaliana* in iron transport. Biochem J 347:749–755, 2000.

107. DP Schachtman, R Kumar, JI Schroeder, EL Marsh. Molecular and functional characterization of a novel low-affinity cation transporter (LCT1) in higher plants. Proc Natl Acad Sci USA 94:11079–11084, 1997.

108. CK Cohen, TC Fox, DF Garvin, LV Kochian. The role of iron-deficiency stress responses in stimulating heavy-metal transport in plants. Plant Physiol 116:1063–1072, 1998.

109. CL Rugh, HD Wilde, NM Stack, DM Thompson, AO Summers, RB Meagher. Mercuric ion reduction and resistance in transgenic *Arabidopsis thaliana* plants expressing a modified bacterial *merA* gene. Proc Natl Acad Sci USA 93:3182–3187, 1996.

110. CL Rugh, JF Senecoff, RB Meagher, SA Merkle. Development of transgenic yellow poplar for mercury phytoremediation. Nat Biotechnol 16:925–928, 1998.

111. SP Bizily, CL Rugh, AO Summers, RB Meagher. Phytoremediation of methylmercury pollution: *merB* expression in *Arabidopsis thaliana* confers resistance to organomercurials. Proc Natl Acad Sci USA 96:6808–6813, 1999.

112. SP Bizily, CL Rugh, RB Meagher. Phytodetoxification of hazardous organomercurials by genetically engineered plants. Nat Biotechnol 18:213–217, 2000.

113. A Läuchli. Selenium in plants: uptake, functions, and environmental toxicity. Bot Acta 106:455–468, 1993.

114. EA Pilon-Smits, S Hwang, C Mel Lytle, Y Zhu, JC Tai, RC Bravo, Y Chen, T Leustek, N Terry. Overexpression of ATP sulfurylase in indian mustard leads to increased selenate uptake, reduction, and tolerance. Plant Physiol 119: 123–132, 1999.

115. AS Murphy, WR Eisinger, JE Shaff, LV Kochian, L Taiz. Early copper-induced leakage of K^+ from *Arabidopsis* seedlings is mediated by ion channels and coupled to citrate efflux. Plant Physiol 121:1375–1382, 1999.

116. AS Murphy, L Taiz. Correlation between potassium efflux and copper sensitivity in 10 *Arabidopsis* ecotypes. New Phytol 136:211–222, 1998.

117. DE Salt, RD Smith, I Raskin. Phytoremediation. Annu Rev Plant Physiol Plant Mol Biol 49:643–668, 1998.

118. J Jungmann, HA Reins, C Schobert, S Jentsch. Resistance to cadmium mediated by ubiquitin-dependent proteolysis. Nature 361:369–371, 1993.

119. S van Nocker, JM Walker, RD Vierstra. The *Arabidopsis thaliana UBC7/13/ 14* genes encode a family of multiubiquitin chain-forming E2 enzymes. J Biol Chem 271:12150–12158, 1996.

120. JA de Knecht, M van Dillen, PLM Koevoets, H Schat, JAC Verkleij, WHO Ernst. Phytochelatins in cadmium-sensitive and cadmium-tolerant *Silene vulgaris*. Plant Physiol 104:255–261, 1994.

121. H Schat, MMA Kalff. Are phytochelatins involved in differential metal tolerance or do they merely reflect metal-imposed strain? Plant Physiol 99:1475–1480, 1992.

122. H Harmens, PR Den Hartog, WM Ten Bookum, JAC Verkleij. Increased zinc tolerance in *Silene vulgaris* (Moench) Garcke is not due to increased production of phytochelatins. Plant Physiol 103:1305–1309, 1993.

123. JAC Verkleij, PLM Koevoets, MMA Blake-Kalff, AN Chardonnens. Evidence for an important role of the tonoplast in the mechanism of naturally selected Zn tolerance in *Silene vulgaris*. J Plant Physiol 153:188–191, 1998.

124. AN Chardonnens, PLM Koevoets, A van Zanten, H Schat, JAC Verkleij. Properties of enhanced tonoplast zinc transport in naturally selected zinc-tolerant *Silene vulgaris*. Plant Physiol 120:779–786, 1999.

Index

Abies grandis, 329
Abiotic elicitors, 82
Abscisic acid, 428
Adenosine triphosphatase, 643
Advanced Technologies Cambridge, 489
Aegilops cylindrica, 610, 613, 618
Aegilops squarrosa, 613
Aeromonas sp., 387
African cassava mosaic begomovirus, 528
Agave sisalana, 28
Aging, diseases of, 207
Agrobacterium, 113, 116–123, 125, 126, 128, 129, 133, 147, 326, 358, 362, 382, 383, 384, 387, 388, 410, 524, 612
Agrobacterium rhizogenes, 79, 94, 96, 143, 146, 147, 331, 351, 357, 358, 361, 388, 611
Agrobacterium tumefaciens, 4, 79, 112, 116–121, 118–121, 127, 150, 328, 361, 362, 385, 387, 388, 390, 413, 588, 611
  marker genes, 117
  Ri plasmids, conjugation systems, 121

[*Agrobacterium tumefaciens*]
  T-DNA, increasing capacity, 121
  Ti-plasmid vectors, 116–118
Agrotis ipsilon, 573
Airlift reactor, 173–176, 183
Ajmalicine, 385
Ajuga pyramidalis, 84, 86
Ajuga reptans, 84, 86
Albumins, 285–286
Alkaloid biosynthetic genes, 351
Alkaloids, 347
  for pharmaceutical applications, 347
  tropane, 29–30
Alkannin, 83
Allergenicity of modified plant, 225
Allium cepa, 338
Allium sativum, 33, 55
Alocasia mycrorrhiza, 583
Alpha amylase inhibitors, insect pest tolerance, 584–585
Alpha-linolenic acids, 312
Alternaria solani, 588
Alternaria longipes, 552
Aluminum, 672–673
Alyssum bertolonii, 146, 153

693

Alyssum lesbiacum, 672
Alyssum montanum, 672
Alyssum tenium, 146
Amaranthus hypochondriacus, 237
Amino acid, chelation, heavy metals,
    672–673
Amino acid content
    barley, 290
    legume proteins, 204–205
    maize, 290
    soybean seed, 290
    wheat, 290
Amino acid metabolic pathways, 236–
    248
Amylopectin structure alteration, 269–
    271
Amylopectin synthesis, 263–264
Amylose, 257
    complement alteration, 269
    synthesis, 263
Ancient art, plant breeding as, 2
Ancient civilizations, decline of, 636
Anise, 335
    as source of flavor component, 337
Anitsense, sense RNAs, RNA-mediated
    protection, 527–528
Anthemis nobilis, 31
Anthocyanin, 80, 84–88
    bioreactor culture, 88
    cell line selection, 86–87
    chemical factors, 87–88
    chemical structure, 85
    commercial production, 165
    cyanidin, 85
    delphinidin, 85
    malvidin, 85
    pelargonidin, 85
    peonidin, 85
    petunidin, 85
    physical environment, 87
    pigments, 12
    recombinant DNA technology, 88
Anthonomonas grandis grandis, 587
Antibiotic resistance genes, 225
Antibodies, transgenic plants for pro-
    duction of, 405

Antibody-derived molecules, 414
Antibody glycosylation, 408–410
Antibody isotype, 413–415
Antimicrobial pathogen tolerance, 549–
    569
    multigene defense mechanisms, 555–
        559
        cell death, 557–562
        constitutive systemic acquired re-
            sistance, 555
        hydrogen peroxide, 555–557
        salicylic acid, 555
    salicylic acid function, programmed
        cell death, 559–562
    single-gene defense mechanisms,
        550–554
        antimicrobial secondary com-
            pounds, 554
        defense peptides, 552–553
        pathogenesis-related proteins, 550–
            552
        ribosome-inactivating proteins,
            553–554
Antineoplastics, 28, 94–100
    camptothecin, 95, 96–97
    Catharanthus alkaloids, 94–96
    chemical structures, 95
    paclitaxel, 95, 98
    podophyllotoxin, 95, 98–100
    taxol, 98
    vinblastine, 95
    vincristine, 95
Antinutritional factors, in developing
    world, 205–208
Antioxidant nutrients, in U.S. popula-
    tion, 206
Antisense, sense RNAs, RNA-mediated
    protection, 527–528
Apomixis, 70
Arabidopsis, 8, 9, 13, 15, 16, 17, 119,
    127, 202, 241, 242, 246, 248,
    265, 271, 292, 293, 307, 311,
    314, 315, 316, 406, 431, 434,
    436, 437, 457, 462, 464, 482,
    552, 553, 555, 557, 559, 560,
    561, 605, 636, 643, 645, 647,

[*Arabidopsis*]
  648, 649, 650, 651, 652, 668,
  670, 671, 674, 675, 676, 677,
  678, 679, 680, 681, 682
Arabidopsis thaliana, 7–14, 8, 13, 146,
  239, 240, 355, 428, 429, 430,
  431, 432, 433, 435, 436, 504,
  508, 520, 644, 647
Arabinogalactan proteins, 460
Aralia cordata, 86, 87, 89
Arbutin, commercial production, 165
Archeological evidence, well-known
  crop species, 2
Arecolin, chemical synthesis, 349
Armoracia rusticana, 152
Aroma, from plants, 323–346
  anise, 335
  cell culture, 331–339
  genomics, 328–331
  hairy root culture, 335
  organ culture, 331–339
  plant enzymes, 328–331
  tissue culture, 331–339
Arsenic, 665
Artemisia annua, 24, 79
Artemisinin, 24–25
"Artificial," aroma compounds, 323
Asclepia syriaca, 310
Ascorbate peroxidase, oxidative stress,
  501–502, 508–509
Aspartate family biosynthetic pathway,
  238
Aspartate kinase genes, 239–240
Aspergillus, 556
Aspergillus fumigatus, 248
Aspergillus nidulans, 652
Assay technologies
  in industrial discovery processes, 50
  cellular assays, 50
  filtration assays, 50
  flashplate, 50
  fluorescence correlation spectros-
    copy, 50
  fluorescence intensity, 50
  fluorescence methods, noncellular
    assays, 50

[Assay technologies]
  fluorescence polarization, 50
  isotropic methods, 50
  luminescence, 50
  quantification of expressed protein,
    50
  reporter gene, 50
  time-resolved fluorescence, 50
  ultraviolet-visible absorbance, 50
Astertripolium, 648
Atropa belladonna, 80, 146, 148, 149,
  152, 355, 358, 362
Atropine, 29
Aulacorthum solani, 586
Automation, in industrial discovery processes, 49–51
  assay technologies, 50
  compound storage system, 49
  data management, 49
  semiautomatic workstations, 49
Auxin, 80
Avena sativa, 355
Axial flow impellers, 181

Bacillus lincheniformis, 387
Bacillus sp., 582
Bacillus stearothermophilus, 268
Bacillus strains, 582
Bacillus subtilis, 272
Bacillus thuringiensis, 363, 572–581,
  588, 600
Bacillus thuringiensis cry genes, insect
  pest tolerance, 572
Bacillus thuringiensis subsp. *kurstaki,*
  573, 575
Bacillus thuringiensis subsp. *tene-
  brionis,* 579
Bacterial ribonuclease barnase, 558
Barley, amino acid content, 290
Benign prostatic hyperplasia, 27,
  36
Berberis koetineana, 354
Berberis stolonifera, 354
Berberis wilsonae, 183
Berkheya coddii, 146
Bertholletia excelsa, 292

Beta-carotene, 18
  biosynthesis, 222
  intake, 206
Beta-sitosterol, cholesterol absorption,
    27
Beta vulgaris, 152, 185
Betacyanins, commercial production,
    165
Betamethasone, 28
Betula, 72
Betula pendula, 72
Bio-diesel, as petrol, 18
Biodiversity, forest, loss of, 74
Bioprospecting, 51–52
Bioreactors, 163–199
  cell, tissue culture types and, 180
  cell-lift fermenter, 183
  cell suspension, 179–184
    impeller types for, 182
    microbial culture, compared, 166
  classification, 169–179
    airlift reactor, 173–176
    bubble column, 173
    fluidized bed reactor, 176
    mechanically driven, 174
    membrane reactor, 176–179
    packed bed reactor, 176
    pneumatically agitated, 175
    rotating drum reactor, 173
    stirred reactor, 173
    tickle bed reactor, 176
  cultures, 80
  eccentric motion stirrer, internal illu-
    mination cage, 188
  economic feasibility, 187–191
  embryogenic cultures, reactors for,
    185–187
  hairy root cultures
    characteristics of, 184
    reactors for, 184–185
  instrumentation, 166–169
  operational strategies, 191–193
  process parameters, 167
  requirements, 164–166
  secondary metabolites, commercial
    production levels, 165

[Bioreactors]
  shoot cultures, reactors for, 185–187
  wave reactor, 190
Biotechnology, defined, 1
Biotic elicitors, 82
Biotin, RDA, 207
Blue light, absorbing cryptochromes, 4
Blumeria graminis f. sp. *tritici,* 556
Boehringer Ingelheim, supplier of tro-
    pane alkaloids, 358
Bomeol, 332
Boragio officinalis, 33, 316
Botrytis cinerea, 429, 431, 554, 588
BPH. *See* Benign prostatic hyperplasia
Brassica, 452
  genome sequence of, 7
Brassica campestris, 387, 602, 604,
    613, 616, 619, 670
Brassica juncea, 243, 669, 681
Brassica napus, 378, 383, 551, 602,
    613, 619
Brassica oleracea, 613
Brassica rapa, 616, 619
Brassinosteroids, herbicide resistance,
    623
Brazillian snapdragon, as source of fla-
    vor component, 337
Brazzein, 326
Bread making, 201
Brochothrix thermosphacta, 387
Bruchus pisorum, 584
Bubble columns, 173, 183
Bupleurum falcatum, 87
Butylidenephthalide, 332

Cadmium, 665–683
  cancer response to, 665
Caenorhabditis, 531
Caenorhabditis elegans, 7, 48
Calcium, 206–207
  environmental stress adaptation, 645
  RDA, 207
Calcium-dependent protein kinases, 432
Calcium signaling, 432–433
Calendula officinalis, 315
Callitris drummondii, 99

Callosobruchus analis, 584
Callosobruchus maculatus, 584
Callus, cell suspension cultures, 78
Calmodulin, 432
Calystegia sepium, 152
Campesterol, 221
Camphene, 332
Camphor, 332
Camptotheca acuminata, 55, 96, 97,
  348, 355
Camptothecin, 45, 95, 96–97
  chemical structure, 350
Capsaicin, 332
Capsicum chinense, 520, 584
Capsicum frutescens, 84, 331, 336, 337,
  338
Carbohydrate biosynthesis modification,
  265–273
  amylopectin structure alteration, 269–
    271
  amylose complement alteration, 269
  nonstarch carbohydrates, 271–273
  simple sugars, in storage organs,
    267–268
  starch quantity alteration, 266–267
Carboxylic acid oxidase, 328
Cardiac glycosides, 25
CaroRx, dental caries prevention, 415–
  418
Carotenoid, 219
  biosynthesis, 222–223
Carthamin, commercial production, 165
Carthamus tinctorius, 185
Carvacrol, 332
Castanospermum australe, 146
Catalase
  gene suppression, antisense strategy,
    556
  oxidative stress, 502
Cathanranthus roseus, indole alkaloid
  production, 373–403
Catharanthus, 94–96
Catharanthus pusillus, 388
Catharanthus roseus, 30, 54, 78, 79, 84,
    86, 88, 94, 95, 147, 152, 354,
    355, 356, 358, 373–403, 587

Catharanthus trichophyllus, 184
Cation diffusion facilitators, sequestration, heavy metals, 674–675
Cell culture, 5–7
  anthocyanin, 84–88
  antineoplastic compounds
    camptothecin, 95, 96–97
    Catharanthus alkaloids, 94–96
    chemical structures, 95
    paclitaxel, 95, 98
    podophyllotoxin, 95, 98–100
    vinblastine, 95
    vincristine, 95
  cell suspension cultures, 6
  cyanidin, 85
  D1 protein, 6
  delphinidin, 85
  dihydroshikonofuran, 93
  geranylhydroquione, 93
  geranylpyrophosphate, 93
  Lithospermum erythrorhizon cell cul-
    tures, 90–91
  m-geranyl-*p*-hydroxybenzoic acid, 93
  malvidin, 85
  manipulated genotypes, 7
  mevalonic acid, 93
  p-hydroxybenzoic acid O-glucoside,
    92
  p-hydroxybenzole acid, 92
  pelargonidin, 85
  peonidin, 85
  petunidin, 85
  phenylalanine, 92
  physical environment, 87
  phytochemical production
    bioreactor cultures, 80
    callus, cell suspension cultures, 78
    immobilized cultures, 78
    organ cultures, 78–80
  public debate on, 7
  recombinant DNA technology, 88
  regulatory mechanism, 91–94
  regulatory mutants, characterization
    of, 6
  resistance mutants, characterization
    of, 6

[Cell culture]
  secondary compounds, 77–109
    antineoplastic compounds, 94–100
    phytochemical production, 78–80
    pigments, 84–94
    secondary metabolite production, 80–84
  secondary metabolite production, 83–84
    abiotic elicitors, 82
    aeration, 83
    alkannin, 83
    anthocyanin, 80
    auxin, 80
    biotic elicitors, 82
    cell density, 83
    cell-to-cell variation, 83
    cytokinin, 80
    elicitors, 82–83
    gibberellin, 80
    light, 83
    medium nutrients, 81–82
    nitrogen, 81
    pH, 83
    phosphate level, 81
    plant growth regulators, 80–81
    rosmarinic acid, 83
    stilbene, 83
    sucrose, 82
    taxol, 83
    temperature, 83
  shikonin, 88–94
  totipotency, 5
  uptake mutants, characterization of, 6
Cell death, microbial pathogen tolerance, 557–562
Cell-lift fermenter, 183
Cell suspension cultures, 6
  potential of, 5
Cell wall, 445–475
  architecture of, 452–456
  assembly of, 457–460
  biosynthesis, 457–460
  components of, 452–456
  fibers of, 465
  functional roles of, 449

[Cell wall]
  functions of, 447–452
  future biotechnology, 466–468
  model, dicot plants, 453
  performance optimization, 460–463
  polymers, 460
  proteins, 459
  timber, 464
  transport in, 450
  wood, 464
Cellulomonas uda, 387
Cellulose microfibrils, 453
Cellulose synthase, 457–458
Cellulose synthetases, 9
Cellulose-xyloglucan network, 454
Cercospora nicotianae, 550
Chalcones, biosynthesis of, 214
Chamaemelum nobile, 31
Chamomile, 31–32
Chamonilla recutita, 217
Chelation, heavy metals, 668–673
  metallothioneins, 669–671
  organic acids, amino acids, 672–673
  phytochelatins, GSH metabolism, 668–669
  proteinaceous chelators, 671–672
Chimeric defense peptides, 553
China, plant breeding, 71
Chinchona officinalis, 354
Chinese herb formulas, in industrial discovery processes, 55
Chinese medicine, 54
Chitinase expression, 550
Chlamydomonas, 265
Chlamydomonas rheinhardtii, 14, 129
Chloroplast
  membrane organization of, 4
  nuclear gene products targeted to, 8
Cholesterol absorption, beta-sitosterol, 27
Chondriome, herbicide resistance, 618–619
Cinchona officinalis, 354, 388
Cinchona sp., 54
Citrobacter freundil, 387
Citrus aurantiifolia, 333

Citrus limon, 333
Citrus paradisi, 333, 337, 338
Citrus sinensis, 333, 337
Citrus tristeza virus, 525
Cladosporium fulvum, 430, 559
Clarkia breweri, 328
Cleanup methods, cost of, 666
Climatic changes, global, 636
Clonal forestry, 75–76
  avoidance of genetic narrowing, 75
Clonally propagated plants, breeding,
    70–71
Coal, replaced by plant raw materials,
    18
Coat proteins, protein-mediated protec-
    tion, 526–527
Cocoa butter, 307
Coconut oil, 306
Codeine, 26, 27
Coleonema album, 333, 337
Coleus blumei, 82
Colon, genomes, 13
Colors, flower, biosynthetic pathway, 3
Compound storage system, in automa-
    tion, 49
Cooking, bioavailability by, 223
Copper, 665–683
Coptis japonica, 80, 83, 374
Corticoids, 28
Cortisone acetate, 28
Corynebacterium, 291
Crataegus azarolus, 40
Crataegus laevigata, 39
Crataegus monogyna, 39
Crataegus nigra, 39
Crataegus pentagyna, 39
Crepis palestrina, 315
Crepis alpina, 315
Cryptochromes, blue light, absorbing, 4
Cryptogein, 558
Cryptomeria, 72
Cucumis melo, 328, 337
Cultural heritage values, in plant breed-
    ing, 59
Culture, cell. *See* Cell culture
Cuphea elegans, 10

Cuphea sp., 307
Cuphea wrightii, 307
Curtovirus, 528
Cuticle, 451
Cyanidin, 85
Cycloarternol, 221
Cycloeucalenol, 221
Cynara scolymus, 273
Cysteine, 236, 290
Cytokinin, 80

Danaus plexippus, 581
Data management, in industrial discov-
    ery processes, 49
Datura, 358
Datura stramonium, 354, 355
Daucus carota, 84, 86, 88, 152
Debate on manipulated genotypes, 7
Defense peptides, 552–553
Defense responses, plants, 519–521
Delphinidin, 85
Dementia, 665
Dental caries prevention, CaroRx, 415–
    418
Deoxyribonucleic acid. *See* DNA
Developed world, nutritional quality of
    food, 206–208
Developing world, nutritional quality of
    food, 204–206
Diabrotica sp., 574, 582
Diacetylmorphine, 26
Dietary fibers, 466
Digitalis, 25
Digitalis lanata, 25, 350
Digitalis purpurea, 25, 54, 84
Digitoxin, 25
Digoxin, 25
Dihydrodipicolinate synthase, 240–242
Dihydroechinofurane, 93
Dioscorea, 28
Dioscorea deltoidea, 336, 338, 340
Diphenol oxidases, genes encoding, 9
Disoma, 333
Disposable reactor systems, 191
Diuretic steroids, 28

Diversa, secondary metabolite production, 165
DNA microarray technology, 11, 249
DNA-mRNA hybridization, 11
DNA transfer methods, 112–113
Docetaxel, 29, 45
Docosahexaenoic acid, 35
Dolichol, 216
Dormancy, herbicide resistance, 622
Doxantha spp., 310
Drechslera teres, 553
Drosophila, 10, 523, 531
Drosophila melanogaster, 7, 48, 509
Drought, breeding for adaptation to, overview, 2
Drought-resistant plants, 600
Drugs. *See also under* specific drug
  isolated from plants, 24–30
    artemisinin, 24–25
    atropine, 29
    beta-sitosterol, 27
    betamethasone, 28
    cardiac glycosides, 25
    codeine, 26, 27
    corticoids, 28
    cortisone acetate, 28
    diacetylmorphine, 26
    diuretic steroids, 28
    diuretic steroids (spironolactone), 28
    docetaxel, 29
    estradiol, 28
    ethylmorphine, 26
    hecogenin, 28
    mestranol, 28
    morphine, 26, 27
    N-allyl-14-hydroxynordihydromorphinone, 27
    N-allylnormorphine, 26
    N-cyclopropylmethyl-14-hydroxynordi-hydromorphinone, 26, 27
    naloxone, 27
    naltrexone, 26
    norethisterone, 28
    opium, 25–27
    oral contraceptives, 28

[Drugs. *See also under* specific drug]
  pholcodine, 26
  progesterone, 28
  scopolamine, 29
  sex hormones, 28
  spironolactone, 28
  steroids, 28
  taxol, 28–29
  testosterone, 28
  tropane alkaloids, 29–30
  vinblastine, 30
  vincristine, 30
Duboisia, 29, 148, 149, 350, 355, 358–365
Duboisia hopwoodii, 358
Duboisia leichhardtii, 29, 358
Duboisia myoporoides, 29, 80, 358
Dunaliella salina, 651
Dwarfing, herbicide resistance, 620, 622

Eccentric motion stirrer, internal illumination cage, 188
Echinacea, 32, 55
Echinacea, 32–33
  polysaccharides, commercial production, 165
Echinacea angustifolia, 32
Echinacea pallida, 32
Echinacea purpurea, 24, 32
Efflux, heavy metals, 676
Eicosapentaenoic acid, 35
Electroporation, 126
Elemicin, 332
Eleucine indica, 606
Elicitors, 82–83
  abiotic elicitors, 82
  alkannin, 83
  biotic elicitors, 82
  rosmarinic acid, 83
  stilbene, 83
  taxol, 83
Enterobacter aerogenes, 387
Enterobacter cloacae, 387
Enterococcus hirae, 676

Environmental stress adaptation, 635–
        664
    calcium, 645
    downstream reactions, protection of,
        639–645
    growth, morphology, changes in,
        649–650
    ion homeostasis, 638
    osmolyte synthesis, 638
    osmotic adjustment, 639–640
    pathway adjustment, 638
    radical scavenging capacity,
        640–641
    signal transduction, 648–649
    stress proteins, 638
    surface properties, 638
    water, ion relations, 641–645
        potassium transporters, channels,
            644
        proton-ATPases, vacuolar pyro-
            phosphatase, 643
        sodium transport systems, 644–
            645
        water channels, 642–643
Epicuticular wax layers, 451
Erwinia amylovora, 272
Erwinia carotovora, 553
Erwinia uredovora, 326
Erysiphe cichoracearum, 431, 557
ESCAgenetics, secondary metabolite
    production, 165
Escherichia coli, 242, 264, 267, 268,
    269, 271, 329, 355, 382, 386,
    387, 408, 415, 417, 669, 680
Eschscholtzia californica, 354, 355
Estradiol, 28
Ethnopharmacology, 23
Ethylmorphine, 26
Etoposide, 98
Eucalyptus, 72, 489
Eucalyptus camaldulensis, 486, 489
Eucalyptus globulus, 489
Eukaryotic genome
    proteins, 7
    sequencing projects, 7
Euphorbia milli, 87, 88, 89

European Union Regulation on Novel
    Foods and Novel Food Ingredi-
    ents, 225
Evaporating light scattering, 54
Evening primrose oil, 33
Expressed sequence tags, 637

Farnesyl diphosphate, 219
Fatty acids
    long chain saturated, 307–310
    medium-chain saturated, 306–307
    monounsaturated, 310–312
    polyunsaturated, 312–314
    short-chain saturated, 306–307
FCS. *See* Fluorescence correlation
    spectroscopy
Female pistil cells, 452
Fermation, developments in, 201
Fibers of cell wall, 465
Field petri dish, 66–67
Fine chemicals. *See* Phytochemicals
Flashplate, 50
Flavonoids, 212–216
    absorption of, 224
    structure of, 213
Flavorless glycosides, 330
Flavors from plants, 323–346
    anise, 335
    aroma biotechnology, research fields,
        324
    cell culture, 331–339
    genetically engineered food plants,
        326–328
    genomics, 328–331
    hairy root culture, 335
    organ culture, 331–339
    plant enzymes, 328–331
    recombinant DNA technology, 326–
        331
    tissue culture, 331–339
FLAVRSAVR tomato, 324, 463
Flower colors, biosynthetic pathway, 3
Fluidized bed reactor, 176, 179
Fluorescence assay technologies, 49
Fluorescence correlation spectroscopy,
    50

Fluorescence intensity, 50
Fluorescence methods, noncellular assays, 50
Fluorescence polarization, 50
Folate, RDA, 207
Folk medicine, 23. *See also* Traditional medicine
Forest
    biodiversity, loss of, 74
    tree population structures, cloning and, 73–74
    tropical, loss of, 74
Forestry
    clonal, 75–76
    tree breeding
        cell, tissue culture in, 72–73
        ecological aspects in, 74–75
Fossils, cells with wall, 460–463
Fragaria ananassa, 86, 87
Fragaria sp., 328
Fragaria x ananassa, 334, 337
Fragrances from plants, 323–346
    anise, 335
    genomics, 328–331
    organ culture, 331–339
    tissue culture, 331–339
Frost resistance, 363
Fusarium oxysporum, 553
Fusarium solani, 553

Galanthus nivalis, 586
Garcinia mangostana, 310
Garlic, 33–35
Gelatin, 294
Geraniol, commercial production, 165
Geranyl, 219
Geranylgeranyl diphosphate, 219
Geranylhydroquinone, 93
Geranylpyrophosphate, 93
Gibberellin, 80, 216
    herbicide resistance, 622–623
Ginger, 37–38
Gingerols, 38
Ginghaosu, 24
Ginkgo, 38
Ginkgo biloba, 38, 55

Ginkolides, 38
Ginseng, 39
    biomass, production of, 164–165
Ginsenoside, 39
    commercial production, 165
Global climatic changes, 636
Globulin storage proteins, 284–285
Glucanase expression, 550
Glutathione S-transferase, 16, 641
Glycine max, 28, 243, 329
Glycine-rich proteins, 9
"Golden rice," recent release of, 18
Golgi apparatus, 459–460
Gossypium arboretum, 82
Gossypium hirsutum, 146
Grapefruit, as source of flavor component, 337
Green revolution, 2, 67–68
Growth regulators, 80–81
    anthocyanin, 80
    auxin, 80
    cytokinin, 80
    gibberellin, 80

Haber-Weiss reaction, 498
Hairy root cultures, 143–161
    Atropa belladonna hairy roots, coculture of, 149
    autotrophy, in plant hormones, 145
    characteristics of, 184
    clonal variation, 147
    cocultures using, 147–148
    doubling times, 146
    fast growth, 145
    foreign proteins, production of, using, 148–151
    genotype stability, 144–145
    growth rates, 146
    heavy metals, 154–155
    phenotype stability, 144–145
    in phytoremediation, 151–156
    properties of, 144–147
    reactors for, 184–185
    secondary metabolites, 145–146
    transformation, species resistant to, 146–147

Hairy root-derived plants, agronomincal performance, 361
Hairy roots, in phytoremediation, 151–156
Hawthorn, 39–40
Heat, breeding for adaptation to, overview, 2
Heavy metal tolerance, 665–683
biotransformation, direct removal, 679–681
chelation, 668–673
metallothioneins, 669–671
organic acids, amino acids, 672–673
phytochelatins, GSH metabolism, 668–669
proteinaceous chelators, 671–672
efflux, 676
hypertolerance, 681–682
mechanisms, 667–681
metal uptake system engineering, 676–679
nonessential metal ions, 678–679
Nramp metal transporters, 677–678
ZIP family of transporters, 677
repair mechanisms, 681
sequestration, 674–675
ABC-type transporters, 674
cation diffusion facilitators, 674–675
Hecogenin, 28
Helianthus tuberosus, 16, 272, 273
Helicobacter pylori, 415
Helicoverpa, 577
Helicoverpa armigera, 576, 577, 578, 580, 583, 584, 586
Helicoverpa punctigera, 576, 577, 584
Helicoverpa zea, 573, 575, 578
Heliothis virescens, 575
Helminthosporium maydis, 74
Hemicellulose, 458
Herbal medicines, 54. *See also under* specific herbal medicine
standardization of, 54
Herbicide metribuzin interference, D1 protein, 6–7

Herbicide resistance, 6, 363, 597–633
assaying, 615–617
gene placement, 618–619
chondriome, 618–619
chromosomal, 618
hybrids, 618
plastome, 618–619
plastome, chondriome, 618–619
transient transgenics, 619
introgression, transgenes, from crops to weeds, 609–618
metabolic resistance, 603–604
mitigation, 622–623
brassinosteroids, 623
dwarfing, 622
gibberellins, 622–623
secondary dormancy, 622
shade avoidance, 623
shattering, 622
preventing, mitigating introgression, 618–623
risks, 612–615
stacked genes, 607–609
target site resistant crops, 604–607
transgenetic mitigation, 619–623
dwarfing, 620
ripening, shattering, 620
seed dormancy, 620
Hibiscus sabdariffa, 86, 87, 88
High-methionine legumes, 291–293
High-performance liquid chromatography, 53
High-throughput-screening, 47
Histidine, 290
History of biotechnology, 2–5
Hodgkin's disease, vinblastine, 30
Hops, as sedative, 23
Hordeum sativum, 243
Hormone levels, signal transduction, 428–429
Host resistance genes, 521–525
Avr genes, 524–525
N gene, 522–523
Rx gene, 523–524

HPLC. *See* High-performance liquid chromatography
Hyalaphora cecropia, 553
Hybanthus floribundus, 146
Hydrocarbon-based polymers, 256
Hydrogen peroxide, 555
  microbial pathogen tolerance, 555–557
Hydrolytic enzymes, 551
Hydroxymethyl-glutathione, 668
Hyoscyamine, 358
Hyoscyamus muticus, 150, 185, 186, 355, 362
Hyoscyamus niger, 150, 355, 358, 362
Hypericin, 40
Hypericum perforatum, 40, 55
Hyperplasia, prostatic, benign, 27, 36
Hypersensitive reaction, 521
Hypersensitive reaction, plant-pathogen interaction, 556
Hypertorin, 40
Hyptis capitata, 146

Illumination cage, eccentric motion stirrer, bioreactor, 188
Immunoglobulin, secretory, structure of, 415
Immunotherapeutic agents, 405–426
  antibody-derived molecules, 414
  antibody glycosylation, 408–410
  antibody isotype, 413–415
  CaroRx, dental caries prevention, 415–418
  cost, protein production, 409
  gene silencing, 410–411
  plantibodies, 413
  production costs, 406–408
  purification, 411–413
  secretory IgA, structure of, 415
  SIgA, 413–415
Impatiens balsamica, 315
Impeller types for cell suspensions, 182
In vitro selection, in cell cultures, 66–67
Indian medicine, 54

Indole alkaloid production, in *Cathanranthus roseus,* 373–403
Indonesia, rice yields, increase in, 202
Industrial, assay technologies, 50
Industrial discovery processes, 45–57
  assay technologies, 50
    cellular assays, 50
    filtration assays, 50
    flashplate, 50
    fluorescence correlation spectroscopy, 50
    fluorescence intensity, 50
    fluorescence methods, noncellular assays, 50
    fluorescence polarization, 50
    isotropic methods, 50
    luminescence, 50
    quantification of expressed protein, 50
    reporter gene, 50
    time-resolved fluorescence, 50
    ultraviolet-visible absorbance, 50
  automation, 49–51
    assay technologies, 50
    compound storage system, 49
    data management, 49
    semiautomatic workstations, 49
  Chinese herb formulas, 55
  current discovery scheme, pharmaceutical industry, 47
  discovery, industrialization of, 46–49
  evaporating light scattering, 54
  high-performance liquid chromatography, 53
  high-throughput-screening, 47
  mass sequencing, 47
  new products, botanical medicines, 55
  pharmaceutical industry, 46
  plant products in, 51–54
  technological breakthroughs, 47
  traditional medicine, 54–55
    Chinese medicine, 54
    herbal-based medicines, 54

[Industrial discovery processes]
United Nations Convention on Biological Diversity, rules of, 52
web sites, natural product companies, 57
INGARD cotton, insect pest tolerance, 577, 578
Insect pest tolerance, 571–595
alpha amylase inhibitors, 584–585
Bacillus thuringiensis cry genes, 572
Bt-corn, 572–575
Bt-cotton, 575–579
Bt-potatoes, 579
digestion, inhibitors of, 583–587
INGARD cotton, 577, 578
lectins, 585–587
nontarget effects, 580–581
oral toxins, 582–583
protease inhibitors, 583–584
secondary metabolites, 587–588
Instrumentation, bioreactors, 166–169
International Breeding Centers, 3
International Rice Research Institute, 71
Iodine, 206–207
RDA, 207
Iodine deficiency, 204–205
Ion homeostasis, environmental stress, 638
Iron, 206–207, 665–683
deficiency of, 18, 666
RDA, 207
Iron-deficient anemia, 204–205
Irrigation, buildup of sodium chloride, other salts, 636
Isochorismate synthase, 429
Isoflavones, biosynthesis of, 217
Isoflavonoids, 212–216
Isoleucine, 290
Isopentenyl diphosphate, 219
Isopentenyl pyrophosphate, biosynthesis of, 218
Isoprenoid pathway, 216–223
secondary metabolites, 219

Jasmonic acid, 428

Juglans regia, 147
Juniperus chinensis, 99

Kibun, secondary metabolite production, 165
Klebsiella, 117, 271
Klebsiella oxytoca, 387
Klebsiella pneumoniae, 387
Klebsiella terrigena, 387

Labeling, with genetic alteration, 203
Laboratory, industrial-scale drug discovery, 49
Lacanobia oleracea, 586
Laccases, genes encoding, 9
Large growth droplet bioreactor system, 185
Larix, 72
Larix decidua, 65, 66
LC-MS. *See* Liquid chromatography-mass spectrometry
LDL. *See* Low-density lipoprotein
Lead, 665
Leaf alcohol, from mint terpene fractions, 330
Lectins, insect pest tolerance, 585–587
Leptinotarsa decemlineata, 579
Lesquerella, 315
Lesquerella fendleri, 315
Leucine, 290
Lignification, cell wall, 449
Lignin genetic engineering, 464, 477–495
alcohol dehydrogenase, down-regulation, 484
biosynthetic pathway, 480–481
CoA reductase, down-regulation, 483–484
coumarate CoA ligase, down-regulation of, 482–483
double transformants, 485
future trends, 487–488
monolignol specific pathway, 481
O-methyltransferases, down-regulation, 479–482
phenylpropanoid pathway, 480

[Lignin genetic engineering]
  potential side effects, 485–487
  socioeconomics and, 488–490
Ligustilide, 332
Limnanthes, 312
Limonene, 332
Linoleic acid, 312
Linseed oil, 35
Linum album, 99
Linum flavum, 99
Linum usitatissimum, 35
Lipid transfer proteins, 553
Lipoxygenases, 330
Liquid chromatography-mass spectrometry, 52
Liriodendron tulipifera, 680
Listeria grayi, 387
Listeria innocus, 387
Lithospermum erythrorhizon, 80, 81, 82, 83, 90–91, 374
Lobelin, chemical synthesis of, 349
Lolium rigidum, 602, 605, 608
Long chain saturated fatty acids, 307–310
Long-term irrigation, buildup of sodium chloride, 636
Lonicera xylosterum L., 388
Lotus corniculatus, 150, 605
Low-density lipoprotein, 27
Luminescence, 50
Lycopersicon esculentum, 14
Lysine, 236, 290
  bacterial genes, 242–243
  biosynthesis, 237–238
  genes, 239–242
  mutants overproducing, 238–239
  overproduction, amino acids, 243
  plant gene transfer, 243–247

M-geranyl-p-hydroxybenzoic acid, 93
Macroscopic fluorescence techniques, 49
Magnifera indica, 329
Male pollen tube, 452
Male sterility cytoplasm, in production of hybrid seed, 74

Malnutrition, 204–205
Malus domestica, 329
Malus pumila, 86
Malvidin, 85
Manduca sexta, 588
MAP. See Mitogen-activated protein
Marker genes, selectable, 114–116
Mass regeneration of specimens, procedures for, 5
Mass sequencing, 47
Mastrevirus, 528
Mathematics, in plant breeding, 59–61
Matricaria chamomilla, 31
Mechanical techniques employed, high-yielding agriculture, 3
Mechanically driven bioreactor, 174
Medicago sativa, 329, 330, 647
Medicine
  botanical, 23, 55, 357–358
  cell wall components, 466
  Chinese, 54
  herbal-based, 54
  Indian, 54
  modern western, 24
  traditional, 23, 54–55
Medium-chain saturated fatty acids, 306–307
Medium nutrients, 81–82
  nitrogen, 81
  phosphate level, 81
  sucrose, 82
Membrane reactor, 176–179
Membrane transport systems, 8
Mentha arvensis, 35, 79, 328
Mentha piperita, 355
Mentha spicata, 35, 328
Mentha x piperita, 35, 338, 339
Menthol, 35–36
Mesembryanthemum, 642, 643, 645, 649, 651, 652
Mesembryanthemum crystallinum, 639, 651
Mestranol, 28
Metabolic excretion, 15
Metabolic flux analysis, 249
Metal hypertolerance, 667

Metal tolerance, 665–683
    biotransformation, direct removal,
        679–681
    chelation, 668–673
        metallothioneins, 669–671
        organic acids, amino acids, 672–
            673
        phytochelatins, GSH metabolism,
            668–669
        proteinaceous chelators, 671–672
    efflux, 676
    hypertolerance, 681–682
    mechanisms, 667–681
    metal uptake system engineering,
        676–679
        nonessential metal ions, 678–679
        Nramp metal transporters, 677–678
        ZIP family of transporters, 677
    repair mechanisms, 681
    sequestration, 674–675
        ABC-type transporters, 674
        cation diffusion facilitators, 674–
            675
Methionine, 236, 290
Methyl esterified, *versus* non-methyl es-
    terified pectin distribution, 450
Methyl jasmonate, 428
Methylene-interrupted double blind,
    314–316
Methylobacterium extorquens, 334
Methylobacterium sp., 334
Metribuzin, 6
Mevalonic acid, shikonin, 93
Microbial pathogen tolerance, 549–
    569
    multigene defense mechanisms, 555–
        559
        cell death, 557–562
        constitutive systemic acquired re-
            sistance, 555
        hydrogen peroxide, 555–557
        salicylic acid, 555
        salicylic acid function, programmed
            cell death, 559–562
    single-gene defense mechanisms,
        550–554

[Microbial pathogen tolerance]
    antimicrobial secondary com-
        pounds, 554
    defense peptides, 552–553
    pathogenesis-related proteins, 550–
        552
    ribosome-inactivating proteins,
        553–554
Micronutrients, 233–253
Micropropagation, 64–66
Microscopic fluorescence technologies,
    49
Miglyol, 339
Mineral fortification of food, 206
Mint oil, 35–36
Mint terpene fractions, 330
Mitochondria
    colon, genomes, 13
    genetic information, 13
Mitogen-activated protein, 9, 433–434,
    647–648
Mitsui Petrochemical Industries, second-
    ary metabolite production, 165
Modern western medicine, focus on iso-
    lated compounds, 24
Molecular markers, phytochemicals,
    364–365
Momordica charantia, 315
Monoterpenes, 219
Monounsaturated fatty acids, 310–312
Morphine, 26, 27
Morteriella alpina, 312, 314
Musk melon, as source of flavor com-
    ponent, 337
Mutants, resistance, characterization of,
    6
Myzus persicae, 586

N-allyl-14-hydroxynordi-hydromorphi-
    none, 27
N-allylnormorphine, 26
N-cyclopropymethyl-14-hydroxynordi-
    hydromorphinone, 26–27
Naloxone, 27
Naltrexone, 26
Narcissus pseudonarcissus, 326

Nattermann, secondary metabolite production, 165
"Natural," aroma compounds, 323
Natural product companies, web sites, 57
"Nature-identical," aroma compounds, 323
Neurospora, 531
Nicandra physaloides, 520
Nickel, 665–683
Nicotiana alata, 584
Nicotiana arabidopsis, 414
Nicotiana benthamiana, 414, 524
Nicotiana clevelandii, 520
Nicotiana edwardsonii, 520
Nicotiana glutinosa, 520, 523
Nicotiana plumbaginifolia, 244, 428, 504, 506, 507, 670
Nicotiana rustica, 84, 355
Nicotiana sylvestris, 242, 245, 520
Nicotiana tabacum, 14, 146, 150, 152, 153, 156, 164, 185, 240, 329, 355, 382, 383, 386, 388, 389, 520, 523, 670
Nilaparvata lugens, 586
Nippon Oil, secondary metabolite production, 165
Nippon Paint, secondary metabolite production, 165
Nippon Shinyaku, secondary metabolite production, 165
Nitrogen, as medium nutrient, 81
Nitto Denko, secondary metabolite production, 165
Noncellulosic polysaccharides, 458
Nonstarch carbohydrates, 271–273
Norethisterone, 28
Noscapin, in opium, 26
Nothapodytes foetida, 97
Nramp metal transporters, metal uptake, 677–678
Nuclear gene products, targeted to chloroplasts, 8
Nucleic acids, defective, interfering, 528–530
Nucleus, genetic information, 13

Nutrients
  absorption of, 223
  medium for, 81–82
    nitrogen, 81
    phosphate level, 81
    sucrose, 82
Nutritional deficiency, 204–205
Nutritional quality of food, 201–231
  amino acid metabolic pathways, 236–248
  antioxidant nutrients, in U.S. population, 206
  aspartate, 238, 239–240
  beta-carotene, 206, 222
  biochemical pathway engineering, 233–253
  campesterol, 221
  carotenoids, 219, 222–223
  chalcones, biosynthesis of, 214
  chemical risk assessment, 226–228
  cycloarternol, 221
  cycloeucalenol, 221
  defined compound, nutritional value, 234–236
  developed world, 206–208
  developing world, 204–206
    antinutritional factors, 205–208
    protein quality, 205
  dihydrodipicolinate synthase, 240–242
  farnesyl diphosphate, 219
  flavonoids, 212–216
    absorption of, 224
    structure of, 213
  genetic engineering, 209
  geranyl, 219
  geranylgeranyl diphosphate, 219
  isoflavones, biosynthesis of, 217
  isoflavonoids, 212–216
  isopentenyl diphosphate, 219
  isopentenyl pyrophosphate, biosynthesis of, 218
  isoprenoid pathway, 216–223
    secondary metabolites, 219

[Nutritional quality of food]
  lysine
    bacterial genes, 242–243
    biosynthesis, 237–238
    content, plant gene transfer, 243–247
    genes, 239–242
    mutants overproducing, 238–239
    overproduction, amino acids, 243
  lysine-rich proteins, 287–289
  micronutrient manipulation, 247–248
  monoterpenes, 219
  obtusifollol, 221
  phenylpropanoid pathway, 210–223
  photosynthetic phyla, sterols, 221
  prolamins, amino acid composition, 287
  quercitin, 224
  recommended dietary allowances, 207
  regulatory issues, 224–228
  rice yields, Indonesia, increase in, 202
  safety issues, 224–228
  secondary metabolites, 210
  seed protein, 283–304
    albumins, 285–286
    barley, amino acid content, 290
    cysteine, 290
    free lysine, 289–291
    globulin storage proteins, 284–285
    high-lysine cereals, 286–291
    high-methionine legumes, 291–293
    histidine, 290
    isoleucine, 290
    leucine, 290
    lysine, 290
    maize, amino acid content, 290
    methionine, 290
    phenylalanine, 290
    prolamins, 285
    soybean seed, amino acid content, 290
    threonine, 290
    tryptophan, 290
    tyrosine, 290
    valine, 290

[Nutritional quality of food]
    wheat, amino acid content, 290
    wheat gluten, viscoelasticity, 294–297
  sitosterol, 221
  soybean globulins
    emulsification properties, 294
    gel formation, 294
  sterols, 219, 220–222
  stigmasterol, 221
  structure, relationship, 223–224
  threonine, 237–238, 238–239
  vitamin C intake, 206
  vitamin E intake, 206
Nutritional supplements, consumption of, 206

Obtusifollol, 221
Odor, sensation of, triggering, 323
Oenothera biennis, 33, 316
Oidium lycopersicon, 555
Oil, modified, from transgenic plants, 305–321
  commercial production, 316
  long chain saturated fatty acids, 307–310
  medium-chain saturated fatty acids, 306–307
  methylene-interrupted double blind, 314–316
  monounsaturated fatty acids, 310–312
  polyunsaturated fatty acids, 312–314
  short-chain saturated fatty acids, 306–307
Oil-derived chemicals, replaced by plant raw materials, 18
Oil palm, 307
Oleic acid, 310
Oligosaccharides, 466
Onobrychis viciaefolia, 144
Ophiorriza pumila, 96, 97
Opium, 25–27
Oral contraceptives, 28
Oral toxins, insect pest tolerance, 582–583

Oregano, as source of flavor component, 337
Organ cultures, 78–80
  root culture, 79–80
  shoot cultures, 79
Organic acids, chelation, heavy metals, 672–673
Organic products, 226
Organization for Economic Cooperation and Development, United Nations, 611
Origanum vulgare, 333, 337
Ormenis multicaulis, 31
Oryza officinalis, 614
Oryza sativa, 601, 614, 647, 651
Osmolyte, synthesis, environmental stress, 638
Osmotic, adjustment, environmental stress, 639–640
Ostrinia nubilalis, 573
Otacanthus coeruleus, 334, 337
Out-crossing plants, breeding hybrid varieties of, 70
Oxalis linalis, 86
Oxidative stress tolerance, 497–516
  antioxidant enzyme levels, 503–509
  ascorbate peroxidase, 508–509
  catalase, 508–509
  defense mechanisms, oxidative stress, 500–502
    ascorbate peroxidase, 501–502
    catalases, 502
    superoxide dismutases, 500–501
  superoxide dismutase levels, 503–508

P-hydroxybenzole acid, 92
P450 enzyme, 16
P450 oxygenases, 10
Packed bed reactor, 176
Paclitaxel, 95, 98, 165
Panax ginseng, 39, 55, 374
Panax pseudoginseng, 39
Panax quinquefolius, 39
Papaver somniferum, 25, 54, 78, 82, 146, 350, 355
Papaya ringspot virus, 525

Paper, production of, 477
Paprika, as source of flavor component, 337
Papyrus Ebers, 33
Parsley, as source of flavor component, 337
Particle bombardment, 121–125
  advantages, disadvantages, 122–123
  reporter genes, 124–125
Pathogen resistance
  classical cross-protection, 525–526
  defense responses of plants, 519–521
  gene silencing, 530–535
    mechanistic aspects of posttranscriptional gene silencing, 532–533
    postranscriptional gene silencing, 531
    suppressors of PTGS, 533–535
    systemic acquired gene silencing, 532
    transcriptional gene silencing, 531
    virus-induced gene silencing, 532
  host resistance genes, 521–525
    Avr genes, 524–525
    N gene, 522–523
    Rx gene, 523–524
  hosts, 518
  interfering replicons, 528–530
    defective interfering nucleic acids, 528–530
    satellites, 530
  multigene defense mechanisms, 555–559
    cell death, 557–562
    constitutive systemic acquired resistance, 555
    hydrogen peroxide, 555–557
    salicylic acid, 555
  nonhosts, 518
  pathogen-derived, 525–530
    classical cross-protection, 525–526
    interfering replicons, 528–530
    protein-mediated protection, 526–527
    RNA-mediated protection, 527–528

[Pathogen resistance]
  protein-mediated protection, 526–527
    coat proteins, 526–527
    dominant negative mutations, 527
  recovery, 518, 521
  resistant plants, 518
  RNA-mediated protection, 527–528
    antisense, sense RNAs, 527–528
  salicylic acid function, programmed
    cell death, 559–562
  sensitive plants, 518
  single-gene defense mechanisms,
    550–554
    antimicrobial secondary com-
      pounds, 554
    defense peptides, 552–553
    pathogenesis-related proteins, 550–
      552
    ribosome-inactivating proteins,
      553–554
  susceptible plants, 518
  sustainability, 535–537
    recombination, 536–537
    resistance breakage, 536
    synergism, 537
    tolerance, 535–536
    transcapsidation, 536
  tolerant plants, 518
  virulence, avirulent virus, 519
  virus, relation between, 520
  virus-specific responses, 519–521
Pathogenesis-related proteins, 550–552
Pectin biosynthesis, 459
Pectinophora gossypiella, 575, 580
Pectins, 466
  as gelling agents, 465
Peganum harmala, 81, 354, 379, 382
Pelargonidin, 85
Peonidin, 85
Perilla frutescens, 83, 86, 87, 89
Perilla sp., 84
Peronospora parasitica, 559
Peroxidases, genes encoding, 9
Pest, insect, tolerance, 571–595
  alpha amylase inhibitors, 584–585
  Bacillus thuringiensis cry genes, 572

[Pest, insect, tolerance]
  Bt-corn, 572–575
  Bt-cotton, 575–579
  Bt-potatoes, 579
  digestion, inhibitors of, 583–587
  INGARD cotton, 577, 578
  lectins, 585–587
  nontarget effects, 580–581
  oral toxins, 582–583
  protease inhibitors, 583–584
  secondary metabolites, 587–588
Petrol
  bio-diesel as, 18
  rape seed oil as, 18
  replaced by plant raw materials, 18
Petrol oil, replaced by plant raw materi-
  als, 18
Petroselinum crispum, 337
Petunia, 3
Petunia hybrida, 81, 117
Petunidin, 85
Pharmaceutical. *See under* specific
  pharmaceutical
Phaseolus vulgaris, 387, 520, 584
Phenolics, 412
Phenylalanine, 92, 290
  synthesis of lignins from, 478–479
Phenylalanine lyase, 356
Phenylpropanoid pathway, 210–223
Philippines, plant breeding, 71
Pholcodine, 26
Phosphate level, as medium nutrient, 81
Phosphoribulokinase, 641
Photoreceptor systems, elucidation of, 4
Photosynthate, 259–261
Photosynthetic phyla, sterols, 221
Photosynthetic processes, elucidation of,
  4
Phylloplane, 451
Physcomitrella patens, 651
Phytoalexins, 12
Phytochemicals. *See also under* specific
  phytochemical
  arecolin, chemical synthesis, 349
  bioreactor cultures, 80
  callus, cell suspension cultures, 78

[Phytochemicals. *See also under* specific phytochemical]
  camptothecin, chemical structure, 350
  cell culture, secondary compounds, 78–80
    bioreactor cultures, 80
    callus, cell suspension cultures, 78
    immobilized cultures, 78
    organ cultures, 78–80
    root culture, 79–80
    shoot cultures, 79
  conventional breeding, 360
  genetic transformation, 361–362
  hairy root-derived plants, agronomincal performance, 361
  immobilized cultures, 78
  lobelin, chemical synthesis of, 349
  medicinal plants, 357–358
  metabolic engineering, 351–356
  molecular markers, 364–365
  organ cultures, 78–80
    root culture, 79–80
    shoot cultures, 79
  piperidine alkaloids, chemical synthesis of, 349
  production of phytochemicals, 348–351
  scopolamine, 358–365
    chemical structure, 350
  taxol, chemical structure, 350
  tissue culture applications, 360–361
Phytolacca americana, 81
Phytomining, 152, 665–683
Phyton Catalytic, secondary metabolite production, 165
Phytophthora, 557
Phytophthora cactorum, 553
Phytophthora cryptogea, 431
Phytophthora infestans, 551, 552, 554, 557
Phytophthora parasitica, 431, 557
Phytophthora sojae, 429
Phytoremediation, hairy roots, 151–156
Pigments, 84–94
  anthocyanin, 84–88
    bioreactor culture, 88

[Pigments]
    cell line selection, 86–87
    chemical factors, 87–88
    chemical structure, 85
    cyanidin, 85
    delphinidin, 85
    malvidin, 85
    pelargonidin, 85
    peonidin, 85
    petunidin, 85
    physical environment, 87
    recombinant DNA technology, 88
  shikonin, 88–94
    biosynthesis, 94
    biosynthetic pathway, 93
    dihydroechinofurane, 93
    dihydroshikonofuran, 93
    engineering biosynthesis, 94
    geranylhydroquione, 93
    geranylpyrophosphate, 93
    Lithospermum erythrorhizon cell cultures, 90–91
    m-geranyl-*p*-hydroxybenzoic acid, 93
    mevalonic acid, 93
    p-hydroxybenzoic acid O-glucoside, 92
    p-hydroxybenzole acid, 92
    phenylalanine, 92
    regulatory mechanism, 91–94
Pilocarpus microphyllus, 350
Pimpinella anisum, 335, 337
Pinus radiata, 489
Pinus sp., 489
Pinus taeda, 489
Piperidine alkaloids, chemical synthesis of, 349
Pistil cells, female, 452
Plagyobotrys arizonicus, 90
Plant cell. *See* Cell
Plantibodies, 413
Plasmid-encoded afflux systems, 676
Plasmodiophora brassicae, 553
Plasmodium falciparum, 24
Plasmodium ovule, 24
Plasmodium vivax, 24

Plastic-lined reactor, 189
Plastids
   genetic information, 13
   mitochondria, colon, genomes, 13
Plastome, herbicide resistance, 618–619
Plodia interpunctella, 580
Plutella xylostella, 580
Pneumatically agitated bioreactor, 175
Podophyllotoxin, 95, 98–100
   commercial production, 165
Podophyllum hexandrum, 99, 146
Podophyllum peltatum, 99
Podophyllum sp., 98
Pollen tube, male, 452
Polyporus umbellatus, 334
Polyunsaturated fatty acids, 312–314
Poppy, opium, cultivation of, 26
Populus, 72
Populus alba, 489
Populus deltoides, 489
Populus nigra, 489
Populus sp., 64
Populus tremula, 73, 489
Populus tremuloides, 73
Porphyridiom cruentum, 312
Posttranscriptional gene silencing, 410
Potassium transporters, 644
Potato virus X, 523
Potato virus Y, 524
Preamylopectin, 265
Process development, 411–413
Processing, bioavailability by, 223
Productivity, agricultural crop plants, 17
Progesterone, 28
Programmed cell death, microbial pathogen tolerance, 559–562
Prolamins, 285
Prostatic hyperplasia, benign, 27, 36
Protease inhibitors, insect pest tolerance, 583–584
Proteinaceous chelators, chelation, heavy metals, 671–672
Protoberberines, commercial production, 165
Proton-ATPases, vacuolar pyrophosphatase, 643

Protoplasts, 125–126
Pseudohypericin, 40
Pseudomonas, 561
Pseudomonas aeruginosa, 673
Pseudomonas mallei, 334
Pseudomonas putida, 559
Pseudomonas syringae pv. *glycinea,* 560
Pseudomonas syringae pv. *tomato,* 435, 553, 557
Puccina graminis, 485
Pueraria lobata, 82
Pulp, production of, 477, 488

Quantitative trait loci, 637
Quercitin glucoside, 224
Quercitin rutinoside, 224

Radical flow impelers, 181
Radical scavenging capacity, environmental stress, 640–641
Radicals, scavenged, 498, 500–502
Ramie, 465
Rape seed oil, as petrol, 18
Raphanus raphanistrum, 602, 614
Rauvolfia serpentina, 354, 378, 386, 387
RDAs. *See* Recommended dietary allowances
Reactive oxygen species, 640
Receptor-ligand interactions, 429–431
Recombinant DNA technology, 88
   flavors, fragrances, 326–331
Recommended dietary allowances, 207
Regulatory issues, in nutritional food quality, 224–228
Regulatory mutants, characterization of, 6
Replicons, interfering, pathogen-derived resistance, 528–530
Restriction fragment length polymorphism, 17
Reverse transcription-polymerase chain reaction, 356
RFLP. *See* Restriction fragment length polymorphism

Rhamnogalacturonan, 459
Rhizoctonia solani, 551, 552, 554
Rhizosphere, 451
Ribes nigrum, 33, 316
Ribonucleic acid. *See* RNA
Ribosome-inactivating proteins, 553–
    554
Ricinus communis, 315
Risk assessment, chemical, food, 226–
    228
RNA-mediated protection, pathogen-de-
    rived resistance, 527–528
Root culture, 79–80
Rosemary, as source of flavor compo-
    nent, 337
Rosmarinic acid, 83
    commercial production, 165
Rosmarinus officinalis, 334, 337
Rotating drum reactor, 173, 179
Rubia tinctorum, 152

Sabal serrulata, 37
Saccharomyces cerevisiae, 7, 10, 397,
    647, 665–683
Safety issues
    foods, 224–228
    with genetic alteration, 203
Salicylic acid, 429, 558
    microbial pathogen tolerance, 555
Salmonella typhimurium, 387, 606
Salt
    breeding for adaptation to, overview,
        2
    buildup of, with irrigation, 636
Salvia officinalis, 329
Santalum album, 183
Satellites, pathogen-derived resistance,
    530
Saturated fatty acids
    long chain, 307–310
    medium-chain, 306–307
    short-chain, 306–307
Saw palmetto, 36–37
Schizosaccharomyces pombe, 669, 670,
    674

Scopolamine, 29, 30, 358–365
    chemical structure, 350
    commercial production, 165
Secondary metabolites, 11–12, 210,
    347–371
    biological factors, 83–84
        cell-to-cell variation, 83
    biosynthetic genes, cloned, 354–355
    commercial production levels, 165
    elicitors, 82–83
        abiotic elicitors, 82
        alkannin, 83
        biotic elicitors, 82
        rosmarinic acid, 83
        stilbene, 83
        taxol, 83
    first production of, 164–165
    formed under stress, 12
    indole alkaloid production, in *Ca-
        thanranthus roseus,* 373–403
    insect pest tolerance, 587–588
    medium nutrients, 81–82
        nitrogen, 81
        phosphate level, 81
        sucrose, 82
    physical factors, 83
        aeration, 83
        cell density, 83
        light, 83
        pH, 83
        temperature, 83
    plant growth regulators, 80–81
        anthocyanin, 80
        auxin, 80
        cytokinin, 80
        gibberellin, 80
Sedanenolide, 332
Seed protein, 293
    2S albumins, 285–286
    barley, amino acid content, 290
    cysteine, 290
    globulin storage proteins, 284–285
    high-lysine cereals, 286–291
        free lysine, 289–291
        lysine-rich proteins, 287–289

[Seed protein]
prolamins, amino acid composition, 287
high-methionine legumes, 291–293
histidine, 290
isoleucine, 290
leucine, 290
lysine, 290
maize, amino acid content, 290
methionine, 290
nutritional quality, 283–304
phenylalanine, 290
prolamins, 285
soybean
amino acid content, 290
emulsification properties, 294
gel formation, 294
threonine, 290
tryptophan, 290
tyrosine, 290
valine, 290
wheat
amino acid content, 290
gluten, viscoelasticity, 294–297
Selenium, 206–207
RDA, 207
Self-pollinated plants, breeding for, 68–69
Selfing plants, hybrid varieties of, breeding, 71–72
Semiautomatic workstations, in industrial discovery processes, 49
Senecio vernalis, 355
Sense RNA, RNA-mediated protection, 527–528
Sequestration, heavy metals, 674–675
ABC-type transporters, 674
cation diffusion facilitators, 674–675
Serenoa repens, 37
Serratia marcescens, 387
Sex hormones, 28
Shattering, herbicide resistance, 620, 622
Shigella flexneri, 117

Shikonin, 88–94
biosynthetic pathway, 93
dihydroechinofurane, 93
dihydroshikonofuran, 93
engineering biosynthesis, 94
geranylhydroquinone, 93
geranylpyrophosphate, 93
Lithospermum erythrorhizon cell cultures, 90–91
m-geranyl-p-hydroxybenzoic acid, 93
mevalonic acid, 93
p-hydroxybenzoic acid O-glucoside, 92
p-hydroxybenzole acid, 92
phenylalanine, 92
regulatory mechanism, 91–94
Shoot cultures, 79
reactors for, 185–187
Short-chain saturated fatty acids, 306–307
Signal peptide sequences, nuclear gene products, 8
Signal transduction, 427–444
calcium signaling, 432–433
environmental stress, 648–649
GTP-binding proteins, 431–432
hormone levels, 428–429
mitogen-activated protein kinases, 433–434
receptor-ligand interactions, 429–431
V transcription factors, 435–437
Silece, 682
Silencing, gene, 530–535
mechanistic aspects of posttranscriptional gene silencing, 532–533
postranscriptional gene silencing, 531
suppressors of PTGS, 533–535
systemic acquired, 532
transcriptional, 531
virus-induced, 532
Silene vulgaris, 681, 682
Simmondsia chinensis, 311
Simple sugars, in storage organs, 267–268
Sitosterol, 221

Snapdragon, Brazillian, as source of flavor component, 337
Sodium chloride, buildup of, with irrigation, other salts, 636
Sodium transport systems, 644–645
Solanum acaule, 520
Solanum aviculare, 146, 152, 336, 338, 340
Solanum nigrum, 152
Solanum stoloniferum, 520
Solanum tuberosum, 244, 384
Somatic embryogenesis, 64–66
Somatic hybridization, 61–62
Sorghum, 245
Sorghum bicolor, 244
Southeast Asia, plant breeding, 71
Soybean globulins
    emulsification properties, 294
    gel formation, 294
Soybean seed, amino acid content, 290
Spirae ulmaria, 54
Spironolactone, 28
Spodoptera litura, 582
St. John's wort, 40–41
Standardized extracts, 37–41
    ginger, 37–38
    gingerols, 38
    ginkgo, 38
    ginkolides, 38
    ginseng, 39
    ginsenoside, 39
    Hawthorn, 39–40
    hypericin, 40
    hypertorin, 40
    pseudohypericin, 40
    St. John's wort, 40–41
Staphyloccoccus epidermidis, 397
Staphylococcus aureus, 397, 676
Starch deposition, 259–265
    amylopectin synthesis, 263–264
    amylose synthesis, 263
    committed pathway, 261–263
    photosynthate, 259–261
Starch quantity alteration, 266–267

Starch structure, 257–259
    amylopectin, 257
    amylose, 257
    properties of starch components, 258
    starch granules, 257–259
State-subsidized breeding, 60
Sterility cytoplasm, in production of hybrid seed, 74
Steroids, 28
Sterol, 216, 219
    biosynthesis, 220–222
Stevia rebaudiana, 187
Stigmasterol, 221
Stilbene, 83
Stirred reactor, 173
Storage carbohydrate properties, variation in, 256
Strawberry, as source of flavor component, 337
Streptalloteichus hindustantus, 117
Streptococcus, 414
Streptococcus mutans, 405, 415–419
Streptomyces hygroscopicus, 117, 573
Streptomyces sp., 587
Strictosidine synthase, from *Catharanthus roseus*, 380–381
Strigs spp, 600
Strobilanthes dyeriana, 87
Substantial equivalence, concept of, 225
Sucrose, as medium nutrient, 82
Sugars, in storage organs, 267–268
Sumitomo Chemical Industries, secondary metabolite production, 165
"Super weeds," 609
Superfund sites, 665
Superoxide, 498
    dismutase, oxidative stress, 500–501, 503–508
Sustainability, 535–537
    recombination, 536–537
    resistance breakage, 536
    synergism, 537
    tolerance, 535–536
    transcapsidation, 536
Sweet orange, as source of flavor component, 337

Symphoricarpus, 379
Synechococcus, 670
Synechocystis, 248, 645

Tabernaemontana pandacaqui, 358, 379, 387
Tabernamontana, 354
Tanacetum pathenium, 55
Taste. *See also* Flavor
   sense of, chemical messengers of, 323
Taxol, 24, 28–29, 45, 83, 98
   chemical structure, 350
   commercial production, 165
Taxotere, 29
Taxus, 348, 374
Taxus baccata, 28, 98
Taxus brevifolia, 28, 29, 98, 355
Taxus canadensis, 98
Taxus chinensis, 98
Taxus cuspidata, 98, 355
Taxus media, 98
Teniposide, 98
Terpenoid indole alkaloid, *Catharanthus roseus,* 375–397
Testosterone, 28
Texture, of fruit, vegetables, 212
TGS. *See* Transcriptional gene silencing
Thalictrum tuberosum, 354
Theobroma cacao, 331
Thermopsis lupinoides, 83
Thermus thermophilus, 268
Thielaviopsis basicola, 431, 557
Thin-layer chromatography, 52
Thionins, 553
Thlaspi arvense, 677
Thlaspi caerulescens, 146, 153, 156, 677
Thlaspi goesingense, 672
Threonine, 290
Thunbergia alata, 310
Thylakoids, membrane organization of, 4
Tickle bed reactor, 176
Timber, cell walls, 464
Time-resolved fluorescence, 50

Topocuvirus, 528
Totipotency, 5
Traditional medicine, 23, 54
Tree breeding
   cell, tissue culture in, 72–73
   ecological aspects in, 74–75
Triazine resistance, 599
Trichoderma viride, 551
Trickle bed reactor, 185
Triticum tauschii, 613
Tropane alkaloids, 29–30
Tropical forests, loss of, 74
Tryptophan, 290
   decarboxylase
      from *Catharanthus roseus,* 379
      expression of, 382–384
Tyromyes sambuceus, 334
Tyrosine, 290

Ubiquinone, 216
Ultraviolet light photoreceptors, 4
Ultraviolet-visible absorbance, 50
Uniform cell, study of, 5
United Nations Convention on Biological Diversity, rules of, 52
Uptake mutants, characterization of, 6

Vaccines, transgenic plants for production of, 405
Vacuolar pyrophosphatase, proton-ATPases, 643
Valeriana officinalis, 55
Valine, 290
Vanilla, as source of flavor component, 337
Vanilla plantifolia, 329, 333, 337
Vanillin, commercial production, 165
Vanillyl alcohol, 332
Vegetarians, nutritional intake, 206
Vernonia galamensis, 315
Verticillum dabliae, 82
Vicia, 293
Vicia narbonensis, 292
Vigna unguiculata, 520
Vinblastine, 30, 94, 95
Vinca rosea, 30

Vincristine, 30, 94, 95
Viral pathogen resistance, 517–548
    defective interfering nucleic acids,
        528–530
    defense responses of plants, 519–521
    gene silencing, 530–535
        mechanistic aspects of posttran-
            scriptional gene silencing, 532–
            533
        postranscriptional gene silencing,
            531
        suppressors of PTGS, 533–535
        systemic acquired gene silencing,
            532
        transcriptional gene silencing, 531
        virus-induced gene silencing, 532
    host resistance genes, 521–525
        Avr genes, 524–525
        N gene, 522–523
        Rx gene, 523–524
    hosts, 518
    nonhosts, 518
    pathogen-derived resistance, 525–530
        classical cross-protection, 525–526
        interfering replicons, 528–530
        protein-mediated protection, 526–
            527
        RNA-mediated protection, 527–528
    protein-mediated protection
        coat proteins, 526–527
        dominant negative mutations, 527
    recovery, 518, 521
    resistant plants, 518
    RNA-mediated protection, antisense,
        sense RNAs, 527–528
    satellites, 530
    sensitive plants, 518
    susceptible plants, 518
    sustainability, 535–537
        recombination, 536–537
        resistance breakage, 536
        synergism, 537
        tolerance, 535–536
        transcapsidation, 536
    tolerant plants, 518
    virulence, avirulent virus, 519

[Viral pathogen resistance]
    virus, relation between, 520
    virus-specific responses, 519–521
Virulence, avirulent virus, 519
Virus, as gene transfer vectors, 128–129
Virus-specific responses, 519–521
Viscoelasticity, wheat gluten, 294–297
Vitamin A, 18, 206–207, 234
    deficiency, 204–205
    RDA, 207
Vitamin B, 206–207
    RDA, 207
Vitamin C, 206
    RDA, 207
Vitamin deficiencies, disease due to,
    204
Vitamin E, 206–207, 234
    RDA, 207
Vitamin K, RDA, 207
Vitis, 86, 89
Vitis labrusca, 331
Vitis vinifera, 81, 82, 86, 87, 88, 554

Wall, cell, 445–475, 478
    architecture of, 452–456
    assembly of, 457–460
    biosynthesis, 457–460
    components of, 452–456
    fibers of, 465
    functional roles of, 449
    functions of, 447–452
    future biotechnology, 466–468
    model, dicot plants, 453
    performance optimization, 460–463
    proteins, 459
    timber, 464
    transport in, 450
    wood, 464
Water, ion relations, environmental
    stress, 641–645
    potassium transporters, channels, 644
    proton-ATPases, vacuolar pyrophos-
        phatase, 643
    sodium transport systems, 644–645
    water channels, 642–643
Wave reactor, 189, 190

Waxes, 311
Western medicine, 24
Wood
  cell walls, 464
  composition of, 478
Workstations, semiautomatic, in industrial discovery processes, 49
World pulp production, amount of, 488

Xanthomonas oryzae pv. *oryzae,* 435

Xenobiochemicals, metabolism of, 14–17
Xyloglucan network, cellulose, 454

Zabrotes subfasciatus, 584
Zinc, 665–683
  deficiency, 666
ZIP family of transporters, metal uptake, 677

group of metals. For instance, fission yeast deficient in phytochelatin synthesis (see later) are Cd and Cu hypersensitive yet grow normally, when compared with wild-type cells, in the presence of elevated Zn concentrations (14). Some Ni-hyperaccumulating plants show the so-called histidine response, which mediates Ni tolerance (15). The sensitivity toward most other heavy metals, however, is not attenuated by the increases in histidine levels.

Five general metal tolerance mechanisms can be distinguished: chelation, sequestration, exclusion, biotransformation, and repair. The first three of these categories constitute the principal elements of the metal homeostasis network. Exclusion includes the regulation of metal uptake into cells. Biotransformation, i.e., reduction and possibly volatilization, appears to be restricted mainly to Hg and the metalloids Se and As.

A. Chelation

Because of their reactivity with N, S, or O, it has been postulated that metal ions are constantly chelated when they are taken up into the cell by specialized or unspecific cation transporters (16–19). By studying the interaction of the Zn/Cu superoxide dismutase, its copper chaperone, and other chelators, it was demonstrated for S. cerevisiae that the intracellular concentration of free copper is indeed limited to less than one atom per cell (20). Specialized chelators exist that function as cytosolic mediators of metal trafficking, the delivery of essential metal ions to organelles and protein complexes.

1. Phytochelatins and GSH Metabolism

The best-known chelators in plant cells are phytochelatins (PCs), small metal-binding peptides that are nontranslationally synthesized from glutathione or glutathione derivatives such as hydroxymethyl-glutathione in a transpeptidation reaction (21–23). PCs of chain lengths up to $n = 11$ are formed in plants and some fungi upon exposure to a range of different metals (24). However, not all of the inducing metals are chelated by phytochelatins. Well characterized is the formation of low-molecular-weight cytosolic PC-Cd complexes, which are transported into the vacuole, where high-molecular-weight complexes are formed on incorporation of S^{2-} (see later). Also, in vitro studies showed that Zn and Cu ions could be transferred efficiently from Zn- or Cu-PCs, respectively, to metal-requiring apoenzymes (25). The importance of PC synthesis was demonstrated by the isolation of the Arabidopsis cad1 mutant (26). This mutant is PC deficient and Cd hypersensitive. The CAD1 gene and other plant and fungal genes encoding phytochelatin synthases (PCSs) have been cloned by positional cloning (27) and based on the screening for Cd tolerance mediating plant complementary